LASER HANDBOOK

VOLUME 3

LASER HANDBOOK

VOLUME 3

edited by

M.L. STITCH

Exxon Nuclear Company Inc.
Richland, Washington, USA

NORTH-HOLLAND PUBLISHING COMPANY
AMSTERDAM · NEW YORK · OXFORD

ISBN 0 444 85271 9

First edition 1979
Reprinted 1986

Published by:

North-Holland Physics Publishing

a division of

Elsevier Science Publishers B.V.
P.O. Box 103
1000 AC Amsterdam
The Netherlands

Sole distributors for the U.S.A. and Canada:

Elsevier Science Publishing Company, Inc.

52 Vanderbilt Avenue
New York, N.Y. 10017
U.S.A.

Printed in The Netherlands

Preface

Volume 3 of the *Laser Handbook* continues in the wake of Volumes 1 and 2 to present a series of articles on laser technology and applications by authors from academic, industrial and government laboratory organizations.

The word "Handbook" is used not in the sense of a collection of tables or formulae, but as in the German "Handbuch" – such as, for example, *Handbuch der Experimentalphysik*. As in earlier volumes, the articles are short, expository monographs by workers in the field primarily for workers in similar or related fields and for advanced students.

Emphasis is placed on technology leading to improved mode characteristics and beam quality, increased efficiency, new wavelength regions, increased power and tunability. These topics have been grouped in a *Laser Technology* section.

In the *Laser Applications* section the applications addressed are either: physical interactions which modify the laser's characteristics, such as obtaining ultrashort pulses and high power second harmonic pulses; or scientific and engineering applications, such as laser induced chemical reactions and isotope separation, pulsed holography, inertial confinement fusion and picosecond spectroscopy. Topics such as the continuing advances in optical communication and material processing have been well covered elsewhere and are not included.

It seems that the maturing of a technology brings together more disciplines. Twenty five years ago I remember how microwave spectroscopy started with physics and microwave electronics and moved more and more into chemistry. Now it is evident from the contents of this volume that what we have been calling "quantum electronics" will contain more and more that is of interest to our colleagues in chemistry and perhaps even those in aeronautical engineering and nuclear engineering.

Most of the authors are at, or recently associated with, leading institutions where the work described has been pursued. The fact that most of the articles originated in the United States and that 2 institutions account for 5 of the articles should be construed not only as a tribute to the excellence of the work at those institutions, but as a matter of convenience. A recent trip to the Soviet Union to visit laboratories and to attend a Laser and Non-Linear Optics Conference held in memoriam to R. V. Khokhlov (who co-authored an article in vol. 2), regent of Moscow State University at the time of his tragic mountaineering death, showed me extensive and impressive research activity in laser related programs. Some of these programs overlapped work discussed in several of the articles in this volume, but it is far more simple and quick to phone a colleague in, say, Los Alamos than write one in Moscow.

For their aid and support in various phases of the incubation, progress and completion of this book I should like to thank: Drs. Willem H. Wimmers, Pres. of North-Holland; Dr. Pieter S. H. Bolman, Senior Physics Editor; Dr. William Montgomery, Physics Editor; Mr. J. R. Boxcer, Desk Editor; and my wife, Sharon Stitch. Dr. J.-C. Diels had the arduous task of preparing the Subject Index and is hereby gratefully acknowledged.

Kennewick, Washington   Malcolm L. Stitch
December, 1978

Contents

Part A   Laser Technology

Part B   Laser Applications

List of Contributors

R.L. Abrams, Hughes Research Laboratories, Malibu, California 90265, USA.

C.D. Cantrell, Los Alamos Scientific Laboratory, University of California, P.O. Box 1663, Los Alamos, New Mexico 87545, USA.

R.E. Center, Mathematical Sciences Northwest, Inc., Bellevue, Washington 98009, USA.

J.J. Ewing, Lawrence Livermore Laboratory, University of California, Livermore, California 94550, USA.

S.M. Freund, Los Alamos Scientific Laboratory, University of California, P.O. Box 1663, Los Alamos, New Mexico 87545, USA.

E.V. George, Advanced Lasers Group, Lawrence Livermore Laboratory, University of California, Livermore, California 94550, USA.

R.A. Haas, Advanced Lasers Group, Lawrence Livermore Laboratory, University of California, Livermore, California 94550, USA.

D. Hon, Laser Systems Division, Aerospace Groups, Hughes Aircraft Company, Culver City, California, USA.

W. Koechner, Hadron Inc., Santa Monica, California 90404, USA.

W.F. Krupke, Advanced Quantum Electronics, Lawrence Livermore Laboratory, University of California, Livermore, California 94550, USA.

H.E. Lessing, Abt. Chemische Physik, Universität Ulm, F.R. Germany.

W.H. Lowdermilk, Lawrence Livermore Laboratory, University of California, Livermore, California 94550, USA.

J.L. Lyman, Los Alamos Scientific Laboratory, University of California, P.O. Box 1663, Los Alamos, New Mexico 87545, USA.

W.H. Steier, Department of Electrical Engineering, University of Southern California, Los Angeles, California 90007, USA.

C.J. Ultee, United Technologies Research Center, East Hartford, Connecticut 06108, USA.

A. von Jena, Abt. Chemische Physik, Universität Ulm, F.R. Germany.

R. Wallenstein, Fakultät für Physik, Universität Bielefeld, Postfach 8640, 4800 Bielefeld, F.R. Germany.

PART A

Laser Technology

R. L. ABRAMS
R. E. CENTER
J. J. EWING
W. H. STEIER
C. J. ULTEE
R. WALLENSTEIN

A1 | Unstable Resonators

WILLIAM H. STEIER

Department of Electrical Engineering,
University of Southern California, Los Angeles, California 90007

Contents

Abstract

The advantages and uses of unstable resonators are reviewed, and the geometrical design formulas are given. The effects of diffraction on the losses and mode patterns of the empty sharp-edged and tapered-edge resonators are discussed. Included is a discussion of when the diffraction effects should be included in the design procedure. The recent computer analyses of loaded resonators are reviewed, and the simpler ray analysis of loaded resonators is outlined. Some of the recent applications of unstable resonators as ring resonators, as frequency-controlled oscillators, and as injection-locked oscillators or regenerative amplifiers are included.

© *North-Holland Publishing Company, 1979*
Laser Handbook, edited by M.L. Stitch

3

1. Introduction

Unstable resonators are now widely and successfully used in many high energy lasers where their properties make them uniquely suited. There have been numerous papers published concerning the properties of these resonators, and they are well understood except perhaps in the case of the design of the optimum resonator for use with a laser medium with significant index and gain inhomogeneities. This review will first summarize the geometrical properties and the design equations based on geometrical optics. The review continues with a discussion of the effects of diffraction and of tapered apertures in order to show in what cases diffraction is important, and when it must be included in the design equations of the empty resonators. The sections on resonators containing gain and index inhomogeneities are to bring the reader up-to-date on the recently published computer techniques and on the emerging ray analysis approaches. The review concludes with some of the recent work on ring unstable resonators and on injection-locked oscillators and power amplifiers employing unstable cavities.

Unstable resonators are uniquely suited for lasers with the following characteristics: (a) the gain volume has a radius-to-length ratio that is relatively large, i.e., 0.1 or larger, and (b) the laser parameters (gain, saturation) are such that the optimum output coupling is relatively large (50% or more). This describes many high energy lasers (CO_2, CO, HF–DF, KrF, etc.) and the unstable resonator has in recent years found extensive use in these systems. The advantages of the unstable resonator were first pointed out by Siegman (1965, 1971, 1974) and by Siegman and Arrathon (1967), whose publications and review articles form the cornerstone of this area of optics. Ananév et al. (1970) and Ananév (1972) have also made significant contributions; in particular, their work provides a good physical insight into the properties of these resonators.

Resonators, both stable and unstable, are traditionally described in terms of g_1 and g_2 where

$$g_1 = 1 - \frac{L}{R_1}, \tag{1}$$

$$g_2 = 1 - \frac{L}{R_2}, \tag{2}$$

L = spacing between mirrors,
R_1, R_2 = radii of curvature of the mirrors,
$\quad$ (R_1, R_2 are positive for concave mirrors).

Stable resonators satisfy the condition

$$0 \leq g_1 g_2 \leq 1, \tag{3}$$

and were the first resonators analyzed and developed for laser use. The modes of the stable resonator are the gaussian–laguerre modes extensively described in the literature (Kogelmk and Li 1966). The lowest order mode is characterized by a 'thread-like' mode volume whose radius-to-length ratio is approximately $(\lambda/2L)^{1/2}$ or on the order of 10^{-3}. This aspect ratio can be increased by using larger radius mirrors, but by no more than a factor of 10 for practical cases. These modes are well confined in the transverse direction and typically have negligible diffraction loss.

In high energy lasers where the radius-to-length ratio of the gain medium is considerably greater than 10^{-3}, it is impractical to attempt to obtain efficient diffraction-limited output with stable resonator. A stable mode volume which fills the gain medium would essentially require flat mirrors. This is an undesirable cavity design because the mode quality of the flat–flat resonator can be severely degraded by optical inhomogeneities of the gain medium. If the cavity is designed with a stable mode volume considerably less than the gain volume, a multi-transverse mode oscillation will result, which is not diffraction-limited. The output beam will essentially diffract with an angle determined by the size of the lowest-order mode, which will be considerably smaller than the total output aperture.

These limitations of the stable resonators are avoided in the unstable resonator. Unstable resonators are those where the $g_1 g_2$ product falls outside the range given above. The modes are not confined in the transverse direction, and the size is determined by some limiting aperture in the system. Hence, they can be designed to fill any mode volume. Because there is no transverse confinement, there is also considerable diffraction loss around the cavity aperture, which can usually be used as the laser output.

2. The geometrical modes and design equations

The first and simplest analysis of the unstable modes is a purely geometrical picture (Siegman 1965, 1971, Kahn 1966, Barone 1967, Bergstein 1968, Streifer 1968). These results give the essential features of the modes and in many cases give a practical set of design equations. Referring to fig. 1, the problem is to find the position of two point sources, which are imaged into one another by the mirrors. That is, waves that leave mirror one and appear to come from S_1, are reflected from mirror two and appear to come from S_2, and vice versa. Hence, on one round trip, the sources are imaged into themselves, and the ray pattern is repeated.

We define a magnification M_{12} which is the geometrical growth of the beam in going from mirror one to mirror two. The geometrical picture assumes uniform transverse beam profiles, and based on this we can calculate a diffraction power

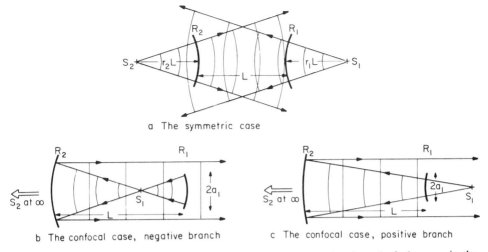

a The symmetric case

b The confocal case, negative branch   c The confocal case, positive branch

Fig. 1. The unstable resonator. S_1 and S_2 are the virtual centers for the spherical waves in the geometrical optics description

coupling δ_1 around mirror one and δ_2 around mirror two determed by M_{12} and M_{21}. Notice that the transverse mode size is determined by the beam filling one or both mirrors, and hence can be increased by increasing the mirror aperture. The total output coupling around both mirrors is fixed by the mirror radii and spacing, but the proportion of this total which is coupled out each end can be adjusted by changing the relative mirror apertures. Table 1 gives the formulas which are basic in the design of the unstable resonator.

A useful parameter in describing unstable resonators is the equivalent Fresnel number which for a symmetric resonator is defined by

$$N_{eq} = \frac{a^2}{\lambda L} \frac{M^2 - 1}{2M}, \tag{4}$$

where

M = round-trip magnification = $M_{12}M_{21}$,

a = radius of the mirror aperture.

For an unsymmetric resonator, N_{eq} can be computed using the equivalence relations developed by Siegman and Miller (1970) and Siegman (1971).

The geometrical ray theory predicts a uniform mode pattern, and the output coupling formulas are based on this assumption. If the more accurate wave analysis is done and diffraction is included, the mode is not uniform and the coupling formula must be modified. It will be shown in the next section that for resonators with circular aperture mirrors, with sharp edges, and when the N_{eq} is low, the diffraction effects will be large. Diffraction will always reduce the coupling below the geometrical values, as discussed in the next section. However, in many practical cases, these conditions are not met (i.e., large N_{eq} or rectangular apertures) and the geometrical coupling formulas are approximately correct and can be used for resonator design.

Table 1.
Basic design equations.

$$g_1 = 1 - \frac{L}{R_1} \qquad\qquad g_2 = 1 - \frac{L}{R_2}$$

1. *Location of virtual point sources*

$$\frac{1}{r_1} = \sqrt{\left(g_1^2 - \frac{g_1}{g_2}\right) + g_1 - 1}$$

$$\frac{1}{r_2} = \sqrt{\left(g_2^2 - \frac{g_2}{g_1}\right) + g_2 - 1}$$

2. *Magnification*

$$M_{12} = \frac{1 + r_1}{r_1} \qquad\qquad M_{21} = \frac{1 + r_2}{r_2}$$

$$M = \text{round-trip magnification} = M_{12}M_{21} = 2g_1g_2 + 2\sqrt{[g_1g_2(g_1g_2 - 1)]} - 1$$

3. *Diffraction coupling*

δ_T = total diffraction coupling (power) per round trip

$$= 1 - \frac{1}{M^2}$$

4. *Confocal resonators – single-ended coupling*

$$\delta_T = \delta_1 = 1 - \frac{1}{M^2} \qquad\qquad \delta_2 = 0$$

Positive branch $g_1 = \dfrac{M+1}{2}, \quad g_2 = \dfrac{M+1}{2M}$

Negative branch $g_1 = \dfrac{1-M}{2}, \quad g_2 = \dfrac{M-1}{2M}$

In figs. 1b and 1c is shown the confocal unstable resonator (Krupke and Sooy 1969), which is a special case that is the most used in practice because it can produce an output from one end only which is collimated. In this case the aperture size of mirror two is made very large, and the transverse mode size is set by the aperture of mirror one. The condition for confocal is

$$g_1 + g_2 = 2g_1g_2, \tag{5}$$

or that $R_1 + R_2 = 2L$. The negative branch is in general avoided because of the internal hotspot, which can lead to breakdown. For confocal resonators (Rensch and Chester 1973)

$$N_{eq} = \frac{M-1}{2} N_1, \tag{6}$$

where $N_1 = a_1^2/\lambda L$, and a_1 = radius of the small aperture.

In all unstable resonators there are one or more limiting apertures, either on the mirrors or internally, which the beam completely fills and which determine the transverse mode size and shape. Note that the mode can be made to fill a rectangular or square cross section gain medium by using appropriately shaped apertures. The beam size and shape can be found by drawing the limiting rays from the two virtual point sources. In all cases the total percentage power coupled out using the geometrical theory is given by

$$\delta_{TOT} \doteq 1 - \frac{1}{(M_{12}M_{21})^2},$$

(7)

which is fixed by R_1, R_2 and L. However, the distribution of coupling around the various apertures depends on the aperture sizes and positions.

It is also a property of unstable resonator modes that the two oppositely traveling waves have different transverse sizes and different radii of phase fronts at any point in the cavity. This fact must be considered when using internal cavity elements such as modulators, etalons or gratings and is contrary to stable cavity modes where the two waves are identical at every point in the cavity.

Since the amount of power intercepted by the apertures is substantial, this is the normal method of coupling out of the unstable resonator rather than using partially transmitting mirrors as is normally done in stable cavities. Several methods have been used for supporting the mirrors to accommodate the power flowing around the mirrors as shown in fig. 2. Perhaps the most effective is the scraper mirror (Krupke and Sooy 1969) in fig. 2c where the output is conveniently directed sidewise out of the cavity; only reflective optics is required, and there is no beam obscuration.

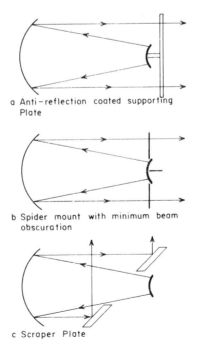

a Anti-reflection coated supporting Plate

b Spider mount with minimum beam obscuration

c Scraper Plate

Fig. 2. Methods of coupling power from the unstable resonator.

3. Wave analysis of the empty cavity—the sharp-edged case

The modes of the empty unstable resonator with diffraction included are found by solving an integral equation derived from Kirchhoff's equation of the general form

$$\gamma_n U_n(x, y) = \int\!\int K(x, y; x', y') U_n(x', y') \, dx' \, dy', \tag{8}$$

where the kernel $K(x, y; x', y')$ accounts for the wave propagation through one round trip in the resonator, including mirror reflectivity. The approach is the same as for the stable resonators, where one searches for eigenfunctions U_n which repeat themselves in each round trip. The diffraction losses which are determined by γ_n are substantially higher in the unstable resonator than in the stable case. Closed form solutions are in general not known, and the solution requires extensive computer calculations using the techniques developed by Fox and Li (1961). Siegman and Miller (1970) introduced the Prony technique to the solution of unstable resonators; this approach can give the losses and mode patterns for several of the modes simultaneously. Recently, the overmoded waveguide approach first used by Weinstein (1969) for stable resonators has been used by Chen and Felson (1973) to analyze unstable resonators.

The results of several computer analyses have been published for the case where the edge of the limiting aperture of the system is sharply defined within the grid limits of the analysis (Seigman and Arrathon 1967, Sanderson and Streifer 1969, Siegman and Miller 1970, Rensch and Chester 1973, Horwitz 1976). As we shall see, this sharp-edge approximation is important in determining the details of the mode patterns and in high Fresnel number systems the sharp-edge approximation may not be realized in practice.

In general, these analyses show the lowest-order mode patterns (maximum γ), with considerable amplitude structure superimposed on the uniform geometrical solution, and a phase front which closely follows the geometrical solution. The number of rings in the amplitude pattern increases, and the height of the rings decreases as the effective Fresnel number of the system increases. There is also a tapering at the edges of the mode. As expected, at high Fresnel numbers the computer solution more closely follows the uniform geometrical solution.

Some typical results from the work of Rensch and Chester (1973), who analyzed a 2-D resonator (i.e., the mirrors are infinite in one transverse direction), are shown in figs. 3 and 4. The results are for a confocal resonator; the upper two parts of the figures show the near-field intensity and phase at the output mirror. The vertical lines mark the edges of the output mirror. The third part shows the far-field intensity as a function of the far-field angle, and the lower part shows the total integrated power within a given beam divergence angle. Fig. 3 is for a resonator with $M = 5$ and $N_{eq} = 0.4$, and fig. 4 for $M = 5$ and $N_{eq} = 4.8$.

It has been shown (Steier and McAllister 1975) that the general shape of the mode for circular apertures can be predicted from the effective Fresnel number

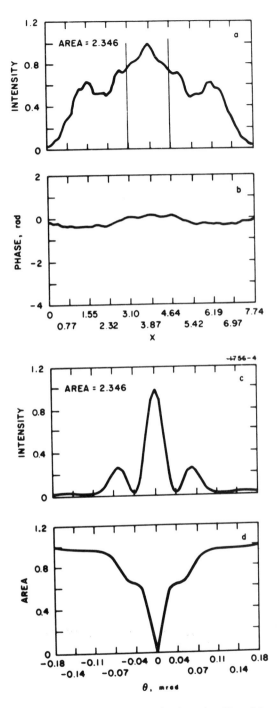

Fig. 3. Normalized near- and far-field intensity distributions for $N_{eq} = 0.4$ and $M = 5$, taken from Rensch and Chester (1973). See text for a description of the various parts.

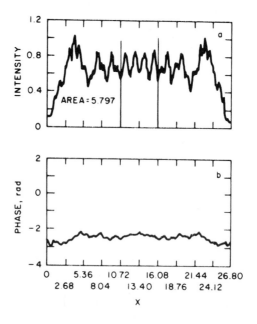

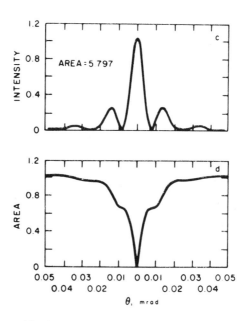

Fig. 4. Normalized near- and far-field intensity distributions for $N_{eq} = 4.8$, $M = 5$, taken from Rensch and Chester (1973). See text for a description of the various parts.

(N_{eff}), which is defined as the Fresnel number of the cavity aperture uniformly illuminated by a point source at the virtual focus. If N_{eff} is odd, there is a relative peak on center; if N_{eff} is even, there is a relative null on center; the number of rings is $(N_{eff} - 1)/2$. For estimating patterns at either mirror in a symmetric resonator

$$N_{eff} = \frac{a^2}{\lambda L} M. \tag{9}$$

For estimating patterns at the output (small) mirror of a confocal resonator

$$N_{eff} = \frac{a_1^2}{\lambda L} \frac{M^2}{M + 1}. \tag{10}$$

In both cases

$$N_{eff} = 2N_{eq} \frac{M^2}{M^2 - 1}, \tag{11}$$

which agrees with the observations of Sziklas and Siegman (1975) on the mode patterns when M is reasonably large.

The lowest-order mode loss $(1 - \gamma_n^2)$ has been shown (Siegman and Arrathon 1967) to oscillate as a function of N_{eq} and to always be less than or equal to the geometrical value. The loss minima occur when $2N_{eq}$ is odd. It has also been shown that this oscillation is due to a periodic crossing of mode losses with a new mode becoming the lowest loss mode at each crossing. For circular sharp-edged apertures this mode loss crossing and interleaving is believed to continue to very high Fresnel numbers. Fig. 5 shows the magnitude and phase of several of the symmetric mode eigenvalues as a function of N_{eq} for an unstable resonator with $M = 5$. The geometrical optics prediction of $|\gamma|$ is $1/M$ or 0.2.

Siegman (1971) has computed the power output coupling at the loss minima and shows that

$$\delta \approx \left(1 - \frac{1}{M^2}\right)^2, \tag{12}$$

for the minima at $2N_{eq}$ of five or larger. Thus for this case of circular sharp-edged mirrors, the coupling is the square of the geometrical value, which demonstrates that diffraction tends to confine the energy and reduce the output coupling.

For large $N_{eq}(>50)$, for noncircular apertures, or for tapered apertures, the effects of diffraction are not as important (see §4), and the resonator can be designed using the geometrical output coupling formula in table 1. Since the power output of a laser is a relatively slowly varying function of output coupling near the optimum coupling, any changes in coupling due to diffraction will not have a large effect on the laser output power.

The analysis does yield higher-order transverse modes with higher diffraction losses. In stable resonators it is not desirable to have a combination of higher-order modes because the entire output aperture is not diffraction-limited. It is

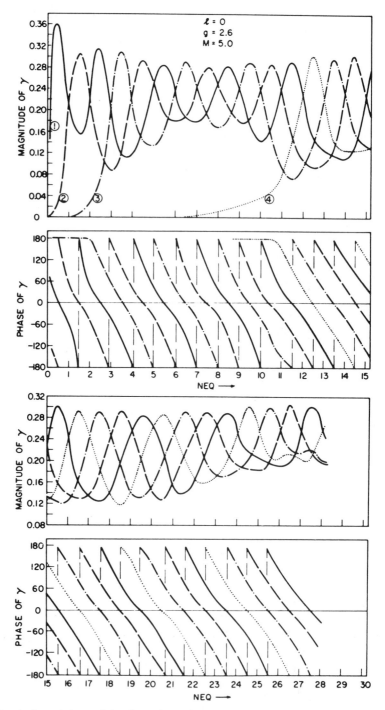

Fig. 5. Magnitude and phase of the eigenvalues of the axially symmetric modes of a circular mirror symmetric unstable resonator with $M = 5$ as a function of N_{eq}. From Siegman and Miller (1970).

not clear if a combination of higher-order unstable resonator modes produces a less-than-diffraction-limited output. Therefore, in practice no attempt is made to control the transverse mode content of the beam.

Siegman (1974) has pointed out another interesting feature of the transverse mode structure which is due to the kernel $K(x, y; y', y')$ being symmetric but not hermitian, and the high loss nature of the modes. The orthogonality relationship is of the form $\iint U_n U_m \, dx \, dy = \delta_{nm}$ and not the usual $\iint U_n^* U_m \, dx \, dy = \delta_{nm}$. This means that in a multi-transverse mode beam, the power cannot be considered as the sum of powers in the individual modes, but some power is carried in the cross-product terms between the modes. The implications of this, if any, are not clear.

Krupke and Sooy (1969), Rensch and Chester (1973) and Horwitz (1976) have also considered the sensitivity of the empty unstable resonator to small mirror misalignment. This misalignment results in the power being coupled more strongly off one side of the coupling mirror and produces an output with a tilted phase front. This results in beam steering in the far field and a fluctuation in output coupling. The steering is similar to that predicted by a geometrical analysis, but the output coupling fluctuations are not simple to predict. The positive branch confocal resonator is more sensitive to mirror misalignment than the negative branch. For small tilt angles, the calculations show that single mode operation is maintained. This has been confirmed in practice where the output beam quality of unstable resonators has not been found to be overly sensitive to mirror alignment. §6 contains some discussion of the effects of mirror tilt on loaded resonators.

4. Far-field pattern of the output

The geometrical fundamental mode near field output for the confocal unstable resonator is a uniform collimated annulus with the ratio between outer and inner diameters given by M. The near-field patterns from the computer solutions show very nearly a collimated phase front and intensity rings superimposed on the annulus. It is important to consider the effect on the far-field intensity that is caused by the annular pattern, the rings and the slight phase deviations.

The far-field intensity pattern $I(\psi)$ (Krupke and Sooy 1969) of a collimated uniform annulus of inner radius ρ_1 and outer radius ρ_2 is

$$I(\psi) = \frac{4I}{\lambda^2 R^2} (\pi \rho_2^2)^2 \left[\frac{J_1(z_2)}{z_2} - \frac{1}{M^2} \frac{J_1(z_1)}{z_1} \right]^2, \tag{13}$$

where I is the intensity in the annulus, R is the distance from the observation plane to the annulus,

$$z_i = \frac{2\pi \psi}{\lambda} \rho_i, \tag{14}$$

and ψ is the angular coordinate measured from the axis of the annulus. The pattern has circular symmetry. It is known that the on-axis intensity for an annulus is reduced by $(1 - 1/M^2)$ from that of a uniform annulus (Siegman and Arrathon 1967). Since the output coupling is also $(1 - 1/M^2)$, it is clear why unstable resonators are only useful in systems which require large output coupling, i.e., 50% or greater. If the laser system is low gain and hence only low output coupling is possible, the output annulus is a narrow ring and much of the far-field energy falls in the side lobes.

The far-field pattern has been measured and agrees well with that of a uniform annulus, even though the near-field intensity pattern shows rings. This is to be expected, since the near-field phase is very close to the plane, and the depth of the ripples is not large. Additional near field diffraction rings are also observed owing to diffraction around the output mirror, but again these seem to have little effect on the far field. Some typical experimental far-field patterns and a comparison to the theory are shown in figs. 6 and 7. Fig. 6 is for a relatively quiet longitudinal discharge CO_2 laser (Krupke and Sooy 1969) and fig. 7 is for a flowing HF laser (Chodzko et al. 1973).

5. Edge-wave analysis—tapered apertures

The deviations of the empty cavity modes from the uniform spherical wave modes of the geometrical analysis can be considered to be due to the diffraction effects at the edges of the mirrors or apertures. The diffraction of light by an aperture can be conveniently described using the concept of an edge wave emanating from the aperture edge. This concept can be obtained from the standard scalar diffraction theory as described by Born and Wolf (1964).

In brief, the edge-wave formalism is obtained by expressing the field amplitude resulting from an illuminated aperture as the sum of a geometrical optics

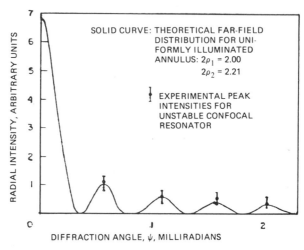

Fig. 6. Theoretical and experimental far-field intensity distributions. From Krupke and Sooy (1969).

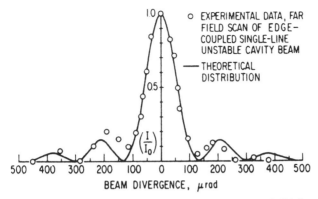

Fig. 7. Far-field intensity distribution of a flowing HF chemical laser. Solid line is the theoretical curve for a uniformly illuminated annulus of 0.94 cm outer radius, 0.66 cm inner radius and a 2.795 μm [$P_2(5)$] wavelength. From Chodzko et al. (1973).

term plus a diffraction term, which is expressed as a line integral around the aperture boundary. The diffracted wave is emitted in all directions.

In the case of the unstable resonator, as Ananév (1972) has pointed out, only that ray which is directed back into the cavity in such a way that it is trapped along the resonator axis should be important in determining the properties of the modes. This is illustrated for a symmetric unstable resonator in fig. 8. The solid lines represent the geometrical rays of the resonator mode that are incident on the sharp edge of the mirror, and the dashed lines represent rays of the resulting sharp edge of the mirror, and the dashed lines represent rays of the resulting edge wave. It is clear that the diffracted rays that are scattered in a direction exactly opposite to that of the ray incident on the mirror edge will, in the geometrical optics picture, be trapped along the resonator axis, whereas the other diffracted waves will be deflected out of the cavity after a few reflections. These rays that are trapped in the geometrical optics picture will, of course, diverge due to diffraction and thus act in part as the source for the mode.

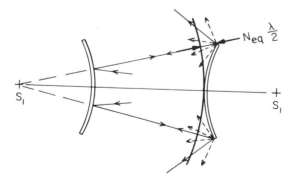

Fig. 8. Symmetric unstable resonator showing the edge-diffracted waves (dashed lines) and the physical significance of N_{eq}.

Thus the edge wave can be considered the feedback mechanism for the unstable resonator. This concept can be useful in gaining a physical insight into the properties of these resonators. For example, the periodicity of the loss with a period of one in N_{eq} can be explained as follows. Ananév and Sherstobitov (1971) have shown that N_{eq} is the number of half-wavelengths between the phase front of the diverging mode and the edge of the mirror, as shown in fig. 8. The diffracted edge ray which is refocused onto the axis is thus $2N_{eq}(\lambda/2)$ out of phase with the on-axis ray of the mode. As this edge ray is repeatedly reflected and focused to the axis, it remains at this phase relative to the center ray. Thus, the feedback interferes with the core ray with a period of one in N_{eq}, and we would expect this periodicity in the mode properties. Why the loss is minimum when the feedback interferes destructively is not clear.

It is obvious from symmetry that the rays from the sharp edge of a *circular* mirror will all arrive on axis in phase. This produces a large diffractive contribution, resulting in modes with large amplitude variations across the radius, large phase deviations from the geometrical optics value and mode losses that vary rapidly with the resonator parameters. It is also clear that altering the shape of the aperture or tapering the edge such that it is no longer sharp can result in destructive interference at the focus, reducing the diffractive contribution to the modes. Ananév and Sherstobitov (1971) and Ananév (1972) have suggested that these destructive interference effects can be introduced by smoothly rolling off the mirror edges, roughing the edges, misalignment or positioning the mirror off center.

Sherstobitov and Vinokurov (1972) and McAllister et al. (1974) have analyzed tapered and shaped apertures using the edge-wave approach and have found that the mode profiles are smoothed both in amplitude and phase and that the loss is more nearly the geometrical loss. The degree of mode improvement is dependent on the exact nature of the tapering (i.e., linear, parabolic, etc.). Typical results are shown in figs. 9 and 10, where the amplitude and phase of the fundamental mode of a symmetric resonator are shown for two tapered apertures and for a sharp aperture. The phase is referred to the geometric phase front. (Fig. 14 in §6 also demonstrates the mode smoothing of a tapered aperture.)

The radius over which the taper extends is critical and brings out an important point for high Fresnel number resonators. Fig. 11 shows the coupling loss for a symmetric resonator with a tapered aperture as a function of the normalized radius at which the taper begins. If the taper extends inward to a radius where the N_{eq} has changed by 0.5 from the N_{eq} at the edge, the coupling loss is very close to the geometrical value. N_{eq} as a function of radius ρ is

$$N_{eq}(\rho) = \frac{\rho^2}{2L\lambda}\left(M - \frac{1}{M}\right). \tag{15}$$

If the tapering extends over a larger area, the loss oscillates but approaches the geometrical loss. At all points in the taper an edge ray is produced whose amplitude depends on the rate of taper and whose phase depends on $N_{eq}(\rho)$. Thus, by correctly tapering the edge over the correct region, the rays will

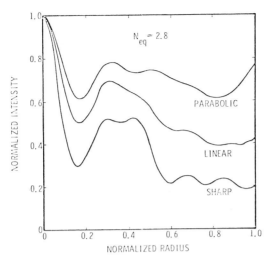

Fig. 9. Normalized mode intensity profiles for symmetric resonators with tapered reflectivity mirrors. $N_{eq} = 2.8$. From McAllister et al. (1974).

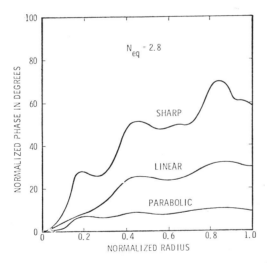

Fig. 10. Normalized mode phase profiles for symmetric unstable resonators with tapered reflectivity mirrors. $N_{eq} = 2.8$. From McAllister et al. (1974).

interfere destructively and the total edge wave can be made zero or very small.

Amplitude tapering of an aperture can be achieved by tapering the dielectric or metallic coatings on the mirrors. Phase tapering can be achieved by rolling off the edges of the mirrors. A mirror with a tapered reflectivity and transmissivity leads to the interesting and possibly useful concept of a combination of diffraction coupling and transmission coupling from the unstable resonator. The diffraction rings in the near field due to the mirror edge may be objectionable in some cases; for example, the damage threshold would be reduced when the

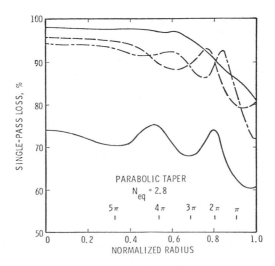

Fig. 11. Single-pass loss for the symmetric ($l = 0$) modes of a symmetric unstable resonator with parabolic amplitude reflectivity tapered mirrors as a function of the normalized radius at which the taper begins. $N_{eq} = 2.8$. Upper curves are for the higher loss symmetric ($l = 1$) modes. From McAllister et al. (1974).

output is used as the pump wave in a nonlinear process. The combination coupling can reduce the amplitude of the rings, but also requires a transparent mirror substrate which can withstand the intensities involved.

From this approach it can also be seen that a mirror with a scalloped edge, or a correctly shaped aperture, will produce a reduced edge wave because of the destructive interference of the rays from the various points on the aperture as the rays are focused to the center. It is also clear why the circular sharp-edged mirror is unique in producing the maximum edge wave on-axis. Indeed, any aperture, i.e., elliptic, square, rectangular, will have significantly less diffraction effects than the circular aperture and the coupling loss will be closer to the geometrical value. Even for a circular aperture, slight misalignment of the mirrors will reduce the total edge-wave effect and change the coupling.

The results shown in fig. 11 agree with those of Ananév and Sherstobitov (1971) which show that the desired taper width is $a/2N_{eq}$, where a is the aperture radius. If N_{eq} at the mirror edge is large, this corresponds to extending the taper over a zone where N_{eq} changes by 0.5.

This leads us to an important point in large N_{eq} systems. For example, if we consider a 3 cm aperture and $N_{eq} = 100$, the aperture edge must be defined to within 1.5×10^{-2} cm for the mode properties to be those of the sharp-edge calculations. Thus, a slightly roughened aperture or slight index inhomogeneities in the gain medium due to shocks or turbulence can reduce the diffraction effects and give a value of the output coupling very near the geometrical value. It thus appears that the geometrical coupling formula (see table 1) is more accurate when N_{eq} is large or when the aperture is noncircular.

6. Unstable resonators filled with an active medium

Any real laser gain medium will show saturation and gain inhomogeneities and in many high power systems will show considerable index inhomogeneities (Clark 1972). In several high power systems the medium is flowing at supersonic velocities, and the medium has index inhomogeneities owing to shock waves. The effects of this saturation and inhomogeneity are not included in the empty cavity analysis and can have a significant effect on the output beam quality of the unstable resonator. Indeed, the problem of analyzing resonators including an active medium and the development of design procedures for these lasers is the major problem remaining in unstable resonators.

The gain saturation tends to smooth the mode profiles and a nonuniform gain due to flow, for example, will tend to distort the beam. The most serious effect is due to index inhomogeneities, since this distorts the output phase profile which reduces the far-field intensity, and at the same time increases the peak intensity at the mirrors and increases the mirror loading. It is known that reducing the cavity magnification M causes an individual ray to remain longer in the cavity, and hence to increase the output phase distortion for a given index inhomogeneity. This can result in situations where the cavity output coupling which results in the maximum output power does not result in the maximum far-field intensity, because of the degradation of the beam quality (Rensch 1974). A design procedure for selecting M and hence the output coupling which does not involve huge amounts of computer time is needed.

In the first published work on loaded resonators (Rensch and Chester 1973), the effects of the medium were confined to a thin sheet at one of the mirrors. This analysis is not sufficient to account for the distributed effects which occur in real lasers. Recently, three techniques for computer analysis of distributed media have been published (Rensch 1974, Siegman and Sziklas 1974, Sziklas and Siegman 1975). In each of them the medium is axially divided into segments in which the gain and index are axially uniform but allowed to vary in the transverse direction; see fig. 12. The computer then propagates a beam across a segment as if in free space, adjusts the resultant E field by the gain and index of

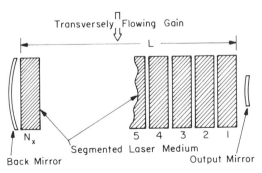

Fig. 12. Schematic showing the laser medium segmented into axially uniform parts for the analysis of loaded unstable resonators.

that segment and propagates through the next segment. The field is thus propagated through the resonator, reflected from the mirror and propagated back through the medium. Gain saturation for a segment is calculated from the sum of the right- and left-hand traveling wave intensities at that segment. If the wave converges to a steady-state solution, that distribution is considered the lowest-order mode of the laser.

The differences between the three papers lies in their technique for free space propagation across the individual segment. Rensch (1974) uses a finite difference algorithm, Siegman and Sziklas use a hermite–gaussian expansion in their first paper (Siegman and Sziklas 1974) and a fast Fourier transform (FTT) method in their second paper (Sziklas and Siegman 1975). In general, the FFT algorithm is preferred for small or moderate Fresnel number cavities, and the finite difference algorithm is preferred for large Fresnel number cavities.

The finite difference algorithm is derived from the slowly varying approximations to the scalar wave equation. To reduce the density of the points in the transverse plane, a variable coordinate system is used to account for the geometrical expansion of the beam. The finite difference approach cannot handle undetermined partial derivatives, and hence a tapered reflectivity mirror is required. The taper was assumed gaussian and extended over 5 to 10% of the diameter. The mode of the *empty cavity resonator* clearly demonstrates the unique properties of the unstable resonator; a typical result is shown in fig. 13 for $N_{eq} = 7.2$. The mode has the square cross section of the mirrors; the mirror reflectivity is tapered, and hence there is no fine structure in the near-field intensity; the near-field phase is relatively flat, and the far-field intensity is a diffraction-limited, well-focused beam. Fig. 14 is a cross section of the near field amplitude for the sharp-edged and tapered apertures, and clearly illustrates the smoothing effect of the taper.

Rensch (1974) considered the flowing CO_2–N_2 laser. His model used four vibrational levels and included gain saturation and density variations due to shockwaves. The results clearly show the necessity for including medium effects in designing the unstable resonator. As shown in fig. 15, the near-field intensity is distorted in the flow direction owing to gain inhomogeneities; the near-field phase is not uniform, and the far-field intensity is severely reduced owing to medium shockwaves. In one example, when the output coupling was decreased from 60 to 40%, the power-out increased from 30 to 48 kW, while the far-field intensity remained unchanged. The reduced coupling magnified the effect of the medium shocks, which degraded the beam quality, causing the increased power output to go into the far-field sidelobes. The higher power-out and enhanced amplitude ripples at the 40% coupling increased the mirror loading by a factor of three. This example demonstrates that when medium inhomogeneities are included, the maximum power-out may not be the optimum operating point.

Computer time is an important consideration if the programs are to be used for design. The finite difference approach typically converged in six to seven round trips, and the computer time decreased for increasing N_{eq}. The computer time decreased by a factor of four when the N_{eq} was increased from 1.25 to 7.2.

NEAR FIELD INTENSITY

EMPTY CAVITY
$N_0 = 60$ ($N_{eq} = 7.2$)
$M = 2.5$ (84 %)

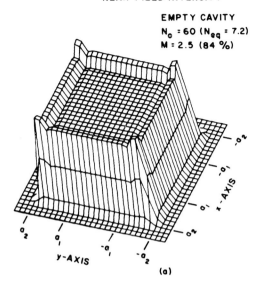

(a)

NEAR FIELD PHASE

EMPTY CAVITY
$N_0 = 60$ ($N_{eq} = 7.2$)
$M = 2.5$ (84 %)

FAR FIELD INTENSITY

EMPTY CAVITY
$N_0 = 60$ ($N_{eq} = 7.2$)
$M = 2.5$ (84 %)

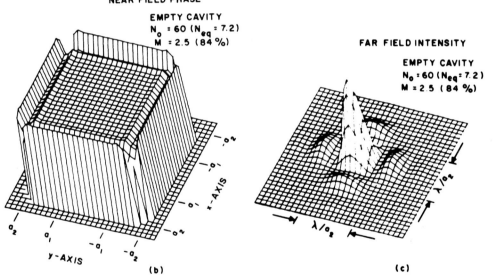

(b)   (c)

Fig. 13. Near- and far-field distributions for an empty, three-dimensional, unstable resonator. From Rensch (1974).

The first result reported by Siegman and Sziklas (1974) used the hermite–gaussian expansion for free space propagation across the axial segments of the resonator. The medium was axially segmented as shown in fig. 12. The hermite–gaussian modes are the free space modes in a Cartesian system and their individual propagation laws are well known (Kogelmk and Li 1966). Their use

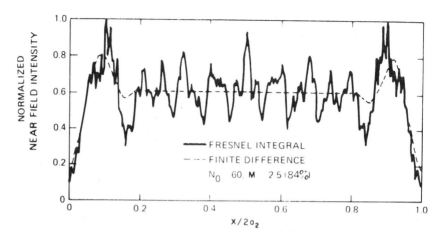

Fig. 14. Comparison of the near-field intensity distribution for sharp-edged (solid line) and the tapered-edged (dashed line) mirrors. $N_{eq} = 7.2$, $M = 2.5$. From Rensch (1974).

has some computational advantages over the Huygens integral. As pointed out, an important consideration in this approach is to choose correctly the family of hermite–gaussian modes (i.e., the confocal parameter) to give the maximum accuracy with the least number of terms.

The results reported for the empty cavity confirm earlier calculations and attest to the accuracy of the method. The gain media model was for the flowing CO_2–N_2 system including shock waves, and the loaded cavity results show the near-field intensity distortion due to the flowing medium and the far-field intensity reduction due to the index inhomogeneities. Results are presented for N_{eq} of 0.5 and 1.5, and the authors point out that in practice this approach is limited to these low Fresnel number systems, and even then it can require considerable computer time.

Sziklas and Siegman (1975) show that the FFT approach is more efficient and reduces the computer time required in some cases For an $N \times N$ grid of transverse data points, both the finite difference algorithm and the hermite–gaussian expansion require a computer effort proportional to N^4. The FFT approach in the same situation requires only $N^2 \log_2 N$ of computer effort. Mode patterns using the FFT program show more mode detail than the hermite–gaussian approach for similar resonators. The FFT method required approximately 30 s per iteration for a loaded resonator and converged in 15 iterations using the IBM 370 Model 168 computer. This compares to the hermite–gaussian approach, which required 1.5 to 2 min per iteration for a coarser grid spacing and hence less mode detail.

A series of calculations is presented for $N_{eq} = 1.5$ which demonstrated the effect of a medium shock wave of various intensities. A shock density strength parameter is defined which is proportional to the single path phase shift through the shock. The parameter equals 1 when the phase shift for one pass for an on-axis ray is 0.2 λ. The near-field amplitude and phase for a density parameter

NEAR FIELD INTENSITY

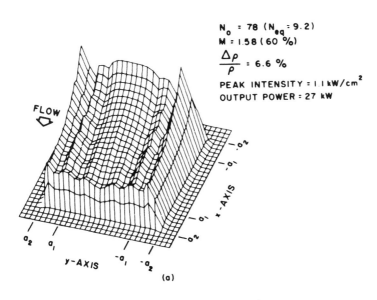

$N_0 = 78 \ (N_{eq} = 9.2)$
$M = 1.58 \ (60 \ \%)$

$\dfrac{\Delta\rho}{\rho} = 6.6 \ \%$

PEAK INTENSITY = 1.1 kW/cm^2
OUTPUT POWER = 27 kW

(a)

NEAR FIELD PHASE

$N_0 = 78 \ (N_{eq} = 9.2)$
$M = 1.58 \ (60 \ \%)$

$\dfrac{\Delta\rho}{\rho} = 6.6 \ \%$

FAR FIELD INTENSIY

$N_0 = 78 \ (N_{eq} = 9.2)$
$M = 1.58 \ (60 \ \%)$

$\dfrac{\Delta\rho}{\rho} = 6.6 \ \%$

PEAK INTENSITY = 700 kW/mrad^2

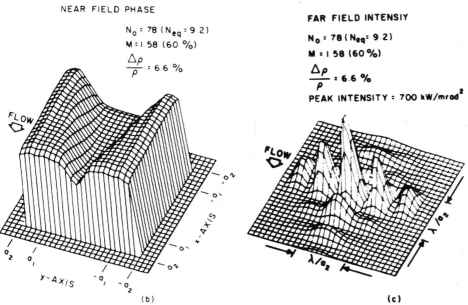

(b)

(c)

Fig. 15. Computed near- and far-field distributions for a CO_2 GDL with shock waves. From Rensch (1974).

of one is shown in fig. 16. The effect on the far-field intensity of various shock intensities is shown in table 2. As the shock density parameter was increased from 0 to 1.5, the power output fell only 6 to 7%, but the peak mirror loading more than doubled. At the same time, the far-field intensity fell to only 15% of the bare resonator value. The far-field intensity ratio i_p is defined as the peak far-field intensity normalized to the peak far-field intensity for the same output power uniformly distributed over the output aperture. Siegman's recent work (Siegman 1977) provides a good physical insight into these results.

Some results are also presented of an attempt to correct for the near-field intensity distortion due to the flowing gain. In general, the near-field output is always larger on the upstream side, and it was thought that a tilt of the mirrors in the direction of the flow may correct for this. The bare resonator results for $N_{eq} = 1.5$ showed a significant decrease in the far-field intensity for tilts greater than 50 μ rad and demonstrated the sensitivity of the positive branch confocal resonator to mirror misalignment, as pointed out by Krupke and Sooy (1969). For the loaded resonator, small tilts of 50 μ rad or less did show a slight increase in the far-field intensity.

An entirely different approach to the analysis of loaded resonators was recently reported by Santana and Felsen (1977), who extended the over-moded waveguide analysis to include the effects of medium index and gain variations.

Karamzin and Konev (1975) have reported a two-dimensional analysis including a saturated flowing gain medium. Their approach uses the slowly varying approximation to the scalar wave equation and a difference equation method of solution. The method appears similar to that of Rensch (1974). No index inhomogeneities were included, but a simple two-level distributed model for the gain medium was included to show the effects of homogeneous saturation. Their results are similar to those discussed earlier.

Based on the analyses for loaded resonators just discussed, it is clear that the medium effects must be included in the design of unstable resonators for high power systems. However, it is also clear that the three approaches require considerable computer time and expense if they are used in the design process of finding the optimum operating conditions. A ray analysis would not be as accurate as the wave approach just discussed, but may use considerably less computer time and may provide a sufficiently accurate description of the medium effects to use in design calculations. We might expect that, when the size of the inhomogeneities is larger than the first Fresnel zone, a ray approach would be justified.

Ananév (1972) and Ananév et al. (1974) have developed a ray technique for including weak index inhomogeneities and gain saturation. It is assumed in this approach that the ray paths are the same in the loaded resonator as in the unloaded resonator, but the phase and amplitude of the rays are calculated with the medium included. That is, the refraction or bending of the rays due to index inhomogeneities is neglected by assuming the inhomogeneities are weak.

The radiation flux $\rho(r)$ at a given distance r from the axis of the resonator can

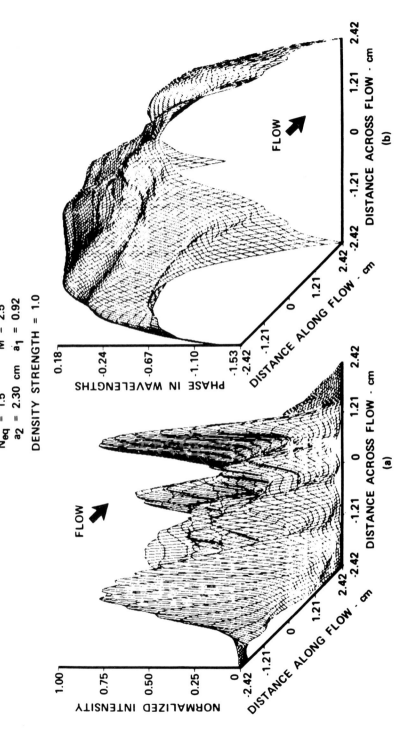

Fig. 16. Computed phase and intensity distributions just inside the output mirror of a confocal loaded unstable resonator. $N_{eq} = 1.5$, $M = 2.5$, normalized shock density strength parameter equal to 1.0. From Sziklas and Siegman (1975).

Table 2.
Calculated results using the FFT method for a loaded $N_{eq} = 1.5$, $M = 2.5$ resonator in which the
normalized strength of the index perturbation due to sidewall shocks was systematically varied. (From
Sziklas and Siegman, 1975).

	Density Strength					
	0	0.25	0.5	1.0	1.5	
Peak internal intensity I_p	11.9	12.5	13.9	19.6	25.5	kW/cm²
Internal power P_{int}	126	123	125	121	116	kW
Reflected power P_{refl}	18.3	17.9	17.2	15.4	13.6	kW
Output coupling	0.852	0.853	0.861	0.872	0.882	
Output power P^a	93.6	91.4	93.6	91.1	87.5	kW
Peak far-field intensity ratio i_p	0.78	0.73	0.58	0.21	0.12	

[a] Transmitted by an aperture of o.d. $D = 4.8$ cm.

be written in terms of the flux on-axis, $\rho(0)$, as

$$\rho(r) = F(r)F\left(\frac{r}{M}\right)F\left(\frac{r}{M^2}\right)\cdots\rho(0), \tag{16}$$

where the function $F(r)$ describes the propagation of the ray from r/M on one
mirror to r on the opposite mirror:

$$F(r) \equiv \frac{R'}{M^2}\exp\left[\int_l k(s)\,ds\right]. \tag{17}$$

The factor $1/M^2$ accounts for the geometrical expansion of the beam during
one pass, and R' is the mirror reflectivity. The medium is described by $k(s)$,
which, if complex, can account for gain and index variations. The ray starts
on-axis with flux density $\rho(0)$, and follows the geometrical path to the point r.
The limiting condition which requires that the gain equal the loss for the on-axis
ray is

$$\lim_{r\to 0} F(r) = F(0) = 1. \tag{18}$$

In Ananév et al. (1974, 1976), the output intensity distribution is computed
using this approach for an homogeneous saturable gain in which the pumping
varies spatially, or the gain medium is flowing. The gain is saturated by the sum
of the fluxes in the two traveling waves. The on-axis intensity is found by an
iteration process, and the intensity distribution is then calculated from eq. 16.
Mirels and Batdorf (1972) have also developed an approach for finding the
on-axis intensity which could be used as a starting point in eq. 16.

Also in Ananév et al. (1974, 1976) the efficiency of a flowing CO_2–N_2 laser is
calculated. The efficiency is defined as the ratio of the number of photons in the
output to the number of excited molecules at the entrance. The goal is to find a

method for designing an unstable resonator for a flowing CW GDL to assure high efficiency and a more uniform output.

Ananév (1972) and Vasilev et al. (1975) use a similar ray approach to calculate the output phase distribution due to weak index inhomogeneities. The examples are for a relatively slowly radially varying index of refraction somewhat different from a typical shock wave inhomogeneity. They show that for the ray method to apply

$$\Delta(r) \ll N_{eq} \pi \left(\frac{r^2}{a^2}\right),$$ (19)

where $\Delta(r)$ is the accumulated phase error at r and a is the radius of the cavity aperture. This condition assures that the geometrical spreading of the beam is much larger than the spreading due to the phase distortions. This expression shows that the ray theory is applicable when N_{eq} is large, which means a high Fresnel number and/or high magnification. It is known that any medium inhomogeneities can seriously affect the output beam quality of a resonator approaching plane parallel ($M \approx 1$, $N_{eq} \approx 0$) since there is little geometrical spreading due to the cavity mirrors.

In one example, Ananév (1972) calculates that the output phase distortion for a cavity with $M = 2$ was 3.5 times larger than the phase shift of a single pass through the index inhomogeneity. This appears consistent with a result of Rensch (1974) using the wave approach; when $M = 1.58$, the output phase distrotion was 3.6 times greater than the phase shift of a single pass through the shock.

Some recent work by Mirels (1976) and by Dreizin and Dykhne (1974) demonstrates the capability of a ray analysis in explaining effects seen in flowing high energy lasers. Yoder and Ahouse (1975) reported the observation of a pulsed output from a cw flowing CO_2 electron-beam-sustainer electric discharge laser using an unstable resonator. This self-oscillation instability is believed to be due to the drop of the on-axis gain below threshold owing to saturation and a subsequent recovery of the gain above threshold as new lasing medium flows into the cavity. Dreizin and Dykhne (1974) published a ray analysis similar in approach to those described above which assumed an excited medium flowing transversely into an unstable resonator but assumed no pumping in the cavity region. The results showed a pulsing behavior under certain operating conditions. Mirels' (1976) analysis included pumping in the cavity region and showed that the output fluctuation amplitude decreased with decreased cavity magnification M and increased pumping rate. These results are consistent with the experimental observations.

7. Unstable ring resonators

The unstable resonator can be made into a ring resonator in the same way as the stable ring resonator. However, the unstable ring resonator has some unique and useful properties (Freiberg et al. 1973).

For example, the unstable ring resonator can support two different intracavity traveling waves of different mode diameters and phase front radii, propagating in opposite directions. One of the traveling waves can be favored by properly placing the gain medium within the ring, so that the mode volume more efficiently overlaps the gain medium. If only one traveling wave is allowed to oscillate, the ring geometry has potential uses as an injection-locked oscillator or as a frequency-controlled or frequency-tuned oscillator. In the case of injection locking, the unidirectional ring requires no isolator to prevent reflection back into the master oscillator. The ring is useful for frequency-controlled lasers since most passive frequency selective elements such as gratings or etalons produce some beam distortion unless illuminated with a collimated beam, and only in the unidirectional ring is this possible.

The confocal asymmetric ring resonator (Freiberg et al. 1973) shown in fig. 17 is of particular interest because for the counter-clockwise wave (CCW), the length of the collimated arm and the mode diameter in that arm can be easily designed to use efficiently the gain medium available. A collimated single-ended output is also possible. Table 3 shows the positions of the virtual centers for the clockwise wave (CW) and the CCW wave, and the magnifications for each wave

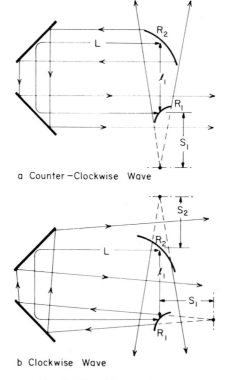

a Counter–Clockwise Wave

b Clockwise Wave

Fig. 17. Unstable ring resonator.

Table 3.
The asymmetric confocal ring resonator.[a]

	Virtual centers		Magnification		
	S_1	S_2	M_{12}	M_{21}	Coupling
CCW	$\dfrac{R_1}{R_2}$	∞	$\dfrac{-R_2}{R_1}$	1	$1 - \left(\dfrac{R_1}{R_2}\right)^2$
CW	$\dfrac{R_1^2 L - R_1 R_2 l_1}{R_2^2 - R_1^2}$	$\dfrac{2l_1^2 - R_2(l_1 + L)}{2(L - l_1)}$	$1 + \dfrac{L(R_2^2 - R_1^2)}{R_1^2 L + R_1 R_2 l_1}$	$1 + \dfrac{2(L - l_1)l_1}{2l_1^2 - R_2(l_1 + L)}$	$1 - \left|\dfrac{R_1}{R_2}\right|^2$

[a] M_{12} = magnification going from mirror 1 to mirror 2.
M_{21} = magnification going from mirror 2 to mirror 1.
Confocal condition $l_1 = \frac{1}{2}(R_1 + R_2)$.

in each arm. Note that the CW wave is not collimated in any portion of the resonator.

Even though the CW and CCW waves have different mode volumes at different points, the geometrical coupling from the cavity for both waves is always equal. Although the total coupling is fixed, as in linear resonators, by the magnification, the distributions of the power out around the various mirrors are different for each wave and can be calculated using ray theory and the positions of the virtual centers.

As in linear resonators, diffraction effects will reduce the coupling when N_{eq} is small and the coupling will vary repetitively as N_{eq} is increased. Freiberg et al. (1974) have shown that N_{eq} depends on the position of the coupling aperture within the ring, and this effect can be used to vary the output coupling when N_{eq} is small.

Freiberg et al. (1974) have shown experimentally that the mode volume effect can be used to discriminate against one wave but not to suppress it completely. They also found that by externally reflecting the output from the undesired wave back into the desired wave, almost complete suppression is possible. The reflected energy must have a phase front radius which is close to that of the desired mode, although not critically so. The frequency and phase of the reflected energy must match those of the desired wave, at least on average. In practice, this was achieved by a piezoelectrically-driven mirror which periodically swept the cavity length through a full wavelength.

8. Frequency control and line selection in unstable resonators

In many of the high power molecular lasers (CO, CO_2, HF–DF), it is necessary for the output to consist of only one, or at most, a selected manifold of the many possible laser transitions. This may be necessary for optimum transmission through the atmosphere or for applications where a selected absorption is desired. The eximer uv lasers (KrF, XeF) and the high pressure CO or CO_2

lasers have sufficient linewidth to be considered for tunable sources. Both the tunability and the line selection require an intracavity frequency selecting element. In placing such an element in an unstable cavity, some care must be taken because the element may introduce astigmatism and because the two internal traveling waves do not have the same divergence at each point in the cavity.

The blazed grating has been used extensively for line selection in stable cavities. The grating is used in the Littrow configuration (the first order reflects back into the incidence beam) to replace one of the cavity mirrors. Any line with sufficient gain can be selected by rotating the grating to the proper angle. Multiple line selection has been achieved by using the grating in the off-Littrow configuration and using several different end mirrors; one for each line.

Recently, Chodzko et al. (1973) and Chodzko (1974) have considered line selection of the HF–DF laser in an unstable cavity using a blazed grating. Fig. 18 shows the Littrow configuration for single line selection for edge coupling using a scraper and for continuous coupling using the zero order of the grating. Fig. 19 shows the off-Littrow configuration for multiple line selection for edge coupling and continuous coupling. Using a plane grating, a confocal resonator is only possible in the off-Littrow configuration.

In the edge-coupled case, the zero-order beams represent power loss and hence it is desirable to have the grating efficiency as high as possible. In the continuously-coupled multi-line selected case, the secondary zero-order beams (the specular reflection of the beams returning from the R_2 reflectors) represent

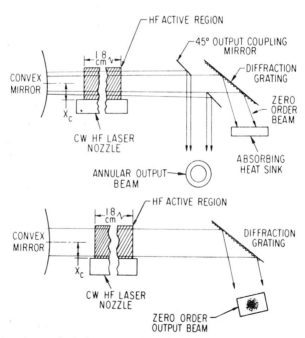

Fig. 18. Schematic of the single-frequency unstable resonator using a grating in the Littrow configuration, (a) edge-coupled, (b) continuously-coupled. From Chodzko et al. (1973).

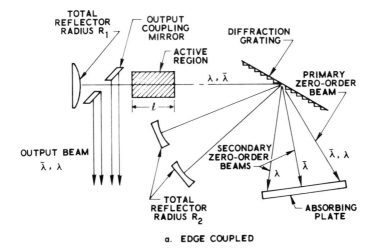

a. EDGE COUPLED

b. CONTINUOUSLY COUPLED

Fig. 19. Schematic of the multi-line selected unstable resonator using a grating in the off-Littrow configuration, (a) edge-coupled, (b) continuously-coupled. From Chodzko (1974).

power loss but the primary zero order is the output coupling. Hence, one must consider what is the optimum grating efficiency to minimize the loss on the secondary zero order but to still give reasonable output in the primary zero order. Chodzko has analyzed this case for both a homogeneous and an inhomogeneous broadened laser line and finds that a rather low grating efficiency (20 to 30%) is appropriate for either case.

The grating in the off-Littrow configuration introduces astigmatism into the beam both in cross section and phase front radius. A spherical beam after reflection into the first order becomes elliptic in cross section with a different phase front radius in the two orthogonal planes.

If

ϕ = angle of the input beam with respect to the grating normal,
ϕ' = angle of the first-order output beam,
a = diameter of the input beam,
a' = diameter of the first-order output beam,
R = radius of the phase front of the input beam in the plane of incidence,
R' = radius of the phase front of the first-order output beam,

then

$$a' = a \frac{\cos \phi'}{\cos \phi}, \tag{20}$$

$$R' = R \left(\frac{\cos \phi'}{\cos \phi} \right)^{2}. \tag{21}$$

This grating distortion causes the output beam from the multi-line selected laser to be elliptic in cross section with a different radius of the phase front in the two orthogonal planes. The amount of distortion, of course, depends on the angle of the grating and the spacing and radii of the cavity mirrors. This astigmatism in the phase front radii should reduce the on-axis far-field intensity. This effect has not been fully analyzed. It can be avoided by operating as close as possible to the Littrow angle or by using grating pairs in a ring resonator configuration (Mann 1977).

Chodzko et al. (1973) and Chodzko (1974) have demonstrated single- and multi-line selection of a cw HF chemical laser using a 600 line per mm grating in an unstable cavity. Near-diffraction-limited output was observed for single-line operation on 10 different lines of HF. Lines at both 2.707 and 2.795 μm were made to oscillate in the multi-line selected laser but with less total power than the free-running laser. The drop in output was due to the additional cavity loss caused by the secondary zero-order beams and the scattering of the grating.

An intracavity etalon could also be considered for frequency narrowing or line selection in an unstable resonator. However, an etalon has a limited aperture for diverging or converging wavefronts, and the beam is collimated in only one direction in a linear unstable resonator. This limited aperture for one of the waves may become a problem in resonators with large magnifications and large cross sections.

9. Power amplifiers and injection-locked oscillators using unstable resonators

In many high energy laser systems it is necessary to use a master oscillator–power amplifier configuration to obtain the maximum energy output, to achieve frequency control, to modulate the output, or to coherently lock together an array of high energy oscillators. To efficiently extract from the amplifier medium requires that the signal inside the amplifier be sufficient to saturate the gain medium. If the amplifier is linear, this may require a pre-amplifier or a rather large signal from the master oscillator. Alternately, a smaller master oscillator signal could be used to effectively saturate a gain medium if some feedback is provided in the amplifier. If the feedback is below the threshold value for the amplifier section to oscillate, this is termed regenerative amplification. If the amplifier section is above threshold and can oscillate as an independent oscillator, this is termed an injection-locked oscillator. Thus a smaller oscillator signal can control an array of amplifier/oscillators. However, the price paid is a limitation on the bandwidth since the frequency of the master oscillator must be within a given bandwidth of the natural frequency of the amplifier/oscillator. The unstable resonator can be effectively used to provide the feedback for the regenerative amplifier or the injection-locked oscillator.

It has been shown (Buczek et al. 1973) that the gain–bandwidth product for regenerative amplification is

$$\left[\frac{P_{MO}}{P_{AO}}\right]^{1/2} \frac{\Delta\omega}{\omega_0} \approx \frac{\sqrt{7}}{2Q}, \tag{22}$$

where

P_{MO} = power injected into the amplifier from the master oscillator,
P_{AO} = power out of the amplifier,
$\Delta\omega$ = frequency difference between the injected signal and the natural frequency of the amplifier,
ω_0 = injected signal frequency,
Q = quality factor of the amplifier resonator
$\approx 4\pi/\gamma)/(L/\lambda)$ for an unstable resonator with γ output coupling.

For the injection-locked oscillator the master oscillator must be within $\Delta\omega$ of the natural frequency of the amplifier, where

$$\frac{\Delta\omega}{\omega_0} = \frac{1}{Q}\tan\left[\frac{P_{AO}}{P_{MO}}\right]^{1/2}. \tag{23}$$

There is an additional effect which must be considered when injection-locking several oscillators with one master oscillator. The phase of the output of each power oscillator depends on the amount of frequency pulling necessary in each case. Hence, if the natural frequency of each power oscillator randomly changes owing to mirror vibrations, the array may remain coherently locked but the phase of each oscillator may randomly change. For a complete review of laser injection-locking see Buczek et al. (1973).

Fig. 20 shows three different possible arrangements using the unstable resonator. The hole coupled arrangement (fig. 20a) uses an isolator to prevent the larger amplified signal which is reflected from the unstable resonator from controlling the frequency of the master oscillator. The size of the coupling hole determines the round-trip loss in the amplifier section and can be used to determine if regenerative amplification or injection-locking is occurring. Ananév and Sherstobitov (1973) have analyzed the hole coupled unstable resonator and have

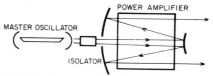

a) Hole injected locked oscillator or regenerative amplifier.

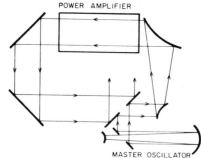

b) Scraper injected locked oscillator or regenerative amplifier.

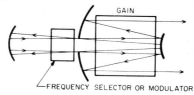

c) Compound Resonator

Fig. 20. Configurations for using the unstable resonator in an injection-locked oscillator or a regenerative amplifier.

derived formulas for computing the round-trip loss given the cavity magnification, N_{eq}, and the size of the coupling hole.

Injection into the center of the unstable resonator is consistent with the concept that the central core acts as the master oscillator and the outer section acts as an amplifier. However, Buczek et al. (1973) observed that the hole coupling was inefficient because of diffraction spreading of the injected power in one pass through the medium. Ananév and Sherstobitov (1973) observed that a small coupling hole did not introduce any significant cavity loss since a relatively low loss antisymmetric mode replaces the symmetric mode as the lowest-order mode.

The scraper injected ring resonator shown in fig. 20b was suggested by Buczek et al. (1973) as being a more efficient injection scheme. The ring resonator requires no isolator if unidirectional oscillations can be maintained. It has been observed that the injected signal plus the reflection back from the output mirror of the master oscillator provided sufficient discrimination against self-oscillation in the wrong direction.

The compound resonator (Little et al. 1976) of fig. 20c essentially uses the same gain medium for the master oscillator and the amplifier. Since there is no isolation this can be viewed as two oscillators with mutual coupling. Hence, the output may tend to be nearer one of the natural frequencies of the larger outer oscillator since this signal will dominate. However, if the hole is large enough to prevent self-oscillation in the amplifier (regenerative amplification) and a line or frequency selecting element is placed as shown, the output will be frequency controlled. The frequency selecting element is not subjected to the entire output power. In a similar way the output could be modulated without the modulator's being capable of handling the entire output power.

All of the configurations in fig. 20 require that at least one of the frequencies of the master oscillator be within the locking range of one of the natural frequencies of the larger amplifier/oscillator. If the longitudinal mode spacings are some small integer multiple of each other, then several modes could be locked simultaneously or several frequencies could see reasonable regenerative gain. This would require some form of frequency control to keep the comb of modes within the locking range. If there is no integer relationship between the mode spacing and there are many longitudinal modes oscillating in the master oscillator, it is likely that at least one mode will be within the locking range of one of the natural frequencies of the amplifier. Hence, even though the mirrors may vibrate, there is always likely to be one or more frequencies which are amplified. If the master oscillator, however, has only one frequency in its output, then some frequency control must be provided to keep this frequency within the locking range of one of the amplifier frequencies.

If the master oscillator is pulsed and the pulse length is less than several round trips in the amplifier, there is no regenerative amplification. The natural frequencies of the amplifier are never established and there is no need for frequency control. The amplifier acts as a multi-path linear amplifier.

The frequency requirements are believed to be similar for the compound resonator in which the master oscillator frequencies are determined by the spacing between the outer two mirrors, and the amplifier frequencies are determined by the spacing between the amplifier mirrors. If a line or frequency selecting element is used and its linewidth covers many longitudinal modes, then no frequency control is required for the reasons just given. The analysis of the mutual coupling in the compound resonator and the frequency pulling to be expected has not been done.

Buczek et al. (1973) have demonstrated hybrid injection-locking of a CO_2 laser with an unstable resonator. In hybrid locking the injected signal is on a different line of CO_2 than the natural self-oscillation of the larger oscillator. The injected signal quenches the self-oscillations and causes the larger oscillator to lase on the injected line. They observed that a cw 200 mW signal on the R(18) line could completely quench the P(18) oscillation in the pulsed oscillator. The 200 mW injected signal controlled a 20 W output of the oscillator. They demonstrated injection-locking of the hole injected linear unstable resonator and the hole injected ring unstable resonator. The measured locking bandwidths agree with the theory.

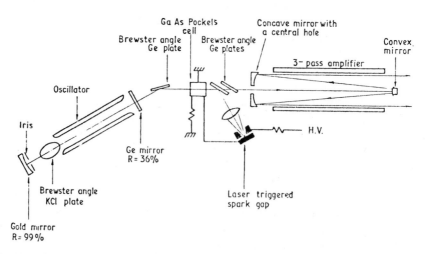

Fig. 21. Experimental configuration for using the unstable resonator as a three-pass amplifier. From Little et al. (1976).

Little et al. (1976) observed linear amplification in a hole injected unstable resonator CO_2 oscillator–amplifier. The injected signal had a pulse length less then one round-trip time in the amplifier, and hence no regenerative amplification was observed and no isolator was necessary. The amplifier acted as a linear three-pass amplifier with an energy gain of 1000 (the single-pass gain was 20). The hole for injection was made large enough to prevent self-oscillation in the amplifier, and care was taken to prevent parasitic oscillations.

Acknowledgement

The author has benefited over the years from discussions and joint research work, particularly in the applications of unstable resonators to high energy lasers, with the staff of the Northrop Research and Technology Center. The author would also like to acknowledge the continued interactions he has enjoyed with his colleagues in laser and optics research at the University of Southern California.

References

Ananév, Yu A., G.N. Vinokurov, L.V. Kovalchuk, N.A. Sventsitskaya and V.E. Sherstobitov 1970, Sov. Phys. – JETP **31**, 420.
Ananév, Yu A. and V.E. Sherstobitov (1971), Sov. J. Quantum Electronics **1**, 263.
Ananév, Yu A. (1972), Sov. J. Quantum Electronics **1**, 565.
Ananév, Yu A. and V.E. Sherstobitov (1973), Sov. Phys. – Tech. Phys. **18**, 640.
Ananév, Yu A., L.V. Kovalchuk, V.P. Trusov and V.E. Sherstobitov (1974), Sov. J. Quantum Electronics **4**, 659.

Ananév, Yu A., V.P. Trusov and V.E. Sherstobitov (1976), Sov. J. Quantum Electronics **6**, 928.

Barone, S.R. (1967), Appl. Optics **6**, 861.

Bergstein, L. (1968), Appl. Optics **7**, 495.

Born, M. and E. Wolf (1964) Principles of Optics (MacMillan, New York).

Buczek, C.J., J. Freiberg and M.L. Skolnik (1973), Proc. IEEE **61**, 1411.

Chen, L.W. and L.B. Felson (1973), IEEE J. Quantum Electronics **QE–9**, 1102.

Chodzko, R.A. (1974) Appl. Optics **13**, 231.

Chodzko, R.A., H. Mirels, F.S. Roehrs and R.J. Pedersen (1973), IEEE J. Quantum Electronics **QE–9**, 523.

Clark, P.O. (1972), AIAA 5th Fluid and Plasma Dynamics Conference, Boston, Mass., June 26–28, Paper No. 72–708.

Dreizin, Yu A. and A.M. Dykhne (1974), JETP Letters **19**, 371.

Fox, A.G. and T. Li (1961), Bell Syst. Tech. J. **40**, 453.

Freiberg, R.J., P.P. Chenausky and C.J. Buczek (1973), Appl. Optics **12**, 1140.

Freiberg, R.J., P.P. Chenausky and C.J. Buczek (1974), IEEE J. Quantum Electronics **QE–10**, 279.

Horwitz, P. (1976), Appl. Optics **15**, 167.

Kahn, W.K. (1966), Appl. Optics **5**, 407.

Karamzin, Yu N. and Yu B. Konev (1975), Sov. J. Quantum Electronics **5**, 144.

Kogelmk, H. and T. Li (1966), Proc. IEEE **54**, 1312; Appl. Optics **5**, 1550.

Krupke, W.F. and W.R. Sooy (1969), IEEE J. Quantum Electronics **QE–5**, 575.

Little, V.I., A.C. Selden and T. Stamatakis (1976), J. Appl. Phys. **47**, 1295.

Mann, M.M.: (1977) Northrop Research and Technology Center, private communication

McAllister, L., W.H. Steier and W.B. Lacina (1974), IEEE J. Quantum Electronics **QE–10**, 346.

Mirels, H. and S.B. Batdorf (1972), Appl. Optics **11**, 2384.

Mirels, H. (1976), Appl. Phys. Letters **28**, 612.

Rensch, D.B. and A.N. Chester (1973), Appl. Optics **12**, 997.

Rensch, D.B. (1974), Appl. Optics **13**, 2546.

Sanderson, R.L. and W. Streifer (1969), Appl. Optics **8**, 2129.

Santana, C. and L.B. Felsen (1977), Appl. Optics **16**, 1058.

Sherstobitov, V.E. and G.N. Vinokurov (1972), Sov. J. Quantum Electronics **2**, 224.

Siegman, A.E. (1965), Proc. IEEE **53**, 277.

Siegman, A.E. and R.W. Arrathon (1967), IEEE J. Quantum Electronics **QE–3**, 156.

Siegman, A.E. and H.Y. Miller (1970), Appl. Optics **9**, 2729.

Siegman, A.E. (1971) Laser Focus **42**, May.

Siegman, A.E. (1974), Appl. Optics **13**, 353.

Siegman, A.E. and E.A. Sziklas (1974), Appl. Optics **13**, 2775.

Siegman, A.E. (1977), IEEE J. Quantum Electronics **QE–13**, 334.

Steier, W.H. and G.L. McAllister (1975), IEEE J. Quantum Electronics **QE–11**, 725.

Streifer, W. (1968), IEEE J. Quantum Electronics **QE–4**, 229 (1968).

Sziklas, E.A. and A.E. Siegman (1975), Appl. Optics **14**, 1874.

Vasilev, L.A., V.K. Kemkin, Yu A. Kalinin and Yu I. Kruzhilin (1975), Sov. J. Quantum Electronics **5**, 27.

Weinstein, L.A. (1969), Open Resonators and Open Waveguides (Golem Press; Boulder, Colorado).

Yoder, M.J. and D.R. Ahouse (1975), Appl. Phys. Letters **27**, 673.

A2 | Waveguide Gas Lasers

RICHARD L. ABRAMS

Hughes Research Laboratories,
Malibu, California 90265

Contents

Abstract

Research and development of waveguide gas laser technology from 1971 to 1977 is reviewed with emphasis on the properties of waveguide laser resonators and CO_2 waveguide lasers. A review of waveguide laser action in other gaseous systems is also presented.

© *North-Holland Publishing Company, 1979*
Laser Handbook, edited by M.L. Stitch

1. Introduction

A waveguide laser resonator can be generally described as a resonator where radiation is transmitted in part by guided wave propagation rather than by free space propagation. A waveguide laser is then a laser which employs a waveguide resonator to provide the necessary feedback to establish oscillation. This is in contrast to a conventional laser where the feedback and resonator modes are established by normal free space propagation, resulting in the well-known gaussian normal modes.

1.1 Historical development

The concept of a waveguide gas laser was first proposed by Marcatili and Schmeltzer (1964), in a paper describing the waveguiding properties of hollow dielectric tubes for long distance optical communications. They pointed out the potential of He–Ne gas laser operation in small bore capillary tubes, citing the well-known increase in laser gain with decreasing diameter demonstrated by Gordon and White (1963). In addition, they showed the existence of an optimum tube diameter for maximum net laser gain, trading off increasing gain against decreasing waveguide transmission with decreasing tube diameter.

Steffen and Kneubuhl (1968) independently introduced the concept of hollow dielectric waveguide modes to explain anomalies observed by Schwaller et al. (1967) on transverse mode frequencies of a far-infrared (FIR) HCN laser. This was, in fact, the first report of a waveguide gas laser although no connection was made with work of Marcatili and Schmeltzer and the full significance of this work was not appreciated by researchers on gas lasers.

Smith (1971) demonstrated the first operation of a waveguide gas laser operating in the visible region of the spectrum. In this beautiful experiment, laser action in He–Ne at 6328 Å was obtained in a 20 cm length of 430 μm diameter capillary tubing using a resonator with external mirrors. This work stimulated a large number of researchers to join the field, resulting in new waveguide gas lasers operating at a wide variety of wavelengths from the visible to the FIR. A large portion of this work was performed with CO_2 waveguide lasers operating near 10 μm, and this will be reviewed in §4.

Much of the waveguide gas laser research employed capillary bore laser discharge tubes and external curved mirrors which served to image the radiation emanating from the capillary tube back onto itself. Thus, the resonators contain regions of free space propagation as well as regions of optical waveguiding. This represented a new type of laser resonator for which there was little intuitive or theoretical understanding. Abrams (1972) attempted to understand the coupling losses of the lowest-order waveguide mode in propagating to a curved mirror and returning back into the waveguide. This initial theory provided a basic understanding of the resonator problem and became an invaluable guide to the design of waveguide resonators with very low loss and a high degree of modal selectivity as shown by Abrams and Chester (1973, 1974). In the meantime, Chester and Abrams (1972), Degnan (1973) and Degnan and Hall (1973) were independently working on similar treatments of the properties of waveguide laser resonators. A review of waveguide resonator theory is included in §2.

In the last few years, a large proliferation in waveguide laser work in a wide variety of systems has been reported. These include lasing in solid state, dye solutions and gaseous media at wavelengths ranging from less than one micrometer to several millimeters. Aside from the CO_2 laser system, the most actively studied waveguide laser systems in recent years have been in the FIR spectral region. In §5, other waveguide gas laser systems are reviewed. Solid-state and dye lasers operating in the waveguide regime will not be covered in this chapter and the reader is referred to the excellent waveguide laser review article by Degnan (1976).

1.2 Unique properties of waveguide lasers

It is reasonable to ask why one wants to operate a laser in the waveguide configuration. This is an excellent question for, as will be shown in §3 on scaling relationships, the power output per unit length and efficiency of most waveguide laser systems is independent of discharge diameter, giving identical performance to conventional lasers. The advantages to be gained include: reduced laser size through smaller transverse dimensions; higher laser gain in Doppler-broadened lasers resulting in the potential for compact low power lasers otherwise not possible; high pressure operation resulting in potential increased frequency tunability in molecular lasers such as CO_2; efficient matching between mode volume and laser excitation region; and finally, excellent mode control through the unique properties of waveguide laser resonators. In the following sections, these advantages will be stressed in discussing the properties of various waveguide laser systems.

2. Waveguide laser resonators

A generalized waveguide laser resonator is illustrated in fig. 1. The waveguide region consists of a hollow dielectric tube which may have arbitrary geometrical cross sections including circular, elliptical, square or rectangular. Only circular waveguides will be treated here.

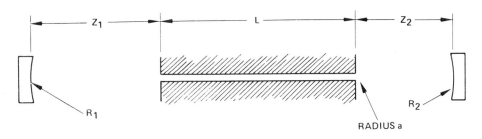

HOLLOW WAVEGUIDE RESONATOR

Fig. 1. Hollow waveguide resonator (Abrams and Chester 1974).

Hollow waveguide *resonator* modes (as opposed to the *waveguide* modes) consist of field distributions that propagate in the waveguide, radiate at the ends of the guide toward a mirror, are redirected and coupled back into the guide. This field distribution must reproduce itself after one round trip through the resonator, both in its amplitude and phase distribution.

In order to appreciate the complexity of the resonator problem it is useful to qualitatively trace this path in terms of the modes of propagation of the hollow waveguide. Consider radiation propagating in a single low-order waveguide mode; it propagates through the guide, suffering some attenuation factor and reaches the end of the guide where it propagates toward a mirror, say R_1, separated by distance Z_1 from the guide. This field distribution of a single waveguide mode then propagates to the mirror and back toward the guide according to the normal laws of diffraction. At the guide, the resulting field distribution then couples to a distribution of waveguide modes, in general different from the original single mode. These modes propagate through the guide encountering their individual phase shifts and attenuations and then radiate toward the opposite mirror, R_2 at a distance Z_2 from the end of the guide. Upon re-entering the guide a new set of mode amplitudes and phases is established which is again, in general, different from the original set. The resonator problem is to find field distributions which reproduce themselves within a constant factor after propagating through round trips of the resonator. These field distributions are the *resonator* modes and the constant factor referred to above is related to the losses of these modes.

There are several techniques and approaches to the solution of this problem. One is to compute the field distributions for a given resonator configuration iteratively for a large number of round trips through the cavity. Steady-state solutions develop after many passes which represent the eigenmodes of the resonator. This technique was developed for waveguide laser resonators by Chester and Abrams (1972) and is discussed in §2.3.1. A second approach is to develop a formulation describing a round trip through the cavity where the field distribution is expressed in terms of an expansion into a finite number of waveguide modes and a propagation matrix derived. Diagonalization of this matrix then results in knowledge of the resonator eigenmodes and their losses. This approach was taken by Abrams and Chester (1973, 1974) and Degnan and Hall (1973) and is discussed in §2.3.2.

In order to approach the above theories, it is necessary to understand the properties of hollow waveguide modes and how they couple to free space. A review of these topics follows in §§2.1 and 2.2.

2.1. Hollow dielectric waveguide modes

The modes of circular hollow dielectric waveguides have been discussed in detail by Marcatili and Schmeltzer (1964). Hollow dielectric waveguide modes in rectangular geometry were considered by Krammer (1976) and Laakman and Steier (1976), and will not be discussed here. The modes of either type of hollow dielectric waveguide differ in nature from the modes of conventional fiber optical waveguide or integrated optical structures. The more conventional dielectric waveguides guide by total internal reflection of optical rays at a dielectric interface where the rays propagate in a core region with a refractive index higher than the surrounding medium. Losses result from imperfect geometries and intrinsic or impurity absorption in the waveguiding materials. In a hollow waveguide, the guiding (or core) region has a lower refractive index than the surrounding media and waveguiding takes place owing to _nearly_ total internal reflection at the dielectric interface (total internal reflection is not possible). Some radiation inevitably leaks into the dielectric medium and radiates. For this reason the modes are often referred to as leaky modes and this results in a finite waveguiding loss, even allowing perfect geometry and no material absorption. This loss is strongly dependent on waveguide dimensions and limits operation of hollow waveguides to highly overmoded structures.

The circular guide supports three basic types of modes: transverse circular electric, whose electric field is to first order tangential to the waveguide surface; transverse circular magnetic, whose magnetic field is to first order tangential to the waveguide surface and whose electric field is radial; and hybrid modes which have both tangential and radial electric and magnetic fields. The field distributions are considerably simplified if we assume

$$ka \gg |\nu| u_{nm}, \tag{1}$$

where $k = 2\pi/\lambda$ is the free space propagation constant, ν is the complex refractive index of the dielectric wall, u_{nm} is the mth root of $J_{n-1}(u_{nm}) = 0$ and n, m are integers labeling the mode. The electric field patterns for the lowest-order mode of each type are shown in fig. 2 along with the approximate expressions for the field, neglecting terms of order u_{nm}/ka. These approximations will be used in the theories described in this chapter and the reader should refer to Marcatili and Schmeltzer (1964) and Stratton (1941) for more accurate functional forms of the modes and additional background.

The lowest-order hybrid mode, EH_{11}, is linearly polarized and closely resem-

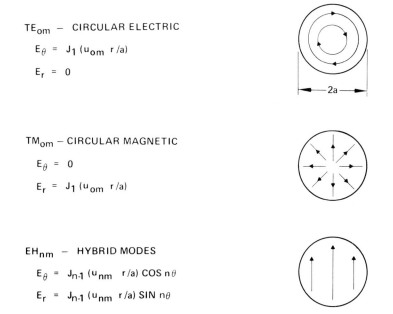

TE$_{om}$ — CIRCULAR ELECTRIC

$E_\theta = J_1 (u_{om}\ r/a)$

$E_r = 0$

TM$_{om}$ — CIRCULAR MAGNETIC

$E_\theta = 0$

$E_r = J_1 (u_{om}\ r/a)$

EH$_{nm}$ — HYBRID MODES

$E_\theta = J_{n-1} (u_{nm}\ r/a)\ \text{COS}\ n\theta$

$E_r = J_{n-1} (u_{nm}\ r/a)\ \text{SIN}\ n\theta$

Fig. 2. Analytic expressions and characteristic electric field patterns for waveguide modes (Abrams and Chester 1974).

bles a TEM$_{00}$ free space gaussian mode. The EH$_{1m}$ modes have circularly symmetric field strengths, are linearly polarized and will clearly be important when considering laser resonators containing polarizing optical elements such as gratings or Brewster windows. Noncircularly symmetric linearly polarized modes can be formed by particular linear combinations of higher-order hybrid modes and have been extensively discussed with reference to waveguide resonators by Degnan and Hall (1973).

Following Marcatili and Schmeltzer (1964), the propagation constant for the various modes is given by

$$\gamma \approx k\left[1 - \frac{1}{2}\left(\frac{u_{nm}\lambda}{2\pi a}\right)^2\left(1 - \frac{i\nu_n\lambda}{\pi a}\right)\right], \qquad (2)$$

where the constant ν_n depends on the type of mode being considered and the refractive index of the guide wall.

$$\nu_n = \begin{cases} \dfrac{1}{(\nu^2-1)^{1/2}} & \text{for TE}_{0m} \text{ modes } (n=0). \\[2ex] \dfrac{\nu^2}{(\nu^2-1)^{1/2}} & \text{for TM}_{0m} \text{ modes } (n=0) \\[2ex] \dfrac{\frac{1}{2}(\nu^2+1)}{(\nu^2-1)^{1/2}} & \text{for EH}_{nm} \text{ modes } (n \neq 0). \end{cases} \qquad (3)$$

The phase and attenuation constants are given by the real and imaginary components of γ,

$$\beta_{nm} = \mathrm{Re}(\gamma) = \frac{2\pi}{\lambda}\left\{1 - \frac{1}{2}\left(\frac{u_{mn}\lambda}{2\pi a}\right)^2\left[1 + \left(\mathrm{Im}\,\frac{\nu_n\lambda}{\pi a}\right)\right]\right\},$$

$$\alpha_{nm} = \mathrm{Im}(\gamma) = \left(\frac{u_{nm}}{2\pi}\right)^2\frac{\lambda^2}{a^3}\,\mathrm{Re}(\nu_n). \tag{4}$$

As the mode number increases, u_{nm} also increases, resulting in a rapidly increasing attenuation constant. Note that for a lossless dielectric wall material, ν_n is real and the phase constant β_{nm} is independent of refractive index.

2.2. Coupling losses

Before considering the complete resonator problem it is useful to gain some understanding of how waveguide modes couple to free space and what mode losses are encountered by an EH_{11} waveguide mode in propagating to an external mirror and back into the waveguide. This calculation also has some utility in predicting the properties of certain waveguide resonators, namely those where the EH_{11} mode coupling loss is low and the guide acts as a perfect EH_{11} mode filter (very long guide length in which higher-order modes are effectively damped and the lowest loss EH_{11} mode predominates). The primary motivation, however, is to gain a physical picture of the coupling problem which will be drawn upon in §§2.3 and 2.4.

2.2.1 Coupling from waveguide to free space

Following Abrams (1972), we consider a hollow dielectric waveguide terminated by flat mirrors normal to the guide axis at the ends of the guide. Clearly, a field distribution or mode in the guide is reflected back on itself at the end points and suffers no coupling loss. Thus the modes of the resonator consist of the usual set of longitudinal modes and, since different waveguide modes have different propagation constants, they are associated with transverse modes or field distributions given by the hollow waveguide modes described in §2.1. The losses of these modes are then simply the losses associated with propagation in the leaky waveguide. If the mirrors are then moved away from the end of the guide, which is the usual laser condition, the modes will suffer additional losses because of imperfect coupling of the radiation back into the guide after propagating to the mirror and back. In this section we calculate these coupling losses as a function of mirror position and mirror radius for the EH_{11} lowest-order linearly polarized mode. In solving the radiation problem at the end of the guide, we will decompose the field amplitude into free space gaussian normal modes. It is emphasized that this is only a mechanism to lend physical insight to the solutions and should not be confused with the modes of a Fabry–Perot resonator. The coupling problem can also be solved by a diffraction integral, as shown by Degnan and Hall (1973).

As the mirrors are moved away from the end of the guide, increasing coupling losses occur as a result of two effects: (1) the field may spread and not re-enter the guiding structure; (2) some of the energy that re-enters the guide is converted to higher-order waveguide modes. To the extent that the second condition occurs, the modes are not truly pure modes and the problem we are really solving is that of a single-mode EH_{11} mode filter in place of a multimode waveguide. However, we shall see that the EH_{11} waveguide mode couples 98% of its energy to the TEM_{00} lowest-order gaussian free space mode for optimum choice of gaussian beam parameters.

We start by choosing our expansion functions, which are the gaussian normal modes described in the review paper by Kogelnik and Li (1966). Only the modes with cylindrical symmetry are required (no θ-dependence) to describe the EH_{11} waveguide mode and we normalize the modes such that

$$\int_0^\infty \psi_p(r)\psi_q(r)2\pi r\, dr = \delta_{pq}. \tag{5}$$

With this normalization, the modes are given at a beam waist by

$$\psi_p = \sqrt{\frac{2}{\pi}}\frac{1}{w_0} L_p\left(\frac{2r^2}{w_0^2}\right)\exp(-r^2/w_0^2), \tag{6}$$

where L_p is the Laguerre polynomial of degree p, and w_0 the $1/e$ radius of the mode amplitude. The free space propagation characteristics of these modes are well understood and are also described by Kogelnik and Li (1966).

The first consideration for choosing a set of modes is to decide what value to take for w_0. Clearly, since we have a complete set of expansion functions, any value is suitable for solving the radiation problem, but some values are more physically meaningful than others and should lead to a more rapidly converging series. Smith (1971) found experimentally that the output of these lasers closely resembles the lowest-order gaussian mode, so it is natural to choose w_0 such that the expansion coefficient A_0 for the lowest-order mode is maximized.

Taking the lowest-order waveguide mode as $E(r) = J_0(ur/a)$, the expansion coefficients are given by ($u = u_{11} = 2.405$ for the EH_{11} mode).

$$A_p = \sqrt{\frac{2}{\pi}}\frac{1}{w_0}\int_0^a J_0\left(\frac{u}{a}r\right)L_p\left(\frac{2r^2}{w_0^2}\right)\exp(-r^2/w_0^2)2\pi r\, dr$$

$$= \sqrt{\frac{\pi}{2}}\,w_0\int_0^K J_0(\xi\sqrt{x})L_p(x)\exp(-x/2)\, dx, \tag{7}$$

where

$$\xi = \frac{u}{\sqrt{2}}\frac{w_0}{a}, \qquad K = 2a^2/w_0^2.$$

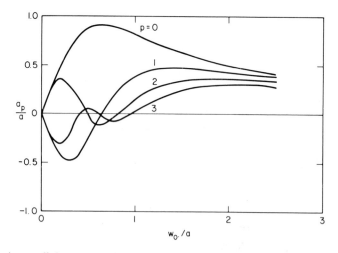

Fig. 3. Expansion coefficients normalized to the waveguide radius versus w_0/a (Abrams 1972).

Now

$$A_0 = \sqrt{\frac{\pi}{2}}\, w_0 \int_0^K J_0(\xi\sqrt{x})\exp(-x/2)\,dx \tag{8}$$

and

$$\frac{\partial A_0}{\partial w_0} = -\sqrt{\frac{\pi}{2}} \int_0^K J_0(\xi\sqrt{x})(1-x)\exp(-x/2)\,dx$$

$$= -\frac{1}{w_0} A_1(w_0), \tag{9}$$

so that at the value of w_0 that maximizes A_0, A_1 is equal to zero. The expansion coefficients normalized to the tube radius a were numerically evaluated as a function of w_0/a for $p = 0$ to 5. The results for $p = 0$ to 3 are shown in fig. 3.

Now we may write the expansion of the EH_{11} field distribution at the end of the waveguide as

$$E(r) = \sum_p A_p(w_0)\psi_p(2r^2/w_0^2), \tag{10}$$

with energy normalized to

$$Q = \int_0^\infty |E(r)|^2\, 2\pi r\, dr = \sum_p |A_p(w_0)|^2, \tag{11}$$

with numerical value

$$Q = 0.84668a^2. \tag{12}$$

The value of w_0/a which maximizes A_0 is $w_0/a = 0.6435 \pm 0.0002$. This is smaller than the value ($w_0/a = 0.728$) calculated by Smith (1971) but is larger than the value ($w_0/a = 0.49$) he determined experimentally. In fact, a more complete treatment of the resonator problem by Chester and Abrams (1972) later predicted $w_0/a = 0.51$ for Smith's resonator, close to the measured value.

Using the value $w_0/a = 0.6435$, the expansion coefficients and $F(p)$, the fraction of total energy contained in the first $p + 1$ terms of the expansion, are calculated and shown in table 1. By including six terms in the expression ($p = 0$ to 5), we are clearly accounting for nearly all the original energy in the waveguide mode. What is surprising is that 98% of the energy is contained in the lowest-order gaussian mode for the optimum value of w_0/a. From here on, this value of w_0/a will be used for our calculations.

Table 1.
Values of A_p and F_p for $w_0/a = 0.6435$ (Abrams 1972).

p	A_p	$F(p)$
0	0.911217	0.9806
1	-5.4×10^{-5}	0.9806
2	-0.11087	0.9952
3	-0.03960	0.9970
4	0.01804	0.9974
5	0.03141	0.9986

2.2.2 Calculation of Coupling Loss

The coupling loss is calculated by propagating the radiation field at the output of the waveguide to the mirror and back and calculating the projection of the resulting field on the EH_{11} mode. At any position Z from the end of the guide we can position a mirror or radius R (fig. 4). It might be expected that optimum coupling at position Z would occur when the radius R is equal to the wavefront radius R' given by

$$R' = B(Z/B + B/Z), \tag{13}$$

where

$$B = \pi w_0^2/\lambda. \tag{14}$$

Fig. 4. Geometry for calculation of coupling loss (Abrams 1972).

For the case when $R = R'$, each gaussian mode is imaged back on the guide end at its waist and the return wave is given by (neglecting constant phase factors)

$$E'(r) = \sum_p A_p \exp(i\phi_p)\psi_p(r), \tag{15}$$

where $\phi_p = 2(2p + 1)$ arctan (Z/B) is the phase shift associated with the pth gaussian mode. It is easily shown that the coupling loss is given by

$$c^2 = 1 - \left| \frac{\int_0^a J_0\left(\frac{u}{a}r\right) E'(r)2\pi r \, dr}{\int_0^a J_0^2\left(\frac{u}{a}r\right)2\pi r \, dr} \right|^2 \tag{16}$$

$$= \frac{2}{Q^2} \sum_{p,q} |A_p|^2 |A_q|^2 \sin^2(\phi_{pq}/2), \tag{17}$$

where $\phi_{pq} = \phi_p - \phi_q$. Note that at $Z = 0$, $\phi_{pq} = 0$, and $c^2 = 0$, confirming the fact that a flat mirror at the end of the guide ($R = \infty$ at $Z = 0$) gives no coupling loss. Also as $Z/B \to \infty$, $\phi_{pq}/2 \to (p - q)\pi$, again $c^2 = 0$. This is equivalent to spherical geometry in a gas laser where all of the transverse modes are degenerate (all modes are in phase) since as $Z \to \infty$, $R \to Z$. One other case of interest is when $Z = B$. Then $R = 2B$, $\phi_{pq}/2 = (p - q)\pi/2$, and all of the even modes return in phase. This is significant for our case because $A_1 = 0$, and 99.5% of the energy is contained in the zeroth and second terms. Note that this does not ensure 99.5% coupling, as all of the odd modes return exactly out of phase and will interfere destructively.

The coupling loss c^2 was calculated numerically as a function of Z/B, where it was assumed $R = R'$ at each value of Z. *The result is* shown as the dashed line in fig. 5. As expected, the best points (lowest coupling losses) are at $Z = 0$, $Z = B$ and $Z = \infty$. At $Z = B$, 1.48% coupling loss is encountered.

We would also like to know what happens to the coupling loss when the mirror radius does not match the curvature of the wavefront i.e., $R \neq R'$. From this we can tell how critical the placement of the mirror is and how far a flat mirror can be placed from the waveguide and still have low coupling loss. This problem is slightly more difficult because the returning wavefront has a radius different from ω_0 and a finite curvature. A model for the calculation has been developed by Abrams (1972) and the results are summarized in fig. 5. The envelope of the family of curves is the dashed line discussed above.

2.2.3 Discussion

The above formalism has been established for calculation of coupling of the EH_{11} mode to free space gaussian modes and for calculating the losses incurred in a waveguide laser geometry (fig. 4). Certain low loss mirror positions have been identified, including $Z < 0.1B$, $Z = B$ or $Z > 10B$. In these cases coupling losses of less than 1.5% are achievable for the proper value of mirror curvature.

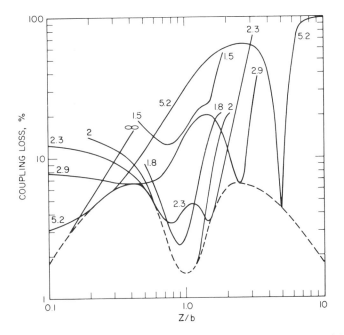

Fig. 5. Coupling loss versus mirror position for the case where the mirror radius matches the curvature of the wavefront (dashed line) and for mirrors of fixed curvature. Each curve is labeled with the value of R/B (Abrams 1972).

One mirror position of some practical importance is that of a flat mirror near the end of the guide. Fig. 5 shows that the loss is less than 2% for $Z/B < 0.11$. Degnan and Hall (1973) have given a theoretical expression for this case, which modified for our notation is

$$\text{Coupling loss} = 0.570(Z/B)^{3/2}. \tag{18}$$

For a CO_2 laser ($\lambda = 10.6 \ \mu\text{m}$) in a 1.5 mm ID waveguide, $B = 6.9$ cm, and the flat mirror should be within 0.74 cm of the end of the waveguide to keep the loss below 2%.

It should be pointed out that other mirror positions and curvatures exist which result in low waveguide resonator losses, where the resonator mode is other than a simple EH_{11} mode. Other modes may be established under these conditions which in fact have even lower coupling losses, but whose field distributions are more complex when described in terms of waveguide modes. These waveguide resonator modes are treated in the §§2.3 and 2.4.

2.3 Waveguide resonator theory

In this section, two approaches to finding the waveguide resonator mode profiles and losses are discussed, an iterative numerical solution and a matrix formulation. The iterative numerical solution is most valuable for finding the lowest-order mode of a few specific waveguide resonator geometries, and this technique is covered in §2.3.2. The more powerful matrix solution, covered in §2.3.3, is most useful when data over a large range of waveguide laser parameters is needed. It has the additional advantage of solving for higher-order modes and their losses, thus yielding valuable information on waveguide laser mode selectivity. In §2.3.1 some symmetry considerations are discussed which allow cataloging of the infinite set of waveguide laser modes into classes, or subsets, according to their azimuthal symmetry.

2.3.1 Resonator mode classes

The waveguide modes are identified by the integers n, m, where n describes the azimuthal symmetry (the rotational symmetry of the field distribution about the tube axis) and m equals the number of maxima and minima in the field along the radial direction between $r = 0$ and $r = a$. It follows from the cylindrical symmetry of the resonator in fig. 1 that modes with different azimuthal symmetry (different values of n) will not be mixed by the interaction of the external mirror. In addition, TE modes will not mix with TM modes because of their orthogonal electric fields. Thus the resonator modes fall into an infinite number of classes, and any resonator mode can be expanded as a linear combination of waveguide modes from a single class. The various classes of resonator modes can be labeled as the following [resonator mode classes, with azimuthal symmetry index n will be identified by the same symbol (TE_{0m}, TM_{0m}, EH_{1m}, etc.) as the set of waveguide modes with the same symmetry; to avoid ambiguity, we will explicitly specify which type of mode we are referring to (either resonator mode or waveguide mode)]:

(1) Circular electric – linear combinations of TE_{0m} modes.
(2) Circular magnetic – linear combinations of TM_{0m} modes.
(3) Linearly polarized hybrid – linear combinations of EH_{1m} modes, resulting in circularly symmetric field strengths.
(4) Higher-order hybrid modes – linear combinations of EH_{nm} modes for fixed values of $n = -1, \pm2, \pm3, \ldots$. The EH_{nm} modes and the $EH_{-|n-2|,m}$ modes are degenerate and can be combined to form linearly polarized noncircularly symmetric modes.

The waveguide modes in each class form a complete orthogonal set of functions on the interval $0 < r < a$, and thus, each class of resonator modes has, in principle, an infinite number of transverse modes. In practice, however, the losses of higher-order modes increase rapidly with mode number, and the approximation made in eq. (1) becomes invalid as u_{nm} increases with values of n, m.

By breaking the resonator modes into classes in the above manner, the problem of solving for the actual resonator modes has been simplified considerably. For example, if we are considering a resonator with Brewster windows, it is usually only necessary to include the linearly polarized hybrid modes (EH_{1m}). Since the guiding loss increases rapidly with mode number, five to ten waveguide modes are usually sufficient to descibe the lowest loss resonator modes.

2.3.2 Iterative approach

In order to solve for the fundamental transverse mode of a waveguide resonator, one can assume a particular radial field distribution at a position within the cavity and propagate repeatedly through the resonator until eventually a field distribution is obtained which reproduces itself on every successive pass (aside from a constant amplitude and phase factor). This amounts to an iterative solution of the electromagnetic field equations to obtain the lowest-order transverse mode in the same spirit of earlier work by Fox and Li (1961). Instead of using the Fresnel integral for the propagation step, Chester and Abrams (1972) followed the procedure described in §2.2 and expanded the field amplitudes in a set of orthonormal modes whose propagation law is well known. Solving for linearly polarized modes of circular symmetry (class #3 of §2.3.1), the linearly polarized EH_{1m} modes are used within the hollow waveguide. In the free space region between the waveguide and the mirror, the gaussian–laguerre free space modes are used. For a given mirror curvature and mirror to waveguide spacing, the radius of the free space modes is chosen so that each mode has its beam waist at the end of the waveguide and a phase front matching the mirror curvature at the mirror position.

Consider a guide of radius a and length L, terminated at one end by a totally reflecting flat mirror as shown in fig. 6. A mirror with radius of curvature R is positioned a distance Z from the other end. Note that this is equivalent to a symmetric resonator of length $2L$ with the same external mirror geometry at both ends. The above geometry is referred to as the half-symmetric configuration.

In terms of these simple expansion functions, the iteration procedure is straightforward. There is a set of complex numbers a_{1i} describing the amplitudes of the free space modes as the they exit from the guide. Propagation to the mirror and back to the guide to obtain amplitudes a_{2i} introduces a complex phase shift $\psi_m = -4m \arctan(\lambda Z/\pi w^2)$ in the mth mode. At the guide the field distribution is re-expressed in terms of the guide modes with complex amplitudes a_{3i}. The conversion from free-mode amplitudes to guide-mode amplitudes simply involves multiplication by a predetermined matrix whose elements are the cross products between individual free modes and individual guide modes, integrated over the tube cross section. The mode amplitudes a_{3i} are propagated within the guide to give amplitudes a_{4i} by using the known phase shifts and losses for these modes within the guide:

$$a_{4m} = a_{3m} \exp[-(u_{1m}\lambda/2\pi a)^2(2L/a)(\nu_n + i\pi a)/\lambda], \tag{19}$$

where u_{1m} is the mth root of $J_0(u) = 0$, and the complex number ν_n is given in eq.

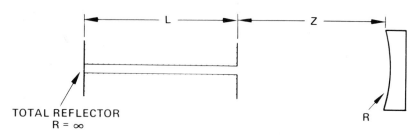

Fig. 6. Half-symmetric resonator geometry (Abrams and Chester 1974).

(3). Another matrix multiplication re-expresses the guide modes in terms of the free space modes, giving a new set of amplitudes a_{1i} to complete one iteration through the resonator.

At any point in the iteration, the total power in the mode is given by the sum of the absolute squares of the mode amplitudes, $\Sigma_i |a_{ki}|^2$. The power loss upon entering the guide, the aperturing loss, is caused by the fact that the entering beam has spread slightly because of diffraction and no longer has zero-field intensity at radii greater than a. The power loss during propagation through the guide, the guide loss, is due to the leaky nature of the guide. There is also an apparent loss upon exiting from the guide because a finite number of free space modes cannot exactly reproduce the field distribution within the guide; this loss is referred to as a truncation loss and a sufficient number of terms had to be taken to insure this was negligible.

In order to demonstrate the validity of the theory, this technique was applied to the waveguide He–Ne laser described by Smith (1971). In order to approximate his experiment the following parameters were used: $\lambda = 6328$ Å, $a = 200\,\mu$m, $L = 20$ cm, mirrors of 30 cm radius of curvature and $\nu_n = 1.5$. With these parameters, the total round-trip loss (coupling loss plus guiding loss) was computed as a function of mirror position for the symmetric resonator, and the results are shown in fig. 7. There are two low loss mirror positions near $Z = R$, at 29.6 and 28.8 cm. Smith reported maximum output for $Z = 29.5$ cm where the loss is 1.0%. For this case, the guiding loss is 0.87% and the coupling loss is only 0.13%. From the known laser gain and laser oscillation bandwidth, Smith estimated his guide-plus-coupling loss at less than 1% per pass, in good agreement with these calculations. At the other low loss position, the loss is 0.95%. The exact mirror locations for low resonator losses are sensitive to the parameters chosen, but in general there always exists a mirror position, near $Z = 30$ cm, where the loss is low ($\sim 1\%$).

The far-field intensity distribution for the Smith laser was also calculated. This is identical to the intensity distribution entering the hollow waveguide and is shown in fig. 8 as a function of far-field angle as well as r/a. The $1/e^2$ intensity radius is $0.51a$, in excellent agreement with the experimental value of $0.49a$ given by Smith. The side lobe, which is down by 26 dB from the peak intensity, is similar to that reported. The number of side lobes and their shapes change for different mirror radii and mirror positions.

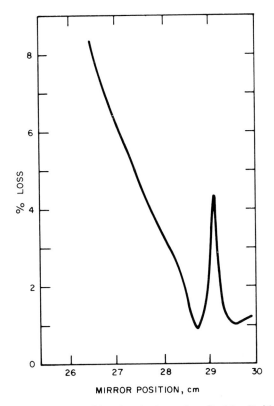

Fig. 7. Coupling loss versus mirror position for the laser described by Smith (1971) (Chester and Abrams 1972).

This technique was applied to several other waveguide laser geometries as described by Chester and Abrams (1972). In general, it was found that the mirror positions at which low coupling loss is achieved are not predictable from any simple wavefront-matching equation. These positions are sensitive to guide length, to a/λ and to mirror radius. As the guide length (or a/λ) is varied, the relative phase shifts of the various guide modes are changed, resulting in significant modification of the field distribution entering the optical waveguide. Low coupling losses are achieved where those relative phase shifts result in modes localized well within the tube radius. Thus, the true situation is much more complicated than the case of a single-mode EH_{11} waveguide filter. It was found, however, that a mirror positioned with z slightly less than R almost always results in low coupling loss. Physically, this results because a small focused spot is formed at waveguide entrance, and there is little aperturing of the beam.

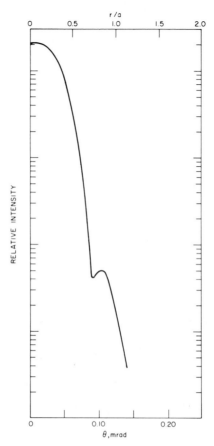

Fig. 8. Predicted far-field radiation pattern for the laser described by Smith (1971) (Chester and Abrams 1972).

2.3.3 Matrix formulation

Degnan and Hall (1972, 1973) presented a theory for waveguide resonators which reduced the calculation to the diagonalization of a complex matrix. They treated the radiation in the free space region from the guide to the mirror by a Fresnel–Kirchhoff diffraction integral, took account of finite mirror aperture and curvature, and then performed a second integral to calculate the return wave to the waveguide aperture. Abrams and Chester (1973) presented and later published (Abrams and Chester 1974) a similar theory in which external mirrors of infinite aperture were assumed. This approximation is appropriate for almost all waveguide resonators of interest and results in a considerable reduction in the complexity of the calculation owing to a contraction of the double Fresnel integral into a single integral. Degnan and Hall considered linearly polarized resonator modes of class 3 and 4, including up to three modes in the expansion.

Chester and Abrams generated a computer program generalized to handle modes of classes 1, 2 and 3 and they included up to 10 waveguide modes in calculating resonator modes and losses in all of these classes.

The technique used in the calculation is discussed with reference to fig. 9 which shows a lens-guide system equivalent to the hollow waveguide resonator of fig. 1. The electromagnetic field in the wave guide is represented by an expansion over an arbitrarily large set of waveguide modes from a single class, as discussed in §2.3.1. The coefficients of the expansion form the components of a column vector describing the mode, which is then easily propagated through the waveguide using the known propagation constants for the various modes. At the end of the guide, the electromagnetic field is numerically evaluated via a transformation matrix onto a finite mesh. The radiation is propagated through the equivalent lens and back onto the waveguide via a Fresnel–Kirchoff diffraction integral, where it is then expanded into waveguide modes via the inverse of the above transformation matrix. This process is repeated until one complete round trip is made through the resonator. Multiplying all the steps together results in an $M \times M$ complex propagation matrix for the waveguide resonator, where M is the initial number of modes used in the expansion. The matrix is then numerically diagonalized. The resulting eigenvectors represent the transverse modes of the resonator, and the eigenvalues yield the corresponding mode losses and the phase shifts experienced by each mode relative to free space propagation.

The first and most tedious step in solving for the waveguide resonator modes is the evaluation of the propagation matrix. As can be seen in fig. 1, there are a large number of parameters necessary to define a waveguide resonator $(a, L, Z_1, Z_2, R_1, R_2,$ and $\nu)$. With such a large number of parameters, it is almost impossible to develop a general set of data or curves that are physically meaningful, and one is left with a mathematical solution that may be applied to any specific problem but that does not lend itself to generating the type of curves useful for design or a general understanding of waveguide resonators. For this reason, the calculations are restricted to symmetric resonators $(R_1 = R_2 = R,$ $Z_1 = Z_2 = Z)$ and the equivalent resonator shown in fig. 6 is solved.

Examination of the components which comprise the propagation matrix reveals that the following dimensionless variables can be used to define the resonator:

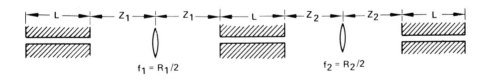

EQUIVALENT LENS – GUIDE SYSTEM

Fig. 9. Lens-guide arrangement equivalent to the resonator shown in fig. 1 (Abrams and Chester 1974).

(a) Fresnel number of the waveguide, $a^2/\lambda L$;

(b) Fresnel number describing mirror position, $a^2/\lambda Z$;

(c) Fresnel number describing mirror curvature, $a^2/\lambda R$;

(d) Waveguide radius, a/λ;

(e) Complex refractive index of wall, ν.

In addition, the following data is specified to define the scope of the computer output:

(i) Type of mode – resonator mode class;

(ii) Dimension of propagation matrix (number of modes);

(iii) Number of mesh points form $0 < r < a$ used in the Fresnel integral.

A computer program was developed that calculates the complex propagation matrix and then finds all the eigenvalues and eigenvectors. The most time-consuming part of the calculation is the generation of the Fresnel integral. For this reason the program is designed to vary the waveguide Fresnel number, while storing the computed Fresnel integral matrix and reusing it. In this manner, waveguide modes and their losses are calculated as a function of waveguide Fresnel number (length) for fixed values of mirror curvature and mirror position.

Once the eigenvalues and eigenvectors are found for a given resonator configuration, the round-trip resonator loss is evaluated from

$$\text{resonator loss} = 1 - |\Lambda|^2, \tag{20}$$

where Λ is the appropriate eigenvalue. Also calculated is the propagation loss through the waveguide for the resonator mode. In this manner the two basic mechanisms for loss in hollow waveguide resonators are calculated. The wave-guiding loss is calculated directly and the difference between resonator and waveguiding loss is a result of aperturing effects on re-entering the waveguide, and this effect is referred to as coupling loss.

The mathematical details of the matrix formulation are covered in the references cited above. In the following numerical results of some specific cases are discussed. It is clear from §2.2 that there are three general cases of interest, namely:

(1) Curved mirrors located a distance $Z = R$;

(2) Curved mirrors located a distance $Z < R$;

(3) Flat mirrors located at $Z = 0$, at the end of the waveguide.

Case 3 has already been discussed in §2.2. Transverse mode selectivity in this case is determined by the relative losses of the waveguide modes.

Case 1 is of general interest because this is the most obvious way (other than Case 3) to make a waveguide resonator. Examination shows that any waveguide mode is imaged back onto itself with some curvature of the wavefront. Some energy is coupled to higher-order waveguide modes because of this curvature, but no energy is lost by aperturing upon re-entering the guide (assuming mirrors of infinite extent). The resonator loss for this configuration is primarily due to waveguide transmission loss, which in general will be higher than the loss for the lowest loss waveguide mode. This is caused by the coupling to higher-order modes. Transverse mode selectivity should not be much better than for Case 3.

Case 2 is potentially the most interesting, for it is possible to choose mirror radii and positions that effectively couple a single waveguide mode back into itself and discriminate against all others. An example of such a mirror position was discussed by Abrams (1972) and in §2.2. It was shown that an EH_{11} mode, in a waveguide of radius a, couples most efficiently to a TEM_{00} mode with $1/e^2$ intensity radius $w_0 = 0.6435a$ and that 98% of the EH_{11} mode energy radiates into this free-space mode. At any position Z from the end of the guide, this gaussian mode has a wavefront with curvature

$$R = Z + B^2/Z, \tag{21}$$

where

$$B = \pi w_0^2/\lambda \tag{22}$$

is a parameter of the gaussian beam. The gaussian beam will be effectively imaged back onto itself if the mirror matches the curvature of the wavefront. The EH_{11} mode will be coupled back into itself very efficiently if we choose $Z = B$ and $R = 2B$. In fact, only 1.5% power loss is suffered propagating to the mirror and back into the EH_{11} mode with the parameters chosen above. Other waveguide modes will not couple as effectively (as shown by Degnan and Hall 1973); thus a resonator of this configuration should have excellent transverse mode selectivity. For this reason, the most extensive numerical calculations were performed for this geometry.

For laser applications, linearly polarized resonator modes are the most common, so the EH_{1m} class of resonator modes is of primary interest. Assuming that the waveguide wall material is a lossless dielectric, the EH_{1m} modes have minimum loss for $\nu = \sqrt{3}$. In this case

$$\text{Re}(\nu) = \sqrt{2}, \qquad \text{Im}(\nu_n) = 0. \tag{23}$$

In all the numerical examples discussed, tha above value for refractive index is assumed. Of course, any other value could be used. Other parameters used in the numerical examples include

$$a/\lambda = 75, \qquad 0.1 < a^2/L < 1.0 \tag{24}$$

These are typical of the parameters used in waveguide lasers. For a 10 μm CO_2 laser, eq. (24) implies a waveguide diameter of 1.5 mm, typical of values used by Abrams and Bridges (1973) in CO_2 waveguide laser experiments. Fresnel numbers $a^2/\lambda L$ for waveguide lasers are less than 1.0, since a conventional laser resonator could be used for the $a^2/\lambda L > 1.0$, and one order of magnitude change in this parameter is covered.

Typical round-trip resonator losses for the above case are shown in table 2. The lowest-order resonator order mode loss (mode no. 1) in the EH_{1m} class has a round-trip loss of 2.6%, while the next mode (mode no. 2) has 73% loss, providing a tremendous margin of mode selectivity as predicted.

Table 2

Round-trip resonator loss for the first four modes of the EH_{1m} Class[a] (Abrams and Chester 1974).

Mode no.	Power loss
1	0.02599
2	0.73019
3	0.99454
4	0.99999

[a] With $a/\lambda = 75$, $a^2/\lambda L = 1.0$, $\nu = \sqrt{3}$, $R = 2B$ and $Z = B$.

The lowest loss resonator mode for the EH_{1m} class is a linear combination of EH_{1m} waveguide modes. The relative energy in each waveguide mode for mode no. 1 (table 2) as it exits from the waveguide is given by the absolute squares of the expansion coefficients and is shown in table 3. The resonator mode is almost a pure EH_{11} waveguide mode, a result of the excellent match of the EH_{11} mode to this particular mirror configuration.

The far-field radiation pattern of the resonator mode is easily computed from the known electric field at the waveguide exit. As an example, the far-field pattern for mode no. 1 of table 2 is shown in fig. 10. The dashed line is the radiation pattern for the TEM_{00} mode with $w_0 = 0.6435a$ and is shown for comparison. Both patterns contain the same power. With the exception of the sidelobes typical of radiation from a finite aperture, the two patterns are similar, having nearly identical divergence.

Table 3

Relative energy in each waveguide mode for mode no. 1 from table 2 resolved into EH_{1m} guide modes at the exit of the waveguide (Abrams and Chester 1974).

m	Relative energy
1	0.998299
2	0.000439
3	0.000464
4	0.000358
5	0.000213
6	0.000133
7	0.000075
8	0.000047
9	0.000025
10	0.000016

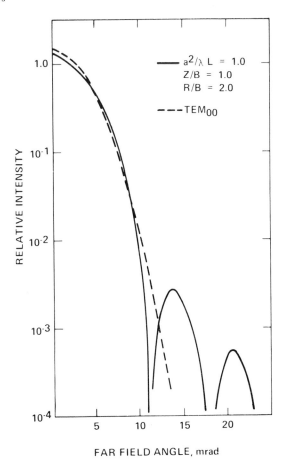

Fig. 10. Far-field radiation pattern for mode no. 1 of table 1 (Abrams and Chester 1974).

It is clear that the resonator configuration described above gives excellent mode selectivity within the EH_{1m} resonator mode class. Using the same resonator parameters, the other mode classes have been considered with Fresnel number of the half-guide from 0.1 to 1.0. The results are depicted in fig. 11. Each of the curves is labeled by its resonator mode class, but only the lowest loss mode of each class is shown. It is clear from these curves that even in the absence of polarizing elements in the cavity, this resonator configuration very strongly favors the lowest loss EH_{1m} resonator mode. The guiding loss for the lowest EH_{1m} resonator mode is very close to the guiding loss for the EH_{11} waveguide mode, reflecting the fact that the resonator mode is nearly a pure EH_{11} waveguide mode. The difference between the EH_{1m} resonator loss and the guiding loss varies between 1.0 and 1.4% over the range 0.1 to 1.0 in Fresnel number. This is just the aperturing loss upon re-entering the waveguide, which varies slightly as the exact mode configuration changes with Fresnel number.

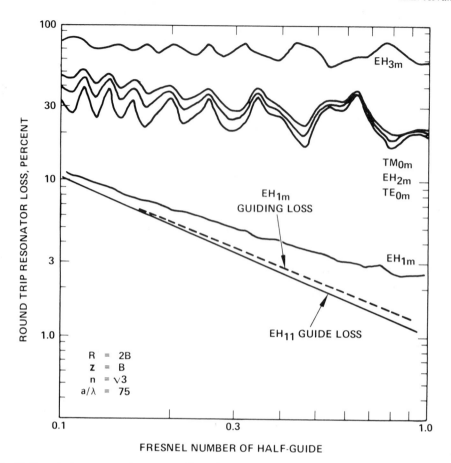

Fig. 11. Round-trip resonator loss versus Fresnel number of the half-guide for the lowest loss mode of each resonator mode class. The dashed line is the guiding loss of the EH_{1m} resonator mode. The guiding loss of the EH_{11} waveguide mode is shown for comparison. For these curves, $R = 2B$, $Z = B$, $\nu = \sqrt{3}$ and $a/\lambda = 75$ (Abrams and Chester 1974).

As an example of the case where $Z = R$ (case no. 1), we take $R = 2B$ (the same as considered above) and $Z = 2B$. All the other parameters were the same. The effect of waveguide length (Fresnel number on the resonator loss for the two lowest-order modes in the EH_{1m} class is shown in fig. 12. The circles and triangles are the two lowest loss modes for $Z = 2B$, whereas the dashed line is reproduced from fig. 11 for the same value of R with $Z = B$. As predicted, the mode selectivity is significantly worse than the case when $Z = B$. The resonator loss behaves much more erratically as the Fresnel number changes, reflecting the fact that there is more contribution from higher-order EH_{1m} modes (the straight lines joining the points are only for illustration). The same calculations were performed for the TE_{0m} and TM_{0m} mode classes with similar results. In general, all mode classes have low loss resonator modes for $R = Z$, but the eigenmodes will, in general, have significant contributions from more than one waveguide

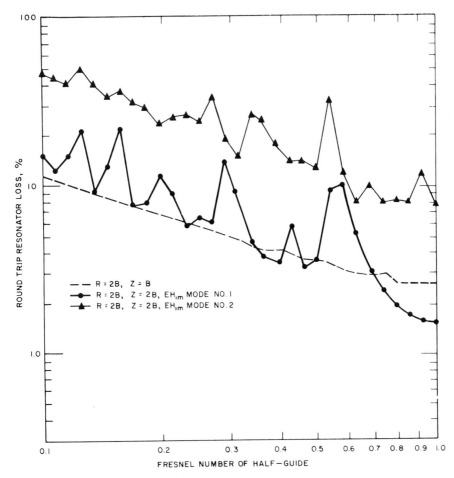

Fig. 12. Round-trip resonator loss versus Fresnel number of the half-guide for the two lowest loss modes of the EH_{1m} class. The circles and triangles are the two lowest loss EH_{1m} modes for $Z = R = 2B$. Other parameters used in the calculation are $a/\lambda = 75$ and $\nu = \sqrt{3}$. The dashed line (from fig. 11) is included for comparison.

mode, resulting in excess guiding losses. However, for large values of $Z = R$ ($\gg B$), the mode mixing becomes negligible and Case 1 becomes equivalent to Case 3.

2.4 Resonator design

Two techniques for computing the modes and losses of hollow waveguide laser resonators have been discussed and a number of numerical examples presented. It is concluded, in general, that a waveguide resonator with low coupling loss may be made by placing a mirror of curvature R a distance $Z = R$ from the end of the waveguide. This configuration results in resonator modes which couple to a number of waveguide modes that may cause excess waveguiding losses (unless $Z = R \gg B$). In addition, the transverse mode selectivity is less than with some other configurations.

If one is interested in optimum mode selectivity and minimum waveguide loss, the theory indicates that a mirror of curvature $R = 2B$ positioned at $Z = B$ is optimum. The resonator mode couples almost exclusively to the EH_{11} waveguide mode resulting in low guide loss and high mode selectivity. When concerned with very high gain lasers, this configuration should be especially useful for transverse mode control.

An interesting extension of this work would be to calculate mode coupling losses for TE_{01} and TM_{01} modes following the same procedure as Abrams (1972) to determine optimum mirror curvature and position. This would lead to resonator designs with high mode selectivity for TE and TM modes in the same manner in which Abrams results led to an EH_{11} mode selective resonator. TE_{01} operation of a ruby laser has been reported by Pohl (1972), but a complex arrangement of intracavity elements was required to force oscillation into this mode. A properly designed waveguide laser resonator could be made to oscillate naturally in this mode.

Finally, in choosing a waveguide resonator design, one should not overlook the simple approach of placing flat mirrors at $Z = 0$ at the end of the guide. The resonator modes are the same as the waveguide modes and mode selectivity is limited by the relative propagation losses. However, the design is simple resulting in a compact structure and this approach has been frequently used successfully.

3. Scaling relations for waveguide lasers

The use of optical waveguiding allows one to extend the operation of gas lasers to smaller diameter discharges without being limited by the usual diffraction losses incurred in free space propagation. In this section we discuss the plasma scaling relationships which allow us to predict how laser parameters such as gain, efficiency and power output will change with decreasing discharge diameter. While these relationships can be used as guidelines to anticipate useful operating ranges of various laser systems, the precise dependence of laser parameters on discharge diameter will be only determined by experiment.

Scaling relationships for gas discharges have been developed for the case where the discharge power causes a negligible rise in gas temperature. These derivations and relations are discussed by von Engel and Steenbeck (1934) and Francis (1956) and were applied to He–Ne lasers by Gordon and White (1963). More recently, they have been modified by Konyukhov (1971) for higher power cylindrical discharges where gas temperature is an important consideration and Doppler-broadened laser transitions are considered. His work was discussed and

Table 4
Similarity relations (Abrams and Bridges 1973).

Quantity	Relation	
Gas temperature T_g	$T_{g2} = T_{g1}$	(1)
Gas number density N	$D_2 N_2 = N_1 D_1$	(2)
Gas pressure p	$p_2 D_2 = p_1 D_1$	(3)
Electron temperature T_e	$T_{e2} = T_{e1}$	(4)
Electron density N_2	$N_{e2} D_2 = N_{e1} D_1$	(5)
Current Density J	$J_2 D_2 = J_1 D_1$	(6)
Current I	$I_2 / D_2 = I_1 / D_1$	(7)
Electric field E	$E_2 D_2 = E_1 D_1$	(8)
DC resistance Z	$Z_2 D_2^2 = Z_1 D_1^2$	(9)
Power input/length P_i/L	$P_{i2}/L = P_{i1}/L$	(10)
Gain coefficient α	$\alpha_2 D_2 = \alpha_1 D_1$	(11a)
	$\alpha_2 = \alpha_1$	(11b)
Saturation flux density S	$S_2 D_2 = S_1 D_1$	(12a)
	$S_2 D_2^2 = S_1 D_1^2$	(12b)
Power output/volume P_0/V	$(P_{o2}/V)D_2^2 = (P_{o1}/V)D_1^2$	(13)
Power output/length P_0/L	$P_{o2}/L = P_{o1}/L$	(14)
Efficiency η	$\eta_2 = \eta_1$	(15)

extended to the case of pressure-broadened lasers by Abrams and Bridges (1973).

Konyukhov showed that if gas temperature and electron temperature are the same at corresponding points in two discharges, then the relationships shown in table 4 hold, where the two discharge diameters are D_1, D_2. Relations (1) to (10) involve only the discharge parameters while the remaining relations also involve the optical interactions. Relations (11) and (12) are given for both Doppler- and pressure-broadened lines and require additional comment. The relations for the Doppler-broadened case (11a), (12a) follow directly from the relation for gas density (2) and the observation by Konyukhov (1971) that the relative populations of all levels remain equal in similar discharges, providing the gas temperature is constant. For Doppler-broadened transitions (11a), a constant linewidth is assumed. At high pressures, the linewidth is proportional to the number density, so that the gain becomes independent of density and hence independent of D (11b). Likewise for the saturation parameter S, the linewidth is broadened by collisions resulting in an additional variation with N (and thus D) giving (12b). Relation (13) assumes a homogeneous interaction, so the power per unit volume is given by αS. This is well satisfied for homogeneously broadened lines, e.g., pressure-broadened, and is approximately true for the well-saturated multimode operation which burns off the top of an inhomogeneously broadened line [see Yariv (1967)]. Relation (14) assumes that the optical mode always fills the entire diameter D, a condition automatically satisfied for waveguide lasers.

It must be emphasized that the relations in table 4 describe *similar* discharges; that is, if the independent parameters of the two discharges are adjusted to satisfy the conditions of table 4, then the remaining parameters should satisfy the relations listed. A convenient choice of parameters is T_g, p and I. The remaining variables may then be considered as dependent characteristics. The scaling laws do *not* tell us how output power or gain vary with gas pressure or current; rather the relations tell us only how the characteristics of two *similar* laser tubes will compare. In fact, according to (14) the output power for tubes of the same length will be identical. Given that optimum conditions have been determined for a particular laser transition and tube geometry, table 4 tells us several interesting things about waveguide lasers. As the tube diameter is made smaller,

(1) the pressure increases,

(2) the current decreases,

(3) the voltage increases,

(4a) the gain increases (for a Doppler-broadened transition),

(4b) the gain remains the same (for a pressure-broadened transition),

(5) the power per unit length remains the same,

(6) the efficiency remains the same.

This assumes that waveguiding losses are negligible, an assumption which eventually becomes invalid at very small diameters.

Another fact self-evident from table 4 is that power per unit volume is not useful as a scaling parameter in the same sense as it is used to characterize TEA lasers. High values of P/V from small diameter lasers cannot be scaled to large volumes. Power per unit length controls the gas temperature and is thus the appropriate quantity for scaling of all gas lasers whose temperature is controlled by wall cooling.

4. CO_2 waveguide lasers

Because of their high efficiency and general usefulness in systems applications, CO_2 waveguide lasers have received considerable attention in research and development. This work has resulted in development of a number of useful waveguide laser devices including a miniature cw laser source, high pressure tunable local oscillator, wideband laser transmitter and high repetition rate waveguide TEA laser. In the following sections, the characteristics of waveguide CO_2 laser devices are discussed along with the development of these specialized devices.

4.1 Characteristics of waveguide CO_2 lasers

Application of the waveguide laser principle to the CO_2 laser system was first accomplished by Bridges et al. (1972) using flowing gas mixtures of CO_2–He–N_2 in a 1 mm diameter ×30 cm long glass capillary tube. They observed substantial increases in gain, output power per unit volume and saturation intensity over conventional CO_2 laser devices, as expected from the scaling relations of §3.1.

Burkhardt et al. (1972) later showed improved operating characteristics in 1 mm diameter ×10 cm long BeO ceramic tubes, where increased thermal conductivity keeps the discharge cooler. Power output of up to 2.4 W and gain coefficients as high as 27 dB/m were realized at high flow rates and −60°C operating temperature. Additional data on characteristics of CO_2 waveguide lasers with flowing gas was presented by Jensen and Tobin (1972) and Degnan et al. (1973).

Motivated by the need for tunable sealed-off lasers in CO_2 laser communications systems, Abrams and Bridges (1973) described the performance characteristics of sealed-off CO_2 waveguide laser devices. The emphasis was on high pressure operation (200–300 Torr) where collision broadening of the CO_2 laser transition exceeds 1 GHz. The particular need was a Doppler tracking laser local oscillator for a heterodyne receiver.

Measurements of laser gain and output power were made over a wide range of gas mixtures, pressure and discharge conditions in a 1.5 mm diameter × 18 cm long water cooled BeO discharge tube. Detailed measurements of gain made for He–CO_2 mixtures are shown in fig. 13 for a fixed discharge current of 3 mA. For mixtures rich in CO_2, the gain peaks at low pressure, while for mixtures lean in CO_2, the peak gain is lower but peaks at higher pressure. It is clear from the data that significant output can be achieved at pressures in excess of 200 torr.

Small amounts of N_2 and Xe were added to selected He–CO_2 mixtures. In general, it was found that $\frac{1}{2}$ part N_2 and $\frac{1}{4}$ part Xe give optimum gain, but only resulting in a 10% increase over the He–CO_2 data. Xe has the advantage of decreasing the discharge potential while N_2 has the opposite effect. Using an optimized mixture of CO_2–Ne–N_2–Xe $= 4:1:\frac{1}{2}:\frac{1}{4}$, the 18 cm device produced 4.0 W of output power at 150 torr of total pressure at an efficiency of 7.1%. This corresponds to a power per unit length of 0.22 W/cm. It is projected that higher efficiency and power per unit length are possible with optimized optics, but these numbers are quite respectable for CO_2 lasers today.

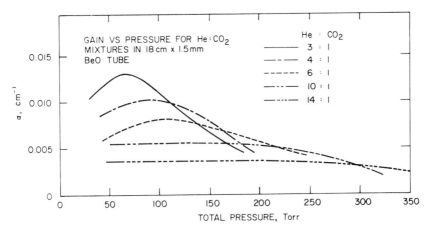

Fig. 13. Gain versus total pressure for various sealed-off He–CO_2 mixtures in an 18 cm × 1.5 mm BeO tube (Abrams and Bridges 1973).

4.1.1 10.6 μm waveguide losses

No discussion of CO_2 waveguide laser characteristics would be complete without mention of waveguide losses. As shown in §2.1, the field loss coefficient for the nm mode is given by

$$\alpha_{nm} = \left(\frac{u_{nm}}{2\pi}\right)^2 \frac{8\lambda^2}{D^3} \, \mathrm{Re}(\nu_n), \tag{25}$$

where u_{nm} is the mth root of J_{n-1} and ν_n is a function of the complex refractive index $\nu = n - ik$. At 10.6 μm, this complex refractive index is quite lossy for most practical waveguide materials, but this does not imply that α_{nm} will be large. In fact, some very large values of n and k can lead to small values of $\mathrm{Re}(\nu_n)$ and thus low loss coefficients. A simple way of picturing this physically is to think of the penetration of the guided mode into the wall material and to note that very high index (n) or conductivity (k) will tend to exclude the field from the dielectric, i.e., the wave is efficiently reflected and thus guided with low loss.

Abrams and Bridges (1973) have evaluated $\mathrm{Re}(\nu_n)$ for typical waveguide laser materials as a function of wavelength and the results are shown in fig. 14. Note that there is a substantial material and wavelength dependence, allowing the possibility of choosing waveguide materials for wavelength control. Using eq.

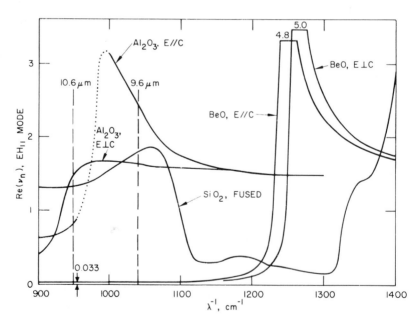

Fig. 14. Refractive-index dependent contribution to the waveguide loss $\mathrm{Re}(\nu_n)$ of the EH_{11} mode shown as a function of frequency for Al_2O_3, SiO_2 and BeO. Curves for Al_2O_3 and BeO show two orientations with respect to the c-axis (Abrams and Bridges 1973).

(25) and the data from fig. 14, the power loss coefficient ($2\alpha_{11}$) is

$$2\alpha_{11} = \begin{cases} 8.6 \times 10^{-5}\,\text{cm}^{-1}, & \text{for BeO} \\ 3.6 \times 10^{-3}\,\text{cm}^{-1}, & \text{for SiO}_2 \end{cases}$$

in a 1 mm diameter waveguide at 10.6 μm. It is clear that BeO is the ideal material for a 10.6 μm waveguide laser, especially in view of the fact it has exceptional thermal conductivity. However, due to its high cost and difficulty in machining, aluminum and silica lasers will also find application.

4.2 Special devices

In this section, a number of specialized CO_2 waveguide laser devices are described.

4.2.1 Miniature sources

A sealed-off miniature laser source only 7.5 cm long, machined out of a block of alumina, and emitting 1.4 W at 6% efficiency was demonstrated by Abrams and Bridges (1973). A photograph of this device is shown in fig. 15, demonstrating that noncircular waveguide geometry can also be considered for waveguide laser operation.

Fry and Brower (1977) extended the data of fig. 13 to lower pressures and higher CO_2:He ratios and showed that gain coefficients as high as 5%/cm were possible in pure CO_2 at 20 torr in 1 mm diameter discharge tubes. Low power laser operation with less than 1 W of discharge power was demonstrated with 40 mW of output power. At slightly higher input power, 9% efficiency was achieved. These results were achieved in a 4 cm long laser, demonstrating that small low power lasers are possible in CO_2 while preserving good efficiency.

Laakman et al. (1977) presented a new concept for waveguide laser excitation using a transverse rf discharge between parallel metal electrodes which also serve as waveguiding walls. cw output of 6 W at 11% efficiency was achieved from a 20 cm long $\times$ 1.5 mm^2 device operating at 10.6 μm.

4.2.2 Tunable local oscillator

The limited tuning range of conventional Doppler-broadened CO_2 lasers (53 MHz) has limited their utility in laser in laser communications, optical radar, pollution detection and spectroscopy. Operation of waveguide CO_2 lasers at high pressure can result in pressure broadening of the laser transition and much larger available gain bandwidth for frequency tuning and other applications. Abrams (1974a) demonstrated a tuning capability of 1.2 GHz on a single vibration–rotation transition in a sealed-off cw waveguide CO_2 laser.

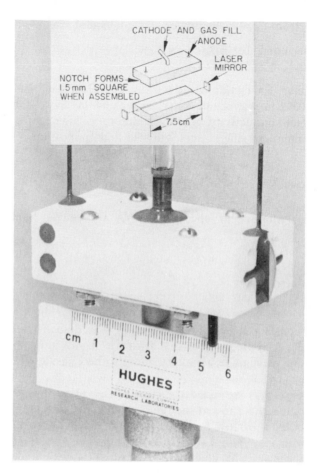

Fig. 15. Photograph and exploded view of 3 in 1.4 W channel laser (Abrams and Bridges 1973).

The tunable laser, shown in fig. 16, was fabricated from four polished BeO slabs, 9.5 cm long, epoxied together to form a 1.0 mm square waveguide. Copper gasketed vacuum flanges are machined to fit over the rectangular outer dimensions of the tube, and are epoxied to the ceramic. All subsequent hardware (mirror mounts, piezoelectric bimorph tuning element) is mounted by means of these mating flanges. Cooling is achieved by conduction to a water cooled aluminum heat sink. Laser output is taken off the zeroth order from the 150 ℓ/mm diffraction grating used in the Littrow configuration for line selection.

Maximum tunability of the laser was found with an $8:1:\frac{1}{2}:\frac{1}{4}$ mixture of He–Co_2–N_2–Xe. The grating was tuned for oscillation on the P(20) $00^\circ1$–$10^\circ0$ transition at $10.59\,\mu m$ and output power was measured versus bimorph voltage (equivalent to cavity length). The results are shown in fig. 17. Maximum tuning was observed to be 1200 MHz at a static fill pressure of 260 torr. Output power at line

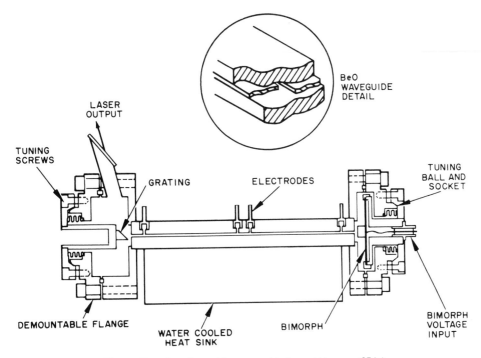

Fig. 16. Drawing of tunable waveguide laser (Abrams 1974a).

center was 80 mW, limited by the small zero-order reflectivity of the grating (0.6%).

This laser was later used by Abrams (1974b) to directly measure line shapes and pressure broadening coefficients for the CO_2 laser transition. A similar tunable waveguide CO_2 laser has been incorporated into a Doppler tracking heterodyne communications system built for NASA by Goodwin (1975) and a flowing gas version of the laser was used by Menzies and Shumate (1977) in airborne heterodyne ozone monitoring experiments.

4.2.3 Wideband power amplifier

The very same considerations that lead to the design of tunable waveguide lasers can be applied to wideband laser power amplifier design. High gain wideband power amplifiers are of interest for CO_2 laser radar transmitters, where the bandwidth is required for FM Doppler radar signals. Klein and Abrams (1975) built a 1.65 mm i.d. × 1 m long waveguide laser amplifier test bed (see fig. 18) and performed a comprehensive analytical and experimental study of the amplifier gain, bandwidth and saturation characteristics.

The device made use of seven separate BeO sections connected together by Kovar flanges with a parallel connection of electrical power and gas flow as

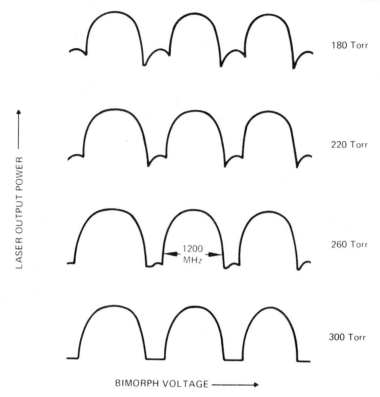

Fig. 17. Tuning characteristics of waveguide laser for several values of total gas pressure. An 8:1:0.5:0.25 mixture of He–CO_2–N_2–Xe was used. Output power was coupled from the zeroth grating order and has a maximum value of 80 mW (Abrams 1974a).

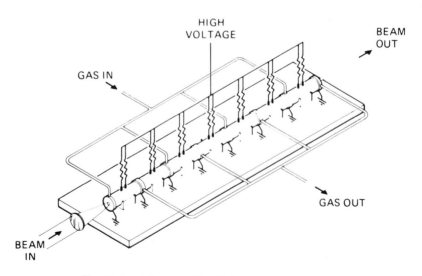

Fig. 18. Amplifier assembly (Klein and Abrams 1975).

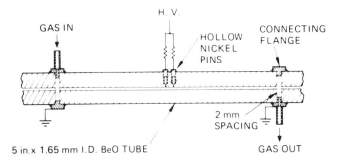

Fig. 19. Detail of electrical and gas flow connections (Klein and Abrams 1975).

shown in fig. 19. Overall transmission of he amplifier was 92% for an EH_{11} mode, which compares with a theoretical value of 97%. The additional loss was due to misalignment at the joints of the segmented system.

Gain measurements were made as a function of pressure and for different gas mixtures with the results shown in fig. 20. Saturation power measurements versus pressure, taken for optimum gas mixtures, are shown in fig. 21. As expected, saturation flux increases with pressure (owing to molecule–molecule

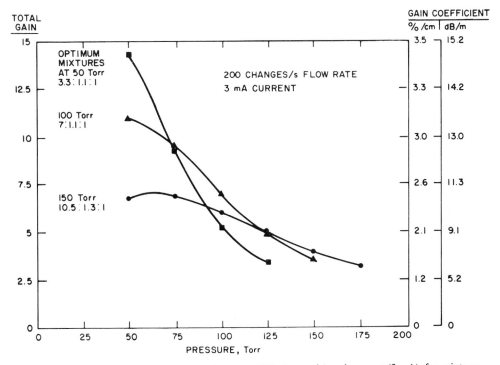

Fig. 20. Gain versus pressure at constant flow rate (200 changes/s) and current (3 mA), for mixtures shown (Klein and Abrams 1975).

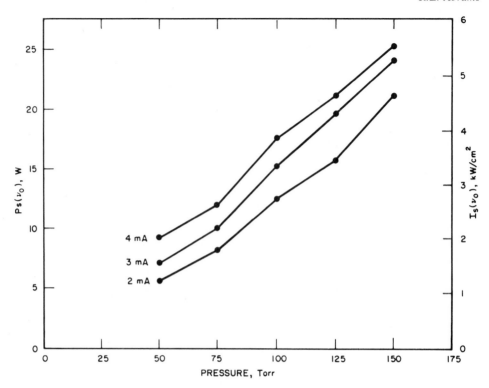

Fig. 21. Saturation power and intensity versus pressure (flow rate = 200 changes/s) (Klein and Abrams 1975).

collisions) and with current (owing to electron–molecule collisions). In a highly saturated amplifier, the product of the gain coefficient and the saturation power gives the power extraction capability. At 3 mA of current, extraction power of 20 W/m at 70% efficiency is possible at an amplifier pressure of 150 torr. For moderate gain, this leads to an amplifier bandwidth of greater than 500 MHz. Significantly improved performance has been achieved by reduced temperature operation of this same amplifier by Tomasetta and Carter (1976) who have implemented this amplifier into a CO_2 laser radar system operated by MIT Lincoln Laboratory.

4.2.4 Wideband transmitter

Modulation of CO_2 laser radiation for wideband communications has best been achieved by coupling modulation. In this scheme, an electro-optic CdTe crystal is placed within the laser cavity where the circulating optical power far exceeds the laser output power. Application of modulating voltage to the modulator results in polarization modulation of the circulating power which can be coupled from the cavity. Using this technique, Kiefer et al. (1972) demonstrated efficient coupling modulation with bandwidths in excess of 200 MHz and Goodwin et al. (1976a) integrated this approach into a 300 MBs CO_2 laser communications demonstration system for NASA.

More recently, Goodwin et al. (1976b) have extended this concept to waveguide CO_2 lasers. Waveguide lasers are ideally suited for intracavity coupling modulation owing to the fact that a guided mode within a square cross-sectioned hollow waveguide can couple very efficiently into an electro-optic modulator rod of the same cross section. A drawing of the sealed-off laser transmitter developed for NASA is shown in fig. 22. The BeO laser structure is surrounded by the Kovar outer cylinder forming the vacuum envelope. The CdTe modulator and housing are precision aligned to the tube bore and are separated by a ZnSe anti-reflection coated window from the vacuum enclosure. A diffraction grating serves to select polarization and laser transition, and also acts as the polarization sensitive output coupler. The laser discharge is in four segments to reduce discharge voltage and increase discharge stability. The laser section is 28.7 cm long while the entire structure measures 40.9 cm.

The laser transmitter performance is summarized in table 5 for a nominal circulating optical power of 30 W and modulation voltage of 60 V rms. Sideband output power was 0.45 W and modulator bandwidth was 400 MHz. Some modulation peaking of the frequency response was observed by $C/2L$ resonances at 300 MHz owing to the incomplete (80%) coupling off of the grating. Complete coupling of the sideband signals from the cavity has been shown to eliminate this problem.

4.2.5 Pulsed waveguide lasers

A number of CO_2 laser applications dictate the use of pulsed output. The requirements vary from the relatively low pulse rates and high energy per pulse required for laser rangefinders and designators to the very short pulse rates required for laser communications. Many of the techniques that have been used for producing pulsed output with conventional CO_2 lasers have been applied to waveguide CO_2 lasers and are discussed here. They include mode locking, Q-switching and gain switching as in transversely excited atmospheric pressure (TEA) lasers.

Mode locking of waveguide CO_2 lasers is an attractive possibility because the large pressure-broadened linewidth allows for the generation of very short pulse widths. Smith et al. (1973a) first demonstrated mode locking of a waveguide CO_2 laser with flowing gas. They produced a cw train of 3 ns wide pulses via intracavity loss modulation at the $C/2L$ frequency using a germanium intracavity acousto-optic modulator. This is similar to the technique developed for TEA laser mode locking developed by Wood et al. (1970). Abrams (1974c) used a similar technique to mode lock a cw sealed-off CO_2 laser, producing the train of

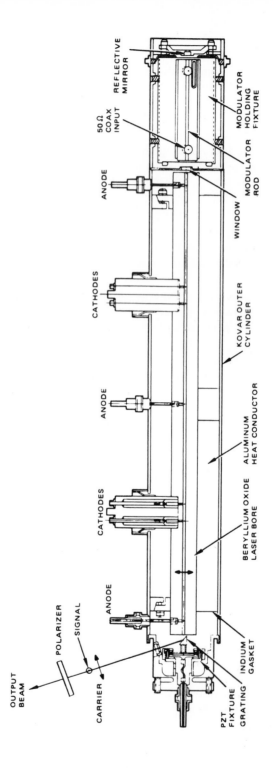

Fig. 22. Waveguide laser transmitter built by the Hughes Aircraft Company for NASA (Goodwin et al. 1976b).

Table 5
Laser transmitter performance.

Parameter	Measured performance
Sideband power	0.45 W
Bandwidth	400 MHz
Laser input power	178 W
Laser input power (excluding ballast)	144 W

2.1 ns wide pulses shown in fig. 23. The pulse repetition rate was 180 MHz and the average power output 0.5 W at a laser pressure of 150 torr. At higher operating pressures continuous trains of 1 ns pulses should be possible.

Q-switching of conventional CO_2 lasers has been useful for production of kilowatt peak powers with pulse widths of 100 to 200 ns. Goodwin et al. (1976b) used intracavity electro-optic loss modulation of a CO_2 waveguide laser to produce 15 to 20 μJ pulses at 10 kHz pulse rates. Typical measured pulse widths and peak power were 110 ns and 120 W respectively. This data was taken with waveguide lasers designed for communications use and normally capable of ~ 1.5 of cw output power with the intracavity modulator. In principle, Q-switched waveguide CO_2 lasers should provide equivalent performance to Q-switched conventional CO_2 lasers.

In the design of waveguide lasers, it was shown in §3.1 that, for conduction cooled lasers, power per unit length is a useful scaling parameter. With mode-locked and cw pumped Q-switched lasers, we can extend this argument to average power per unit length. In the case of TEA lasers, however, energy per unit volume is a more useful scaling parameter. Waveguide TEA lasers therefore can be expected to have lower output energy than conventional TEA lasers simply because of their smaller volume. Their advantages are high gain, small size, lower operating voltage and high prf capability when compared with conventional CO_2 TEA lasers.

Operation of a waveguide TEA laser was first demonstrated by Smith et al. (1973b). They observed laser action up to 1 atm of pressure and pulse rates of up to 4 kHz. Peak output power was ~ 100 W in a 150 ns gain switched spike reminiscent of conventional TEA laser operation. They found operational stability increased at high prf due to pre-ionization of the laser medium by the previous pulse, a phenomenon thoroughly studied recently by Jain et al. (1977). Wood et al. (1975) further exploited this effect in a 1 mm^2 waveguide amplifier achieving small signal gains greater than 5%/cm over the 100 to 760 torr range. These gain measurements versus volumetric input energy were shown to be equivalent to previous data on cold cathode conventional TEA laser measurements. Further development was later reported by Smith et al. (1976). They demonstrated prf as high as 40 kHz. In the 1 mm$^2 \times$ 10 cm BeO device shown in fig. 24, quasi-continuous operation was achieved at pressures up to $\frac{1}{3}$ atm with an

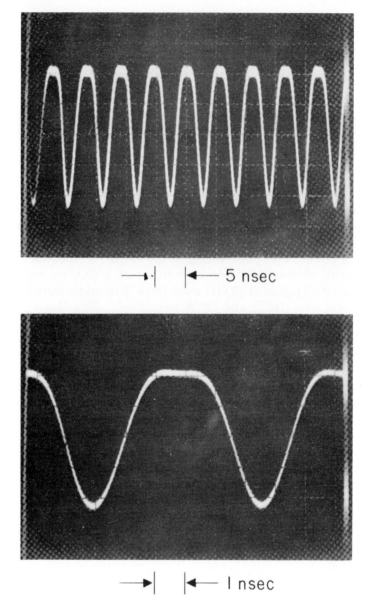

Fig. 23. Mode-locked laser pulses from cw waveguide CO_2 laser (Abrams 1974).

average output power of 1.5 W. The active volume of this laser was only 0.1 cm³. Operation up to 100 kHz prf at 1 atm is projected where tunability could exceed 4 GHz. With the projected characteristics, the laser would be a useful source for spectroscopy and pollution detection, and for military applications such as rangefinders and designators.

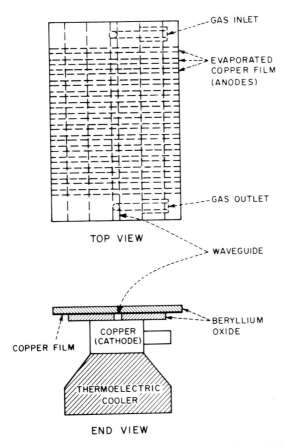

Fig. 24. Waveguide TE laser construction (Smith et al. 1976).

Walker and Rickwood (1976) demonstrated 0.1 mJ output pulses with peak power of 1.2 kW from an extremely simple waveguide TEA structure. The device was fabricated from printed circuit board, glass and aluminum plates and 1 kΩ resistors. No windows were required as the gas simply exited from the open ends. Further simplification was demonstrated by Gibson et al. (1977) where a semiconducting silicon slab was used to replace the resistively loaded cathodes with comparable performance. Other novel waveguide CO_2 laser structures for tactical military applications have been recently reported. Mocker and Willenbring (1977) and Bua and Rudko (1977) discussed waveguide TEA lasers for rangefinders and designators.

A major constraint in the design of high prf waveguide TEA lasers is the necessity for a high rate of gas flow. This flow must either be longitudinal or through holes provided in the waveguiding walls. Papayoanou and Fujisawa (1975) used porous BeO waveguide walls to enable transverse flow in cw waveguide CO_2 lasers, and this technique should also be applicable to TEA lasers. A novel type of low loss parallel plate waveguide which would allow transverse gas flow was proposed and demonstrated by Nishihara et al. (1974). This guide consists of slightly curved parallel plates of metal or dielectric which continually refocuses and confines the radiation. Losses as low as 0.02 dB/m were predicted and measured for a copper guide with a separation of only 0.5 mm, a value six times less than the corresponding circular waveguide. This type of structure was subsequently used by Nishihara et al. (1976) to operate a TEA waveguide laser. This approach offers the possibility of extremely efficient gas flow as well as very low waveguide loss.

5. Other waveguide gas lasers

In this chapter, we have placed heavy emphasis on the properties of CO_2 waveguide lasers. The reason is that far more research has been done in this system than any other and it is expected that important applications will follow. In this section, progress and developments in other waveguide laser systems are briefly reviewed.

5.1 Noble gases

As discussed in §1.1, Smith (1971) built the first visible wavelength waveguide laser, a He–Ne laser operating at 6328 Å. A specially selected 20 cm × 430 μm length of capillary bore tubing formed the waveguide and was placed between two 30 cm radius of curvature mirrors positioned 29.5 cm from the waveguide ends to form a low loss resonator. The primary features of the waveguide He–Ne laser are:
(1) Higher gain due to the inverse dependence of gain on discharge diameter.
(2) Higher operating pressure (~7 torr) resulting in nearly homogeneous broadening of the He–Ne laser transition. This causes the laser to operate in the highly desired single mode condtition owing to strong mode competition effects.
(3) Excellent transverse mode control is achieved via the unique properties of waveguide resonators.
Smith and Maloney (1973) demonstrated some unique features of the He–Xe laser system at 3.5 μm operated in the waveguide geometry. This transition exhibited extraordinary gain coefficients, approaching 1000 dB/m in a 250 μm bore. Single isotope He^3–Xe^{136} was used with a combination of dc and rf excitation. Measured gain versus pressure for dc discharge excitation is shown in fig. 25. An amazing property of such high gain lasers is the strong dispersion associated with the gain, resulting in pronounced mode pulling effects. It was found that single frequency oscillation was obtained with a frequency always within $\frac{1}{100}(C/2L)$, independent of the setting of the cavity spacing.

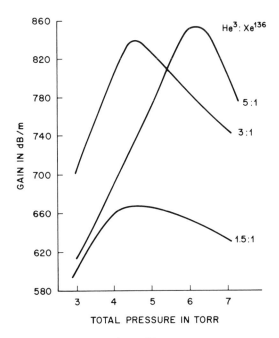

Fig. 25. Gain of a 250 μm- bore dc-excited He^3 – Xe^{136} laser as a function of gas pressure and mixture (Smith and Maloney 1973).

5.2 Molecules

All of the molecular lasers which had been successfully operated in conventional resonators should be operable in the waveguide configuration. The electrically excited CO laser is a good candidate because of its high efficiency and cw operation. Yusek and Lockhart (1973) obtained cw operation in sealed-off glass capillaries at a wall temperature of 205 K. They achieved 300 mW output from a 2 mm diameter discharge at an optimum operating pressure of 65 torr.

Asawa (1974) reported measurements on a 2 mm dia × 14 cm long sealed-off BeO tube cooled by dry ice to 210 K. He achieved 1.1 W output at 5.7% efficiency in a single mode. Optimum pressure was found to be 80 torr (close to the value reported above) with a CO–He–Xe–N$_2$ mixture of 1:8:1:1. He also found that the waveguide results were consistent with previous sealed-off CO laser measurements, fitting a scaling relationship of $pD = 160$ torr mm.

Evidence that the waveguide laser principle is quite general was given by Bua and Rudko (1977). Operation of miniature waveguide TEA lasers with a number of molecular and atomic gases in the near-infrared was demonstrated. In addition to CO$_2$, lasing was accomplished in HF(2.8 μm), DF(3.8 μm), HBr(4.2 μm), CH$_3$I(3.04 μm), ArF(1.69 μm), Xe(2.03) μm) and Ar(1.25, 1.28 μm). They found that output power and efficiency increased with anode to cathode distance owing to the increased room for fan-out of the discharge. At the most efficient operating point, the spacing was 4 mm, where the lasers are not really waveguide lasers but just small TEA lasers. Millijoule outputs were obtained with DF at 3.8 μm in these devices. Thus we see that although laser action was obtained in the waveguide configuration, for many transitions optimized device performance did not occur in the waveguiding region.

5.3 Far-infrared waveguide lasers

The far-infrared (FIR) spectral region from 20 μm to 2 mm is of potential importance for spectroscopy, plasma diagnostics, radiometry and radar, but has not been fully exploited partly owing to the lack of efficient high power sources. Chang and Bridges (1970) discovered optically pumped FIR lasing in CH_3F, and subsequently hundreds of FIR sources have been found by optical pumping of rotational transitions in molecular systems. Partial tabulations of these transitions can be found in articles by Chang (1974), Yamanaka (1976) and Rosenbluh et al. (1976). Inversion is generally achieved by resonant or near-resonant absorption of a strong pump laser (CO_2 or N_2O) followed by lasing on a pure rotational transition. Pumping with 10 to 50 W of CO_2 laser power has resulted in 1 to 10 mW of FIR laser output.

Optically pumped FIR laser action in a waveguide configuration was first achieved by Hodges and Hartwick (1973). A CO_2 laser was used to pump CH_3OH in a hollow 1 cm diameter × 70 cm long metal FIR waveguide. The pump and FIR output were both hole coupled to the resonant cavity formed by flat metal mirrors terminating the hollow tube. CW output at power levels greater than 1 mW was achieved with a 4 W pump laser.

Rate equation models of FIR lasers have been given by Tucker (1974), Yamanaka et al. (1974), Henningsen and Jensen (1975) and DeTemple and Danielewisc (1976). The photon conversion efficiency is limited by the slow vibrational relaxation rate of the lower laser level. Wall diffusion has been shown to increase this rate, as shown by Hodges et al. (1976a). Chang and Lin (1976) further increased this rate and the FIR output by adding buffer gases to the laser. The combination of small diameter waveguide geometry and use of buffer gas has been suggested by Hodges et al. (1976b) to be optimum for efficient high power FIR lasers.

In general, optically pumped FIR lasers have operated with efficiencies less than 1% of the theoretical maximum value, which is shown by Hodges et al. (1976b) to be given by

$$\eta_{max} = \frac{1}{2}(\nu_{FIR}/\nu_p), \tag{26}$$

where ν_{FIR}/ν_p is the ratio of the FIR frequency to the pump frequency. Hodges et al. (1976b) were able to pay careful attention to output coupling, pump beam absorption and spatial distribution through use of waveguide laser geometry. A summary of their high power, high efficiency results is given in table 6. They

Table 6
CW performance data for 38 mm diam dielectric waveguide lasers.

Output power [mW]	Wavelength [μm]	Diam. × length [mm × m]	Molecule	Pump line [9.6 μm]	Pump power [W]	Efficiency [%]
100	71	38 × 1.1	CH_3OH	P(34)	30	4.8
150	118	38 × 1.1	CH_3OH	P(36)	27	13.1
15	496	38 × 1.1	$C^{13}H_3F$	P(20)	30	5.0
22	496	38 × 1.1	$C^{12}H_3F$ + hexane	P(20)	30	7.3
10	1222	38 × 1.1	$C^{13}H_3F$	P(32)	30	8.3
400	118	38 × 2.0	CH_3OH	P(36)	60	15.7
40	496	38 × 2.0	$C^{12}H_3F$	P(20)	70	5.7

[A]Output power and efficiency are listed for single end only. Output coupling at 71 μm is provided by 6 mm diam. hole in plane metallic reflector. Remaining data were taken using Si hybrid output coupler with 12 mm diam. hole. The efficiency is the ratio of the measured power to the theoretical maximum output power expressed by eq. (26) (Hodges et al. 1976b).

achieved 150 mW output at 118 μm in CH_3OH and 15 mW at 496 μm in CH_3F from a 1 m long waveguide laser with 30 W of CO_2 laser power. A 2 m long FIR laser of the same type produced 400 mW at 118 μm and 40 mW and 496 μm. Efficiencies as high as 20% of theoretical and data on additional laser transitions were reported by Hodges et al. (1977). A key component for achieving efficient operation at wavelengths greater than 100 μm was a novel output mirror. The mirror consisted of a silicon window dielectric coated for high reflectivity at 10 μm and overcoated with 5000 Å of gold, masked to leave an uncoated coupling hole in the center of the mirror. The dielectric coating reflects the 10 μm pump and transmits the FIR efficiently. It is projected that a number of FIR sources with as much as 0.1 to 1 W cw power may be possible with a 100 W CO_2 laser pump.

A review of optically pumped waveguide lasers has been given by Yamanaka (1977). He discusses FIR attenuation in waveguides for various materials, the theory of FIR lasers, output coupling, new lines, polarization and modes, and power and stability, as well as applications. Kneubuhl (1977) has given a review of waveguide structures for FIR lasers.

Stark tuning of FIR lasing molecules has proven useful for both tuning the absorption near useful pump laser transitions and for modulation of the FIR laser output. Fetterman et al. (1973) found new FIR laser transitions in NH_3 by Stark tuning and were able to significantly enhance FIR laser output by Stark modulation of the absorption frequency, thereby reducing the effects of hole burning in the pump absorption. These techniques could easily be applied to waveguide FIR lasers. Tobin and Jensen (1977) showed that sizeable modulation effects could be achieved with small ac Stark fields in CH_3OH and CH_3F waveguide FIR lasers. If laser pumped FIR lasers are to be used in practical communications, radar or tactical military systems, simple modulation techniques such as Stark tuning may be important.

6. Conclusions

Waveguide gas laser research has enjoyed considerable support in universities and industrial and government research laboratories since 1971 following publication of the first He–Ne waveguide laser results by Smith (1971). Exciting research has been conducted on the characteristics of waveguide lasers in various laser systems and on the properties of waveguide laser resonators.

The application of waveguide lasers to useful systems will depend, of course, on some improvement in performance or new capability not offered by the corresponding conventional device. Potential improvements offered by waveguide laser technology will be in the areas of size, weight, cost and efficiency. New capabilities will be features such as high pressure operation for tunability and mode control, longer lifetime, low operating voltages for TEA devices, excellent transverse mode control unique to waveguide lasers, compatibility with waveguiding modulator structures, high gain and power per unit volume, and very low prime power devices. It will be the exploitation of these features and others yet to be discovered that will pave the way for waveguide laser applications. The CO_2 waveguide laser has already been chosen for its tunability in spectroscopic and communications applications, and it is anticipated that miniature TEA lasers and waveguide FIR lasers will also be used in applications where their unusual properties offer systems advantage.

References

Abrams, R.L. (1972), IEEE J. Quantum Electronics **OE–8**, 838.
Abrams, R.L. (1974a), Appl. Phys. Letters **25**, 304.
Abrams, R.L. (1974b), Appl. Phys. Letters **25**, 609.
Abrams, R.L. (1974c), Wideband Waveguide CO_2 Lasers, in: Laser Spectroscopy – Proceedings of an International Conference on Laser Spectroscopy, Vail, CO, 1973 (R.G. Brewer and A. Mooradian, eds., Plenum, New York), 263–272.
Abrams, R.L. and W.B. Bridges (1973), IEEE J. Quantum Electronics **OE–9**, 940.
Abrams, R.L. and A.N. Chester (1973), Spring Meeting of Optical Society of America, Denver, CO, paper ThD16.
Abrams, R.L. and A.N. Chester (1974), Appl. Optics **13** 2117.
Asawa, C.K. (1974), Appl. Phys. Letters **24**, 121.
Bridges, T.J., E.G. Burkhardt and P.W. Smith (1972), Appl. Phys. Letters **20**, 403.
Bua, D.P. and R.I. Rudko (1977), IEEE/OSA conference on Laser Engineering and Applications, Washington, D.C., paper 6.9.
Burkhardt, E.G., T.J. Bridges and P.W. Smith (1972), Opt. Commun. **6**, 193.
Chang, T.Y. (1974), IEEE Trans. Microwave Theory Tech. **MTT–22**, 983.
Chang, T.Y. and T.J. Bridges (1970), Opt. Commun. **1**, 423.
Chang, T.Y. and C. Lin (1976), J. Opt. Soc. Amer. **66**, 362.
Chester, A.N. and R.L. Abrams (1972), Appl. Phys. Letters **21**, 576.
Degnan, J.J. (1973), Appl. Optics **12**, 1026.
Degnan, J.J. (1976), Appl. Phys. **11**, 1.
Degnan, J.J. and D.R. Hall (1972), IEEE International Electron Devices Meeting, Washington, D.C.
Degnan, J.J. and D.R. Hall (1973), IEEE J. Quantum Electronics **QE–9**, 901.

Degnan, J.J., H.E. Walker, J.H. McElroy and N. McAvoy (1973), IEEE J. Quantum Electronics **QE–9**, 489.

DeTemple, T.A. and E.J. Danielewisc (1976), IEEE J. Quantum Electronics **QE–12**, 40.

Fetterman, H.R., H.R. Schlossberg and C.D. Parker (1973), Appl. Phys. Letters **23**, 684.

Fetterman, H.R., C.D. Parker and P.E. Tannenwald (1976), Opt. Commun. **18**, 10.

Fox, A.G. and T. Li (1961), Bell Syst. Tech. J. **40**, 453.

Francis, G. (1956), Handbuch Der Physik, Glow Discharge at Low Pressure (Springer-Verlag; Berlin), 53–195.

Fry, S.M. and R.E. Brower (1977), IEEE/OSA Conference on Laser Engineering and Applications, Washington, D.C., paper G.6.

Gibson, A.F., K.R. Rickwood and A.C. Walker (1977), Appl. Phys. Letters **31**, 176.

Goodwin, F.E. (1975), Opto-Mechanical Sybsystem of a $10\,\mu$m Wavelength Receiver Terminal, National Aeronautics and Space Administration Contract NAS 5–21859.

Goodwin, F.E., T.A. Nussmeier, L.S. Stokes and E.J. Vourgourakis (1976a). Experimental Definition Space Shuttle Laboratory – Laser Data Relay Link $10.6\,\mu$m Experiment, National Aeronautics and Space Administration Contract NAS 5–20018.

Goodwin, F.E., D.M. Henderson, J. Wilderson, and A. Reiss (1976b), 10 Micrometer Transmitter, National Aeronautics and Space Administration Contract NAS 5–20623.

Gordon, E.I. and A.D. White (1963), Appl. Phys. Letters **3**, 199.

Henningsen, J.O. and H.G. Jensen (1975), IEEE J. Quantum Electronics **QE–11**, 248.

Hodges, D.T. and T.S. Hartwick (1973), Appl. Phys. Letters **23**, 252.

Hodges, D.T., J.R. Tucker and T.S. Hartwick (1976a), Infrared Phys. **16**, 175.

Hodges, D.T., F.B. Foote and R.D. Reel (1976b), Appl. Phys. Letters **29**, 662.

Hodges, D.T., F.B. Foote and R.D. Reel (1977), IEEE J. Quantum Electronics **QE–13**, 491.

Jain, R.K., O.R. Wood, P.J. Maloney and P.W. Smith (1977), Appl. Phys. Letters **31**, 260.

Jensen, R.E. and M.S. Tobin (1972), Appl. Phys. Letters **20**, 408.

Kiefer, J.E., T.A. Nussmeier and F.E. Goodwin (1972), IEEE J. Quantum Electronics **QE–8**, 173.

Klein, M.B. and R.L. Abrams (1975), IEEE J. Quantum Electronics **QE–11**, 609.

Kneubuhl, F.K. (1977), J. Opt. Soc. Amer. **67**, 959.

Kogelnik, H. and T. Li (1966), Appl. Optics **5**, 1550.

Konyukhov, V.K. (1971), Sov. Phys. – Tech. Phys. **15**, 1283.

Krammer, H. (1976), IEEE J. Quantum Electronics **QE–12**, 505.

Laakman, K.D. and W.H. Steier (1976), Appl. Optics **15**, 1334.

Laakman, K.D. et al. (1977), unpublished.

Marcatili, E.A.J. and R.A. Schmeltzer (1964), Bell Syst. Tech. J. **43**, 1783.

Menzies, R.T. and M.S. Shumate (1977), IEEE/OSA Conference on Laser Engineering and Applications, Washington, D.C., paper 15.10.

Mocker, H.W. and G.R. Willenbring (1977), IEEE/OSA Conference on Laser Engineering and Applications, Washington, D.C., Paper 6.7.

Nishihara, H., T. Inoue and J. Koyama (1974), Appl. Phys. Letters **25**, 391.

Nishihara, H., T. Mukai, T. Inoue and J. Koyama (1976), Appl. Phys. Letters **29**, 577.

Papayoanou, A. and A. Fujisawa (1975), Appl. Phys. Letters **26**, 158.

Pohl, D. (1972), Appl. Phys. Letters **20**, 266.

Rosenbluh, M., R.J. Temkin and K.J. Button (1976), Appl. Optics **15**, 2635.

Sehwaller, P., H. Steffen, J.F. Moser and F.K. Kneubuhl (1967), Appl. Optics **6**, 827.

Smith, P.W. (1971), Appl. Phys. Letters **19**, 132.

Smith, P.W. and P.J. Maloney (1973), Appl. Phys. Letters **17**, 259.

Smith, P.W., T.J. Bridges, E.G. Burkhardt and O.R. Wood (1973a), Appl. Phys. Letters **21**, 470.

Smith, P.W., P.J. Maloney and O.R. Wood (1973b), Appl. Phys. Letters **23**, 524.

Smith, P.W., C.R. Adams, P.J. Maloney and O.R. Wood (1976), Opt. Commun. **16**, 50.

Steffen, H. and F.K. Kneubuhl (1968), Phys. Letters **27A**, 612.

Stratton, J.A. (1941), Electromagnetic Theory (McGraw-Hill; New York and London), 524.

Tobin, M.S. and R.E. Jensen (1977), IEEE J. Quantum Electronics **QE–13**, 481.

Tomasetta, L.R. and G.M. Carter (1976), private communication.

Tucker, J.R. (1974), Proceedings of International Conference on Submillimeter Waves, Atlanta, GA, IEEE Cat. 74CH0 856–5 MTT, p. 17.

von Engel, A. and M. Steenbeck (1934), Electrische Gasentlandungen, Ihre Physic und Tecknik (Springer-Verlag, Berlin).

Walker, A.C. and K.R. Rickwood (1976), J. Phys. **E9**, 432.

Wood, O.R., R.L. Abrams, and T.J. Bridges (1970), Appl. Phys. Letters **17**, 376.

Wood, O.R., P.W. Smith, C.R. Adams and P.J. Maloney (1975), Appl. Phys. Letters **27**, 539.

Yamanaka, M. (1976), Rev. Laser Eng. (Japan) **3**, 253.

Yamanaka, M. (1977, J. Opt. Soc. Amer. **67**, 952.

Yamanaka, M., Y. Homma, A. Tanaka, M. Takada, A. Tanimoto and H. Yoshinaga (1974), Jap. J. Appl. Phys. **13**, 843.

Yamanaka, M., R.J. Temkin and R.J. Button (1976), Appl. Optics **15**, 2635.

Yariv, A. (1967), Quantum Electronics (Wiley; New York), 243.

Yusek, R. and G. Lockhart (1973), IEEE/OSA Conference on Laser Engineering and Applications, Washington, D.C., paper 16.1.

A3 | High-Power, Efficient Electrically-Excited CO Lasers

R. E. CENTER

Mathematical Sciences Northwest, Inc.,
Bellevue, Washington 98009

Contents

Abstract

A review is presented of the development of efficient high-power, electrically-excited CO lasers. The review includes a description of the basic physics underlying the electrically-excited CO laser with emphasis on the effect of the molecular anharmonicity on the relaxation behavior and on the inversion mechanism. Also included is a discussion of the vibrational excitation of CO by electron impact with examples of predictions for the partitioning of electrical energy in several typical gas mixtures. These sections are followed by an outline of the development of kinetic models for the CO electric discharge laser and a derivation of simple scaling relations and performance curves of direct use to the laser designer. Because of the temperature sensitivity and desirability of uniform low temperature operation, a brief description is given of gas thermal conditioning techniques. Finally, specific examples are presented of various electrical-optical configurations with a discussion of the performance characteristics which have been achieved in practical devices.

1. Introduction

This chapter is concerned with the practical development of efficient high-power, electrically-excited CO lasers that emit in the infrared near $5\,\mu$m owing to vibration–rotation transitions in the ground electronic state of the CO molecule. With the advent of electron-beam sustained discharge techniques, it has been possible to obtain cw powers in the range of hundreds of kilowatts and pulse energies of the order of kilojoules at electrical efficiencies (defined by the ratio of optical output/electrical input) in the range from 20 to 50%. This chapter discusses the basic physical concepts pertinent to the CO electric discharge laser (EDL) as well as some of the practical aspects and it includes descriptions of the development of several basic types of high-power/energy CO EDLs.

Legay and Legay-Sommaire (1964) appear to have been the first to suggest the possibility of laser action in the CO molecule utilizing the vibrational–rotational transitions in the $X^1\Sigma^+$ electronic ground state of CO. The first experimental observation of laser action on these vibration–rotation transitions in CO appears to have been made by Patel and Kerl (1964), who observed laser oscillations in the wavelength range from 5.0 to $5.4\,\mu$m, using a pulsed discharge in low-pressure CO. The observed transitions corresponded to P-branch rotational transitions in the 6–5, 7–6 and 10–9 vibrational bands of the ground state of CO. Shortly after these first pulsed discharge experiments, continuous wave (cw) lasing action was achieved by Legay-Sommaire et al. (1965) and Patel (1965), both experiments employing mixing techniques. Legay-Sommaire et al. (1965) excited N_2O in a high-frequency discharge and subsequently mixed in CO. In a similar experiment, Patel (1965) excited N_2 in a discharge and mixed in CO. He was able to obtain cw oscillation on 143 transitions between 5.0 and $6.2\,\mu$m, the wide wavelength range being obtained by varying the temperature of the mixing region from 15 to $-78°$C.

Two other methods of excitation of CO were explored in the next few years. These were the CO chemical laser and the CO gasdynamic laser (GDL). Pollock (1966) observed stimulated emission on over 30 vibration–rotation transitions of the ground electronic state of CO in the region 5.1 to $5.7\,\mu$m, following flash photolysis of CS_2 and O_2 mixtures. In later experiments, Gregg and Thomas (1968) using similar flash photolysis found 270 new lasing lines, including the R-branch of seven vibrational transitions. Since then, a number of both pulsed and cw CO chemical lasers have been reported in the literature. Hancock and

Smith (1971) have shown that the reaction

$$O + CS \rightarrow CO^\dagger + S + 75 \text{ kcal/mole}$$

is responsible for producing the vibrationally-excited $CO^\dagger$ in the flash photolysis CS_2–O_2 chemical lasers. It has been shown that the vibrationally-excited CO product in the above reaction contains more than 80% of the available energy. This high vibrational energy yield suggests the possibility of high chemical efficiency. However, it is offset by the electrical energy input to the flashlamp or electrical discharge required to dissociate the CS_2 and/or O_2 in the first step of the reaction sequence. In recent years, the cw CO chemical laser (Wittig et al. 1971, Jeffers and Ageno 1975) has been developed to quite high specific output powers, on the order of 70 kJ/lb and chemical efficiencies of 97%. However, the total electrical efficiency of such systems is much less than 1%, and they rely on the dissociation of O_2 in a discharge with subsequent injection of CS_2. Scaling to high cavity pressures has not been achieved experimentally owing to the difficulty of efficient generation of O-atoms at high pressures. This system is therefore not considered any further in the present context of high electrical efficiency CO lasers.

A brief mention should be made of the gasdynamic CO laser (GDL). McKenzie (1970) reported the first experimental observation of stimulated emission at 5 μm from the supersonic expansion of CO which had been compressed to high stagnation temperature and pressure behind a reflected shock wave in a shock tube. By the use of large area ratio nozzles, up to an area ratio of 2700, CO–N_2 mixtures were expanded to very low translational temperatures, and over 100 W was extracted in a 2 ms pulse. The thermal efficiency in this experiment and a similar one by Watt (1971) was on the order of 0.5%, comparable with the thermal efficiency in the CO_2 GDL. Klosterman (1975) was later able to extract over 100 kW of power by scaling to large stagnation pressures, of the order of 10 000 psi, and stagnation temperatures of 2000 K. He observed cw oscillation for the order of 2 ms in this experiment. Although the GDL does not fall within the context of this chapter, it is mentioned here because it is a practical means of achieving very high power in a CO laser.

During the latter part of the 1960s, the electrical and GDL CO_2 lasers were in the forefront of high-power research and development efforts in gas lasers. Little emphasis was given to the CO laser during this period, because of lack of understanding of the basic physics in the device and the very modest performance of experimental CO lasers in comparison to the CO_2 laser. Considerable impetus was given to the research efforts in CO lasers following reports by Osgood and Eppers (1968) and Osgood et al. (1969) of high-power, high-efficiency operation of a direct discharge-excited CO laser operating at liquid-nitrogen temperature. In these early experiments, 20 W of cw power were obtained at 9% electrical efficiency and peak powers of 7.7 kW were obtained in a Q-switched system. In later experiments (Osgood et al. 1970), they obtained output powers of 95 W at an electrical efficiency of over 20%. This performance was comparable to that obtainable from CO_2 discharge lasers.

Shortly thereafter, Bhaumik et al. (1970), using Xe as an additive in a directly-excited electrical CO laser obtained efficiencies of 40% with liquid-nitrogen cooling. The effect of gas composition and temperature was investigated extensively by Bhaumik et al. (1970, 1972), who demonstrated the rather remarkable performance of 70 W cw output power at an electrical efficiency of 47% at liquid-nitrogen temperature (Bhaumik et al. 1972), and 25 W cw output and a total electrical efficiency of 17% at room temperature (Bhaumik 1970). These experiments were the forerunner to the demonstration of completely sealed-off CO lasers (Freed 1971, Sequin et al. 1972) operating at temperatures between −90 and +30°C, with output powers of up to 70 W and performance comparable to that of sealed-off CO_2 lasers.

Another contribution to the development of the CO laser arose from extensive kinetic modelling efforts in the early 1970s. In contrast to the analysis of the CO_2 laser system, which uses the concepts of vibrational relaxation among harmonic oscillators, it was found necessary to include the effects of anharmonicity in the interpretation of the vibrational–relaxation process, as first suggested by Treanor et al. (1968). The inclusion of anharmonic terms was crucial to the understanding of the temperature sensitivity of the laser experiments and the multivibrational level relaxation processes leading to partial inversions over the observed very wide wavelength range.

A major experimental development in laser technology occurred in 1971 with the development of the externally-sustained discharge technology using an electron beam as the source of ionization (Daugherty et al. 1972, Fenstermacher et al. 1972). This proved to be the key to scaling the electrical discharge excitation to high average power and high pulse energy laser systems. The concept of an externally-sustained discharge was applied initially to the CO_2 laser (Daugherty et al. 1972, Fenstermacher et al. 1972), but its obvious extension to the CO laser was effected soon thereafter. Mann et al. (1974) measured 153 J pulse output in a cryogenically cooled device with an electrical conversion efficiency of 63%, a value which has not been achieved in any other CO EDL. Single-pulse energies of 1500 J have been measured (Boness and Center 1977) at an electrical efficiency of approximately 36%. These electron-beam sustained discharge techniques have also been applied to quasi-cw operation, with powers of the order of 100 kW and electrical efficiencies up to 30% (Jones et al. 1974, Jones and Byron 1974, Thompson et al. 1975, Klosterman and Byron 1977).

These experimental performance developments place the electrically-excited CO laser among the most important of the potentially high-power gas laser systems. These include the HF/DF chemical and electrical/chemical systems operating at 2.7 and 3.5 μm, and the CO_2 laser operating at around 9.4 and 10.6 μm. Apart from its potential high-power applications, the large number of discrete lasing transitions available in the CO EDL lead to some unique applications for pollution monitoring, since the CO laser overlaps the 5.3 μm absorption band of NO and the corresponding stretch mode in N_2O (Menzies 1971, Kreuzer et al. 1972). At the same time, however, the CO laser transitions also overlap the 6.3 μm bending mode of H_2O vapor, which is a possible interference in the atmospheric transmission of CO lasing transitions. Bhaumik (1972) and Rice (1974) have demonstrated the use of an intracavity water vapor cell to perform spectral line selection to enhance the atmospheric transmittance of the CO EDL output.

The above short historical summary of the CO laser has necessarily been restricted to some of the major developments in the CO laser field. The interested reader is referred to an extensive review of the earlier experimental work published by Sobolev and Sokovikov (1973a). The remainder of this chapter is devoted to a discussion of the basic aspects of efficient electrically-excited CO lasers. The following sections describe first the basic physics underlying the electrically-excited CO laser, the vibrational excitation of CO by direct electron impact and an outline of the kinetic modelling, including the dominant scaling parameters. This is followed by a brief description of gas thermal conditioning techniques and a description of various device configurations, including specific examples and a discussion of their operating characteristics which have been achieved to date.

2. Basic physics of vibrational relaxation mechanism

2.1 Relaxation time scales

There are three basic collisional phenomena in the relaxation of the CO molecule in the electrically-excited laser. These are the processes of vibration–vibration (V–V), vibration–translation (V–T) and electron excitation of vibration (e–V) energy exchange. (For the present discussion it is assumed that the translational and rotational degrees of freedom are in equilibrium so that vibration–rotation (V–R) or vibration–rotation–translation (V–R–T) energy exchange processes are indistinguishable from V–T processes.) In general, the V–V relaxation rates are many orders of magnitude faster than the V–T rates, i.e., $\tau_{V-V} \ll \tau_{V-T}$. This is exemplified in the CO molecule which exhibits the lowest V–T relaxation rate constants of all the common diatomic molecules (Millikan and White 1963). Under typical discharge excitation conditions, the e–V excitation rates are small compared with the V–V relaxation rates, and, in the absence of stimulated emission, radiative decay processes are very slow and usually can be neglected. Because of these different time scales, electrical energy transferred into the vibrational mode of the CO molecule will be redistributed amongst the vibrational levels to produce a state of vibrational quasi-equilibrium before the V–T processes (or any other decay processes) become important.

2.2 Relaxation of anharmonic oscillators

Treanor et al. (1968) appear to have been the first to have considered the relaxation behavior of anharmonic oscillators such as CO when the vibrational mode is out of thermal equilibrium with the translational mode. For the conditions of low energy loss from the system, they deduced non-Boltzmann vibrational distributions with overpopulation of the higher vibrational levels as compared with the equivalent Boltzmann distribution for the same vibrational energy content. This overpopulation of the higher levels is a consequence of detailed balancing among the anharmonic energy levels under conditions in which the translational temperature is small compared with the characteristic vibrational temperature of the gas.

The overpopulation in the high levels of the vibrational anharmonic oscillator can be qualitatively understood by considering the exchange between two oscillators with only two active vibrational states, as shown in fig. 1. Detailed balancing requires that the forward and reverse rate constants for the vibrational exchange

$$A^\dagger + B \underset{k_r}{\overset{k_f}{\rightleftharpoons}} A + B^\dagger \tag{1}$$

be related by

$$k_f = k_r \exp(\Delta E/kT). \tag{2}$$

Here, T denotes the translational temperature, k_f and k_r are the forward and the reverse rate constants respectively, ΔE represents the energy defect between the two oscillators and k is the Boltzmann constant. Under conditions of quasi-vibrational equilibrium, the net forward and reverse reaction rates are equal and it follows that

$$(E_A/T_A) - (E_B/T_B) = \Delta E/T, \tag{3}$$

where T_A and T_B are the vibrational temperatures of species A and B respectively. For $T < T_A$, it is clear that $T_B > T_A$ and oscillator B with the smaller energy spacing is pumped by oscillator A. It should be noted that the energy

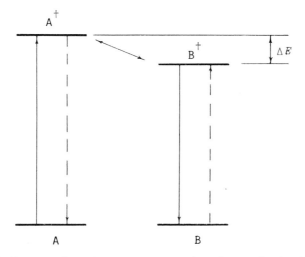

Fig. 1. Energy exchange between molecules with only two vibrational levels.

defect in each collision must be taken up by the thermal bath at the translational temperature.

This simple example of the two-level system illustrates the physical principles of the vibrational pumping effect in the anharmonic oscillator in which the anharmonicity causes the energy level spacing to decrease with increasing vibrational level. Thus, an overpopulation of the higher vibrational levels in an anharmonic oscillator can occur under thermal nonequilibrium conditions when the system is cold translationally but not vibrationally.

In the absence of energy deactivation processes, Treanor et al. (1968) showed that the vibrational quasi-equilibrium in an anharmonic oscillator is described by

$$N_v/N_{v+1} = \exp\{[(\omega_e - 2\omega_e x_e)/k\theta_1^*] - 2v\omega_e x_e hc/kT\}, \tag{4}$$

where ω_e and $\omega_e x_e$ are the usual molecular constants, θ_1^* is the vibrational temperature characterizing the zeroth and first vibrational levels and N_v is the population density in the vth vibrational level. The above result depends solely upon detailed balancing and leads to a total inversion in the high vibrational levels. In practice, collisional deactivation by V–T processes and radiative decay lead to energy drain out of the upper vibrational levels and preclude the total inversion predicted by the Treanor result. However, in the low vibrational levels, the non-Boltzmann nature of the Treanor distribution is exhibited under thermal nonequilibrium conditions. An example of such a distribution is shown in fig. 2 taken from Caledonia and Center (1971) which compares the experimental measurements of Joeckle and Peyron (1970) with the predicted vibrational distribution function for two different V–V transition probabilities. The main point of this figure is the qualitative agreement between the experimental observation and the kinetic modelling prediction of increasing vibrational temperature with increasing vibrational levels up to the levels where collision and radiative decay effects dominate.

2.3 Inversion process

The basic inversion mechanism in the CO laser can now be understood with respect to the non-Boltzmann vibrational distribution in the anharmonic oscillator. The preferential transfer or pumping vibrational energy up the vibrational ladder leads to high local vibrational temperatures. Although a total inversion (negative vibrational temperature) is unlikely, it is possible to have an inversion between two rotational levels for positive vibrational temperature under conditions of high local vibrational temperature and low rotational/translational temperature. This is called a partial inversion and occurs on a P-branch transition for which $v, J \rightarrow v - 1, J + 1$, where v, J are the vibrational and rotational quantum numbers respectively. It can be readily shown (e.g., Center and Caledonia 1971b) that the small-signal gain is optimized near the condition $T_r/\theta_r = 2(T_v/\theta_v)^2$. Here, T_v, θ_v are the local vibrational temperature and characteristic vibrational energy level spacing respectively, and T_r, θ_r are the rotational temperature and characteristic rotational level spacing respectively. In general,

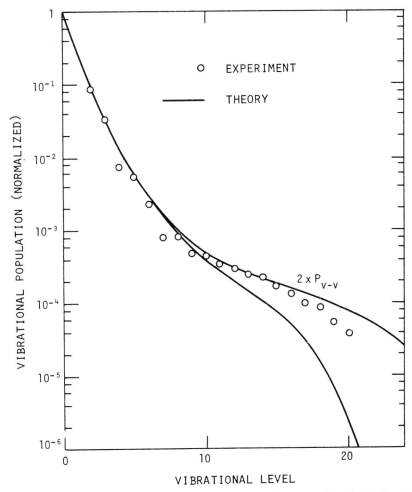

Fig. 2. Comparison of experimental (○) CO vibrational distribution with distribution function prediction (from Caledonia and Center 1971).

the rotational and translational degrees of freedom are in thermal equilibrium so that $T_r = T$. Since $\theta_v/\theta_r \simeq 10^3$, the above thermal condition implies a small translational temperature and a large vibrational temperature for maximum gain. Thus, it is possible to obtain useful gain on the high vibrational levels of the anharmonic CO molecule.

The experimentally observed sensitivity to low translational temperature can also be explained in terms of this anharmonic oscillator model. The over-population in the high vibrational levels is a strong function of the translational temperature as shown by eqs. (3) and (4). Lowering the translational temperature leads to an increase in the local vibrational temperature among high lying levels and, consequently, an increase in the partial inversion gain coefficient.

From the above discussion it can be seen that it is possible to obtain simultaneous partial inversions between many pairs of vibrational levels. Thus, one could expect laser oscillation on a multitude of vibrational–rotation transitions. This is observed experimentally as described in § 1. The high energy conversion efficiency of the electrical CO laser can also be explained qualitatively. First, there is an inherently high quantum efficiency in the laser since the available energy on any discrete vibration–rotation transition is large compared with the anharmonic defect (characterized by $2\omega_e x_e$) and the rotational energy difference in the partial inversion. For CO, $2\omega_e x_e/\omega_e$ is approximately equal to 0.013 and this leads to an expected quantum efficiency greater than 90%. Second, since the energy rejected to the thermal bath is small compared to the characteristic vibrational energy, it is possible to recycle vibrational energy through the molecule many times before significantly increasing the thermal bath temperature. This leads to a potentially high total efficiency. In practice, a certain amount of energy has to be invested in the vibrational mode of the CO molecule to reach threshold conditions, as discussed in the following sections, and a high total electrical efficiency can only be achieved if the vibrational energy is cycled through the molecule many times after reaching threshold conditions. This requirement will become evidence in discussion of practical systems in § 6, where the choice must be made between separation of the excitation region from the optical extraction region and coincidence of these regions.

At low gas densities where pressure broadening effects are small, there is sufficient spectral separation of the vibration–rotation transitions such that transitions with positive gain, namely the P-branch transitions, do not overlap R-branch transitions, which would be an absorption in the absence of a total vibrational inversion. Thus, it is possible to extract lasing energy simultaneously on many vibration–rotation transitions, which is characteristic of CO laser experiments. However, at high gas densities, the collisional broadening effect may be large enough such that resonant self-absorption by overlapping lines may become important. This was first pointed out by Lacina and McAllister (1975b) and observed experimentally by Boness and Center (1975). This resonant self-absorption, which is also discussed in § 3 on kinetic modelling, will probably preclude the development of a continuously tunable high-power CO laser which was originally proposed by Basov et al. (1971).

3. Vibrational excitation of CO in an electric discharge

The CO EDL relies on electron impact processes to excite the vibrational mode of CO. The maximum electrical efficiency of the laser is limited to the conversion efficiency of electron impact excitation in the electric discharge. Fortunately, the vibrational excitation cross section in CO is very large, near 2 eV, as a result of a resonance process (Schultz 1964, 1976). N_2, which is isoelectronic with CO, exhibits similar behavior with respect to electron impact. At small electron energies, the vibrational excitation cross section (Schultz 1964, Ehrhardt et al. 1968) dominates the inelastic processes, as shown in fig. 3, which

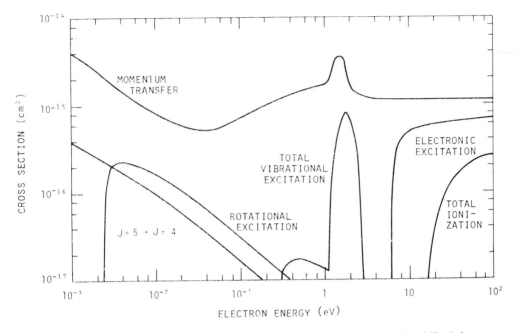

Fig. 3. Electron impact excitation cross-section data for CO (after Hake and Phelps 1967). Only two rotational levels are included in this figure.

also includes cross sections for rotational excitation, electronic excitation (Hake and Phelps 1967) and total ionization (Rapp and Englander-Golden 1965). From the curves of fig. 3 it can be inferred that the large vibrational cross section can act as a barrier to prevent electrons from reaching energies above about 2 eV. This has been demonstrated by calculations of the electron energy distribution function which is obtained numerically by solving the Boltzmann transport equation. Such calculations, for example by Hake and Phelps (1967) and Nighan (1970, 1972), show that the electron distribution function is sharply cut off towards the higher electron energies, which results in very efficient excitation of the vibrational levels of CO.

The partitioning of electrical energy into the different degrees of freedom can be calculated as a function of E/N (the ratio of the electric field to the gas density) from the electron distribution function and the cross-section distribution for each degree of freedom. An example of such calculations is shown in fig. 4 for CO–He and CO–Ar mixtures. The use of He as a diluent is expected to lead to increased gas heating, as indicated in fig. 4 because of its large cross section for energy transfer by elastic collisions with electrons. Such calculations are essential in designing discharge experiments, since they indicate the range of E/N values for efficient energy transfer to the vibrational mode. Although not shown in fig. 4, detailed calculations also can be used to predict the partitioning of energy amongst the individual vibrational levels. This is not expected to be of primary importance in the CO laser since the electron excitation rates are

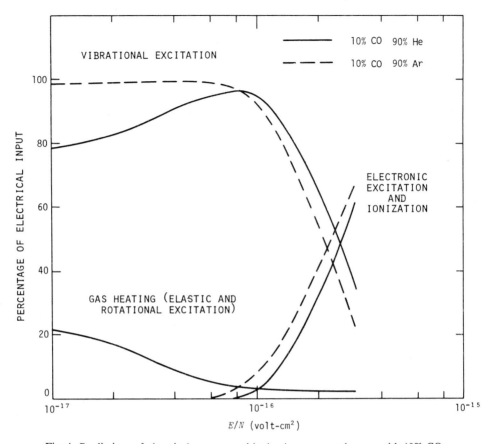

Fig. 4. Predictions of electrical energy partitioning in two gas mixtures with 10% CO.

generally slow compared with the V–V rates and the vibrational distribution is usually dominated by intramolecular transfer of vibrational energy.

The results shown in fig. 4 have been obtained neglecting superelastic collisions in which electron impact with vibrationally-excited molecules results in energy being transferred to the electrons. These superelastic collisions can become important and increase the high energy tail in the distribution function at high vibrational temperatures. The energy partitioning predictions are thus expected to be quantitatively correct only for vibrationally cold gases (e.g., at the start of the discharge). Since the local vibrational temperature among high vibrational levels of CO can be of the order of 10 000 K, the results of distribution function calculations neglecting superelastic collisions should only be considered qualitative for a vibrationally-excited gas mixture. Even for this case such calculations give a useful first-order estimate of the appropriate discharge parameters. It should also be pointed out that the rotational excitation by electron impact is usually treated by a continuous function (Hake and Phelps 1967). More recent calculations have shown that there is a large rotational excitation cross section in the 1 to 2 eV electron energy region (Chandra and Burke 1973).

3.1 Discharge techniques

There are two basic discharge techniques which have been applied to the CO EDL. These are the self-sustained positive column discharge (both pulsed and cw) and the externally-sustained discharge with either an electron beam or uv ionization as the source of ionization. In the self-sustained discharge, the electron loss by ambipolar diffusion to the walls and by volume processes (e.g., recombination and attachment) is balanced by electron impact ionization. This uniform low-pressure discharge is restricted to pd products of the order of 10 Torr cm to avoid discharge instabilities. Here, p is the pressure and d is the smallest dimension. The electron temperature required to maintain this self-sustained discharge is typically several electron volts and is a factor of 2 to 3 larger than necessary for the most efficient vibrational excitation. Of course, the variation of energy partitioning with electron energy can be modified by altering the gas mixture, e.g., by the inclusion of a low ionization potential material such as Xe (Bhaumik et al. 1970). Thus, it is possible to improve and tailor the energy partitioning to the particular discharge technique employed.

In order to operate a discharge at the optimum E/N (or electron temperature) for vibrational excitation, it is necessary to provide an external source of ionization. This can be seen with reference to fig. 4 from which it is evident that the rate of electronic excitation and ionization is extremely small under conditions of maximum partitioning/conversion into the vibrational degree of freedom. An example of such an external source is the high voltage electron beam which was developed by Daugherty et al. (1972) and Fenstermacher et al. (1972) for application to the CO_2 EDL. In this technique, the external source of ionization provides the necessary electron density (conductivity) while the discharge maintains the optimum value of E/N. This technique is scalable to pd products of 10^4 Torr cm and larger, as demonstrated in CO_2 and CO EDLs (see § 6 for examples).

It is clear that the actual discharge operating conditions depend upon the particular discharge technique that is to be used and may be limited by the development of instabilities which result in plasma constriction into an arc. The precise causes of plasma constriction in electrically-excited CO lasers are not well understood, and the subject is beyond the scope of the present chapter. The interested reader is referred to a review article by Nighan (1976). It is mentioned here simply to warn the reader that arcing phenomena can have serious consequences in high-power laser applications and can limit the maximum electrical input power density.

3.2 Power density scaling

Approximate scaling of the discharge power density with the electrical parameters and the gas density can be obtained as follows. The input power density can be written

$$P = jE = en_e v_d E, \tag{5}$$

where j and E are the current density and electric field respectively, e is the electron charge, and n_e and v_d are the electron density and electron drift velocity respectively. The drift velocity can be expressed in terms of the value of E/N by balancing the force on the electron due to the electric field with the rate of momentum change due to collisions with the neutral gas:

$$eE = m_e v_d N k_M, \tag{6}$$

where m_e is the electron mass, k_M is the rate constant for momentum transfer and N is the gas density. It follows that

$$P \sim n_e N (E/N)^2. \tag{7}$$

The range of E/N is restricted by the requirement of high vibrational excitation efficiency as described above and by discharge instability problems at high electron energies. The benefits of high neutral gas density and high electron density are obvious in going to high power input densities. This illustrates another advantage of the externally-sustained discharge technique which, in principle, is not limited by gas density in comparison with the pd limit of the self-sustained discharge. Afterglow measurements in high-pressure CO discharges (Center 1973) have indicated that dissociative recombination is the dominant volumetric electron loss mechanism. Thus, for an external source of ionization of strength S, electrons/cm^3 s^{-1}, the electron density n_e varies as $\sqrt{(S/\alpha)}$, where α is the rate constant for dissociative recombination.

Typical values for the parameters specifying the input power density are: $N \sim 2.5 \times 10^{18} - 2.5 \times 10^{19}$ particles/cm^3, n_e approximately 10^{12} to 10^{13} electrons/cm^3 and E/N approximately 10^{-16} V cm^2. Thus, the power input density can vary from approximately 100 W/cm^3 to 100 kW/cm^3.

3.3 Mixture effects

The use of diluents in the mixture has already been alluded to in preceding paragraphs in terms of modifying the electron distribution function in tailoring the excitation to fit the particular discharge technique. However, there are also other reasons for sometimes requiring the use of diluents; e.g., monatomic gases provide a lower temperature than diatomic gases in supersonic expansion sometimes used for gas conditioning (see § 5). The monatomic diluent also provides a useful thermal bath to limit the temperature rise due to heating in the vibrational relaxation process as described in § 2. In general, the heavier monatomic gases play no role in the discharge excitation because they have small momentum transfer cross sections at electron energies of interest (Frost and

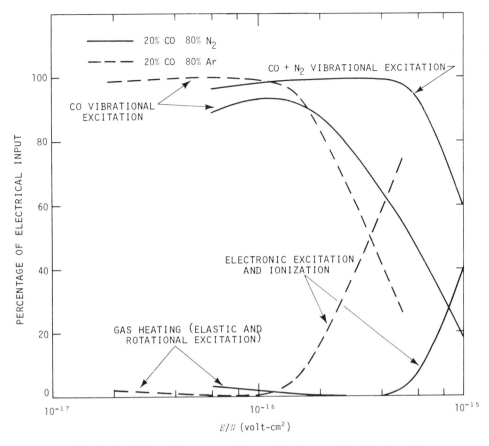

Fig. 5. Predictions of energy partitioning CO mixtures diluted in Ar or N_2.

Phelps 1964), and the inelastic processes have very high threshold energies, well beyond the tail of the distribution function. The use of helium with its large momentum transfer cross section can lead to some translational heating in the discharge, as already seen in fig. 4.

Another diluent that is sometimes useful is N_2, since its presence tends to enhance the CO vibrational excitation rate by electron impact as compared with the rate in pure CO. This is due to a modification of the electron energy distribution function in the presence of N_2, whose vibrational excitation cross sections are displaced to higher energy than those of CO. An example is shown in fig. 5 which compares the electrical energy partitioning for 20% CO dilute in N_2 with that for 20% CO dilute in Ar. It is clear that there is preferential excitation of the CO with respect to the N_2. Also notice that the optimum E/N value is larger with the use of N_2 as a diluent rather than Ar. Thus, at a given electron density the power input density can be increased by adding N_2 to the mixture. Distribution function calculations indicate that for CO/N_2 mole fraction ratios greater than 0.2, most of the discharge energy goes into the CO, as evident in fig. 5. However, for much smaller mole fractions, the discharge preferentially excites the N_2 vibrational mode which can subsequently pump the CO vibrational mode by intramolecular transfer. This can be used to advantage to modify the vibrational distribution function in the CO and, in particular, to preferentially excite the lower vibrational levels and weight the lasing spectra towards the shorter wavelengths. An example of this mixture effect was demonstrated by Djeu (1973) in the development of a CO discharge laser to oscillate on the lowest vibrational transition, i.e., $v = 1 \rightarrow v = 0$.

4. Kinetic modelling

Detailed kinetic models of the CO electric discharge laser have been developed by many authors to predict the laser performance parameters, such as total efficiency, small-signal gain, power spectral distribution and energy partitioning in the internal modes of the gas mixture (Center and Caledonia 1971a, Rich 1971, Rockwood et al. 1973, Lacina et al. 1973, Sobolev and Sokovikov 1973b, Hall and Eckbreth 1974, Center and Caledonia 1975, Lacina and McAllister 1975a and Pistoresi and Nelson 1975). These models have been carried out to various levels of sophistication but all are based on solutions to the kinetic master equation which includes V–V and V–T energy transfer processes, electron impact excitation, and spontaneous and stimulated emission. The electron excitation terms are derived from solutions to the electron transport equation as described in § 3, and the model usually includes the thermodynamic and fluid dynamic conservation equations. The inclusion of superelastic electron–molecule collisions (in which the electrons gain energy at the expense of the CO vibrational energy) depends on the degree of sophistication of the model (Pistoresi and Nelson 1975).

4.1 Scaling parameters and performance curves

A detailed review of these kinetic models and their validity is beyond the scope of the present chapter. However, one can derive some simple scaling relations and general performance curves which permit approximate performance predictions to be made. This is of obvious use to the CO EDL designer. Furthermore, the general performance curves may be applied to both pulsed and flowing cw CO lasers by use of the standard time-to-distance transformation.

The expected scaling with gas density N and electron density n_e can be readily derived for specific gas mixtures and electron distribution functions. Since only binary collisional processes take part, the kinetic master equation can be nondimensionalized and written in terms of the particle mole fractions (Center and Caledonia 1975) provided the optical transitions are pressure broadened (which applies for pressures above approximately 20 torr). It follows that the electron density scales with the gas density N, and both the power density P and the optical flux ϕ scale with N^2. Similarly, the time scales inversely with the gas density N, so that the energy per particle (tP/N) is independent of the gas density. For pure CO or CO mixtures with monatomic gases, the conversion efficiency from electrical to CO vibrational energy is almost 100%, as described in the previous section. This is also approximately true for mixtures containing N_2 with the CO/N_2 ratio greater than 0.2. Thus, the vibrational distribution function in the CO molecule and the laser performance are determined by the vibrational energy per CO molecule, which becomes the critical scaling factor at any given translational temperature.

In practice, the electron temperature (or the electron distribution function) will be weakly dependent on the CO vibrational content owing to superelastic collisions. Therefore, at constant input power density, the rate constants for vibrational excitation by electron impact will vary somewhat with time (or distance). As a result, one may expect a small departute from the above predicted linear scaling with vibrational energy per CO molecule. Using a simplified model, Lacina and McAllister (1975a) were able to show that the time to reach threshold scaled almost inversely with the CO vibrational energy density, as shown in fig. 6. The actual slope of these curves is approximately $-5/6$ and is almost independent of the threshold loss (output coupling). A slope of unity would correspond to a constant ratio of threshold time to vibrational energy content. It should be pointed out that these predicted threshold times are in good agreement with the experimental data (Lacina et al. 1973, Boness and Center 1977). The magnitude of the vibrational energy density necessary to

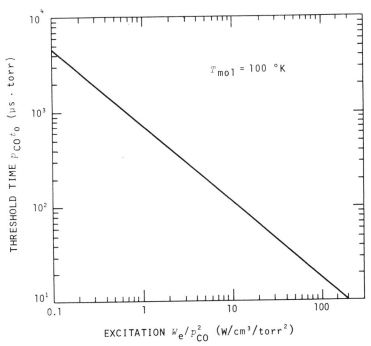

Fig. 6. Scaled threshold time $p_{CO}t_0$ as a function of the excitation parameter W_e/p_{CO^2} (after Lacina and McAllister 1975).

reach threshold conditions varies by less than a factor of 2 in the translational temperature range of 100 to 200 K. The threshold energy density at 100 K is typically of order 0.1 eV per CO molecule.

Lacina and McAllister (1975a) were also able to derive generalized power and energy conversion efficiencies as a function of the parameter

$$p_{CO}t[W_e/p_{CO^2}]^{5/6},$$

where W_e is the electrical input power density. An example of these generalized curves is shown in fig. 7 for a translational temperature of 100 K. If one neglects the 5/6 power dependence, the performance curves would scale with the CO energy density which, in fig. 7, would be in units of joules per liter per torr of CO. These performance curves assume constant power input density to the CO vibrational mode and neglect the small rise in translational temperature, resulting from the anharmonic defect in the intramolecular transfer of vibrational energy.

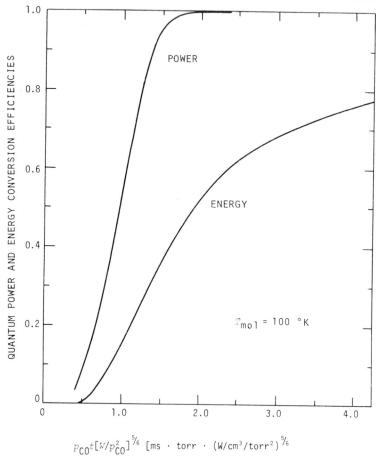

Fig. 7. Generalized power and energy conversion efficiencies normalized by the steady-state values (from Lacina and McAllister 1975).

Despite these simplifications, the calculations are in good agreement with the results obtained by more extensive computer calculations.

The kinetic modelling calculations are expected to be quite insensitive to mixtures with inert gases since they do not influence the V–V relaxation process and only have a small effect on the electron distribution function for a fixed ratio of the electric field to the CO molecular density. On the other hand, the use of N_2 in the gas mixture can modify the vibrational distribution function owing to the intramolecular transfer of vibrational energy between N_2 and CO. The electron distribution function is also modified in the presence of N_2, but both these effects are small for CO–N_2 fractions greater than approximately 0.2, so that the performance predictions such as those shown in fig. 7 can be applied to a wide range of CO mixtures.

As described in § 2, the vibrational distribution function and the small-signal gain are strongly dependent on the translational temperature, with the result that the overall laser performance is expected to be temperature dependent. Specifically, the laser energy conversion efficiency decreases with increasing temperature, and the power spectral distribution shifts to higher vibrational and rotational levels with increasing temperature. Fig. 8, taken from Center and Caledonia (1975), shows the energy partitioning between the vibrational mode and the translational degrees of freedom as a function of the CO vibrational energy density for three different initial translational temperatures. The increase in the translational temperature is due to the conversion of the anharmonic defect into translational energy through the V–V collisional process. The increase in the stored vibrational energy, as a function of translational temperature, is necessary to counterbalance the strong temperature dependence of the small-signal gain. Fig. 8 clearly indicates the decrease in overall conversion efficiency with increasing translational temperatures.

4.2 Limitations of existing modelling predictions

Although the kinetic modelling calculations agree qualitatively with experimentally observed laser performance, there are serious discrepancies in the detailed performance comparisons. These include the absolute value of the overall efficiency as well as the small-signal gain and the detailed power spectral distribution. While total energy conversion efficiencies in the range of 40 to 60% have been achieved at cryogenic temperatures (Mann et al. 1974, Boness and Center 1977), the observed efficiency at room temperature has not exceeded 6% in large volume uniformly excited systems (Center and Boness 1974) and has only reached 17% in wall-stabilized discharges with the addition of Xe and Hg in the gas mixture (Bhaumik 1970). Furthermore, the predicted power spectral distribution is usually downshifted by one to two vibrational levels.

These discrepancies may be due in part to errors in the assumed V–V and V–T relaxation processes among the high vibrational levels of CO for which no experimental data exist. For example, V–T processes are expected to be rela-

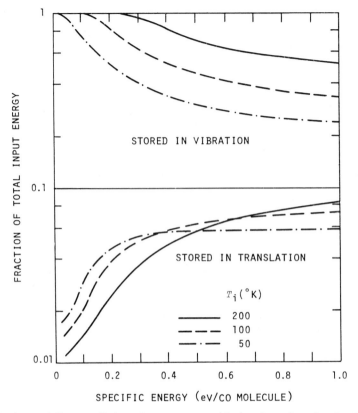

Fig. 8. Kinetic modelling predictions for energy partitioning into the vibrational and translational/rotational modes for three different initial translational temperatures (from Center and Caledonia 1975).

tively unimportant in the low vibrational levels, based on the limited available data, but may be grossly underestimated in the high vibrational levels. Since the vibrational distribution is weighted towards the higher vibrational levels with increasing translational temperature, the V–T deactivation might be expected to become more important at the high translational temperatures. Similar arguments apply to the effect of superelastic electron molecule collisions in the high vibrational levels. Calculations by Pistoresi and Nelson (1975) have indicated that such superelastic processes may contribute significantly to the vibrational de-excitation of the high vibrational levels, and hence may limit the laser performance at high translational temperatures as compared with low temperatures.

Another discrepancy between the predicted and observed performance concerns the small-signal gain and consequently the variation of saturation intensity with the output coupling (or the threshold loss coefficient). The magnitude of the experimentally observed small-signal gains is smaller than predicted by kinetic modelling (Boness and Center 1975) and the optimum output coupling is significantly less than predicted. Deactivation by superelastic electron–molecule collisions and V–T processes would certainly decrease the small-signal gain, and it is clear that more accurate modelling calculations are needed to understand and predict detailed laser performance.

Since the externally-sustained discharge technique can be scaled to high pressures, one might expect continuous tunability (Basov et al. 1971) in the 5 to 6 μm region as a result of line broadening for CO lasers operating at densities of 10 amagats (Basov et al. 1971).[1] However, it was observed in the early cryogenic experiments at moderate densities, of the order of $\frac{1}{2}$ amagat, that certain transitions were always missing in the output spectra. Lacina and McAllister (1975a) attributed these spectral discrepancies to resonant self-absorption of the missing transitions by overlapping transitions. The extent of the overlap and the population distribution determine the degree of absorption, which has generally been neglected in the theoretical models of the CO laser performance. Subsequent small-signal gain measurements by Boness and Center (1975) verified this overlapping absorption effect which resulted in anomalously low small-signal gain coefficients in several transitions originating from the sixth and seventh vibrational levels. Lacina (1975) has shown that the inclusion of resonant self-absorption in the kinetic calculations results in a shift in the spectral distribution towards higher vibrational levels under high density operation but has little effect on the overall energy conversion efficiency.

4.3 Summary

The vibrational energy content per CO molecule is clearly the dominant scaling parameter at any specific translational temperature. The existing kinetic modelling codes give qualitative agreement with the experimental data and are therefore useful for first-order design calculations. However, there are serious discrepancies in the predicted spectral distribution and the variation of laser performance with translational temperature, and this is probably the major limitation in the application of the modelling predictions. The resolution of these discrepancies will probably require careful consideration of the effects of superelastic electron–molecule collisions and a better understanding of the variation of the V–V and V–T collisional transfer probabilities among the high vibrational levels.

5. Thermal gas conditioning requirements

The experimental observations and theoretical predictions discussed in the preceding sections indicate the desirability of low temperature operation, with temperatures of the order of 100 K or less, in order to achieve overall electrical

---

[1] As discussed in this section, the number density is of more fundamental significance than pressure in the kinetic scaling. A convenient unit for the number density is the amagat. 1 amagat = 2.69×10^{19} particles/cm^3 and corresponds to Loschmidt's number.

efficiencies in excess of 20%. This leads to the experimental requirement to provide both the low initial temperature in the gas stream as well as the ability to remove the waste heat for repetitively pulsed or cw operation.

5.1 Gas cooling techniques

The simplest technique is to use diffusion cooling to the walls of a cooled discharge chamber. This technique was used in the early wall-stabilized discharge devices described in the introduction. Wall-stabilized discharges are limited in the power input density by the rate of radial heat conduction and by the discharge stability, which depends on ambipolar diffusion of electron–ion pairs to the wall. Both these processes have characteristic timescales which vary as ρR^2, where ρ is the gas density and R the tube radius. However, the diffusion coefficients for ambipolar diffusion are typically an order of magnitude greater than that for heat diffusion. The maximum input power to an electrical glow discharge device can be readily estimated by balancing the power input to the discharge with the radial heat conduction to the walls. For a long cylinder such that $L/R \gg 1$,

$$P = 2\pi\kappa RL(2.4\Delta T/R),$$

where P is the electrical input power, κ the gas thermal conductivity, ΔT the allowable temperature difference between the centerline and the wall and L is the tube length. The electrical input power is typically limited to the order of 1 kW/m of discharge tube length for a gas mixture dilute in He (to benefit by the high thermal conductivity of the He).

By precooling the gas in a heat exchanger and flowing the gas through the discharge cavity, it is possible to avoid the power limit imposed by transverse heat conduction. This approach has been used with both wall-stabilized and volumetrically scalable discharges as described in § 6. Of course, there is still an upper bound to the product ρR^2 in order to limit the ambipolar diffusion time and, hence, maintain a stable discharge. This limit obviously does not apply in the case of volumetrically scalable discharges such as the externally-sustained discharge. The flow velocity through the discharge region is usually subsonic when the gas is precooled, with the minimum flow velocity being determined by the waste head load, as described below. There is an additional limitation on the flow velocity imposed in the case of precooled gas flowing in a warm wall discharge region. In this case, the flow velocity must be fast enough to produce the thin thermal boundary layers on the walls of the discharge region and so to thermally isolate the discharge region from the warm walls. These thermal boundary layers (as opposed to recirculation zones) are desirable because they concentrate the density change in a narrow distance. Estimates of a boundary-layer thickness for forced and free convection flows and either laminar or turbulent boundary layers may be readily calculated (Szewczyk 1964, Wilks 1973, Woodroffe 1975).

The use of a supersonic expansion upstream of the optical cavity is an alternate approach to gas cooling. The gas temperature T is determined by the flow Mach number M and is given by

$$T_0/T = 1 + \tfrac{1}{2}(\gamma - 1)M^2, \tag{8}$$

where T_0 is the stagnation temperature and γ is the ratio of the specific heats. For typical laser mixtures which are diluted in a monatomic gas, a Mach number of about 3 is needed to cool the gas to 100 K or less with the gas initially at room temperature.

The supersonic expansion eliminates the need for a heat exchanger and a cryogenic coolant supply. However, it introduces an added complexity to the design of the discharge region when this is located downstream of the supersonic expansion. Since any disturbance in the flow will generate compression and expansion waves, these waves result in variations in the gas density which in turn may alter the value of E/N in the discharge region and lead to nonuniform excitation. It is also desirable to use rapidly expanding nozzles in order to restrict the thickness of the supersonic boundary layer on the sidewalls.

The limit to the input power in the case of the wall-stabilized discharge has already been described above. For any flow velocity, there is a corresponding upper bound to the electrical input power density in the case of the volumetrically scalable discharge to limit the temperature rise resulting from the system inefficiency, i.e., the waste heat. When the electrical discharge and optical cavity are coincident, the temperature gradient can readily be calculated by balancing the gas heating with the input power density, described in § 3, i.e.,

$$NUC_p(dT/dx) \sim fNn_e(E/N)^2, \tag{9}$$

where U is the flow velocity, dT/dx is the gradient of the translational temperature in the flow direction, f is the fraction of electrical energy converted to thermal energy at the position x and C_p is the specific heat at constant pressure. (In supersonic flow it is necessary to use C_v, the specific heat at constant volume.) For a given gas mixture it follows that $U\,dT/dx \sim fn_e(E/N)^2$ for cw operation, and increasing the electrical power density requires shorter flow transit times through the discharge region. This applies to both cw and pulse excitation for pulse lengths that are comparable with or larger than the gas transit time through the cavity. For short pulse discharges with pulse lengths appreciably less than the gas transit times, $dT/dt \sim fn_e(E/N)^2$, and the pulse energy is limited by the maximum allowable temperature rise, while the average power for repetitively-pulsed systems is limited by the flow velocity as above.

Another consideration which may influence the power handling capacity in flowing gas discharges has to do with the changes in the flow velocity resulting from heat input to the constant area flow. Heat addition leads to acceleration of subsonic flows and deceleration of supersonic flows (Shapiro 1953), thereby tending to choke the flow. Even without choking, there may be a considerable density and pressure change resulting from the heat addition, and this would lead to a variation in the value of the applied E/N (and the corresponding electron temperature) in the flow direction. This effect of heat addition is illustrated in fig. 9 which shows the variation in Mach number and density with heat addition for

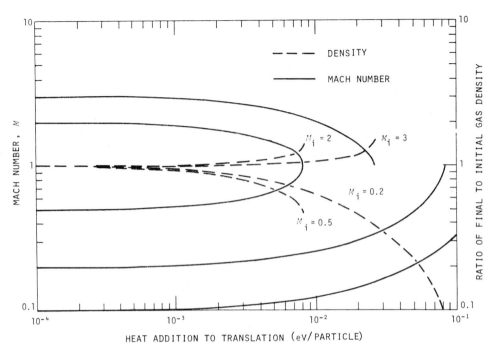

Fig. 9. Variation in Mach number and gas density with heat addition into the translational mode for 20% diatomic gas in a monatomic diluent. M_i is the initial Mach number.

an initial translational temperature of 80 K and a typical ratio of specific heats, $\gamma = 1.59$, corresponding to 20% diatomic gas in a monatomic diluent. The heat addition is plotted in eV/gas particle for comparison with the threshold energy densities of approximately 0.1 eV/CO molecule, as described in § 4.

One problem in estimating the maximum electrical power or energy input concerns the uncertainty in the fraction of electrical energy, f, which appears as translational energy. The magnitude of this fractional conversion to the translational degrees of freedom also depends on whether the electrical excitation is coincident with the optical extraction region or not. Under optimum discharge excitation conditions, as described in § 3, the direct conversion of electrical to translational energy is at most a few percent except in the case of He as a diluent, in which case the conversion may be as high as 10%. The remainder of the energy is stored in the vibrational mode of CO, and the rate of conversion to translational energy is determined by V–T deactivation from the high vibrational level, the rejection of the anharmonic defect in the intramolecular V–V collisions and the rotational defect in the stimulated P-branch transitions. Based on the kinetic modelling calculations, the latter two terms account for almost 10% of the electrical input power. Thus, for a coincident cavity, the total conversion of electrical to translational energy is expected to be on the order of 20% or less. The corresponding number in case of noncoincidence between the excitation region and the optical cavity depends on the particle transit time between the discharge region and the optical V–T deactivation of the population in the high vibrational levels. This leads to a larger value for f than in the coincident cavity.

Although the expected conversion of electrical to translational energy is less than 20% within the cavity region, there is considerable vibrational energy in the gas as it flows out of the cavity. The minimum value of the energy content per CO molecule exiting the cavity is simply the energy density required to reach the threshold gain conditions. In practice, of course, the specific energy exiting the cavity is considerably larger because of inefficiencies in the system, including the increase in the threshold density resulting from gas heating effects. Since the overall conversion efficiency from electrical input energy to lasing energy is no more than approximately 50%, some 30 to 40% of the input electrical energy remains in the vibrational mode of CO but must finally appear as translational energy somewhere downstream of the cavity. This large gas heating will tend to choke the flow and must be taken into account in designing the flow system.

5.2 Optical quality

In order to achieve high optical quality in the laser beam, it is necessary to limit the density inhomogeneities (which lead to phase variations) because of the linear dependence of the index of refraction on the gas density. These phase variations reduce the intensity in the far field, and it can be shown that (Born and Wolf 1964)

$$I/I_0 = \exp[-(\Delta\psi)^2],\tag{10}$$

where I and I_0 are the actual intensity and the ideal intensity respectively, and $\Delta\psi$ is the average phase variation over the aperture of the beam.

For ordered disturbances, the average phase variation can be written as (Sutton 1969)

$$\Delta\psi = (2\pi/\lambda)(n - 1)(\Delta\rho/\rho_0)L,\tag{11}$$

where $\Delta\rho$ is the density disturbance, ρ_0 is the density at S.T.P., λ is the wavelength and L is the optical path length through the gas medium. An example of such ordered disturbance would be the boundary layer on the optical windows isolating the active discharge region from the surrounding environment. Woodroffe (1975) has derived approximate expressions for the effect on the optical path length of a boundary layer in the case of both forced and free convection. Such calculations are used to compute the integral $\int_0^\delta (\rho - \rho_\infty)/\rho_\infty \, dy$ where δ is the boundary-layer thickness, ρ_∞ is the free-stream gas density and y is the distance normal to the windows. This density change defines the appropriate phase shift. Similar calculations can be readily made for supersonic boundary layers. Sutton (1969) has analyzed the effective phase shift resulting from random fluctuations, e.g., turbulent fluctuations, through the optical medium. In this case, the above expression for the phase shift is modified by a parameter Λ/L, where Λ is the integral scale size associated with the turbulent spectrum.

The average phase shift $\Delta\psi$ must be minimized in order to approach the ideal intensity for a diffraction-limited beam. Thus, to achieve an I/I_0 of 0.9 requires that the average density disturbance $\Delta\rho/\rho$ be of the order 10^{-2} to 10^{-3}, depending upon the gas density and the diluent, with the upper limit being set by the use of He as the diluent.

Another form of ordered density disturbance is due to the presence of expansion and compression waves, or shock waves, resulting from physical disturbances in supersonic flow. For example, extreme care must be taken in eliminating discontinuities in the wall surface at the electrode junctions when using the supersonic expansion method of cooling the gas. Even then it is possible to generate a transverse shock wave owing to discharge-induced disturbances in the boundary layer.

Large density changes can result from the heat addition to the constant area flow, as discussed in § 5.1. The density change is much larger in subsonic flow than in supersonic flow for the same heat input, and this is illustrated in fig. 9. For supersonic flow or subsonic flow with small heat addition, the density profile is almost linear in the streamwise direction. Only the deviation from linearity is a problem, since a linear gradient acts as a wedge and the resulting beam steering can be corrected easily by tilting one of the optical elements. However, with large heating rates in subsonic flows leading to density changes $\Delta\rho/\rho$ of order 0.1 or greater, the density gradient becomes nonlinear and it becomes difficult to correct the associated beam steering effects.

An additional optical interference problem arises in the case of transversely-excited pulsed lasers owing to the generation of heat shock waves at the electrodes. The shock waves result from the rapid deposition of energy in the gas on a time scale that is short compared with the acoustic transit time between the electrodes. Since these electrode-generated disturbances can lead to density changes $\Delta\rho/\rho$ of order unity, the pulse length for good optical homogeneity is limited to a time short compared with the acoustic transit time between the electrodes. It is also necessary to estimate the clearing time t_c for the acoustic disturbances to damp out and allow the gas to return to its undisturbed condition. This clearing time defines the maximum repetition frequency of t_c^{-1}. The usual technique to minimize t_c is to use a choked orifice plate upstream of the discharge and downstream acoustic attenuation. The choked orifice plate reflects all disturbances travelling upstream. A semi-empirical formula for the pulse clearing time is

$$t_c = l/M(1 - M)a, \tag{12}$$

where l is the distance from the choked orifice plate to the downstream edge of the discharge, M is the Mach number and a is the speed of sound. Basically, this formula defines the sum of the time for an expansion wave to travel from the downstream edge of the discharge to the orifice plate plus the time to flush the gas from the orifice plate to the downstream edge of the discharge.

5.3 Gas purity

The requirements on gas purity for high-power, high-efficiency CO lasers are more stringent than in other infrared laser systems. The overall system efficiency relies upon the slow V–T deactivation of the CO vibrational energy. It is therefore necessary to eliminate any gas impurity that may be an effective V–T or V–V collision partner with the CO. An example of this would be the elimination of CO_2, which could cause rapid decay of vibrational energy from the CO as a result of intramolecular V–V transfer between the CO and CO_2, followed by rapid V–T deactivation of the CO_2.

Another potential impurity effect concerns the possible presence of electronegative gases which may readily dissociatively attach electrons and thereby limit the discharge current density. Some examples are water and iron pentacarbonyl, the latter being a common impurity in commercial CO supplies. It is usually necessary to limit the iron pentacarbonyl to less than a few ppm, which can be done by use of a gas filter such as activated alumina (Center 1974).

6. Cavity configurations – examples and operating conditions

There are basically 16 modes of operation for the electric CO laser, as shown in table 1. These can be subdivided into four broad categories, i.e., closed- versus open-cycle operation and coincidence of the electrical excitation with the optical extraction region versus noncoincidence. Checkmarks in table 1 indicate those modes of operation which are deemed feasible for high-power and high-electrical-efficiency operation. The basis for this restricted selection is discussed below, and examples of experimental lasers are described for the selected mode of operation under open-cycle conditions. These examples are limited to devices with average power of more than 100 W or pulse energies in excess of 1 J. For low-power CO EDLs, e.g., conventional and transversely-excited discharge lasers, the reader is referred to previous reviews by Mann (1976) and Sobolev and Sokovikov (1973a).

Table 1
Modes of operation for cavity design.

	Open cycle		Closed cycle	
	cw	Pulsed	cw	Pulsed
Noncoincident				
Subsonic				
Supersonic	$\sqrt{}$		$\sqrt{}$	
Coincident				
Subsonic	$\sqrt{}$	$\sqrt{}$	$\sqrt{}$	$\sqrt{}$
Supersonic	$\sqrt{}$	$\sqrt{}$	$\sqrt{}$	$\sqrt{}$

Static gas systems have been excluded from table 1 since they are severely restricted to low average power because of the rate of heat diffusion to the walls, as discussed in § 5. As of 1977, there does not appear to have been any experimental development of the closed-cycle CO EDL in contrast to the closed-cycle electric CO_2 laser, which has been developed for commercial application up to cw power levels of the order of 10 kW. Probably the major engineering factor to be considered in the development of the closed-cycle CO EDL is the thermal conditioning needed to maintain low gas temperature required in the discharge cavity. Thus, the relative cost of gas conditioning is far higher for CO than for CO_2, and efficient closed-cycle CO lasers will probably be restricted to higher average power levels than for CO_2 lasers. Another engineering factor which has not been considered in the design of the closed-cycle CO EDL concerns gas poisoning. This can be due to gaseous impurities inherent in the system and/or impurities generated by ion–molecule chemistry as a result of ions generated in the electric discharge region.

6.1 Electrical–optical configurations

The two basic electrical–optical configurations are defined by coincidence or noncoincidence between the electrical discharge and the optical cavity. A fundamental disadvantage of the noncoincidence configuration is the potential loss of vibrational energy during the transit time from the discharge region to the optical cavity. This loss will be mainly due to V–T deactivation of high vibrational levels which have been populated by the V–V equilibration. As a result, there can be a large conversion of the input electrical energy to translational energy before the gas reaches the discharge region, and this gas heating will tend to choke the flow, as discussed in the previous section. It, therefore, only becomes practical to consider noncoincident cavity configurations in supersonic flows in which the transit time between the discharge and the optical cavity can be restricted to the order of 1 ms or less. Also, pulsed excitation is not compatible with the noncoincident configuration since the optical pulse length is determined by the flow time through the cavity rather than by the excitation pulse length. For these reasons, noncoincident excitation is probably restricted to cw supersonic flow configurations, as indicated in table 1.

Perhaps the most important advantage of the supersonic noncoincident mode of operation over the coincident cavity is the possibility to avoid flow disturbances created by the discharge and by discontinuities at the electrode boundaries. Thus, in the noncoincident configuration, the discharge electrode geometry can be optimized for uniform gas excitation, while the flow/medium uniformity is separately determined by the supersonic expansion channel geometry. It is also easy to correct for boundary-layer growth by a small expansion of the flow in the optical cavity without affecting the electric field distribution in the discharge.

Another approach to the noncoincident mode of operation is the mixing laser in which one species is excited in the electric discharge and is subsequently mixed with the CO in the optical cavity where it transfers its energy to the CO vibrational mode. N_2 is the ideal candidate for this mixing laser since its behavior under discharge excitation is similar to that of CO, with which it is isoelectronic. In addition, the N_2 vibrational spacing is slightly larger than that of CO so that the transfer from N_2 to CO is exothermic ($169 \, cm^{-1}$). However, the rate constant for the vibrational energy transfer between N_2 and CO is quite small (Stephenson and Mosburg 1974), and the transfer process is only effective for cw operation at high gas densities.

6.2 Examples of coincident cavities

6.2.1 Subsonic cw

One of the earliest experimental CO EDLs exploiting the benefits of forced convection flow was that of Kan and Whitney (1972). They showed that there could be a considerable increase in the average power over that obtainable with diffusion cooling to the wall. Fig. 10 is a schematic drawing of their forced convection flow CO laser using precooled gas mixtures flowing at approximately 10^4 cm/s through the tube. They were able to extract 400 W cw power at 20% electrical efficiency from a 1 m long, 2.5 cm diameter bore laser. This system cannot be scaled volumetrically since the discharge was limited by ambipolar diffusion to the walls, but the experiment was effective in removing the power limit imposed by thermal diffusion. Under slow flow conditions, the output power was only 35 W.

In order to develop a volumetrically scalable CO EDL, Falk and Kennedy (1976) applied an electron-beam sustained discharge to a 1ℓ subsonic flow CO

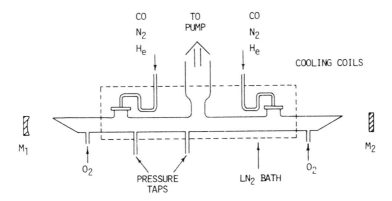

Fig. 10. Schematic drawing of forced-convective CO laser (from Kan and Whitney 1972).

laser which is shown schematically in fig. 11. The gas mixtures were precooled and flowed into the laser, being uniformly distributed across the flow area by the flow screen. Both electrodes were actively cooled to liquid-nitrogen temperature, and a Lexan thermal liner was used to isolate the flow from the outer structure and thereby permit core flows with static temperatures of approximately 100 K. Most of the experimental measurements were made with cavity flow speeds of order 3×10^3 cm/s (i.e., Mach 0.1) and laser performance was measured in run times extending to 16 s.

Typical performance data for both subsonic devices are summarized in table 2. The maximum output power in the 1ℓ electron-beam sustained discharge was 1100 W as measured with closed cavity calorimetric mirrors, and the maximum electrical efficiency in the core flow was 18%. The electrical efficiency is defined as optical power out divided by electrical power deposited in the gas discharge. This performance is considerably short of that predicted by kinetic modelling for a uniformly excited system. Several experimental factors were thought to have contributed to the difference between predicted and measured performance. These include unexplained gas heating and low energy loading per CO particle, typically 0.3 eV/CO. The experiments indicated low efficiency for the conversion of electrical to vibrational energy in the discharge, possibly as a result of large nonuniformities in the ionization and energy deposition. Much higher energy loadings have been achieved in other electron-beam sustained discharges as discussed in the following sections.

6.2.2 Subsonic pulsed

The highest electrical efficiency and specific energy have been achieved by the use of pulsed electron-beam sustained discharges operating at cryogenic

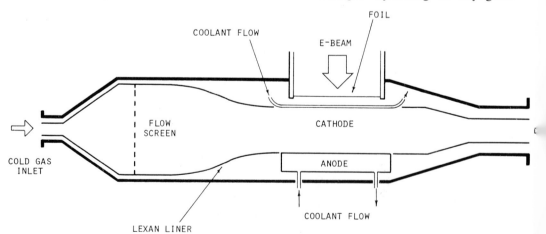

Fig. 11. Schematic drawing of subsonic CO laser using electron-beam sustained discharge excitation (from Falk and Kennedy 1976).

Table 2
Subsonic cw performance summary.

Inlet temperature [K]	Gas density [ama]	Velocity [cm/s]	Gas mixture	Specific electrical input energy		$10^{16} E/N$ [V cm^2]	Laser power [W]	Electrical efficiency [%]	Spectral range [μm]
				[kJ/lb]	[eV/particle]				
80[a]	0.12	$\sim 10^4$	He–CO–N$_2$–O$_2$ 370:14:6:0.4	450	~ 1.0	~ 0.6	400	20	5.02–5.72
100[b]	0.24	3.4×10^3	He–CO 0.93:0.07	170	0.26	0.54	1120	14	Not reported

[a]Kan and Whitney (1972).
[b]Falk and Kennedy (1976).

temperatures. The first experimental results were reported by Mann et al. (1972, 1974) for a 1 m long, 2ℓ volume system. They reported pulse energies up to 200 J and a maximum efficiency of 63%. This energy conversion efficiency has not been achieved in any of the larger CO EDLs. In later experiments in a 10ℓ device (McAllister et al. 1975, Mann 1976), they obtained output pulse energies of 500 J at a maximum efficiency of 43%. The highest energy output in a subsonic pulse device has been reported by Center (1974), who extracted 1700 J from a 20ℓ device at a total electrical efficiency of 20%. More recent experiments by Boness and Center (1977) in a modified 16ℓ device with improved thermal uniformity have resulted in a laser output of 1560 J and 35% overall electrical efficiency. These performance data and the corresponding experimental conditions are summarized in table 3.

Different gas thermal conditioning techniques were used in the experiments of Mann et al. (1974) and Mann (1976) as compared to those of Center (1974) and Boness and Center (1977). The former experiments (Mann et al. 1974, Mann 1976) employed a slow flowing system in which the gas was precooled to liquid-nitrogen temperature and flowed through an isothermal discharge cavity, also cooled to liquid-nitrogen temperature. A cross section of this system is illustrated in fig. 12 and shows precooled gas flowing into the discharge region through a porous screen of the liquid-nitrogen-cooled anode. The gas flow is transverse to the optical axis and is exhausted through the cathode screen. The gas velocity in the discharge region is of order 10 cm/s. This isothermal cavity can provide a very high degree of uniformity in the gas temperature (to the order of 0.1°C) but necessitates very complex thermal engineering to allow for differential contraction as the system is cooled down to liquid-nitrogen temperature. In particular, the cavity windows, CaF_2, present a large design problem because of differential contraction. As shown in fig. 12, the output windows were

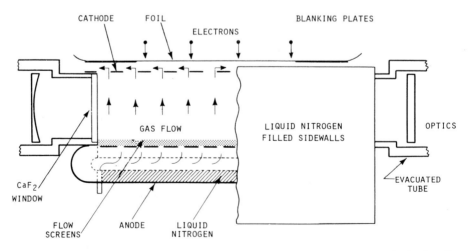

Fig. 12. Schematic view of isothermal CO discharge cavity using electron-beam ionization (from McAllister et al. 1975).

Table 3
Summary of subsonic pulsed CO experimental results.

Temperature [K]	Gas density [ama]	Gas mixture [CO-N₂-Ar]	Excitation pulse length [μs]	Specific electrical input energy [J/ℓ ama]	Specific electrical input energy [eV/particle]	$10^{16} E/N$ [V cm²]	Laser energy [J]	Electrical efficiency [%]	Spectral range [μm]
80[a]	0.45	1:0:10	100	270	0.69	0.5	153[f]	63	4.94–5.40
80[b]	0.8	1:6.7:10	100	200	0.82		500[f]	42	Not reported
90[c]	0.5	1:5:0	60	560	0.65	0.81	1560[g]	35	5.04–5.39
90[c]	0.5	1:0:5	40	332	0.39	1.9	1010[g]	38	5.02–5.34
295[d]	0.5	1:5:0	80	1170	1.36	2.2	700[f]	6	Not measured
295[e]	0.1	1:6:7 CO:N₂:He	15	250	0.81	1.3	50	2	5.41–5.82

[a]Mann et al. (1974).
[b]McAllister et al. (1975); also Mann (1976).
[c]Boness and Center (1977).
[d]Center and Boness (1974).
[e]Basov et al. (1976).
[f]Stable resonator, multimode output.
[g]Unstable resonator, near-diffraction-limited.

thermally isolated by use of an evacuated double-window structure. Near-diffraction-limited performance, at a relatively low laser energy density of 20 J/ℓ ama, has been demonstrated by McAllister et al. (1975) in the 10ℓ system using an unstable resontator with 75% output coupling.

A completely different design philosophy was used by Center (1974) and Boness and Center (1977). Here, the primary aim was to develop a room temperature discharge cavity capable of operating with cryogenic gas flow, and yet able to maintain good density and uniformity in the gas. The use of a room temperature cavity avoids materials compatibility problems, mentioned above, resulting from differential contraction and expansion. However, the room temperature environment introduces thermal boundary layers on the walls, which can affect the optical homogeneity. The final design used in the 16ℓ device (Boness and Center 1977) is shown in fig. 13. The basic concept was to introduce

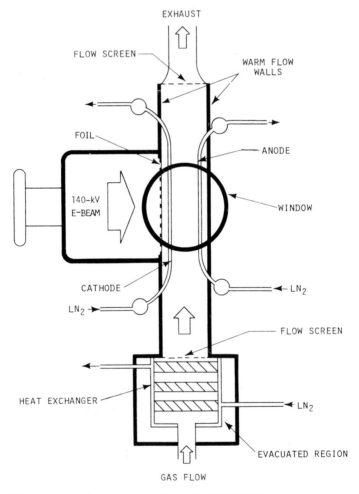

Fig. 13. Schematic drawing of cryogenic flow pulsed CO EDL in room temperature cavity.

the gas through a porous heat exchanger and to allow free and forced convection boundary layers to be established on room temperature side and end walls. The flow was directed vertically upwards to ensure that buoyancy effects were favorable and to avoid recirculation. By the use of controlled boundary layers, the density change due to the difference in temperature between the room temperature walls and the cryogenic gas is concentrated in a narrow region and isolates the main cold flow from the warm walls. Boundary-layer calculations (Woodroffe 1975) had indicated that medium flow requirements for a near-to-diffraction-limited beam could be met with flow velocities greater than 100 cm/s. The gas velocity in the experiments was about 400 cm/s and near-field diffraction patterns have indicated beam quality better than 1.5 times the diffraction-limited performance at pulse energies of 1000 to 1500 J.

Table 3 also includes some representative data for room temperature operation. The maximum energy and efficiency are those reported by Center and Boness (1974) using the 20ℓ electron-beam sustained discharge. The measured efficiency of 6% is far below that expected on the basis of the kinetic model predictions (Rockwood et al. 1973), which have suggested the possibility of 50% conversion efficiency at input energy densities of the order of kJ/ℓ ama. Thweatt et al. (1974) have reported experiments using a high current density, cold cathode electron beam as the external ionization source. Although they were also able to achieve input electrical energy densities in excess of 1 kJ/ℓ ama, the maximum conversion efficiency was only 2%. The large discrepancy between theory and experiment is poorly understood but may be due in large part to de-excitation of the CO by superelastic electron collisions. Such de-excitation processes become progressively more important in the higher vibrational levels associated with room temperature operation and at the high electron densities associated with pulsed operation. These results again emphasize the dramatic improvement in the electrical efficiency that can be achieved by operating at low translational temperatures.

6.2.3 *Supersonic* cw CO *lasers*

The supersonic cw CO laser excited by an electron-beam sustained discharge has received a lot of attention recently because of the potential benefits of the uniform cooling which can be achieved in a carefully designed supersonic expansion. Also, as indicated in § 5, the streamwise density variation due to gas heating is much smaller in supersonic flow than in subsonic flow, with resulting improvement in the optical quality which should be achievable in the supersonic flow. Both the small scale and large scale experiments to date have simulated cw operation by long pulse excitation of the gas with optical extraction over 3 to 5 cavity flow times.

The first electron-beam stabilized supersonic CO EDL was reported by Jones et al. (1974) and Jones and Byron (1974). This device had an active discharge volume of 0.5ℓ. Klosterman and Byron (1977a, b) subsequently performed detailed parametric measurements using a modified flow system. Thompson et al. (1975) have reported measurements from a similar device of 0.9ℓ volume. Experimental measurements in a 4.5ℓ device have been reported by Plummer et al. (1975) and by Plummer (1977), and a 39ℓ system is currently being studied by O'Brien (1977). Table 4 lists some of the representative performance characteristics.

As shown in this table, power output of the order of 100 kW can be obtained readily from a half-liter system with an overall efficiency of the order of 20 to 40%. Significantly higher output power may be expected from the large volume device currently under investigation by O'Brien (1977). Substantial improvements in the laser performance have been demonstrated by small variations in the gas mixture and optimization of the input power density distribution by use of a multi-element ballasted electrode. The addition of small mole fractions of H_2 or O_2 (typically 0.01) has resulted in increases of up to 50% in the specific output energy and output laser power over that achieved with the same mixture without the additives. Further improvement in the laser efficiency can be obtained by preventing the discharge from running downstream of the cavity (Klosterman and Byron 1977a, b).

All of these devices have in common a Ludwieg tube storage reservoir to provide a quasi-steady supersonic flow, obtained by expansion of the gas through a minimum length supersonic nozzle. Typically, the gas is expanded to a temperature of approximately 60 to 80K, with a corresponding flow Mach number of 3 to 4, depending upon the diluent and the effective ratio of the specific heats, γ. A high Mach number is desirable to maximize the allowable heat input without choking the flow. Furthermore, the laser kinetics improve with decreasing translational temperature, as discussed in § 3. However, there is an upper bound to the expansion which is set by condensation of the CO, which typically limits the expansion to translational temperatures above 50 K. The electron-beam current density in these experiments is usually limited to $1 \, mA/cm^2$ to simulate the foil heating limit for true cw operation (Daugherty 1976).

A schematic drawing of the 0.5ℓ device used in the parametric studies of Klosterman and Byron (1977a, b) is shown in fig. 14. Gas is supplied by the Ludwieg tube and is expanded through the supersonic nozzle to produce a quasi-steady supersonic flow of approximately 10 ms duration. The flow is initiated by rupturing a diaphragm which separates the Ludwieg tube from the supersonic nozzle. A dual-foil assembly, initially described by Thompson et al. (1975), is used as the electron-beam window with one of the foils acting as a flush discharge electrode backed by a support structure. The discharge channel is twice as wide (20 cm) as the nominal electron beam aperture (10 cm × 10 cm) to avoid discharge arcing through the warm boundary layer on the sidewalls. A significant design innovation in this apparatus is the use of a screen nozzle in place of the conventional slit nozzle or grid nozzle array. The application of screen nozzles to gasdynamic and electric discharge lasers was first described by Russell et al. (1975). In the application to the supersonic CO EDL, the screen

Table 4
Supersonic cw CO EDL experimental data.

Temperature (K)	Gas density (ama)	Gas mixture	Particle flow time [μs]	Discharge duration [μs]	Specific electrical input energy [kJ/lb]	[eV/CO]	$10^{16}E/N$ [V cm^2]	Laser power [kW]	Electrical efficiency [%]	Spectral range [μm]
60[a]	0.4	10:45:45 CO-N$_2$-Ar-He-O$_2$	150	300	84	0.43	0.47	50	18.5	4.98–5.19
64[b]	0.3	10:5:74:10:1 CO-N$_2$-Ar-He-O$_2$	187	360	83	0.65	0.51	79	36	5.00–5.32
64[b]	0.3	5:10:40:41:1 CO-N$_2$-Ar-He-O$_2$	150	425	152	1.53	0.60	75	35	4.94–5.26
75[c]	0.34	1:9 CO-Ar	300	1000	38	0.32		60	10	5.09–5.55
85[d]	0.32	1:4:5 CO-N$_2$-Ar	240	1000	152	1.15	0.59	117	18	Not reported
50[e]	0.19	78:91.7:0.4 CO-He-O$_2$		cw[f]	125	0.22		0.94	14	4.88–5.63

[a]Jones and Byron (1974).
[b]Klosterman and Byron (1976).
[c]Plummer et al. (1975).
[d]Thompson et al. (1975).
[e]Daiber and Thompson (1977).
[f]Noncoincident optical cavity. See text.

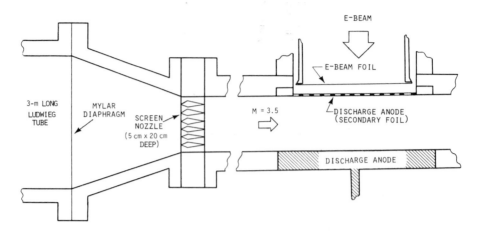

Fig. 14. Schematic drawing of 0.5ℓ supersonic cw CO EDL.

nozzle has an inherent advantage over the conventional slit nozzle of minimizing the required distance between the nozzle throat and the discharge region. This, in turn, leads to thinner boundary layers over the discharge electrodes and consequently more efficient use of the discharge and optical volume.

Although the supersonic expansion provides an initially uniform low temperature medium, severe disturbances can be generated in the flow field as a result of the interaction with the discharge. For example, gas heating of the boundary layer in the discharge region can lead to disturbances which propagate upstream in the boundary layer and change the flow characteristics. Jones (1977) has made observations of large axial density gradients in quasi-cw operation at low power loadings, corresponding to a CO vibrational energy density of approximately 0.3 eV/CO molecule. Density profile measurements by Klosterman and Byron (1977a, b) have shown that there is significantly more boundary-layer heating on the cathode than on the anode. This leads to an increase in the static pressure and to the possibility of separation of the cathode boundary layer. Thus, it is important to take account of the boundary-layer growth in designing a discharge channel geometry. The warm boundary layer on the sidewall also provides a possible path for electrical breakdown of the gas as a result of the increased E/N in the boundary layer. It is, therefore, necessary to limit the boundary-layer growth and to maintain a sufficient distance between the sidewalls and discharge electrodes to prevent the possibility of arc breakdown. In a true cw device, the value of E/N could be kept constant by cryogenic cooling of the sidewalls.

Another flow-related problem investigated by Klosterman and Byron (1977a, b) was the nonuniform power loading distribution in the discharge, resulting from expansion and compression waves generated at the edge of the discharge region. The flow outside the discharge region sees no discharge heating and therefore experiences compression waves due to the pressure rise in the discharge region. Corresponding expansion waves move into the discharge region, which lead to local increases of E/N and highly nonuniform power input density. Current density variations of a factor of 4 have been observed as a result of these expansion waves. This discharge current distribution can be externally controlled to some extent by the use of a variable resistance ballasted multi-element electrode. However, there is a clear need for further investigation of this problem in order to achieve true volumetric scaling.

The gas heating leads to a streamwise temperature gradient in both the subsonic and supersonic cw EDL. Since the vibrational relaxation process and the small-signal gain are strongly dependent on the translational temperature, the gas heating will cause a shift in the optical gain distribution in the streamwise direction. This can have a significant impact on the power extraction since transitions which are inverted in the low temperature gas near the inlet to the discharge region may change to absorption near the downstream edge of the optical cavity. Large streamwise variations in the gain coefficient have been observed experimentally by Klosterman and Byron (1977a, b), and it is clear that different rotational lines will be optimum for the upstream and downstream portions of the cavity.

6.2.4 Supersonic pulsed CO lasers

The pulsed supersonic CO EDL combines the advantages of the supersonic flow for gas conditioning with the spatially uniform discharge excitation obtained when the pulse length is short compared with the gas flow time through the cavity. The most extensive measurements obtained to date are those of Jones (1977), using a $2.5\,\ell$ cavity ($6.5\,cm \times 11\,cm \times 35\,cm$). The gas supply system is similar to that described in the previous section. The only difference in the electron-beam sustained discharge between pulsed and cw operation is the possibility to increase the electron-beam current density during the pulse while maintaining the average current density of $1\,mA/cm^2$ set by foil heating limits. Some single-pulse performance data are listed in table 5. The pulse length was limited to less than $45\,\mu s$ to minimize performance degradation resulting from the propagation into the cavity of strong acoustic waves which are generated at the discharge electrodes. The short pulse length is to be compared with the particle flow time of the order of 200 to $300\,\mu s$.

One surprising result from these pulse performance measurements is the improvement in the electrical efficiency with the substitution of Ne for Ar. It is not known whether this is due to a kinetic effect or related to the decreased refractive index of the Ne, with the consequent improvement in the optical medium quality. It may also be associated with the lower condensation temperature of Ne compared to Ar, which allowed higher density operation with Ne diluent (see table 5). The use of Ne as the major diluent imposes a penalty of lower electron density and lower power input density resulting from the decreased stopping power of the Ne for high energy electrons. Furthermore, the economics of the use of Ne as a diluent in a CO laser system would indicate the

Table 5

Pulsed supersonic CO EDL experimental data.

Temperature [K]	Gas density [ama]	Gas mixture	Particle flow time [μs]	Excitation pulse width [μs]	Specific electrical input energy		$10^{16}E/N$ [V cm^2]	Laser energy [J]	Electrical efficiency [%]	Spectral range [μm]
					[J/ℓ ama]	[eV/CO]				
65[a]	0.4	15:75:10 CO–Ar–He	230	44	420	0.65	0.8–0.4	118	28	4.89–5.01
65[a]	0.5	10:80:10 CO–Ne–He	170	34	210	0.49	0.58–0.50	82	31	4.88–4.99
65[a]	1	10:80:10 CO–Ne–He	170	35	200	0.47	0.42–0.37	162	32	4.88–5.06

[a]Jones et al. (1977).

desirability of closed-cycle operation because of the relatively high cost of Ne in comparison with Ar.

Although the boundary layer problems in the cw supersonic laser are some-what mitigated in the pulsed supersonic device, there remains some uncertainty about the minimum clearing time of boundary-layer-induced disturbances. This clearing time may limit the frequency for repetitively-pulsed operation. Except for this uncertainty, the repetition frequency can be estimated from the inverse of the flow transit time through the cavity [see eq. (12)]. Thus, one might expect high average power operation from a repetitively-pulsed supersonic laser with electrical efficiencies in excess of 30%. Further experimental work is necessary to establish limitations to the repetition frequency as well as to investigate the optical homogeneity of the laser medium.

6.3 Noncoincident supersonic cw CO EDL

This mode of excitation was initially investigated by Kan et al. (1972) and subsequently scaled to higher pressures by Rich et al. (1975). Further scaling to larger discharge and cavity volumes has been reported by Daiber and Thompson (1977). Self-sustained discharges were used in these experiments with aerody-namic stabilization to permit operation in helium-rich mixtures at stagnation pressures of the order of 1 atm. In all cases, the discharge is located in the subsonic plenum immediately upstream of the supersonic nozzle. A schematic diagram is shown in fig. 15, and the reader is referred to Daiber et al. (1976) for discussions of the aerodynamically-stabilized glow discharge. They have repor-ted detailed measurements of the efficiency of the CO vibrational excitation in this self-sustained glow discharge. The measured excitation efficiency was in the range from 30 to 80% and was significantly less than expected in an externally-sustained discharge with independent control of the value of E/N.

Representative data for the noncoincident experiment of Daiber and Thomp-son (1977) is shown in table 4. The output power was extracted from an optical volume of 0.19ℓ. To date, the best performance data in a noncoincident laser is

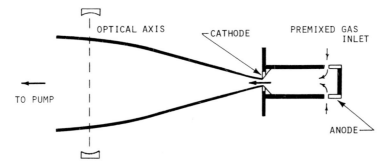

Fig. 15. Schematic diagram of noncoincident supersonic CO EDL (after Rich et al. 1975 and Daiber and Thompson 1977).

significantly less than achievable in the coincident cw supersonic device. There are two basic reasons for this. First, the discharge excitation efficiency is low, as mentioned above, and the energy density was usually limited to approximately 0.2 eV/CO molecule. Secondly, there is a significant loss of vibrational energy during the expansion. This energy loss amounted to some 33% of the expected vibrational energy content, and the cause of the energy loss has not been determined. These losses severely limit the overall efficiency obtainable from a noncoincident excitation device and illustrate the major problem in this approach. The power spectral distribution in the noncoincident configuration is significantly different from that in a coincident device, as indicated by comparison of the data in table 4. This is a result of the lower translational temperature achievable in the noncoincident mode of operation, since most of the gas heating occurs upstream and can be taken into account by suitable design of the supersonic nozzle.

7. Conclusions

The basic physical phenomena describing the efficient electrically-excited CO laser are well understood. In particular, the relaxation behavior of the anharmonic oscillator has been carefully investigated under conditions of thermal nonequilibrium. The potentially high electrical efficiency of the laser is seen to result from several properties inherent to the CO molecule, including (a) very slow energy loss by V–T collisions, (b) rapid redistribution of vibrational energy in the vibrational manifold, (c) high quantum efficiency due to the small anharmonic defect and small rotational energy spacing, (d) ability to recycle vibrational energy many times through the molecule and (e) the high efficiency for electrical excitation of the vibrational mode.

The important scaling parameters have been identified, and the vibrational energy per CO molecule is clearly the dominant parameter. This is verified by the detailed kinetic modelling calculations which have been shown to be in qualitative agreement with experimental results. In addition, the kinetic models can quantitatively predict the lasing threshold conditions of energy density and the corresponding time to reach threshold. However, these kinetic models are incomplete and fail in their quantitative estimates of the lasing spectral distribution and performance variation with translational temperature. This latter defect is a serious limitation to the use of the detailed kinetic models in the design of practical lasing systems. Clearly, more development is needed to explain the very limited electrical efficiency observed in room temperature CO EDLs.

The development of efficient cryogenically cooled CO EDLs has been experimentally demonstrated. This includes the application of the scalable externally-ionized discharge techniques. Several gas thermal conditioning techniques have been investigated in subsonic and supersonic flows. By far, the simplest systems involve the flow of cryogenic gas through a warm (room temperature) cavity. This is a logical choice in the supersonic CO laser by virtue of the expansion cooling of the gas. The pulsed CO EDLs, both subsonic and supersonic, are truly scalable in volume and have yielded electrical efficiencies in the range of 30 to 60%. A question still remains regarding the volumetric scalability of the cw supersonic CO EDL because of interactions between the flow and the discharge. This uncertainty does not apply to the noncoincident supersonic laser using a subsonic excitation region, but the efficiency of those systems is clearly limited.

One final comment concerns the possible future development of the closed-cycle CO EDL. The practicality of the efficient closed-cycle CO laser will depend on the engineering development of efficient cryogenic loops and a better understanding of the effect of translational temperature in the laser performance. In addition, it will be necessary to identify the possible buildup of impurities and to maintain continuous purification of the gas supply to limit the mass flow of makeup gas.

Acknowledgments

The author gratefully acknowledges private communication and technical discussions with many colleagues, including M.J.W. Boness, E.L. Klosterman, T.G. Jones and B.B. O'Brien. Particular thanks go to S.R. Byron for a critical review and helpful discussions of this manuscript.

References

Basov, N.G., E.M. Belenov, V.A. Danilychev and A.F. Suchkov (1971), JETP Letters **14**, 375.
Basov, N.G., V.A. Danilychev, A.A. Ionin, O.M. Kerimov, I.B. Kovesh, A.F. Suchkov, B.M. Urin and M.V. Khasenov (1976), Sov. J. Quantum Electronics **6**, 619.
Bhaumik, M.L. (1970), Appl. Phys. Letters **17**, 188.
Bhaumik, M.L. (1972), Appl. Phys. Letters **20**, 342.
Bhaumik, M.L., W.B. Lacina and M.M. Mann (1970), IEEE J. Quantum Electronics **QE-6**, 575.
Bhaumik, M.L., W.B. Lacina and M.M. Mann (1972), IEEE J. Quantum Electronics **QE-8**, 150.
Boness, M.J.W. and R.E. Center (1975), Appl. Phys. Letters **26**, 511.
Boness, M.J.W. and R.E. Center (1977), J. Appl. Phys., to be published.
Born, M. and E. Wolf (1964), Principles of Optics, 2nd edn. (Pergamon Press; Oxford).
Caledonia, G.E. and R.E. Center (1971), J. Chem. Phys. **55**, 552.
Center, R.E. (1973), J. Appl. Phys. **44**, 3538.
Center, R.E. (1974), IEEE J. Quantum Electronics **QE-10**, 208.
Center, R.E. and M.J.W. Boness (1974), AVCO Everett Research Laboratory Technical Report, Contract N00014-72-C-0030, May.
Center, R.E. and G.E. Caledonia (1971a), Appl. Phys. Letters **19**, 211.
Center, R.E. and G.E. Caledonia (1971b), Appl. Optics **10**, 1795.
Center, R.E. and G.E. Caledonia (1975), J. Appl. Phys. **46**, 2215.
Chandra, N. and P.G. Burke (1973), J. Phys. B, Atom. Molec. Phys. **6**, 2355.
Daiber, J.W. and H.M. Thompson (1977), IEEE J. Quantum Electronics **QE-13**, 10.
Daiber, J.W., H.M. Thompson and T.J. Falk (1976), IEEE J. Quantum Electronics **QE-12**, 686.

Daugherty, J.D. (1976), Principles of Laser Plasmas (G. Bekefi, ed.; Wiley; New York), Chapt. 9.

Daugherty, J.D., E.R. Pugh and D.H. Douglas-Hamilton (1972), Bull. Amer. Phys. Soc. **17**, 3.

Djeu, N. (1973), Appl. Phys. Letters **23**, 309.

Ehrhardt, H., L. Langhans, F. Linder and H.S. Taylor (1968), Phys. Rev. **173**, 222.

Falk, T.J. and J.J. Kennedy (1976), Air Force Weapons Laboratory Report AFWL-TR-76-275, August.

Fenstermacher, C.H., M.J. Nutter, W.T. Leland and K. Boyer (1972), Bull. Amer. Phys. Soc. **17**, 399.

Freed, C., (1971) Appl. Phys. Letters **18**, 458.

Frost, L.S. and A.V. Phelps (1964), Phys. Rev. **136**, A1538.

Gregg, D.W. and S.J. Thomas (1968), J. Appl. Phys. **39**, 4399.

Hake, R.D. and A.V. Phelps (1967), Phys. Rev. **158**, 70.

Hall, R.J. and A.C. Eckbreth (1974), IEEE J. Quantum Electronics **QE-10**, 580.

Hancock, G. and I.W.M. Smith (1971), Trans. Faraday Soc. **67**, 2586.

Jeffers, W.Q. and H.Y. Ageno (1975), Appl. Phys. Letters **27**, 227.

Joeckle, R. and M. Peyron (1970), J. Chim. Physique **67**, 1175.

Jones, T.G. (1977), private communication.

Jones, T.G. and S.R. Byron (1974), Mathematical Sciences Northwest, Inc., Report No. 74-128-1, Contract F29601-73-A-0039-0003, December.

Jones, T.G., S.R. Byron, A.L. Hoffman, B.B. O'Brien and W.B. Lacina (1974), AIAA Paper 74-562, Palo Alto, California.

Kan, T. and W. Whitney (1972), Appl. Phys. Letters, **21**, 213.

Kan, T., J.A. Stregack and W.S. Watt (1972), Appl. Phys. Letters **20**, 137.

Klosterman, E.L. (1975), Mathematical Sciences Northwest, Inc., Final Technical Report 75-123-1, Contract N00014-24-C-0258, March.

Klosterman, E.L. and S.R. Byron (1976), 29th Annual Gaseous Electronics Conference, October. Klosterman, E.L. and S.R. Byron (1977a) Air Force Weapons Laboratory Report, AFWL-TR-76-298.

Klosterman, E.L. and S.R. Byron (1977), submitted to J. Appl. Phys.

Kreuzer, L.B., N.N. Kenyon and C.K.W. Patel (1972), Science **177**, 347.

Lacina, W.B. (1975), IEEE J. Quantum Electronics **QE-11**, 297.

Lacina, W.B. and G.L. McAllister (1975a), IEEE J. Quantum Electronics **QE-11**, 235.

Lacina, W.B. and G.L. McAllister (1975b), Appl. Phys. Letters **26**, 86.

Lacina, W.B., M.M. Mann and G.L. McAllister (1973), IEEE J. Quantum Electronics **QE-9**, 588.

Legay, F. and N. Legay-Sommaire (1964), Compt. Rend. Hebd. Séances Acad. Sci. **A259**, 99.

Legay-Sommaire, N., L. Henry and F. Legay (1965), Compt. Rend. Hebd. Séances Acad. Sci. **A260**, 3339.

Mann, M.M (1976), AIAA J. **14**, 549.

Mann, M.M., W.B. Lacina and M.L. Bhaumik (1972), IEEE J. Quantum Electronics **QE-8**, 617.

Mann, M.M., D.K. Rice and R.G. Eguchi (1974), IEEE J. Quantum Electronics **QE-10**, 682.

McAllister, G.L., V.G. Dragoo and R.G. Eguchi (1975), Appl. Optics, **14**, 1290.

McKenzie, R.L. (1970), Appl. Phys. Letters **17**, 462.

Menzies, R.T. (1971), Appl. Optics **10**, 1532.

Millikan, R.C. and D.R. White (1963), J. Chem. Phys. **39**, 3209.

Nighan, W.L. (1970), Phys. Rev. **A2**, 1989.

Nighan, W.L. (1972), Appl. Phys. Letters **20**, 96.

Nighan, W.L. (1976), Principles of Laser Plasmas (G. Bekefi, ed.; Wiley; New York), Chapt. 7.

O'Brien, B.B. (1977), private communication.

Osgood, R.M. and W.C. Eppers (1968), Appl. Phys. Letters **13**, 409.

Osgood, R.M., E.R. Nichols and W.C. Eppers (1969), Appl. Phys. Letters **15**, 69.

Osgood, R.M., W.C. Eppers and E.R. Nichols (1970), IEEE J. Quantum Electronics **QE-6**, 145.

Patel, C.K.N. (1965), Appl. Phys. Letters **7**, 246.

Patel, C.K.N. and R.J. Kerl (1964), Appl. Phys. Letters **5**, 81.

Pistoresi, D.J. and D.J. Nelson (1975), Supersonic CO Laser Code, Air Force Weapons Laboratory Technical Report TR-75-256.

Plummer, M.J. (1977), private communication.

Plummer, M.J., J.W. Wagner and W.J. Glowacki (1975), IEEE J. Quantum Electronics **QE-11**, 700.

Pollock, M.A. (1966), Appl. Phys. Letters **8**, 237.

Rapp, D. and P. Englander-Golden (1965), J. Chem. Phys. **43**, 1464.

Rice, D.C. (1974), Appl. Optics **13**, 2812.

Rich, J.W. (1971), J. Appl. Phys. **42**, 2719.

Rich, J.W., R.C. Bergman and J.A. Lordi (1975), AIAA J. **13**, 95.

Rockwood, S.D., J.E. Brau, W.A. Proctor and G.H. Canavan (1973), IEEE J. Quantum Electronics **QE-9**, 120.

Russell, D.A., S.E. Neice and P.H. Rose (1975), AIAA J. **13**, 593.

Schultz, G.J. (1964), Phys. Rev. **135**, A988.

Schultz, G.J. (1976), Principles of Laser Plasmas (G. Bekefi, ed.; Wiley; New York), Chapt. 3.

Sequin, H.G., J. Tulip and B. White (1972), Appl. Phys. Letters **20**, 436.

Shapiro, A.H. (1953), The Dynamics and Thermodynamics of Compressible Fluid Flow (Ronald Press; New York).

Sobolev, N.N. and V.V. Sokovikov (1973a), Sov. J. Quantum Electronics **2**, 305.

Sobolev, N.N. and V.V. Sokovikov (1973b), Sov. Phys. – Uspekhi, **15**, 350.

Stephenson, J.C. and E.R. Mosburg (1974), J. Chem. Phys. **50**, 3562.

Sutton, G.W. (1969), AIAA J. **9**, 1737.

Szewczyk, A.A. (1964), Trans. ASME, J. Heat Transfer **86**, 50¹.

Thompson, J.E., B.B. O'Brien, C.G. Parazzoli and W.B. Lacina (1975), 28th Annual Gaseous Electronics Conference, Rolla, October.

Thweatt, W.L., G.W. Sullivan and R.F. Weber (1974), 27th Annual Gaseous Electronics Conference, October.

Treanor, C.E., J.W. Rich and R.G. Rehm (1968), J. Chem. Phys. **48**, 1728.

Watt, W.S. (1971), Appl. Phys. Letters **18**, 487.

Wilks, G. (1973), Int. J. Heat and Mass Transfer **16**, 1958.

Wittig, C., J.C. Hassler and P.D. Coleman (1971), J. Chem. Phys. **55**, 5523.

Woodroffe, J.A. (1975), AIAA J. **13**, 942.

A4 | Excimer Lasers

J. J. EWING

University of California,
Lawrence Livermore Laboratory,
Livermore, California 94550

Contents

Abstract

The general class of lasers operating in the ultraviolet and vacuum ultraviolet spectral regions commonly called excimer lasers is reviewed here. The general characteristics of these laser media are discussed along with the technology required to excite these lasers. The spectroscopy of this class of molecular species is reviewed. The spectral characteristics of absorbers generated in the media in the course of generating laser gain are discussed. Nonsaturating losses in these media limit the efficiency and scalability of these devices. The kinetic mechanisms for producing and quenching these excited molecular species are reviewed and tabulations of pertinent kinetic data are given. The technologies associated with electrically exciting these species are discussed.

© *North-Holland Publishing Company, 1979*
Laser Handbook, edited by M.L. Stitch

1. Introduction

The class of short wavelength molecular gas lasers broadly known as excimer lasers is a subject of large and growing interest. The use of the term excimer laser for our purposes implies lasers which operate on electronic transitions of molecules, up to now diatomic, whose ground state potential energy curve is essentially repulsive. Excimer lasers thus generally operate on a continuum band caused by a molecular excited state emitting to the repulsive lower level (see fig. 1). Population inversion and hence gain naturally occur for bound–free transitions by the rapid dissociation of the ground state molecules. A more detailed description of excimer spectroscopy and some special cases are given in §3 of this chapter.

Because of the broad bandwidth of their continuum transitions, excimers have inherently small cross sections for stimulated emission. Large excited state populations are therefore necessary for reasonable gain. Collisional processes among the excited states thus become important. Excited state densities sufficient for excimer laser action have been produced by using intense excita-

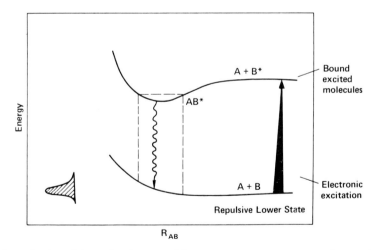

Fig. 1. Schematic plot of the molecular potential energy curves, energy versus internuclear separation, R_{AB}, of an excimer or exciplex species. The molecule AB is unbound with respect to $A + B$ and hence dissociates. Excited molecules AB* can be formed from the excited atoms. AB* emits a continuum band with the transition terminating on an unstable portion of the (AB) energy curve, leading to rapid dissociation into $A + B$.

tion by electron beams, discharges and optical pumping. Efficient channels have been discovered for the rapid production of the excimer emitting states. Various processes such as electron–molecular ion recombination, positive ion–negative ion recombination, electron impact excitation and chemical reaction of excited atoms have all been found to be important kinetic mechanisms for producing excimer excited states.

It is now clear that efficient excimer lasers offering high average power can be built. The variety of visible, ultraviolet and vacuum ultraviolet wavelengths currently demonstrated, as well as modest tunability, make these laser sources of interest in advanced photochemical processing concepts, such as isotope separation. These developments are based on a rapidly maturing understanding of the fundamental physics of the media. It is the purpose of this chapter to introduce to the reader both the microscopic media issues and technological aspects of excimer lasers.

An overview of excimer lasers describing their history, emission properties, and excitation requirements is given in § 2. A summary of spectroscopic nomenclature using various excimers for examples is discussed in § 3. Also in § 3 the spectroscopy of specific absorbers that have been found to compete with the gain and energy extraction in excimer laser media is discussed. Kinetic processes in excimer laser media are described in § 4 starting with the pure rare gases and concentrating on the important case of rare gas–halogen molecule mixtures. In § 5 the technology and physics of electron-beam and discharge sources are described along with the optical components and tuning techniques of excimer lasers.

2. Overview of excimer lasers

2.1 History

A synopsis of the history of excimer lasers follows, highlighting the first laser demonstration of the various media. Demonstrated excimer lasers are listed in table 1 by wavelength. For brevity, many papers in supporting research and technology are not discussed in detail in this chapter. Identification of excimer emission begins with Hopfield (1930a, b) who discovered He_2 emission in the extreme ultraviolet. The classical work of Finkelnburg (1938) describes Hg_2, Cd_2 and other metal vapor excimer emissions that were discovered in close relation to atomic resonance lines. Other rare gas excimer emissions were later found by Tanaka and Zelikoff (1954a, b) and Tanaka (1955) who studied the emission spectra of Xe_2, Kr_2 and Ar_2, and developed useful lamps for their production. Excimer emissions, and indeed the derivation of the name itself, are also well known in condensed phase organic materials (Stevens and Hutton 1960, Forster 1975). The word 'excimer' rigorously refers to an excited species made by combination of two identical moieties, atoms or molecules. Excited complexes

Table 1
Excimer laser wavelengths.

Species	Wavelength [nm]	Reference
Ar_2	126	Hughes et. al. (1974)
Kr_2	146	Hoff et. al. (1973)
Xe_2	172	Kohler et. al. (1972)
ArCl	175	Waynant (1977)
ArF	193	Hoffman et. al. (1976)
KrCl	222	Murray and Powell (1976)
KrF	248	Ewing and Brau (1975b)
XeBr	282	Searles and Hart (1975)
XeCl	308	Ewing and Brau (1975b)
XeF	351, 353	Brau and Ewing (1975a)
XeO	540	Powell et. al. (1974)
ArO	558	Powell et. al. (1974)
KrO	558	Powell et. al. (1974)

not in this category are also called an 'exciplex' (Birks 1967). This more accurate but less alliterative name is also found in the excimer laser literature.

The concept of using a bound–free electronic transition to provide gain was first suggested by Houtermans (1960) in the same year as the first demonstration of an actual laser by Maiman (1960). A general analysis of the gain in an excimer system based on differing electronic and translational temperatures was given by Leonard et al. (1963). Early work concentrated on the bound–free continuum band of $H_2(a^3\Sigma_g^+ - b^3\Sigma_u^+)$, familiar in discharge lamps. Later work by Palmer (1970) demonstrated the impracticality of gain from the H_2 continuum because of competing continuum absorption from electronically excited states. The broad 490 nm continuum common in Hg discharges was also investigated as an excimer laser candidate by Carbone and Litvak (1968). Spectrocopic assignment of this band is still uncertain, and recent work by Drullinger et al. (1977) suggests it may be an Hg_3 rather than Hg_2 feature. Once again, actual probe measurement by Hill et al. (1973) found absorption rather than gain in electron-beam-excited, high pressure Hg at 488 nm. All this work emphasizes the difficulty of achieving practical gain on extremely broad excimer transitions and also the possibility of unexpected medium absorptions.

First demonstration of an excimer laser was made by Basov et al. (1971) who showed stimulated emission of Xe_2 near 170 nm. Intense excitation was provided by electron-beam pumping of cryogenic liquid Xe. Extension of the Xe_2 laser to the gas phase was made by Kohler et al. (1972) using electron-beam pumping of Xe at several atmospheres pressure. Gas phase rare gas excimer lasers based on Kr_2 at 146 nm and Ar_2 at 126 nm were later demonstrated by Hoff et al. (1973) and Hughes et al. (1974) respectively. Although the fluorescence efficiency of rare gas excimers pumped by electron beams was found to be very high, the reported laser efficiencies were disappointingly low. Kinetic models, the most extensive developed by Werner et al. (1977), again predicted the laser efficiency is limited by excited state absorption. Rhodes (1974) has reviewed the early work on rare gas excimer lasers.

The addition of oxygen-bearing impurities to electron-beam-excited rare gases at high pressure produced lasing on XeO, KrO and ArO (Powell et al. 1974, 1975). These excimers emit near 558 nm and are a molecular version of the weak $O(^1S)-O(^1D)$ auroral transitions. Lasing occurred only after the current pulse and was severely limited by low gain. Probe measurements showed large medium absorption during the pumping followed by recovery to small gain only in the afterglow.

A major step in the excimer laser field was initiated by the discovery of an entirely new class of excimers, the rare gas monohalides. In studies of product yields of Ar metastable reactions, Golde and Thrush (1974) identified an ArCl continuum band at 170 nm. They attributed the emission to an ionic excited state of argon monochloride, Ar^+Cl^-, emitting to the repulsive ArCl ground state. Velazco and Setser (1975) observed an entire class of ultraviolet emissions produced by Xe metastable reactions with halide molecules at low pressures. They identified the series as XeF, XeCl, XeBr and XeI. The spectroscopy of these excimers was also studied by Ewing and Brau (1975a), based on observation of the emission bands of XeI produced in high pressure mixtures excited by an electron beam. Using the model of an ionic excited state emitting to a covalent lower state, they accurately predicted the wavelengths of the most intense bands for the potential rare gas halide emitters (Brau and Ewing 1975b).

Demonstration of laser action followed rapidly. Searles and Hart (1975) produced weak lasing from XeBr while Brau and Ewing (1975a) shortly thereafter produced stronger lasing from XeF. Electron-beam pumping of near-atmospheric-pressure gas mixtures at current densities of order 100 A/cm^2 was used in both cases. The pumping threshold for these lasers was clearly much lower than previous excimer lasers. A host of rare gas halide laser demonstrations followed (see table 1). Of these the KrF laser (Ewing and Brau 1975b) was found to be the best, having a medium efficiency of 15% as reported by Bhaumik et al. (1976) and Brau and Ewing (1976). Shortly thereafter single pulse energies of about 100 J were reported for both KrF and ArF lasers (Hoffman et al. 1976). High energy outputs with medium efficiencies of 10% have also been observed for pulse durations of about 1 μs (Hunter 1977, Bhaumik 1977, Rokni et al. 1978a, b, c).

The basic mechanisms by which excitation could efficiently channel into the rare gas halide excited states were pointed out in the early papers. However, the accurate determination of all of the relevant reaction rate coefficients is still a subject of research. § 4 and the review by Brau (1978a) tabulate some of the pertinent rate coefficients. The mechanism of electron attachment to produce negative ions which rapidly combine with positive ions producing the ion pair excited states of rare gas halides was a novel excited state production mechanism which had not been previously discussed in a laser context (Brau and Ewing 1975a).

The use of electron-beam-controlled discharges operating in long pulses to excite rare gas halide lasers was demonstrated by Mangano and Jacob (1975). Ultimately higher powers and higher laser efficiencies may be possible by this technique. XeF and XeCl also have been excited by this technique (Mangano et al. 1976, Levatter et al. 1978). Excitation of XeF by use of a simple discharge was achieved by Burnham et al. (1976a). System efficiencies of ~1% and pulse energies of 100 mJ on XeF and KrF were demonstrated with a uv preionized discharge approach (Burnham et al. 1976b). Despite their limited pulse length (≤50 ns) and efficiency (~1%), such simple discharge lasers have quickly become the method of choice for convenient laboratory applications, and have been extended to other rare gas halides with good results (Sze and Scott 1978).

The problem of competing medium absorption has also arisen with the rare gas halide lasers. Champagne and Harris (1977) indirectly inferred significant medium loss from the XeF laser output dependence on mirror coupling. Hawryluk et al. (1977) directly measured KrF gain superimposed on broadband absorption. Discussion of various absorbing species is given in §3.2. Several

Table 2
Some proposed excimer laser species.

Species	Wavelength	Reference
a) Metal complexes akin to rare gas excimers/large binding in excited state		
Hg_2	335	Schlie et. al. (1976)
		Komine and Byer (1977)
Hg_2 or Hg_3	485	Hill et al. (1973)
		Drullinger et al. (1977)
HgCd	470	McGeoch and Fournier (1978)
Cd	480	Drullinger and Stock (1978)
HgZn	455	Eden (1978)
b) Weakly bound metal complexes/atomic line wings		
Metal–rare gases	—	Phelps (1972)
KXe	820	Palmer (1976)
TlXe	680	Cheron et. al. (1976)
HgXe	270	Lam et. al. (1978)
HgTl	459, 656	Drummond and Schlie (1976)
c) Polyatomic complexes		
XeOH	234, 340	Hutchinson (1978)
$HgNH_3$	350	Callear and Connor (1972)
d) Other rare gas halides		
Triatomics	—	Lorents et. al. (1978)
Kr_2F	400	Hunter et. al. (1978)
ArXeF	460	Hunter et. al. (1978)
XeI and other broad bands	254, 320	Ewing and Brau (1975a, c)
		Brau and Ewing (1975b)
		Kligler et al. (1978)

reviews of rare gas halide laser development have been prepared (Ewing and Brau 1976, Rokni et al. 1978b, Brau 1978a).

Work on possible metal vapor excimer lasers has continued from the earliest Hg_2 studies. Recently positive gain measurements have been reported for the 335 nm Hg_2 band (Schlie et al. 1976), and the 470 nm HgCd band (McGeogh and Fournier 1978). Other experiments have, however, not confirmed these results (Komine and Byer 1977). Recent laser demonstrations by Parks (1977a, b) of HgCl and Schimitschek et al. (1977) of HgBr give considerable encouragement to the field. Although these molecules are not excimers, they are kinetically similar to rare gas halides and are excited by similar techniques. The problems of temperature nonuniformity leading to optical distortions for large index-of-refraction media, as well as component failure from corrosion, have clearly slowed progress in the metal excimer area.

Table 2 lists some of the potential excimer lasers that have been suggested in the literature but have not yet been made to lase. There are obviously many directions for future research.

2.2 Some general media features

Fig. 1 shows the schematic potential curves for the hypothetical excimer molecule AB. Collision of ground state A and B atoms occurs along a repulsive potential curve. However, if the electrons in one atom, say B, are excited, creating a state labeled B*, B* may combine with atom A to produce an excited state of the molecule AB*. Binding energies, relative to excited atoms, for such excited state complexes range from 0.1 eV for many rare gas–metal atom combinations, to about 5 eV for rare gas monohalides. There is typically a manifold of potential curves that can be derived from A + B*. Higher lying states, and the molecular ion AB$^+$, are often important in determining whether or not absorption processes originating in the excimer's upper laser level AB* will lower the efficiency of a particular laser.

The bandwidth of the excimer radiation depends on both the excited state potential energy curve and the ground state potential energy curve, as well as the distribution of excited states among the various vibrational and rotational states of the upper laser level. The majority of the excimer lasers demonstrated to date operate under conditions such that the excited states of the lasing molecules are deeply bound (rare gas oxides being the exception), and are vibrationally relaxed to temperatures below 1 000 K. Under such circumstances the emission bandwidth is determined by the effective width of the minimum region of the AB* curve and the slope of the lower level potential energy curve in the region of the minimum of the AB* potential curve. For example, the rare gas excimer excited state minima are in a region of strong repulsion of the lower state. Thus their excimer emission has a broad bandwidth, $\Delta\lambda \sim 20$ nm. The rare gas monohalides have sharp continua $\Delta\lambda \sim 2$ nm, corresponding to transitions terminating on a very flat portion of the lower potential energy curve.

The excited state of the excimer has vibrational and rotational energy levels associated with it. Many of these states may be populated under lasing conditions. Usually the emission spectra of each individual vibrational level overlap in such a manner that the emission band is homogeneous. Stimulated emission can then extract all of the excited states, independent of vibrational and rotational distributions.

Most excimer species also have excited electronic states that lie close and sometimes below the upper laser level. These states act as reservoirs for excitation that can funnel through the upper laser level. The reservoir states may also have gain or loss at the laser frequency. The role of these states is still being explored at this time.

The principal novelty of an excimer laser system is the characteristic mechanism for removal of the lower laser level via ground state dissociation. This naturally prevents bottlenecking, filling up of the lower laser level and termination of inversion. The quantum ratio, defined to be the ratio of the laser photon energy to the energy of the excimer precursor, can also be very high, since the lower laser level is only excited translationally. Hence excimer lasers are potentially very efficient. The principal efficiency limitations are due to the efficiency with which the excimer excited state can be made, and the efficiency with which it can be extracted in the presence of small absorptions typically present in these media.

Since ground state excimer molecules have very small concentrations, it is clear that the practical laser excitation mechanism for the upper excimer state must be indirect. For many cases of interest, the channeling of excitation into excimer upper levels is rapid and occurs with high branching ratios. Initial excitation can be to ions or atomic excited states. Consider the example of pure Ar excited and ionized by an intense electron beam source. Initially ions, Ar^+, and excited atoms, Ar^*, Ar^{**}, are formed. Ar^* denotes the lowest atomic excited state, and Ar^{**} the other high lying excited states. The Ar^+ ions combine with neutral Ar to form the molecular ion Ar_2^+, viz.,

$$Ar^+ + 2\,Ar \rightarrow Ar_2^+ + Ar. \tag{1}$$

The dimer ions are neutralized by electrons to form excited atoms, Ar^* and Ar^{**}. The Ar^{**} species are channeled by collisions and radiation to form Ar^*. At high pressures Ar^* combines with Ar via three-body collisions to form Ar_2^*. This process requires a time of approximately $100\,\text{ns}/p^2$ where the density p is given in amagats. The Ar_2^* excimers thus created decay either by fluorescence or stimulated emission near 130 nm, or are destroyed by self-quenching and superelastic electron collisions, or their energy may be transferred to other components in the gas. Examples of such energy transfer lasers are the $Ar-N_2$ laser (Searles and Hart 1974, Ault et al. 1974) and the rare gas oxide lasers which operate at high pressure of rare gas with lesser amounts of oxygen-bearing molecules present. It has been found that, in pure rare gases, production of excimer molecules can proceed with unit yield, i.e., every initial ion or excited state can be channeled through the excimer molecules (Turner et al. 1975, Lorents 1976).

Bimolecular reactions of excited rare gas atoms with some halogen-bearing compounds have been shown to produce rare gas halide excited states with branching ratios of unity (Velazco et al. 1976). The rate coefficients are sufficiently large that excitation can be channeled from rare gas metastables into rare gas halide excited states in a time of order 25 ns for typical laser mixtures. Thus at low pressures, excitation of rare gases in the presence of halogens leads to rare gas halide emissions rather than rare gas dimer emissions.

In rare gas–halogen mixtures, electron attachment followed by ion recombination is also very important. For example, the attachment of an electron by F_2

$$e + F_2 \rightarrow F^- + F \tag{2}$$

occurs in about 10 ns for F_2 pressures of 1 Torr. The negative ions thus formed recombine with positive ions by the Thompson process (Thompson 1924), well known to be characterized by very large rate coefficients. Thus the process

$$Ar^+ + F^- + Ar \rightarrow ArF^* + Ar \tag{3}$$

yields the ArF excimer in a time of 10 ns or less at atmospheric pressure for the ionization levels of interest in laser media (Brau and Ewing 1975a). This process, like the reactions of the excited atoms with halogen molecules, short circuits the production of the rare gas excimer. Production of the excited state of the rare gas halide proceeds with unit yield also (Rokni et al. 1977a).

The indirect excited state production and kinetic channeling in real excimer lasers is somewhat more complicated than the general outline given here, owing to the variety of reactant and product quantum states possible, the quenching of excited states and the typical use of more complicated mixtures.

2.3 Pumping requirements

Since excimer lasers operate on broadband transitions, they have small stimulated emission characteristics. This requires larger pumping rates than atomic line lasers to produce net practical gain. The net excited state production rate and corresponding power deposition required for a given gain in an excimer species dictate the excitation technology.

The small-signal gain per unit length in the absence of loss is given by

$$g_0 = \sigma_s N^*, \tag{4}$$

where σ_s is the stimulated emission cross section and N^* the instantaneous excited state number density in the upper laser level. It is presumed in this simple analysis that N^* is the population of all vibrational and rotational levels in the emitting state and that σ_s describes the gain cross section with respect to this distribution. The lower level population can be assumed to be zero for a first approximation. The cross section depends on the wavelength, radiative lifetime and bandwidth according to

$$\sigma_s = \frac{\lambda^2}{8\pi} \frac{1}{\tau_r} g(\nu). \tag{5}$$

The normalized, fluorescence lineshape-factor is given by $g(\nu)$. Expression (5) depends only on the Einstein relationship between spontaneous and stimulated emissions, as discussed, for example, by Yariv (1967), and does not require a detailed knowledge of the excimer emission process. Excimer gain cross sections are typically in the range 10^{-18} to 10^{-16} cm^2. For comparison, electronic transition lasers on allowed bound–bound molecular transitions have σ_s in the range 10^{-17} to 10^{-14} cm^2, and allowed atomic transitions have σ_s in the range $10^{-14} - 10^{-11}$ cm^2.

The steady-state inversion density N^* is related to the production rate of excimers R^* by

$$N^* = fR^*\tau_u, \tag{6}$$

where τ_u is the lifetime of the excited state with respect to radiative and collisional decay. f is the instantaneous fraction of excimer molecules in the excimer electronic states that can undergo stimulated emission. Thus $1-f$ is the fraction of excimer species in reservoir electronic states. The steady state is reached for pumping times long compared to the slowest dominant kinetic step, and is usually achieved in a time short compared to laser pump times in most excimer systems.

The power that must be deposited per unit volume to achieve a given R^*, and hence N^*, is given by

$$P_D = E'R^*(\eta_B \eta_P)^{-1}, \tag{7}$$

where E' is the energy of the primary excitation, η_B is the branching ratio for creation of excimer species from initial excitation and η_p is the efficiency with which an initial excitation is created. η_B is usually unity for excimer systems of interest, but of course can be less if other species can be formed. η_p is dependent on the definition of initial excitation and the value chosen for E'. Combining (4), (6) and (7), one obtains the power deposition required for a given small-signal gain g_0 as

$$P_0/g_0 = E'/\tau_u \eta_p \eta_B f \sigma_s. \tag{8}$$

Fig. 2 gives a plot of the power deposition required for 10% cm^{-1} for several excimer systems of interest. Losses are not accounted for in this figure. In making this comparative plot we have chosen E' to be the energy of the atomic metastable of the background gas that is primarily excited, e.g., Ar* in mixtures that are predominantly Ar, and chosen $\eta_p = 60\%$ for pure e-beam excitation (Peterson and Allen 1972, Lorents 1976). For comparison, with electric discharge excitation η_p can be as high as 80% (Mangano and Jacob 1976), but can also be degraded to a considerably lower value, as shall be discussed in § 3.4. η_B is unity in the cases cited, excepting the rare gas oxides. Values of σ_s and τ_u are from

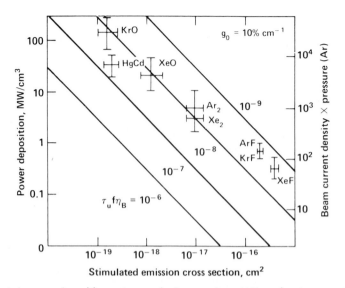

Fig. 2. Estimated power deposition rates required to produce $10\%\,cm^{-1}$ gain on various excimer systems, characterized by their stimulated emission cross sections.

best literature values or estimated ranges. As mentioned, the role of reservoir states is still under investigation, but for this graph we take a value of $f = 0.5$ for rare gas halide lasers and $f = 0.25$ for rare gas excimers.

It is interesting to note the power loading requirements shown in fig. 2. An e-beam-excited KrF laser will have a gain of order $10\%\,cm^{-1}$ for power depositions of order 0.5 to $1\,MW/cm^3$, dependent on pressure, which determines τ_u. This corresponds to conditions achieved with an electron beam of about $200\,A/cm^2$ exciting atmospheric pressure Ar. In comparison, the rare gas excimers require deposition rates about 10 times higher. This is usually achieved with a somewhat greater electron-beam intensity and high pressure operation. The rare gas oxides, because of their very low effective $\sigma_s f$ product and low η_B, are very low gain systems. However, their long excited state lifetimes make it reasonable to excite them with available e-beams and finite pressures.

Discharge pumping rates appropriate for exciting rare gas halides are obtained at discharge current densities of order $30\,A/cm^2$ at electric field strengths of order $10^4\,V/cm$.

An additional requirement on the length of the pumping pulse for an efficient laser oscillator is mandated by the finite build-up time of laser radiation within the resonator. The laser radiation grows, starting from the small fraction of fluorescence directed along the cavity axis. As a rule of thumb, an exponentiation time τ_e given by

$$\tau_e \sim 20/\sigma_s N^* c \tag{9}$$

is required to produce an optical power sufficient to saturate the medium. This criterion is equivalent to that commonly used for amplifier superfluorescence, noting $c\tau_e$ defines a length. For our example of 10% cm^{-1} net gain, the required growth time is about 6 ns. For low-gain cases the build-up time can be a significant limitation.

3. Excimer spectroscopy

3.1 Nomenclature

We first review the basic physical properties of diatomic molecules in order to understand the labeling of excimer electronic states. This section attempts only to explain the frequently encountered electonic nomenclature. No attempt is made to predict molecular potential curves or transition moments. The reader is referred to texts such as Herzberg (1950) for a more detailed treatment of this complex subject. There are many similarities of molecular spectroscopic labeling to those applying for atomic states. Thus we briefly review the labeling of atomic states. The complete rotational symmetry of an atom implies that its total angular momentum J is a good quantum number. For simplicity we neglect nuclear contributions, and J is the sum of orbital and spin angular momenta, denoted as L and S. Because an atom is symmetric with respect to inversion through the nucleus, its energy is invariant by this operation. The atomic wave function must either be unchanged or change sign by this operation. States are accordingly labeled as having even or odd parity, as well as their J label, an integer or half integer. Parity and angular momentum are always good quantum numbers for atoms and lead to absolute combination rules for optical transitions between atomic states. For instance, for electric dipole radiation, by far the strongest type, the strict selection rules are

$$\Delta J = 0, \pm 1, 0 \not\leftrightarrow 0, \tag{10}$$

and parity must change.

The spin angular momentum S and the orbital angular momentum L couple to each other by a magnetic interaction. The coupling is strongly dependent on the atomic number, being more important for heavier elements, and weak for elements in the first two rows. This trend in the strength of coupling of spin to orbital motion is followed in molecules as well, introducing some complexity in excimer nomenclature. In the lighter atoms L and S are approximately good quantum numbers, since energy splittings due to electrostatic interactions are much larger than those due to magnetic interactions. As a result, there exists a broadly used atomic nomenclature that is based on S, L and J. The key components of a so-called 'L–S' designation are the spin multiplicity of a state, $2S + 1$ and the orbital angular momentum of the state, L, as well as J. An alphabetical symbol is used to denote L. The capital letters S, P, D, F, represent orbital angular momenta of a state of 0, 1, 2, 3,... A 'term symbol' is built around this letter with the spin degeneracy, $2S + 1$, written as a left-hand superscript, the value of J for a given state as a right-hand subscript and a right-hand superscript 'o' for states of odd parity. States with $S = 0$ are called singlets, $S = \frac{1}{2}$ doublets, $S = 1$ triplets, etc., according to the corresponding value of $2S + 1$. For light atoms the electric dipole optical selection rules are

$$\Delta S = 0, \qquad \Delta L = 0, \pm 1, \tag{11}$$

as well as the strict rules previously given.

The complete specification of an atomic state is given by the atomic orbitals which are occupied, together with the angular momentum term symbol. For instance, the low lying states of the O atom derive from the configuration of electrons $1s^2 \, 2s^2 \, 2p^4$, and have the term symbols $^3P_2$, $^3P_1$, $^3P_0$, $^1D_2$ and $^1S_0$. The ^3P states are split only by $300 \, \text{cm}^{-1}$, a small energy compared to the larger energy gap to the ^1D and ^1S states or states of other configurations. All of these low lying states have the same parity, even. Thus transitions among these states are electric dipole forbidden. Excitation of a single electron to higher energy configurations produces a number of states which may have the proper parity and angular momentum to couple by electric dipoles to the low lying states.

The labeling of the electronic states of diatomic molecules is governed by analogous rules. A diatomic molecule has the symmetry operations of rotations about the internuclear axis as well as reflections through any plane containing this axis. Hence the rigorously good quantum numbers of a diatomic molecule are its total angular momentum component along the axis and its parity with respect to reflections through the internuclear plane. The angular momentum along the axis is called Ω, and must be integer or half-integer in units of $\hbar$. It can be shown, for example, Herzberg (1950), that for cases where $\Omega \neq 0$, states of the same Ω value that are either positive or negative with respect to reflection, both have the same energy. A molecular term symbol can be based on the Ω value. Such electronic angular momentum states are labelled as: $0^+, 0^-, \frac{1}{2}, 1, \frac{3}{2}$, etc., with no $+$ or $-$ label superscripted for $\Omega \neq 0$.

Homonuclear diatomic molecules have an additional symmetry: reflection through a plane perpendicular to the internuclear axis midway between the atoms. Electronic states which are unchanged by this reflection are labeled g, those that change sign are labeled u. The g or u label appears as a right-hand subscript after the Ω value, as 0_g^+ or 1_u. The German words 'gerade' and 'ungerade' for even and odd are the origin of the g and u label. These symmetries for homonuclear molecules place restrictions on the allowed values of the rotational quantum number of the nuclear pair according to the Pauli principle. Since rotational spectra can rarely be seen for excimer molecules, usually this degree of freedom is neglected. This is fortunate, since the coupling of electronic spin and angular momentum together with the nuclear rotation leads to great richness in molecular spectroscopy.

The selection rules for electric dipole radiation are given by

$$\Delta \Omega = 0, \pm 1, \qquad 0^+ \nleftrightarrow 0^-, \tag{12}$$

and for homonuclear molecules

$$g \nleftrightarrow g, \qquad u \nleftrightarrow u.$$

In the case of molecules containing light atoms, the spin and orbital angular momenta of the electrons may only be weakly coupled. The total spin angular momentum is then an approximately good quantum number. The interaction of primary importance in such molecular systems is the coupling of electronic orbital motion to the internuclear axis. In analogy to the atomic nomenclature, a molecular nomenclature exists for this special case. This nomenclature is broadly used, even when only approximate and in cases where it can be quite imprecise. The projection of the electronic angular momentum on the internuclear axis is called Λ. Term symbols are based on the capital Greek letters Σ, Π, Δ, ... depending on $\Lambda = 0, 1, 2$, the most frequently encountered situations. The symbol also shows the spin multiplicity as in atoms, as a left-hand superscript, the $+$ or $-$ symmetry, and g or u symmetry when appropriate. Examples are $^1\Sigma_g^+$ for the ground state of He_2 or analogous rare gas pairs, and $^2\Sigma^+$ and $^2\Pi$ for the lowest energy states of rare gas halides. The Ω values are sometimes added as right-hand subscripts or in parenthesis. Selection rules for this case are given by

$$\Delta S = 0,$$
$$\Delta \Lambda = 0, \pm 1,$$
$$+ \nleftrightarrow -,$$
$$g \nleftrightarrow g, \; u \nleftrightarrow u. \tag{13}$$

The angular momentum states of a molecule can be simply correlated with the L–S atomic limits in this approximate coupling scheme. The states that result from the collision of various angular momentum states with an atom in an S state, for example, any rare gas ground state atom, ground state Hg or ground state alkali, are as follows: $S + S$ gives a Σ molecular state, $S + P$ gives a Σ and a Π molecular state, and $S + D$ gives Σ, Π and Δ states. An atom in a P state, for example an excited alkali or an excited Hg or inert gas atom, colliding with another P state atom gives Σ, Π and Δ states also. To date, the important excimer systems have states of such atomic parenthood. Analogous rules for correlating Ω values to J values exist.

The electronic spin, designated by the spin degeneracy, is of great importance because of the $\Delta S = 0$ electric dipole selection rule. For excimer molecules with an even number of electrons the important states are singlets and triplets. The important states for molecules with an odd number of electrons are usually doublet and quartet states having either a net spin of one half or three halves. The rules for combining spin angular momenta in going from two atoms to a molecule are as follows: singlet + singlet gives a singlet molecular state; singlet + doublet a doublet; singlet + triplet a triplet; doublet plus doublet a singlet and a triplet; and so on, the higher multiplicities typically not important in the excimers studied to date.

In many cases molecules can have several different electronic states having the same angular momentum designation. Whereas the labeling for atoms is clear since electron configurations are also given, this is not always practical in molecules. It is common practice with diatomic molecules to label the ground electronic state the X state and the higher states of the same spin as the ground state by A, B, C, etc., in order of their energy. Higher states with different spin than the ground state are labeled a, b, c, etc. In other cases for historical reasons or author preference, as we shall see for KrF, different labelings may be found in the literature.

It is important to note that the state designations based on J or Ω values are rigorous but not always utilized. Moreover, as an atom or molecule is electronically excited, the validity of an approximate L–S or Λ–S state designation decreases. Thus, for a molecule such as KrF the Σ and Π labels for the ground and first excited states are quite accurate, but for the upper laser level and its nearby states the Ω-based labels more accurately reflect the true nature of the state.

3.2 Excimer emission spectra

We describe here the emission spectra of three prototype excimer systems: the rare gas continua, the KrF emission spectra and the emission spectra of rare gas-metal vapor complexes. These are each prototypes for physically similar excimer systems as characterized by spectral shapes, excited state binding energies and bond nature.

The rare gas molecular fluorescence bands observed at high pressure are quite broad, $\Delta\lambda \sim 20$ nm. The stimulated emission cross section is of order 10^{-17} cm^2 for these transitions, based on the bandwidths and radiative lifetimes of the radiating states (see table 3). The rare gas excimer continua and the corresponding potential energy curves serve as prototypes for other excimers derived from the collision of two identical closed shell configuration atoms, such as Hg_2, Cd_2 and Mg_2, and in a limited number of cases some mixed excimers such as HgCd. The most bound excited states of these species are bound by energies of order 1 eV with respect to the lowest excited atomic state, a ^3P state with $J = 2$ for rare gases, and $J = 0$ for Hg and its analogs. This excited state binding is considerably larger than that found in excimer excited states of rare gases with nonidentical species, such as ArXe or XeNa. Because the binding is primarily

Table 3
Radiative lifetimes of rare gas excimers [s][a].

Ne$_2$	$1_u, 0_u^-$('triplet')	5.1(-6)	Leichner et. al. (1975a, b)
Ar$_2$	$1_u, 0_u^-$('triplet')	2.9(-6)	Keto et. al. (1976)
	0_u^+('singlet')	4.2(-9)	Keto et. al. (1976)
Kr$_2$	$1_u, 0_u^-$('triplet')	2.8(-7)	Quigley and Hughes (1978a, b)
		3(-7)	Leichner and Erickson (1974)
Xe$_2$	$1_u, 0_u^-$('triplet')	1(-7)	Keto et. al. (1976)
	0_u^+('singlet')	6(-9)	Keto et. al. (1976)

[a] See also references cited in referenced papers.

due to a 'covalent' or electron exchange interaction, the potential wells in the excited state are similar to those of ground state covalent molecules. They are relatively narrow in radial extent, $\Delta R \sim 2\,\text{Å}$. This is in marked contrast to the rare gas halide upper levels as shall be discussed shortly. The value of R at the excited state potential minima for all such excimers is less than the characteristic value of the onset of strong repulsion in the lower states, producing the broad bandwidth of emission.

A second important point in these systems is that the lowest excited excimer state is not the upper laser level. The upper laser level is an $\Omega = 0_u^+$ state. Lying below this state is a reservoir state, $^3\Sigma_u$ or $\Omega = 1_u$, 0_u^- for the rare gases. This 'triplet' state reservoir, whose emission is spin forbidden, has profound effects on the efficiency and extraction kinetics of these lasers. The optical coupling of the reservoir state to the ground state is weak (see table 3). Photoionization absorption of the upper laser level and the reservoir states decreases the attainable gain for a given total excited state density that is shared between the upper laser level and reservoir state. Note the radiative lifetimes of the emitting component of the 'triplet' reservoir state 1_u become considerably shorter for the heavier excimers owing to the increased spin–orbit interaction.

Fig. 3 shows schematic potential curves for KrF, a typical rare gas mono-halide, and a spectroscopic prototype for the laser transition in mercury halides. Since the ground state atoms are $Kr(^1S)$ and $F(^2P)$, two low lying potential curves derive from the collision of Kr with F, a $^2\Sigma$ state and a $^2\Pi$ state. The $^2\Sigma(\Omega = \frac{1}{2})$ state lies lower than the $^2\Pi(\Omega = \frac{1}{2}, \frac{3}{2})$ states because the $^2\Sigma$ state has one electron

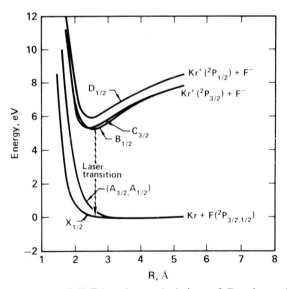

Fig. 3. Potential energy curves of KrF based on calculations of Dunning and Hay (1978). The relative positions of the $B(\frac{1}{2})$ and $C(\frac{3}{2})$ states are still the subject of research. The ionic excited state Kr^+F^- is characterized by an extremely long range interaction at high energy, thus ion recombination can feed these states at a high rate.

and one electron hole from the halogen atom aligned along the internuclear axis, while the $^2\Pi$ states place the unpaired halogen electron and the electron hole off-axis and two electrons on-axis.

The parentage of the excited states of the rare gas monohalides is not that of the lowest lying atomic excited states. The upper laser levels are *ion pair states*, sometimes called charge transfer states, that are bound ion pairs, Kr^+F^-. The coulombic potential curves for these charge transfer states drop below the potential curves coming from $Kr^* + F$ at very large internuclear separation. The range of the ionic potential curves is very large owing to the long-range nature of a coulomb force. The states emanating from $Kr^* + F$ are several electron volts higher in energy. The interaction of the halide negative ion, a ^1S state, with the rare gas positive ion, $^2P_{3/2}$ and $^2P_{1/2}$ states, gives a $^2\Sigma(\frac{1}{2})$ state and a $^2\Pi(\frac{3}{2})$ and $^2\Pi(\frac{1}{2})$ state. The two $\Omega = \frac{1}{2}$ states' wave functions are strongly mixed for the heavier rare gases such as Kr and Xe. Unlike the rare gas excimer lasers, it had generally been thought that the $(\frac{1}{2})$ state, the parent of the laser transition, lay below both the $^2\Pi$ states which would act as reservoirs, rapidly drained by exothermic collisions into the upper laser level. Recent measurements imply that the $\Omega = \frac{3}{2}$ state is slightly lower in energy than $\Omega = \frac{1}{2}$ for several species, (Kliger et al. 1978, Brashears and Setser 1978, Kotts and Setser 1978). The proximity of the $\Omega = \frac{3}{2}$ state to the upper laser level, a few kT energy difference, implies that the fractional population of the upper laser level is only $\frac{1}{2}$ in the small-signal intensity regime.

As in Xe_2, the excited state angular momentum nomenclature is governed by the strong effects of spin–orbit coupling, hence the importance of the Ω nomenclature. Several different authors have used slightly different names for these states in KrF and analogous species. Table 4 compares the names used for the different states by various writers.

The spectra of the main bands of KrF are shown in fig. 4. As an example of how vibrational excitation within an electronic excited state can change the spectral shape, fig. 4 shows the spectrum of KrF taken at 2 atm pressure and at very low pressure. The excited KrF*($\ddagger$) molecule is initially formed with both electronic excitation, superscripted (*) designation, and vibrational and rota-

Table 4
Rare gas halide state labellings: KrF as an example.

Parenthood	Energy (eV)	Brau and Ewing (1975a, b)	Hay and Dunning (1977)	Tellinghuisen et. al. (1976a, b, c)
Kr + F	0.1	$X^2\Sigma(\frac{1}{2})$	$I(\frac{1}{2})$	$X_{1/2}$
	1	$A^2\Pi(\frac{1}{2},\frac{3}{2})$	$II(\frac{1}{2}), I(\frac{3}{2})$	$A_{1/2,3/2}$
Kr^+	5	$B^2\Sigma(\frac{1}{2})$	$III(\frac{1}{2})$	$B_{1/2}$
	5	$C^2\Pi(\frac{3}{2})$	$III(\frac{3}{2})$	$C_{3/2}$
	5.7	$D^2\Pi(\frac{1}{2})$	$IV(\frac{1}{2})$	$D_{1/2}$

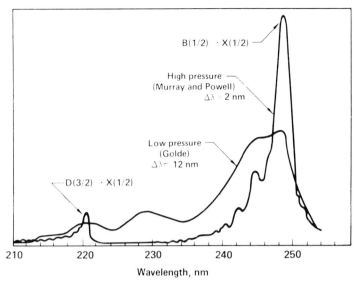

Fig. 4. High and low pressure emissions of KrF. Vibrational relaxation sharpens the KrF* emission band. Low pressure data of Golde (1975) and high pressure spectra provided by J. Murray and H. Powell, Lawrence Livermore Laboratory.

tional excitation, signified by the ($\ddagger$) notation. The vibrationally excited KrF* molecules emit at shorter wavelengths and in a broader wavelength distribution. The initial broad excited state distribution produced by the elementary feeding reactions is narrowed to the high pressure spectrum by the collisional relaxation of the vibrationally and rotationally excited molecules. The rare gas halide excited species also produce very broad emission bands at longer wavelengths owing to transitions to the strongly repulsive $^2\Pi$ states, presumably with the $^2\Pi(\Omega = \frac{3}{2}, \Omega = \frac{1}{2})$ states as upper levels. At shorter wavelengths than the intense main band is a sharp transition terminating on $^2\Sigma(\frac{1}{2})$ from the $^2\Pi(\frac{1}{2})$ ionic state. The sharp high pressure emission bands have fluctuations in intensity as a function of wavelength. This is caused by emission from the low lying vibrational levels of KrF* above $v = 0$ and $v = 1$.

Fig. 5 shows the calculation of the emission spectra of several v states and how they overlap to form the KrF main band at high buffer pressures. The low pressure emission spectrum has been synthesized in a similar manner (Tamagaki and Setser 1977), and the XeI, KrF and XeBr continuum emission bands at high pressure have been analyzed as well (Tellinghuisen et al. 1976a). The stimulated emission constant for the high pressure laser band is about 2×10^{-16} cm^2 for KrF, and is expected to be about this value for the other species.

The XeF and XeCl molecules differ somewhat from the other rare gas halides in that the low levels are slightly bound (Brau and Ewing 1975c, Tellinghuisen et al. 1976b, c, Smith and Kobrinsky 1978, Monts et al. 1979). The laser transitions are bound–bound in these species, with complicated rotational and isotopic effects. It is also thought that the lowest excited state in XeF is not the upper laser level, $B(\Omega = \frac{1}{2})$, but rather the $C(\Omega = \frac{3}{2})$ state (Kliger et al. 1978, Brashears and Setser

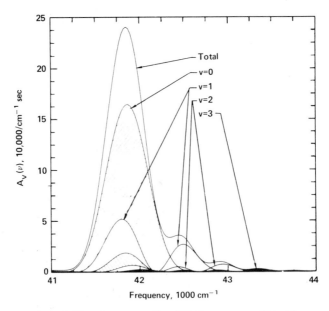

Fig. 5. Emission rates from KrF* low v states, $T = 4000$ K, courtesy of N. Winter and T. Rescigno, Note how the bands of the low states overlap emission from $v = 0$. The calculated frequency is shifted from experimental, since this synthetic spectrum is based on the a priori calculated energies rather than experiment.

1978). This is probably due to the configuration interaction of the $B(\frac{1}{2})$ and ground $X(\frac{1}{2})$ states that leads to binding in the ground state and a slightly higher energy of $B(\frac{1}{2})$ than may be otherwise anticipated.

The rare gases and rare gas halides exemplify excimer lasers with substantial excited state binding. However, it has been proposed that the molecular bands of weakly bound excimers could be useful for laser gain. Such emissions are observed in the wings of atomic resonance lines buffered with rare gases. Analysis of this option was given by Phelps (1972) and gain and loss coefficient expressions were derived by Hedges et al. (1972). A review of the concept and experiments is given by Gallagher (1978). These continua correspond to electronic transitions from excited states that are primarily unbound to lower states that are strongly repulsive. Thus these excimer bands are superpositions of bound to free emissions with primarily free–free emissions. Many metal atoms in collision with rare gases have excited state binding energies of 100 to 1 000 cm, too small an energy to recombine a substantial fraction of the excited states into excimer molecules. However, when the excited atom radiates in the presence of large pressures of buffer gas, a significant fraction of the emission can occur in these continuum bands which are in the red wings of the atomic lines. Of the many systems proposed which would fall in the class of weakly bound excimers, only the rare gas oxide lasers have been demonstrated. These are a special case of an excimer laser originating from weakly bound excited states in which the lower level is collisionally quenched (Murray and Rhodes 1976), and the free atom transition is electric dipole forbidden. A detailed description of the collisionally induced rare gas oxide spectra has been given by Julienne (1978). The XeO emission not only has bound–free and free–free components, but bound–bound and free–bound emissions, since the lower laser level has a predissociating, bound segment.

Fig. 6 shows the calculated gain and absorption profiles for the typical case of the TlXe complex. Two laser transitions can originate from the weakly bound $Tl^*(7\,^2S_{1/2})+Xe$ excited state. Repulsive potentials derive from the ground $Tl^2P_{1/2}$ state and the low lying $Tl6^2P_{3/2}$ state. Measurements of the profiles of wing radiation in high pressure Xe doped with Tl (Cheron et al. 1976), coupled with an assumed excited state distribution and translational temperature, and the analysis of Hedges et al. (1972) yield the plots of gain and loss versus wavelength. The calculated gains are of order $0.1\%\,cm^{-1}$ for excited state densities corresponding to a temperature of about 1 eV. Gain exceeds loss only for emission shifted several hundred angstroms away from the resonance lines. The ratio of gain to loss is small for the transition terminating on the ground state, $\lambda \sim 420$ nm. This fact would make this laser inefficient even if laser action could be achieved, as shall be discussed. The calculation of course is an upper limit on the gain in that it does not include effects due to absorptions by other excited states, ions, or windows present in real systems. Greater gains can be obtained if greater excimer excited state densities are achieved. This can happen if the excited state well depth is larger. Complexes of Hg* with Xe and Tl* with Hg may have this feature and thus prove useful for laser gain (Drummond and Schlie 1976, Lam et al. 1978).

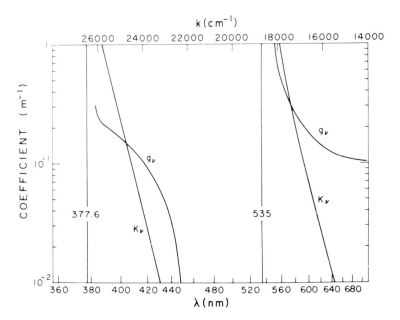

Fig. 6. Gain and absorption profiles for TlXe, courtesy of A. Gallagher.

3.3 Absorbers in excimer media

Absorbers seriously impact the performance of excimer lasers. Excimer lasers utilize electronic excitation and ionization to produce the upper laser level. Most of the ions and excited species formed in the kinetic channeling process absorb in the ultraviolet. The excimer species themselves may absorb at the wavelength of interest. Ground state species, both precursors such as F_2 in a rare gas fluoride laser, or the lower laser level itself, can also be important absorbers. The efficiency, extractable intensities and laser scalability are all linked to the amount of absorption created in the medium along with gain.

Consider first a general medium which is characterized by a per cm gain coefficient g and loss γ. The gain g is the product of the excimer upper level number density and the stimulated emission cross section σ. The gain is dependent on the intensity in any given volume element, and drops as the intensity exceeds the saturation intensity $I_s = h\nu/\sigma\tau$, where τ is upper level lifetime including effects due to collisions. Thus

$$g(I) = g_0/(1 + I/I_s), \tag{14}$$

where g_0 is the small-signal gain coefficient. The absorption γ is often divisible into two types of loss, those which saturate in the same manner as the excited state, and those that saturate at much higher intensities, so-called nonsaturable loss. Thus the loss depends on intensity as

$$\gamma(I) = \gamma_0/(1 + I/I_s) + \gamma_{ns}. \tag{15}$$

The species which contribute to γ_0 will be the upper laser level itself and absorbing species formed by quenching of the upper laser level, i.e., species whose density is linked to that of the upper laser level. γ_{ns} is the nonsaturating loss coefficient. In an amplifying medium the growth in intensity is given by

$$dI/dx = (g(I) - \gamma(I))I. \tag{16}$$

The presence of nonsaturating loss limits the intensity that can be drawn from an amplifier, and that maximum intensity is given, setting $dI/dx = 0$, by

$$I = I_s\{[(g_0 - \gamma_0)/\gamma_{ns}] - 1\}. \tag{17}$$

A loss-free amplifier can amplify to arbitrary intensity, given sufficient gain. Clearly, to maximize intensity achievable, one desires as large as possible a ratio of $(g_0 - \gamma_0)/\gamma_{ns}$. In the presence of a nonsaturating loss, however, the maximum intensity can only be approached at the cost of decreased efficiency.

The local extraction efficiency, defined as the ratio of excited states being extracted by a photon field from a unit volume to the rate of production of those excited states, is given simply by η_x (Brau 1978a, b):

$$\eta_x = I(g - \gamma)/(N_0^*/\tau_0)h\nu, \tag{18}$$

when N_0^* and τ_0 are excited state densities and lifetimes in the absence of a stimulating field. In the loss-free limit, the local extraction efficiency is simply

$\eta_x = I/(I + I_s)$, and can be made close to unity as long as the stimulated rate well exceeds the collisional and spontaneous losses. In the limit of all losses being saturable, the extraction efficiency is $(I/(I + I_s))(1 - \gamma_0/g_0)$, i.e., a product of the kinetic extraction efficiency and a term expressing the ratio of saturating loss to saturating gain. In this case the local extraction efficiency can be high if $I \gg I_s$ and γ_0/g_0 is small. For realistic cases nonsaturating losses are always present. The efficiency is optimized by choosing an optimum stimulating intensity (Brau 1978a). For losses that are totally nonsaturable, the local extraction efficiency is maximized at $\eta_x(\mathrm{max}) = 1 - 2(\gamma_{ns}/g_0)^{1/2} + \gamma_{ns}/g_0$. Thus, $\eta_x(\mathrm{max})$ approaches unity for $\gamma_{ns}/g_0 \to 0$. For typical situations, however, $\gamma_{ns}/g_0 \sim 0.1$ and $\eta_x(\mathrm{max})$ is of order 50%.

In a large volume laser or amplifier the intensity varies along the medium gain length, and depends on the input intensity or feedback. Expression (18) may be integrated to find the maximum efficiency for an entire amplifier gain length. This maximum efficiency is achieved by varying the laser cavity output coupling or the amplifier input intensity. Analysis shows that extraction efficiencies of order 50% require that $\gamma_{ns}L$ be less than about 1.0 and the gain/loss ratio be greater than about 10. Increased efficiency can be achieved with lower loss lengths and/or greater ratios of $(g_0 - \gamma_0)/\gamma_{ns}$. The input intensity needed to produce an optimum amplifier extraction efficiency of order 50% is in the range 0.5 to 1.0 I/I_s. Plots of amplifier extraction efficiency versus input intensity for various $\gamma_{ns}L$ values are shown in a separate chapter in this volume (Krupke et al. 1978). The case of a laser is slightly different in that laser flux can travel in two directions, and for the important cases of unstable resonator geometries or high output couplings, the cavity flux is spatially varying in at least one dimension and often in two dimensions. In such circumstances estimates of device extraction efficiency are best made by computer solution of the coupled kinetic and optical equations. Again it is found that optimum medium efficiency is obtained for low values of $\gamma_{ns}L$ and high gain to loss ratios. In the limit of high output coupling, the optical extraction efficiency drops to 50% for loss lengths of order unity. As with amplifiers, increasing the gain (and loss) length can yield higher output intensities, but at the cost of lower medium extraction efficiencies for $\gamma_{ns}L$ in the range 0.5 to 2.0 (Rokni et al. 1978c, Rigrod 1978).

Absorbers important in excimer lasers can be classed in terms of charged particles or neutrals. Charged particle absorptions of importance are positive molecular ion photodissociation, photodetachment and photodissociation of negative ions, and at very high electron densities and long wavelengths electron inverse bremsstrahlung. Absorption bands in neutral molecular species often mimic ionic absorption, as we shall see. Neutral absorbers of importance can be either excited atoms or molecules, or ground state molecules, including absorption out of the lower level of the excimer. We discuss each in this order.

Diatomic, and possibly polyatomic, positive ions play an important role in the energy channeling kinetics of all excimers. They can also absorb in the visible and ultraviolet. Fig. 7 shows calculations of rare gas dimer ion absorption cross sections in the uv for Ne_2^+, Ar_2^+, Kr_2^+ and Xe_2^+ for two vibrational tem-

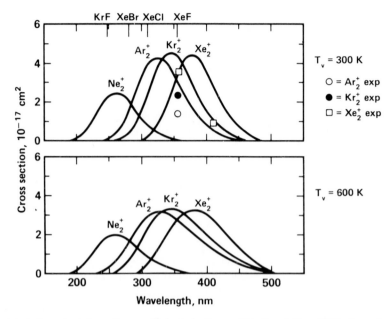

Fig. 7. Calculated dimer ion absorption coefficients in the uv (Wadt and Hay 1978). Experimental points are also shown (Vanderhof 1978).

peratures of the dimer rare gas ion molecule (Wadt and Hay 1978). The Ar_2^+ absorption has also been calculated, separately (Stevens et al. 1977). Measurements of Ar_2^+ absorption in molecular beam experiments in the visible (Moseley et al. 1977) and ultraviolet (Vanderhof 1978) are found to be in reasonable agreement, within a factor of 2, with the calculated curves. The rare gas dimer absorbers are most important in the rare gas halide systems, since the absorption bands formed from the typical buffers, Ne and Ar, overlap many of the laser frequencies. Weak positive ion absorption may also be important for pure rare gases. However, extrapolating these absorption curves into the vuv is probably speculative because of uncertainties in the dependence of loss on vibrational temperature, which dramatically changes wing absorbance. Absorption due to mixed ions, such as $ArXe^+$ and $ArKr^+$, has been postulated as important (Rokni et al. 1978, Bender and Winter 1978). Triatomic positive ions, such as Xe_3^+, have been postulated as kinetically important, but absorption spectra have not been estimated (Werner et al. 1977).

The photodetachment of negative ions is important in the rare gas halide lasers since negative halide ions are rapidly formed. It is probably not important in rare gas excimers, except in that oxide or hydroxide ions can be formed by attachment to oxygen or water impurities. Fig. 8 shows the absorption profiles of F^-, Cl^-, Br^- and I^-. The heavier negative ions have absorption cross sections which are quite large, and detract significantly from laser performance of the relevant lasers.

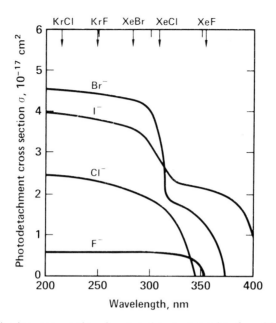

Fig. 8. Halide negative ion cross sections for photodetachment, taken from Mandl (1970) and Rothe (1969).

Inverse bremsstrahlung absorption in the uv is small at the electron number densities typically encountered. However, at 10.6 μm the absorption at densities of order 10^{17} electrons/cm^3 is large enough that electron number densities in rare gas plasmas have been probed (Zamir et al. 1974).

Excited atoms typically are photoionized by the short wavelengths encountered in excimer lasers. Calculations of absorption coefficients for rare gas metastables have now been performed by several authors (Hyman 1977, McKann and Flannery 1977, Hazi and Rescigno 1977). It is generally agreed that the cross section for photoionizing these species is low in the near ultraviolet, exhibiting minima near 3 eV, and always being less than about 10^{-19} cm^2 in the wavelength regions of interest. Hyman has shown, however, that the cross sections for higher lying rare gas excited states are much larger, $\sigma \sim 10^{-18}$ cm^2 (Hyman 1977). The largest unknown in modeling these systems is the ratio of metastable rare gas excited states to the higher lying states that absorb strongly.

Absorption due to photoionization of rare gas excimers was recognized early. Predictions of the absorption by Ar$_2^*$, Kr$_2^*$ and Xe$_2^*$ in the far uv were made using a hydrogenic model (Lorents et al. 1974). The photoionization cross sections for any species in this model are simply a function of the energy above threshold. Cross sections of order 10^{-18} cm^2 were predicted for the 170 to 150 nm region, viz.: Xe$_2^*$ at 173 nm, $(1-4) \times 10^{-18}$ cm^2; Kr$_2^*$ at 147 nm, $(0.7-3) \times 10^{-18}$ cm^2; and Ar$_2^*$ at 128 nm, $(0.4-1.8) \times 10^{-18}$. This calculational method probably overestimates slightly the absorption in the far uv, based on recent state-of-the-art ab initio calculations (Rescigno et al. 1978). The ab initio calculations also show that in the spectral region near threshold, important contributions due to the molecular nature of the absorber come into play. Strong autioionizing peaks with shapes and cross sections similar to Ar_2^+ absorption were found. Thus, in the near uv, Ar_2^* and other rare gas dimers should absorb with a cross section and wavelength dependence very similar to that for the rare gas dimer ions. The Ar_2^* peak absorption near 320 nm is expected to be about 10^{-16} cm^2.

Calculations of other relevant excited states, such as high lying halogen atom states or higher lying states of excimers themselves have not been performed. Presumably the hydrogenic prescription given by Lorents et al. (1974) is as good an estimate as any in this situation.

For the case of rare gas halides, quenching of excited states can yield an absorbing species, viz., $KrF^* + Kr + Ar \rightarrow Kr_2F^* + Ar$ (Lorents et al. 1977, Rokni et al. 1977a, b, c). These excited species are also ion pair states, viz., $Kr_2^+F^-$ (Lorents et al. 1978). Calculations of the absorption by Ar_2F^* and Kr_2F^* have been made (Wadt and Hay 1978). It has been found that the dominant feature is essentially the same transition that leads to the intense rare gas dimer ion absorption and the calculated Ar_2^*, 320 nm absorption. Thus the Kr_2F^* absorption at 248 nm can be safely assumed to have the same cross section as that of Kr_2^+ at 248 nm. Mixed ion species such as ArKrF should absorb as $ArKr^+$. The absorption by Kr_2F in a KrF laser mixture can easily be as large as the nonsaturable losses due to ions and excited states at low excitation levels (Hawryluk et al. 1977). In fact at sufficiently high pressures and excitation rates, absorption by this species should far exceed the nonsaturable losses. Although gain measurements have been made on vigorously excited high pressure KrF (Bradford et al. 1977), gain/loss measurements and saturation behavior have not yet been studied.

Ground state absorption is typically a nonsaturable loss. The KrF and XeF laser bands overlap the weak F_2 absorption (Stuennenberg and Vogel 1956), while XeCl laser is overlapped by the strong Cl_2 band (Seery and Britton 1964). The F_2 absorption, $\sigma = 1.5 \times 10^{-20}$ cm^2 at 248 nm and 6.6×10^{-21} cm^2 at 351 nm, is small, about 0.2/m for typical laser mixtures. The XeCl laser, however, only operates efficiently in the absence of Cl_2, $\sigma(308 \text{ nm}) = 1.7 \times 10^{-19}$ cm^2, and positive dimer absorbers such as Ar_2^+. The other source of ground state absorption is that due to the excimer species themselves. For the excimer lasers demonstrated to date, excepting XeCl and XeF for the moment, ground state excimer absorption is a minor effect. An analysis of this effect for the rare gas excimer lasers has been given by Rhodes (1974), and experimental measurements in Xe and Kr show its minor importance (Gerardo and Johnson 1974). A general analysis has been given (Hedges et al. 1972) which is applied in many of the papers dealing with weakly bound metal–rare gas excimers. For the rare gas halides, ground state absorption is minor because the absorption/emission frequency change with internuclear separation is large, and the transition moment for these charge transfer bands drops precipitously as the internuclear separation is increased.

For the two special cases, XeF and XeCl, in which the lasing transition is bound–bound, the degree of ground state absorption is not yet quantified since the rate coefficients for vibrational relaxation and dissociation in the lower level are not yet known. Because of the large binding of the XeF $X^2\Sigma^+$ state, $1\ 100\ \text{cm}^{-1}$ (Tellinghuisen et al. 1976b, 1978), the rate of dissociation of the lower laser level can be slow compared to the rate of excited state stimulation. The variation of XeF laser efficiency with ground state absorption has been discussed (Finn et al. 1978, Rokni et al. 1978c).

4. Excimer laser kinetics

This section describes the kinetic mechanisms and tabulates pertinent reaction rate coefficients for three genre of excimer lasers: e-beam-excited rare gases, e-beam-excited rare gas halides, and electric-discharge-excited rare gas halides. The mechanisms by which excitation channels into upper laser levels are fairly well understood. Rate coefficients for many of the reactions leading to excited states or quenching the upper levels have been measured or estimated. Some controversies and uncertainties still exist regarding aspects of kinetic models for these lasers, however. The kinetic pathways in other proposed excimer laser systems have received considerably less attention, but many of the generalities discussed here apply. An extensive tabulation of some of the relevant kinetic data has been prepared (McDaniel et al. 1977). The recent reviews of Rokni et al. (1978c) and by Brau (1978b) also contain extensive discussions and rate data compilations.

4.1 E-beam-excited rare gases

The kinetics of pure rare gases represent the simplest starting point for describing other excimer lasers. Argon will be used in the example here, but the kinetic description pertains to Kr and Xe as well. Extensive modeling of the rare gas systems has been performed, but a number of key uncertainties remain (Werner et al. 1977, Gerardo and Johnson 1974, Lorents 1976). Rare gas excimer lasers have only been excited using electron beams, although discharge produced fluorescence has been observed (Huber et al. 1976). E-beam excitation utilizes a beam of high energy electrons, $0.2 \lesssim E \lesssim 2\ \text{MeV}$, to ionize and excite the gas. The beam is formed in a vacuum diode and injected into the gas through a thin metal or plastic foil. Details of the technology of e-beam production are given in §5.1. As the high energy electrons traverse the gas, they slow and are scattered by collisons with the gas, resulting in ionization and secondary electron formation, and a smaller amount of direct excitation. Using Ar as an example, the first pumping step is

$$e' + Ar \rightarrow Ar^+ + e + e', \tag{19}$$

where primary electrons are symbolized by e'. The rate of power deposition in the gas is related to the rate of ion pair formation by

$$P_i = (eW)_i R_i,$$
(20)

where $(eW)_i$ is the energy loss per ion pair produced and R_i is the net rate of primary ionization and excitation by the e-beam. The power deposition is also proportional to the e-beam current density J_{eb}, given in A/cm², and the medium stopping power S, expressed in V/cm. The stopping power S is proportional to the gas density. Tables of stopping powers have been prepared (Berger and Seltzer 1964), and fig. 9 shows a plot of the mass density normalized stopping power S/ρ, in MeV cm²/g, for He, Ar and Xe. For cases of interest S/ρ is in the range 1 to 2 MeV cm²/gm. The net rate of producing ions and electrons in the excited medium is thus given by

$$R_i = \rho J_{eb}(S/\rho)/eW_i.$$
(21)

For fixed J_{eb}, the ionization rate is larger in heavier rare gases such as Xe because of the greater mass density and lower eW_i. Some typical values of eW_i are given in table 5.

For the heavy elements, Ar, Kr, Xe, Hg, etc., the stopping power tables of Berger and Seltzer underestimate the net rate of ionization. This is due to the effects of scattering of the beam in the gas. This scattering is due to the dominant importance of collisions with the gas nuclei rather than electrons. As a result, large angle scattering events become more pronounced for the heavier elements and a correction for increased deposition in the range of 2 to 3 must be

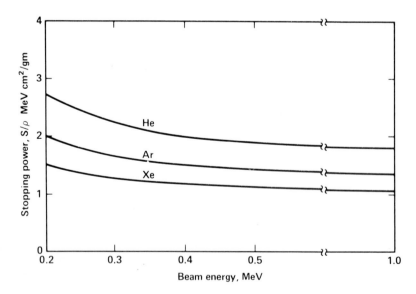

Fig. 9. The mass density normalized stopping power of several rare gases. S/ρ is not strongly dependent on species or beam voltage.

Table 5
Energy to ionize rare gases.[a]

	eW_i	Ionization potential
He	42	24.6
Ne	36	21.6
Ar	26	15.8
Kr	24	14.0
Xe	22	12.1

[a] From Christophorou (1971). See also Fano (1973).

applied (Hart and Searles 1976). The correction for scattering of broad area beams is not strongly dependent on voltage in the range V greater than 300 keV.

Once the ion–electron pair is formed in the gas the excited medium begins to lose energy in two parallel channels, one for electrons and one for ions. The secondary electrons are created with a kinetic energy of order the ionization potential of the gas. These electrons cool by collisions with the rare gas. Both momentum transfer collisons and inelastic collisions are important. The inelastic collisions produce rare gas excitation, with the total number of excited states formed being about 20% of the total ion/secondary electron yield. Thus for the case of Ar the net energy expenditure for producing one ion or excited state is only 22 eV rather than 27 eV, the value of eW_i (Lorents 1976, Peterson and Allen 1972, Elliott and Greene 1976). The secondary electrons reach a steady-state temperature of around 1 eV in pure rare gases. Addition of more than a few percent of molecular species such as N_2 will lower this electron temperature to less than 0.2 eV.

At pressures of interest, the rare gas ion recombines in a three-body process to form a molecular dimer ion, losing roughly 1 eV per recombination into gas heating:

$$Ar^+ + 2\,Ar \rightarrow Ar_2^+ + Ar. \tag{22}$$

The termolecular ion recombination rate coefficients are of order $2 \times 10^{-31}\,\mathrm{cm}^6\,\mathrm{s}^{-1}$ for the rare gases (see table 6, and also Brau 1978a, b). The decay time of Ar^+

Table 6
Three-body rare gases dimer ion formation rate coefficients: $Rg^+ + Rg + M \rightarrow Rg_2^+ + M$.

Ion formed Rg_2^+	Third body M	Rate coefficient [cm⁶/s]	Reference
He_2^+	He	6.8(−32)	Smith and Copsey (1968)
Ne_2^+	Ne	7.2(−32)	Gaur and Chanin (1964)
	He	3(−31)	Veach and Oskam (1970)
Ar_2^+	Ar	2.1(−31)	Liu and Conway (1974)
Kr_2^+	Kr	5(−32)	Werner et. al. (1977)
Xe_2^+	Xe	2(−31)	Vitols and Oskam (1973)

into Ar_2^+ is about 12 ns at 1 atm pressure and decreases as P^{-2}. Assuming the presence of no negative ions, the dimer ions can recombine with electrons, or at sufficiently high pressure convert a fraction of their density to triatomic ions:

$$Ar_2^+ \begin{cases} + e \quad \nearrow Ar(^{**}) + Ar \\ \\ + 2\,Ar \searrow Ar_3^+ + Ar. \end{cases} \tag{23}$$

The electron dimer ion recombination rate coefficients at electron temperatures of order 1 eV are of order $10^{-7}\,cm^3\,s^{-1}$, and are larger for lower electron temperatures (see table 7). Rate constants for triatomic ion formation and subsequent recombination are not independently known, but their importance in laser modeling has been pointed out (Werner et al. 1977). For ion–electron number densities of order $3 \times 10^{15}/cm^3$, pertinent for an ionization rate of $10^{24}\,cm^{-3}\,s^{-1}$, P_i of about 5 MW/cm^3 and gains on Ar_2^* of about 5% cm^{-1}, the molecular ion is neutralized on a time scale of order 3 ns.

Molecular ion–electron recombination yields a variety of electronically excited rare gas atoms, denoted as Ar** above. The excited states thus formed can relax to the lowest electronically excited states, denoted as Ar*, by collisions or radiation. In the presence of electrons at $10^{15}/cm^3$, the Ar* and Ar** states maintain a population ratio of about 3 to 1. Radiation decay to the ground state is very slow because of radiation trapping at the high densities utilized in excimer lasers.

The rare gas excited states, coupled together by electron collisions, feed the excimer molecular states by three-body recombination

$$Ar^* + Ar + M \to {}^{1,3}Ar_2^* + M, \tag{24}$$

where M is any third body, typically the same rare gas. Slight variations of recombination rate with the atomic quantum state have been found. The excimer can be formed as either a singlet state 0_u^+, the upper laser level, or a triplet state, 1_u, 0_g^-, which acts as a kinetic reservoir and possible absorber. The recombination from atoms into the excimer upper level is one of the slow steps in the

Table 7
Rare gas dimer ion–electron recombination rate coefficients (Shiu and Biondi 1978).[a]

Rare gas ion	Recombination coefficient at 300 K = T_e	Electron temperature Dependence
Ne_2^+	$1.7 \times 10^{-7}\,cm^3/s$	$T_e^{-0.43}$
Ar_2^+	8.5×10^{-7}	$T_e^{-0.67}$
Kr_2^+	1.6×10^{-6}	$T_e^{-0.55}$
Xe_2^+	2.3×10^{-6}	$T_e^{-0.72}$

[a] See also Brau (1978b) for extensive comparison.

kinetic chain, having typical recombination rate coefficients in the range $(1-5) \times 10^{-32}$ cm^6 s^{-1} (see table 8). At 1 atm of Ar, where recombination into the excimer is somewhat slower than that for Xe, the recombination time is 250 ns, and decreases as P^{-2}. Thus, for excitation/ionization rates of 10^{24} cm^{-3} s^{-1} and pressures of 1 and 10 atm, atomic excited state densities of 2×10^{17} cm^{-3} (1 atm) and 2×10^{15} cm^{-3} (10 atm) are expected. Assuming an effective cross section of 10^{-19} cm^2 for photoionization of the various states of Ar* and Ar** that are populated, the optical loss due to atoms is about 2% cm^{-1} at low pressure.

Penning ionization reactions such as

$$Ar^* + Ar^* \rightarrow Ar^+ + e + Ar \tag{25}$$

are a loss process of atomic excited states. Rate constants for this process are large, 5×10^{-10} cm^3 s^{-1} (see Werner et al. 1977, for example). This bimolecular excited state loss rate can easily exceed the recombination rate into excimers at high excitation levels, $Ar^*/Ar > 4 \times 10^{-4} P$ (atm). Pressures greater than a few atmospheres minimize atomic Penning decay and force all of the excitation into the excimer species.

The rare gas excimer states can either radiate or be quenched. Important loss processes are electron superelastic collisions and Penning processes yielding Ar_2^+. Mutual ionization lowers the overall efficiency of producing radiation from Ar_2^*. The importance of this excimer self-quenching process cannot be diminished by increasing pressure, as in the case of Ar* Penning processes. Thus a rollover in excimer production efficiency with increased deposition has been observed at high pressures (Powell and Murray 1977).

The radiative decay times for the singlet and triplet excimer states are considerably different, as was shown earlier in table 3. Under conditions where the lower triplet state is collisionally coupled to the radiating singlet state, the kinetic lifetime of the coupled excimer excited state is longer than that of the singlet state and shorter than that of the triplet. Values of collisional cross coupling rates by electrons and rare gases have been extracted for limited cases (Keto et al. 1974, 1976). Only electrons can couple the states rapidly, with mixing rate constants of order 10^{-7} to 10^{-6} cm^3/s inferred (Werner et al. 1977). At

Table 8

Three-body excimer formation rate coefficients: $Rg^* + Rg + M \rightarrow Rg_2^* + M$[a]

Excimer formed Rg_2^*	Third body M	Rate coefficient [cm^6/s]	Reference
Ne_2^*	Ne	6×10^{-34}	Leichner et. al. (1975a, b)
Ar_2^*	Ar	1.1×10^{-32}	Chen et. al. (1978)
Kr_2^*	Kr	2.6×10^{-32}	Tracy and Oskam (1976)
Xe_2^*	Xe from $^3P_1$	4×10^{-32}	Leichner et al. (1975b)
	Xe from $^3P_2$	7.7×10^{-32}	Barber et. al. (1975)
	Ar	2.2×10^{-32}	Gleason et. al. (1977)
	He	1.4×10^{-32}	Rice and Johnson (1975)

[a]See also Brau (1978b) for extensive comparison.

electron densities of order $10^{15}/cm^3$, ratios of 'triplets' to 'singlets' in the range of 3:1 to 10:1 are estimated. Thus for an ionization rate of $10^{24}\,cm^{-3}\,s^{-1}$ and an effective total upper state lifetime of order 20 ns, the total steady-state excimer number density is $2 \times 10^{16}\,cm^{-3}$. The population of the upper laser level is in the range 2×10^{15} to $5 \times 10^{15}/cm^3$, depending on excited state mixing rates, and the reservoir triplet state has a population of between 1.5 and $1.8 \times 10^{16}\,cm^{-3}$. Thus the estimated gain is 2 to $5\%\,cm^{-1}$ due to the singlets, and the loss is of order $1.5\%\,cm^{-1}$. Net gains of 1/2 to $3.5\%\,cm^{-1}$ are thus expected. Detailed measurements of gain to loss ratios in rare gases have not yet been reported. The losses are sufficiently large that efficient extraction of long gain length rare gas excimers appears doubtful. Such losses probably explain the low laser efficiencies observed to date in pure rare gases. A quantitative knowledge of excimer excited state density ratios and coupling rates remains a large unknown in these systems.

The rare gas excimer cycle is completed by radiation in the vuv and dissociation of the lower state. Because the lower state potential curve is strongly repulsive, dissociation is very rapid, $10^{-13}\,s$. About 1 eV of heat is added to the gas per initial excitation. Because all of the kinetic processes above allow for excitation to funnel through the excimer states with unit yield, the fluorescence efficiencies of these species are measured to be very high, 50%, in the regimes of low total excitation rate, (Turner et al. 1975).

Rare gas mixtures have also been utilized to produce excimer fluorescence or laser action (Hoff et al. 1973). The kinetics in these situations is virtually the same: excitation channels down from ions into excimers with high efficiency. The initial ratios of ionization depend on the relative densities and stopping powers of the mixture. Thus in the case of Xe-He mixtures with Xe mole fractions of 3% or more, the primary excitation is of the more dense Xe, $\rho \sim 6\,mg/cm^3$ at 1 atm, rather than He, $\rho \sim 0.2\,mg/cm^3$ at 1 atm. However, in Ar-buffered Xe, the primary ionizing event is the creation of Ar^+ ions. In such mixtures the light ion recombines with another atom, since the atomic charge transfer rates are low compared to the net three-body recombination rate. The dimer ions formed can recombine as in pure rare gases or charge exchange in a

Table 9
Rare gas ion charge transfer rates.

Positive ion	Atom	Charge transfer rate constant[cm³/s]	Reference
He_2^+	Ar	2×10^{-10}	Bohme et. al. (1970)
He_2^+	Kr	2×10^{-11}	Bohme et. al. (1970)
Ne_2^+	Ar	$< 5 \times 10^{-14}$	Bohme et. al. (1970)
Ne_2^+	Kr	$< 10^{-13}$	Johnson et. al. (1978)
Ne_2^+	Xe	$< 10^{-13}$	Johnson et. al. (1978)
Ar_2^+	Kr	$(6 \pm 2) \times 10^{-10}$	Johnson et. al. (1978)
Kr_2^+	Xe	2×10^{-10}	Kebarle et. al. (1967)

Table 10
Rare gas excitation transfer rate coefficients.

Excited species	Atom	Rate coefficient [cm^3/s]	Reference
Ar*	Kr	1.4×10^{-11}	Bourene and LeCalve (1973)
	Xe	2×10^{-10}	Bourene and LeCalve (1973)
Ar$_2^*$	Kr	1.5×10^{-9}	Gedanken et. al. (1972)
	Xe	4×10^{-10}	Gleason et. al. (1977)

few cases (see table 9). Any energy that funnels into the lighter atoms or excimers can be rapidly transferred to a heavier rare gas, by reactions such as

$$Ar_2^* + Xe \rightarrow Xe^* + 2\,Ar \tag{26}$$

(see table 10).

Xe pressures of order 10 Torr are sufficient to capture excitation in Ar$_2^*$ with high efficiency. Detailed models of rare gas excimer lasers utilizing mixtures have not yet been fully developed.

4.2 E-beam-excited rare gas halides

The kinetic pathways of rare gas halide lasers are significantly different and are somewhat more complex, owing to the fact that tripartite mixtures of two rare gases and a halogen source tend to give the best laser performance. At least three entirely different pumping mechanisms are important in feeding the upper levels. As with rare gas excimers, the cross coupling of the upper laser level with the nearby excited states needs further research. Excited state quenching is very important. Since Ar is used as a buffer in many cases and since ArF is a promising laser in its own right, the discussion centers first on Ar-F$_2$ mixtures.

For an e-beam-excited ArF laser, F$_2$ to Ar ratios are typically 1% or less. Thus, the primary electrons ionize and excite Ar to initiate the kinetic chain. At pressures above 2 atm, the Ar$^+$ can rapidly recombine to form Ar$_2^+$. However, the use of electron-attaching halogen compounds opens up an entirely new channel of energy flow. Secondary electrons undergo rapid dissociative attachment reactions with most halogen compounds, viz.,

$$e + F_2 \rightarrow F + F^-. \tag{27}$$

Rate constants for these processes vary considerably with electron temperature and species, and a review has been given (Caledonia 1975). For F$_2$ and $T_e \sim 1$ eV, the attachment rate constant is about 2×10^{-9} cm^3 s^{-1}, and electrons attach at a frequency of 2×10^8 s^{-1} for typical F$_2$ densities (Chen et al. 1977). A recent paper compares all of the F_2 attachment data (Schneider and Brau 1978). For ionization rates less than about 2×10^{23} ions/cm^3 s, electron attachment is faster than electron–diatomic ion recombination, even at high pressures where all of the positive charge is carried by Ar_2^+. Positive–negative ion recombination proceeds very rapidly by the three-body processes

$$Ar^+ + F^- + Ar \rightarrow ArF^{*\ddagger} + Ar \tag{28}$$

or

$$Ar_2^+ + F^- + Ar \rightarrow ArF^{*\ddagger} + 2\,Ar, \tag{28}$$

where $ArF^{*\ddagger}$ denotes the ionic excited states of the ArF with large amounts of vibrational excitation. Because of the exceptionally long range of the Coulomb well characterizing the ion pair states of rare gas halides, these three-body rate constants are about six orders of magnitude larger than those pertaining to recombination along neutral covalent potential curves, such as those for pure rare gases. At one atmosphere pressure three-body ion–ion recombination occurs on a time scale of less than 1 ns for negative/positive ion densities of order 10^{15}/cm^3. The rate of three-body ion–ion recombination decreases at high pressures because diffusion becomes rate limiting. Expressed as a pressure dependent recombination coefficient, one finds a rollover in recombination rate at pressures above ~ 1 atm as discussed by McDaniel (1964) and shown in fig. 10 (taken from Wadehra and Bardsley 1978). Calculations over a more extended pressure range have also been presented (Flannery and Yang 1978a, b). The ion that the F^- recombines with depends on pressure. At pressures above 2 atm, F^- recombination with diatomic ions, Ar_2^+, is important. It has been found that the product of this recombination is also ArF* (Rokni et al. 1977a, b, c). The recombination coefficients for F^- with dimer ions differ slightly from those for atomic ions, as shown in fig. 10. The F^- density in steady state scales as $J_{eb}^{1/2}$. The corresponding absorption is a nonsaturable loss.

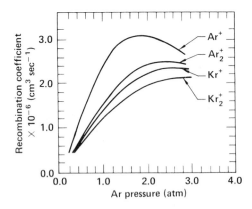

Fig. 10. Effective bimolecular recombination rate coefficient for F^- recombination in Ar buffer.

A second channel for production of ArF* is the reaction of atomic excited states or rare gas excimers with F_2:

$$Ar^* + F_2 \rightarrow ArF^* + F$$
$$Ar_2^* + F_2 \rightarrow ArF^* + 2\,Ar. \tag{29}$$

Such reactions are common in all of the rare gas–halogen systems, and the bimolecular rate constants are quite large (see table 11). For typical F_2 densities, Ar* reacts in a time of about 16 ns. In analogy to the similar bound state chemistry of alkali–halogen reactions (Ewing and Brau 1976), these reactions are called 'harpooning' reactions, because they proceed through a triatomic 'charge transfer' intermediate, which decays into the ArF excited state. Application of the harpooning model suggests that the more highly excited states, Ar**, with lower ionization potentials, should have even larger reactive rate constants. In a few select cases, the molecular excited state yields are high. Yields less than unity are more common and are attributable to predissociation of the intermediate or product rare gas halide into other electronic excited species.

Reactions of Ar_2^* with F_2 are characterized by large rate constants and yields of ArF* (Chen et al. 1978). At sufficiently high pressures ($\geqslant 2$ atm) any Ar* present recombines into Ar_2^* faster than it reacts with F_2. Because of the low ionization potential of Ar_2^* the 'harpooning' mechanism applies. Formation of excited species such as $Ar_2^+F^-$ is possible, both for $Ar_2^+ + F^-$ recombination and $Ar_2^* + F_2$ reaction. However, this triatomic excimer species is apparently formed primarily by ArF* quenching, as will be discussed shortly.

Four reactions forming ArF* have been mentioned so far. There are analogous reactions that can form other rare gas halide excited states in similar mixtures. Recombination of electronically excited halogen atoms could also yield excited rare gas halides. The pressure–excitation regimes in which the different major mechanisms are important are compared in table 12. At low excitation levels the

Table 11
Rare gas metastable–halogen species reaction rate coefficients.

Rare gas	Halogen	Branching ratio to rare gas halide	Rate coefficient [cm³/s]	Reference
Ar*	F_2	0.56	6(−10)	Kolts and Setser (1978)
	Cl_2	0.5	7(−10)	Gundel et. al. (1975)
Kr*	F_2	1.0	7(−10)	Velazco et. al. (1976)
	OF_2	1	5(−10)	Velazco et. al. (1976)
	NF_3	0.57	9(−11)	Velazco et. al. (1976)
	Cl_2	0.9	7(−10)	Velazco et. al. (1976)
Xe*	F_2	1.0	7(−10)	Velazco et. al. (1976)
	OF_2	0.9	6(−10)	Velazco et. al. (1976)
	NF_3	1.0	9(−11)	Velazco et. al. (1976)
	Cl_2	1.0	7(−10)	Velazco et. al. (1976)
	Br_2	1.0	6(−10)	Velazco and Setser (1975)

Table 12
Regimes of ArF* formation mechanisms.

| Pressure | Deposition rate | |
	Low	High
Low	$Ar^+ + F^- + M$ $Ar^* + F_2$ (Direct Ar*)	$Ar^+ + F^- + M$ $Ar^* + F_2$ (Some Ar* from Ar_2^+)
High	$Ar_2^+ + F^- + M$ $Ar_2^* + F_2$ (Direct Ar_2^*)	$Ar_2^+ + F^- + M$ $Ar_2^+ + e \rightarrow Ar^*$ $Ar_2^* + F_2$ (Ar_2^* from Ar_2^+)

Ar* that reacts is *only* that formed directly, rather than by recombination of Ar^+ with electrons via Ar_2^+. At higher pressures, but low total ionization/excitation rates, both the Ar^+ and Ar^* chemistry convert into the chemistry appropriate to dimer ions or excimers. The high pressure regime applies for the positive ions, and atoms/excimers for pressures greater than about 2 atm. At high excitation levels increased pressure changes the relative importance of ion–ion recombination and ion–electron recombination.

In this situation no one particular mechanism truly dominates. Hence it is essential to computer model the kinetic system to adequately estimate the densities of all of the relevant species and their absorption, especially in the range of current densities of 30 to 1 000 A/cm^2 and pressures over 2 atm. Very few of the laser experiments reported to date have been performed in a regime where only one of the relevant pumping mechanisms mentioned above have been truly dominant.

The kinetic cycle in Ar–F_2 mixtures is completed with excited state decay. ArF* decay has radiative, bimolecular quenching or termolecular quenching components. Table 13 lists some of the ArF* decay rate constants which have

Table 13
ArF*(B) decay rate coefficients.

Radiative lifetime [s], for low v, J states

4(−9)(calc.)	Michels (1977)
4(−9)	Chen et. al. (1978)
4(−9)(calc.)	Dunning and Hay (1978)

Bimolecular quenching [cm^3/s]

Ar: 9(−12)	Rokni et. al. (1978b)
Kr: 1.6(−9)	Rokni et. al. (1978b)
Xe: 4.5(−9)	Rokni et. al. (1978b)
F_2: 1.9(−9)	Rokni et. al. (1978b)
1.9(−9)	Chen et. al. (1978)

Termolecular quenching [cm^6/s]

2 Ar: 4(−31)	Rokni et. al. (1978b)
5.3(−31)	Chen et. al. (1978)

been measured or calculated. Quenching by Kr is extremely fast and apparently produces the heavier KrF* molecule. The importance of this displacement reaction will be discussed shortly as a formation mechanism in three-part mixtures. Quenching by F_2 is quite rapid, a similar feature in all rare gas halides. Electron superelastic quenching has been suggested as an important excited state decay mechanism at high deposition rates (Ewing and Brau 1976). No calculational estimates exist for such processes. A measurement for XeF(B) quenching by electrons has been reported, giving a rate coefficient of 3×10^{-7} cm^3/sec (Jacob 1978). The discovery of the importance of termolecular quenching in rare gas halide lasers was a fundamental breakthrough in the understanding of these lasers (Mangano et al. 1977, Lorents et al. 1977, 1978). For ArF*, three-body quenching has a rate coefficient of 5×10^{-31} cm^6 s^{-1}. The quenching decay time at one atmosphere pressure roughly equals the radiative lifetime. Three-body quenching dominates the quenching at higher pressures. The product of three-body quenching is the ionic triatomic species Ar_2F^*. Ar_2F^* can decay by radiation, $\tau_r \sim 180$ ns, or by bimolecular quenching by F_2 with a quenching rate constant 2×10^{-10} cm^3 s^{-1} (see table 14).

Ar_2F^* is calculated to absorb strongly around 250 nm (Wadt and Hay 1978), but the absorbance in the wing of this band in the ArF* spectral band is much smaller and subject to greater calculational error. The strong absorber for ArF radiation should be F^- with any Ar^{**}, and Ar_2^* also contributing. Under conditions where ArF will have a small-signal gain of 10% cm^{-1} at 1 atm pressures, the loss at 193 nm should be about 0.3% cm^{-1}. Gain/loss measurements have not been reported. In the near uv, absorption is dominated by Ar_2^+ and species which mimic it, Ar_2F^* and Ar_2^*. Near 3 000 Å, one expects losses of order 12% cm^{-1} that could be saturated, and 4% cm^{-1} nonsaturable loss due to Ar_2^+ and Ar_2^* for atmospheric pressure Ar–F_2 mixtures.

The heavier rare gas halide lasers, KrF, XeF, KrCl, XeCl and XeBr, usually are operated with tripartite mixtures. In e-beam-pumped lasers Ar or Ne are the

Table 14
Triatomic rare gas monofluoride kinetic properties.

Ar_2F	Radiative lifetime [ns]	
	= 128(calc.)	Wadt and Hay (1978)
	= 185 ± 45	Chen and Payne (1978)
	Two-body quenching [cm^3/s]	
	$F_2 : 2 \pm 0.5 \times 10^{-10}$	Chen and Payne (1978)
Kr_2F	Radiative lifetime [ns]	
	= 132(calc.)	Wadt and Hay (1978)
	181 ± 12	Quigley and Hughes (1978b)
	Two-body quenching [cm^3/s]	
	$F_2 : 4.3 \pm 0.4 \times 10^{-10}$	Quigley and Hughes (1978b)
	$Kr < 2 \times 10^{-14}$	Quigley and Hughes (1978b)

buffer gases encountered to date. Because of the Ar_2^+ absorption, Ne is preferred for the longer wavelength XeCl, XeF lasers. E-beam ionization thus produces Ar^+ or Ne^+ ions with some Kr^+ or Xe^+ formed directly, as well as some direct excitation of atomic excited states of electrons. As in the case of mixtures of rare gases, parallel energy funneling chains are set up with cross linking of the chains possible at several points. Fig. 11 schematically shows this for the case of Ar–Kr F_2 mixtures, omitting minor energy channels such as direct production of Ar* by the e-beam. In this example, using an Ar buffer, it is currently thought that the important cross linking steps connecting excitation in the Ar^+–ArF* chain with the analogous chain in Kr are the following:

Charge transfer: $Ar_2^+ + Kr \rightarrow Kr^+ + 2\,Ar,$ (30)

Displacement: $ArF^* + Kr \rightarrow KrF^* + Ar,$ (31)

Energy transfer: $Ar_2^* + Kr \rightarrow Kr^* + 2\,Ar.$ (32)

A detailed accounting of the energy flow channeling requires computer model-ing, since it is a sensitive function of the mixture ratios, energy deposition rate and pressure. As with pure Ar–F_2 mixtures, low deposition rates and low pressures cause the ion recombination mechanism to dominate. Ar^+ and Kr^+ rapidly convert to the corresponding fluorides. Measurements by Rokni et al.

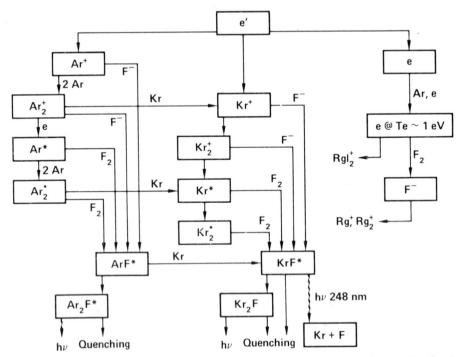

Fig. 11. Dominant energy flow kinetic pathways in e-beam-excited Ar–Kr–F_2 mixtures. Despite the large number of potential paths shown, and others not shown, virtually all of the initial ionization can be made to flow through KrF*.

(1978a) suggest that the displacement by Kr of the Ar^+ in ArF* occurs with an exceptionally large bimolecular rate constant, $2 \times 10^{-9}\,cm^3\,s^{-1}$, using a 4 ns radiative lifetime for ArF*. However, displacement may occur for ArF* molecules which have a very high vibrational temperature and hence a longer lifetime. The details of this important step are not yet fully understood. At low deposition rates and higher pressures, the Ar^+ is combined into Ar_2^+, which then rapidly charge transfers with the abundant Kr, producing Kr^+. Krypton ions then funnel energy down a Kr^+–KrF* cycle entirely analogous to that described for Ar^+– ArF*. At high pressures and deposition rates sufficiently large to have electron densities in excess of about 10^{-3} the Kr density, energy corresponding to Ar_2^+ excitation can funnel through Ar* and Ar_2^*. Directly produced Ar* of course funnels through these species as well. The energy transfer cross linking [eq. (32)] is primarily important at low mole fractions of the second rare gas and high e-beam current densities.

Despite the complicated kinetic chains that three-part mixtures entail, production of KrF* and other rare gas halides in such mixtures can occur with high yield. A major limitation on efficiency is collisional quenching and absorption. Quenching rate constants for KrF and XeF have been measured and are compared in tables 15, 16 and 17. Superelastic electron quenching and excited state mixing rate constants need to be determined.

Table 15

KrF*(B) decay rate coefficients.

Radiative lifetime [s]	
6.5(−9)(calc.)	Hay and Dunning (1977)
6.8(−9)	Eden et. al. (1978)
9(−9)	Burnham and Searles (1977)
9(−9)	Quigley and Hughes (1978a)
Bimolecular quenching [cm³/s]	
Ar 1.8(−12)	Eden et. al. (1978)
Kr 8.6(−12)	Eden et. al. (1978)
F₂ 4.8(−10)	Eden et. al. (1978)
5.7(−10)	Quigley and Hughes (1978a)
7.8(−10)	Jacob et. al. (1978)
Termolecular quenching [cm⁶/s]	
+ 2Ar 9(−32)(calc.)	Shui (1977)
1(−31)	Eden et. al. (1978)
7(−32)	Rokni et. al. (1978a)
+ 2Kr 3(−31)	Quigley and Hughes (1978a)
1(−30)	Eden et. al. (1978)
5(−31)(calc.)	Shui (1977)
7(−31)	Mangano et. al. (1977)
+ Kr + Ar 6(−31)	Rokni et. al. (1978a)

Table 16
XeF*(B) decay rate coefficients.

Radiative lifetime [s]

12(−9)(calc.)	Dunning and Hay (1978)
19(−9)	Fisher and Center (1978)
16(−9)	Eden and Searles (1977)
13(−9)	Ewing et. al. (1977)
18(−9)	Burnham and Harris (1977)
14(−9)	Eden and Waynant (1978a, b)

Termolecular quenching [cm^6/s]

+ 2Ne 4(−41)	Rokni et. al. (1978b)
+ 2Ar 2(−32)	Rokni et. al. (1977a, b, c)
7(−32)	Eden and Waynant (1978a)
+ Xe + Ne 8(−31)	Rokni et. al. (1978b)
+ 2Xe 2(−2)	Eden and Waynant (1978b)
+ Xe + Ar 3(−31)	Rokni et. al. (1977a, b, c)

Other than parent halogens, the important absorbing species produced in this and analogous cycles in other rare gas halide mixtures are believed to be the positive dimer ions, negative halide ions, triatomic rare gas halides and excited states of the atoms. The importance of these absorbers depends on deposition rate and pressure. At low current densities F^-, Ar_2^+, Kr_2F^* and Ar_2F^* are important absorbers at 248 nm. However, at high current densities the importance of Kr_2F^*, Ar_2F^* and excited atoms dominates, since these species

Table 17
XeF*(B) energy sharing/quenching rates [cm^3/s].[a]

Sharing into C($\frac{3}{2}$) state	Brashears and Setser (1978)
He 2(−12)	
Ne 7(−13)	
Ar 5(−11)	
Xe 5(−11)	
NF$_3$ 5(−11)	
F$_2$ 2(−11)	

Quenching of B($\frac{1}{2}$) or coupled B/C.

	Brashears and Setser (1978)	Fisher and Center (1978)	Eden and Waynant (1978a, b)	Rokni et. al. (1978b)
He	<2(−13)	2(−12)	4(−13)	—
Ne	<7(−14)	<(−13)	7(−13)	small
Ar	<5(−12)	—	5(−12)	—
Xe	6(−11)	6(−11)	3(−11)	3(−11)
F$_2$	2(−10)	1(−10)	4(−10)	3(−10)
NF$_3$	3(−11)	3(−12)	3(−11)	—

[a] See also Brashears et. al. (1977), Rokni et. al. (1977a, b, c).

steady-state density scales as J_{eb} while the density of ions scales as $J_{eb}^{1/2}$. The low excitation rate gain/loss measurements of Hawryluk et al. (1977), $J_{eb} = 6$ A/cm^2, were dominated by ion absorption, while those of Bradford et al. (1977), $J_{eb} \approx 500$ A/cm^2, were dominated by excited state absorption.

The kinetics of XeF lasers entail a few subtleties not shared by other rare gas halide lasers. First, because of the bound–bound nature of the transition, lower level bottlenecking can occur. The details of lower vibrational level filling and subsequent vibrational redistribution and dissociation have not yet been quantified. Secondly, it appears that the XeF C(3/2) excited state lies considerably lower in energy than the B(1/2) upper laser level (Kligler et al. 1978, Kolts and Setser 1978). The cross relaxation rates between the C and B states under laser conditions, electrons present, have not yet been resolved. Finally, the background absorption in XeF lasers can be decreased by the use of neon as a diluent, (Champagne and Harris 1977, Rokni et al. 1978a, b, c). The collisional energy channeling processes in such mixtures are not yet fully quantified although a plausible mechanism has been proposed (Rokni et al. 1978a, b, c). The intermediate species NeF* and F*, formed by reaction of Ne* + F$_2$, or the Ne$^+$ + F$^-$ + M ion channel, have sufficient energy to ionize the Xe atom. See also the recent work of Huestis et al. (1978).

4.3 Discharge excited rare gas halides

The rare gas monohalides can be excited by electrical discharges as well. The kinetic chain is biased more toward neutral chemistry. Ion recombination is important as well, but electron–atom and electron–molecule processes dominate considerations of efficiency and discharge stability. Discharges can be self-sustained or sustained by an external ionization source such as a spark source of ultraviolet light or an electron beam. The kinetic regime is slightly different in each case. Typically the self-sustained and uv sustained discharges operate in a short pulse, $\leqslant 100$ ns, the pulse length presumably limited by discharge instabilities. Tripartite mixtures containing He as the buffer are normally used in these lasers. The laser volumes which have been excited for the short pulse lasers tend to be small, typically less than 1 ℓ. The e-beam-controlled discharge technique has been used to excite much larger volumes, and the mixtures used are similar to those used for pure e-beam pumping. The e-beam-controlled discharge technique has also yielded longer pulse durations. In the short pulse self-sustained or uv sustained discharge lasers, all of the excitation energy is pumped into the gas by the discharge current. In the long pulse e-beam-controlled discharges, the relativistic electrons from the e-beam produce a substantial fraction of the excited states as well as producing the volumetric ionization of the discharge medium. For such devices one often sees reference to a 'discharge enhancement factor', the excited state production by the discharge current ratioed to that provided by the e-beam ionizer. For stable long pulse rare gas halide discharges, the discharge enhancement factor tends to be low, in the range 1.5 to 3.5 (Jacob and Mangano 1976a, b, Brown and Nighan

1978). Thus, the laser pumping kinetics include important contributions due to the ionizing electron beam, as described earlier.

A discharge excited rare gas halide laser differs kinetically from an e-beam-excited system in its additional reliance on low energy electrons to excite the rare gas atoms in the mixture, for example,

$$e + Kr \rightarrow Kr^* + e. \tag{33}$$

The excited states so produced, Kr^* and Kr^{**}, and Ar^* and Ar^{**}, are reacted away by the halogen to produce excited states as discussed previously. The details of the electron energy distribution function, the efficiency with which excited states of atoms are produced, and the ratio of highly excited rare gas atomic states to the lower energy metastables have been the subject of several computer models (Jacob and Mangano 1976a, b, Long 1977, Nighan 1978a, b). The various models agree that for conditions of low power deposition by the discharge, the efficiency with which excited states are produced is very high, 90%. However, for larger steady-state populations of rare gas excited atoms, the efficiency of producing excited states decreases, as illustrated in fig. 12. The fraction of energy deposited in the gas that results in rare gas atom excitation decreases with increased excited state density because of two-step excitation and ionization:

$$Kr^* + e \rightarrow Kr^{**} + e, \tag{34}$$

$$Kr^*, {}^{**} + e \rightarrow Kr^+ + 2e. \tag{35}$$

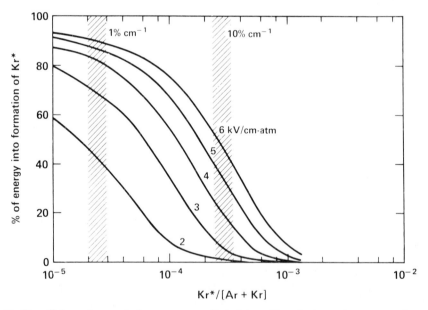

Fig. 12. The efficiency for producing rare gas excited states with a discharge drops as one increases the excitation level. High gains come at lowered efficiency. Based on calculations by J. Jacob and J. Mangano.

These reactions have low threshold energies compared to excitation of ground state atoms. Cross sections similar to those of alkali atoms should pertain. The excitation step, (33), utilizes electrons in the high energy tail of the distribution, while (34) and (35) require the more numerous lower energy electrons. Thus, under higher gain situations, larger Kr*, the efficiency of producing Kr* drops. On fig. 12 are also shown the Kr*–Kr + Ar ratios required for approximate 1 and $10\% \, cm^{-1}$ small-signal gains due to KrF in a 2 atm mixture of Ar–Kr–F_2 containing $0.3\% \, F_2$. $1\% \, cm^{-1}$ gain requires an excitation rate of 0.7×10^{23} excitations per $cm^3 \, s$, and correspondingly has an excited state mole fraction of 2.5×10^{-5}. Somewhat higher gains may in fact be observed when KrF* formation from Ar*/ArF* is taken into account, since the relative importance of the Kr* and Ar* reactive channels is both a function of applied field and fractional excitation (Long 1977). As we shall see momentarily, discharge stability limits the achievable gain to about $3\% \, cm^{-1}$ for such a mixture. Discharges that are unstable, such as the short pulse rare gas halide lasers, can operate with higher gains, albeit for shorter times and possibly with lower intrinsic efficiencies.

It has been pointed out that at lower excited state relative densities, Kr*–(Kr + Ar) $< 3 \times 10^{-5}$, that the effects of electron–electron collisions become more important in determining the electron distribution function than are the effects of superelastic and excited state ionization collisions. Calculations accounting for this suggest that the efficiencies of excited state production are higher than those shown in fig. 12, especially for the lower values of E/N (Long 1977). Experiments on small-scale e-beam-stabilized discharges appear to fit this model (Bradford et al. 1976).

A second major difference between a discharge and a pure e-beam-excited rare gas halide laser is the possible instability in the volumetric discharge approach. Over a wide range of the parameter space of excitation rate–number density, the discharge tends to rapidly become an arc. The microscopic phenomenon leading to this is thought to be the local increase in the number density of electrons due to electrons ionizing excited states. Thus as one produces electrons, the net rate of producing more electrons increases. Fortunately the rare gas halide lasers have large stimulated emission cross sections compared to other excimers and do not require very large steady-state densities of easily ionized atoms or molecules. Moreover, halogen attachment reduces the growth rate of electrons, thus delaying the onset of an arc. Fig. 13 shows calculations of the net rate of electron density exponentiation by collisions with excited rare gas atoms for various excitation ratios and discharge electric field strengths. For small-signal gain coefficients of 3 to $10\% \, cm^{-1}$ these times are of order to 2–25 ns. The F_2 present in the mixture however effectively 'buffers' the growth of electron number density by attachment. For F_2 pressures in the range 2 to 4 Torr and electron temperatures of a few volts, the electron avalanching rate can be balanced by the more rapid attachment. A criterion for a 'stable' discharge whose electron number density is constant in time has been derived for rare gas–halogen mixtures (Daugherty et al. 1976). A stable electron number density can be attained if the net rate of excess ionization (excess over

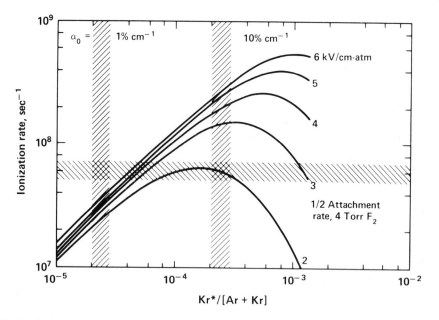

Fig. 13. The ionization rate of rare gas metastables increases with excited state density. Although attachment can stabilize electron density growth, it does so only for a limited region of efficient gain production. (Adapted from Jacob and Mangano.)

any external controlling ionization source such as an electron beam) is one-half the rate of electron attachment. Fig. 13 shows the approximate upper limit on ionization rates, and consequently excitation ratios and electric field strength for a 4 Torr F_2 discharge laser mixture. This suggests, by comparison with fig. 12, that net gains of ~2% cm^{-1} should be achievable with excited state production efficiencies that are high, ~80%. The low gains and the losses due to $F_2(\sim0.2\%\ cm^{-1})$ and discharge created absorbers, suggest that efficient extraction in a large device may be difficult to achieve since the ratio of g_0/γ_{ns} is lower than 10. A discussion of the gains achievable for various electron number densities and discharge field strengths has been given, Brau (1978a, b) arriving at the same conclusion: small-signal gains in stable discharges will be low, and extraction efficiency may suffer. A theory which accounts for F_2 depletion also predicts low gains for efficient stable discharges (Nighan 1978b). Gains in the region of 3 to 10% are achievable but correspond to inefficient excited state production, operation in an unstable regime, or use of even more F_2 with consequently increased photoabsorption by this species. The short pulse, uv preionized discharge lasers operate at high gains and high electric field strengths in what is apparently an unstable regime.

A series of papers by Nighan and Brown (Nighan 1978a, b, c, Brown and Nighan 1978) has theoretically considered and experimentally measured discharge instability onset times over a wide range of parameters. This work has pointed out the importance of F_2 dissociation by discharge chemistry in determining discharge stability times. F_2, which serves to buffer the growth of electron number density by attachment, dissociates during the discharge pulse. Direct electron impact dissociation reactions with rare gas excited states, attachment of secondary electrons and collisional quenching of rare gas halide excited states all act to decrease the F_2 number density, hence producing discharge instability when the F_2 density falls below that required to balance two-step ionization. The efficiency of excited state production is also decreased by the decreasing F_2 densities, since the electron density rises, and concurrently more power is dissipated in the electron metastable reactions of two-step excitation and ionization.

The optical gain and loss properties of e-beam-controlled discharges in rare gas halide lasers need to be studied as no measurements of gain and loss and their saturation have been reported. The important absorbers, other than F_2, should be atomic and molecular excited states rather than ions. Of the atomic excited states the absorption due to states above the rare gas metastables, Ar** and Kr**, should be most important. Calculations show that the relevant cross sections are about two orders of magnitude larger than those for Ar* and Kr* (Hyman 1977). However, the ratio of Ar** to Ar* that one predicts depends on the details of the electron distribution and its coupling to the atoms being pumped. Modeling calculations for typical discharge mixtures suggest this ratio could be as high as 20%. Thus a gain of about $3\% \, cm^{-1}$ on KrF will have a nonsaturating Kr** and Ar** loss of about $0.2\% \, cm^{-1}$.

Publications dealing with the accurate modeling of simple discharges using uv preionization have not been as extensive (Greene and Brau 1978). The role of negative halide ions in these systems as sources of easily ionized electrons has been discussed (Hsia 1977). As discussed in § 5.2, a uv preionized laser utilizes an array of uv sources to ionize the gas before the main discharge voltage is applied. Rapid attachment converts these electrons into halide ions in a very short time. However, negative ion densities of order $10^{11}/cm^3$ can persist for tens of microseconds. When the main discharge field is applied, electrons are collisionally released from the negative ion providing the electrons to excite the medium.

5. Excimer laser device technology

Excimer lasers require the integration of several technologies. The most important are those involving excitation techniques, materials technology and optics. In principle, the hardware required to construct an excimer laser is simply an extension of that utilized in infrared lasers. However, the requirements of excimer lasers are usually more stringent. A few comparisons are cited here. The e-beams used to either directly excite an excimer laser or to control a discharge in a rare gas–halogen mixture must run at higher current densities than those needed for CO_2 or CO lasers. The discharge circuitry required for an efficient, fast pulse discharge rare gas halide laser is less inductive than that used in similar pulsed infrared lasers. The ultraviolet wavelengths and high cavity fluxes required stress the current state-of-the-art in mirrors and optical components. The chemical purity requirements for rare gas excimer lasers are rigorous. The F_2 compatibility problems of ArF, KrF and XeF lasers are similar to those encountered in pulsed HF lasers. However, F_2-compatible laser systems are not necessarily compatible with Cl and Br donors.

The following sections discuss the most important uv excimer laser technologies: e-beams, electric discharge technologies and optics related areas. The current practical limitations in each of these areas are highlighted. There has not previously existed a strong requirement to improve performance in some of the 'problem' areas of these technologies. Thus research and development in coming years could overcome several of the performance limits mentioned.

5.1 E-beams for excimer lasers

The technology of generating pulsed e-beams in the 1 to 1 000 A/cm^2 range is a pivotal part of excimer laser design. Direct electron-beam pumping is the most general technique for exciting excimer lasers. E-beam pumps in the 0.2 to 2.0 MeV range have been used to excite small and large aperture lasers. Overall laser efficiencies of several percent are possible with this pump technique. Broad area electron sources have also been used to control large volume electric discharges. In single shot or low pulse repetition frequency devices, the electron beam and its ancillary hardware are the major cost items of an excimer laser. Several companies supply complete electron-beam systems for laser experiments. It is usually found that any existing combination of beam accelerating voltage, total energy and e-beam aperture is of limited use in a new situation. Thus most new e-beams for excimer lasers tend to be modifications of existing systems or custom built to the user's specifications. A useful survey article on e-beam-ionized laser principles has been prepared (Daugherty 1976).

Broad area beams of high energy electrons can be produced in a variety of ways. Because of the high current densities required, the approach most frequently encountered is the use of a diode configuration with a cold cathode as an electron source. Thermionic electron emitters, commonly used in e-beam-ionized infrared lasers, typically produce current densities of less than 1 A/cm^2. Thus they are of limited utility in excimer lasers. Extending the useful current density range is the subject of some current development.

The relevant components of a cold cathode electron gun suitable for use in an excimer laser are shown schematically in fig. 14. The cold cathode electron gun consists of a cathode which is pulse charged to high voltage, an anode and a thin foil separating the laser mixture from the diode. The diode is evacuated typically to pressures of order 10^{-4} Torr. A high voltage insulator feed-through connects a pulsed high voltage source to the diode. Since the pressure differential across the foil is typically one atmosphere or more, the electron-beam foil is supported on a solid structure. This is typically an array of slots, often referred to as a hibachi. The foil and support can also be used as the anode. The cathode is an extended

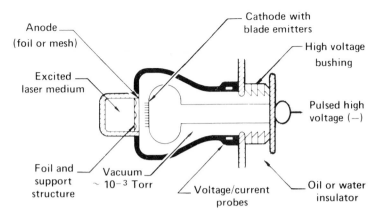

Fig. 14. Schematic diagram of a cold cathode e-beam-excited excimer laser.

metal structure pulse charged to high voltage. The entire structure can emit electrons, but the use of field-enhancing techniques in the region of desired emission can practically limit spurious emission. Such field enhancement is achieved by having a much smaller spacing between cathode and anode in the region where emission is desired, and by using sharp structures such as metal razor blades, arrays of needles, notched carbon and so forth to enhance the electric field at the cathode. The metal surface areas where emission is undesirable are polished very smooth. Electric field strengths of less than 100 kV/cm are recommended for the nonemitting surfaces, and local field strengths of over 500 kV/cm are recommended for a rapid start of the emission process in the desired emitting regions.

A variety of e-beam-pumped laser geometries have been utilized. Fig. 15 shows some of the more frequently discussed layouts. By far the most commonly used to date is the transverse geometry. This arrangement has the drawback of low e-beam energy utilization. Its simplicity is favored for small-scale experiments since power deposition rates can be readily calculated and measured. It has the simplest mechanical design. Improved beam energy utilization can be obtained by use of magnetic guide fields to bend the beam so it can excite a medium longitudinally. This technique was used to produce the first single pulse 100 J excimer laser (Hoffman et al. 1976). It has the drawback that scaling to larger apertures requires energy intensive magnetic fields. The longitudinal geometry is of limited utility unless the beam voltage is such that the distance needed to efficiently stop the electron beam is smaller than the characteristic optical absorption depth for a medium which has a gain profile overlapped by background losses. The cylindrical converging geometry has been used successfully to produce uniformly excited gain media, both for simple rare gas excimer lasers (Bradley et al. 1974) and rare gas halide lasers (Bhaumik et al. 1976). Total system efficiencies of 1% have been achieved with this geometry for KrF. Experience with two-sided transverse geometries has been limited to date.

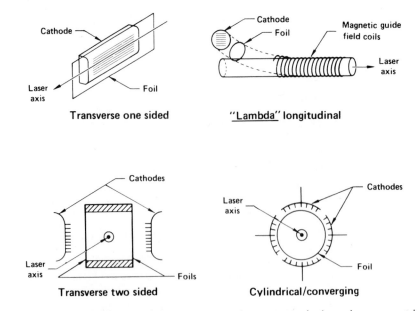

Fig. 15. Schematic diagram of the most commonly encountered e-beam laser geometries.

Calculations suggest that reasonably uniform deposition can be achieved with this approach. Such a laser is intrinsically flow compatible.

In order to efficiently utilize e-beam energy, the dimensions of the device along the direction of beam transport should be a large fraction of the electron penetration depth at the voltage of the beam electrons. A plot of the electron range in one atmosphere of various rare gases is given in fig. 16. As mentioned

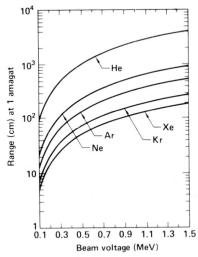

Fig. 16. Berger and Seltzer range for rare gases at 1 atm. Efficient beam utilization requires the dimension of beam transport in the gas be of order 30% of the range for the heavier gases.

previously, the energy deposition rate is increased, and the beam penetration depth is decreased by the importance of beam scattering for high-Z materials such as Ar, Kr or Xe. A number of three-dimensional energy deposition codes have been developed to treat this effect (Colbert (1973, Hart and Searles 1976). The approximate distance into the gas that the electrons penetrate is the range in which they are scattered $90°$ from the initial trajectory. This length is called the transport mean free path (Spencer 1955, Daugherty 1976), λ_T, and is a function of e-beam voltage, scaling roughly linearly with the voltage above 300 keV. The product of electron transport mean free path, mass density and atomic number is roughly a constant for all elements, dependent on beam voltage, $\lambda_T \rho (Z + 1) = f(V)$. At about 400 keV, $\lambda_T \rho (Z + 1)$ is approximately 1 g/cm^2 for the heavier elements. For Ar at 1 atm and 400 keV, λ_T is roughly 25 cm.

Three important phenomena govern the actual performance of a cold cathode electron gun for an excimer laser driver. The current density that can be drawn is space charge limited and an upper limit on the current density is given by Child's Law. The pulse duration is limited by a phenomenon called diode closure as well as foil heating. The total areas excited depend on a phenomenon called beam pinch.

The maximum current density drawn in a space charge limited diode is governed by the relationship (Child 1911)

$$J_{eb} = 2.3 \times 10^3 \, V^{3/2}/d^2 \tag{36}$$

where V is the applied voltage in megavolts, d the anode–cathode spacing in cm and J_{eb} is the current density in A/cm^2. The total current drawn in the diode is not entirely delivered to the laser cell. Losses are encountered by shadowing of the electrons by foil support structures, the finite transmissions of the anode and foil, and the flow of current in regions of the diode that are not desired. In very well designed systems ~80% of the beam current can be utilized, although coupling efficiencies of 30 to 50% are more typical.

The total current that is drawn in the diode of a real electron-beam system usually varies during the pumping pulse. This is so because both the applied voltage, the total emitting area of the cathode and the anode cathode spacing vary in time. The voltage applied to the diode can vary because of the circuit characteristics of the high voltage source. The impedance of the gun can also vary with time. Any electron emission in the diode from parts of the cathode that take longer to begin to emit electrons, for example, regions at lower field strengths, will cause increased current flow over that expected. These two effects can be minimized by good design. An intrinsic effect observed in cold cathode electron guns is called diode closure, and corresponds to a time variation of anode–cathode spacing d. Diode closure is caused by the motion of a plasma that is formed near the surface of the cathode. The beam electrons are drawn out of this plasma and then accelerated in the diode. The plasma cloud created at the cathode surface drifts toward the anode at a velocity between 1 and 3×10^6 cm s^{-1}, (Bugaev et al. 1969, Friedman and Ury 1972, Ahlstrom et al. 1972). To insure a beam current growth of less than 20% over the pulse duration, the effective diode spacing can decrease by only 10%. Thus, the pulse duration is limited by diode closure by

$$\tau_p \lesssim 1.6 \ V^{3/4} J_{eb}, \tag{37}$$

where τ_p is in μs, V in MV and J_{eb} in A/cm^{-2}. Thus for a typical baseline case, a 600 keV, 200 A/cm^2 pulse can be drawn for $\tau_p \sim 75$ ns, growing to 240 A/cm^2 at the end. If one requires a given energy fluence, and also requires the beam current variation to be less than some fixed amount, say 20%, then one prefers to work with an electron beam of higher voltage.

Foil heating is a limitation on pulse duration in two regimes, very high current density single shot experiments, and in average power applications. Except in situations where beam pinching is extreme or when an arc forms in the diode, single shot foil heating is not of importance in excimer lasers. A concise analysis of the heat loading delivered to foils in repetitively pulsed e-beam-pumped lasers has been given (Daugherty 1976). Since roughly 10 W/cm^2 can be dissipated by an Al foil with a temperature increase of about 200°C, corresponding to a decrease in tensile strength of a factor of 2, a pulse repetition frequency approaching 100 Hz should be achievable at interesting current densities. Repetitively pulsed excimer lasers that have been built to date have operated well below foil heating limits.

The phenomenon of beam pinch limits the area of any single electron beam diode. The current that flows in the diode and elsewhere in the device produces a magnetic field. The field due to the beam current density is given by

$$B = 0.5h(\mu_0 J_{eb}) \tag{38}$$

where h is the beam height in cm. The field so generated causes the electrons furthest from the beam center to deflect toward the center of the beam, to pinch. After the beam has traversed the diode gap, it makes an angle relative to the normal of the anode plane. If $\theta = 90°$, the beam is fully pinched. The achievable beam height h is related to the anode–cathode spacing and the applied voltage on the gap by (Schlitt 1978)

$$h = \alpha \ \sin \theta (d/V^{1/2}). \tag{39}$$

The factor α is dependent on geometry:

$$\alpha \cong \frac{12}{(\ln(2l/h) + 1)}. \tag{40}$$

For e-beam geometries of interest for exciting lasers, α varies from about 2 for long thin geometries to about 7 for square geometries. The scaling of the sine of the pinch angle with the quantity $V^{1/2}(d)^{-1}$ has been experimentally verified (Schlitt 1978). The predictions of the formula 39 are accurate to 25% for the long rectangular geometries studied. Combining the pinch angle expression with Child's law yields

$$h \cong 48\alpha \ \sin \theta \ V^{1/4}/\sqrt{J_{eb}}. \tag{41}$$

Using $\theta = 30°$ as an acceptable deviation from normal incidence, and taking $V = 0.5\,\text{MeV}$ one finds the relationship

$$h \lesssim 20\alpha/\sqrt{J_{eb}}. \tag{42}$$

An analogous relationship has been derived (Daugherty 1976), giving the achievable beam height for a 300 keV electron beam approximately as

$$h \lesssim 88/\sqrt{J_{eb}}. \tag{43}$$

For laser geometries of interest, e-beam heights of order 10 cm at 100 A/cm^2 and 30 cm at 10 A/cm^2 appear achievable. If for a particular combination of cathode material and pulse duration the full space charge limited current is not achieved, the beam height available may increase, usually at the cost of a shorter time to diode closure because of decreased anode–cathode spacing.

Magnetic fields have been used to control beam pinch for both longitudinal and transverse geometries. The applied field strength is such that the self-pinching field is weaker than that due to the externally applied magnetic field, typically several kilogauss. The control of diode pinch with a guide field has the corollary effect of controlling beam spread in the laser media. The beam spread is due to the scattering of beam electrons by the laser media, since the transport mean free path is usually much greater than the anode cathode spacing except for Xe (or Hg) lasers operating at pressures of about 10 atm. For apertures even larger than that obtainable with reasonable magnetic fields, some form of diode modularization is required. Several laboratories are currently investigating methods for practically achieving such modularized e-beam sources.

An important aspect of e-beam laser technology is the choice of foil and foil support structure. Low-Z foils, for example, Al or plastics are usually used. For high temperature operation or larger thermal loadings by the e-beam, super-strength alloys such as stainless steel, Havar, or Inconel have also been used. Fig. 17 and 18 show a plot of the temperature dependence of the tensile strength and density normalized tensile strength of some common foil materials. The strength is density normalized since e-beam stopping and scattering is dependent on the materials density and atomic number. Beam scatter in the foil is more pronounced in the foils made of the heavier elements. Scatter in the foil causes the electrons to emerge from the foil into the gas at a nonnormal angle. This has the effect of further reducing the effective range of electrons into the gas, and increasing the total beam energy deposited in the foil. It is thus preferred to limit the foil thickness to a small fraction of the electron mean free path in the foil. The transport mean free path for the various foils is shown in fig. 18 as well.

The foil support structures are usually designed to have an open to total area ratio of order 90%. A support structure having slots of width 0.6 to 1.2 cm with solid ribs of one-tenth that dimension, 0.6 to 1.2 mm, constructed of 1 cm thick Al will support pressure differentials of 1 atm over a total support height of 10 cm. A 12 μm Al foil is required to support 1 atm across this slot width. Stronger support and foil materials allow use of a thinner support structure, larger e-beam aperture and higher pressure differential. Because of the presence

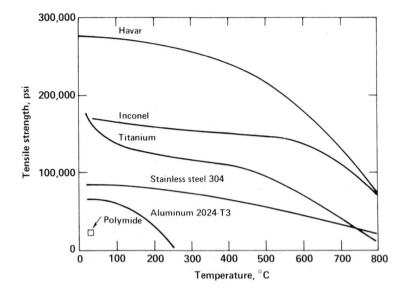

Fig. 17. Tensile strengths of common foil materials. (Courtesy of J.C. Swingle.)

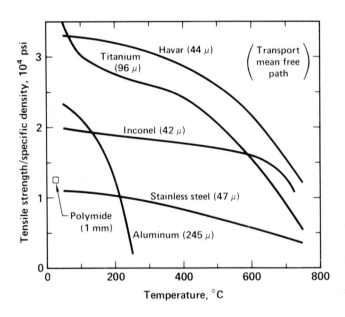

Fig. 18. Density normalized tensile strength of foils. Beam scatter and deposition in the foil are dependent on foil density.

of transverse velocity components in the electron beam and beam pinch effects, a higher transmission of the beam through the support structure is achieved by minimizing the support thickness.

The foil is usually vacuum sealed to the foil support structure by an O-ring. In very high temperature systems, this is impossible because of the lack of suitable O-ring materials. In such cases foils may be welded or brazed to the support. Alternatively, heat pipe technology could be brought to bear. This practical problem, and its fairly expensive solutions, is one of the more vexing limitations of proposed high temperature e-beam-excited excimer systems.

A separate part of electron beam technology for an excimer laser is the selection of the pulsed power source used to impress the accelerating potential across the diode. These approaches have been quite varied. A number of manufacturers offer high voltage pulse generators, either separately or as part of an actual e-beam system. The basic concept typically used is to store energy in a capacitor array, and switch the voltage, possibly with some voltage multiplication onto the diode. For efficient generation of short pulses, with fast rise times, an intermediate pulse forming network or pulse forming line is often used between the high voltage source and the diode. The most commonly encountered energy storage/voltage multiplication component is a Marx generator, although L–C circuits and directly charged coax cables have also been used.

5.2 Electric discharge technology

The only excimer laser systems that have been successfully excited over laser threshold by discharge techniques to date are those utilizing halogen-bearing compounds in the laser mixture. Pure rare gas excimer discharges have been researched (Basov et al. 1973, Huber et al. 1976). A number of proposed metal vapor excimer mixtures are thought to be discharge compatible. As discussed previously, the number density of electrons in a discharge pumped excimer laser can rapidly multiply because excited atoms and molecules which have low ionization potentials are ionized by the secondary electrons. Any multiplication of electrons will lower the discharge plasma impedance. This changing plasma impedance can result in inefficient coupling of electrical energy into the laser medium. Thus the principle technology issue in discharge pumped excimers is the technique used to avoid problems associated with electron multiplication.

Two generic techniques have been used to accommodate this basic problem: to control very carefully the ionization rates with an e-beam and thus be able to use long discharge pulse durations, or to work in a patently unstable regime and switch the voltage off before arc formation, so-called fast discharges. The fast discharges have been operated both without any source of preionization (Burnham et al. 1976a, b, Wang et al. 1976) and with weak preionization uv spark sources (Burnham et al. 1976b). The successful discharge pumping experiments have used transverse rather than longitudinal discharges. Of these approaches, the e-beam-controlled technique has yielded the longest pump pulse durations and promises the highest specific energy densities and the largest excitation volumes. However, the simpler discharge schemes readily yield pulse energies in the range of 0.01 to 1 J on time scales of 10 to 50 ns. The simplicity of the non-e-beam-ionized approaches makes them the technology of choice for useful energies of less than 1 J. Future technology advances could raise this energy to the 5 J/pulse regime. Pulse repetition frequencies of commercial discharge rare gas halide lasers have been limited to less than 100 Hz. Recently, a 40 W, kHz pulse repetition frequency KrF laser has been demonstrated (Fahlen 1978). No repetitively pulsed e-beam-controlled discharge excimer laser has yet been reported.

Aside from the simplicity of the small-scale discharge lasers, the motivation for discharge pumping lies in its potentially higher system efficiency and the larger pulse repetition frequency potentially available. The efficiency of discharge pumping is potentially higher because the effective excited state production efficiency can be higher than for e-beams. Maintaining discharge stability with high extraction efficiency is the key technology issue. The potentially higher pulse repetition frequency is based on the fact that in pure e-beam pumping all of the pump energy deposited in the gas must pass through the thin foil separating the laser mixture from the vacuum diode. The foil suffers a resultant heat loading in an average power mode. For a similar tolerable time averaged heat loading on the foil, the e-beam-controlled discharge approach can result in a higher pulse repetition frequency and presumably higher average power system. For discharge enhancements that have been observed, overall efficiencies of order 10% may be possible. High medium efficiencies have been observed in small volume e-beam-controlled discharges (Bradford et al. 1976). Fig. 19 shows a schematic diagram of a typical electron-beam-controlled rare gas halide laser. Laser action has been achieved on KrF (Mangano and Jacob 1975), XeF (Mangano et al. 1976) and more recently on XeCl, (Levatter et al. 1978) in such devices. Large volume, high energy e-beam-controlled discharges have also been built, but not fully documented in the literature (Rokni et al. 1978c). Similar design considerations apply to these devices as discussed above for pure e-beam-pumped lasers. However, since the e-beam only provides a fraction of the energy, the total current in the e-beam diode is decreased, with typical values

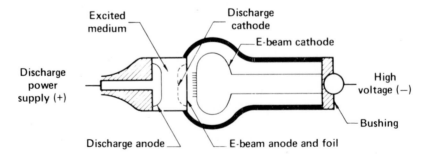

Fig. 19. Schematic diagram of an e-beam-controlled discharge laser.

for laser discharge control of about 10 A/cm² e-beam current density. The discharge currents that are used are in the range 10 to 30 A/cm² for stable discharges (Jacob and Mangano 1976b, Brown and Nighan 1978). As with pure e-beam-pumped lasers, pinching of the e-beam due to the magnetic fields associated with current flows limits the aperture of the device. In a discharge, however, the larger discharge currents in the laser medium can cause the e-beam to pinch in the medium even when self-pinching does not occur in the diode. Previous experience with CO_2 lasers has shown that modularization of the discharges at an aperture size smaller than the medium pinching limit is required (Daugherty 1976). For typical rare gas halide laser discharge currents, this corresponds to a discharge width of order 20 cm. As with pure e-beam pumping, magnetic guide fields could be used to counteract this effect.

The complexity of e-beam-controlled discharges is not merited for smaller pulse energies useful for many laboratory experiments. In such cases fast pulse discharges are of more use. Laser action in ArCl, KrCl, XeCl, ArF, KrF, XeF and XeBr is readily achieved. Fig. 20 shows a schematic diagram of a typical fast pulse uv excimer laser. The first lasers of this sort used no independent source of volumetric preionization. These lasers require a very fast discharge rise time and low circuit impedance. Near atmospheric pressures of He with ~10% of appropriate rare gas and 1% of halogen source are utilized. Electric field strengths of order 10 to 20 kV/cm atm provide both breakdown ionization and discharge pumping. Because of the finite chemical reaction time, the laser pulses last longer than the discharge pump pulse. Discharge pulse durations in excess of 10 ns at discharge spacings of order 0.5 to 1.0 cm at these applied voltages will lead to arc formation if a preionizing scheme is not used. The fast pulse discharge circuitry most frequently encountered utilizes a parallel plate extended capacitor Blumlein geometry, discrete capacitor Blumlein geometries or rapidly charged cable arrays. The nonpreionized discharge lasers operate as small

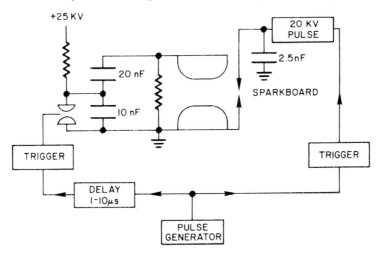

Fig. 20. Schematic diagram of a fast pulse discharge laser, courtesy of R. Burnham.

volume devices, ~25 cm³/m of active laser length. As such, the single pulse energies are limited to about 10 mJ. Operation at kilohertz pulse repetition frequencies has been achieved (Wang 1978).

To produce single pulse energies in the range 0.1 to 1 J, a larger volume discharge is required. To provide discharge uniformity in the larger scale device, some form of preionization is required. UV preionization is a most convenient approach to ionization of a small volume, although other sources such as low current e-beams could be used. Fig. 20 shows a schematic of such a system. The key features are a fast pulse discharge system, as used in the nonpreionized devices, and some source of uv preionization. Commercial devices use an independent capacitor array and high voltage pulser to provide preionization. A time delay is then provided between ionizer turn-on and the fast pulse discharge. Alternatively the ionizing array can be driven by the same capacitors used to drive the discharge. The ionizers can be made from simple automotive spark plugs, and can be positioned either as shown in fig. 20 or behind the discharge electrodes. If the ionizer is placed behind the discharge electrodes, a direct, transverse gas flow path is available, and a higher prf can be achieved. As with the nonpreionized devices, a premium is placed on a rapid rise time on the discharge voltage pulse, hence the use of low inductance circuitry. Closely coupled Blumlein capacitor arrays, low inductance cables and pulse forming networks have also been used (Sarjeant et al. 1977, 1978).

The use of uv preionization allows for an increase in laser volume, and laser operating pressure as well. Typically laser volumes in excess of 100 cm³/m of laser length can be excited by the discharge for discharge pulse times of up to 50 ns. The increase in laser volume and the increased time that the discharge pumps the rare gas mixture allows the device energies to rise to well above 100 mJ/pulse. Pressure scaling of XeF, ArF and KrF lasers has lead to single pulse energies in excess of 500 mJ with laser volumes of order 300 cm³/meter of active length. If high pressure is used, the device requires careful laser cell design, consistent with the high pressures used, as well as concommitantly higher voltage applied across the laser medium to achieve the requisite operating electric field strengths. A pulse forming network with low inductance multi-channel spark gap coupling to the discharge electrodes has been described (Sarjeant et al. 1978). Applied voltages of 110 kV at 5 atm and a discharge gap of 2.5 cm were used. The approach produces fast rise time rectangular voltage pulses of constant E/N. Independent control of voltage pulse duration and applied E/N yield optimum performance over wide pressure and composition ranges. Alternatively, larger discharge electrode spacing can be used (~5 cm) with gas pressures in the 1 to 2 atm regime to achieve energies of order 1 J (Goldhar and Swingle 1978, Watanabe et al. 1978). A paper comparing the characteristics of such lasers has recently been prepared by Sze and Loree (1978).

5.3 Optics issues for excimer lasers

The optical resonators and tuning techniques used with excimer lasers are not different in principle from any other laser system. A few significant aspects of this technology are highlighted here. Tuning of an excimer laser was first achieved for the Xe_2 laser on the broad band centered at 172 nm (Bradley et al. 1975). For operation in the vuv they found that a high quality suprasil quartz prism inserted in the cavity would allow the Xe_2 laser to be tuned over a 20 nm range. Prism tuning rather than grating tuning is preferred for $167 < \lambda < 200$ nm because high quality quartz will have greater dispersion for less insertion loss in this range. Tuning of the intense lasing bands in ArF, KrF and XeF has been achieved (Kudryavtsev and Kuzmina 1977, Barker and Loree 1977, Goldhar et al. 1977, Loree et al. 1978). With either prisms or diffraction gratings as the tunable element, the laser bandwidth decreases from about 0.3 nm to about 0.05 nm for the ArF and KrF excimers. A tuning range of about 2 nm is found. For the bound–bound XeF transition the insertion of a dispersive element can result in the selection of and oscillation on a particular narrow line. For a laser pulse duration longer than the vibrational relaxation time in the upper level, the energy normally emitted on other lines in the XeF laser can be funneled into the selected transition. However, experiments with short pulse uv preionized discharges in XeF suggest that only states of the same initial vibrational parentage (but different rotational levels) can be extracted in a 30 ns pulse (Goldhar et al. 1977). For the homogeneous ArF and KrF bands, there is a little loss of laser power when the laser is tuned near the peak of the gain curve, for bandwidths of order 0.05 nm. Tuning into the wings of the KrF or ArF gain profiles or inserting more lossy or more selective tuning elements in the cavity results in considerably decreased laser power for short pulse devices. This limits the utility of a multiple etalon tuned laser to achieve both high power and narrow linewidth. However, an injection-locked oscillator–amplifier scheme has been used (Goldhar and Murray 1977) to achieve linewidths of 0.1 cm^{-1}, on the 248 nm KrF band with high power. The oscillator produced only 10 μJ in the narrow linewidth. It is used to injection-lock a second laser that produces of order 100 mJ in the narrow band, essentially the same energy as when operated broadband. The relative timing of the two lasers is assured by powering both discharges from the same pulsed power supply through cables of appropriately chosen lengths.

As mentioned above, the tuning of the XeF system is limited because of its bound–bound nature. The practical available tuning range in KrF and ArF is limited by absorbers. For ArF the O_2 Schumann–Runge band overlaps the lasing transition (see fig. 21). Thus, the laser mixture and laser cavity external to the active medium should be kept free of oxygen. Usually a dry nitrogen purge will suffice to remove this absorption. An intrinsic absorption at 1 930 Å has been observed in ArF, (Loree et al. 1978). The nature of this absorber is unknown but appears to be due to impurities. In KrF a wide absorption at 2 488 Å has been observed in tuning and gain experiments (Hawryluk et al. 1977, Loree et al.). It has been attributed in part to absorption by the CF_2 radical (Goldhar and Murray 1977). This radical can form by electron attachment to fluorocarbons formed by reaction of fluorine oxidizers with impurities such as pump oil.

The choice of optical resonator components for an excimer laser is usually

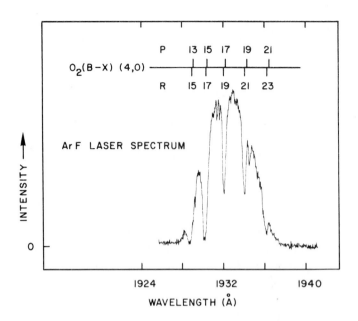

Fig. 21. ArF laser transition showing O_2 absorption. Such lasers must be purged of air. (Courtesy of R. Burnham.)

determined by the use anticipated and the size of the device. For laser demonstration and laser optimization and kinetics experiments, simple stable resonator cavities have primarily been used. However, in any situation where either high beam brightness or minimum beam divergence is required and high gain and high Fresnel number characterize the laser source, an unstable resonator is preferred (Siegmann 1974). The Fresnel number is given by $F = a^2/\lambda L$, where λ is the excimer laser wavelength, a the aperture radius and L the mirror separation. For even the small simple discharge devices, F is of order 400, a very high Fresnel number. Diffraction-limited operation of KrF and XeF lasers has been reported (McKee et al. 1977) for a negative branch unstable resonator. Positive branch confocal resonators have also been utilized. These resonators have the advantage of no focal points in the cavity, yield a collimated beam and fill the medium quite well.

The choice and reliability of reflective materials for uv excimer lasers depend strongly on the wavelength of interest. Long wavelength reflectors, $\lambda > 300\,\text{nm}$, are quite reliable and will safely operate at high power loadings. In the mid uv range, reasonable reflectivities ($\sim$99%) and damage tolerance are possible. Further development needs to be done, however. At the very short vuv wavelengths, solid Al reflectors are often used, with reflectivities of order 90%. Dielectric coatings are also available on MgF_2 substrates for short wavelengths, or on quartz for $\lambda > 170\,\text{nm}$. These have reflectivities of order 95% and absorptions of order 2%. The energy and power loadings that can be tolerated for such coatings in the vuv are low, typically less than $20\,\text{MW/cm}^2$. Russian research with LaF$_3$-based dielectric coatings has yielded high reflectivity vuv mirrors capable of withstanding power loadings up to 100 MW/cm^2 (Danilychev et al. 1976). Time-dependent mirror damage was a key element of some of the early Xe$_2$ excimer laser kinetics codes (Johnson and Gerardo 1974, Gerardo and Johnson 1975).

Acknowledgements

The author wishes to thank many co-workers at several laboratories whose discussions, figures and publications preprints proved valuable. Of specific benefit was H.T. Powell, who contributed to §§ 2 and 3, R.A. Haas who discussed media modeling, and L. Schlitt, L. Pleasance and J. Goldhar who discussed device technology. Prof. D. Setser, C.A. Brau and R.E. Center kindly provided manuscripts before publication. The help of Ms. P. Anderson in preparing this manuscript is acknowledged.

References

Ahlstrom, H.G., G. Inglesakis, J.F. Holzrichter, T. Kan, J. Jenson and A.C. Kolb (1972), Appl. Phys. letters **21**, 492.
Ault, E.R. (1975), Appl. Phys. Letters **26**, 619.
Ault, E.R., M.L. Bhaumik and N. Olson (1974), IEEE J. Quantum Electronics **10**, 624.
Ault, E.R., R.S. Bradford, Jr. and M.L. Bhaumik (1975), Appl. Phys.
Barber, A., N. Sadeghi and J.C. Pebay-Peyroula (1975), J. Phys. **B8**, 1776.
Basov, N.G., V.A. Danilychev, Y.M. Popov and D.D. Khodkevich (1970), JETP Letters **12**, 329.
Basov, N.G. V.A. Danilychev and Y.M. Popov (1971), Sov. J. Quantum Electronics **1**, 18.
Basov, N.G., E.M. Belehov, V.A. Danilichev, O.M. Kerimov, I.B. Kovsh, A.S. Podsosonnyi and A.F. Suchov (1973), Sov. Phys. – JETP **37**, 58. Letters **27**, 413.
Bender, C.F. and N. Winter (1978), Appl. Phys. Letters **33**, 29.
Berger, J.J. and S.M. Seltzer (1964), Nuclear Science Series Report No. 10, NAS-NRC Pub.-113 National Academy of Sciences, Washington, D.C.
Bhaumik, M.L. (1977), unpublished.
Bhaumik, M.L., R.S. Bradford and E.R. Ault (1976), Appl. Phys. Letters **28**, 23.
Birks, H.B. (1967), Nature **214**, 1187.
Bohme, D.K., N.G. Adams, M. Mosesman, D.B. Dunkin and E.E. Ferguson (1970), J. Chem. Phys. **52**, 5094.
Bradford, R.S., Jr., W.B. Lacina, E.R. Ault and M.L. Bhaumik (1976), Opt. Commun. **18**, 210.
Bradford, R.S., Jr., W.B. Lacina, E.R. Ault and M.L. Bhaumik (1977), Electronic Transition Lasers II (J.I. Steinfeld, ed.; MIT Press; Cambridge, Mass.).
Bradley, D.J., D.R. Hull, M.H.R. Hutchinson and M.W. McGeoch (1974), Opt. Commun. **11**, 335.
Bradley, D.J., D.R. Hull, M.H.R. Hutchinson and M.W. McGeoch (1975), Opt. Commun. **14**, 1.
Brashears, Jr., H.C. and D.W. Setser (1978), Appl. Phys. Lett. **33**, 821.
Brashears, Jr., H.C., D.W. Setser and D. Desmarteau (1977), Chem. Phys. Letters **48**, 84.
Brau, C.A. (1978A), Rare Gas Halogen Excimers, in: Excimer Lasers (C. Rhodes, ed.; Springer-Verlag; Berlin) to be published.
Brau, C.A. (1978b), in: High Power Lasers and Applications (K. Kompa and H. Walther, eds.; Springer-Verlag; Berlin).
Brau, C.A. and J.J. Ewing (1975a), Appl. Phys. Letters **27**, 435.

Brau, C.A. and J.J. Ewing (1975b), J. Chem. Phys. **63**, 464D.

Brau, C.A. and J.J. Ewing (1976), Electronic Transition Lasers (J.I. Steinfeld ed.; MIT Press; Cambridge, Mass.).

Brown, R.T. and W.L. Nighan (1978), Appl. Phys. Letters **32**, 730.

Bugaev, S.P., G.A. Mesyats and D.I. Proskvrovskii (1969), Sov. Phys.–Dokl. **14**, 605.

Burnham, R. and N. Djeu (1976), Appl. Phys. Letters **29**, 707.

Burnham, R. and N.W. Harris (1977), J. Chem. Phys. **66**, 2742.

Burnham, R. and S.K. Searles (1977), J. Chem. Phys. **67**, 5967.

Burnham, R., N.W. Harris and N. Djeu (1976a), Appl. Phys. Letters **28**, 86.

Burnham, R., F.X. Powell and N. Djeu (1976b), Appl. Phys. Letters **29**, 30.

Callear, A.B. and J.H. Connor (1972), Chem. Phys. Letters **13**, 245.

Carbone, R.J. and M.N. Litvak (1968), J. Appl. Phys. **39**, 2413.

Champagne, L.F. and N.W. Harris (1977), Appl. Phys. Letters **31**, 513.

Chen, C.H. and M.G. Payne (1978), Appl. Phys. Letters **32**, 358.

Chen, C.H., M.G. Payne and J.P. Judish (1978), unpublished.

Chen, H.L., R.E. Center, D.W. Trainor and W.I. Fyfe (1977), Appl. Phys. Letters **30**, 99.

Cheron, B., R. Scheps and A. Gallagher (1976), J. Chem. Phys. **65**, 326.

Cheshnovsky, O., B. Raz and J. Tortner (1973), J. Chem., Phys. **59**, 330.

Child, C.D. (1911), Phys. Rev. **32**, 492.

Christophorou, L.G. (1971), Atomic and Molecular Radiation Physics (Wiley-Interscience; New York).

Colbert, H.M. (1973), unpublished.

Danilychev, V.A., V.A. Dolgikh, O.M. Kerimov, S.I. Sagitov and D.B. Stavrovskii (1976), Sov. J. Quantum Electronics 989.

Daugherty, J.D. (1976), Laser Plasmas (G. Bekefi, ed.; Wiley; New York).

Daugherty, J.D., J.A. Mangano and J.H. Jacob (1976), Appl. Phys. Letters **28**, 581.

Drullinger, R.E. and M. Stock (1978), J. Chem. Phys. **68**, 5294.

Drullinger, R.E., M.M. Hessel and E.W. Smith (1977), J. Chem. Phys. **67**, 5656.

Drummond, D. and L.A. Schlie (1976), J. Chem. Phys. **65**, 3454.

Eden, J.G. (1978), Opt. Commun. **25**, 201.

Eden, J.G. and S.K. Searles (1977), Appl. Phys. Letters **30**, 387.

Eden, J.G. and R.W. Waynant (1978a), Opt. Letters **2**, 14.

Eden, J.G. and R.W. Waynant (1978b), J. Chem. Phys. **68**, 2850.

Eden, J.G., R.W. Waynant, S.K. Searles and R. Burnham (1978), Appl. Phys. Letters **32**, 733.

Elliott, C.J. and A.E. Greene (1976), J. Appl. Phys. **47**, 2946.

Ewing, J.J. and C.A. Brau (1975a), Phys. Rev. **A12**, 129.

Ewing, J.J. and C.A. Brau (1975b), Appl. Phys. Letters **27**, 350.

Ewing, J.J. and C.A. Brau (1975c), Appl. Phys. Letters **27**, 557.

Ewing, J.J. and C.A. Brau (1976), High Efficiency UV Lasers, in: Tunable Lasers and Applications (A. Mooradian et al., ed.; Springer-Verlag; Berlin).

Ewing, J.J., J.C. Swingle and A. Szoke (1977), unpublished.

Fahlen, T.S. (1978), J. Appl. Phys. **49**, 455.

Fano, U. (1973), Ann. Rev. Nucl. Sci. **13**, 1.

Finkelnburg, W. (1938), Kontinuierliche Spektren (Springer; Berlin).

Finn, T.G., L.J. Palumbo and L.F. Champagne (1978), Appl. Phys. Letters **33**, 148.

Fisher, C.H. and R.E. Center (1977), Appl. Phys. Letters **31**, 106.

Fisher, C.H. and R.E. Center (1978), J. Chem. Phys. **69**, 2011.

Flannery, M.R. and T.P. Yang (1978a), Appl. Phys. Letters **32**, 327.

Flannery, M.R. and T.P. Yang (1978b), Appl. Phys. Letters **32**, 356.

Forster, T. (1975), The Exciplex (M. Gordon and W.R. Ware, eds.; Academic Press; New York).

Friedman, M. and M. Ury (1972), Rev. Sci. Instr. **43**, 1659.

Gallagher, A. (1978), to be published.

Gaur, J.P. and L.M. Chanin (1964), Phys. Rev. **182**, 167.

Gedanken, A., J. Jortner, B. Raz and A. Szoke (1972), J. Chem. Phys. **57**, 3456.

Gerardo, J.B. and A.W. Johnson (1974), Phys. Rev. **A10**, 1204.

Gerardo, J.B. and A.W. Johnson (1975), Appl. Phys. Letters **26**, 582.
Gleason, R.E., T.D. Bonifield, J.W. Keto and G.K. Walters (1977), J. Chem. Phys. **66**, 1589.
Golde, M.F. (1975), J. Molec. Spectrosc. **58**, 261.
Golde, M.F. and B.A. Thrush (1974), Chem. Phys. Letters **29**, 485.
Goldhar, J. and J. Murray (1977), Opt. Letters **1**, 199.
Goldhar, J. and J.C. Swingle (1978), unpublished.
Goldhar, J., J. Dickie, L.P. Bradley, and L.D. Pleasance (1977), Appl. Phys. Letters **31**, 677.
Greene, A.E. and C.A. Brau (1978), IEEE Quant. Electron. **QE-14**, 951.
Hart, G.A. and S.K. Searles (1976), J. Appl. Phys. **47**, 2033.
Hawryluk, A.M., J.A. Mangano and J.H. Jacob (1977), Appl. Phys. Letters **31**, 164.
Hay, P.J. and T.H. Dunning, Jr. (1977), J. Chem. Phys. **66**, 1306.
Hazi, A.U., and T.N. Rescigno (1977), Phys. Rev. **A16**, 2376.
Hedges, R.E.M., D.L. Drummond and A. Gallagher (1972), Phys. Rev. **A6**, 1519.
Herzberg, G. (1950), Spectra of Diatomic Molecules (Van Nostrand; New York).
Hill, R.M., D.J. Eckstrom, D.C. Lorents and H.H. Nakano (1973), Appl. Phys. Letters **23**, 373.
Hoff, P.W., J.C. Swingle and C.K. Rhodes (1973), Appl. Phys. Letters **23**, 245.
Hoffman, J.M., A.K. Hayes, and G.C. Tisone (1976), Appl. Phys. Letters **28**, 538.
Hopfield, J.J. (1930a), Phys. Rev. **35**, 1133.
Hopfield, J.J. (1930b), Phys. Rev. **35**, 784.
Houtermans, F.G. (1960), Helv. Phys. Acta **33**, 933.
Huber, E.R., L.R. Jones, E.V. George and R.M. Lerner (1976), IEEE J. Quantum Electronics **12**, 353.
Huestis, D.L., R.M. Hill, H.H. Nakans and D.C. Lorents (1978), J. Chem. Phys. **69**, 5133.
Hughes, W.M., J. Shannon and R. Hunter (1974), Appl. Phys. Letters **24**, 488.
Hunter, R.O. (1977), unpublished.
Hunter, R.O., J. Oldenettel, C. Howton and M.V. McCusker (1978), J. Appl. Phys. **49**, 549.
Hutchinson, M.H.R. (1978), Paper X-9, Tenth International Quantum Electronics Conference, unpublished.
Hsia, J. (1977), Appl. Phys. Letters **30**, 101.
Hyman, H. (1977), Appl. Phys. Letters **31**, 14.
Ishchenko, V.N., V.N. Lisitsyn and A.M. Razhev (1977), Opt. Commun. **21**, 30.
Jacob, J.A. (1977), Appl. Phys. Letters **31**, 252.
Jacob, J.H. and J.A. Mangano (1976a), Appl. Phys. Letters **28**, 724.
Jacob, J.H. and J.A. Mangano (1976b), Appl. Phys. Letters **29**, 467.
Jacob, J.H., M. Rokni, J.A. Mangano and R. Brochu (1978), Appl. Phys. Letters **32**, 109.
Johnson, A.W. and J.B. Gerardo (1974), J. Appl. Phys. **45**, 867.
Johnson, R., J. Macdonald and M.A. Biondi (1978), J. Chem. Phys. **68**, 2991.
Julienne, P.S. (1978), J. Chem. Phys. **68**, 32.
Keto, J.W., R.E. Gleason, Jr. and G.K. Walters (1974), Phys. Rev. Letters **33**, 1365.
Keto, J.W., R.E. Gleason, Jr., T.D. Bonifield, G.K. Walters and F.K. Soley (1976), Chem. Phys. Letters **42**, 125.
Kligler, D., H.H. Nakano, D.L. Huestis, W.K. Bischel, R.M. Hill and C.K. Rhodes (1978), Appl. Phys. Letters **33**, 39.
Kohler, H.A., L.J. Ferderber, D.L. Redhead and P.J. Ebert (1972), Appl. Phys. Letters **21**, 198.
Kohler, H.A., L.J. Ferderber, D.L. Redhead and P.J. Ebert (1974), Phys. Rev. **A9**, 768.
Kohler, H.A., L.J. Ferderber, D.L. Redhead and P.J. Ebert (1975), Phys. Rev. **A12**, 968.
Kolts, J.H. and D.W. Setser (1978), J. Phys. Chem., **82**, 1766.
Komine, H. and R.L. Byer (1977), J. Chem. Phys. **67**, 2536.
Krupke, W.F., R.A. Haas and E.V. George (1978), Advanced Lasers for Fusion, this volume.
Kudravtsev, Y.A. and N.P. Kuzmina (1977), Appl. Phys. **13**, 107.
Lam, L.K., A. Gallagher and R. Drullinger (1978), J. Chem. Phys. **68**, 4411.
Lacina, W.B. and D.B. Cohn (1978), Appl. Phys. Letters **32**, 106.
Leichner, P.K. and R.J. Ericson (1974), Phys. Rev. **A9**, 251.
Leichner, P.K., J.D. Cook and S.J. Luerman (1975a), Phys. Rev. **A12**, 2501.
Leichner, P.K., K.F. Palmer, J.D. Cook and M. Thieneman (1975b), Phys. Rev. **A13**, 1787.
Leonard, D.A., J.C. Keck and M.M. Litvak (1963), Proc. IEEE **51**, 1785.

Levatter, J.I., J.H. Morris and S.C. Lin (1978), Appl. Phys. Letters **37**, 630.
Liu, W.F. and D.C. Conway (1974), J. Chem. Phys. **60**, 784.
Long, W.H., Jr. (1977), Appl. Phys. Letters **31**, 391.
Loree, T.R., K.B. Butterfield and D.L. Barker (1978), Appl. Phys. Letters **32**, 171.
Lorents, D.C. (1976), Physica, **826**, 19.
Lorents, D.C., D.J. Ekstrom and D. Huestis (1974), SRI Report MP73-2, unpublished.
Lorents, D.C., R.M. Hill, D.L. Huestis, M.V. McCusker and H.H. Nakano (1977), in: Electronic Transition Lasers II (J.I. Steinfeld, ed.; MIT Press; Cambridge, Mass.).
Lorents, D.C., D.L. Huestis, M.V. McCusker, H.H. Nakano and R.M. Hill (1978), J. Chem. Phys. **68**, 4657.
Maiman, T.H. (1960), Nature **187**, 493.
Mandl, A. (1970), Phys. Rev. **A3**, 251.
Mangano, J.A. and J.H. Jacob (1975), Appl. Phys. Letters **27**, 495.
Mangano, J.A., J.H. Jacob and J.B. Dodge (1976), Appl. Phys. Letters **29**, 426.
Mangano, J.A., J.H. Jacob, M. Rokni and A. Hawryluk (1977), Appl. Phys. Letters **31**, 26.
McCann, K.J. and M.R. Flannery (1977), Appl. Phys. Letters **31**, 594.
McDaniel, E.W. (1964), Collision Phenomena in Ionized Gases (Wiley; New York).
McDaniel, E.W., M.R. Flannery, H.W. Ellis, F.L. Eisele, W. Pope and T.G. Roberts (1977), Technical Report H-78-1, US Army Missile Research and Development Command, Redstone Arsenal, Alabama, unpublished.
McGeoch, M.W. and G.R. Fournier (1978), J. Appl. Phys. **49**, 2659.
McKee, T.J., B.P. Stoicheff and S.C. Wallace (1977), Appl. Phys. Letters **30**, 278.
Mehr, F.J. and M.A. Boindi (1968), Phys. Rev. **176**, 322.
Monts, D.L., L.M. Zïurys, S.M. Beck, M.G. Liverman and R.E. Smalley (1979), to be published.
Moseley, J.T., R.P. Saxon, B.A. Huber, P.C. Cosby, R. Abouaf and M. Tadjeddine (1977), J. Chem. Phys. **67**, 1659.
Murray, J.R. and H.T. Powell (1976), Appl. Phys. Letters **29**, 252.
Murray, J.R. and C.K. Rhodes (1976), J. Appl. Phys. **47**, 5041.
Nighan, W.L. (1978a), Appl. Phys. Letters **32**, 297.
Nighan, W.L. (1978b), Appl. Phys. Letters **32**, 424.
Nighan, W.L. (1978c), IEEE J. Quant. Electron. **QE-14**, 714.
Nygaard, K.J., S.R. Hunter, J. Fletcher and S.R. Foltyn (1978), Appl. Phys. Letters **32**, 351.
Palmer, A.J. (1970), J. appl. Phys. **41**, 438.
Palmer, A.J. (1976), J. Appl. Phys. **47**, 3088.
Parks, J.A. (1977a), Appl. Phys. Letters **31**, 192.
Parks, J.A. (1977b), Appl. Phys. Letters **31**, 297.
Patterson, E.L., J.B. Gerardo and A.W. Johnson (1972), Appl. Phys. Letters **21**, 293.
Peterson, L.R. and J.E. Allen (1972), J. Chem. Phys. **56**, 6068.
Phelps, A.V. (1972), Tunable Gas Lasers Utilizing Ground State Dissociation, JILA Report 110, Joint Institute for Laboratory Astrophysics, Boulder, Colorado.
Philbrick, J., F.J. Mehr and M.A. Biondi (1969), Phys. Rev. **181**, 271.
Powell, H.T. and J. Murray (1977), Lawrence Livermore Laboratory, Laser Program Annual Report UCRL-50021-76, unpublished.
Quigley, G.P. and W.M. Hughes (1978a), Appl. Phys. Letters **32**, 627.
Quigley, G.P. and W.M. Hughes (1978b), Appl. Phys. Letters **32**, 649.
Rescigno, T.N., A.V. Hazi and A.E. Orel (1978), J. Chem. Phys. **68**, 5283.
Rhodes, C.K. (1974), IEEE J. Quantum Electronics **QE–10**, 153.
Rice, J.K. and A.W. Johnson (1975), J. Chem. Phys. **63**, 5235.
Rigrod, W.W. (1978), IEEE J. Quantum Electronics **QE-14**, 377.
Rokni, M., J.H. Jacob, J.A. Mangano and R. Brochu (1977a), Appl. Phys. Letters **31**, 79.
Rokni, M., J.H. Jacob, J.A. Mangano and R. Brochu (1977b), Appl. Phys. Letters **30**, 458.
Rokni, M., J.H. Jacob, J.A. Mangano and R. Brochu (1977c), Appl. Phys. Letters **32**, 223.
Rokni, M., J.H. Jacob and J.A. Managno (1978a), Appl. Phys. Letters **32**, 622.
Rokni, M., J.H. Jacob and J.A. Mangano (1978b), Phys. Rev. **A16**, 2216.

Rokni, M. Mangano, J.A. Jacob, J.H. and J.C. Hsia (1978c), IEEE J. Quantum Electronics **QE-14**, 464.
Rothe, D.E. 1969, Phys. Rev. **177**, 93.
Rothe, D.E. and R.A. Gibson (1977), Opt. Commun. **22**, 265.
Sarjeant, W.J., A.J. Alcock and K.E. Leopold (1977), Appl. Phys. Letters **30**, 635.
Sarjeant, W.J., A.J. Alcock and K.E. Leopold (1978), IEEE J. Quantum Electronics **QE-14**, 177.
Schimitschek, E.J., J.E. Celto and J.A. Trias (1977), Appl. Phys. Letters **31**, 608.
Schlie, L.A., B.O. Guenther and R.D. Rathge (1976), Appl. Phys. Letters **28**, 393.
Schlitt, L. (1978), Lawrence Livermore Laboratory, Laser Program Annual Report UCRL-50021-76, unpublished.
Schneider, B.I. and C.A. Brau (1978), to be published.
Searles, S.K. and G.A Hart (1974), Appl. Phys. Letters **25**, 79.
Searles, S.K. and G.A. Hart (1975), Appl. Phys. Letters **27**, 243.
Shiu, V.H. (1977), Appl. Phys. Letters **31**, 50.
Shiu, Y.J. and M.A. Biondi (1977), Phys. Rev. **A16**, 494.
Shiu, Y.J., M.A. Biondi and D.P. Sipler (1977), Phys. Rev. **A15**, 494.
Siegman, A.E. (1974), Appl. Optics **13**, 353.
Smith, A.L. and P.C. Kobrinskey (1978), J. Mol. Spectrosc. **69**, 1.
Smith, D. and M.J. Copsey (1968), J. Phys. **B1**, 650.
Stevens, B. and E. Hutton (1960), Nature **186**, 1045.
Stevens, W.J., M. Gardner, A. Karo and P. Julienne (1977), J. Chem. Phys. **67**, 2860.
Sutton, D.G., S.N. Suchard, O.L. Gibb and C.P. Wang (1976), Appl. Phys. Letters **28**, 522.
Sze, R.C. and T. Loree (1978), IEEE J. Quant. Electron. **QE-14**, 944.
Sze, R.C. and P.B. Scott (1978), Appl. Phys. Letters **32**, 479.
Tamagaki, K. and D.W. Setser (1977), J. Chem. Phys. **67**, 4370.
Tanaka, Y. (1955), J. Opt. Soc. Amer. **45**, 710.
Tanaka, Y. and M. Zelikoff (1954a), Phys. Rev. **93**, 933.
Tanaka, Y. and M. Zelikoff (1954b), J. Opt. Soc. Amer. **44**, 254.
Tellinghuisen, J., A.K. Hays, J.M. Hoffman and G.C. Tisone (1976a), J. Chem. Phys. **65**, 4473.
Tellinghuisen, J., G. Tisone, J.M. Hoffman and A.K. Hays (1976b), J. Chem. Phys. **64**, 4796.
Tellinghuisen, J., J.M. Hoffman, G.C. Tisone and A.K. Hays (1976c), J. Chem. Phys. **64**, 2484.
Tellinghuisen, P.C., J. Tellinghuisen, J.A. Coxon, J.E. Velazco and D.W. Setser (1978), J. Chem. Phys. **68**, 5187.
Thompson, J.J. (1924), Phil. Mag. **47**, 337.
Tisone, G.C., A.K. Hays and J.M. Hoffman (1975), Opt. Commun. **15**, 188.
Tracy, C.J. and H.J. Oskam (1976), J. Chem. Phys. **65**, 1666.
Turner, C.E. (1977), Appl. Phys. Letters **31**, 659.
Turner, C.E., P.W. Hoff, J. Taska and L.G. Schlitt (1975), Lawrence Livermore Lab. Report UCRL-50021-75.
Vanderhof, J.A. (1978), J. Chem. Phys. **68**, 3311.
Veach, G.E. and J.H. Oskam (1970), Phys. Rev. **A2**, 1422.
Velazco, J.E. and D.W. Setser (1975), J. Chem. Phys. **62**, 1990.
Velazco, J.E., J.H. Kolts and D.W. Setser (1976), J. Chem. Phys. **65**, 3468.
Vitols, A.P. and H.J. Oskam (1973), Phys. Rev. **A8**, 1860.
Waderha, J.M. and J.N. Bardsley (1978), Appl. Phys. Letters **32**, 76.
Wadt, W.R. and P.J. Hay (1978), J. Chem. Phys. **68**, 3850.
Wadt, W.R., D.C. Cartwright and J.S. Cohen (1977a), Appl. Phys. Letters **31**, 672.
Wadt, W.R., D.C. Cartwright and J.S. Cohen (1977b), Appl. Phys. Letters **31**, 672.
Wang, C.P. (1978), Appl. Phys. Letters **32**, 300.
Wang, C.P. (1976), Rev. Sci. Instr. **47**, 92.
Wang, C.P., H. Mirels. D.G. Sutton and S.N. Suchard (1976), Appl. Phys. Letters **28**, 326.
Watanabe, S., S. Shiratori, T. Sato and H. Kashiwagi (1978), Appl. Phys. Letters **33**, 141.
Waynant, R.W. (1977), Appl. Phys. Letters **30**, 234.
Werner, C.W., E.V. George, P.W. Hoff and C.K. Rhodes (1977), IEEE J. Quantum Electronics **13**, 769.
Yariv, A., (1967), Quantum Electronics, (Wiley; New York), 264.
Zamir, E., C.W. Werner, W.P. Lapatovich and E.V. George (1975), Appl. Phys. Letters **27**, 56.

A5 | Chemical and Gasdynamic Lasers

C. J. ULTEE

United Technologies Research Center, East Hartford, Connecticut 06108

Contents

Abstract

After a general discussion of molecular and chemical lasers, individual chemical laser systems are discussed in detail. Electrically and photochemically initiated pulsed HF and HCl lasers including HF chain reaction lasers and electrical and combustion driven cw HF and HCl are reviewed in some detail, with emphasis on recent work. CO lasers using electrically produced oxygen atoms and CO flame lasers requiring no external power source are discussed. Other chemical or pseudo-chemical laser systems, including iodine and other photodissociation lasers are also reviewed. Although the gasdynamic laser is not usually considered a chemical laser, it is included in this discussion because of its similarity to combustion chemical lasers. Pertinent references to the literature are provided.

© *North-Holland Publishing Company, 1979*
Laser Handbook, edited by M.L. Stitch

1. Introduction

The chemical production of light was certainly one of the earliest practical applications of chemistry. From the cavemen's fire to the gas lamps at the turn of the century, the nature of the light produced changed very little. It was essentially the black (grey) body radiation emitted from solids heated by the combustion process, with only minor contributions from molecular and atomic spectra. The nature and intensity of the light emitted from conventional chemical sources can be characterized by a temperature which is very nearly identical to the equilibrium temperature in the combustion zone. Thus, the energy liberated by chemical reactions is usually obtained as translational energy (heat). Detailed consideration of the formation of new species in simple chemical reactions suggests, however, that the reaction energy can initially occur in the newly formed molecule in the form of internal energy, most likely in vibrational or rotational modes. Under the usual conditions of pressure and temperature, this energy is rapidly converted by collisional processes to translational energy. Experimentally, certain reactions when studied at low pressures emit light suggestive of temperatures (excitation) far in excess of the gas temperature in the reaction zone. Such radiation is referred to as chemiluminescence. Of particular interest to early chemical laser development is the work by Polanyi and co-workers on the infrared chemiluminescence of bimolecular exchange reactions (see, for example, Charters and Polanyi 1962, Polanyi 1973). In an atmospheric H_2–Cl_2 flame, the infrared emission is typically characteristic of temperatures near 2500 K, and arises from a Boltzmann distribution of vibrational and rotational populations. However, Polanyi (1960, 1961), and Charters and Polanyi (1962), in a detailed study of the $H + Cl_2$ flame at low pressure, observed vibrational spectra characteristic of 7350 K, simultaneously with rotational spectra indicating temperatures of 500 K. This work clearly showed that the HCl in such flames is not in equilibrium and is initially produced in excited vibrational states. Polanyi (1961) and Penner (1961) first suggested that this type of chemical excitation could be used to produce an infrared laser.

Operation of a laser requires some means for obtaining a population inversion (see also below), i.e., $N_u/g_u > N_l/g_\ell$, where N_u and N_l are the populations in an upper and lower state and g_u and g_ℓ their respective degeneracies. In the above context a chemical laser may be defined as a laser in which the population inversion results from the direct production of the excited states (N_u) by a chemical reaction. Historically, the definition of a chemical laser has included the production of excited states by photolysis (photo-chemistry), transfer lasers and sometimes gasdynamic lasers. In a more practical sense, a chemical laser could be defined as a laser requiring no electrical power input or generation. Most chemical lasers reported in the literature fall short of the concept of the purely chemical laser, i.e., a laser requiring no external power input. In many cases the required laser 'reagents' such as atoms are not readily storable and have to be prepared in situ by electric discharges, photolysis, electron bombardment, etc. Exceptions are flame lasers such as the CS_2-O_2 lasers, certain pulsed chain reaction lasers and combustion type lasers. From an operational viewpoint the combustion gasdynamic laser is similar to a combustion chemical laser, although its excitation mechanism is purely thermal. For completeness, a brief discussion of photodissociation lasers and gasdynamic lasers is included in this review. More recently, a new type of laser has been reported which results from the reaction of an *excited* rare gas, such as Kr^*, with a reagent, for example, F_2, to form a bound energetic molecule $(KrF)^*$ which has an unstable lower state. Since in these 'excimer' lasers electronic excitation of one of the reagents is a prerequisite, these lasers have been (admittedly somewhat arbitrarily) excluded from the present discussions.

Several criteria can be formulated for suitable reactions for chemical lasers. The reactions must obviously have the required exothermicity. Vibrational excitation to $HF(v = 1)$ corresponds to 11 kcal/mole. Reaction energies vary widely, but for simple gas phase reactions, excluding atom recombinations, values up to 50 kcal are common, making it rather easy to meet the energy requirements for vibrational lasers operating in the infrared region. On the other hand, visible and ultraviolet lasers would require excitation above 40 kcal/mole, limiting the choice of reactions with sufficient energy. Aside from this energy consideration, there are some reaction rate criteria for successful chemical lasers. If the reaction product is formed in several excited states, as for example in the case of HF (see fig. 1), then the rate of formation into an upper state must in general be larger than the rate into a lower state (see following section for a more precise definition of the state distribution requirements for molecular lasers), and also the absolute rate must be fast enough to compete with losses due to spontaneous emission and collision induced relaxation. In the case of infrared lasers the spontaneous emission rate is in the range of 10 to $10^3 \, s^{-1}$ and usually insignificant compared to the collisional relaxation rates, which have rate constants of the order 10^9 to $10^{12} \, cm^3 \, mole^{-1} \, s^{-1}$ for the hydrogen halides. Expressing the reaction rate in the usual form $k = A \exp(-E_a/RT)$, we would expect to find that chemical lasers require exothermic reactions with low activation energies (E_a) and high frequency factors (A). Again, these conditions are generally met by simple reactions involving atoms or free radicals. The requirements for visible lasers with electronic excitation are similar. As pointed out above, the energy needed is higher, and although collisional losses can be lower, the radiative losses tend to be in the 10^6 to $10^9 \, s^{-1}$ range. In addition, the large number of available vibrational and rotational levels in a molecular electronic state makes it difficult to obtain a large population excess in any given

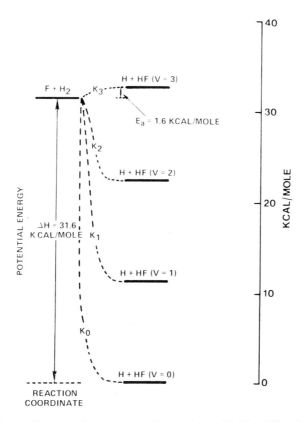

Fig. 1. Energy diagram and reaction coordinates for the $F + H_2 \rightarrow HF + H$ reaction.

state. Much effort has been expended in an attempt to find reactions which could lead to visible chemical lasers through direct electronic excitation. Some metal oxidation reactions such as $Ba + N_2O \rightarrow BaO^* + N_2$ showed considerable promise (see, for example, Jones and Broida 1974). However, more recent work by Dickenson and Zare (1974) suggests that even in these systems the chemical reaction produces vibrationally excited ground state molecules. The observed chemiluminescence apparently results from vibrational to electronic energy transfer. It is, therefore, not surprising that to date no chemical visible laser has been reported.

Although chemical lasers can be classified by reaction types (see, for example, Pimentel and Kompa 1972) this classification is not particularly convenient, since a given laser species generally can be generated by more than one reaction. In the discussion following, the more important chemical lasers and their pumping reactions are discussed individually. Since at least two monographs on chemical lasers (Gross and Bott 1972, Kompa 1973) and several bibliographies (Dobratz 1972, Arnold and Rojeska 1973, Wiswall et al. 1973) have recently appeared, no attempt has been made to include an exhaustive review of the older literature.

Emphasis here has been on a representative description of both small and large cw and pulsed devices, with some bias towards work carried out in the author's laboratory.

2. Partial inversions

The energies of a diatomic molecule in the harmonic oscillator–rigid rotator approximation are given by

$$E(v, J) = \omega(v + \tfrac{1}{2}) + BJ(J + 1), \tag{1}$$

where ω is the vibrational frequency in cm^{-1} (band center), B the rotational constant, cm^{-1}, v the vibrational quantum number and J the rotational quantum number.

For heterogeneous diatomic molecules vibration–rotation transitions are allowed for $\Delta v = 1$ and $\Delta J = \pm 1$, and the line positions $\omega(v, J)$ are then given by

$$\omega(v, J) = E(v + 1, J + 1) - E(v, J), \quad \text{R-branch}, \tag{2}$$

$$\omega(v, J) = E(v + 1, J - 1) - E(v, J), \quad \text{P-branch}. \tag{3}$$

Substitution of (1) in eqs. (2) and (3) gives

$$\omega(v, J) = \omega + 2B(J + 1), \quad J = 0, 1, 2, 3, 4, \ldots,$$

and

$$\omega(v, J) = \omega - 2BJ, \quad J = 1, 2, 3, 4.$$

Thus, the spectrum of a diatomic molecule will appear as two sets of equispaced lines a distance $2B$ apart and centered about ω. In the actual spectra there is a slight convergence of the lines owing to the interaction of rotation and vibration. Similarly, the spacing of vibrational levels decreases with increasing v owing to the anharmonicity. Fig. 2 is a spontaneous emission spectrum from HF showing these effects. For precise calculations of the frequencies a more complete expression involving Dunham coefficients should be used (see, for example, Herzberg 1950, Proch and Wanner 1971). Table 1 gives ω and B values for diatomic laser molecules. HF and DF, because of their low molecular weight, have a large B value compared to most other molecules, in particular, those not involving hydrogen.

A consequence of the rotation–vibration structure of molecules is the concept of partial population inversion, first pointed out by Polanyi (1961). An examination of various relaxation processes shows that vibration to translation (V–T) and rotation (V–R) energy transfer is generally much slower than either vibration–vibration (V–V), rotation–rotation (R–R) and rotation–translation (R–T) transfer. This suggests the possibility of a kind of pseudoequilibrium situation where the vibrational levels satisfy the Boltzmann equation with a temperature

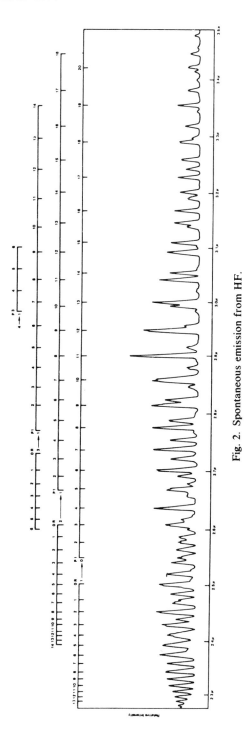

Fig. 2. Spontaneous emission from HF.

Table 1
Vibrational and rotational constants for chemical
laser molecules.

	$\omega[\mathrm{cm}^{-1}]$	$B[\mathrm{cm}^{-1}]$		$\omega[\mathrm{cm}^{-1}]$	$B[\mathrm{cm}^{-1}]$
HF	3958	20.9	HBr	2559	8.5
DF	2907	11.0	DBr	1841	4.3
HCl	2886	10.6	CO	2143	1.9
DCl	2091	5.4	OH	3570	18.9

T_v:

$$N_v = \frac{N}{Q_v} \exp\left[-\omega\left(v+\frac{1}{2}\right)hc/\,k\,T_v\right], \tag{4}$$

while rotational populations similarly follow a Boltzmann equation characterized by a temperature T_R:

$$N_{vJ} = \frac{N_v(2J+1)}{Q_R} \exp(-J[J+1]Bhc/\,k\,T_R), \tag{5}$$

where N_v is the number of molecules in a vibrational state, N the total number of molecules N_{vJ} the number of molecules in a rotational state J within the v level, Q_v and Q_R the vibrational and rotational partition functions, and h, c and k have their usual meaning. In addition, the rotational levels, because of the fast R–T processes, will be in equilibrium with the translational temperature. The two 'T' parameters defined by these equations are usually, and somewhat unfortunately, referred to as vibrational and rotational temperatures. Furthermore, in the case of vibrational populations the term T_v is also used to define, via the Boltzmann equation, the population ratio of any two levels, thus leading to the term 'negative temperatures' to designate population inversions. Neither T_v or T_R are, of course, temperatures in the true thermodynamic sense; they merely characterize the Boltzmann distribution over vibrational and rotational levels respectively.

Assuming a system characterized by T_v and T_R, the condition for gain

$$N_u/g_u > N_d/g_\ell$$

can be written as

$$N_{vJ}/g' > N_{vJ}/g,$$

where the ' indicates the upper state. Substitution of the appropriate Boltzmann equations (4) and (5) gives

$$[\exp[-\omega hc(v' + \tfrac{1}{2})/k\, T_v]]\,[\exp[-Bhc\, J'(J' + 1)/k\, T_R]]$$
$$> [\exp[-\omega hc(v + \tfrac{1}{2})/k\, T_v]]\,[\exp[-Bhc\, J(J + 1)/k\, T_R]]. \tag{6}$$

Taking logarithms and setting $\Delta v = 1$, i.e., $v' = v + 1$:

$$(-\omega hc/k\, T_v) - (Bhc\, J'(J' + 1)/k\, T_R) > (-Bhc\, J(J + 1)/k\, T_R) \tag{7}$$

or

$$\omega/T_v + BJ'(J' + 1)/T_R < BJ(J + 1)/T_R. \tag{8}$$

For P-branch transitions with $J' = J - 1$, eq. (8) gives

$$T_R/T_v < 2JB/\omega, \tag{9}$$

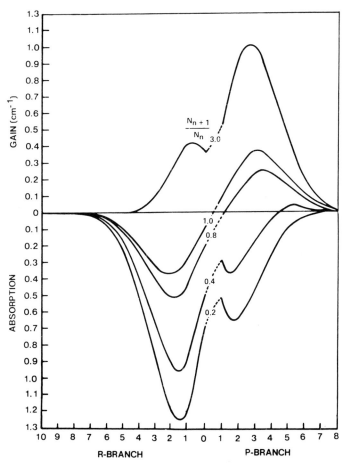

Fig. 3. Relative gain and absorption for HF. Inversion as indicated. Temperature 300 K. Pressure 1 Torr.

and for the R-branch transitions with $J' = J + 1$, eq. (8) leads to

$$T_R/T_v < -2(J + 1)B/\omega. \tag{10}$$

Since J, B and ω are always positive quantities, lasing on the P-branch can occur with T_v and T_R positive, i.e., without an actual inversion in either vibrational or rotational population. It is only required that T_v is higher than T_R by the indicated ratio. This result is a consequence of the fact that P-branch transitions have a higher J value in the lower state. It should be clear from the above derivation that an inversion, of course, exists between the populations of the actual levels involved in the transition and the term partial inversion refers to the vibrational levels only. For the R-branch a 'negative' T_v or total vibrational inversion is required for lasing. Under most conditions, molecular lasers have been found to oscillate only on P-branch transitions, even when operated with an intercavity grating to overcome the normally lower gain in the R-branch transitions. Thus, the above treatment appears to be a reasonable approximation for most chemical lasers. Deviations in terms of non-Boltzmann distributions in both rotation and vibration have been observed in pulsed lasers, and close to the nozzle in supersonic flow cw chemical lasers, and will be discussed later. Fig. 3 shows gain calculations for HF for a constant $T_R = 300$ K and various ratios of vibrational populations. For ratios smaller than or equal to one, i.e., with positive 'temperatures', gain occurs on the P-branches only. In accordance with eq. (9), the gain becomes positive at lower Js as the inversion increases. Vibrational population ratios above one, indicating negative 'temperatures' or actual vibrational inversions, provide gain on both R- and P-branches.

3. HF and DF lasers

3.1 Introduction

The HF chemical laser was first reported by Kompa and Pimentel (1967) and independently by Deutsch (1967a). Since this early work, the HF laser has by far become the most investigated chemical laser. Most HF lasers are based on the reaction $F + H_2 \rightarrow HF^* + H$. More than 60% of the reaction energy is released as vibrational excitation, making the HF laser an excellent candidate for scale-up to high powers for various applications. The large rotational constant (see table 1) makes it relatively easy to satisfy the partial inversion criterion. As a result, the HF laser is experimentally one of the easiest systems to get going. The laser output is in the 2.6 to 3.3 μm region of the infrared spectra. Unfortunately, this wavelength region coincides with strong atmospheric water vapor absorption. As a result of this, the corresponding DF laser, with an output in the 3.6 to 4.2 μm region, has also been of considerable interest for applications where atmospheric transmission is important. In spite of its favorable factors, the HF system is not without problems. In any laser system the net production of excited states is a balance between the pumping rates and the various loss processes. For the

Table 2

Representative values of hydrogen halide relaxation rates $HX(v = 1) + M \rightarrow HX(v = 0) + M$ [k in cm^3/mole s at 300 K] (selected from Cohen and Bott 1976).

Deactivator (M) Molecule	HF	DF	HCl	DCl	HBr	DBr	H$_2$	D$_2$
HF	1×10^{12}	6.5×10^{11}	–	–	–	–	1.9×10^{10}	1.5×10^9
DF	6×10^{11}	3.9×10^{11}	–	–	–	–	–	–
HCl	–	–	1.6×10^{10}	–	–	–	3.2×10^9	–
DCl	–	–	–	4.6×10^9	–	–	–	–
HBr	–	–	–	–	1.1×10^{10}	–	–	–
DBr	–	–	–	–	–	3.2×10^9	–	–

hydrogen halides and for HF in particular the vibrational relaxation rates are the highest known. Table 2 gives representative values. Relaxation rates for higher levels are even faster, as are V–V rates. The relaxation processes, as well as that part of the heat of reaction not going into vibrational energy, raise the translational and rotational temperature, resulting in a reduction of the gain. Efficient operation of a chemical laser, therefore, requires careful choice and control of temperature, pressure, flow rates, composition, photon flux, etc., to maximize the available radiative power. Considering the complexities of the chemical and physical processes and the lack of detailed data on the rates and their temperature dependences, this is indeed a formidable task. The problem has been approached in several ways. The easiest approach is that of operating the laser in a pulsed mode. The reagents can be premixed and most of the conditions can be preset, and at least some power extraction can take place before the build-up of relaxation processes, temperature, pressure, etc., becomes overwhelming. Once lasing is obtained, the operating parameters can then be readily optimized for peak laser performance. This approach has led to efficient pulsed lasers, culminating in the development of high power chain reaction pulsed lasers. The other approach has been to produce the lasing medium in a flowing stream, using aerodynamic processes to obtain the desired conditions and to remove waste HF and heat. This process had led to combustion driven multikilowatt cw HF and DF lasers. In the following discussion of HF lasers we will start with a discussion of the pumping reactions, followed by a description of pulsed and cw HF lasers.

3.2 Pumping reactions

The most important pumping reaction for the HF chemical laser is

$$F + H_2 \rightarrow HF + H, \; \Delta H = -31 \; \text{kcal/mole}. \tag{11}$$

In spite of a considerable number of studies, the absolute rate of this reaction is poorly known. Bott and Cohen (1976) have reviewed the literature and concluded that the rate must be considered uncertain by an order of magnitude. Most laser models use a rate constant of 10^{13} cm^3 mole^{-1} s^{-1} with an activation energy of 1,6 kcal/mole. The rate constant for the reaction $F + D_2 \rightarrow DF + D$ is about half this value. As indicated in fig. 1, reaction (11) can produce HF in different vibrational states; for each vibrational state characterized by the quantum number v, a reaction rate constant k_v can be specified. Two techniques have been developed to determine the product or energy distribution: chemiluminescence from a flow system, and gain measurements in pulsed chemical lasers. The first method was pioneered and developed by Polanyi and co-workers. In the most successful form of this method the reagents are introduced as diffuse crossed beams in a reaction vessel. Product molecules are removed rapidly by condensation on the liquid-nitrogen-cooled vessel walls to avoid relaxation effects. Pressures are typically in the 10^{-5} Torr range to further minimize relaxation. The cell is provided with multiple reflection mirrors to enhance collection of infrared radiation. Fig. 4 shows the cell used for $F + H_2$

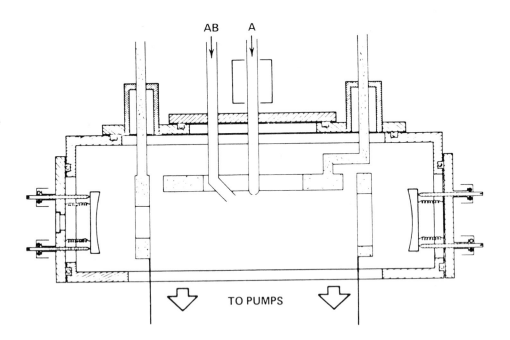

Fig. 4. Schematic drawing of reaction vessel for chemiluminescence measurements. Atomic species are formed in a microwave cavity (A), the molecular reagent is introduced through AB. Dotted portions are liquid-nitrogen cooled. The radiation is collected by the gold coated mirrors at the ends of the vessel and passes out through a sapphire window (Polanyi and Woodall 1972).

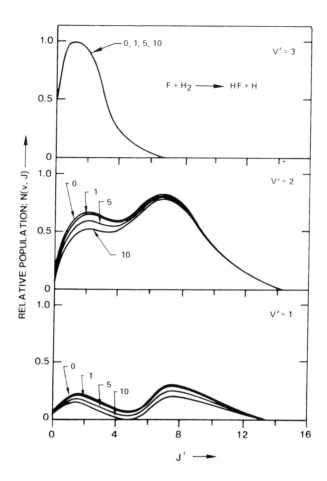

Fig. 5. Experimental rotation distribution, labeled zero, corrected for radiational relaxation on the basis of various contact times. Curve 1: $\tau = 4 \times 10^{-5}$, curve 5: $\tau = 2 \times 10^{-4}$ and curve 10: $\tau = 4 \times 10^{-4}$ s. The curves illustrate the small effect of radiational relaxation (Polanyi and Woodall 1972).

studies by Polanyi and Woodall (1972). From an analysis of the spectra, corrected for radiation and rotational relaxation, the population distributions are determined. Fig. 5 shows results obtained for $F + H_2 \rightarrow HF + H$ and the effect of the radiation corrections. The radiation corrections depend on the residence time τ of the molecule in the cone-of-sight of the spectrometer. Fig. 5 shows corrections for various values of τ and demonstrates that these corrections have only a small effect. The rotational distributions in fig. 5 show some evidence of relaxation. The maximum at high J values for $v = 1$ and 2 represents the initial product distribution. The second maximum at lower J values results from relaxation processes. As more energy goes into vibration ($v = 3$) the rotational maximum shifts to lower J values. This distribution is typical of that observed in many reactions. The second method, developed by Pimentel and co-workers, uses the gain or threshold times in a pulsed laser to determine population ratios. In the equal gain method the medium temperature is varied until the delay time between the flashlamp and laser pulse appearance is equal for two transitions. For this condition the gain equations for the two transitions are numerically equal and the vibrational population ratio can be calculated, assuming rotational equilibrium and equal linewidths. Linewidth differences can be corrected for if pressure broadening coefficients are known. In the zero gain modification of this method, a separate oscillator is used to determine the variation of gain with temperature in an 'amplifier'. The point of zero gain, found by interpolation, defines the zero gain temperature which is used to define the population ratios. Under the proper conditions the times involved are sufficiently short to avoid vibrational relaxation effects. The method does not give any information about rotational distributions. Also, in view of evidence of rotational nonequilibrium in lasers, it is not clear that the rotational equilibrium requirement is easily met. The fluorescence experiment seems to provide more detail and reliability for product distribution determinations. Details of these methods can be found in the references for the individual reactions. A detailed experimental study of the product distribution of the $F + H_2 (D_2)$ reaction was made by Polanyi and Woodall (1972). They obtain $k_1 = 0.31$ and $k_3 = 0.47$ relative to $k_2 = 1$. Other studies have given similar values (Krogh and Pimentel 1972, Jonathan et al. 1971a, b). Since the experimental studies all use emission to determine the relative concentrations, and since the $v = 0$ state does not emit, no data on k_0 are available. Theoretical calculations by Wilkins (1972) indicate $k_0 = 0$. Fig. 6

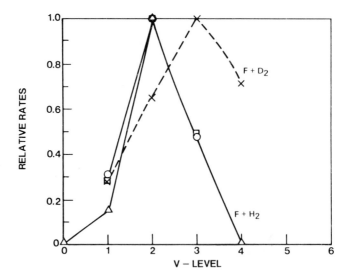

Fig. 6. Relative rates of formation of HF and DF in vibrational levels: — relative rates for $F + H_2 \rightarrow HF(v) + H$; ○ Polanyi and Woodall (1972); ∅ Anlauf et al. (1970); □ Jonathan et al. (1971a, b); △ Wilkins (1972); --×-- relative rates for $F + D_2 \rightarrow DF(v) + D$ (Polanyi and Woodall 1972).

summarizes the data for the $F + H_2$ reaction. In the case of DF the lower value of ω makes it possible to excite higher levels. Fig. 6 also shows the product distribution for this reaction. Both reactions convert about 66% of the available reaction energy into vibrational energy. A high conversion of reaction energy into vibration is important, not only from the overall efficiency standpoint, but also because the remaining energy appears as heat, raising the temperature of the laser medium. Aside from reducing the gain, this heat release can be extremely detrimental in supersonic flow devices. A number of product distributions have been reported for other reactions and are shown in fig. 7. Methane and presumably other hydrocarbons (Parker and Pimentel 1968) are similar to H_2 in their product distribution. More recent data on aliphatic hydrocarbons (Chang and Setser 1973, Duewer and Setser 1973, Johnson et al. 1973, Kim and Setser 1973, Kim et al. 1974, Parker 1975, Bogan and Setser 1976) give essentially identical results for CH_4 and similar results for other hydrocarbons, with the highest rate into the $v = 2$ state. Aromatic hydrocarbons tend to give maximum rates into $v = 1$. Moehlman and McDonald (1975) obtained $k_1 : k_2 : k_3 = 0.53 : 0.38 : 0.09$ for benzene and $k_1 : k_2 : k_3 = 0.52 : 0.34 : 0.14$ for toluene. The reaction of F atoms with HI is of interest in that it allows excitation up to $v = 6$, thus making the $F + HI$ reaction a rich source of HF radiation (Jonathan et al. 1971a, b). However, the corrosive nature of HI in mechanical vacuum pumps appears to have discouraged extensive work on this reaction. Another reaction of considerable interest is the $H + F_2 \rightarrow HF$ reaction. The exothermicity of this reaction is 98 kcal/mole, sufficient to populate the $v = 10$ level. The population distribution resulting from this reaction has been studied by Jonathan et al.

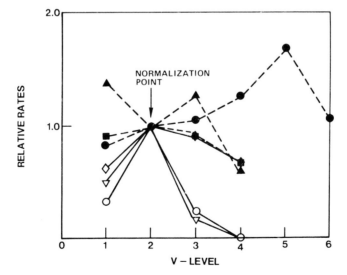

Fig. 7. Relative rates of HF(v) formation; $\bigcirc$ $F + CH_4$, Jonathan et al. (1971b); ∇ $F + HCl$, Jonathan et al. (1971b); $\diamondsuit$ $F + HBr$, Jonathan et al. (1971b); $\bullet$ $F + HI$, Jonathan et al. (1971b); $\blacksquare$ $F + H_2S$, Chang et al. (1971); $\blacktriangle$ $F + H_2O_2$, Chang et al. (1971).

Table 3
Relative rate constants for the reaction $H + F_2 \xrightarrow{k_v} HF(v) + F$.

	k_0	k_1	k_2	k_3	k_4	k_5	k_6	k_7	k_8	k_9	k_{10}
Jonathan et al. (1972)	0.04	0.09	0.11	0.13	0.45	0.90	1.00	0.45	0.20	0.04	0.04
Polanyi and Sloan (1972)	–	0.12	0.13	0.25	0.35	0.78	1.00	0.40	0.26	<0.16	–

(1972) and Polanyi and Sloan (1972), and is shown in table 3. About 50% of the reaction energy results in vibrational excitation. No study has been reported of the $D + F_2$ reaction. The reactions $F + H_2 \rightarrow HF + H$ and $H + F_2 \rightarrow HF + F$ together provide a chain mechanism for the H_2–F_2 reaction, which after initiation would not require a separate source of atoms. Although considerable evidence exists from pulsed laser work that both reactions can participate in the HF chemical laser, no *efficient* cw chain laser has been demonstrated. In part, the problem lies in the exothermicity of the reaction. Even with 50% conversion of the reaction energy into vibrational energy, the remaining energy leads to a fast heating of the medium, which in turn leads to increased reaction rates, thus resulting in a thermal run-away with detrimental results in both pulsed and cw devices. Although no detailed product distributions have been measured for other reactions, lasing has been obtained in mixtures containing F atoms and saturated hydrocarbons (Parker and Pimentel 1968, 1969, Kompa et al. 1969, Brus and Lin 1971, Green and Lin 1971a, b, Jacobson and Kimbell 1971a, b, 1973, Jacobson et al. 1973, Pearson et al. 1973, Robinson et al. 1973, Hon and Novak 1974, Obara and Fujioka 1974a, b, Patterson et al. 1974, Gerber and Patterson 1974, Patterson and Gerber 1975), olefins (Pearson et al. 1973), aromatic hydrocarbons (Pearson et al. 1973, Obara and Fujioka 1975), halogen-substituted hydrocarbons (Parker and Pimentel 1968, 1970, 1971, Brus and Lin 1971, Rice and Jensen 1972), and inorganic hydrides (Gregg et al. 1971, Pearson et al. 1973). HF lasers also have been pumped by elimination reactions such as

$$CH_3 + CF_3 \rightarrow CH_3CF_3^* \rightarrow HF^* + CH_2CF_2, \tag{12}$$

studied by Berry and Pimentel (1968), and

$$CH_3 + N_2F_4 \rightarrow CH_3NF_2^* + NF_2, \tag{13}$$

$$CH_3NF_2^* \rightarrow HF^* + CH_2NF, \tag{14}$$

$$CH_3NF_2^* \rightarrow HF^* + HCN, \tag{15}$$

studied by Padrick and Pimentel (1971, 1972a, b) and by Brus and Lin (1971). For reaction (12), gain in 2–1 and 1–0 transitions was observed in accordance with the product distribution observed by Clough et al. (1970) who report $N_1 : N_2 : N_3 = 1 : 0.43 : 0.13$. Similarly for reactions (13 to 15), only $v = 1$–0 emission is observed. Apparently the reaction energy is these reactions is distributed among the vibrational modes of both products. Since the polyatomic molecules have many more modes, only a small amount goes into the HF. A similar

Table 4
O-atom insertion reactions producing vibrationally excited HF.

	ΔH [kcal/mole]	N_2/N_1	N_1/N_0
$O(^3P) + CHF \rightarrow HFCO \rightarrow HF^* + CO$	-181		
$O(^3P) + CHF_2 \rightarrow HF_2CO \rightarrow HF^* + FCO$	-103		
$O(^3P) + CH_2F \rightarrow H_2FCO \rightarrow HF^* + HCO$	-114		
$O(^1D) + CHF_3 \rightarrow CF_3OH \rightarrow HF^* + F_2CO$	-155	0.8	
$O(^1D) + CH_2F_2 \rightarrow CF_2HOH \rightarrow HF^* + FHCO$	-148	0.8	
$O(^1D) + CH_3F \rightarrow CH_2FOH \rightarrow HF^* + H_2CO$	-143		0.35
$O(^1D) + CH_2ClF \rightarrow ClFHCOH \begin{array}{l}\rightarrow HF^* + HClO \\ \hookrightarrow HCl + HFCO\end{array}$	-145 -151	0.8 –	– –
$O(^1D) + CHCl_2F \rightarrow Cl_2FCOH \begin{array}{l}\rightarrow HF^* + Cl_2CO \\ \hookrightarrow HCl + ClFCO\end{array}$	-154 -160	0.8 –	– 0.5
$O(^1D) + CHClF_2 \rightarrow ClF_2COH \begin{array}{l}\rightarrow HF^* + ClFCO \\ \hookrightarrow HCl + F_2CO\end{array}$	-160 -168	0.8 –	– –

situation appears to exist in addition reactions such as

$$CH_2 = CH_2 + NF \rightarrow CH_2 - CH_2 \rightarrow CH_3C \equiv N + HF^*,$$
$$\underset{\substack{\diagdown \diagup \\ N \\ F}}{}$$

which in spite of its high reaction energy, $\Delta H = -116$ kcal/mole, leads only to $v = 1$–0 lasing (Padrick and Pimentel 1972a, b).

Vibrationally excited HF has also been produced by insertion reactions followed by HF elimination. Lin (1971, 1972, 1973) and Lin and Brus (1971) studied the reaction of O atoms with fluorinated methanes. The observed population ratios are given in table 4. $O(^1D)$ atoms were prepared from uv photolysis of O_3, $O(^1P)$ from uv photolysis of SO_2. The exothermicities for these reactions are very high, capable of exciting HF to very high levels if all the energy ended up in HF rather than distributed among all reaction products. From the laser output, however, it appears that again only a small fraction of the reaction energy goes into HF. Other insertion–elimination reactions only gave weak gain as reported by Roebber and Pimentel (1973) for $CF_3H + CH_2 \rightarrow [CF_3CH_3] \rightarrow CF_2 - CH_2 + HF^*$ and by Poole and Pimentel (1975) for $CF_3H + NH \rightarrow CF_3NH_2 \rightarrow CF_2 = NH + HF^*$. In both cases production of HF^* in the lower levels only was observed. Vibrational excited HF has also been produced by photoelimination reactions such as $HFCO + hv(>1650 \text{ A}) \rightarrow HFCO^* \rightarrow HF^* + CO$. A small fraction of the energy goes into HF excitation. Only $v = 2$–1 and 1–0 lasing was observed (Klimek and Berry 1973). Other photoelimination reactions for HF are given in table 10 along with those for HCl.

3.3 Pulsed HF and DF Lasers

The first discharge initiated pulsed HF laser was reported by Deutsch (1967a) followed shortly by a flashlamp initiated HF laser reported by Kompa and

Pimentel (1967). Both initiation techniques have been used extensively to study pulsed HF lasers. Arnold and Rojeska (1973), Wiswall et al. (1973), Kompa (1973) and Suchard and Airey (1976) have tabulated and summarized pulsed HF laser work up to 1973. Initiation by flashphotolysis generally gives a well-defined experiment resulting in a single pumping reaction, usually $F + H - R$. Unfortunately, F_2 absorption is low in the ultraviolet region and not very well matched with most flashlamp outputs. UF_6 (Parker and Pimentel 1968, 1970, 1971, Kompa et al. 1968, Dolgov-Savel'ev et al., 1970); MoF_6 (Dolgov-Savel'ev et al. 1970, Hess 1971a, b, Chester and Hess 1972), F_2O (Gross et al. 1968), WF_6 (Gensel et al. 1970b, Kompa and Wanner 1972), SbF_5 (Kompa et al. 1969), XeF_4 (Kompa et al. 1969), IF_5 (Gensel et al. 1970a), ClF_3 (Pan et al. 1971, Krogh and Pimentel 1972), and N_2F_4 (Gregg et al. 1971, Brus and Lin 1971) have been used to improve the initial F-atom yield in flash photolysis. All these compounds as well as F_2, are very reactive and difficult to handle. Electric discharge initiation has the advantage that it works with almost any fluorine-containing compound; it can readily be varied over a wide range of conditions, it can be done at high repetition rates giving quasi-cw outputs and generally is easier to handle experimentally. It has the big disadvantage that after the discharge, the medium is poorly defined in terms of chemical composition, temperature, initiation volume, etc. Thus, flashlamp initiation is better suited for basic studies of kinetics, relaxation, product distribution, etc., while discharge initiation is more convenient for development of lasers as a source of coherent radiation.

Jensen and Rice (1971) used a flashlamp to trigger an explosion in ClN_3–NF_3–H_2 and ClN_3–SF_6–H_2 mixtures. SF_6 and NF_3 have no appreciable absorption in the spectral region of the lamps used, and by themselves, showed no laser emission. Addition of 12 to 25% of ClN_3 led to explosion with sufficient heat generation from the reaction

$$ClN_3 \rightarrow \tfrac{1}{2} Cl_2 + \tfrac{3}{2} N_2, \quad \Delta H = -93 \text{ kcal/mole},$$

to lead to thermal dissociation of SF_6 and NF_3. Suchard and Pimentel (1971) observed DF overtone emission from a flashlamp pumped N_2F_4–CO_4 mixture. Stimulated emission on the $v = 3 \rightarrow 1$ band was observed when $\Delta v = 1$ lasing was blocked with a filter in the cavity. Krogh and Pimentel (1972) observed both HCl and HF emission from flashlamp pumped mixtures of ClF and ClF_3 with H_2. They conclude that the reaction $H + ClF \rightarrow HCl + HF^*$ leads to vibrational excitation in both HF and HCl (see also the discussion of HCl lasers).

The initial work on electrically pulsed lasers by Deutsch (1967a) was followed by a similar study (Ultee 1970) in a longitudinal pulsed discharge laser, which first reported the use of SF_6 as a source of fluorine. This same study also showed the high gain (approximately 60%/cm) available from a transversely pulsed discharge. SF_6 has become the most used and most convenient source of fluorine atoms in pulsed lasers. Jacobson and Kimbell (1971a, b, 1973, also see below) obtained the best performance with SF_6–He–C_3H_8 or –C_2H_6 mixtures. Other workers prefer SF_6 with H_2; some operate without He, depending on the type of discharge, its geometry, flow rates, etc. In the discharge laser, the SF_6 (or other fluorine compounds) are dissociated by electron bombardment. With SF_6 the main reactions are probably $SF_6 + e^- \rightarrow SF_5 + F + e^-$ (Ahearn and Hannay 1953) and $SF_6 + e^- \rightarrow SF_5^+ + F + 2e^-$ (Asundi and Craggs 1964). H_2 or hydrocarbons are always used in much smaller quantities and any H atoms formed by the discharge do not play a significant role in the laser, since the reaction $H + SF_6 \rightarrow HF + SF_5$ has a high activation energy and is too slow to contribute to HF* formation during the laser pulse. The formation of F_2 from fluorine atom recombination is unlikely to occur to any extent during the laser pulse. It is too slow to compete with the $F + R - H$ (or H_2) reaction, and no evidence of a significant contribution from the $H + F_2 \rightarrow HF + F$ reaction has been observed in pulsed lasers not containing F_2 or other chain-carrying fluorine compounds. Most of the early electrical discharge pulsed lasers involved longitudinal discharges, with the optical axis, the gas flow and the discharge all coaxial. A convenient small pulsed HF or DF laser of the coaxial type has been described by Ultee (1971a, b, 1972). The laser tube is constructed of 0.25 in o.d. Pyrex tubing (fig. 8). The small inner dimension of the tube provides a well-defined path for the discharge and appears to be critical for good pulse reproducibility. The windows are sapphire with the C-axis and Brewster angle in the proper orientation for operation with a grating. Multiline laser operating was obtained with a Cu mirror

Fig. 8. Small pulsed HF laser and power supply. The laser tube is on the right between the mirror and iris. The grating on the left completes the cavity.

with a radius of curvature anywhere from 1 to 10 m and an Irtran 4 flat. Short cavity lengths are preferred to avoid absorption of HF lines by atmospheric water vapor. In particular, the $P_2(5)$ and to a lesser extent the $P_2(1)$ and $P_2(2)$ are strongly absorbed, and may be suppressed with too much of the optical path in air. For use as a single line laser the mirror is replaced by a 300 or 600 lines/mm grating blazed at 3.0 μm.

The electric discharge pulse is supplied by discharging a small capacitor through a 50:1 pulse transformer. Most of the data reported here were obtained with approximately 0.06 J, 25 kV pulses; increasing the power input to 0.2 J did not affect the general characteristics of the laser. The main current pulse is 0.1 μs wide with a peak current of about 0.75 A (see fig. 9). The spectrum typically consists of the P(2) to P(17) lines of the 1–0 transition, the P(2) to P(9) lines of the 2–1 transition, and P(4) and/or P(5) of the 3–2 transition. The 2–1 and 1–0 transitions are dominant, and the 3–2 transition occurs only weakly. The average output on all lines under these conditions is approximately 20 mW. Table 5 shows typical results. The single line powers are somewhat lower because of the necessity of using optical stops to prevent multiple reflections and simultaneous operation on more than one line. In addition, the power in the 1–0 transition is decreased because these levels are no longer pumped by the 2–1 transition. The individual pulses are approximately triangular in shape with a width of 1 μs (see fig. 9). The 2–1 transitions occur 0.15 to 0.20 μs after the peak of the current pulse. The 0–1 pulses have a 0.4 to 0.5 μs delay time. For the 1–0 P(4) line at 2.639 μm the energy is 11 μJ/pulse with peak power of 15 W; for the 2–1 P(4) line 22 μJ/pulse and 22 W peak power are obtained. Pulse repetition rates of 100 pps for 2–1 and 40 pps for 0–1 transitions were possible with a 21 ℓ/min pump. With larger laboratory pumps increased pulse rates are readily obtained.

Although coaxial-type discharges work well with short discharge lengths for low power pulses, the electronegativity of SF_6 readily leads to the formation to filamentary discharges. Multiple-pin-type discharge lasers can be operated at higher pressures and can be scaled-up to larger powers. The multiple-pin discharge configuration also lends itself readily to fast or transverse flow

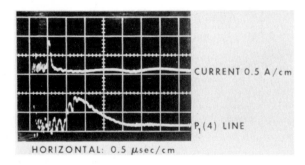

Fig. 9. Current and laser pulse.

Table 5
Typical single line output from HF probe laser.

Transition	Wavelength [μm]	Average power [mW]
1–0 P(1)	2.551	< 0.1
P(2)	2.579	0.10
P(3)	2.608	0.40
P(4)	2.640	0.40
P(5)	2.673	0.27
P(6)	2.707	0.44
P(7)	2.744	< 0.1
2–1 P(2)	2.696	0.47
P(3)	2.727	1.0
P(4)	2.760	1.35
P(5)	2.795	1.35
P(6)	2.832	0.82
P(7)	2.870	0.27
P(8)	2.911	0.10
P(9)	2.954	< 0.1

configurations for high repetition rates. Also, the fast rise times obtainable with this type can lead to very high gains (Ultee 1970). A typical multiple-pin-type laser tube used in the author's laboratory is shown in fig. 10. Jacobson and Kimbell (1971a, b, 1973) have investigated the operation of a transverse type of laser in detail with SF_6, C_2F_6, CF_4 and C_4F_8 with H_2, CF_4, C_2H_6, C_3H_8 and C_4H_{10}. Lasing was observed with all combinations. Pulse energies of 960 mJ were

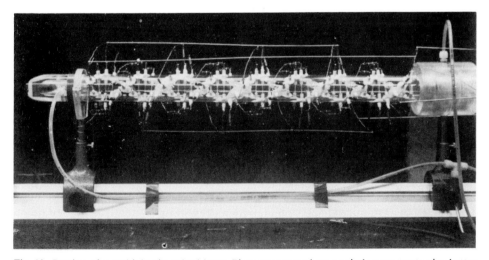

Fig. 10. Portion of a multiple-pin pulsed laser. Pins are arranged on a spiral geometry to give better uniformity. Each pin is individually ballasted.

observed at low pressure, with pulse widths of 0.7 μs; operation was possible at atmospheric pressure with reduced output. Superradiance has been reported with these lasers (Jacobson and Kimbell 1971a, b, 1973, Marcus and Carbone 1971, Goldhar et al. 1971). Operation at R-branch lines, indicative of a total vibrational inversion, has also been observed (Marcus and Carbone 1971, Green and Lin 1971a, b, Wood and Chang 1972). Pummer and Kompa (1972) have reported pulse energies of 1.2 J with peak powers of 3 MW from a 1.8 ℓ volume with a simple, spark-gap-triggered capacitor discharge. The output spectrum consisted of 1–0 P(3)–P(7), 2–1 P(3)–P(16) and the 3–2 P(2)–P(9) transitions, except for $P_1(10)$ and $P_2(10)$ which for unknown reasons were not observed. Lasing is observed over a wide pressure range with pulse lengths from 200 ns to 35 μs, depending on the gas pressure. An improved version (Pummer et al. 1973) of this laser operating with 114 Torr of SF_6 and 6 Torr of H_2 has given pulse energies of 11 J, peak powers >100 MW, with an overall (wall plug) efficiency of 4%. A scaled-down version of this laser with a small Marx bank pulse generator makes a convenient medium power pulsed laser and has been used extensively in the author's laboratory. Wenzel and Arnold (1972) have used a double discharge technique to obtain 3.5 J (10.6 J/ℓ) pulse energies. A high repetition, small transversely excited laser has been reported by Jacobson et al. (1973). Using SF_6–C_3H_8 at 45 Torr in a 10 cm long active volume and a pin-to-plate discharge configuration they obtained an average power of 7.2 W with pulse repetition rates of 1100 Hz. With multiline operation, $P_1(4)$–$P_1(9)$, $P_2(2)$–$P_2(9)$ and $P_3(3)$–$P_3(9)$ were observed. Only $v = 2 \rightarrow 1$ and $v = 1 \rightarrow 0$ lines were observed with single line operation. Single line average powers of 160 MW were observed for several lines. Voignier and Gastaud (1974) with a double discharge initiated pulsed laser using Rogowski electrodes obtained 1 J (17 J/ℓ) with an electrical efficiency of 4%. Pressures used were SF_6 = 250 Torr, H_2 = 20 Torr and He = 280 Torr. Superradiance was observed for P_2(2, 3, 5 and 6). Deka et al. (1976) report 1 J, 3 MW pulses with near-diffraction-limited quality from a transversely excited laser using an unstable oscillator. Peak radiance was 8×10^{12} W cm^{-2} sr^{-1}. Schilling and Decker (1976) have reported an oscillator–amplifier configuration for the generation of 25 to 3 ns pulses with powers in the 1 to 10 MW range. Getzinger et al. (1976, 1977) obtained, in a laser similar to that reported by Wenzel and Arnold (1976), 1 to 5 ns pulses from transversely excited SF_6–C_2H_6 mixtures using electro-optic gating. An interesting recirculating DF laser has been reported by Fradin et al. (1975). A scaled-down version of a multiple-pin configuration reported by Pummer and Kompa (1972) was used with small recirculating pump and an absorbent bed with NaOH and zeolite (molecular sieve 5A) was used to remove HF or DF. The laser will operate for periods up to an hour with an average power of 0.1 W (10 pps) with only a 20% power loss. Continuous operation could be obtained by replenishing D_2 and SF_6 at the rate of 2.7×10^{-4} and 1.3×10^{-4} mole/min respectively.

In most of the pulsed laser studies, the use of H_2 or hydrocarbons limited the laser output to the $v = 3$–2, 2–1 and 1–0 transitions. Operation with other reagents can expand the wavelength range and provide additional wavelengths for many experiments. The reaction $F + HI$ has an exothermicity of 64.5 kcal/mole, sufficient for excitation of the sixth vibrational level. Fluorescence measurements on this reaction show that vibrational levels up to $v = 6$ are populated with good efficiency (Johnathan et al. 1971a, b). Mayer et al. (1973) have operated a pulsed laser with HI and report observation of the $P_4(3, 4, 5)$, $P_5(6)$ and $P_6(4, 5)$ lines in addition to the usual P_3, P_2 and P_1 lines. Greiner (1975) in a similar study with SF_6 and HI in a transversely pulsed laser observed $P_4(4-9)$, $P_5(4-9)$ [$P_5(8)$ missing] and $P_6(5, 6)$. Computer modeling of electric discharge lasers using SF_6 and H_2 has been described by Lyman (1973).

Although the multiple-pin-type lasers overcome some of the problems of longitudinal configurations, they are eventually also limited in efficiency and performance by arc formation in the discharge. Also, from the studies of Pummer et al. (1973) it appeared that only the initial part of the input pulse is effective in generating the laser pulse; the remainder is wasted. Electron-beam (e-beam) initiation, therefore, offers several advantages. Extremely fast rise times can be obtained, relatively large volumes can be initiated homogeneously and current densities can be regulated independently of the gas properties. The first reported use of e-beam initiation for chemical lasers was by Gregg et al. (1971) for N_2F_4–H_2, N_2F_4–B_2H_6 and NF_3–H_2. Pan et al. (1971) used IF_7–H_2 mixtures. In later work, Robinson et al. (1973) reported 60 J pulses with a 250 ns half width, from an e-beam pumped SF_6 (120 Torr)–C_2H_6 (20 Torr) mixture with a 0.5 ℓ volume. Dovbysh et al. (1974) obtained 24 mJ from 20 cm^3 with SF_6–H_2 (6:1) mixtures at a total pressure of 3 atm. Recent studies of nonchain e-beam lasers are summarized in table 6. The specific energies listed in the table are those given by the authors or calculated from their data. They may or may not include corrections for that part of the volume not initiated by the electron beams. The original papers should be consulted for details. A similar problem exists for quoted electrical efficiencies, which can be based on power input from the wall plug, e-beam power absorbed by the gas, or anything in between.

In all the above lasers the laser output is essentially proportional to the number of F atoms directly produced by the discharge or the flashlamp, i.e., there is little or no contribution from a chain reaction. Since the dissociation energy of SF_6 is high, these lasers have a low electrical efficiency (<10%) and in a practical sense are not much different from a conventional nonchemical discharge laser. There is a second class of HF lasers which is much closer to the ideal chemical laser. This is the H_2–F_2 chain reaction laser. The mechanism of the H_2–F_2 chain reaction is not entirely clear. After initiation by the reaction

$$F_2 \xrightarrow{h\nu(e.d.)} 2F$$

the chain propagates via the reactions

$$F + H_2 \rightarrow HF + H,$$

$$H + F_2 \rightarrow HF + F.$$

Table 6
Pulsed nonchain e-beam chemical HF lasers.

Composition	Pressures	Input	Output	Volume	Reference
N_2F_4–H_2 N_2F_4–B_2H_6 NF_3–H_2	10–250 Torr	1.4 MeV 400 J	– 30 ns	131 cm³ –	Gregg et al. (1971)
$1F_7$–H_2 6:1	150 Torr	0–5 MeV 12 J	8.3 mJ	12 cm³ 0.7 J/ℓ	Pan et al. (1971)
SF_6–C_2H_6 6:1	140 Torr	0.6 MeV 5 kJ	60 J 250 ns	131 cm³ 460 J/ℓ	Robinson et al. (1973)
SF_6–H_2 6:1	to 3 atm	2–4 MeV ~1 kJ	24 mJ –	20 cm³ 1.2 J/ℓ	Dovbysh et al. (1974)
SF_6–C_2H_5 5:1	20 Torr	0.3 MeV 225 J	5.5 J 125–30 ns	15.7 ℓ 0.35 J/ℓ	Patterson et al. (1974)
SF_6–H_2	450 Torr		0.2 J	1 ℓ 0.2 J/ℓ	Gross and Wesner (1974)
SF_6–H_2 8:1	600 Torr	50 kV 900 J	250 mJ 200 ns	244 cc 1.0 J/ℓ	Aprahamian et al. (1974)
SF_6–C_2H_5 10:1	440 Torr	2 MeV 5.5 kJ	228 J 55 ns	36 ℓ 6.3 J/ℓ	Gerber and Patterson (1974)
SF_6–C_2H_5 10:1	330 Torr	2 MeV 5.5 kJ	380 J 23 ns	26.3 ℓ 14.1 J/ℓ	Patterson and Gerber (1975)
SF_6–H_2 2:1	to 2 atm	0.1 KeV 4–6 J	0.2 J –	120 cm³ 1.7 J/ℓ	Ponomarenko et al. (1975)
SF_6–C_2H_6 10:1	398 Torr		8 J, 116 J from osc.–amplif. 30 ns	2.7 ℓ 3 J/ℓ	Getzinger et al. (1976)
SF_6–H_2 8:1	1.75 atm	0.15 MW	0.85 J	124 cm³ 6.8 J/ℓ	Khapov et al. (1976)

The chain can be terminated by recombination reactions

$$H + H + M \rightarrow H_2 + M,$$
$$F + F + M \rightarrow F_2 + M,$$
$$F + H + M \rightarrow HF + M.$$

The experimental observation that a small initial F atom concentration can carry the reaction to completion, as well as the observation of a second explosion limit, indicates that a chain branching step exists. Such a step produces more free radicals or atoms than it uses, and thus provides a counterpart to the chain terminating steps. Sullivan et al. (1975) have reviewed the available experimental data and conclude from computer modeling that the branching reactions $HF^*(v \geq 4) + F_2 \rightarrow 2F + HF$ and $H_2^* + F_2 \rightarrow H + F + HF$ both must occur in order to explain the experimental data.

Although initial attempts of cw operation of the chain laser were unsuccessful, pulsed operation was reported at an early date by Russian workers (Batovskii et al. 1969, Basov et al. 1969a, b). Mixtures of H_2 and F_2 were initiated by flash photolysis, and lasing was observed from the $v = 6, 5, 4, 3, 2, 1$ levels. Many studies of the H_2-F_2 chain laser initiated by electric discharges, flash lamps, laser photolysis and e-beams have been reported. Detailed observations vary greatly from laboratory to laboratory, probably owing to difficulties of mixing H_2 and F_2 and the occurrence of appreciable reaction between H_2 and F_2 prior to initiation. Also, since in many cases the initiation of $F_2-H_2-O_2$ mixtures appears to be wall dominated, different vessel materials, as well as impurities, doubtlessly played an important role. There appear to be only incomplete explosion limit data for H_2-F_2 mixtures. Generally it is assumed that the diagram for a given H_2-F_2 ratio appears qualitatively like that in fig. 11. Stable mixtures (slow reaction) can be

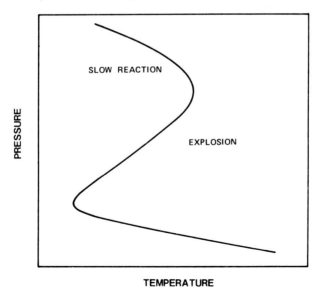

Fig. 11. Explosion limits for H_2-F_2 mixtures.

obtained under conditions represented by the left side of the diagram. Initiation of the reaction can be achieved by changing pressure or temperature across the boundary. O_2 has been widely used to stabilize H_2–F_2 mixtures. Truby (1976) has reported 'stable' mixtures up to 11 atm total pressure at room temperature (see fig. 12). However, the effects of surface conditions on the reported explosion limits raise doubt that, even at these pressures, the observed limits are the classical third limit. Similarly, Greiner (1973, 1974) reported explosions with mixtures which earlier in the same equipment had been stable, again indicating that significant surface effects can occur.

Table 7 summarizes results obtained by electric discharge, e-beam and photo-initiation of $H_2(D_2)$–F_2 mixtures. Highest chemical and electrical efficiencies have been obtained with e-beam initiation. Aprahamian et al. (1974) reported pulse energies of 50 J (12 J/ℓ) for equimolar mixtures of F_2 and H_2 with some SF_6 at a total pressure of 240 Torr. Similar mixtures at 420 Torr in a smaller volume gave 11 J at 46 J/ℓ. Modeling of the experiment predicted 1 kJ/ℓ atm. Gerber et al. (1974) have reported 2.3 kJ (96 J/ℓ) using H_2–F_2–O_2–SF_6 mixtures. Electrical efficiencies of 200% were observed. Mangano et al. (1975) obtained 5 J of 51 J/ℓ with a 1 atm F_2–H_2–He–O_2 = 30:8:61:1 mixture. Their best performance was obtained with the highest percentage fluorine. The electrical efficiency *based on the effective* volume was 875%. Pulse lengths varied from 1 to 10 μs. The best performance reported so far (Gerber and Patterson 1975) is 4.2 kJ (130 J/ℓ) at 1240 Torr with F_2–H_2–O_2–SF_6 mixtures excited by a 2 MeV, 50 kA 70 ns fwhm e-beam pulse. A magnetic field was used to guide the electrons into the laser

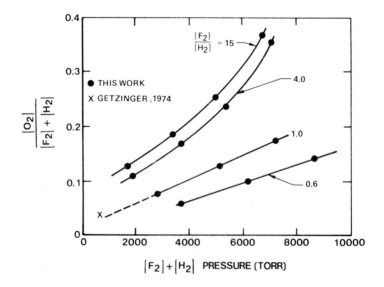

Fig. 12. The O_2–[F_2 + H_2] ratio required to achieve stability as a function of F_2 + H_2 pressure for various values of F_2 to H_2. Explosions occur at and below each curve (Truby 1976).

Table 7
Pulsed H_2–F_2 chain lasers.

Initiation	Pressure	F₂	H₂	O₂	Other	Energy out	Specific energy	Chemical efficiency [%]	Comments	Reference
Electric discharge	—	1	1					0.2		Batovskii et al. (1969)
Electric discharge and photolysis	—								No details	Basov et al. (1969a, b)
Electric discharge									No details	Burmasov et al. (1969)
Photolysis	120 Torr	1	1		He 30				Output spectra at 20 Torr	Dolgov-Savel'ev et al. (1970)
Photolysis	50 Torr	3	1		He 10	4.4 mJ				Hess (1971a)
Electric discharge	4 Torr	9	1		He 60				HF and DF spectra–DF–CO_2 transfer results	Basov et al. (1971)
Photolysis	128 Torr	3	1		He 60	8.8 mJ	0.1 J/ℓ	0.8	With MoF_6–H_2–He = 1:1:53, 12% chem. efficiency	Hess (1971b)
Photolysis	50 Torr	1	1		He 60				Output spectra at 50 torr. Appearance sequence of lines	Suchard et al. (1971)
Photolysis	1.5 atm	1	1		N_2 17	9.6 mJ		0.07	Also DF–CO_2 transfer laser at 350 mJ	Wilson and Stephenson (1972)
Photolysis	40 Torr	1	1			0.1 J	2.1 J/ℓ	~0.3	Initiation at 100 K, output spectra	Dolgov-Savel'ev et al. (1972)
Photolysis	500 Torr	1	1		He 40, MoF_6 0.25	65 mJ	3.3 J/ℓ	0.8	Range of pressures and compositions	Hess (1972)
Photolysis	50 Torr	1	1		He 40				Comparison: experiment and theory	Suchard et al. (1972a, b)
Electric discharge	120 Torr	1	1	0.06	He 10	10.6 mJ	0.9 J/ℓ	0.3	Double discharge	Parker and Stephens (1973)
	36 Torr	1	1	0.25	He 10	16.1	1.3 J/ℓ	2	Appearance sequence for individual lines	

Table 7 contd.

Initiation	Pressure/composition				Energy out	Specific energy	Chemical efficiency [%]	Comments	Reference
	F_2	H_2	O_2	Other					
Electric discharge	1	0.7	0.08	He 15	73 mJ	0.1 J/ℓ	0.07	Electrical efficiency 115%	Kerber et al. (1973a, b)
Photolysis	1	0.5		He 40				Lasing observed from high v levels. Temporal sequence of lines	Suchard (1973a. b)
Photolysis	7	2	1	He 1 SF$_6$ 1	0.83 J	21 J/ℓ	0.8	FWHM 250 ns	Greiner (1973)
E-beam	1	1		SF$_6$ 1.3	50 J	12 J/ℓ	0.5	HF and DF. DF $\frac{1}{2}$ to $\frac{1}{3}$ the HF output	Aprahamian et al. (1974)
	1	1		1.5 SF$_6$ 1	11 J	46 J/ℓ	1.3		
E-beam	3.6	1	1.4	1	2.3 kJ	96 J/ℓ	3.3	Electrical efficiency 178%; chemical efficiency at lower output; electrical efficiency up to 200% observed	Gerber et al. (1974)
E-beam	2.7	3	0.5		12 J	16 J/ℓ	0.25	FWHM 60 ns. 14 J with 100 Torr SF$_6$ added	Greiner et al. (1974)
Electric discharge	1	2						HF and DF output spectra	Batovskii and Gur'ev (1974a)

Pressure figures (Torr) per row: 120 Torr, 50 Torr, 560 Torr, 240 Torr, 420 Torr, 700 Torr, 620 Torr, 6 Torr.

Electric discharge	240 Torr	1	0.23	0.08	He 12	0.15 J	0.24 J/ℓ	0.01	Electrical efficiency 144%	Whittier and Kerber (1974)
Photolysis	90 Torr	1	1		N_2 10, He 8				Time resolved spectral data	Nichols et al. (1974)
Photolysis	1.1 atm	1	1			8 J	80 J/ℓ	4	R-lines observed. 10% of output in far ir	Chen et al. (1974)
Photolysis	250 Torr	10	3	5		36 J	8 J/ℓ			Batovskii and Gur'ev (1974b)
E-beam	1 atm	30	8	1	He 61	5.3 J	51 J/ℓ	3	Electrical efficiency 875%	Mangano et al. (1975)
Electric discharge	300 Torr	4.5	D_2 2	6	He 87				DF output spectra. DF–CO_2 transfer laser	Turner and Poehler (1976)
E-beam Preionization	1 atm	6	3	0.6	He 54, Ar 37	2.5 J	42 J/ℓ	6.3	Electrical efficiency 148%	Hofland et al. (1976)
E-beam	1240 Torr	5	1.25	1.5	SF_6 2	4.2 kJ	130 J/ℓ	3.5	Chemical efficiencies up to 11%; electrical efficiencies up to 180%	Gerber and Patterson (1975)
Photolysis	1 atm	10	1.2	0.7	A 88	292 J	22.8 J/ℓ	8.1	DF 144 J	Nichols et al. (1976)

volume (see fig. 13). Fig. 14 shows the effect of SF_6 addition. Chemical efficiencies up to 11% and electrical efficiencies up to 180% are also reported (see fig. 15).

Flash photolysis initiation has produced 292 J (22.8 J/ℓ) at an efficiency of 8.1% (Nichols et al. 1976). Highest specific energy in flash photolysis was reported by Chen et al. (1974), who obtained 8 J in a 0.1 ℓ volume. In addition to the usual P-branch lines, Chen et al. (1974) also report the observation of several R-lines, indicating the presence of a total vibrational inversion. They also found that 10% of the total energy output falls in the range above 15 μm, presumably owing to pure rotational transitions (see below). The spectral distribution observed by various authors is shown in table 8. Again there appears to be a wide diversity of results. Early Russian workers reported observations of lasing from levels up to $v = 6$, but generally no lasing from the $v = 1$ state. This lack is probably due to prereaction of their uninhibited F_2–H_2 mixtures, which leads to strong absorption of the 1–0 transition by ground state HF. A number of later studies failed to observe radiation from higher levels, probably because of coupling optimization for maximum power, which lead to cavity Qs too low for observation of the low gain transitions. Wide distributions over J values have been observed, with a general trend to lower J values for the higher vibrational transitions. This is in agreement with the distributions expected from the chemical reactions. As more energy is deposited in the vibrational modes, the fraction going into rotation becomes lower, leading to narrow, low J distributions in the higher vibrational states. Rotational equilibration rates are slow (Hinchen 1975, 1976) and the observed spectral distribution shows clearly that nonequilibrium distributions are present even in the high pressure lasers. The

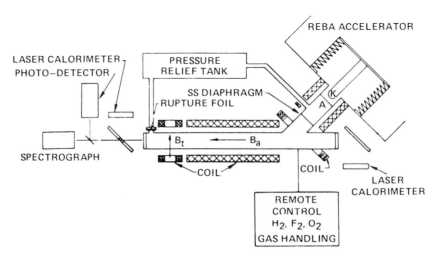

Fig. 13. Schematic drawing of electron beam pumped HF laser (Gerber and Patterson 1976).

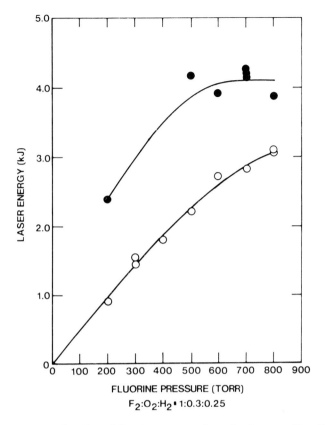

Fig. 14. Laser energy as a function of fluorine pressure for a fixed composition. The lower curve represents mixtures without SF_6; the upper curve is for mixtures with 100 Torr SF_6 added (Gerber and Patterson 1976).

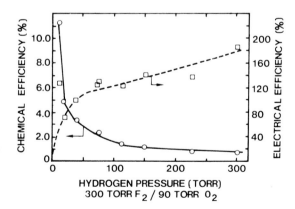

Fig. 15. Chemical and electrical efficiencies as a function of hydrogen pressure with 300 Torr F_2 and 90 Torr D_2 (Gerber and Patterson 1976).

Table 8
Spectral distribution of pulsed chain HF lasers.

Vibrational levels	P(J)	cm⁻¹	(1)	(2)	(3)	(4)	(5)	(6)	(7)	(8)	(9)	(10)	(11)	(12)
1–0	1	3920.5												
	2	3877.8												
	3	3833.8				x		x					x	x
	4	3788.3				x		x			x		x	x
	5	3741.6				x		x	x		x	x	x	x
	6	3693.5				x		x	x		x	x	x	x
	7	3644.2				x		x	x	x	x	x		x
	8	3593.8				x		x	x		x	x	x	x
	9	3542.2			x			x	x		x	x	x	x
	10	3489.6						x	x	x	x	x		
	11	3436.0								x	x			
	12	3381.5						x			x			
	13	3326.1						x			x			
	14	3269.8						x						
	15	3212.8									x*			
2–1	1	3749.9				x							x	R(0)
	2	3708.8				x		x	x				x	
	3	3666.4			x	x	x	x	x				x	x
	4	3622.6	x		x	x	x	x	x		x	x	x	x
	5	3577.5	x		x	x	x	x	x	x	x	x	x	x
	6	3531.2	x		x	x	x	x	x	x	x	x	x	x
	7	3483.7	x		x	x	x	x	x	x	x	x		x
	8	3435.0	x		x		x	x	x	x	x		x	x
	9	3385.3	x		x		x	x	x		x			x
	10	3334.5	x	x			x	x			x			x
	11	3282.8	x		x		x	x			x			x
	12	3230.1	x	x	x		x	x			x			x
	13	3176.6		x		x	x	x			x			x
	14	3122.3									x*			
	15	3067.3											x	x*
3–2	1	3584.0				x					x			
	2	3544.5				x					x			
	3	3503.6			x	x		x			x			
	4	3461.4		x	x	x		x	x		x			x
	5	3418.0	x	x	x	x	x	x	x	x	x		x	x
	6	3373.3		x	x	x	x	x	x	x	x		x	x
	7	3327.5		x	x	x	x	x	x	x	x			x
	8	3280.6				x	x	x	x		x			x
	9	3232.6				x	x	x	x		x			
	10	3183.6				x					x			
	11	3133.6				x					x			
	12	3082.8				x					x			
	13	3031.1												
	14	2978.6												
	15	2925.4												
4–3	1	3422.3				x			x					R(1)
	2	3384.3				x			x				x	
	3	3344.9				x			x	x				
	4	3304.3							x					x
	5	3262.5			x			x	x	x				x
	6	3219.4			x			x	x					x
	7	3175.2			x		x	x	x		x			x

Table 8 contd.

Vibrational levels	P(J)	cm⁻¹	(1)	(2)	(3)	(4)	(5)	(6)	(7)	(8)	(9)	(10)	(11)	(12)
	8	3130.0	x	x			x	x			x			
	9	3093.6	x					x	x		x			
	10	3036.3						x			x			
	11	2988.1												
	12	2939.0												
	13	2889.0												
	14	2838.3												
	15	2786.9												
5–4	1	3264.2							x					R(0)
	2	3227.6					x		x					
	3	3189.8							x					x
	4	3150.7	x	x					x		x			x
	5	3110.4	x	x				x	x					x
	6	3068.9	x	x				x	x					x
	7	3026.3	x	x				x	x		x			
	8	2982.6		x				x	x		x			
	9	2937.9		x					x		x			
	10	2892.2									x			
	11	2845.6												
	12	2798.1												
	13	2749.8									x			
	14	2700.8									x			
	15	2651.0									x*			
6–5	1	3109.1							x					x
	2	3074.0							x					x
	3	3037.6							x		x			x
	4	2999.9	x						x					x
	5	2961.1	x	x				x			x			x
	6	2921.1		x				x						
	7	2880.0		x										
	8	2837.9		x										
	9	2794.7												
	10	2750.6									x			
	11	2705.6												
	12	2659.7												
	13	2612.9												
	14	2565.5												
	15	2517.3												

* Additional lines were observed.

[1] Basov et al. (1969a, b)
[2] Dolgov-Savel'ev et al. (1970)
[3] Basov et al. (1971)
[4] Suchard et al. (1971)
[5] Dolgov-Savel'ev et al. (1972)
[6] Parker and Stephens (1973)
[7] Suchard (1973a, b)
[8] Aprahamian et al. (1974)
[9] Gerber et al. (1974)
[10] Batovskii and Gurev (1974a, b)
[11] Nichols et al. (1974)
[12] Chen et al. (1974)

temporal development of the laser pulse is extremely complex. Fig. 16 shows the time resolved spectroscopy of a pulsed HF laser as observed by Suchard (1973a, b). The simultaneous lasing of several J lines in a given vibrational band is an indication that rotational equilibration is incomplete.

The chain reaction HF laser has a total chemical energy release of 66.7 kcal/per mole of HF formed. This is equivalent to 12.5 kJ/ℓ for an equimolar mixture of H_2 and F_2 at 1 atm. Of this energy, 7 kJ/ℓ goes into vibration. The highest chemical efficiency observed is 11.4%, obtained at 400 Torr with a F_2–H_2 ratio of 30 (50 J/ℓ atm). The reasons for these low efficiencies are not completely understood. Modeling and comparisons between theory and experimental results have been reported by Hess (1971a, b), Dolgov-Savel'ev et al. (1972), Kerber et al. (1972), Suchard et al. (1972a, b) and Hough and Kerber (1975). The complete modeling of the pulsed chain laser is difficult because of the changing composition and temperatures during the pulse. Since the reaction and relaxation kinetics and the line broadening coefficients are not well known over the ranges of temperatures and composition present, modeling has met with only limited success. The laser pulse is much shorter than the reaction time, probably as a result of the increasing temperature and relaxation as HF is generated by the mixture. The vibrational relaxation appears particularly severe for the higher vibrational levels, and is postulated to be responsible for the fact that the most

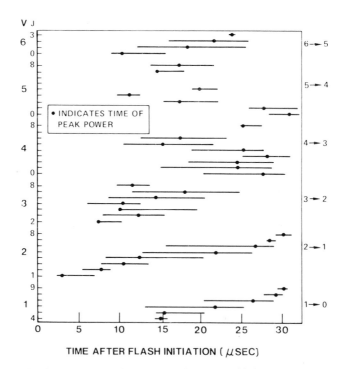

Fig. 16. Time resolved spectroscopy of the observed laser transitions of H_2–F_2–He 0.5 : 1 : 40 mixture. Total pressure 50 Torr, output coupling 10% (Suchard 1973a, b).

energy is contained in the 3–2, 3–1 and 1–0 lines. Although chain branching via $HF(\nu \geq 4) + F_2 \to 2F + HF(\nu = 0)$ has also been suggested for the low energy in the higher transitions, present evidence tends to favor the relaxation as the main loss channel (Chen et al. 1974).

3.4 Pulsed DF–CO₂ transfer lasers

The high relaxation rates of HF and DF favor low pressure operation. Although premixed pulsed lasers can operate at high pressure, the pulse lengths tend to be short ($<1 \mu$s) and more importantly, lasing is terminated long before the reaction between H_2 and F_2 is completed. Finally, the HF or DF laser is essentially a two-level system, always operating with substantial lower-level populations. One of the early DF laser studies (Gross 1969) circumvented these problems by transferring the DF excitation to CO_2 and extracting the optical power at 10.6μm. The DF vibrational frequency is close to the CO_2 001 level (see fig. 41 in the GDL section), resulting in a fast energy transfer:

$$DF(\nu) + CO_2(000) \to DF(\nu - 1) + CO_2(001).$$

Stephens and Cool (1972) have measured the rate at 350 K and obtained 3.8×10^{12} cm³/mole s for $v = 1$, compared to 3.9×10^{11} cm³/mole s for DF($v = 1$) relaxation. The best 'match' is obtained for DF $v = 8$, $J = 0$; thus the rate for higher v's is even faster (Airey and Smith 1972). The important point is that effective DF transfer to CO_2 (001) occurs even though the partial inversion of the DF is insufficient to allow lasing at 3.5μm. The complete description of the transfer system is complex. Kerber et al. (1973a, b) have made a detailed analysis of the DF–CO₂ transfer laser. In addition to the DF to CO_2 transfer reaction, the vibrational deactivation of CO_2 by DF is an important factor not present in the nontransfer CO_2 laser. They predict chemical efficiencies in the 6–8% range with 10.6μm pulse lengths up to 260μs for a 50 Torr mixture.

Poehler et al. (1972, 1973) have observed a chemical efficiency of 15%, based on the active volume, for 5 J pulses of 30μs duration in a D_2–F_2–CO_2–He = 1:1:6:19 mixture at 1 atm. Pulse duration and energy were a strong function of pressure and were in reasonable agreement with predictions from a kinetic model. Suchard et al. (1972a, b) obtained 2.8 J (9.6 J/ℓ) from D_2–F_2–CO_2–He = 0.33:1:8:10 at 0.5 atm with a chemical efficiency of 5%. Other pulsed DF–CO₂ transfer lasers have been reported by Basov et al. (1971), Wilson and Stephenson (1972), Poehler and Walker (1973) and Turner and Poehler (1976).

3.5 CW HF and DF lasers

In the pulsed chemical lasers the reagents are premixed at the desired concentrations and temperature, and each laser pulse is essentially obtained with a fresh mixture. Both heat and ground state HF are removed between pulses with the spent gas. Ground state HF is particularly detrimental because of the extremely fast V–T and V–V rates. Operation of cw lasers requires, therefore, as a minimum, fast flow to remove both HF and heat. Since the reaction rates of necessity have to be fast, the reagents generally are mixed in the optical cavity region. Again, mixing must be fast, or the reaction stream will be out of the optical region before the laser reaction has taken place. Also, the HF* concentration must be built up fast enough for the laser to reach threshold in competition with the relaxation processes. The mixing generally is the rate-determining step and tends to set the maximum operating pressure of the cw mixing type lasers. Early attempts to obtain lasing from a diffusion hydrogen fluorine flame met with failure. In 1969 Airey and McKay observed 1.8 ms laser pulses by mixing HCl with F atoms obtained in a reflected shock and expanded through a supersonic nozzle. The pulse length was orders of magnitude longer than the pulses obtained in typical flashlamp or discharge pulsed lasers and was limited only by the flow time of the shock tube.

The first truly cw HF laser was demonstrated by Spencer et al. (1969) using arc heated N_2 to dissociate SF_6 and a supersonic mixing configuration. Since that time many cw HF lasers have been reported. They can be divided into two groups: (1) hybrid cw chemical lasers, requiring nonchemical energy input and (2) pure chemical HF lasers.

In the high power hybrid chemical lasers F atoms are obtained by high temperature dissociation of F_2 or SF_6 with an arc or arc heated gas. The gas mixture is then expanded through a supersonic nozzle to lower the temperature without significant recombination of the fluorine atoms. The H_2 or other reactant is then mixed into the supersonic stream in or just prior to the optical region. For lower powers it is possible to obtain the F atoms with a low pressure, low temperature discharge. In this case little or no cooling is required and much slower subsonic flows can be used. Because of the slower flows, these lasers generate lower powers and require only modest pumping speeds. They are ideally suited for a variety of laboratory applications. Gross and Spencer (1976) have given detailed descriptions of the arc driven supersonic diffusion lasers and have given an excellent summary of the literature through 1974. The present discussion will be primarily concerned with the small discharge cw HF lasers and with more recent work on the pure chemical or combustion HF lasers.

As discussed earlier, the operation of a cw chemical laser requires the experimental definition of a set of conditions such that the rate of excited state production exceeds the losses due to various relaxation processes. In practice it has turned out to be very difficult to obtain the required conditions by a cw discharge of premixed reagents. In general, the problem appears to be due to difficulties of simultaneously tailoring the discharge to the proper e/p and electron temperature with fluorine-containing gases, and preventing the build-up of HF and discharge heating of the medium. The only premixed cw HF laser reported has been described by Buczek et al. (1970). It uses a transverse electric discharge maintained in a fast (8 m/s) premixed SF_6, H_2 and He flow. The fast flow removes heat and HF. To avoid the 'blowing out' of discharge in the flow

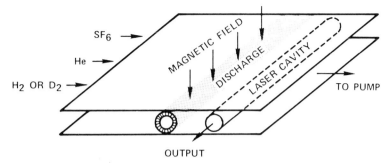

Fig. 17. Magnetically stabilized dc discharge HF laser (Buczek et al. 1970).

direction a weak magnetic field is used to maintain the discharge parallel to the optical axis and optimize the distance between discharge and optical axis (see fig. 17). Using a 300 ℓ/s pump, 800 mW of HF laser output was obtained.

Other approaches have relied on F production from SF_6 in either rf (or microwave) or dc discharges. RF discharges have been used by Cool et al. (1970), Stephens and Cool (1971), Rosen et al. (1973), Glaze and Linford (1973), Gagne et al. (1974, 1975), Conturie et al. (1976) and Bertrand et al. (1977). DC discharges have been used by Hinchen and Banas (1970), Hinchen (1974) and Proch et al. (1975). The most convenient and best characterized laser is the one first described by Hinchen (1974), which has been used extensively for gain, absorption and other measurements. Fig. 18 shows a diagram of the laser. The fluorine atoms are produced in a water-cooled Pyrex tube 55×2.5 cm i.d. The cathode consists of ten 10×0.03 cm nickel rods individually ballasted with 400 kΩ resistors. The anode is 5×1.9 cm i.d. copper tube. A dc discharge of approximately 10 kV, 125 mA is maintained across the electrodes. The mixing section of the laser consists of a smooth transition from the cylindrical cross section of the discharge to the rectangular mixing flow channel. The flow channel

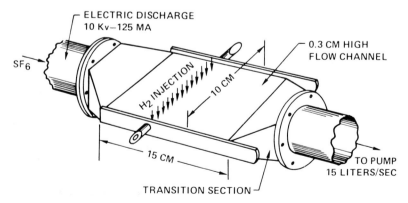

Fig. 18. Electric discharge HF laser (Hinchen 1974).

is a single electrodeposited nickel piece with a 0.3×10 cm i.d. cross section. Hydrogen or deuterium is introduced through a series of 0.025 cm diameter holes in the top and bottom of the channel. The whole nickel mixing and flow section is water cooled. No attempt has been made to passivate the walls. However, the discharge through SF_6 produces elemental sulfur, and much of the metallic surface is covered with a thin layer of sulfur. Although the sulfur deposition can be minimized by addition of a small amount of O_2 (see below) it is not completely eliminated. The sides of the mixing channel consist of flat copper pieces with Brewster's angle window holders. The side plates are movable in order to optimize the optical cavity axis with respect to the mixing zone. A small He purge flow keeps the window channels free from ground state HF. SF_6 and He are introduced into the discharge section through a glass wool plug wedged between the ten cathode pins. This apparently smoothes out the flow across the tube cross section and greatly increases the discharge and laser stability. Optimum gas flows for maximum HF multimode power were: SF_6, 2.4 mmole/s; He, 5.8 mmole/s; and H_2, 0.8 mmole/s. A small amount of O_2 (0.5 mmole/s) will decrease sulfur deposition and enhance 2–1 lines. Optimization of individual lines can be achieved by small changes in the gas flows. Although cooling and relaxation phenomena obviously play a role, the details are not well understood. Pressures in the mixing region are from 6 to 20 Torr, depending on the detailed gas flows. The laser is pumped with two 7 ℓ/s mechanical pumps.

For DF similar flows were used. However, in the case of DF, addition of O_2 did not increase the power. The optical cavity for multimode output consisted of a total reflector and a 75% reflecting output mirror. Total power was 1.3 W on HF, 1.0 W on DF. For single line operation, the cavity is made up of a 600 lines/mm Littrow mounted blazed grating and a 2 m radius partially transmitting mirror mounted on a piezoelectric crystal. The mirrors are mounted in an Invar rod structure independent of the laser body. The short optical length of the laser allows a mirror spacing such that the longitudinal mode spacing ($c/2L$) is larger than the Doppler width of the laser lines, about 380 Hz for HF. Higher-order transverse modes are suppressed by intracavity irises. Fig. 19 shows a typical beam profile taken 70 cm from the mirror exit surface. Heterodyne experiments using two of these lasers indicated frequency stabilities of ~ 5 parts in 10^8 in a typical laboratory environment. For a pneumatically isolated laser, Fabry–Perot interferometer measurements gave frequency stabilities of 1 part in 10^8 over a 150 ms time span, 3 parts in 10^8 over a 660 ms time span (Hinchen and Freiberg 1976). Free running, the laser has a long-term amplitude stability of several percent. In order to improve on this the laser can readily be locked to the minimum of the Lamb dip and excellent long-term stability can be achieved. Considerable higher power is obtained from the above laser by increasing the flow through the laser. Proch et al. (1975) used a similar subsonic configuration. F atoms are produced from SF_6 in Ar with two 1 m $\times 5$ cm i.d. electric discharges. The flow from the discharges expands into a rectangular 40×1.5 cm flow channel, which is connected to pumps with a total capacity of 1250 ℓ/s. H_2 or D_2 is injected through a movable spray bar. The laser operates typically at 2

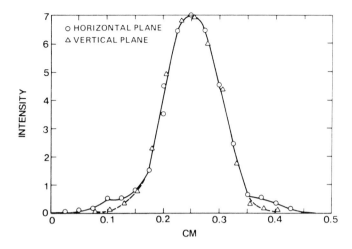

Fig. 19. Beam profile of $P_1(4)$ line from a small electric discharge HF laser (Hinchen 1974).

to 3 Torr in the cavity, with a linear flow velocity of 2×10^4 cm/s. Output power is 40 W for HF, 16 W for DF on the usual transitions. A more elaborate approach to the mixing has been described by Glaze and Linford (1973). They use a water-cooled 1×14 cm nozzle block with 47 individual fluorine ducts and 46 0.1×1 cm H_2-injector ducts. The output power is 10 W with a pump capacity of 95 ℓ/s.

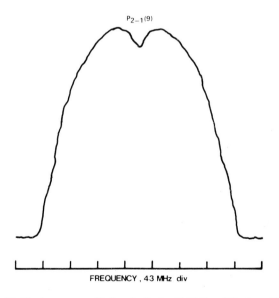

Fig. 20. Tuning curve with Lamb dip for $P_2(9)$ line (Hinchen 1974).

In general, the power output of these lasers scales approximately with the pump capacity. Exceptions are the longitudinal configuration of Cool et al. (1970), which had much slower flow and generated only 50 mW with a 200 ℓ/s pump, and the configuration of Glaze and Linford (1973) discussed above.

The use of electric discharges, arcs, or electric heaters for cw chemical lasers is inefficient and essentially reduces the chemical laser to a special type of electrically pumped laser. In addition, the scale-up of these types of lasers to high power levels is often difficult, if not impossible. Since all HF lasers require either F- or H-atoms, and since neither of these species have any long-term chemical stability at room temperature, a chemical generation process is required. Cool et al. (1970) used the low pressure, low temperature reaction

$$NO + F_2 \rightarrow NOF + F,$$

to generate the required F atoms. Although lasing on HF was observed when the F–NOF stream was mixed with H_2 in subsonic longitudinal lasers, the laser output obtained was disappointingly low, probably because the combined result of the reactions

$$F_2 + NO \rightarrow NOF + F, \qquad F + NO + M \rightarrow NOF + M,$$

failed to produce sufficient F atom concentrations. No further development has taken place using this approach.

3.6 Combustion HF and DF lasers

A much more successful method for a purely chemical laser was developed by Meinzer (1970) and led to a scalable cw HF laser. Meinzer (1970) used combustion of H_2 and F_2 with excess F_2 to generate F atoms. The H_2–F_2 combustion is particularly suitable because of the low dissociation energy of F_2 and the high heat of formation of HF and DF. Two problems have to be overcome. The high temperature combustor mixture containing the F atoms must be cooled to temperatures where advantage can be taken of partial conversion, and secondly the deleterious relaxation and absorption effects of the HF formed in the combustor must be circumvented. The first problem is again overcome by expansion of the combustor mixture through a supersonic nozzle. Nozzle contours are chosen to provide the proper nozzle exit temperatures and pressures for given combustor exit temperatures, pressures and compositions. The second problem is minimized by choosing a different isotopic species for the combustor fuel, i.e., D_2 in the combustor for a HF laser. Although the deactivation of HF ($v = 1$) is similar with HF and DF as collision partners (see table 2), reactions of the type $HF(v = 2) + HF(v = 0) \rightarrow 2HF(v = 1)$ are very fast, and thus the large ground state population coming from the combustor would relax the higher levels to $v = 1$, while effective lasing from the $v = 1$ levels is prevented by the ground state absorption. Both of these problems are avoided by the use of D_2 as the combustor fuel. Fig. 21 shows the temperatures and compositions for D_2–F_2 mixtures with various F_2–D_2 ratios. With molar mixture

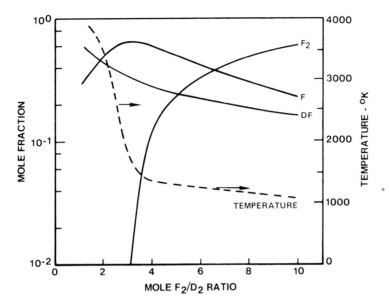

Fig. 21. Composition and ideal temperature for combustion of various F_2–D_2 mixtures at 1 atm pressure.

ratios close to 3 almost all of the excess fluorine is dissociated into F atoms, with ideal (adiabatic temperatures) between 1500 and 1800 K. Concentrations and temperatures can be further modified by addition of He or N_2 as a diluent. The combustion products are generally referred to as the primary flow. The low recombination rate of fluorine (Ultee 1977) allows the expansion of this mixture through the primary nozzles without serious F atom loss. Once expanded into the laser cavity, the secondary fuel, generally H_2, is mixed in from a second series of nozzles for H_2 injection added to the nozzle block. For operation of a DF laser the primary and secondary fuel are interchanged, H_2 is used in the combustor and D_2 in the laser cavity. The requirement of mixing the secondary fuel into the supersonic primary stream has led to the development of configurations made up of many small primary and secondary nozzles. Fig. 22 shows a diagram of a simple nozzle configuration used by Meinzer. Fig. 23 shows a photograph of this matrix hole type nozzle. A small amount of F_2 present in the primary stream leads to the chain reaction and to HF generation in high vibrational levels, resulting in a visible $\Delta v = 3$ overtone emission. Fig. 24 shows this emission in the flow field of this type of laser. Although not important in generating laser power, this emission can provide detailed information about the flow field.

Since the original work on combustion lasers, a large amount of development has been carried out on these types of lasers. Complex nozzles have been developed and multikilowatt output powers have been observed. Unfortunately,

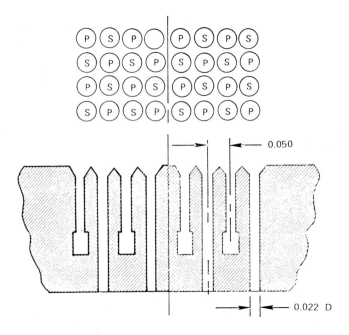

Fig. 22. Combustion chemical laser nozzle diagram (Meinzer 1970).

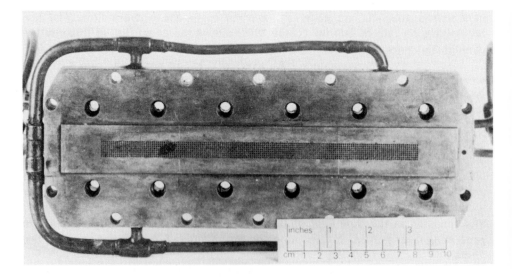

Fig. 23. Typical front view matrix hole type nozzle. The nozzle bolts onto the combustor which feeds the primaries. The alternate secondary holes are fed from a manifold.

Fig. 24. HF overtone emission.

little of this work has been published other than in internal company or government reports, which are not always readily available.

The combustion chemical laser involves a number of complex fluid dynamic, chemical kinetic and relaxation processes, as well as the quantum mechanical interaction between the medium and the radiation field. Modeling of the complete laser is obviously a matter of a judicious choice of approximations with respect to both the processes and the various rates. A detailed description is beyond the scope of this article. Gross and Emanuel (1976) recently reviewed the gasdynamics of mixing in chemical lasers. Emanuel (1976) has reviewed numerical modeling, and Chester and Chodzko (1976) have summarized the optical aspects. There has been considerable interest recently in the effects of rotational nonequilibrium in chemical lasers. Chemical laser models including rotational relaxation effects have been reported by Sentman (1975, 1976), Hough and Kerber (1975), Skifstad and Chao (1975), Ben-Shaul et al. (1976) and Hall (1976). The model by Hall appears to be the most complete with detailed rotational relaxation rates for individual levels based on experimental work by Hinchen and Hobbs (1976). Without finite rotational relaxation rates the models predict lasing on one rotational line at a time in each vibrational manifold with a shift to higher J levels as lasing progresses. This is in strong contrast with experimental observations, which almost always consist of simultaneous lasing on several P-branches. Inclusion of finite rotational rates, therefore, leads to a more realistic prediction of the output spectrum. These models also allow a much better assessment of available power for operation on single or selected lines. An additional feature of the finite rotational rates model is the prediction of very high gain on pure rotational lines, either early in the pulse for low pressure pulsed lasers or close to the nozzle exit plane in combustion lasers. Although rotational lines have been reported for some pulsed lasers Wood et al. 1971, Wood and Chang 1972, Cuellar et al. 1974, Deutsch 1967a, b), no effort has apparently been made to observe the long wavelength rotational lines in cw lasers.

4. The HCl and DCl chemical laser

4.1 Introduction and product distributions

Although the first chemical laser was a HCl laser reported by Kasper and Pimentel (1965), the HCl laser has received relatively little attention. In most respects HF and HCl are quite similar; the main difference in the chemistry of the two systems lies in the lower bond energy of HCl, resulting generally in lower reaction exothermicities for HCl-forming reactions. In particular, in the H_2–Cl_2 chain reaction, only the reaction

$$H + Cl_2 \rightarrow HCl + Cl, \quad \Delta H = -45 \text{ kcal/mole},$$

leads to vibrational excitation. The reaction

$$Cl + H_2 \rightarrow HCl + H, \quad \Delta H = +1 \text{ kcal/mole},$$

produces only HCl in the ground state. Thus, although the H_2–Cl_2 chain reaction has been used for flashlamp pumped HCl lasers, the system is inherently inefficient. Known pumping reactions for HCl are listed in table 9. The product

Table 9
Product distributions for HCl lasers.

Reaction	$-\Delta H$ kcal/mole	v_{max}[a]	Product distribution $k_1 : k_2 : k_3 : k_4 : k_5 \ldots$	Reference
$H + Cl_2 \rightarrow HCl + Cl$	45	6	$0.21 : 0.95 : 1.00 : 0.23$	Menard-Bourcin et al. (1975), Anlauf et al. (1972)
$D + Cl_2 \rightarrow HCl + Cl$	46	8	$0.1 : 0.3 : 1.00 : 0.9 : 0.06$	Anlauf et al. (1972)
$H + SCl_2 \rightarrow HCl + SCl$	46	6	$0.53 : 0.72 : 1.00 : 0.83 : 0.25$	Heydtman and Polanyi (1971)
$Cl + HI \rightarrow HCl + I$	43	4	$0.22 : 0.35 : 1.00 : 0.74$	Maylotte et al. (1972)
$Cl + DI \rightarrow DCl + I$	32	5	$0.08 : 0.14 : 0.35 : 0.73 : 1.00 : 0.05$	Maylotte et al. (1972)
$Cl + HBr \rightarrow HCl + Br$	16	1(2)	$1.0 : 0.4$	Maylotte et al. (1972)
$H + S_2Cl_2 \rightarrow HCl + S_2Cl$			$1.1 : 1.2 : 1.0 > 0.05 > 0.06$	Johnson et al. (1970)

[a] Maximum v level based on ΔH; value in parentheses is for DCl.

distribution for the $H + Cl_2 \rightarrow HCl + Cl$ reaction has been measured by several investigators using the same methods described under HF. The most recent results are those of Menard-Bourcin et al. (1975), and are given. These results are in good agreement with the measurements by Polanyi and co-workers (Anlauf et al. 1972). Available data for the other reactions are also given. The $Cl + HI$ and $Cl + DI$ reactions, in particular, appear to be a rich source of HCl and DCl radiation. Although the efficiency of vibrational energy production is high, 71% for the $Cl + HI$ reaction, most experimental devices have produced low powers with unknown or low efficiencies. HCl photoelimination lasers have received considerable attention. In the photoelimination process, the uv absorption by a polyatomic molecule is followed by elimination of a molecular fragment. In particular, there has been a detailed study of HCl photoelimination lasers resulting from chloroethylene photolysis at wavelengths above the quartz cut-off (1550 Å). Table 10 shows the results of these studies. The photoelimination process, in spite of its apparent simplicity, is rather complex. The absorption can be either directly into the fragment continua or can involve intermediate bound states. In the latter case, if the lifetime of the excited state is long compared to molecular vibration times ($\sim 10^{-13}$ s) the excess energy can be distributed in the intermediate molecule, and the product distribution is determined by the intermediate excited state configuration and by any interaction of the separating fragments. From the results obtained with isomeric chloroethylenes it is clear that both $\alpha\alpha$ (Cl and H from the same C atom) and $\alpha\beta$ (Cl and H from different C atoms) elimination can occur. Thus, in the case of $CH_2 = CCl_2$, HCl can only be formed by $\alpha\beta$ elimination; in the case of $C(CH_3)_2 =$

Table 10
Product yields for photoelimination lasers.

			N_4/N_3	N_3/N_2	N_2/N_1	N_1/N_0	Reference
$CH_2 = CHF$	HF*	$P_2(5-9)P_1(3-10)$					Berry and Pimentel (1969)
$CH_2 = CF_2$	HF*	—					
$CH_2 = ClF$	HF*, HCl*	—					
$CHF = CHCl$	HF*	—					
$CHCl = CCIF$	HF*						
$CH_2 = CHCl$	HCl*	$P_3(5-10)P_2(4-11)P_1(4-12)$					
$CH_2 = CHCl$	HCl*	$P_4(5-8)P_3(5-10)P_2(5-10)P_1(4-11)$	0.78	0.91	0.90	0.86	Berry (1974)
$CH_2 = CDCl$	HCl*	$P_4(5-9)P_3(5-9)P_2(4-10)P_1(4-10)$	0.86	1.05	0.91	0.80	Molina and Pimentel (1972)
$CH_2 = CCl_2$	HCl*	$P_4(6-8)P_3(5-10)P_2(4-10)P_1(4-10)$		0.85	1.08		Berry and Pimentel (1970)
$CH_2 = CCl_2$	HCl*	$P_4(5-9)P_3(4-10)P_2(4-11)P_1(4-12)$	0.83	1.02	0.97	0.75	Berry (1974)
cis CHCl = CHCl	HCl*	$P_3(6-9)P_2(5-9)P_1(5-10)$			0.71	0.76	Berry and Pimentel (1970)
cis CHCl = CHCl	HCl*	$P_3(6-8)P_2(5-9)P_1(5-10)$		0.72	0.83	0.79	Berry (1974)
trans CHCl = CHCl	HCl*	$P_3(6-9)P_2(6-10)P_1(6-10)$			0.70		Berry and Pimentel (1970)
trans CHCl = CHCl	HCl*	$P_3(6-9)P_2(6-9)P_1(6-10)$		0.62	0.65	0.69	Berry (1974)
CHCl=CCl₂	HCl*	$P_3(6-8)P_2(5-9)P_1(5-10)$		0.72	0.74	0.69	Berry (1974)

$CH_2 = CF_2$, $CHF = CHCl$ and $CHCl + CCIF$ have been reported to yield HF*; $CH_2 = CFCl$ yields both HF and HCl; $CH(CH_3) = CHCl$, $CH_2 = C(CH_3)Cl$ and $C(CH_3)_2 = CHCl$ yield HCl*. No details have been reported (Berry and Pimentel 1969).

CClH only $\alpha\alpha$ elimination can occur, assuming that $\alpha\gamma$ elimination is highly unlikely. In the case of $CCl_2 = CHF$, where both HCl* and HF* are observed, both $\alpha\alpha$ and $\alpha\beta$ elimination must take place. From a detailed theoretical and experimental study, Berry (1974) concludes that for the chloroethylenes the photoelimination process involves an intermediate excited state with a short ($\leq 10^{-13}$ s) lifetime, resulting in a nonstatistical HCl product distribution. The excitation in photoelimination seems to be higher than that observed in elimination reactions following radical recombination, perhaps resulting from the fact that photochemical activation is more selective than radical recombination. These studies, although perhaps only of secondary interest with respect to lasers per se, demonstrate the extensive contribution that chemical laser studies have made to our understanding of chemical dynamics on a molecular level. The photoelimination product distributions, for example, imply that in the reverse reactions, i.e., the addition of HCl to chloroacetylene, the reaction products depend on the HCl excitation, opening up the real possibility of sterochemical control in chemical synthesis.

The following discussion of HCl lasers is divided into three parts: flash photolysis and electric discharge initiated pulsed lasers; cw lasers using electric power to generate Cl or H atoms; and combustion-type HCl chemical lasers.

4.2 Pulsed HCl and DCl lasers

The initial HCl laser (Kasper and Pimentel 1965) was essentially a low pressure H_2–Cl_2 explosion. As pointed out above, this system is inherently inefficient in that at least half the HCl is produced in the ground state, leading to absorption of the 0–1 transitions and increased relaxation. Corneil and Pimentel (1968) have made a detailed study of this system over a wide variation of pressure and composition. Output powers up to 70 mJ were observed for flashlamp inputs up to 3000 J. Only 2–1 transitions were observed, in agreement with other flash photolysis results reported in table 11. A more convenient way to initiate the Cl_2–H_2 chain reaction is by an electric discharge. Deutsch (1967a, b) using a 2 m longitudinal discharge tube obtained lasing from H_2–Cl_2 mixtures on a large number of transitions. In this case H atoms can be produced directly by the discharge, and the reaction $H + Cl_2 \rightarrow HCl + Cl$ can take place without prior production of ground state HCl in the $Cl + H_2 \rightarrow HCl + H$ reaction. Other workers on similar discharges have observed 1–0 emission (Henry et al. 1968, Bourcin et al. 1970). The transverse electric discharge technique was applied to H_2–Cl_2 mixtures by Wood and Chang (1972). They obtained 0.3 to 15 μs pulses with kilowatt peak powers on 0–1 to 4–3 transitions for HCl and DCl. Similar results have been observed by others. Both flash photolysis and electric discharge initiation have been used for Cl_2–HI mixtures. While in the flash photolysis the lasing is most likely due to the $Cl + HI$ reaction, the electric discharge can lead to a $H + Cl_2 \rightarrow HCl + Cl$ pumping reaction as well. In a study of a transverse discharge using chlorine-containing hydrocarbons and HI, Taylor et al. (1974) concluded that lasing under these conditions is mainly due to reactions of H

Table 11
Pulsed HCl lasers.

Excitation	Reaction	Composition	Pulse energy	Spectral distribution	Reference
Flash photolysis	$H + Cl_2$	$H_2-Cl_2 = 2:1$ $P = 15$ Torr	—	Assignment incorrect, see Airey (1970)	Kasper and Pimentel (1965)
Pulsed electric discharge	$H + Cl_2$ $D + Cl_2$	—	—	10 HCl lines 24 DCl lines	Deutsch (1967a)
Flash photolysis	$Cl + HI$	$P = 6-40$ Torr		$P_1(9-13)$ $P_2(5, 6, 7)$ $P_3(4, 5)$	Airey (1967)
Pulsed electric discharge	$H + Cl_2$	$H_2-Cl_2 = 10:1$ $P = 1.6$ Torr $D_2-Cl_2 = 1.5:0.4$ $P = 1.9$ Torr		*HCl:* $P_2(4-10)$ $P_3(4-9)$ *DCl:* $P_2(5-9)$ $P_3(4-11)$ $P_4(5-11)$ $P_5(6-9)$	Deutsch (1967b)
Pulsed electric discharge	$Cl + HI$	$Cl_2 = 1.2$ Torr $HI = 0.2-0.4$ Torr	3×10^{-4} J 0.5 mJ/ℓ	$P_2(5-13)$ $P_3(5-12)$	Moore (1968)
Flash photolysis	$H + Cl_2$	Wide range of compositions; pressures up to 90 Torr	70 nJ	$P_2(4-10)$	Corneil and Pimentel (1968)
Flash photolysis	$D + Cl_2$	Pressures up to 230 Torr	—	$P_2(7, 8, 9)$	
Pulsed electric discharge	$H + Cl_2$ $H + NOCl$	$Cl_2 = 0.07-0.25$ Torr $H_2 = 0.6-1.5$ Torr $NOCl = 0.2$ Torr $H_2 = 1-3$ Torr	— —	$P_1(6-11)$ $P_2(5-13)$ $P_3(5-10)$ $P_1(4-10)$ $P_2(4-10)$ $P_3(4-10)$ $P_4(5-9)$	Henry et al. (1968)
Flash photolysis	$H + Cl_2$	$Cl_2 = 15$ Torr $H_2 = 40$ Torr	—	P_2	Basov et al. (1969a, b)
Pulsed electric discharge	$H + NOCl$	$NOCl = 0.05$ Torr $H_2 = 0.35$ Torr	—	$P_1(1-10)$ $R_1(0-3)$ $P_2(4-8)$ $P_3(5, 6)$	Henry et al. (1969)
Flash photolysis	CH_2CHCl $HCl + C_2H_2$	$C_2H_3Cl-Ar = 1.50$ $P = 50$ Torr	—	$P_1(4-12)$ $P_2(4-11)$ $P_3(5-10)$	Berry and Pimentel (1969)
Flash photolysis	$Cl + HBr$	6–8 Torr	0.1 J	$P_1(4-14)$	Airey (1970)
Pulsed electric discharge	$H + Cl_2$	$P = 46$ Torr		$P_1(5-9)$ $P_2(5-9)$ $P_3(5-8)$	Bourcin et al. (1970)
Pulsed electric discharge	$H + Cl_2O$	$Cl_2 = 0.5$ Torr $H_2 = 5$ Torr			Lin (1970)

Table 11 contd.

Excitation	Reaction	Composition	Pulse energy	Spectral distribution	Reference
Pulsed electric discharge	$H + Cl_2$	$Cl_2 = 6$ Torr $H_2 = 14$ Torr $He = 30$ Torr		$P_1(4-8)$ $P_2(4-8)$ $P_3(4-9)$ $P_4(4-7)$	Wood and Chang (1972)
	$D + Cl_2$	$Cl_2 = 8$ Torr $D_2 = 5$ Torr $He = 7$ Torr		$P_1(8-13)$ $P_2(6-10)$ $P_3(3-10)$ $P_4(4-11)$	
Pulsed electric discharge	$H + Cl_2$	$Cl_2 = 3$ Torr $H_2 = 47.5$ Torr $He = 36.5$ Torr		$P_1(5-11)$ $P_2(4-10)$ $P_3(4-9)$	Burak et al. (1972)
Flash photolysis	$H + Cl_2$	$H_2-ClN_3 = (5-10):$ $(95-90)$		$P_2(4-12)$	Rice and Jensen (1972)
	$H + ClN_3$	$P = 20$ Torr		$P_3(7)$	
Flash photolysis	$H + ClF$	$ClF_x-H_2-Ar = 1:1:20$ $P = 60$ Torr		$P_2(4-7)$ $P_1(3, 4)$	Krogh and Pimentel (1972)
	$H + ClF_3$			$P_3(5, 6)$ $P_2(4-7)$ $P_1(3-8)$	
Pulsed electric discharge	$H + SOCl_2$ $H + SO_2Cl_2$ $H + Cl_2CNCl$ $H + ClCN$	$RCl = 0.4-0.6$ Torr $H_2 = 2.8$ Torr $He = 7-16$ Torr		$P_1(7-12)$ $P_1(5-12)$ $P_1(6-10)$ $P_1(6-9)$	English et al. (1972)
Pulsed electric discharge	$H + CCl_4$ CCl_3H etc.	$RCl = 1-2$ Torr $HI = 0.6$ Torr $He = 70-80$ Torr		$P_1(4-5)$ $P_2(3-6)$ $P_3(4-6)$	Taylor et al. (1974)
Flash photolysis	$Cl + H_2S$	$H_2S-Cl_2-He = 1:10:40$ $P = 40$ Torr		$P_1(6-13)$	Coombe et al. (1975)
	$Cl + H_2Sc$	$H_2Sc-Cl_2-He = 1:10:40$ $P = 50$ Torr			
	$Cl + PF_3H_2$	$PF_3H_2-Cl_2-He = 3:18:110$		$P_1(7-12)$ $P_2(7)$	

atoms with the chlorocarbons. English et al. (1972) observed lasing from electrically pulsed mixtures of H_2 with $SOCl_2$, SO_2Cl_2, Cl_2CNCl and $ClCN$. Manuccia et al. (1975) in a similar study found that $CCl_4 + HI + He$ mixtures in a transverse electric discharge gave 4.5 times the output obtained with H_2-Cl_2-He mixtures in the same laser. Gorshkov et al. (1971) report gain measurements in a flashlamp pumped H_2-Cl_2 HCl laser. From their data they conclude that V–V rate constants at 300 K are of the order 1 to 3×10^{12} cm^3/mole s for the 0–1, 2–1 and 0–1, 3–2 exchange and 3 to 9×10^{12} for the 3–2, 2–1 exchange. These values are in reasonable agreement with those reported by others (Ridley and Smith 1972, Hopkins and Chen 1972, Burak et al. 1972, Leone and Moore 1973) and somewhat lower than those for HF. An extensive study was made of the $Cl + HBr \rightarrow HCl^* + Br$ reaction by Airey (1970). The main characteristics of the laser pulses were successfully modeled. Peak powers of 12 kW/cm² were observed. As in the case of HF, there are indications that in HCl rotational equilibrium is not maintained during the lasing pulse. Indeed, pure rotational lasing has been observed for HCl (Deutsch 1967a, b, c). In spite of the many similarities of the HCl and HF systems, the performance of pulsed HCl lasers has been well below that obtained with HF–DF systems. There appears to be no good explanation for this other than the fact that the HCl laser has not been the subject of the same intensive study that HF has received.

4.3 CW HCl *lasers*

As in the case of pulsed HCl lasers, cw HCl lasers have been developed to a much lesser extent than the corresponding HF lasers. In spite of the fact that reaction rates for HCl formation are equivalent to those for HF, and the relaxation rates considerably slower, cw HCl lasing is generally more difficult to achieve. Table 12 shows the reported cw HCl lasers. The first reported cw laser by Naegeli and Ultee (1970) used a 0.5×15 cm flow channel similar to that used later by Rosen et al. (1973). Cool et al. (1970) used a longitudinal flow channel. Glaze et al. (1971) used a nozzle to obtain supersonic flow for the primary stream. This configuration appears reasonably efficient and resulted in 70 mW

Table 12
CW HCl lasers.

Reaction	Pressure [Torr]	Power	Pump capacity [CFM]	Transition	Reference
Cl + HI	2	50 mW	500	$P_2(4-6)$	Naegeli and Ultee (1970)
Cl + HI	21	12 mW	450	$P_1(4-7)$ $P_2(4-7)$	Cool et al. (1970)
H + Cl$_2$	45	1 mW	450	$P_1(4-6)$	
Cl + HI	3	70 mW	80	$P_1(7)$ $P_2(4, 5, 6, 9, 10)$	Glaze et al. (1971)
Cl + HI	6	1.65 W	1800	P_1 P_2 P_3	Rosen et al. (1973)
Cl + HBr	6	100 mW	1800	P_1	
H + Cl$_2$	1	200 mW	2350	$P_1(4-8)$ $P_2(4-8)$ $P_3(6, 7)$	Menard-Bourcin et al. (1975)
H + ClF	4.5	300 W	5000	$P_1(9-12)$ $P_2(5-12)$ $P_3(5-11)$	Meinzer (1974)
2NO + ClO$_2$ $\rightarrow$ 2NO$_2$ + Cl Cl + HI $\rightarrow$ HCl* + I	2.4	13 W	1860	—	Arnold et al. (1977)

power output with 80 CFM pumping capacity. Menard-Bourcin et al. (1975) used 10 individual H_2 discharges leading into an approximately 22 cm wide cavity, pumped with a 2350 CFM pump. In agreement with the product distribution studies (table 9) the Cl + HI reactions appear to provide the highest laser output. No detailed parameter studies (gain or population measurements) of the cw HCl have been reported. The generally poor performance is probably due to a number of problems such as the slow mixing rate of HI and Cl (compared to F and H_2), inefficient discharges for Cl atom production, etc. In addition the HI, Cl_2 and HCl mixtures are much more corrosive to pumps and pump oil, requiring therefore, extensive trapping in the vacuum lines. In addition to the extra cost and operating complexity, this also tends to limit the flow rate well below that expected from the pump capacity.

In HCl lasers discussed so far, an electric discharge was used to produce Cl- or H-atoms. The use of arcs or other high temperature sources to produce Cl atoms has not been reported. The production of Cl atoms in a H_2–Cl_2 combustion process is much more difficult than in the case of H_2–F_2, because of the higher dissociation energy of Cl_2 (59 kcal/mole) compared to F_2 (38 kcal/mole) and the lower heat of formation of HCl compared to HF (22 versus 65 kcal/mole). Meinzer (1974) reported a combustion-type laser based on the H + ClF → HCl + F reaction. Chlorine and excess F_2 are reacted in a combustor to form ClF and F. Fig. 25 shows a typical composition–temperature plot for a F_2–Cl_2 ratio of 2:1. In practice, a small amount of H_2 is added to the combustor flow to aid ignition and to provide independent temperature control. At temperatures near 1600 K it is possible to obtain primarily ClF and F as combustion products. Using the

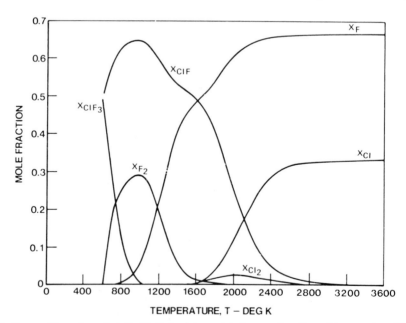

Fig. 25. Composition of reacted F_2–Cl_2 2:1 mixture at 1 atm as a function of temperature (Meinzer 1974).

same type nozzles as for HF, this mixture is expanded and H_2 is mixed in the optical cavity region. Vibrationally excited HCl (and HF) are produced by the chain reaction

$$F + H_2 \rightarrow HF^* + H, \tag{16}$$

$$H + ClF \rightarrow HCl^* + F, \tag{17}$$

$$H + ClF \rightarrow HF^* + Cl. \tag{18}$$

Neither the rates of reaction (17) and (18) nor their product distributions are known. In modeling the laser, reasonable agreement is obtained by giving reactions (17) and (18) a rate of $1.2 \times 10^{14} \exp(-2400/RT)$ cm^3/mole s and a product distribution of

$$k_1 : k_2 : k_3 : k_4 = 0.5 : 1 : 0.75 : 0.25.$$

The details of the chemistry in this laser are poorly known, and the effects of other reactions involving the atomic species and of energy transfer and relaxation processes are largely unknown. Depending on the flow rate, composition and pressure, either HCl or HCl and HF lasing can be obtained. HF lasing occurs earlier in the reaction stream in accordance with the chemistry (see fig. 26). Also, HCl lasing persists at higher cavity pressures because of slower vibrational relaxation. Fig. 27 shows typical performance as a function of primary H_2 flow, the effect being due to optimization of the combustor process.

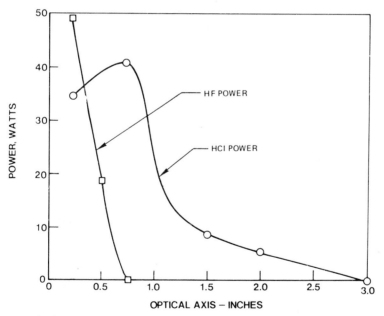

Fig. 26. Power distribution of a HCl–HF combustion laser as a function of the optical axis position. The distance is measured from the nozzle exit plane. Primary flows (combustor): He = 0.10, $F_2 = 0.0585$, $Cl_2 = 0.027$, $H_2 = 0.016$. Secondary flow: $H_2 = 0.3$, all in mole/s. Combustor pressure 293 Torr, cavity pressure 3.0 Torr (Meinzer 1974).

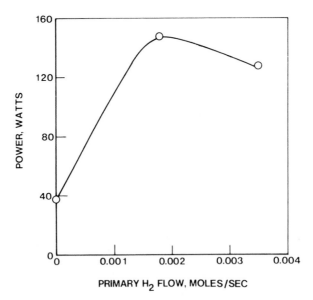

Fig. 27. Power variation with H_2 combustor flow in HCl combustion laser. H_2 combustor flow as indicated. Other flows: combustor; He = 0.055, F_2 = 0.119, Cl_2 = 0.0652 mole/s; secondary flow: H_2 = 0.594 mole/s. Combustor pressure = 2.6–3.3 Torr (Meinzer 1974).

A different reaction sequence was used by Arnold et al. (1977). They used a chain reaction between NO and ClO_2 to produce Cl atoms by the reaction sequence

$$k$$

$$NO + ClO_2 \rightarrow NO_2 + ClO, \qquad 2.1 \times 10^{11}\ cm^3/mole\ s, \qquad (19)$$

$$ClO + NO \rightarrow NO_2 + Cl, \qquad 1.0 \times 10^{13}\ cm^3/mole\ s, \qquad (20)$$

$$Cl + ClO_2 \rightarrow 2\ CLO, \qquad 3.6 \times 10^{13}\ cm^3/mole\ s, \qquad (21)$$

followed by

$$Cl + HI \rightarrow HCl^* + I, \qquad 1.0 \times 10^{14}\ cm^3/mole\ s. \qquad (22)$$

HCl laser emission was observed in a transverse configuration with a $1.5 \times 14\ cm$ flow channel. NO was mixed with ClO_2, HI was then added downstream. In this manner the reaction sequence (19) to (21) can proceed, producing the maximum Cl-atom concentration, without competition from the faster $Cl + HI$ reaction. With flow rates (in standard cm^3/min) of: He, 58 000; ClO_2, 650; NO, 1700; and HI, 1360, HCl multiline power of 4 W was observed. The linear flow rate was $1.6 \times 10^4\ cm/s$ at 2.4 Torr. With a larger $40\ cm \times 1\ cm$ laser and flow rates (cm^3/min) of: He, 103 600; ClO_2, 1500; NO, 4500; and HI, 1950, 13 W was obtained corresponding to a 8% chemical efficiency. The linear flow rate in this cavity was $2.2 \times 10^4\ cm/s$ at a total pressure of 2.41 Torr.

After providing the initial chemical laser, there has been little development of

HCl. In a relatively short time, the interest in high efficiency, high power lasers has channeled most research work into the HF–DF system. It is clear, however, that in spite of lower potential for high power application, the HCl laser and the HBr and CO lasers as well, will continue to provide a wealth of information leading to a better understanding of chemical kinetics, energy transfer and relaxation processes.

5. The HBr and DBr laser

The final member in the hydrogen halide chemical laser series has received little attention. The lower heat of dissociation of HBr (88 kcal/mole) limits the number of reactions with sufficient exothermicity. Only two reactions are known to produce excited HBr:

$$H + Br_2 \rightarrow HBr + Br, \quad \Delta H = -41 \text{ kcal/mole},$$

and

$$Br + HI \rightarrow HBr + I, \quad \Delta H = -15 \text{ kcal/mole}.$$

The reaction $Br + H_2 \rightarrow HBr + H$ is endothermic by 17 kcal/mole and has a similar activation energy, making it too slow to carry an effective chain reaction in H_2 and Br_2. Anlauf et al. (1968) have studied the product distribution of the $H + Br_2$ reaction and find $k_2 : k_3 : k_4 : k_5 = 0.148 : 1.00 : 0.982 : 0.216$. Airey et al. (1967) give $k_3 : k_4 : k_5 : k_6 = 1.0 : 0.64 : 0.19 : 0.05$. Deutsch (1967a, b) first observed HBr and DBr stimulated emission from a longitudinal discharge through mixtures of Br_2 and $H_2(D_2)$. The HBr emission occurred on the $P_4(5\text{–}8)$, $P_3(4\text{–}9)$, $P_2(4\text{–}9)$ and $P_1(4\text{–}9)$ transitions covering a wavelength from 4.1 to 4.0 μm. For DBr the observed transitions were $P_5(6\text{–}9)$, $P_4(5\text{–}11)$, $P_3(5\text{–}8)$ and $P_2(8)$ from 6.3 to 5.8 μm. In both cases the lines occurred in pairs, owing to the Br^{81} and Br^{79} isotopes present in approximately equal abundance.

Wood and Chang (1972) using H_2–Br_2–He and D_2–Br_2–He mixtures in a 179 cm long pulsed transverse discharge obtained strong (5 kW) HBr and DBr emission at pressures up to 550 Torr. The observed lines were similar to those reported by Deutsch (1967a, b), except that for DBr the 1–0 transitions were also observed. Similar results were obtained by Burak et al. (1972). They also observed strong CO_2 emission at 10.6 μm from the transfer reaction: $HBr(v = 1) + CO_2(000) \rightarrow HBr(v = 0) + CO_2(001)$. Oodate et al. (1974) report an enhancement in a transversely excited HBr laser by addition of SF_6. No significant changes in pulse delays or duration of the individual lines occurred. HF lasing occurred simultaneously. Apparently the addition of SF_6 results in additional H-atom formation through the reaction

$$F + H_2 \rightarrow HF + H.$$

Inoue and Tsuchiya (1974) have discussed some of the chemical and relaxation processes in the HBr laser in detail. Cool et al. (1970) failed to observe cw lasing from the $H + Br_2$ reaction, although transfer to CO_2 was observed in the same equipment (Cool and Stephens 1970).

6. CO lasers

6.1 Introduction

The CO chemical laser is the only major nonhydrogen halide chemical laser developed to date. Most of the work reported so far has been based on the pumping reaction

$$CS + O \rightarrow CO^* + S, \qquad \Delta H = -80 \pm 5 \, kcal/mole.$$

The CS radical is usually produced in situ by the reaction $CS_2 + O \rightarrow CS + SO$. Thus the laser requires O atoms for its operation. No successful chemical means for generating large quantities of O atoms have been found and, consequently, no cw CO combustion type mixing lasers have been developed. However, recently lasing has been observed from premixed flames, opening up the possibility of future development of 'flame lasers'. The high exothermicity of the reaction (23) leads to excitation of CO up to $v = 15$; this, coupled with a small rotational constant, results in an extremely rich output spectrum, covering the wavelength region from 4.6 to 5.7 μm. Unfortunately many of the lines coincide with atmospheric absorption bands which limit the use of this laser. An important difference between the CO and hydrogen halide laser results from the tremendous difference in relaxation rates. The CO self-relaxation rate constant is $1.86 \times 10^5 \, cm^3/mole \, s$ (compared with $HF = 1 \times 10^{12} \, cm^3/mole \, s$), and is insignificant in the loss processes. Direct vibration–translational relaxation by other species also tends to be slow and the relaxation losses generally take place by V–V processes involving species such as CS_2 ($2.8 \times 10^{11} \, cm^3/mole \, s$). Because of the low V–T rates, V–V processes play a dominant role in restructuring the CO population distribution. CO V–V rates vary with vibrational quantum number, reaching a maximum $1.4 \times 10^{12} \, cm^3/mole \, s$ for $v = 3$ and decreasing to $2.8 \times 10^{10} \, cm^3/mole \, s$ for $v = 11$ (Powell 1973). With the slower relaxation and pumping rates, and the possibility of some contribution from chain reactions, the lasing process tends to last longer in pulsed lasers, and will extend over larger distances in the cw lasers. As a result of the V–V dominance, different CO lasers, including electrically pumped CO lasers, tend to have a similar population distribution (Treanor distribution) and lead to very similar output spectra.

Kompa (1973) and Bronfin and Jeffers (1976) have summarized the pulsed CO laser work up to 1973. In the following discussion we are primarily concerned with more recent developments, with emphasis on cw devices.

6.2 Population distributions

The first chemical CO laser was obtained by flash photolysis of CS_2–O_2 mixtures (Pollack 1966a, b, c). After some early speculation that the reaction $O_2 + CS \rightarrow$

CO + SO was responsible for the excited CO formation (Arnold and Kimbell 1969, Rosenwaks and Yatsiv 1971), it was soon established that the reaction $CS + O \rightarrow CO^* + S$ was the main reaction responsible for the production of vibrationally excited CO. With a few exceptions all chemical CO lasers are based on this reaction. The exothermicity of this reaction is given as 75 to 85 kcal/mole. The 85 kcal/mole value is sufficient to excite CO to $v = 15$. Hancock and Smith (1971) first measured the product distribution. Additional work based on cw and pulsed laser gain or output characteristics has been reported by Hancock et al. (1971), Foster (1972), Tsuchiya et al. (1973), Powell and Kelley (1974), and Djeu (1974). Considering the complexities of the system, the uncertainties inherent in the laser methods and differences in temperatures, pressures, etc., of the experiments, the results are in reasonable agreement. All studies give a maximum rate into $v = 13$, except those of Djeu (1974), who obtained a maximum population at $v = 12$. Above $v = 13$ all investigators show a steep decrease in rate, with a zero rate for states above $v = 19$. For the lower levels there is some disagreement. Powell and Kelley (1974) and Tsuchiya et al. (1973) obtained higher rates for the states below $v = 6$ than do Hancock et al. (1971) and Djeu (1974). Fig. 28 shows the distribution obtained by combining and averaging the above results. Kelley (1976) has recently suggested that the discrepancies below $v = 6$ are due to contributions in varying degrees of the reaction $CS_2 + O \rightarrow CO^* + S_2$. Using a model based on surprisal analysis correlation, he concludes that the $CS + O \rightarrow CO^* + S$ distribution smoothly goes to zero from its maximum at $v = 13$. Lasing has also been observed from the reaction

$$O + CSe \rightarrow CO^* + Se, \qquad \Delta H = -118 \text{ kcal/mole,}$$

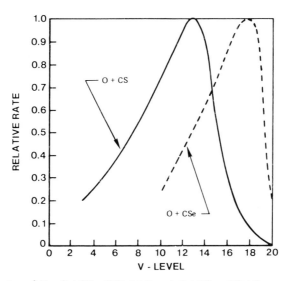

Fig. 28. Relative rates of the $O + CS \rightarrow CO(v) + S$ and $O + CSe \rightarrow CO + Se$ reactions. See text for sources.

in pulsed lasers (Wittig and Smith 1972, Rosenwaks and Smith 1973). The energy of the reaction is sufficient to excite CO to the $v = 20$ level. Emission from the $v = 20 \rightarrow 19$ to $9 \rightarrow 8$ bands has been observed, with lasing appearing first on the $19 \rightarrow 18$ to $16 \rightarrow 15$ bands. The absence of lasing from levels $v \leq 9$ is probably due to CO relaxation by the reaction $CO(v \leq 9) + Se(^3P_2) \rightarrow CO + Se(^3P_1)$. Excitation of $Se(^3P_1)$ requires $1989\,cm^{-1}$, close to the CO vibrational energy. Fig. 28 shows the product distribution observed by Wittig and Smith (1972). The behavior of the $O_2 + CSe_2$ system appears generally similar to the $O_2 + CS_2$ systems.

Several other reactions of O atoms lead to excited CO: Lin and Bauer (1970) obtained CO lasing from pulsed or cw discharges of mixtures of $C_3O_2 + O_2$. In the $C_3O_2 + O_2$ systems the CO formation apparently results from reaction sequence

$$O + C_3O_2 \rightarrow CO_2 + C_2O, \qquad \Delta H = -38\,kcal/mole,$$

$$O_2 + C_2O \rightarrow 2CO^* + O, \qquad \Delta H = -85\,kcal/mole,$$

$$O + C_3O_2 \rightarrow 3CO^*, \qquad \Delta H = -115\,kcal/mole,$$

with the second and third reaction producing the excited CO. Excitation up to $v = 11$ was observed, indicating that the energy was not evenly distributed among the CO produced, which would have limited excitation to $v = 7$ and 6 respectively. Clough et al. (1970a, b) earlier observed infrared emission from these systems and suggest that the $O(^3P) + C_3O_2 \rightarrow 3CO$ produces CO excited up to $v = 8$.

In the reaction of the $C_3O_2 + O_3$, flash photolysis leads to excited $O(^1D)$ atoms. In this case Lin and Brus (1971) suggest that the reaction

$$O(^1D) + C_3O_2 \rightarrow 3CO^*, \qquad \Delta H = -160\,kcal/mole,$$

is responsible for the observed lasing. Photolysis of a $O_3-C_3O_2-Ar = 1:1:30$ mixture at 70 Torr with a 600 J flash gave emission from $v = 13-12$ to $v = 5-4$. Since the times appear too short for extensive V–V pumping, these results also indicate an unequal energy distribution among the three CO molecules produced. Pukhal'skaya et al. (1975) in a similar study come to the same conclusion. They also observed excitation up to CO ($v = 13$) and concluded that the energy must be primarily deposited in one of the three CO molecules. The $C_3O_2 + O_3$ appears inefficient compared to the $CS_2 + O_2$ system, probably because of many competing reactions and deactivation processes. Also, since carbon suboxide is difficult to synthesize and store, these lasers are primarily of interest because of their chemical kinetics.

Other reactions such as

$$O(^3P) + CN \rightarrow CO^* + N, \qquad \Delta H = -74\,kcal/mole,$$

$$O(^1D) + CN \rightarrow CO^* + N, \qquad \Delta H = -119\,kcal/mole,$$

$$O(^3P) + C_2H_2 \rightarrow CO^* + CH_2, \qquad \Delta H = -48\,kcal/mole,$$

$$O(^3P) + CH_2 \rightarrow CO^* + 2H, \qquad \Delta H = -73 \text{ kcal/mole},$$
$$O(^3P) + CH \rightarrow CO^* + H, \qquad \Delta H = -176 \text{ kcal/mole},$$
$$O(^3P) + CHF \rightarrow CO^* + HF^*, \qquad \Delta H = -181 \text{ kcal/mole},$$

have been used by Lin and co-workers (Brus and Lin 1972, Lin 1973) to pump CO lasers. Little detail is available on the laser output and the product distributions. Clough et al. (1970a, b) studied the infra-emission of the reactions

$$O + C_2H_2 \rightarrow CO + CH_2, \quad \Delta H = -48 \text{ kcal/mole},$$
$$O + CH_2 \rightarrow CO + 2H, \quad \Delta H = -73 \text{ kcal/mole},$$

and observed excitation up to $v = 14$. For the $O + CH$ reaction, $v = 33$ is observed. These observations at 0.1 to 2 Torr do not necessarily correspond to the original product distribution. Lin et al. (1972), using a fast flow system, observed excitation up to $v = 14$ with a non-Boltzmann distribution. Again, however, at their pressure and observation time, 1.40 Torr and 100 μs, considerable Treanor pumping could have occurred and no good information is available on the product distribution from these reactions. Apparently none of the above reactions can compete with CS_2 in efficiency (see below), and CS_2 has become the dominant fuel for CO lasers.

6.3 Pulsed CO lasers

Early pulsed CO lasers were obtained by flash photolysis of CS_2–O_2 mixtures. The flash photolysis $CS_2 \rightarrow CS + S$ is followed by reactions $S + O_2 \rightarrow SO + O$ and $SO + O_2 \rightarrow SO_2 + O$, which produce the O atoms for the pumping reaction $O + CS \rightarrow CO^* + S$. Gregg and Thomas (1968) made a detailed study of the temporal sequence in which the laser lines occurred after the initiation. Using a grating in the cavity and a wide variety of CS_2–O_2 and CS_2–O_2–air mixtures at pressures up to 18 Torr, they observed 250 lines covering $v = 16$–15 down to $v = 1$–0, with J values from 6 to 33. In addition R-branch lines were observed for $v = 15$–14 to $v = 9$–8 with J values between 2 and 28. The occurrence of the R-branch lines is indicative of a total vibrational inversion. Jacobson and Kimbell (1970) used a transverse electric discharge with CS_2–$O_2 = 2:25$ at 10 Torr with 20 Torr He and obtained 25 mJ. Ahlborn et al. (1972) developed a fast flow transverse discharge laser operating on CS_2 and O_2 at 70 Torr. They obtained 0.1 mJ pulses at ~100 pps. Pulse lengths were ~70 μs. Tiee et al. (1975), using flash photolysis of CS_2–O_2 mixtures, obtained 240 mJ/pulse. Addition of N_2O and O_3 increased the energy to 360 mJ/pulse. Bashkin et al. (1976), also using CS_2–O_3 mixtures, obtained 0.28 J/pulse with a specific power of 4.5 J/ℓ. Gordon et al. (1976) obtained 0.5 J/pulse with CS_2–O_2–$N_2O = 1:24:5$ and an input energy of 2.2 kJ. Ahl and Birns (1976) have reported 115 mJ/pulse (1 J/ℓ) from a double discharge transversely excited system through CS_2–O_2 mixtures at 35 Torr. Apparently no e-beam-excited CS_2–O_2 lasers have been reported.

6.4 CW CO *mixing lasers*

All reported cw CO chemical mixing lasers are based on the reactions

$$O + CS_2 \rightarrow CS + SO, \tag{23}$$

$$O + CS^* \rightarrow CO + S. \tag{24}$$

Some participation of the chain reaction via $S + O_2 \rightarrow SO + O$ and $SO + O_2 \rightarrow SO_2 + O$ is also possible, particularly in the longitudinal configurations, depending on the flow velocities and temperatures. Even without the chain contribution, it is generally not possible to calculate the chemical efficiency, since the O-atom concentrations are unknown. In table 13 the efficiencies listed are the authors' estimates. Other characteristics of the lasers are given as reported. The flow rates and pressures are generally for the highest power listed. A number of authors have reported cw lasers obtained by discharging mixtures of CS_2 and O_2. Although the laser output in part is due to reactions (23) and (24), the possibility of some direct electrical excitation of CO cannot be excluded. These reports have, therefore, been excluded from table 13.

The first cw CO laser was reported by Wittig et al. (1970a, b). He and O_2 were passed through a 60 Hz 12 kV, 50 mA (neon sign transformer) discharge. The flow passed into a 10 mm i.d tube, was mixed with CS_2 in the cavity region and pumped out with a 500 ℓ/min pump. The active length was 15 cm. Observed power was 100 mW. Suart et al. (1970a, b) in a similar longitudinal configuration obtained 2.3 W using a microwave discharge to generate O atoms. In this same laser addition of CO_2, N_2, Ar, O_2 and NO_2 to the laser decreased the power, OCS, CO and N_2O enhanced the power output. Addition of 1 ℓ/min CO increased the power from 30 to 300 mW (see table 13). Lecuyer and Legay-Sommaire (1970) also obtained 0.3 W in a longitudinal configuration. The same laser had an output of 2.3 W when a mixture of O_2, CS_2 and He was discharged. The spectral output of all the above lasers is similar and covers $v = 12$–11 to $v = 8$–7 with J values from 8 to 17 (see table 13). A detailed study of this type of laser has been reported by Suart et al. (1972). In a longitudinal configuration similar to that reported previously (Suart et al. 1970a, b) but with improved mixing, they obtained 23 W. O atoms were measured by titration with NO_2 and other species by mass spectroscopy. Assuming complete conversion of CS_2 to CS they report a chemical efficiency of 0.92%. The effect of cold CO is explained as at least in part due to transfer of CO excitation in high levels to lower levels where it is less susceptible to deactivation by O_2 present in excess. The above lasers provide a convenient source of modest power at CO wavelengths and require only a limited pumping capacity. They are, however, inherently inefficient in that the optical path passes through CO with varying degrees of excitation and at different temperatures, which gives rise to absorption and high cavity losses. Higher power output and lasing on many more transitions can be obtained with transverse flow laser geometries. Jeffers and Wiswall (1970)

Table 13

CW CO chemical mixing lasers.

O-atom source	Flow rates [mmole/s]				Pressure [Torr]	Pump capacity [ℓ/min] Linear flow [cm/s]	Power	Output spectra	Comments	Reference
	O_2	CS_2	He	Other						
Pulsed capacitor discharge 3 J input	5.2	0.2	23	—	—	5000 ℓ/min	100 mW		Longitudinal configuration. Quasi-cw	Wittig et al. (1970a)
Microwave discharge	2.7	0.76	—	—	4	8490 ℓ/min	few watts	P_{10}, P_9 P_8(8–13)	Longitudinal configuration	Suart et al. (1970a)
60 Hz discharge 12 kJ. 60 mA	—	—	—	—	10–16	500 ℓ/min	1 mW	P_8(12–15) P_9(11–16) P_{10}(11–17) P_{11}(12, 12–16) P_{12}(12–15)	Longitudinal configuration	Wittig et al. (1970b)
60 Hz electric discharge 50 mA	1.8	0.18	21	x	10–40	1132 ℓ/min 10⁴ cm/s	70 mW	P_{13}(9–17) to P_4(9–17)	Transverse flow. Note increase in power on cold CO addition	Jeffers and Wiswall (1970)
Microwave discharge	2.2	0.3	10.1	CO 1.4	2.8	1.4×10^4 cm/s	300 mW	P_7(9–12) P_8(8–12) P_9(9–13) P_{10}(9–13) P_{11}(8–12)	Longitudinal geometry. Addition of O_2, NO_2, N_2, CO_2 and Ar decreased power. OCS, N_2O and CO increased power	Suart et al. (1970b)
	1.34	0.31	25.3	—	14	3000 ℓ/min	0.3 W	P_8, P_9, P_{10} P_{11}, P_{12}	Longitudinal configuration: also reports on cw discharge through CS_2. O_2 He mixtures	Lecuyer and Legay-Sommaire (1970)
60 Hz discharge		0.6			$O_2 = 1.8$ He = 9.3		0.25 W	11–10 to 8–7		Wittig et al. (1971)
60 Hz discharge for O. Microwave discharge for CS	2.3	0.1	4.8		7.2	1500 ℓ/min	10 mW 100 µW overtone	v = 16–14 14–12 to 2–0		Sadie et al. (1972)
Microwave discharge	4.7	0.76	45	CO 2.3	9	33 960 ℓ/min	2.3 W	P_7(9–12) P_8(8–14) P_9(9–14) P_{10}(8–14) P_{11}(8–12)	Longitudinal configuration. includes chemiluminescence measurements and modeling	Suart et al. (1972)
Electric arc	8.4	2.9	2.56	Ar 88.2	40		15 W	v = 14–18 to 3–2	Supersonic expansion 5 cm optical length	Boedeker et al. (1972)

Table 13 contd.

O-atom source	Flow rates [mmole/s] O₂	CS₂	He	Other	Pressure [Torr]	Pump capacity [ℓ/min] Linear flow [cm/s]	Power	Output Spectra	Comments	Reference
60 Hz discharge	4.2	0.3	18	CO 2.3		2×10^4 cm/s	1.5 W	with CO: $P_7(6\text{–}15)$, $P_8(2\text{–}21, 4\text{–}11)$, $P_9(2\text{–}21, 3\text{–}15)$, $P_{10}(3\text{–}19)$, $P_{11}(9\text{–}13)$	Transverse geometry R-lines indicate total vibrational inversion in the presence of CO	Jeffers (1972)
Microwave discharge	9.9	0.3	14.7	CO 2.6	4.3	6.1×10^3 cm/s	1 W	$P_{13}\text{–}P_5$	Description of transverse flow laser. Product distribution of CS + O reaction	Foster (1972)
60 Hz discharge for O. Microwave discharge for CS	2.5	0.08	12.2		3.15	3×10^4 cm/s	2.15 W		Transverse flow. CS produced by microwave discharge	Jeffers et al. (1973)
60 Hz discharge 800 W	9.7	0.5	32		5.4	14 150 ℓ/min	4.5 W	$P_3(8\text{–}10)$, $P_4(8\text{–}15)$, $P_5(8\text{–}16)$, $P_6(8\text{–}12)$, $P_6(14\text{–}17)$, $P_7(8\text{–}12)$, $P_7(15, 17)$, $P_8(8\text{–}10, 12, 14\text{–}16)$, $P_9(9\text{–}17)$, $P_{10}(9\text{–}17)$, $P_{11}(8,9,11,13,16,18)$, $P_{12}(10, 16)$, $P_{13}(12,13,15,17,18)$, $P_{14}(11, 14)$	Transverse flow. Detailed gain measurements for various locations	Ultee and Bonczyk (1974)
Microwave discharge 1 kW	5.6	0.6	20	N₂O 3.1	~5	5×10^3 cm/s	28.8 W[a]		Transverse flow. Mass spectroscopy O-atom titration. Effect of N₂O, CO and Ar addition	Jeffers and Wiswall (1974)
Thermal source for CS	7.3	1.4	65		10.6		84 W[a]			Jeffers and Ageno (1975)
60 Hz discharge	13.6	0.71	32.6	Ar 7.5, H₂ 0.007	5.3		17.0 W	$P_{14}\text{–}P_1$	Transverse flow (see Jeffers 1974). 0–1 emission observed	Jeffers et al. (1976b)
Microwave discharge	11.2	0.6	26.9	Ar 7.1, H₂ 0.04, N₂O	5.0				Transverse flow. Mainly effect of additives on gain	Jeffers et al. (1976c)
Microwave discharge	6.5	1.05	44.1	15.2	10.8	51 000 ℓ/min	94.4 W[a]	12–11 to 3–2	Thermal CS source	Jeffers et al. (1976a)

reported the first cw subsonic CO chemical mixing laser with an output of 70 mW. The output spectrum, in contrast with longitudinal flow lasers, included transitions down to $v = 4$–3. In later studies, Jeffers and Wiswall (1973) report 22 W. Similar lasers have been reported by Foster (1972) and by Ultee and Bonczyk (1974). Fig. 29 shows a typical example of this class of lasers. A mixture of O_2 and He is discharged and the resulting He–O_2–O mixture is passed into the laser channel and mixed with CS_2. The mixing channel dimensions are chosen to provide near-optimum linear flow velocities for the system. CS_2 is injected through two spray bars each with forty-seven 0.013 in. diameter holes which inject the flow perpendicular to the mainstream to promote rapid mixing. The optical windows are mounted on movable sidewalls so that the distance between the CS_2 injection and the optical cavity axis can be optimized. Fig. 30 shows the flow channel with the usual sidewall replaced by Plexiglass. The O–CS_2 flame and the SO + O continuum are clearly visible. Gain for this system was measured with a stabilized line-selective electric discharge CO laser. Fig. 31 shows the development of gain in the flow direction for several lines. Fig. 32 shows the corresponding laser power. Both of these figures show that a

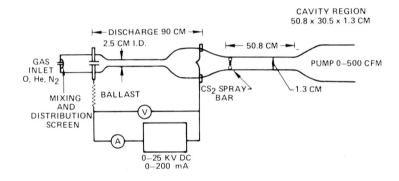

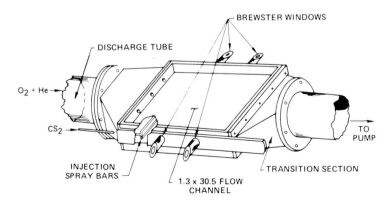

Fig. 29. CO mixing laser (Ultee and Bonczyk 1974).

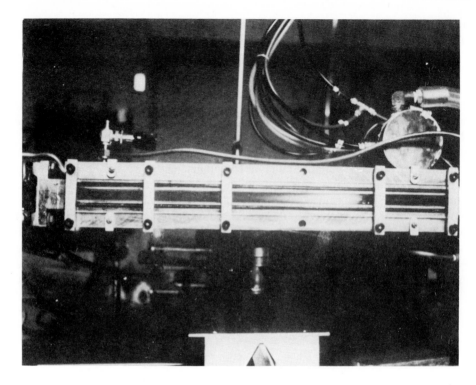

Fig. 30. SO + O$_2$ continuous emission in CO laser flow channel.

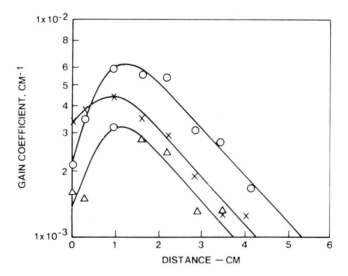

Fig. 31. Gain as a function of distance in a CO mixing laser. O$_2$ = 1.17 mole/min, He = 1.69 mole/min, CS$_2$ = 0.028 mole/min. Pressure = 5.90 Torr (Ultee and Bonczyk 1974).

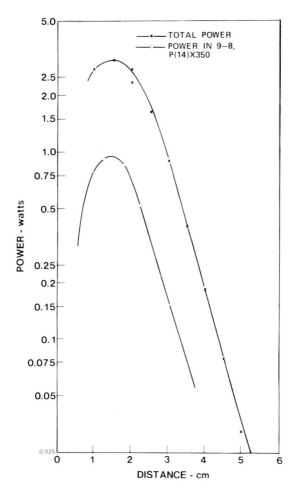

Fig. 32. Power versus distance between optical axis and CS_2 injection. $O_2 = 1.17$ mole/min, He = 1.69 mole/min, $CS_2 = 0.028$ mole/min. Pressure = 5.9 Torr (Ultee and Bonczyk 1974).

time (distance) of 2 to 3×10^{-4} s is required to reach maximum gain. This time corresponds to the time necessary for mixing on a molecular scale and reaction. The subsequent decay is due to the combined relaxation processes. Fig. 33 shows gain measurements over a large number of bands. Maximum gain observed was 6×10^{-3} cm^{-1}. Somewhat lower gain has been reported by Jeffers et al. (1976a, b, c). A small-scale supersonic chemical mixing laser was developed by Boedeker et al. (1972). Argon was heated in an arc heater, mixed with O_2 and exhausted through a slit nozzle. The expansion freezes the O-atom recombination and results in a supersonic flow with a low translational temperature. CS_2 is added through a pair of slot injectors (see fig. 34). A typical output spectrum from this device is shown in fig. 35. With a near-confocal, closed optical cavity the resulting power was 15 W, corresponding to a chemical

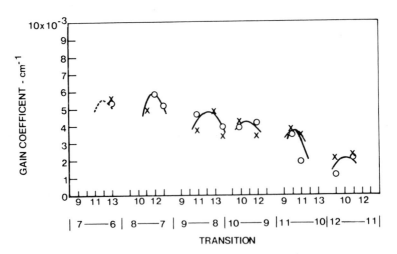

Fig. 33. Gain coefficient for a number of transitions. See fig. 32 for conditions (Ultee and Bonczyk 1974).

efficiency of ~6%. In the same equipment 34 W was obtained for unspecified conditions.

6.5 CO flame lasers

The observation of lasing from a CS_2–O_2 explosion initiated by flash photolysis or electric discharge suggests the possibility of lasing from a low pressure CS_2–O_2 flame. Lasing obtained from discharging CS_2–O_2–He and C_2H_4–O_2–He mixtures at pressures and with diluent concentrations close to those for which a stable combustion could be maintained also indicated that lasing from a sustained flame could be possible. Foster and Kimbell (1970) first observed a complete population inversion in a CS_2–O_2 flame. With CS_2–O_2 ratios of 1:3.9 at 0.5 Torr, the observed emission spectra with distributions corresponded to

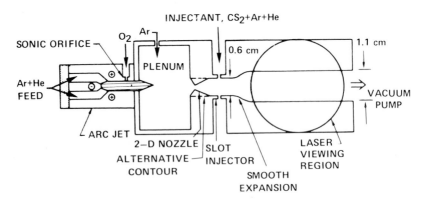

Fig. 34. Small supersonic CO laser (Boedeker et al. 1972).

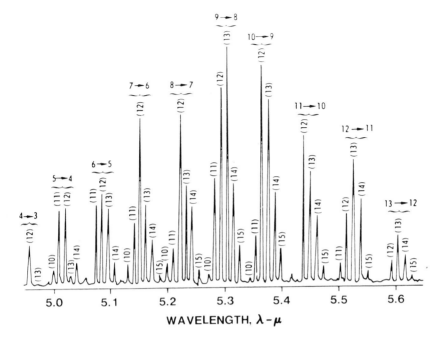

Fig. 35. Spectral distribution of supersonic CO laser output (Boedeker et al. 1972).

$T_v = 25\,000$ K for $v = 5$–10. The addition of N_2 to the flame resulted in a complete population inversion up to $v = 9$. Djeu et al. (1971) observed gain in the 10–9, 9–8, 8–7 vibrational bands of a similar flame operation with 0.9 and 6.0 Torr of CS_2 and O_2 respectively. The observed gain was 2% over a 120 cm path. Gain was also observed by Ultee (1971a, b) in a similar system operating with CS_2 and O_2 at 0.1 and 5 Torr respectively. Pilloff et al. (1971) were the first to obtain stimulated emission from a free-burning CS_2–O_2 flame. The burner was 60 cm long and consisted of twenty-four mm o.d. parallel tubes with fifty 1 mm holes through which CS_2 and O_2 were (separately) fed. The system was pumped by a 300 CFM vacuum pump. Lasing was a sensitive function of the CS_2–O_2 ratio and total pressure. The output was 1 mW, consisting of the $P_8(11)$, $P_9(12)$ and $P_{11}(10)$ lines. The addition of small amounts of OCS (0.1 Torr) doubled the observed power. Searles and Djeu (1971) further developed this system. With a 12 in ribbon burner the gases were mixed between the burner surface and 80 mesh screen. Lasing was observed with O_2–CS_2 ratios of 34 to 1. With CS_2 and O_2 alone 19 mW was obtained. The addition or substitution of He for O_2 decreased the output, indicating that the excess O_2 served to increase the reaction rates rather than act as a coolant. Addition of N_2O, CO, CO_2, SO_2, SF_6 and N_2 increased the output power. Best performance was obtained with CS_2, 3.9 mmole/s; O_2, 110 mmole/s; and N_2O, 7.6 mmole/s. The output power with these flows was 0.6 W. The highest power from a CS_2–O_2–N_2O flame has been reported by Linevski and Carabetta (1973) using a 16×1.5 in Meker type burner with sideplates. They obtained 25 W from a $CS_2–N_2O–O_2 = 1:2.5:10$ mixture operating at 24 Torr. Lasing takes place in the 600 to 1000 K region of the flame. The output spectra consisted of $P_{11}(20–15)$, $P_{10}(20–16)$, $P_9(21–13)$, $P_8(22–13)$, $P_7(22–12)$, $P_6(22–12)$, $P_5(19–15)$ and $P_4(20–15)$. The effect of additives to either the $CS_2–O_2$ or $CS_2–O_2–N_2O$ flame in their laser was similar to that reported by Searles and Djeu (1971). CO_2 and CO were most effective in increasing the output from the $CS_2–O_2$ flame. Addition of other gases to the $CS_2–O_2–N_2O$ flame laser decreased the output. Addition of fuels such as C_2N_2, C_2H_2 and hydrocarbons terminated lasing. The same burner was used to investigate low pressure flames of acetylene (C_2H_2), ethylene (C_2H_4), methane (CH_4), cyanogen (C_2N_2), methyl acetylene (CH_3CCH), dimethyl ether (CH_3COCH_3), ethylene oxide

(CH–CH) and methyl alcohol (CH_3OH). No lasing was observed with any of these fuels. A typical $CS_2–O_2$ flame laser is shown in fig. 36.

The chemistry of the $CS_2–O_2$ flame is extremely complex. Howgate and Barr (1973) selected 24 chemical reactions and have reviewed the status of the rate data and their temperature dependence. There appears to be little agreement among different investigators and for many reactions rate data are missing completely. The main reactions, with the rate constants in cm^3 $mole^{-1}$ s^{-1},

Fig. 36. $CS_2–O_2$ flame laser. The burner is horizontal just below the bright region. The glow is due to the $SO + O_2$ continuum.

selected by Howgate and Barr (1973) are

$$CS_2 + O \rightarrow CS + SO, \qquad 5 \times 10^{13} \exp(-1.9/RT),$$
$$O + CS_2 \rightarrow CO + S, \qquad 2.4 \times 10^{14} \exp(-2/RT),$$
$$S + O_2 \rightarrow SO + O, \qquad 1.0 \times 10^{12},$$
$$SO + O_2 \rightarrow SO_2 + O, \qquad 3.5 \times 10^{11} \exp(-6.5/RT).$$

Fig. 37 shows a plot of these rate constants over the temperature range from 300 to 1000 K. For a self-supporting flame the last two reactions are required to maintain an adequate O-atom supply. Also the last reaction requires a reasonably high temperature and O_2 pressure to become significant. Howgate and Barr (1973), Wittig et al. (1971), Suart et al. (1972), Lilenfeld et al. (1975) and Lilenfeld and Jeffers (1976) have modeled various $O + CS_2$ lasers with considerable success. No attempt has been made to model the flame laser.

The initial high specific energy, the chemical efficiency and the promise of high pressure operation made the CO chemical laser an attractive candidate for a high power, high efficiency laser. In addition, the required fuel (CS_2) and oxidizer

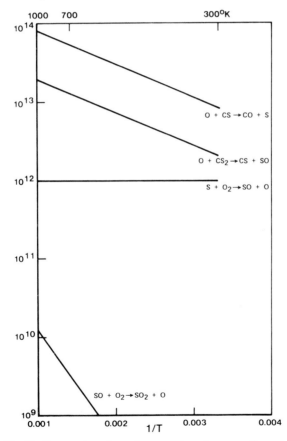

Fig. 37. Temperature dependence of flame laser reaction rates.

were inexpensive and relatively safe compared to those for HF. Failure to find a practical solution for the O-atom requirement has, however, led to a decreasing interest. Use of electrical energy is inefficient in that it requires an energy expenditure of at least 119 kcal ($O_2 \rightarrow 2O$) for a return of part of 85 kcal ($CS + O \rightarrow CO + S$). With the normal inefficiencies inherent in both steps the overall electrical efficiency becomes too low to make the chemical CO laser competitive with electrically excited CO or CO_2 lasers.

7. OH chemical laser

The chemical excitation of OH radicals has been of considerable interest because of the presence of hydroxyl emission bands in the night sky air glow. These bands consist of overtone emission sequences with $\Delta v = 4$, 5 and 6 from vibrationally excited OH in its electronic ground state. Heaps and Herzberg (1952) suggested the reaction

$$H + O_3 \rightarrow OH^+ + O_2 \qquad \Delta H = -77 \text{ kcal/mole},$$

as a possible source. Several investigations of this reaction confirmed that it could lead to vibrationally excited OH. These studies were carried out at relatively high pressures, and as a result the observed spectra include the results of perturbations due to the fast V–T and V–V processes. The first attempt to obtain the nascent energy partition of this reaction was by Anlauf et al. (1968). Working at pressures in the 3.3×10^{-4} Torr range, four orders of magnitude below previous workers, they obtained a distribution peaking at $v = 9$, the maximum accessible level. In a later refinement of this study, Charters et al. (1971) at 1×10^{-4} Torr, obtained $k_9 = 1.00$, $k_8 = 0.8$, $k_7 = 0.4$, $k_6 < 0.4$, all relative to k_9. The first stimulated OH emission was reported by Callear and Van den Bergh (1971). They flash photolyzed $O_3 + H_2$ mixtures at pressures from 1 to 10 Torr with an O_3–H_2 ratio of 1:10, and observed seven lines in the $v = 3$–2, 2–1 and 1–0 vibrational bands at 2.1 to 3.3 μm. The addition of He had no effect. Vietzke et al. (1971) obtained OH emission from O_3–H_2–He mixtures in the 3–2 and 2–1 vibrational bands using a 2 m long longitudinal pulsed discharge. Wauchop et al. (1974), using a transverse electric discharge through O_3–H_2–He mixtures, similarly observed lasing on the 3–2 and 2–1 vibrational bands. The output of OH was about 10^{-2} times that observed for SF_6–H_2 in the same laser. The OH laser's failure to produce the emission corresponding to the product distribution apparently results from the unavoidable presence of O_2 resulting from the initiation

$$O_3 \xrightarrow{h\nu} O_2 + O(^1D).$$

The O_2 is a fast relaxant for OH and leads to deactivation of OH in the higher vibrational states. Ducas et al. (1972) have observed lasing in pure rotational transitions of OH and OD radicals at 44 lines in the $v = 0$, 1 and 2 vibrational levels of OH covering wavelengths from 509 to 815 cm^{-1}. Seventeen lines were observed for OD in the $v = 0$ level at wavelengths in the 492 to 552 cm^{-1} range. Lasing was obtained from both longitudinal and transverse discharges through

H_2-O_2-SF_6 mixtures. The processes, chemical or otherwise, responsible for the excitation are unknown. Basov et al. (1972) and Searles and Airey (1973) have observed 10.6 μm CO_2 radiation from flashlamp pumped mixtures of O_3, D_2 and CO_2. Searles and Airey (1973) also observed transfer from OD to N_2O. No cw lasing has been reported from OH.

8. Miscellaneous lasers

8.1 Metal oxidation lasers

Explosion of metal wires or films in O_2 and F_2 has led to long wavelength laser emission (Rice and Jensen 1973, Rice and Beattle 1973, Dekoker and Rice 1974, Rice 1975). Thin wires or film of the metal supported on glass tubes are electrically exploded inside a laser tube filled with F_2 or O_2. A 0.4 to 3 μs laser pulse is observed during the latter part of the wire explosion. Lasing has been observed for Li (F_2), C (F_2 and O_2), Mg (F_2), Al (F_2), Ti (F_2, NF_3 and O_2), Fe (F_2), Ni (F_2), Cu (F_2), Pt (F_2), Au (F_2), U (F_2), U (O_2), Zr (O_2, F_2), V (F_2, O_2), Zn (F_2), Mo (F_2, O_2), Ag (F_2), Ta (F_2, O_2) and W (F_2, O_2). Little is known of the nature of the lasing species or their excitation. It is assumed to be a diatomic molecule produced by the reaction $M + O_2 \rightarrow MO + O$ or $M + F_2 \rightarrow MF + F$. No detailed spectroscopic measurements have been made. The wavelengths are generally between 10 and 24 μm, and presumably result from vibration–rotation transitions of the species MO or MF, although transient ions or even triatomic species cannot be ruled out at this time. In the case of B and Al, lasing has been observed in a quasi-cw mode by vaporizing B powder in a shock tube and expanding the resulting vapor with Ar through a series of nozzles and reacting it with F_2 (Johnson et al. 1976). Al was present as an impurity. Lasing observed between 6.1 and 13 μm was postulated to be due to BF, that observed in the 12.8 to 20 μm region is thought to arise from AlF. The pulse duration was in the millisecond range and the laser presumably would be continuous if the metal vapor flow could be maintained. A more detailed description of the exploding wire technique has been given by Jensen (1976).

8.2 CO–N_2O flame laser

A CO–N_2O flame laser has been reported by Benard et al. (1973). Walker et al. (1973) investigated a Na-catalyzed CO-N_2O flame:

$Na + N_2O \rightarrow NaO + N_2$, $\qquad \Delta H = -25.4 \, kcal/mole$,

$NaO + CO \rightarrow CO_2 + Na$, $\qquad \Delta H = -61.9 \, kcal/mole$.

At pressures in the 3.5 to 8 Torr range they observed strong emission from CO, CO_2 and N_2O hot bands in addition to strong Na D-line chemiluminescence. Somewhat surprisingly only N_2O 001 $\rightarrow$ 100 lasing was observed by Benard et al. (1973). The excitation mechanism and the reason for the lack of gain on CO_2 transitions are unknown.

9. Photodissociation lasers

In photodissociation lasers the excited molecule or atom results from the decomposition of an optically excited parent molecule. This process can lead to very large inversions if the photochemical dissociation produces only excited state species. A considerable number of photodissociation lasers are known. Only one, the iodine laser, has been the subject of extensive study and has been developed into a high power laser system. The Asterix III iodine laser system has produced 500 J pulses with a maximum power of 1 TW. A complete discussion of photodissociation lasers is beyond the scope of this review. Some of the more interesting aspects of these lasers will be discussed here to illuminate our current level of understanding of these systems. For a more detailed treatment of the I laser the reader is referred to a recent review by Hohla and Kompa (1976).

9.1 The iodine laser

The first chemical laser (Kasper and Pimentel 1965) was the I laser obtained by photodissociation of trifluorometryl iodide:

$$CF_3I + h\nu(2700 \text{ Å}) \rightarrow CF_3 + I(^2P_{1/2}),$$
$$I(^2P_{1/2}) \rightarrow (I(^2P_{3/2}) + h\nu(13\ 150 \text{ Å}).$$

A large number of other iodine compounds have been investigated and found to yield I*–I ratios sufficient for lasing. Hydrocarbons and fluorinated hydrocarbons (Kasper and Pimentel 1965, Hohla and Kompa 1972), alkyl arsine and phosphine iodides (Birich et al. 1974) have been used with varying degrees of success. Recently, Davis et al. (1976) re-examined a number of compounds and concluded that 1–C_3F_7I gave the best performance (see fig. 38). In agreement with this conclusion, CF_3I or C_3F_7I are used by most investigators. The reasons for this are evident from table 14. In terms of fractional I* yield (I*/I + I*) on photolysis, CH_3I, CF_3I and the higher perfluoroalkylates and fluorinated hydrocarbons are about equivalent. However, the rate constant for deactivation by the reactions

$$RI + I^* \rightarrow RI + I \rightarrow R + I_2,$$

is about 2 to 3 orders of magnitude smaller for the perfluoroalkylades than for the hydrogen compounds. The quenching rate constant of I* by I_2 is 2.2×10^{13} cm^3/mole s (Burde et al. 1975), considerably larger than that for most other molecules. The recombination reaction of CF_3 radicals is very fast but does not affect the laser operation, other than competing with reactions such as $CF_3 + I \rightarrow CF_3I$ or $CF_3 + I_2 \rightarrow CF_3I + I$ which regenerate the starting material. The recombination of iodine atoms,

$$I + I + M \rightarrow I_2 + M,$$

although slow, is important in that I_2 is an extremely effective quencher for I*

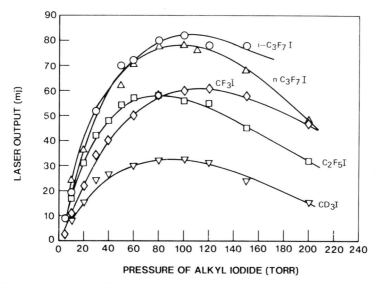

Fig. 38. Pressure dependence of the I photodissociation laser energy for various alkyl iodides, using fast flash excitation (Davis et al. 1976).

atoms. Build-up of I_2 in a repetitively pulsed laser leads to a drastic decrease in laser output (DeMaria and Ultee 1966, Fuss and Hohla 1976). Thus, at first sight, the chemical and kinetic processes appear rather simple, consisting of recombination of the CF_3 and I radicals and quenching of $I(^2P_{1/2})$ by CF_3I, CF_3, I_2 and

Table 14
Yield of I* and collisional deactivation.

	I* yield[a]	Deactivation rate constant[b] [cm^3 molecule^{-1} s^{-1}]
HI	0.10	1.5×10^{-13}
CH_3I	0.92	5.7×10^{-13}
CD_3I	0.99	1.8×10^{-14}
C_2H_5I	0.69	6.1×10^{-13}
$n\text{-}C_3H_7I$	0.67	3.0×10^{-10}
$i\text{-}C_3H_7I$	<0.10	6.3×10^{-13}
$n\text{-}C_4H_9I$	0.82	—
$s\text{-}C_4H_9I$	<0.10	—
$i\text{-}C_4H_9I$	0.69	—
$t\text{-}C_4H_9I$	<0.10	—
CF_3I	0.91	3.5×10^{-16}
C_2F_5I	<0.98	1.1×10^{-16}
$n\text{-}C_3F_7I$	<0.99	8×10^{-16}
$i\text{-}C_3F_7I$	0.90	3×10^{-16}

[a] Donohue and Wiesenfeld (1975a, b).
[b] Selected from a compilation by Davis et al. (1976).

C_2F_6 or the higher alkyls. These processes do indeed appear to describe the laser during the early part of the pulse. However, lasing can be observed after the flashlamp pulse (Gensel et al. 1971). Other investigators (Andreeva et al. 1969, 1971, DeWolf-Lanzerotti 1971, Zalesskii 1972, Hohla and Kompa 1972) have also suggested that additional pumping mechanisms may be present. Quantum yields larger than one (Gensel et al. 1971) also point to formation of I* by processes other than direct photolysis. Most often the reaction

$$CF_3 + CF_3I \rightarrow C_2F_6 + I*$$

is suggested as responsible, although the possible presence of F-atoms (Berry 1972) and pyrolysis during the photolysis makes other reactions also possible. In contrast to the above, Palmer and Gusinow (1974a, b) in a series of experiments with CF_3I were unable to find any evidence for delayed gain or lasing. Gusinow et al. (1973) have suggested that diffusion effects may be responsible. Thus, in spite of their apparent simplicity, the detailed chemistry and physics of the I* laser still remain uncertain at this time.

Detailed analysis and modeling have been published by Hohla and Kompa (1972), O'Brien and Bowen (1969, 1971), Zalesskii (1970), Zalesskii and Venediktov (1969), Zalesskii and Krupenikova (1973), Filyukov and Karpov (1972), Kuznetsova and Maslev (1974), Pirkle et al. (1975), Andreeva et al. (1969), Turner and Rapagni (1973) and Franklin (1973). Skorobogatov et al. (1974) have included fast CF_3 reactions and revised the rate constants to give a better representation of the chemistry. Zalevsskii (1975) has modeled the CF_3I laser with emphasis on the effects of pyrolysis and I_2 formation. Recent interest in the I laser has been stimulated by the possible use of this laser for controlled nuclear fusion. Early work by DeMaria and Ultee (1966) demonstrated the potential of this system. They obtained 65 J/pulse with peak powers of 10^5 W for 1.5 ms from flash photolysis of 15 Torr of CF_3I. Basov et al. (1973) reported 50 J in 5 ns, Hohla et al. (1975) reported 60 J in less than 1 ns and, more recently, Antonov et al. (1975) reported 720 J in 60 μs from C_3F_7I at 0.01 atm in a 152 ℓ volume. Pulses of 500 J in 500 ps have been reported from Asterix III at the Max Planck Institut für Plasma Physik (Hohla 1977). The Asterix III system has been described by Brederlow et al. (1976). The system consists of a mode-locked oscillator and four amplifiers of increasing diameter. The oscillator and amplifiers are pumped by Xe flashlamps. The laser medium consists of C_3F_7I with added Ar and CO_2 to broaden the emission line and provide for optimum energy storage and extraction. Fig. 39 shows the configuration and typical conditions. The design goals of this system are 1 kJ in 1 ns.

9.2 Other photodissociation lasers

The electronic structure of F, Br and Cl atoms is similar to that of I atoms, but with smaller $^2P_{1/2}-^2P_{3/2}$ separations. For F the separation is 404 cm^{-1}, for Cl 881 cm^{-1}, and for Br 3685 cm^{-1} or 24.7, 11.3 and 2.7 microns, respectively. Only in the case of Br have efforts to obtain a laser operating on this transition been

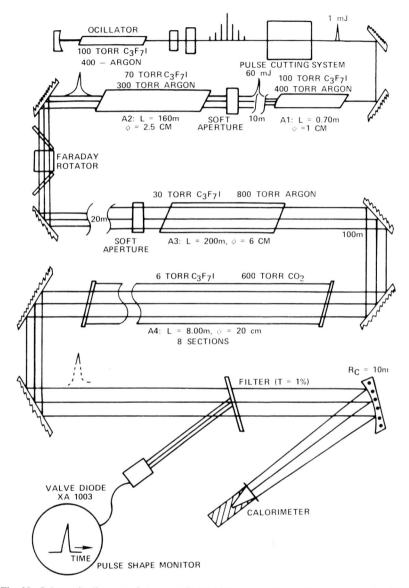

Fig. 39. Schematic diagram of the Asterix III high power I laser (Brederlow et al. 1976).

successful. Giuliano and Hess (1969, 1970) and Dudkin et al. (1970) have reported lasing from flash photolysis of IBr. Campbell and Kasper (1971) have reported lasing from CF_3Br. The observed power and efficiencies have been much lower than those for CF_3I.

Another photodissociation laser of considerable interest is the CN laser. Basco et al. (1963) had observed vibrationally excited CN in the flash photolysis of C_2N_2. Shuler et al. (1965) in an early assessment of possible chemical lasers suggested the possibility of stimulated emission from this system. Lasing from CN was first observed by Pollack (1966b). C_2N_2 at 15 to 20 Torr was flash photolyzed with a 2000 J pumping pulse. Emission was obtained at $P_4(9, 10, 11)$ at 5.18 to 5.20 μm. West and Berry (1974) have made a detailed study of the flash photolysis of a number of CN compounds. They obtained electronic A $^2\Pi_{3/2}$ $v = 0 \rightarrow X\,^2\Sigma^+$ $v = 2$ laser emission by photolysis of CH_3CN, C_2N_2 HCN, BrCN, CF_3CN and C_2F_5CN. The most intense emission was obtained with methyl isocyanate. The observed emission consisted of A $^2\Pi_{3/2}$ $v = 0 \rightarrow X\,^2\Sigma^+$ $v = 1$ Q(5–21) and the A $^2\Pi_{3/2}$ $v = 0 \rightarrow X\,^2\Sigma^+$ $v = 2$ Q(4–16) P(8–16) lines (see fig. 40) covering wavelengths from 1.09 to 2.02 μm. Ground state vibrational lasing was observed for $(CN)_2$, CF_3CN, BrCN and C_2F_5CN photolysis, but not with CH_3CN and HCN, apparently due to fast exchange and abstraction reactions. Emission was observed at $P_5(7–15)$, $P_4(5–15)$, $P_3(6–14)$ and $P_2(8, 9)$ for $(CN)_2$ (10 Torr + 40 Torr Ar). CF_3CN gave similar output with fewer J lines in the higher transitions but additional $P_2(10–31)$ lines. BrCN only lased on 4–3 and 3–2 transitions. CF_3CN did not lase on 2–1 transitions. Fig. 40 shows the CN potential energy level diagram with laser transitions. The authors suggest that the mechanism

$$RCN \xrightarrow{\ h\nu\ } R + CN^* \left(A\,^2\Pi_L, v = 0 \right),$$

$$CN^* \left(A\,^2\Pi_L, v = 0 \right) + M \rightarrow CN^+ \left(X\,^2\Sigma^+, v = 4 \right) + M,$$

is consistent with the experimental observations. The $(CN)_2$ and IBr lasers are of particular interest in that they form reversible systems, regenerating the parent compound after photodissociation. This allows an essentially closed system, as opposed to CF_3I and others where continuous flow is required.

Another photodissociation laser of interest is the NO laser first reported by Pollack (1966a, b, c). NOCl at 1 Torr with 1000 Torr. He was flash photolyzed. Lasing was observed on vibrational transition in the ground state. The observed lines between 6 and 6.3 μm were $P_6(16)$, $P_7(9, 10, 12, 14, 15)$, $P_8(7, 9, 11, 15, 16, 19)$, $P_9(19)$ of the $^2\Pi_{1/2}$ state and $P_7(14)$ and $P_8(11)$ of the $^2\Pi_{3/2}$ state. (NO has a doubled ground state split by 121 cm^{-1}.) The laser pulse lasts 4 μs and occurs before the flashlamp pulse reaches its maximum. The reasons for this quenching are unknown. Giuliano and Hess (1967) report similar observations. Deutsch (1966) reports NO lasing from an electric discharge through 5.8 Torr He with 0.1 Torr NOCl. Sixty lines were observed in the 11–10 through 6–5 vibrational levels with J values from 6 to 15. The pumping mechanism is unknown; it could involve the same excited electronic state of NOCl as in the flash photolysis, since direct electric discharge pumping of NO has not been reported. Lasing in the alkali metals has been obtained by photo-dissociation of the dimers. K_2, Rb_2 and Cs_2 on photodissociation give electronically excited K, Rb and CS atoms (Sorokin and Lankard 1969, 1971). It is of interest to note that the K laser was discussed in the original suggestions for an optical (as opposed to microwave) laser by Schalow and Townes in 1958!

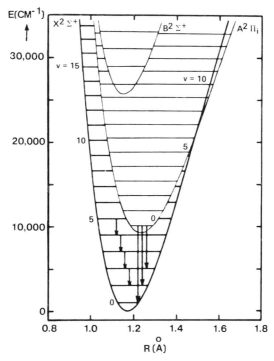

Fig. 40. Potential energy functions for CN and observed transitions (West and Berry 1974).

10. Gasdynamic lasers

In the mid 1960s the main candidate for high power cw lasers was electrically pumped CO_2. Simple scale-up of longitudinal N_2–CO_2 lasers led to outputs in the multikilowatt range with lasers which were hundreds of feet in length. Thus, although high powers could be obtained, the lasers became very cumbersome. The discharge pumping of CO_2, remarkably efficient, is limited by simultaneous heating of the gas which thermally populates the lower level, thus destroying the inversion. Cooling by diffusion to the walls eventually sets a limit on the power density achievable. A new approach which eventually circumvented this difficulty was based on the ideas of Basov and Oraevskii (1963), Hurle and Herzberg (1965) and Konyukhov and Prokhorov (1966). These workers suggested that a rapid temperature change in a gasdynamic process could create nonequilibrium populations among vibrational or electronic energy levels with relaxation rates too slow to follow the rapid heating or cooling. These ideas lead to the concept of gasdynamic lasers.

In the gasdynamic lasers a gas mixture at high temperature is expanded rapidly through a supersonic nozzle. The high temperature gas or gas mixture can be obtained in a number of ways; common methods are heating by electric arc, shock or combustion. Anderson (1976) has recently published a monograph on gasdynamic lasers and given a detailed discussion of the theory and modeling of the various types. The present discussion will only provide sufficient detail to indicate the differences and similarities between GDLs and combustion chemical lasers. Most GDL lasers reported have been with CO_2. Other gases such as CO (McKenzie 1972) and CS_2 (Gavrekov et al. 1976) have been used in GDLs; however, they are not amenable to heating in a combustion process. Only combustion GDLs will be considered here, thus limiting the discussion to CO_2. Fig. 41 shows the pertinent energy levels of CO_2 and N_2. The CO_2 laser operates between the 001 (ν_3) level at 2349 cm^{-1} and 100 (ν_1) and 020 (ν_2) levels at 1388 and 1286 cm^{-1} respectively. Because of Fermi resonance, the V–V processes between the ν_1 and ν_2 modes are very fast and these modes can be considered in 'local' equilibrium. They relax through V–T processes of ν_2. Relaxation rates for ν_3 are generally slower. The pertinent relaxation rates are shown in table 15. As can be seen from the table the relaxation rate of the ν_3 level, although slower than that of the ν_1 level, is still fast, particularly in the presence of H_2O. As in the electric CO_2 lasers, N_2 can also be used in GDLs as a vibrational energy storage reservoir. The V–T rates for N_2 are extremely slow, and the fast exchange rate between N_2 ($\nu = 1$) and CO_2 (001) creates an equilibrium between these two modes. The result is that the effective relaxation time for ν_3 in an

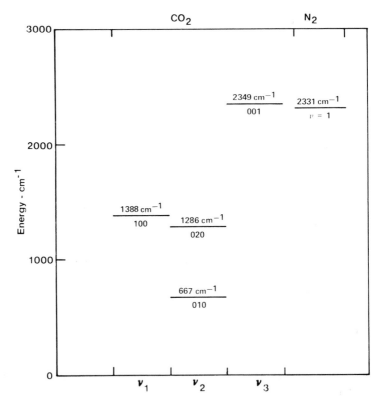

Fig. 41. CO_2–N_2 energy level diagram.

Table 15
CO_2 and N_2 relaxation rates.

	$k[\text{cm}^3\text{mole}^{-1}\text{s}^{-1}]$
$CO_2(001) + CO_2 \rightarrow CO_2(000) + CO_2$	2.4×10^9
$CO_2(001) + H_2O \rightarrow CO_2(000) + H_2O$	1.8×10^{11}
$CO_2(010) + CO_2 \rightarrow CO_2(000) + CO_2$	4.8×10^9
$CO_2(010) + H_2O \rightarrow CO_2(000) + H_2O$	6.0×10^{12}
$CO_2(010) + N_2 \rightarrow CO_2(010) + H_2O$	6.0×10^8
$N_2(v = 1) + N_2 \rightarrow N_2 + N_2$	$< 6 \times 10^3$
$N_2(v = 1) + H_2O \rightarrow N_2 + H_2O$	1.2×10^9

Data taken from a recent compilation by Lewis and
Trainor (1974).

N_2–CO_2 mixture is much longer than in pure CO_2. During expansion of a heated
gas through a supersonic nozzle the translational energy, i.e., the kinetic energy
due to the random motion of the molecules, is changed into directed kinetic
energy, i.e., the kinetic energy due to the motion in the flow direction. As a result,
the temperature defined by the random kinetic energy of the molecules
decreases, and the rotational temperature also decreases. The vibrational temperature, because of the slow conversion of vibrational to translational energy,
cannot follow the static temperature, and depending on the relaxation rate of the
particular level, it will stay between the stagnation (i.e., the combustor) temperature and the static temperature. Thus, when a N_2–CO_2 mixture at high
temperatures is expanded and cooled through a supersonic nozzle, the lower
CO_2 level populations tend to be characterized by a vibrational temperature
close to the translational temperature, while the N_2 vibrational temperature
remains frozen, i.e., close to the temperature prior to the expansion. In practice,
the difference between the upper and lower temperature is emphasized by
addition of H_2O as a 'catalyst'. The H_2O increases the relaxation rate of the
lower levels, keeping them close to equilibrium, and prevents build-up of
molecules in the lower laser level. The required high temperature, high pressure
gas mixture for CO_2 GDLs is generally obtained by either arc heating or
combustion. In the arc driven laser, N_2 is heated, and CO_2 and H_2O or He are
added at their optimum concentrations. In a combustion GDL, fuel is burned to
provide both the thermal energy and the CO_2 and H_2O. Since the combustion
fuel, diluent, temperature and combustion products are all interrelated, much
work has been carried out to define the optimum conditions. Gerry (1970)
describes a high power cw GDL using either CO or $(CN)_2$ as the fuel. The
combustion takes place at 15 to 20 atm; N_2 addition keeps the temperature at
~ 1400 K. The use of hydrogen-less fuels has the advantage that the H_2O
concentration can be independently varied by addition of small amounts of H_2.
Typically the combustion products have a composition of 91% N_2, 8% CO_2 and
1% H_2O. This mixture is expanded through an array of two-dimensional nozzles
to generate a Mach 4 supersonic flow at pressures of 0.1 atm and 300 to 400 K
gas temperatures. Fig. 42 shows typical conditions in the combustor of a GDL

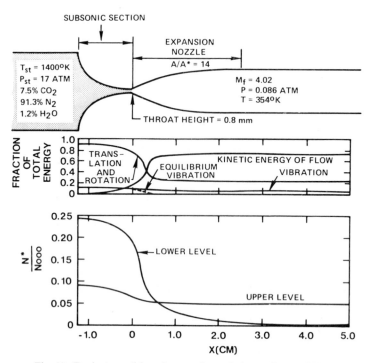

Fig. 42. Typical conditions in a gasdynamic laser (Gerry 1970).

just prior to entry into the nozzle. This mixture is generally completely equilibrated; the N_2 and CO_2 level populations can be calculated with the Boltzmann distribution equations. At 1400 K these populations as fractions of the total population of that species are

$$N_2(v = 2) = 0.01, \quad N_2(v = 1) = 0.08,$$
$$CO_2(001) = 0.015, \quad CO_2(100) = 0.04,$$
$$CO_2(020) = 0.14, \quad CO_2(010) = 0.17, \quad CO_2(000) = 0.17.$$

Fig. 42 shows the vibrational population for the upper and lower level of CO_2 upon expansion through the nozzle. Since the lower level relaxes much faster than the upper level (coupled with the N_2) an inversion results and is maintained for a considerable distance downstream.

Early GDLs generally used CO or C_2N_2 as fuels in order to allow independent control of the H_2O content. Many fuel–oxidizer combinations are more exothermic than $CO–O_2$; however, their combustion leads to too much H_2O to be useful. C_2N_2, although better than CO, is not convenient because of its very high toxicity. Pallay and Zovko (1973) have given an extensive review of GDL fuels, including solids.

There have been few detailed publications on combustion driven gasdynamic lasers. Meinzer (1972) has described results obtained with a small cw combustion

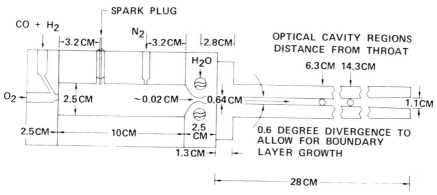

Fig. 43. Small combustion gasdynamic laser (Meinzer 1972).

GDL. The combustor and laser are shown schematically in fig. 43. The laser is operated by burning CO and H_2 with O_2, giving CO_2 and H_2O. N_2 is added as a diluent, to make a composition of CO_2–H_2O–N_2 = 22.1 : 5 : 73. With a total flow rate of 0.7 mole/s and a 2.4% transmitting mirror, laser power of 50 W was obtained. Gain in this GDL was measured in the 10 cm direction with a small stabilized CO_2 probe laser operating on a single frequency. Fig. 44 shows gain versus CO_2–H_2O ratio for a fixed N_2 concentration. Fig. 45 shows a ternary diagram mapping gain as a function of H_2O, CO_2 and N_2 concentration. These figures show that the gain is a sensitive function of the water content; maximum gain is generally obtained with water contents less than 10 mole %, although low gain can be observed with combustion products containing up to 30% water. At very low water concentrations the lower level is not fully deactivated, and gain again decreases. The low optimum water concentration limits the use of hydrocarbon fuels. Meinzer (1972) observed lasing with ethylene but not with

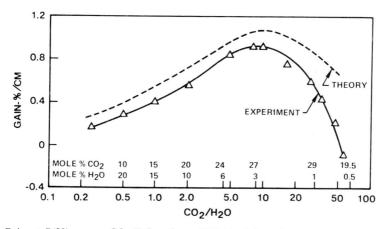

Fig. 44. Gain at P(20) versus CO_2–H_2O ratio at 70% N_2, 6.3 cm from nozzle throat. Total flow 0.6 mole/s, combustor pressure 12 atm, area ratio 30 (Meinzer 1972).

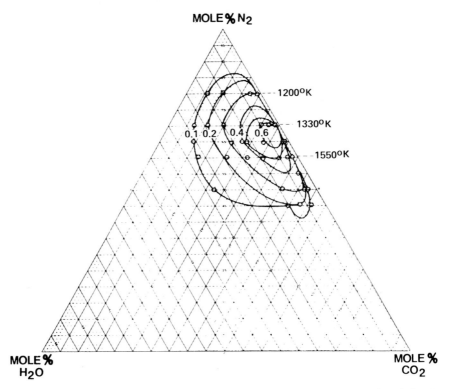

Fig. 45. Gain variation with combustion product composition. Total flow 0.3 mole/s, gain in % cm⁻¹ on P(20), 6.3 cm from nozzle throat stagnation pressure 2.7 to 7.7 atm (Meinzer 1972).

methane, in agreement with the data in fig. 45. Fig. 46 shows gain as a function of CO flow. For a fixed N_2 and O_2 flow the CO variation results in a combustion temperature variation, also given in fig. 46. From a gain versus wavelength (fig. 47) plot the rotational temperature can be estimated. This temperature was in good agreement with the temperature obtained from gasdynamic measurements.

The efficiency of the gasdynamic CO_2 laser is essentially determined by the vibrational population that can be achieved in a combustion process. Thus, one would expect a steady increase in laser power with increasing combustion temperature. However, aside from limitations set by available fuels and materials, the collisional deactivation also increases with temperature, and some optimum temperature should be reached around 3000 K. Highest power reported for a combustion gasdynamic laser is 60 kW for a 14 kg/s flow (Gerry 1970). Typical efficiencies achieved with CO_2 GDLs are in the 0.5 to 2% range, without much hope of large improvements in the near future. This low efficiency, or equivalently the low available laser energy per unit weight of fuel compared to combustion chemical lasers, makes the GDLs less attractive for high power applications. The GDL, however, is inherently a much simpler device than a chemical HF laser. The GDL requires no mixing in the laser cavity and, hence,

MEASUREMENT AT 6.3 CM FROM THROAT OF 2D COPPER NOZZLE

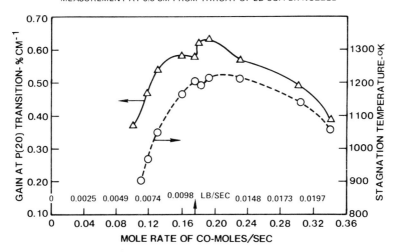

Fig. 46. Gain on P(20) and stagnation temperature as a function of CO flow. Other flows: $N_2 =$ 0.738 mole/s, $H_2 = 0.003$ mole/s, $O_2 = 0.089$ mole/s. Measurements 6.3 cm from throat (Meinzer 1972).

can use larger and less complex nozzles that the combustion driven chemical lasers. In addition, the chemical reaction releases considerable heat in the supersonic flow, which in turn can cause inhomogeneities affecting beam quality. Finally, GDLs can operate at pressure levels which allow recovery to pressure of 1 atm with a simple diffuser, thus eliminating the need for mechanical

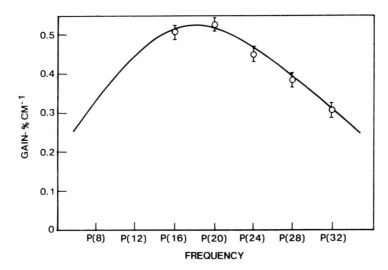

Fig. 47. Representative gain versus wavelength dependence. Distance from throat 14.3 cm. The solid line corresponds to a rotational temperature of 331 K. The temperature determined from gasdynamic measurements is 320 K.

pumps or ejectors. The choice of laser for a particular application thus becomes a trade-off of many factors and it appears that both GDLs and chemical lasers will find their place among high power lasers.

References

Ahearn, A.J. and N.B. Hannay (1953), J. Chem. Phys. **21**, 119.
Ahl, J.L. and W.R. Binns (1976), IEEE J. Quantum Electronics **QE–12**, 26.
Ahlborn, B., P. Gensel and K.L. Kompa (1972), J. Appl. Phys. **43**, 2487.
Airey, J.R. (1967), IEEE J. Quantum Electronics **QE–3**, 208.
Airey, J.R. (1970), J. Chem. Phys. **52**, 156.
Airey, J.R., P.D. Pacey and J.C. Polanyi (1967), Eleventh Symposium on Combustion, The Combustion Institute, Pittsburgh, PA.
Airey, J.R. and S.F. McKay (1969), Appl. Phys. Letters **15**, 401.
Airey, J.R. and I.W.M. Smith (1972), J. Chem. Phys. **57**, 1669.
Anderson, J.D., Jr. (1976), Gasdynamic Lasers (Academic Press; New York).
Andreeva, T.L., V.I. Malyshev, A.I. Maslov, I.I. Sobelman and V.N. Sorokin (1969), JETP Letters **10**, 271.
Andreeva, T.L., S.V. Kutnetsova, A.I. Maslov, I.I. Sobelman and V.N. Sorokin (1971), Sov. Phys.–JETP Letters **13**, 449.
Anlauf, K.G., D.H. Maylotte, P.D. Pacey and J.C. Polanyi (1967), Phys. Letters **24A**, 208.
Anlauf, K.G., R.G. MacDonald and J.C. Polanyi (1968), Chem. Phys. Letters **1**, 619.
Anlauf, K.G., D.S. Horne, R.G. MacDonald, J.C. Polanyi and K.B. Woodall (1972), J. Chem. Phys. **57**, 1561.
Antonov, A.V., N.G. Basov, V.S. Zuev, V.A. Katulin, K.S. Korol'kov, G.V. Mikhailov, V.N. Netemin, F.A. Nikolaev, V. Yu. Nasach, O. Yu. Petrov and A.V. Shelobolin (1975), Sov. J. Quantum Electronics **5**, 123.
Aprahamian, R., J.H.S. Wang, J.A. Betts and R.W. Barth (1974), Appl. Phys. Letters **24**, 239.
Arnold, G.P. and R.G. Wenzel (1973), IEEE J. Quantum Electronics **QE–9**, 491.
Arnold, S.J. and G.H. Kimbell (1969), Appl. Phys. Letters **15**, 351.
Arnold, S.J. and H. Rojeska (1973), Appl. Optics **12**, 169.
Arnold, S.J., K.D. Foster, D.R. Snelling and R.D. Suart (1977), Appl. Phys. Letters **30**, 637.
Asundi, R.K. and J.D. Craggs (1964), Proc. Phys. Soc. (London) **83**, 611.
Basco, N., J.E. Nicholas, R.G.W. Norrish and W.H.J. Vickers (1963), Proc. Roy. Soc. **272A**, 147.
Baskin, C.P., C.F. Bender, R.R. Lucchese, R.R. Bauschlicher, Jr., H.F. Schaefer (1976), J. Molec. Struct. **32**, 125.
Basov, N.G. and A.N. Oraevskii (1963), Sov. JETP **17**, 1171.
Basov, N.G., V.V. Gromov, E.L. Koshelev, E.P. Markin and A.N. Oraevskii (1969a), JETP Letters **9**, 147.
Basov, N.G., L.V. Kulakov, E.P. Markin, A.I. Nikitin and A.N. Oraevskii (1969b), JETP Letters **9**, 375.
Basov, N.G., V.T. Galochkin, V.I. Igoshin, L.V. Kulakov, E.P. Markin, A.I. Nikitin and A.N. Oraevskii (1971), Appl. Optics **10**, 1814.
Basov, N.G., A.S. Bashkin, V.I. Igoshin, A.N. Braevskii and N.N. Yuryshev (1972), JETP Letters **16**, 389.
Basov, N.G., L.E. Goluber, V.S. Zuev, V.A. Katulin, V.N. Netemin, V.Y. Nasach, O.Y. Nasach, and A.L. Petrov (1973), Kvarit. Elektron **6**, 116.
Batovskii, O.M. and V.I. Gur'ev (1974a), Sov. J. Quantum Electronics **4**, 380.
Batovskii, O.M. and V.I. Gur'ev (1974b), Sov. J. Quantum Electronics **4**, 801.
Batovskii, O.M., G.K. Vasil'ev, E.F. Makarov and V.L. Tal'roze (1969), JETP Letters **9**, 200.
Benard, D.J., R.C. Benson and R.E. Walker (1973), Appl. Phys. Letters **23**, 82.

Ben-Shaul, A., K.L. Kompa and V. Schmailze (1976), J. Chem. Phys. **65**, 1711.
Berry, M.J. (1972), Chem. Phys. Letters **15**, 269.
Berry, M.J. (1974), J. Chem. Phys. **61**, 3114.
Berry, M.J. and G.C. Pimentel (1968), J. Chem. Phys. **49**, 5190.
Berry, M.J. and G.C. Pimentel (1969), J. Chem. Phys. **51**, 2274.
Berry, M.J. and G.C. Pimentel (1970), J. Chem. Phys. **53**, 3453.
Bertrand, L., J.M. Gagne, B. Mongeau, B. LaPointe, Y. Conturie and M. Moisan (1977), J. Appl. Phys. **48**, 224.
Birich, G.N., G.I. Drozd, V.N. Sorokin and I.I. Struck (1974), Sov. Phys. – JETP Letters **19**, 27.
Boedeker, L.R., J.A. Shirley and B.R. Bronfin (1972), Appl. Phys. Letters **21**, 247.
Bogan, D.J. and D.W. Setser (1976), J. Chem. Phys. **64**, 586.
Bourcin, F., J. Menard and L. Henry (1970), Compt. Rend. Hebd. Séances Acad. Sc. **B270**, 494.
Brederlow, G., K.J. Witte, E. Fill, K. Hohla and R. Volk (1976), IEEE J. Quantum Electronics **QE–12**, 152.
Bronfin, B.R. and W.Q. Jeffers (1976), Handbook of Chemical Lasers (R.W.F. Gross and J.F. Bott, eds.; Wiley; New York).
Brus, L.E. and M.C. Lin (1971), J. Phys. Chem. **75**, 2546.
Brus, L.E. and M.C. Lin (1972), J. Phys. Chem. **76**, 1429.
Buczek, C.J., R.J. Freiberg, J.J. Hinchen, P.P. Chenausky and R J. Wayne (1970), Appl. Phys. Letters **17**, 514.
Burak, I., Y. Noter, A.M. Ronn and A. Szöke (1972), Chem. Phys. Letters **13**, 322.
Burde, D.H., R.A. McFarlane and J.R. Wiesenfeld (1975), Chem. Phys. Letters **32**, 296.
Burmasov, V.S., G.G. Dolgov'Savel'ev, V.A. Polyakov and G.M. Chumak (1969), JETP Letters **10**, 28.
Callear, A.B. and H.E. Van den Bergh (1971), Chem. Phys. Letters **8**, 17.
Campbell, J.D. and J.V.V. Kasper (1971), Chem. Phys. Letters **10**, 436.
Chang, H.W. and D.W. Setser (1973), J. Chem. Phys. **58**, 2298.
Charters, P.E. and J.C. Polanyi (1972), Disc. Faraday Soc. **33**, 107.
Charters, P.E., R.G. MacDonald and J.C. Polanyi (1971), Appl. Optics, **10**, 1747.
Chen, H.L., R.L. Taylor, J. Wilson, P. Lewis and W. Fyfe (1974), J. Chem. Phys, **61**, 3066.
Chester, A.N. and R.A. Chodzko (1976), Handbook of Chemical Lasers (R.W.F. Gross and J.F. Bott, eds.; Wiley; New York).
Chester, A.N. and L.D. Hess (1972), IEEE J. Quantum Electronics **QE–8**, 1.
Clough, P.N., J.C. Polanyi and R.T. Taguchi (1970), Can. J. Chem. **48**, 2919.
Clough, P.N., S.E. Schwartz and B.A. Thrush (1970b), Proc. Roy. Soc. (London) **A317**, 575.
Cohen, N. and J.F. Bott (1976), Handbook of Chemical Lasers (R.W.F. Gross and J.F. Bott, eds.; Wiley; New York).
Conturie, Y., J.M. Gagne and L. Bertrand (1976), Rev. Physique Appl. **11**, 421 (1976).
Cool, T.A. and R.R. Stephens (1970), J. Chem. Phys. **52**, 3304.
Cool, T.A., R.R. Stephens and J.A. Shirley (1970), J. Appl. Phys. **41**, 4038.
Coombe, R.D., A.T. Pritt, Jr. and D. Pilysovich (1975), Chem. Phys. Letters **35**, 345.
Corneil, P.H. and G.C. Pimentel (1978), J. Chem. Phys. **49**, 1379.
Cuellar, J.H. Parker and G.C. Pimentel (1974), J. Chem. Phys. **61**, 422.
Davis, C.C., R.J. Pirkle, R.A. MacFarlane and G.J. Wolga (1976), IEEE J. Quantum Electronics **QE–12**, 334.
Deka, B.K., P.E. Dyer and D.J. James (1976), Optics Commun **18**, 462.
Dekoker, J.G. and W.W. Rice (1974), J. Appl. Phys. **45**, 2770.
DeMaria, A.J. and C.J. Ultee (1966), Appl. Phys. Letters **9**, 67.
Deutsch, T.F. (1966), Appl. Phys. Letters **9**, 295.
Deutsch, T.F. (1967a), Appl. Phys. Letters **10**, 234.
Deutsch, T.F. (1967b), IEEE J. Quantum Electronics **QE–3**, 419.
Deutsch, T.F. (1967c), Appl. Phys. Letters **11**, 18.
Deutsch, T.F. (1971), IEEE J. Quantum Electronics **QE–7**, 174.
DeWolf-Lanzerotti, M.Y. (1971), IEEE J. Quantum Electronics **QE–7**, 207.

Dickenson, C.R. and R.N. Zare (1974), Chem. Phys. **7**, 361.

Djeu, N. (1974), J. Chem. Phys. **60**, 4105.

Djeu, N., H.S. Pilloff and S.K. Searles (1971), Appl. Phys. Letters **18**, 538.

Dobratz, B.M. (1972), Chemical Lasers: An Overview of the Literature, 1960–1971, Report UCRL-51285 TID-4500, UC-4, Lawrence Livermore Laboratory.

Dolgov-Savel'ev, G.G., V.A. Polyakov and G.M. Chumak (1970), Sov. Phys. – JETP **31**, 643.

Dolgov-Savel'ev, G.G., V.F. Zharov, Yu. S. Neganov and C.M. Chumak (1972), Sov. Phys. – JETP **34**, 34.

Donahue, T. and J.R. Wiesenfeld (1975a), Chem. Phys. Letters **33**, 176.

Donahue, T. and J.R. Wiesenfeld (1975b), J. Chem. Phys. **63**, 3130.

Donovan, R.J. and D. Husain (1971), Annual Report Chem. Soc. (London) **A68**.

Dovbysh, L.E., N.I. Zavada, A.T. Kazakevich, A.A. Karpikov, S.P. Melnikov, I.V. Podmoshenskii and A.A. Sinyanskii (1974), JETP Letters **20**, 183.

Ducas, T.W., L.D. Geoffrion, R.M. Osgood, Jr. and A. Javan (1972), Appl. Phys. Letters **21**, 42.

Dudkin, V.A., I.N. Knyazev and V.I. Malyshev (1970), Kratk. Soobshch Fix. **32**.

Duewer, W.H. and D.W. Setser (1973), J. Chem. Phys. **58**, 2310.

Emanuel, G. (1976), Handbook of Chemical Lasers (R.W.F. Gross and J.F. Bott, eds.; Wiley; New York).

English, J.R. III, H.C. Gardner, R.W. Mitchell and J.A. Merritt (1972), Chem. Phys. Letters **16**, 180.

Filyukov, A.A. and Ya. Karpov (1972), Sov. Phys. – JETP **35**, 63.

Foster, K.D. (1972), J. Chem. Phys. **57**, 2451.

Foster, K.D. and G.H. Kimbell (1970), J. Chem. Phys. **53**, 2539.

Fradin, D.W., P.P. Chenausky and R.J. Freiberg (1975), IEEE J. Quantum Electronics **QE–11**, 631.

Franklin, R.D. (1973), Technical Report ARL-73-0071, Wright-Patterson Air Force Base, Dayton, Ohio.

Fuss, W. and K. Hohla (1976), Opt. Commun. **18**, 427.

Gagne, J.M., S.Q. Mah and Y. Conturie (1974), Appl. Optics **13**, 2835.

Gagne, J.M., L. Bertrand, Y. Conturie, S.Q. Mah and J.P. Monchalin (1975), J. Opt. Soc. Amer. **65**, 876.

Gavrekov, V.F., A.P. Dronov, V.K. Orlov and A.K. Piskunov (1976), JETP Letters **23**, 595.

Gensel, P., K.L. Kompa and J. Wanner (1970a), Chem. Phys. Letters **7**, 538.

Gensel, P., K.L. Kompa and J. Wanner (1970b), Chem. Phys. Letters **5**, 179.

Gensel, P., K. Hohla and K.L. Kompa (1971), Appl. Phys. Letters **18**, 48.

Gerber, R.A. and E.L. Patterson (1974), IEEE J. Quantum Electronics **QE–10**, 333.

Gerber, R.A. and E.L. Patterson (1975), J. Appl. Phys. **47**, 3524.

Gerber, R.A., E.L. Patterson, L.S. Blair and N.R. Greiner (1974), Appl. Phys. Letters **25**, 281.

Gerry, E.T. (1970), IEEE Spectrum 7, 51.

Getzinger, R.W., N.R. Greiner, K.D. Ware, J.P. Carpenter and R.G. Wenzel (1976), IEEE J. Quantum Electronics **QE–12**, 556.

Getzinger, R.W., K.D. Ware, J.P. Carpenter and G.L. Schott (1977), IEEE J. Quantum Electronics **QE–13**, 97.

Giuliano, C.R. and L.D. Hess (1967), J. Appl. Phys. **38**, 4451.

Giuliano, C.R. and L.D. Hess (1969), J. Appl. Phys. **40**, 2428.

Giuliano, C.R. and L.D. Hess (1970), J. Quantum Electronics **QE–6**, 186.

Glaze, J.A. and G.J. Linford (1973), Rev. Sci. Instr. **44**, 600.

Glaze, J.A., J. Finzi and W.F. Krupke (1971), Appl. Phys. Letters **18**, 173.

Goldhar, J., R.M. Osgood, Jr. and A. Javan (1971), Appl. Phys. Letters **18**, 167.

Gordon, M.S., P.M. Saatzer and R.D. Koob (1976), Chem. Phys. Letters **37**, 217.

Gorshkov, V.I., V.V. Gromer, V.I. Igoshin, E.L. Keshelev, E.P. Markin and A.N. Oraevskii (1971), Appl. Optics **10**, 1781.

Green, W.H. and M.C. Lin (1971a), J. Chem. Phys. **54**, 3222.

Green, W.H. and M.C. Lin (1971b), IEEE J. Quantum Electronics **QE–7**, 98.

Gregg, D.W. and S.J. Thomas (1968), J. Appl. Phys. **39**, 4399.

Gregg, D.W., B. Krawetz, R.K. Pearson, B.R. Schleicher, S.J. Thomas, E.B. Huss, K.J. Pettipiece,

J.R. Creighton, R.E. Niver and Y.L. Pan (1971), Chem. Phys. Letters **8**, 609.

Greiner, N.R. (1973), IEEE J. Quantum Electronics **QE–9**, 1123.

Greiner, N.R. (1975), IEEE J. Quantum Electronics **QE–11**, 844.

Greiner, N.R., L.S. Blair and P.F. Bird (1974), IEEE J. Quantum Electronics **QE–10**, 646.

Gross, R.W.F. (1968), J. Chem. Phys. **48**, 3821.

Gross, R.W.F. (1969), J. Chem. Phys. **50**, 1889.

Gross, R.W.F. and J.F. Bott (1972), Handbook of Chemical Lasers (Wiley; New York).

Gross, R.W.F. and G. Emanuel (1976), Handbook of Chemical Lasers (R.W.F. Gross and J.F. Bott, eds.; Wiley; New York).

Gross, R.W.F. and J.D. Spencer (1976), Handbook of Chemical Lasers (R.W.F. Gross and J.F. Bott, eds.; Wiley; New York).

Gross, R.W.F. and F. Wesner (1974), Appl. Phys. Letters **23**, 559.

Gross, R.W.F., N. Cohen and T.A. Jacobs (1968), J. Chem. Phys. **48**, 3821.

Gusinow, M.A., J.K. Rice and T.D. Padrick (1973), Chem. Phys. Letters **21**, 197.

Hall, R.J. (1976), IEEE J. Quantum Electronics **QE–12**, 453.

Hancock, G., C. Morley and I.W.M. Smith (1971), Chem. Phys. Letters **12**, 193.

Heaps, H.S. and G. Herzberg (1952), Z. Phys. **133**, 48.

Henry, A., F. Bourcin, I. Arditi, R. Charneua and J. Menard (1968), Compt. Rend. Hebd. Séances Acad. Sci. **B267**, 616.

Henry, A., I. Arditi and L. Henry (1969), Compt. Rend. Hebd. Séances Acad. Sci. Paris, **B268**, 1245.

Herzberg, G. (1950), Spectra of Diatomic Molecules (Van Nostrand; Princeton, NJ).

Hess, L.D. (1971a), Appl. Phys. Letters **19**, 1.

Hess, L.D. (1971b), J. Chem. Phys. **55**, 2466.

Hess, L.D. (1972), J. Appl. Phys. **43**, 1157.

Heydtman, H. and J.C. Polanyi (1971), Appl. Optics **10**, 1738.

Hinchen, J.J. (1974), J. Appl. Phys. **45**, 1818.

Hinchen, J.J. (1975), Appl. Phys. Letters **27**, 672.

Hinchen, J.J. and C.M. Banas (1970), Appl. Phys. Letters **17**, 386.

Hinchen, J.J. and R.J. Freiberg (1976), Appl. Optics **15**, 459.

Hinchen, J.J. and R.H. Hobbs (1976), J. Chem. Phys. **65**, 2732.

Hirose, Y., J.C. Hassler and P.D. Coleman (1973), J. Quantum Electronics **QE–9**, 114.

Hofland, R., A. Ching, M.L. Lundquist and J.S. Whittier (1976), IEEE J. Quantum Electronics **QE–10**, 781.

Hohla, K. (1977), Paper Presented at the 4th Colloquium on Electronic Transition Lasers, June 20–24, Munich, Germany.

Hohla, K. and K. Kompa (1972), Z. Naturforsch **27a**, 938.

Hohla, K. and K.L. Kompa (1976), Handbook of Chemical Lasers (R.W.F. Gross and J.F. Bott, eds.; Wiley; New York).

Hohla, K., G. Brederlow, W. Fuss, K.L. Kompa, J. Raeder, R. Volk, S. Witkowski and K.J. Wille (1975), J. Appl. Phys. **46**, 808.

Hon, J.F. and J.R. Novak (1974), Appl. Phys. Letters **24**, 202.

Hopkins, M.M. and H.I. Chen (1972), Chem. Phys. Letters **17**, 500.

Hough, J.J.T. and R.L. Kerber (1975), IEEE J. Quantum Electronics **QE–11**, 699.

Howgate, D.W. and T.A. Barr (1973), J. Chem. Phys. **59**, 2815.

Hurle, I.R. and A. Hertzberg (1965), Phys. Fluids **8**, 1601.

Inoue, G. and S. Tsuchiya (1974), Jap. J. Appl. Phys. **13**, 1421.

Jacobson, T.V. and G.H. Kimbell (1970), J. Appl. Phys. **41**, 5210.

Jacobson, T.V. and G.H. Kimbell (1971a), J. Appl. Phys. **42**, 3402.

Jacobson, T.V. and G.H. Kimbell (1971b), Chem. Phys. Letters **8**, 309.

Jacobson, T.V. and G.H. Kimbell (1973), IEEE J. Quantum Electronics **QE–9**, 173.

Jacobson, T.V., G.H. Kimbell and D.R. Snelling (1973), IEEE J. Quantum Electronics **QE–9**, 496.

Jeffers, W.Q. (1972), Appl. Phys. Letters **21**, 267.

Jeffers, W.Q. and H.Y. Ageno (1975), Appl. Phys. Letters **27**, 227.

Jeffers, W.Q. and C.E. Wiswall (1970), Appl. Phys. Letters **17**, 67.

Jeffers, W.Q. and C.E. Wiswall (1973), Appl. Phys. Letters **23**, 626.

Jeffers, W.Q. and C.E. Wiswall (1974), J. Quantum Electronics **QE-10**, 860.

Jeffers, W.Q., H.Y. Ageno and C.E. Wiswall (1976a), J. Appl. Phys. **47**, 2509.

Jeffers, W.Q., H.Y. Ageno, C.E. Wiswall and J.D. Kelley (1976b), Appl. Phys. Letters **29**, 242.

Jeffers, W.Q., C.E. Wiswall and H.Y. Ageno (1976c), IEEE J. Quantum Electronics **QE-12**, 693.

Jensen, R.J. (1976), Handbook of Chemical Lasers (R.W.F. Gross and J.F. Bott, eds.; Wiley; New York).

Jensen, R.J. and W.W. Rice (1971), Phys. Chem. **76**, 805.

Johnson, R.L. and D.W. Setser (1969), Chem. Phys. Letters **3**, 207.

Johnson, R.L., M.J. Perona and D.W. Setser (1970), J. Chem. Phys. **52**, 6372.

Johnson, R.L., K.C. Kim and D.W. Setser (1973), J. Phys. Chem. **77**, 2499.

Jonathan, N., C.M. Melliar-Smith and D.H. Slater (1971a), Molec. Phys. **20**, 93.

Jonathan, N., C.H. Melliar-Smith, S. Okada, D.H. Slater and D. Timlin (1971b), Molec. Phys. **22**, 561.

Jonathan, N., S. Okuda and D. Timlin (1972), Molec. Phys. **24**, 1143.

Jones, C.R. and H.P. Broida (1974), J. Chem. Phys. **60**, 4369.

Kasper, J.V.V. and G.C. Pimentel (1964), Appl. Phys. Letters **5**, 231.

Kasper, J.V.V. and G.C. Pimentel (1965), Phys. Rev. Letters **14**, 352.

Kasper, J.V.V., J.H. Parker and G.C. Pimentel (1965), J. Chem. Phys. **43**, 1827.

Kelley, J.D. (1976), Chem. Phys. Letters **41**, 7.

Kerber, R.L., G. Emanuel and J.S. Whittier (1972), Appl. Optics **11**, 1112.

Kerber, R.L., N. Cohen and G. Emanuel (1973a), IEEE J. Quantum Electronics **QE-9**, 94.

Kerber, R.L., A. Ching, M.L. Lundquist and J.S. Whittier (1973b), IEEE J. Quantum Electronics **QE-9**, 607.

Khapov, Yu.I., A.G. Ponomarenko and R.I. Soborikhin (1976), Opt. Commun. **18**, 466.

Klimek, D.E. and M.J. Berry (1973), Chem. Phys. Letters **20**, 141.

Kim, K.C. and D.W. Setser (1973), J. Phys. Chem. **77**, 2493.

Kim, K.C., D.W. Setser and C.M. Bogan (1974), J. Chem. Phys. **60**, 1837.

Kompa, K.L. (1973), Fortschritte der Chemische Forschung **37**, Springer-Verlag, Berlin, p.3.

Kompa, K.L. and G. Pimentel (1967), J. Chem. Phys. **47**, 857.

Kompa, K.L. and J. Wanner (1972), Chem. Phys. Letters **12**, 560.

Kompa, K.L., J.H. Parker and G.C. Pimentel (1968), J. Chem. Phys. **49**, 4257.

Kompa, K.L., P. Gensel and J. Wanner (1969), Chem. Phys. Letters **3**, 210.

Kompa, K.L., J. Wanner and P. Gensel (1970), IEEE J. Quantum Electronics **QE-6**, 185.

Konyukov, W.K. and A.M. Prokhorov (1966), JETP Letters **3**, 286.

Krogh, O.D. and G.C. Pimentel (1972), J. Chem. Phys. **56**, 969.

Kuznetsova, S.V. and A.I. Maslev (1974), Sov. J. Quantum Electronics **3**, 468.

Lecuyer, A. and N. Legay-Sommaire (1970), Compt. Rend. Hebd. Séances Acad. Sci. **B271**, 1212.

Leone, S.R. and C.B. Moore (1973), Chem. Phys. Letters **19**, 340.

Lewis, P.F. and D.W. Trainor (1974), AVCO Everett Research Laboratories Report AMP422.

Lilenfeld, H.V. and W.Q. Jeffers, (1976), J. Appl. Phys. **47**, 2520.

Lilenfeld, H.V., R.F. Webbink, W.Q. Jeffers and J.D. Kelley (1975), J. Quantum Electronics **QE-11**, 660.

Lin, M.C. (1970), Chem. Phys. Letters **7**, 209.

Lin, M.C. (1971), J. Phys. Chem. **75**, 3642.

Lin, M.C. (1972), J. Phys. Chem. **76**, 1425.

Lin, M.C. (1973), Int. J. Chem. Kin. **5**, 173.

Lin, M.C. and S.H. Bauer (1970), Chem. Phys. Letters **7**, 223.

Lin, M.C. and L.E. Brus (1971), J. Chem. Phys. **54**, 5423.

Linevski, M.J. and R.A. Carabetta (1973), Appl. Phys. Letters **22**, 288.

Lyman, J.L. (1973), Appl. Optics **12**, 2736.

Mangano, J.A., R.L. Limpaecher, J.D. Daugherty and F. Russell (1975), Appl. Phys. Letters **27**, 293.

Manuccia, T.J., J.A. Stregack and W.S. Watt (1975), IEEE J. Quantum Electronics **QE-11**, 921.

Marcus, S. and R.J. Carbone (1971), IEEE J. Quantum Electronics **QE-6**, 493.

Mayer, S.W., D. Taylor and M.A. Kwok (1973), Appl. Phys. Letters **23**, 434.

Maylotte, H.M., J.C. Polanyi and K.B. Woodall (1972), J. Chem. Phys. **57**, 1547.
McKenzie, R.L. (1972), Phys. Fluids **15**, 2163.
Meinzer, R.A. (1970), Int. J. Chem. Kin. **2**, 335.
Meinzer, R.A. (1972), AIAA J. **10**, 388.
Meinzer, R.A. (1974), Paper Presented at the VIIIth International Quantum Electronics Conference, San Francisco, CA.
Menard-Bourcin, F., J. Menard and L. Henry (1975), J. Phys. Chem. **63**, 1479.
Moehlmann, J.G. and J.D. McDonald (1975), J. Chem. Phys. **62**, 3061.
Molina, M.J. and G.C. Pimentel (1972), J. Chem. Phys. **56**, 3988.
Moore, C.B. (1968), IEEE J. Quantum Electronics **QE-4**, 52.
Naegeli, D.W. and C.J. Ultee (1970), Chem. Phys. Letters **6**, 121.
Nichols, D.B., K.H. Wrolstad and J.D. McClure (1974), J. Appl. Phys. **45**, 5360.
Nichols, D.B., R.B. Hall and J.D. McClure (1976), J. Appl. Phys. **47**, 4026.
Obara, M. and T. Fujioka (1974a), Jap. J. Appl. Phys. **13**, 675.
Obara, M. and T. Fujioka (1974b), Appl. Phys. Letters **25**, 656.
Obara, M. and T. Fujioka (1975), Jap. J. Appl. Phys. **14**, 1183.
O'Brien, D.E. and J.R. Bowen (1969), J. Appl. Phys. **40**, 4767.
O'Brien, D.E. and J.R. Bowen (1971), J. Appl. Phys. **42**, 1010.
Oodate, H., M. Obara and T. Fujioka (1974), Appl. Phys. Letters **24**, 272.
Padrick, T.D. and G.C. Pimentel (1971), J. Chem. Phys. **54**, 720.
Padrick, T.D. and G.C. Pimentel (1972a), J. Phys. Chem. **76**, 3125.
Padrick, T.D. and G.C. Pimentel (1972b), Appl. Phys. Letters **20**, 167.
Pallay, B.G. and C.T. Zovko (1973), AIAA Paper No. 73–1233.
Palmer, R.E. and M.A. Gusinow (1974a), J. Appl. Phys. **45**, 2174.
Palmer, R.E. and M.A. Gusinow (1974b), IEEE J. Quantum Electronics **QE-10**, 615.
Pan, Y.L., C.E. Turner, Jr. and K.J. Pettipiece (1971), Chem. Phys. Letters **10**, 577.
Parker, J.H. (1975), Int. J. Chem. Kin. **7**, 433.
Parker, J.H. and G.C. Pimentel (1968), J. Chem. Phys. **48**, 5273.
Parker, J.H. and G.C. Pimentel (1969), J. Chem. Phys. **51**, 91.
Parker, J.H. and G.C. Pimentel (1970), IEEE J. Quantum Electronics **QE-6**, 175.
Parker, J.H. and G.C. Pimentel (1971), J. Chem. Phys. **55**, 857.
Parker, J.V. and R.R. Stephens (1973), Appl. Phys. Letters **22**, 450.
Patterson, E.L. and R.A. Gerber (1975), IEEE J. Quantum Electronics **QE-11**, 642.
Patterson, E.L., R.A. Gerber and L.S. Blair (1974), J. Appl. Phys. **45**, 1822.
Pearson, R.K., J.O. Cowler, G.L. Hermann, D.W. Gregg and J.R. Creighton (1973), IEEE J. Quantum Electronics **QE-9**, 879.
Penner, S.S. (1961), J. Quant. Spectrosc. Radiative Transfer **1**, 163.
Pilloff, H.S., S.K. Searles and N. Djeu (1971), Appl. Phys. Letters **19**, 9.
Pimentel, G.C. (1973), J. Chem. Phys. **58**, 1270.
Pimentel, G.C. and K.L. Kompa (1972), Handbook of Chemical Lasers (Wiley; New York).
Pirkle, R.J., C.C. Davis and R.A. MacFarlane (1975), Chem. Phys. Letters **36**, 305.
Poehler, T.O. and R.E. Walker (1973), Appl. Phys. **22**, 282.
Poehler, T.O., M. Shandor and R.E. Walker (1972), Appl. Phys. Letters **20**, 497.
Poehler, T.O., J.C. Pirkle, Jr and R.E. Walker (1973), IEEE J. Quantum Electronics **QE-9**, 83.
Polanyi, J.C. (1960), Proc. Roy. Soc. (Canada) **54C**, 25.
Polanyi, J.C. (1961), J. Chem. Phys. **34**, 347.
Polanyi, J.C. (1963), J. Quant. Spectrosc. Radiative Transfer **3**, 471.
Polanyi, J.C. (1967), Phys. Letters **24A**, 208.
Polanyi, J.C. and J.J. Sloan (1972), J. Chem. Phys. **57**, 4988.
Polanyi, J.C. and K.B. Woodall (1972), J. Chem. Phys. **57**, 1574.
Pollack, M.A. (1966a), Appl. Phys. Letters **8**, 237.
Pollack, M.A. (1966b), Appl. Phys. Letters **9**, 94.
Pollack, M.A. (1966c), Appl. Phys. Letters **9**, 230.
Ponomarenko, A.G., R.I. Solonkhin and Yu.I. Khapov (1975), Sov. Phys.–Dokl. **20**, 282.

Poole, P.R. and G.C. Pimentel (1975), J. Chem. Phys. **63**, 1950.

Powell, H.T. (1973), J. Chem. Phys. **59**, 4937.

Powell, H.T. and J.D. Kelley (1974), J. Chem. Phys. **60**, 2191.

Proch, D. and J. Wanner (1971), Tables of Vibrational Rotational Transitions in Diatomic Molecules Pertinent to Chemical Lasers, Max-Planck Institut für Plasma Physik, 1PP LV/17.

Proch, D., H. Pummer, K.L. Kompa and J. Wanner (1975), Rev. Sci. Instr. **46**, 1101.

Pukhal'skaya, G.V., G.P. Zhitneva, S.Ya. Pshezhetskii, A.L. Evseenko and N.V. Sikeeva (1975), Sov. J. Quantum Electronics **2**, 1701.

Pummer, H. and K.L. Kompa (1972), Appl. Phys. Letters **20**, 356.

Pummer, H., W. Breitfeld, H. Wedler, G. Klement and K.L. Kompa (1973), Appl. Phys. Letters **22**, 319.

Rice, W.W. (1975), IEEE J. Quantum Electronics **QE–11**, 689.

Rice, W.W. and W.H. Beattle (1973), Chem. Phys. Letters **19**, 82.

Rice, W.W. and R.J. Jensen (1972), J. Phys. Chem. **76**, 805.

Rice, W.W. and R.J. Jensen (1973), Appl. Phys. Letters **22**, 67.

Rice, W.W., W.H. Beattle, R.C. Oldenborg, S.E. Johnson and P.B. Scott (1976), Appl. Phys. Letters **28**, 444.

Ridley, B.A. and I.W.M. Smith (1972), J. Chem. Soc. Faraday Trans. **68**, 123.

Robinson, C.P., R.J. Jensen and A. Kolb (1973), IEEE J. Quantum Electronics **QE–9**, 963.

Roebber, J. and G.C. Pimentel (1973), IEEE J. Quantum Electronics **QE–9**, 201.

Rosen, D.I., R.N. Sileo and T.A. Cool (1973), IEEE J. Quantum Electronics **QE–9**, 163.

Rosenwaks, S. and S. Yatsiu (1971), Chem. Phys. Letters **9**, 266.

Rosenwaks, S. and I.W.M. Smith (1973), Trans Faraday Soc. **69**, 1416.

Sadie, F.G., P.A. Büger and O.G. Malan (1972), J. Appl. Phys. **43**, 2906.

Schalow, A.L. and C.H. Townes (1958), Phys. Rev. **112**, 1940.

Schilling, P. and G. Decker (1976), Infrared Phys. **16**, 103.

Searles, S.K. and J.R. Airey (1973), Appl. Phys. Letters **22**, 282.

Searles, S.K. and N. Djeu (1971), Chem. Phys. Letters **12**, 53.

Searles, S.K. and N. Djeu (1973), IEEE J. Quantum Electronics **QE–9**, 116.

Sentman, L.H. (1973), Chem. Phys. Letters **18**, 493.

Sentman, L.H. (1975), J. Chem. Phys. **62**, 3523.

Sentman, L.H. (1976), Appl. Optics **15**, 744.

Sentman, L.H. and W.C. Solomon (1973), J. Chem. Phys. **59**, 89.

Shuler, K.E., T. Carrington and J.L. Light (1965), Appl. Optics Chem. Laser Suppl. **87**.

Skifstad, J.G. and C.M. Chao (1975), Appl. Optics **14**, 1713.

Skorobogatov, G.A., V.M. Tret'yak and V.S. Komarov (1974), Sov. Phys. – Tech. Phys. **19**, 495.

Sorokin, R.P. and J.R. Lankard (1969), J. Chem. Phys. **51**, 2929.

Sorokin, R.P. and J.R. Lankard (1971), J. Chem. Phys. **54**, 2184.

Spencer, D.J., T.A. Jacobs, H. Mirels and R.W.F. Gross (1969), Int. J. Chem. Kin. **1**, 493.

Stephens, R.R. and T.A. Cool (1971), Rev. Sci. Instr. **42**, 1489.

Stephens, R.R. and T.A. Cool (1972), J. Chem. Phys. **56**, 5214.

Suart, R.D., G.H. Kimbell and S.J. Arnold (i970a), Chem. Phys. Letters **5**, 519.

Suart, R.D., S.J. Arnold and G.H. Kimbell (1970b), Chem. Phys. Letters **7**, 337.

Suart, R.D., P.H. Dawson and G.H. Kimbell (1972), J. Appl. Phys. **43**, 1022.

Suchard, S.N. (1973a), J. Chem. Phys. **58**, 1269.

Suchard, S.N. (1973b), Appl. Phys. Letters **23**, 68.

Suchard, S.N. and J.R. Airey (1976), Handbook of Chemical Lasers (R.W.F. Gross and J.F. Bott, eds.; Wiley; New York).

Suchard, S.N. and G.C. Pimentel (1971), Appl. Phys. Letters **18**, 530.

Suchard, S.N., R.W.F. Gross and J.S. Whittier (1971), Appl. Phys. Letters **19**, 411.

Suchard, S.N., A. Ching and J.S. Whittier (1972a), Appl. Phys. Letters **21**, 274.

Suchard, S.N., R.L. Kerber, G. Emanuel and J.S. Whittier (1972b), J. Chem. Phys. **57**, 5065.

Sullivan, J.H., R.C. Feber and J.W. Starner (1975), J. Chem. Phys. **62**, 1714.

Taylor, D., S.W. Mayer and S.N. Suchard (1974), IEEE J. Quantum Electronics **QE–10**, 389.

Tiee, J.J., C.R. Quick, Jr., C.D. Harper, A.B. Peterson and C. Wittig (1975), J. Appl. Phys. **46**, 5191.

Truby, F.K. (1976), Appl. Phys. Letters **29**, 247.

Tsuchiya, S., N. Nielsen and S.H. Bauer (1973), J. Phys. Chem. **77**, 2455.

Turner, R. and T.O. Poehler (1976), J. Appl. Phys. **47**, 3038.

Turner, C. and N.L. Rapagni (1973), UCRL Report 50021-73-1, Lawrence Livermore Laboratory, Livermore, CA.

Ultee, C.J. (1970), IEEE J. Quantum Electronics **QE–6**, 617.

Ultee, C.J. (1971a) Rev. Sci. Instr. **42**, 1174.

Ultee, C.J. (1971b), Appl. Phys. Letters **19**, 535.

Ultee, C.J. (1972), IEEE J. Quantum Electronics **QE–8**.

Ultee, C.J. (1977), Chem. Phys. Letters **46**, 366.

Ultee, C.J. and P. Bonczyk (1974), J. Quantum Electronics **QE–10**, 105.

Vietzke, E., H.I. Schiff and K.H. Welge (1971), Chem. Phys. Letters **12**, 429.

Voignier, F. and M. Gastaud (1974), Appl. Phys. Letters **25**, 649.

Walker, R.E., B.F. Hochheimer, T.W. Wang and J.C. Creeden (1973), Chem. Phys. Letters **20**, 528.

Wauchop, T.S., H.I. Schiff and K.H. Welge (1974), Rev. Sci. Instr. **45**, 653.

Wenzel, R.G. and G.P. Arnold (1976), IEEE J. Quantum Electronics **QE–8**, 26.

West, G.A. and M.J. Berry (1974), J. Chem. Phys. **61**, 4700.

Whittier, J.S. and R.L. Kerber (1974), IEEE J. Quantum Electronics **QE–10**, 844.

Wilkins, R.L. (1972), J. Chem. Phys. **57**, 912.

Wilson, J. and J.S. Stephenson (1972), Appl. Phys. Letters **20**, 64.

Wiswall, C.E., D.P. Ames and T.J. Menne (1973), IEEE J. Quantum Electronics **QE–9**, 181.

Wittig, C. and I.W.M. Smith (1972), Appl. Phys. Letters **21**, 536.

Wittig, C., J.C. Hassler and P.D. Coleman (1970a), Appl. Phys. Letters **16**, 117.

Wittig, C., J.C. Hassler and P.D. Coleman (1970b), Nature **226**, 845.

Wittig, C., J.C. Hassler and P.D. Colemen (1971), J. Chem. Phys. **55**, 5523.

Wood, O.R. and T.Y. Chang (1972), Appl. Phys. Letters **20**, 77.

Zalesskii, V.Yu. (1972), Sov. Phys.–JETP **34**, 474.

Zalesskii, V.Yu. (1975), Sov. J. Quantum Electronics **4**, 1009.

Zalesskii, V.Yu. and T.I. Krupenikova (1973), Opt. Spectrosc. **30**, 439.

Zalesskii, V.Yu.and E.I. Moskalev (1970), Sov. Phys. – JETP **30**, 1019.

Zalesskii, V.Yu.and A.A. Venedektov (1969), Sov. Phys. – JETP **28**, 1104.

A6 | Pulsed Dye Lasers

Fakultät für Physik, Universität Bielefeld,
48 Bielefeld, West Germany

Contents

Abstract

Construction and performance of pulsed dye lasers are discussed by reviewing the physics and technology of dye laser devices excited by flashlamps or powerful pulsed laser light sources. The review includes basic properties and experimental results obtained for dye laser oscillators, dye amplifiers, oscillator-amplifier systems and injection-locked dye lasers. In addition some important applications of dye lasers such as mode-locking and frequency mixing in nonlinear optical materials are considered.

© *North-Holland Publishing Company, 1979*
Laser Handbook, edited by M.L. Stitch

1. Introduction

Since the first report on laser action in organic dyes by Sorokin and Lankard and by Schäfer, Schmidt and Volze in 1966, dye laser physics and technology have developed very rapidly. Important factors in this fast advance were certainly the considerable interest in the quantum electronics of this new type of laser and the unique applications in many fields of physics, chemistry or biology. Most applications for dye lasers depend on their remarkable capability to generate intense coherent light which is frequency tunable continuously over a wide range of the optical spectrum. Presently dye lasers optically pumped by intense flashlamps or powerful pulsed lasers operate from the near ultraviolet (340 nm) to the near infrared (1200 nm). With excitation by the visible or ultraviolet output of gas ion lasers, continuous dye laser emission has also been generated throughout the visible and the near infrared. A considerable extension of the accessible wavelength range in both directions towards the vacuum ultraviolet and the far infrared is possible by frequency mixing of dye laser radiation in nonlinear optical materials. Owing to the high spectral intensity of generated dye laser pulses, efficiencies up to 50% have been obtained for such nonlinear frequency conversion. The output energy of pulsed dye lasers presently ranges from a few microjoules to more than 10 joules per pulse and the peak power lies between some milliwatts and more than 10 megawatts. Continuous wave (cw) dye lasers generate an output of a few milliwatts to some ten watts. An average power output up to more than hundred watts is obtained with flashlamp-pumped high repetition rate dye laser devices. Besides wide tunability and high output power, narrow spectral width of the dye laser light is of great importance for many applications. The coherent emission of pulsed and cw dye lasers can be narrowed quite easily to bandwidth between 1 and 10 pm with wavelength-selective cavity elements such as prisms or Fabry–Perot-type interferometers. With multiple dispersive cavity elements the linewidth of pulsed lasers has been reduced to less than 0.04 pm or 30 MHz (at 600 nm). Even narrower linewidths down to less than 1 MHz can be generated with frequency controlled cw dye lasers. Although narrow bandwidth, wideband tunability and high output power are very desirable in most applications, these requirements are somewhat contradictory for laser construction and operation. Dye laser oscillator–amplifier systems, as developed more recently, can, however, provide all these essential properties to a large extent. In addition to these advantageous properties, dye lasers are extremely versatile for the generation of short light pulses. Flashlamp-pumped devices can provide pulses with durations ranging from a microsecond to several milliseconds. Laser-pumped dye lasers operate at pulselengths from 1 to 200 ns. Extremely short pulses down to less than 1 ps can be obtained by mode-locking.

In the following sections construction and performance characteristics of pulsed dye lasers will be discussed by reviewing important results which have been obtained so far in dye laser research. The review will start with a short outline of dye properties and the analysis of dye lasers. A more detailed description of the physics and technology of laser- and flashlamp-pumped dye lasers is then followed by a discussion of the essential properties of special dye laser devices such as dye laser amplifiers, oscillator-amplifier systems, and, injection-locked dye lasers. The contribution will conclude with a review of basic results which have been reported in important fields such as mode-locking of dye lasers and the generation of tunable coherent light in the infrared and ultraviolet by frequency mixing of dye laser radiation in nonlinear optical materials. Finally, future prospects for new dye lasers, as, e.g., vapor phase systems, will be considered.

2. Principles of dye laser operation

2.1 Dye properties

Operation and construction of organic dye lasers depend on the photophysical and chemical properties of the dye solutions, which are used as laser medium. The invention of dye lasers therefore triggered a thorough investigation of the chemical and spectroscopic properties of suitable laser dyes, which are large organic molecules with conjugated double bonds. The results of this research have led to a more detailed understanding of the connections between molecular structure and those spectroscopic properties which make a dye suitable for use as an efficient laser medium. As a consequence, numerous new and efficient laser dyes have been synthesized. The structural and spectroscopic properties of efficient laser dyes have been discussed in detail in several review articles, as, for example, in the excellent summaries given by Schäfer (1973) and Drexhage (1973). Since a detailed review on dye properties is also given by Schäfer in Vol. 1 of this handbook, the present chapter will only summarize those spectroscopic dye properties which are most essential for laser operation.

The spectral behavior of most dyes can be described in zero-order terms by means of a level diagram, which consists of a series of singlet and triplet states (fig. 1). Owing to the numerous vibrational and rotational sublevels the electronic singlet and triplet states are broad quasi-continuous bands. Strong, broadband absorption of visible or ultraviolet light excites the molecules from the ground state S_0 to a level B' of the first excited singlet state S_1. The excitation may also populate levels B" in the S_2 state. The excited states B' and B" are transfered

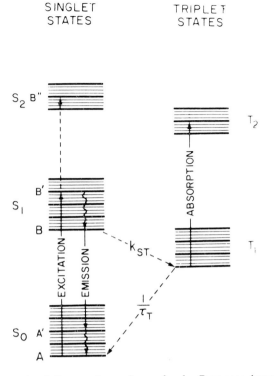

Fig. 1. Schematic energy level diagram for a dye molecule. Processes important for dye laser operation are indicated by arrows.

rapidly by radiationless transitions into the state B. The B state is the upper laser level, and laser emission results from the stimulated transition of B to A', a higher lying vibration–rotation state of S_0. Populated levels A' thermalize rapidly into the lowest vibration–rotation levels A of S_0. Molecules in the excited state B may decay by several competing processes. Besides stimulated laser transitions B → A', most of the B-state molecules normally return to levels A' by spontaneous emission, which causes intense broadband fluorescence. In addition to these radiative processes, nonradiative decay may transfer the B state into the S_0 ground state (internal conversion) or into T_1, the lowest triplet state (inter-system crossing). Molecules in the B state which decay by spontaneous radiative or nonradiative processes are lost for stimulated laser emission. Thus the ratio of the rate for stimulated decay and the rate of the fluorescent decay, together with the nonradiative transition rates, is an important factor for the efficiency of dyes, when used as laser medium. In good laser dyes the fluorescent quantum yield Φ is close to one. For a 10^{-3} M/ℓ solution of rhodamine 6G in water, for example, $\Phi = 0.92$. The rate for internal conversion to the electronic ground state is negligibly small and only a few percent of the excited molecules decay via intersystem crossing into T_1. Transitions from T_1 to S_0, however, are spin forbidden and the corresponding transition rate constants range from 10^3 to

$10^7 s^{-1}$, depending upon the chemical environment of the dye molecule. Molecules such as O_2 or triplet quenching agents like cyclooctatetraene tend to enhance the $T_1 \rightarrow S_0$ process and thus shorten the triplet lifetime τ_T. Conversely, τ_T may be rather long if no triplet quenching molecules are present in the dye solution.

In all cases the transition $T_1 \rightarrow S_0$ is slow compared to $S_1 \rightarrow S_0$, which has a rate constant τ_1^{-1} for spontaneous decay of about $2 \times 10^8 s^{-1}$. Therefore the triplet state acts as a trap for excited molecules and reduces the number of molecules available for the laser process. The accumulation of molecules in the lowest triplet state T_1 leads to further losses for the laser process. This is due to the large cross section for excitation into higher excited triplet states T_2. The absorption spectrum of those transitions overlap the $S_1 \rightarrow S_0$ fluorescence band. Consequently the absorption $T_1 \rightarrow T_2$ produces a large optical loss at wavelengths for which laser emission is probable. This loss may inhibit or halt the lasing process even if only a relatively small fraction of the molecules is accumulated in T_1.

The spectral dependence of the absorption cross section $\sigma_A(\lambda)$ for $S_0 \rightarrow S_1$ transitions is displayed in fig. 2 for the dye rhodamine 6G, which is one of the most efficient laser dyes. Compared to $\sigma_A(\lambda)$ the fluorescence spectrum $E(\lambda)$ is shifted towards longer wavelengths. The energy difference between absorption and emission is taken up by the radiationless processes $B' \rightarrow B$ and $A' \rightarrow A$. For dye laser operation this separation between the fluorescence and singlet absorption spectra has the important consequence that unexcited dye solution is transparent for spontaneous and stimulated fluorescence. Fig. 2 also displays the cross section $\sigma_E(\lambda)$ for stimulated emission, which is given by Yariv (1967)

$$\sigma_E(\lambda) = E(\lambda)\lambda^4/8\pi\tau_1 cn. \tag{1}$$

In this equation, n is the index of refraction, c the speed of light and $E(\lambda)$ the fluorescence line shape, normalized such that $\int_0^\infty E(\lambda)\, d\lambda = \Phi$. Owing to the strong fluorescence $E(\lambda)$, rhodamine 6G exhibits, like other dyes, large cross sections for stimulated fluorescence. The triplet absorption cross section $\sigma_T(\lambda)$ for the process $T_1 \rightarrow T_2$ is also included in fig. 2. The shape of $\sigma_T(\lambda)$ was measured by Morrow and Quinn (1973) for a solution of rhodamine 6G in ethanol. The comparison of the triplet absorption of other dyes in different solvents indicates, however, that the triplet spectra would not differ greatly if rhodamine 6G were dissolved in water or ethanol.

2.2 Dye laser analysis

In principle, a dye laser consists of an optically excited dye solution placed inside an optical laser resonator. Many aspects of the characteristic performance of such lasers can be deduced from the optical dye properties which were summarized in the previous section. An adequate tool for the analysis of the excited dye laser medium and the laser process are the rate equations which

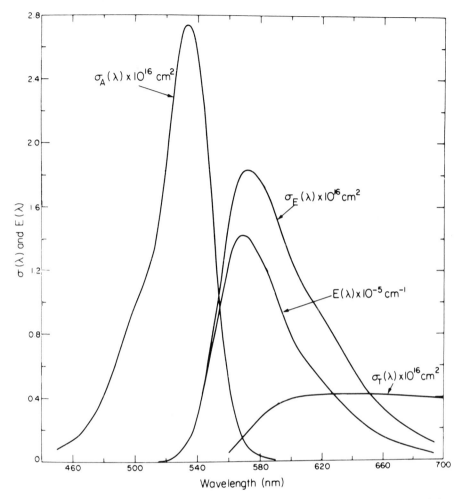

Fig. 2. Spectral dependence of singlet absorption $\sigma_A(\lambda)$ and emission $E(\lambda)$, as measured for a $10^{-4}\,M/\ell$ solution of rhodamine 6G in water with 2% Ammonyx LO added for deaggregation. The stimulated emission cross section $\sigma_E(\lambda)$ is calculated according to eq. (1). The triplet absorption $\sigma_T(\lambda)$ was measured for a solution in ethanol.

describe the balance of pump light absorption, the number of molecules in the ground and excited states and the emitted photons which are generated by spontaneous or stimulated fluorescence. This fluorescence is reduced by various losses in the laser medium and the laser resonator. According to the excitation and de-excitation processes which were discussed in § 2.1, the rate equation for the population density N_1 of the first excited singlet level S_1 is given by

$$\frac{dN_1}{dt} = N_0(\sigma_{AP}I_P + \sigma_{AL}I_L) - N_1\tau_1^{-1} - N_1\sigma_E I_L - N_1(\sigma'_{AP}I_P + \sigma'_{AL}I_L) + N_2\tau_2^{-1}. \qquad (2)$$

In this equation the absorption cross sections for absorption of pump light I_p and laser light I_L by the ground state S_0 with concentration N_0 are designated as σ_{AP} and σ_{AL} respectively. The excited state S_1 decays with a rate constant τ_1^{-1}, which includes the radiative decay constant τ_{1R}^{-1}, the nonradiative (internal conversion) $S_1 \rightarrow S_0$ relaxation rate K_{SS} and the singlet–triplet cross-over rate K_{ST}. τ_1 is thus given by $\tau_1^{-1} = \tau_{1R}^{-1} + K_{SS} + K_{ST}$. Molecules in the S_1 state are also removed with a cross section σ_E by the process of stimulated emission. σ'_{AP} and σ'_{AL} are the cross sections for absorption of pump and laser radiation by the singlet state S_1 with concentration N_1 to the singlet state S_2 with population N_2. The (internal conversion) decay time from S_2 to the S_1 state is given by τ_2, which is of the order of 10^{-11} s (Lin and Dienes 1973). Molecules in the S_2 state will therefore return immediately to S_1, and unless the pump and laser intensities are very high there is no effective change of the population of S_1 due to the $S_1 \rightarrow S_2$ absorption. In most practical cases the contribution of the last three terms of eq. (2) can therefore be neglected and this equation simplifies to

$$\frac{dN_1}{dt} = N_0(\sigma_{AP}I_P + \sigma_{AL}I_L) - N_1\tau_1^{-1} - N_1\sigma_E I_L. \tag{3}$$

Although the lifetimes of higher lying triplet states T_2 have not been measured, it is known from qualitative data and theoretical considerations (Liu and Kellog 1969) that those states decay nonradiatively with lifetimes of the order of 10^{-11} s or shorter. As for the singlet state S_1, since absorptive depopulation of T_1 can be neglected, the population and depopulation rate of this level is given by

$$\frac{dN_T}{dt} = N_1 K_{ST} - N_T \tau_T^{-1}. \tag{4}$$

N_T and τ_T are the population density and decay time of T_1, respectively. The population densities N_0, N_1 and N_T are connected by the relation

$$N_0 + N_1 + N_T = N = \text{constant.} \tag{5}$$

The photon density I_L of the laser light, which propagates along the z-axis of the laser cavity, is increased by stimulated emission and reduced by the absorption processes $S_0 \rightarrow S_1$, $S_1 \rightarrow S_2$ and $T_1 \rightarrow T_2$. Thus the net rate at which I_L increases along the z-axis is given by

$$\frac{dI_L}{dz} = \frac{n}{c}\frac{dI}{dt} = (N_1\sigma_E - N_0\sigma_{AL} - N_1\sigma'_{AL} - N_T\sigma^T_{AL})I_L. \tag{6}$$

The rate equations (2) through (6) form the basic set of relations which permit a rather detailed analysis of the performance of dye lasers. In the past, several authors have derived solutions of these equations under various approximations for pulsed laser systems excited by flashlamps or pulsed lasers and for cw dye laser devices. A first rate equation analysis was given by Sorokin et al. (1967) for a laser-pumped dye laser. Assuming a gaussian shaped pumping pulse, the authors solved the rate equations for the population N_1 of S_1 and the number of photons in the resonator cavity. The result was used to predict the shape of the laser output pulse. In their experiment, the dye was pumped by the intense short pulse of a ruby laser. The dye laser is excited to threshold in a time of 10 to 50 ns, which is short compared to the characteristic time for intersystem crossing, typically 10^{-3} to 10^{-7} s. Thus the accumulation of molecules in the triplet state is negligible. This justifies the neglect of triplet states in the analysis of this type of dye laser. For the long pulses of flashlamp-excited lasers, however, this approximation is not applicable. In a first analysis of such systems, Schmidt and Schäfer (1967) took the triplet state population into account and solved by computer coupled rate equations for the description of the population of the upper laser level and the triplet state. The dynamic effects of triplet absorption upon flashlamp-excited laser emission have also been considered by Sorokin et al. (1968) by adding to the rate equations of their first analysis, a third equation describing the population of the triplet states. From the results of these calculations (displayed in fig. 7 through 8, on pp. 390 and 391 of Vol. 1) it became evident that triplet effects can terminate the laser pulse even before the excitation reaches its maximum. Because the triplet losses are time dependent they strongly decrease the efficiency with increased duration of the pump light pulse. In addition, triplet effects change the emission wavelength of the dye laser during the laser pulse, as calculated by Bass and Steinfeld (1968) and Weber and Bass (1969), and displayed in fig. 9 on p. 392 of Vol. 1. As a consequence, efficient dye laser operation requires flashlamp pulses with fast rise time and short duration.

The analysis of the triplet state kinetics provided a useful insight into the importance of triplet state populations and their control for the laser process. From the triplet state rate equation (4) Sorokin et al. (1968) determined the rise time of the exciting pulse which is required to generate laser action in a dye medium when triplet state effects are important. If no triplet quenching is present ($\tau_T \to \infty$), the generation of laser emission in a dye medium with typical values of $K_{ST} = 2 \times 10^7 \, s^{-1}$ and $\sigma_E / \sigma_{AL}^T = 10$ necessitates a flashing pulse which produces the critical inversion population N_{1C} in a time of less than 10^{-6} s. N_{1C} is the population of S_1 which is sufficient for balancing laser gain and intrinsic optical losses. Eq. (4) may also be used to get an idea of the longest laser pulse which can be obtained from a solution with unquenched triplet state population. Above laser threshold the population of S_1 is fixed at N_{1C} and does not increase significantly with excitation (Smith and Sorokin 1966). If critical inversion is reached at the time $t = 0$ the population of N_T is given by eq. (4) as approximately $N_T(t) = N_{1C} K_{ST} t$. Losses due to triplet state absorption are given by $N_T \sigma_{AL}^T$ and lasing will be possible only if the stimulated fluorescence exceeds the triplet losses, $N_{1C} \sigma_E \geq N_T \sigma_{AL}^T$. At the peak of σ_E (fig. 2) the ratio σ_E / σ_{AL}^T equals approximately 10. Therefore the net gain will be reduced to zero as soon as $N_T \approx 10 N_{1C}$ and the longest laser pulse duration t_{max} is expected to be of the order of $10 K_{ST}^{-1}$. With an intersystem crossing rate of $K_{ST} = 2 \times 10^7 \, s^{-1}$, the duration of the laser pulse should be less than $t_{max} = 5 \times 10^{-7}$ s. This value is consistent with the observations in dye laser systems of uncontrolled triplet state population (Sorokin et al. 1968, Snavely 1969).

These considerations indicate that the population N_T of T_1 must be limited if long pulse or cw operation of a dye laser is to be achieved. The improvement in dye laser performance which can be achieved with reduction of N_T was shown theoretically by Keller (1970) through solution of the rate equations of Sorokin et al. (1968) after adding to the equation for the triplet state population density a term N_T/τ_T accounting for the deactivation of the triplet state. Obviously N_T can be controlled by K_{ST} or τ_T. The maximum allowable values of K_{ST} and τ_T are estimated using eq. (4). In steady state, the time derivative of N_T vanishes and

$$N_T = N_{1C}K_{ST}\tau_T. \tag{7}$$

As obtained above, N_T/N_{1C} has to be smaller than 10 for laser operation. The maximum allowable value for $K_{ST}\tau_T$ is therefore on the order of 10. For long pulse and cw operation $K_{ST}\tau_T$ has to be kept below this upper limit. This may be achieved by quenching the triplet state concentration rapidly enough by chemical additives and by a rapid flow of the dye through the excited region. Molecules of O_2 are known to quench triplet states of dye molecules. At the same time, however, O_2 also enhances K_{ST} and quenches the fluorescence of the dye (Schäfer and Ringwelski 1973). It is therefore more advantageous to quench triplet molecules by energy transfer to added molecules with a lower lying triplet state, which can act as an acceptor. Using this scheme, Pappalardo et al. (1970a, b) added cyclooctatetraene as an acceptor molecule and obtained output pulses from a rhodamine 6G dye laser as long as $500 \mu s$. These long pulses demonstrated that the quenching action of cyclooctatetraene reduced the triplet lifetime below the steady-state value required for cw laser operation. This compound is still one of the most effective triplet quenchers known for rhodamine 6G, although a number of others have been tested and several were found to be rather efficient (Marling et al. 1970, 1971).

As was the case for the analysis of the dependence of dye laser operation on triplet state effects and triplet quenching, rate equations are very useful for the description of performance properties of dye lasers as they depend on loss and gain at and above threshold, and the effects of system parameters upon tuning and threshold. Gain, losses, threshold conditions and power output of long pulse and cw dye lasers have been studied in detail by Snavely and Peterson (1968), Snavely (1969), Peterson et al. (1971), Pike (1971), Jacobs et al. (1973), Tuccio and Strome (1972), Strome and Tuccio (1973) and Snavely (1973).

The increase of laser light I_L of wavelength λ along the laser axis is given according to eq. (6) by

$$\frac{dI_L}{dz} = I_L[N_1\sigma_E(\lambda) - N_0\sigma_{AL}(\lambda) - N_T\sigma_{AL}^T(\lambda)]. \tag{8}$$

In eq. (8) the absorption of laser light by $S_1 \rightarrow S_2$ transitions has been neglected. Following the analysis of Peterson et al. (1971) and Snavely (1973), the gain of the active medium may be defined as $g(\lambda) = (1/I_L)(dI_L/dz)$. The total gain $G(\lambda)$ for a round trip through the active medium of length d and the laser resonator with mirror reflectances $R_1(\lambda)$ and $R_2(\lambda)$ is obtained by integrating $g(\lambda)$ along the light path:

$$G(\lambda) = 2d[N_1\sigma_E(\lambda) - N_0\sigma_{AL}(\lambda) - N_T\sigma_{AL}^T] - \ln[R_1(\lambda) R_2(\lambda)]. \tag{9}$$

At laser threshold, $G(\lambda) = 0$. Taking into account the relationship in eq. (5) and in eq. (7), which contains the equilibrium triplet approximation as introduced by Peterson et al. (1971), the threshold condition can be written as

$$N_{1C}\gamma(\lambda) - \sigma_{AL}N + r(\lambda) = 0, \tag{10}$$

where

$$r(\lambda) = -(1/2d) \ln[R_1(\lambda) R_2(\lambda)]$$

$$\gamma(\lambda) = \sigma_E(\lambda) + \sigma_{AL}(\lambda) + K_{ST}\tau_T[\sigma_{AL}(\lambda) - \sigma_{AL}^T(\lambda)].$$

The laser system parameters are contained in $r(\lambda)$. The term $\gamma(\lambda)$ depends only on spectrophotometric parameters of the dye solution and relates the rate at which photons are produced to the population of the upper laser level. Thus in eq. (10) the critical inversion N_{1C} is given as a function of wavelength and system parameters.

$$\frac{N_{1C}}{N} = \frac{1}{\gamma(\lambda)} \left(\sigma_{AL} - \frac{r(\lambda)}{N}\right).$$

This equation describes a critical inversion surface and provides interesting details of the performance of dye lasers (Peterson et al 1971). The critical inversion surface has its intrinsic or self-tuned form, if $r(\lambda)$ is nondispersive. Such a self-tuned critical inversion surface, calculated for the dye rhodamine 6G, is displayed in fig. 3. With increasing excitation intensity the ratio N_1/N increases until N_1/N reaches the critical inversion surface. At this value of N_1/N, laser oscillation starts and N_1 is fixed at the value N_{1C}. Therefore the critical inversion is constrained to lie along the valley of the surface, as is indicated in fig. 3 by the dashed line. The wavelength of the laser oscillation can be tuned by adjusting the mirror reflectivity (which changes r) and dye concentration N over the whole range given by the extremes of the dashed curve. With N fixed, an increase in r causes the laser to oscillate at a shorter wavelength, while, for an increase in N with r fixed, the laser wavelength is shifted to a longer wavelength. This dependence of the output wavelength on r and N is consistent with the early experimental results obtained by Sorokin et al. (1968) and Schäfer et al. (1966). With dispersive optical elements in the laser cavity, the laser may operate at points other than those indicated by the dashed line. The dispersive laser cavity modifies the shape of the critical inversion surface. Such a modified surface (an example is given by Snavely 1973) never lies at any point below the intrinsic surface, displayed in fig. 3. From the critical inversion surface, the excitation power required to reach laser threshold, and its dependence on laser system parameters, can be estimated for steady-state conditions (Peterson et al. 1971, Snavely 1973). The calculation of the performance of practical laser devices, however, requires an analysis of the gain above threshold in a more

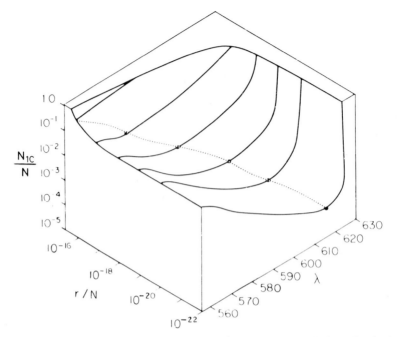

Fig. 3. The relative excited singlet state population N_{1C}/N at threshold, calculated for rhodamine 6G (with $K_{ST}\tau_T = 1.0$) as a function of lasing wavelength λ (nm) and normalized external loss r/N (cm²/dye molecules). The iso-loss lines represent the intersections of constant loss planes with the critical inversion surface. With a nondispersive cavity, the laser operates at points along the dashed curve (Peterson et al. 1971).

realistic laser model which takes into account the longitudinal and radial dependence of the population of the upper laser level in the excited medium. The very
detailed investigations of the operation of longitudinally excited cw dye lasers
(Pike 1971, Tuccio and Strome 1972, Snavely 1973) analyze the gain above
threshold, the output power, the efficiency, the tuning behavior and the dependence of these characteristics on system parameters and excitation power. The
precise knowledge of these dependences is an important basis for optimization
of cw laser design and operation. Since the pump sources of cw dye lasers are of
comparatively low power, efficient operation of a cw dye laser requires a very
careful design and construction to minimize optical losses. In cw lasers, the
mechanical and optical tolerances are therefore rather severe. In pulsed lasers,
the tolerances for design and operation are relaxed since pulsed excitation
sources are capable of producing light pulses greatly in excess of the intensity
necessary to reach laser threshold. The laser analysis, although extremely useful
for design and operation of cw devices, is of rather limited value for the
optimization of pulsed dye lasers pumped by flashlamps or pulsed lasers. This is
mainly due to the many quantitative uncertainties in the temporal and spatial
variation of the inversion density and of loss mechanisms, including excited
state absorption of pump and laser light. In addition, the powerful pump pulses

often produce uncontrolled changes of the optical properties of the dye medium, as, for example, local time dependent variations of the refractive index, which strongly influence laser performance and output power.

In most theoretical dye laser models, excited singlet state absorption $(S_1 \rightarrow S_2)$ is neglected for pump and laser light. However, excited singlet state absorption of the pump and laser light can have important effects upon the output power in high intensity cw or pulsed dye lasers. A relatively complete dye laser model, which includes excited singlet and triplet absorption for both pump and laser light, was studied by Teschke et al. (1976). From derived expressions for gain and output power, the authors concluded that excited state absorption of the laser light reduces the effective stimulated emission cross section and thus reduces gain and output efficiency in the same way as triplet absorption of laser light. The excited state absorption of the pump changes the dye transmission and transfers part of the pump into thermal heating of the dye solution. Since gain and efficiency are reduced, excited state absorption can substantially alter design parameters of high power pulsed or cw dye lasers, as for example in the manner of optimum outcoupling.

In pulsed dye lasers the lasing parameters such as gain, output power, emission wavelength and spectral bandwidth vary during the laser pulse. To analyze the time and wavelength dependence of these parameters the spectral variation of the photon flux–molecule interaction has to be included in the solution of the appropriate rate equations. Considering this spectral dependence, Atkinson and Pace (1973) computed the spectral distribution of a tuned and untuned flashlamp-pumped dye laser and its temporal evolution during the laser pulse. Their model calculations were performed for the line narrowing of a dye laser with a broadband filter and Fabry–Perot etalon as wavelength-selective intracavity elements, and clearly showed that most of the line narrowing occurs before the onset of laser action. The rapid decrease obtained for the rate of narrowing was attributed to saturation of the laser transition. A detailed investigation of the temporal evolution of the spectral distribution of untuned, tuned and injection-locked pulsed lasers was performed by Meyer and Flamant (1976) and by Juramy et al. (1977). Juramy et al. computed the spectro-temporal properties of the laser pulse from the solutions of coupled rate equations which included the time and wavelength dependence of excited state populations, molecular gain, and, of the loss coefficients of the cavity. Experimentally, spectral sweeps of untuned laser emission have been observed for dye laser pulses longer than 10 ns in laser-pumped dye lasers (Bass and Steinfeld 1968, Farmer et al. 1968, Gibbs and Kellock 1968, Neporent and Shilov 1971, Richardson et al. 1977) as well as flashlamp-pumped laser systems (Furumoto and Ceccon 1968, Bradley and O'Neill 1969, Ferrar 1969a). Depending on the device, red or blue shifts were observed and sometimes even a reversing sweep was detected during the laser pulse. These sweeps were interpreted as due to the spectral dependence of the singlet state emission and absorption cross sections and to triplet–triplet absorption. Computations of Juramy et al. (1977), displayed in fig. 4, demonstrate a spectral redistribution during the laser pulse. A redistri-

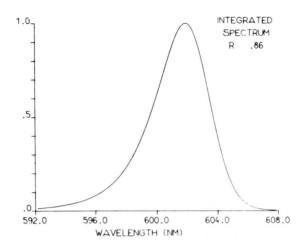

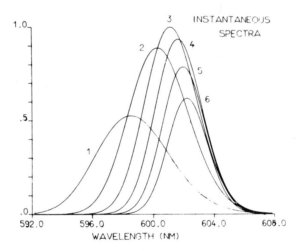

Fig. 4. Spectro-temporal evolution of the broadband laser output computed for a flashlamp-pumped rhodamine 6G dye laser. The upper part displays the time-integrated spectrum, the lower shows instantaneous spectra computed for 1.6 μs (curve 1), 2.6 μs (2), 3.6 μs (3), 4.6 μs (4), 5.6 μs (5) and 6.6 μs (6) after the onset of the pump light pulse (Juramy et al. 1977).

bution can occur even in the absence of any transient species or losses. In the absence of triplet losses, a red spectral sweep occurs during the whole laser pulse. In the case of an etalon-tuned laser the results of Juramy et al. show a line narrowing together with a progressive sweep of the emission (fig. 5). The rate of narrowing was shown to decrease without saturation of the laser transition. As seen in fig. 5, even with an intracavity tuning element a significant spectral sweep of the laser wavelength can occur. With increasing spectral width of the wavelength-selective element the time dependent evolution of the tuned laser

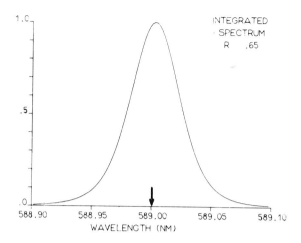

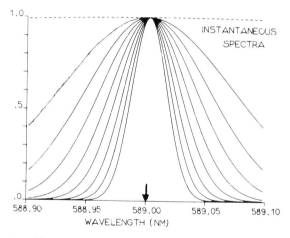

Fig. 5. Time-integrated and instantaneous spectrum of the output of a flashlamp-pumped rhodamine 6G dye laser, operated with an intracavity Fabry–Perot of 6 nm free spectral range and finesse of 5. The dotted line indicates the passive single-pass transmission of the etalon, with maximum transmission at 589 nm indicated by arrows. The instantaneous spectra are computed at 1 μs, 1.4 μs, 2 μs, 2.8 μs, 4 μs, 5.6 μs, 8 μs and 11 μs after the onset of the pump light pulse. The intensity of each curve is normalized to unity. Besides progressive narrowing, these spectra clearly show a progressive red shift of the central laser wavelength as a function of time (Juramy et al. 1977).

will tend to the behavior of an untuned device. The output wavelength of a tuned dye laser may be shifted with respect to the center wavelength of the wavelength-selective intracavity element. The magnitude of this shift (which is towards the wavelength at which lasing would occur without tuning element) is determined by the spectral width of the tuning element, by the gain characteristics of the excited dye medium and by cavity design parameters (Singer et al. 1976).

The optimization of pulsed high gain dye laser systems strongly depends on the spatial distribution and temporal development of the population and photon densities. Gassmann and Weber (1971) studied a theoretical model of a flash-lamp-pumped high gain dye amplifier and calculated the gain in the limit of weak pumping as a function of dye concentration and amplifier geometry. Signal amplification and amplified spontaneous emission in transversely laser-pumped dye laser systems are analyzed by Ganiel et al. (1975) in theory and experiment. Their results describe the temporal and spatial dependence of the excited state population densities and amplified spontaneous emission photon fluxes on pumping rate, spectral narrowing and gain saturation. The gain characteristics and performance of these dye laser amplifiers were analyzed as a function of input signal intensity and pumping rate. Details of these investigations will be discussed in § 4.

In addition to their application in dye laser analysis, rate equations have been used in numerous studies of the performance of special dye laser systems. Ganiel et al. (1976) considered the behavior of a dye laser system in which the wavelength of a powerful pulsed dye laser is controlled by an injected narrow-band signal. In their study, the temporal evolution of the spectral components of the output, the tuning of the laser output as the injected wavelength is tuned across the gain curve of the dye laser, and its dependence on the injected intensity are calculated from a general set of rate equations. With linearized rate equations, Lin (1975a) analyzed transient phenomena like the initial spiking and relaxation oscillations in organic dye lasers. Based on the results of these studies Lin and Shank (1975) and Lin (1975b) describe a general technique for the generation of subnanosecond dye laser pulses by controlling the transient relaxation oscillations. Using this simple method, tunable laser pulses of 0.6 to 0.9 ns duration can be generated throughout the visible spectral range (for details see § 10.1).

Since the rate equation model only contains the balance between exciting photons, excited molecular states and stimulated photons it cannot account for coherent phenomena occurring during laser operation such as, for example, spatial hole burning which has been investigated by Pike (1974), Hertel and Stamatovicz (1975) and Marowsky and Kaufmann (1976). Finally it should be noted that a quantum mechanical treatment of the dye laser, which is based on a six-level model to account for triplet losses, has been published by Baczynski et al. (1976).

3. Construction and performance

In organic dye lasers the upper laser level has a lifetime of typically 4 to 8 ns. Therefore, the laser medium cannot store energy for more than a few nanoseconds and consequently the single-pass gain is approximately proportional to the instantaneous intensity of the exciting pump light. For this reason, the generation of gain sufficient for dye laser oscillations requires pump light which is of high intensity rather than high energy. Today, sufficiently high excitation rates are obtained with flashlamps and pulsed or cw lasers. Alternative methods of excitation, such as discharge or electron-beam pumping of dye vapors, have attractive aspects but have not yet been successful.

3.1 Laser excited pulsed dye lasers

A pump light source which is very suitable for the excitation of pulsed dye lasers is the nitrogen laser. This laser can deliver pulses of typically 1 to 2 MW peak power and 5 to 10 ns duration with a repetition rate up to 100 Hz. The short wavelength (337 nm) of the laser light permits the excitation of a large number of laser dyes operating at wavelengths in the near ultraviolet, the visible and even in the near infrared. Owing to its simple design and low operating costs, the N_2 laser is a very reliable and economic light source for dye laser excitation. Besides the N_2 laser, Nd:YAG laser systems as newly designed (Herbst et al. 1977), and already commercially available, seem to be particularly attractive for this purpose. Since the resonator configuration of this laser design is of the unstable type, the single-mode power density inside the Nd:YAG laser rod is relatively low. This feature reduces considerably the possibility of radiation damage of optical laser components and facilitates operation. With such an oscillator followed by a single-stage amplifier, Herbst et al. (1977) generated output energies of more than 700 mJ in 10 to 15 ns long pulses at a repetition rate of 10 to 30 Hz. Since the long laser wavelength at 1.06 μm is not suitable to pump dye solutions the laser output has to be converted to shorter wavelength by frequency doubling (532 nm) or tripling (355 nm). The radiation at 532 nm is of great advantage for pumping dye lasers operating in the yellow, red or near-infrared region of the spectrum. The third harmonic at 355 nm, on the other hand, is very suitable for dyes lasing in the blue and the green. With appropriate nonlinear optical crystals like KDP, KD*P, CD*A and RDP, the frequency conversion can be very efficient. Using a Nd:YAG laser system with 50 MW pulse power, Kato (1975), for example, reported an efficiency close to 40% (20 MW) for the generation of the second harmonic and almost 20% (10 MW) for the third harmonic.

For several applications of laser-pumped dye lasers, pulses of longer duration are highly desirable. Pulses of 0.1 to 1 μs can be generated with Xe-ion lasers (Schaerer 1975). Xe lasers which have been developed so far, generate peak powers of typically 0.1 to 80 kW at repetition rates of more than 100 Hz. The most intense lines of the laser output are located in the blue, green and red. Thus, Xe-laser-pumped dye lasers (Hänsch et al. 1973, Levenson and Eesley 1976, Schaerer, 1975, Harper and Gundersen 1974) are operated only in the yellow and red part of the spectrum. Pulsed laser systems, which are very promising for dye laser pumping, are the KrF and XeF excimer lasers. The short wavelength of about 250 nm (KrF) and 350 nm (XeF) is very useful for dye laser operation in the visible and even in the near ultraviolet down to 347 nm (Rulliere et al. 1977). With KrF lasers, pulses of more than 5 J energy with 100 MW peak power have been reported by Tisone et al. (1975). For the XeF laser, output pulses as long as 100 ns with 80 mJ (Ault et al. 1975) and 1μs with 300 mJ of energy (Champagne et al. 1977), have been reported. Presently, pulse energy, pulse length and repetition rate are limited mainly by the capability of the excitation source used for fast discharge pumping or electron-beam excitation. The present intense research on excimer lasers promises, however, the development of systems with rather ideal properties for dye laser pumping since they possess short wavelength, high peak power, long pulse length and high repetition rate.

In laser-pumped dye lasers, the laser medium can be excited in either a longitudinal or transverse pumping arrangement. Which of these pumping schemes is more appropriate depends mainly on the beam profile of the pump laser. For pump sources with an extended beam profile which is more or less homogeneous and of nearly rectangular cross section (typical for transversely excited N_2 and excimer lasers), the transverse pumping scheme is more adequate. In this scheme the laser light is focused by a cylindrical lens through a side window into the cell which contains the dye solution. The focusing parameter and dye concentration are adjusted so that the excited region of the dye solution forms a pencil-shaped volume with nearly circular cross section on the inner cell wall, parallel to the axis of the dye laser resonator. Since only a rather thin layer near the inner wall of the dye cell is excited by the pump light, diffraction usually causes a substantial angular spread of the dye fluorescence, which emerges from the cell in the direction of the dye laser axis. For pump lasers with beams of circular intensity distribution, such as the Nd:YAG laser or the Xe-ion laser, the longitudinal pumping scheme appears more appropriate. In such arrangements, the exciting laser beam enters the dye laser resonator through one of the cavity mirrors which is of low reflectivity for the exciting pump light. Inside the resonator, cavity pump and laser beam are aligned to be collinear. In very short active dye media it is sufficient if laser and pump beam overlap under a small angle. Therefore the pump beam may enter the dye laser cavity by the side of one laser mirror. Concentration of the dye and the length of the cell have to be adjusted so that the excitation power is nearly uniform in the whole excited volume. The longitudinal scheme has the advantage of an almost perfect overlap of pumped dye volume and dye laser beam. This results in efficient power extraction and low beam divergence. The schemes of different arrangements useful for laser-pumped dye lasers are displayed in Vol. 1, p. 398. Some important aspects will be discussed also in the following sections.

The usual way to generate laser radiation of narrow bandwidth is to isolate a single axial mode of the laser resonator. In long pulse flashlamp-pumped or cw dye lasers this can be achieved with wavelength selectors of low dispersion, such as prisms and low finesse Fabry–Perot interferometers, owing to the multiple passes of the light inside the cavity. In laser-pumped dye lasers, however, the number of light passes is very limited during the short excitation time of typically 5 to 10 ns. Therefore the linewidth of the laser radiation will not be much smaller than the bandwidth of any spectral filter inserted into the laser cavity. For this reason, frequency-selective elements of low dispersion are rather inefficient and an alternative way has to be chosen in the construction of laser-pumped pulsed dye lasers. In dye solutions pumped by powerful pulsed lasers a single-pass gain of more than 30 dB can be achieved easily over a path length of a few millimeters (Shank et al. 1970, Hänsch et al. 1971). Thus, rather high losses can be tolerated in the laser cavity, and very narrowband spectral filters, as for example high-order diffraction gratings together with high finesse Fabry–Perot interferometers are very suitable wavelength-selective elements.

According to this philosophy, several approaches have been reported to generate tunable emission from laser-pumped pulsed dye lasers. Bradley et al. (1968a) placed a dye cell into a laser cavity which consisted of a mirror with 70% reflectivity and an echelle grating with 300 lines/mm. Pumping the dye solution near-longitudinally with 20 MW of the giant pulse output of a ruby laser produced dye laser pulses of 2 MW peak power and 0.4 nm spectral width. The insertion of a tilted Fabry–Perot interferometer with 75 pm free spectral range and 65% reflectivity between the dye cell and the output mirror reduced the linewidth to less than 50 pm. The arrangement of this dye laser device has, however, the disadvantage that the small diameter of the laser beam, of the order of one millimeter, covers only a few lines at the surface of the grating. This seriously reduces the spectral resolution. In addition, the high power density of the laser light at the small illuminated grating area may damage the reflecting metal film. These difficulties are overcome by a cavity design reported by Myers (1971) for a dye laser pumped transversely by a N_2 laser (fig. 6). In

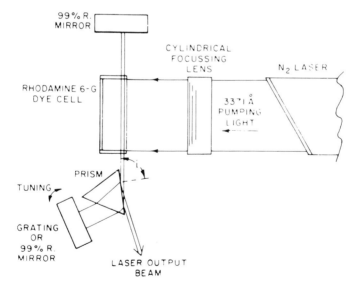

Fig. 6. Schematic cavity design of a dye laser pumped transversely by a pulsed N_2 laser (Myers 1971).

this dye laser, a combination of prism and grating spectrally narrows the generated dye laser emission. The prism is positioned at very high angle of incidence, close to 90°. The reflection at the prism face serves as an outcoupling device which is variable between 20% reflection at 80° of incidence to 100% reflection at 90°. Passing the prism, the laser beam is considerably expanded with a simultaneous reduction in beam divergence. With a laser system of this design Myers generated tunable dye laser emission from a 10^{-3} M/ℓ solution of rhodamine 6G in ethanol with a spectral width of about 0.09 nm. In a similar laser scheme, Stokes et al. (1972) inserted an additional intracavity Fabry–Perot into the expanded beam between prism and grating. With this system they generated tunable radiation of typically 10 to 30 kW peak power in the wavelength range 350 to 730 nm. The linewidth of the laser emission was reduced to about 10 pm.

A dye laser design which has been used very successfully for the generation of reliably tunable radiation was reported by Hänsch (1972). The basic elements of this design are displayed in fig. 7. A short dye cell of only 10 mm length is side-pumped by the output of a N_2 laser. The pump light is focused at the inner cell wall to a line of near-circular cross section of about 0.15 mm width. The dye solution is transversly circulated with a flow speed of the order of 1m/s. Despite the high pump energy densities (up to several J/cm³) no noticable thermal optical distortion is observed during the short excitation of 5 to 10 ns. Thermal schlieren

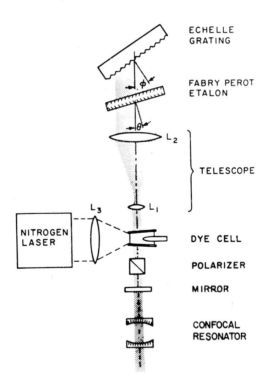

Fig. 7. Basic components of a narrowband tunable dye laser (Hänsch 1972).

effects occur with a substantial time delay after the absorption of the pump light, and the medium returns to optical homogeniety before the arrival of the next pump pulse. The laser pulse duration is short compared to the intersystem crossing time constant K_{ST}^{-1} of typically 10^{-7} s. Thus triplet effects are without any importance in this type of laser.

The dye cell is placed in an optical cavity formed by a plane dielectric-coated mirror of low reflectivity and a high-order echelle grating. The optimum of the reflectivity of the outcoupling mirror depends mainly on the gain of the excited dye and ranges from 4 to about 30%. In practical systems of this design the initial diffraction-limited angular divergence of the light which emerges from the dye cell is about 2.4 mrad. A collimating telescope, consisting of the two lenses L_1 and L_2 enlarges the diameter of the light beam of about 0.2 mm to approximately 10 mm. Simultaneously, it reduces the divergence to less than 0.05 mrad. This reduction of the beam divergence greatly increases the spectral resolution of the grating. The resolution is determined by the deflection angle which displaces the back-reflected light at the position of the active dye volume by an amount which prevents it from being fully transmitted and further amplified. It turns out that this deflection angle must be at least equal to the beam divergence (Hänsch 1972). For the combination of a high-order echelle grating and a high quality telescope with a magnification factor of 50, the estimated single-pass resolution is of the order of 8 to 20 pm (Hänsch 1972, Lawler et al. 1976). In agreement with these results the generated laser light had a measured bandwidth of typically 10 to 50 pm. At pump power levels of 100 kW this simple laser system has an optical conversion efficiency of the order of 10 to 20%. The bandwidth can be reduced by inserting a tilted Fabry–Perot interferometer into the collimated beam. With a finesse of 30 and a free spectral range of 0.5 to 1 cm^{-1} this spectral filter narrows the linewidth by more than a factor of 10 to less than 1 pm. The losses in such a filter, however, reduce the efficiency considerably, and the remaining output power is typically 1 to 5 kW.

To achieve a very narrow linewidth in the megahertz region, the laser output can be passed through a confocal interferometer acting as an ultra-narrowband pass filter. Owing to the small free spectral range of 2 GHz and the high finesse of almost 200 which can be achieved with this type of optical filter, the bandwidth of the transmitted radiation is expected to be of the order of 10 MHz. Unfortunately, in such narrowband interferometric filters considerable losses in intensity have to be accepted and the peak power of the output is usually less than 1 W. The spectral filtering not only introduces a substantial loss of intensity but also reveals large spectral fluctuations within the laser bandwidth, which are transformed into amplitude fluctuations of the filtered radiation. The pulse heights of the dye laser pulses filtered by a confocal resonator were measured by Curry et al. (1973) and exhibit to a good approximation an exponential distribution (fig. 8). This can be understood from the fact that the dye medium initially acts as a source of amplified spontaneous emission, exhibiting gaussian statistics. The wavelength selector permits feedback only for photons in a limited frequency band $\Delta\omega$ near the frequency ω. Radiation within the band-

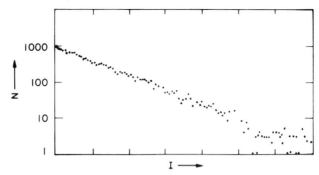

Fig. 8. Pulse height distribution of dye laser pulses generated by a system of the design displayed in fig. 7 (Curry et al. 1973).

width $\Delta\omega$ is further amplified in multiple passes through the gain cell. Owing to the short pulse duration τ, the bandwidth $\Delta\omega$ corresponds to Z distinguishable frequency intervals, i.e., $Z = \Delta\omega\tau$.

As long as the amplifier is not saturated, the amplitude distribution of the amplified light remains gaussian and the photon numbers in each of the Z frequency intervals exhibit an exponential distribution. The fluctuations of the total unfiltered laser output are an average over the Z intervals and are thus effectively reduced. As shown by Curry et al. (1973), the remaining amplitude fluctuations of the total intensity of the dye laser oscillator are expected to be of the order of ±40%, if the gain is not saturated. Under actual operating conditions, however, partial saturation of the gain reduces these fluctuations to only 5%. Owing to the homogeneous broadening of the laser gain, saturation does not affect very substantially the statistics of each individual frequency interval. Thus the isolation of a single frequency channel by an external spectral filter results in an almost exponential amplitude distribution of the transmitted light.

Although dye lasers of the design displayed in fig. 7 have been very successful for many applications, especially in atomic and molecular spectroscopy, an increase in efficiency and output power is very desirable, in particular at narrowband operation. A more intense output can be obtained, of course, by pumping the dye solution with more powerful pump laser radiation as is now available from N_2 or Nd:YAG lasers. Pumping a dye laser longitudinally with 10 MW third-harmonic pulses of a Nd:YAG laser, Kato (1975) obtained tunable dye laser radiation of 1 to 2 MW pulse peak power in the wavelength range 430 to 510 nm. With a repetition rate of 20 pps, and 8 to 10 ns pulses, the average power was of the order of 200 to 400 mW. The dye laser system was operated with a diffraction grating together with a beam-expanding telescope. These wavelength-selective elements reduced the linewidth to about 0.2 nm. With the 24 MW second-harmonic output of the same Nd:YAG system Kato (1976) operated a carbazine-122 dye laser with an efficiency of 44%. This dye laser contained no frequency-selective elements and delivered pulses of 14 nm spectral width and more than 10 MW peak power at the operating wavelength of 700 nm. In another example, Lotem (1973) generated tunable radiation in the near infrared (1.0 to 1.14 μm) with an output power in excess of 2 MW by transverse pumping suitable dyes with a 25 MW ruby laser. In this dye laser, the output was taken from the zeroth order of the grating which, together with a totally reflecting mirror, formed the laser cavity.

Although high pulse power can be obtained in this way, the bandwidth of the generated radiation is too large to be useful for many applications. To achieve a narrower bandwidth additional spectral filters have to be inserted into the laser cavity. The optical losses in those filters will reduce efficiency and output power considerably. Therefore the development of a laser scheme with improved efficiency remained very important. Such an improvement in efficiency has been achieved for narrowband dye laser systems by exciting a narrowband but lossy dye laser oscillator with only a fraction of the available pump light, and by using the remainder to boost the laser intensity afterwards in a travelling wave dye laser amplifier. Construction and performance of such systems will be discussed in detail in § 5.

3.2 Flashlamp-pumped dye lasers

The principles of design, time behavior and spectra of flashlamp-pumped dye lasers have been reviewed by Schäfer in Vol. 1, p. 397 ff. Therefore, this section will summarize important aspects of design and performance and discuss in more detail recent achievements in construction and operation of this type of dye laser.

In general, the flashlamp-pumped dye laser consists of the dye cell, the flashtube as pump light source and a reflector. For most efficient utilization of the pump light it is advisable that the dye cell and flashtube be about the same length. The reflector, surrounding lamp and cell, is used to focus the pump light emitted by the extended flashtube into the absorbing dye solution. The reflector can be of the imaging type (e.g., an elliptical cylinder whose focal lines determine the position of the linear lamp and dye cell tube) or of the close coupling type (e.g., a circular cylinder). The latter is advantageous, for example, if the dye cuvette is pumped by several linear flashlamps or a single helical shaped flashtube.

In flashlamp-pumped dye lasers, triplet effects have to be minimized for best output power and efficiency. As discussed above this can be achieved by a fast rise time and short duration of the flashlight pulse. In different laser designs a great variety of flashlamps and electrical discharge circuits have been used to generate very fast high power light pulses.

In most laser systems the flashlamps are operated by connecting a low inductance capacitor, loaded to a certain voltage, repetitively to the electrodes of the lamp. The switching element is usually a triggered spark gap or a fast thyristor. In this way the lamp can be operated at a voltage which is much higher than the self-firing voltage. The excess voltage ensures a rapid build-up of the plasma of the discharge, which results in a fast rise time and high peak power.

Rise times of 200 to 300 ns have been obtained with commercial Xe-filled linear flashlamps of 8 cm length and 5 mm bore diameter, driven by a circuit of a spark gap and a capacitor (0.3 μF/20 kV) of especially low inductance (Schmidt 1970). Helical flashlamps usually have somewhat longer rise times. For comparison, a helical tube of 8 cm length and 13 mm inner helix diameter gives a rise time of about 0.5 μs. Pulses with very fast rise times and extremely high power output are obtained in low inductance coaxial lamps. In such lamps the cylindrical plasma surrounds the active dye medium, which is contained in the inner tube of the lamp. This type of flashlamp was used by Sorokin and Lankard (1967) and by Schmidt and Schäfer (1967) in their successfully operated flashlamp-pumped dye laser systems. With a spark gap in series to the lamp, Furomoto and Ceccon (1969) achieved rise times of typically 140 ns in an improved version of such coaxial lamps. By adjusting the pressure to an optimum value they obtained a uniform discharge over the whole cross section. This ensured a very uniform excitation of the dye solution. A uniform illumination and heating of the dye liquid by the flashlamp pulse is of particular importance because nonuniform absorption of the pump light may induce small thermal gradients which severely degrade the optical quality of the long, extended dye medium (Gavronskaya et al. 1977). In order to reduce such optical distortions, laser operation requires rapid mixing and exchange of the dye liquid by a fast turbulent flow which may be either transverse or longitudinal in respect to the laser beam. In the design of the dye circulating system special care has to be taken that it does not contain any material which may give off chemicals or other impurities and thus spoil the absorptive and fluorescent behavior of the dye solution. Materials which have proven to be useful for this purpose include glass, Teflon and stainless steel.

As mentioned before, triplet effects are important phenomena for the operation of a flashlamp-pumped dye laser. The accumulation of dye molecules in the triplet state during the excitation pulse causes quenching of the laser action by triplet state losses, and laser action is terminated before the pumping pulse decreases back to the threshold value. With triplet quenching agents added to the dye solution triplet losses are reduced drastically. Consequently rise time requirements of the pump pulse can be relaxed and much longer laser pulses are obtainable. In an oxygen-saturated solution of rhodamine 6G in methanol, Snavely and Schäfer (1969) generated laser pulses of more than 100 μs duration. Adding cyclooctatetraene as a triplet quenching agent to rhodamine 6G, Pappalardo et al. (1970a) obtained pulses up to 500 μs from a pumping pulse of about 600 μs. Long pulse dye laser emission of 50 to 60 μs duration throughout the visible was obtained by Marling et al. (1971) with different dyes by adding various triplet quenching agents. These results indicate that early laser termination by the triplet state problem can be overcome.

It turns out, however, that additional effects arise which terminate the laser output, in particular during high power excitation. In the investigation by Snavely and Schäfer (1969), early termination of laser action appeared despite the presence of oxygen as a triplet quencher. This was explained by thermal lensing effects which distort the optical light path in the dye solution. These effects were shown to become important after 20 to 30 μs and did explain the experimental result of 40 to 50 μs pulse width rather well. In the experiments where this difficulty did not appear (Pappalardo et al. 1970a),the pumping pulse was relatively long with a rather slow rise time. Typically, a few hundreds of joules were discharged in hundreds of microseconds. Thermal effects can be overcome even in high power systems by a sufficiently short duration of the discharge pulse. Pulse lengths of less than 10 μs are usually short enough to avoid this problem. A further cause of early laser termination is the shock waves generated by the high power flashlamps (Ewanizky et al. 1973, Blit et al. 1974). If a shock wave propagates into the dye liquid it destroys the optical homogenity of the laser medium. This, of course, has detrimental effects on the operation of the laser. Blit et al. (1974) describe two mechanisms which are responsible for the generation of such waves. First, a mechanical shock wave is generated in the flashtube and propagates into the dye liquid. There it creates inhomogenities in the refractive index through pressure gradients. This kind of shock wave generation is expected in particular in the coaxial design of flashtube and dye laser cell. In addition, it was found that a substantial part of the flash energy which is emitted in the infrared and absorbed by the solvent of the dye solution can cause a shock wave in the liquid. This effect can be reduced if the flashlamps are surrounded by an infrared absorbing medium like water. In fact, in most systems the flashlamps are surrounded by water which is circulated through a cooling system in order to remove the heat from the flashtubes. The infrared light will be absorbed completely in the water and will not reach the dye solution. Since the velocity of shock waves generated in water is about 1600 m/s (Blit et al. 1974), the shock disturbance will typically reach the laser medium in 6 to 10 μs after the onset of the laser pulse. In fast laser systems, both pump and laser pulse have already terminated by this time.

Surrounding the flashlamp by water has the additional advantage that the pump light can be filtered by various additives. This can be an important technique for reducing the photochemical decomposition of the dye molecules which is caused primarily by the strong ultraviolet component of the flashlamp light (see fig. 17, p. 404 in Vol. 1).

The dependence of laser dye stability on laser parameters like flash frequency, fluid filtration effects, input energy per flash, dye concentration and mirror reflectivity effects has been investigated, for example, by Fletcher et al. (1977) and Fletcher (1977a, b). The effect of dye deterioration on laser performance has been discussed in detail by Hammond (1977).

An example for the design of a dye laser head which can be considered as rather typical for a flashlamp-pumped dye laser is displayed schematically in fig. 9. This laser head, developed by Jethwa and Schäfer (1974), consists of two linear flashlamps of 5 cm arc length and 3 mm bore in a double elliptical reflector. The lamps are cooled with a 0.75 M/ℓ solution of $CuSO_4$ in water which serves to cool the lamps and to absorb the harmful ultraviolet and infrared part of the pump radiation. As mentioned before, these spectral components are mainly responsible for the photochemical decomposition of the dye, for the generation

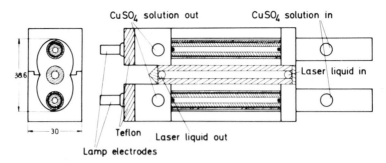

Fig. 9. Cross section of the laser head of a flashlamp-pumped dye laser (dimensions are in mm)
(Jethwa and Schäfer 1974).

of thermal gradients and for shock wave formation in the solvent of the dye
which would prematurely terminate the laser action. The 73 mm long quartz dye
cell has a 4 mm inner diameter. Its inner surface is matted to prevent whispering
modes along the cell walls. The dye cell is sealed by a 90° roof top prism at one
end and a parallel quartz window at the other end. To the solution of $2 \times 10^{-4}\,M/\ell$ of rhodamine 6G in water are added 5% ammonyx as a deagregating
agent and 0.05% cyclooctatetraene as a triplet quencher. The dye liquid flowed
through a 1 μm millipore Teflon filter to remove any scattering centers which
would quench the laser output. The laser cavity, formed by the roof top prism
and an outcoupling mirror with a dielectric coating of 83%, had a length of
130 mm. For excitation a low inductance capacitor of 10 μF is charged to
typically 1.8 kV and discharged through the flashlamps. The flashlamps were
operated in the simmering mode in which a constant current of approximately
30 mA provides preionization in the flashlamp. Owing to this preionization the
operating voltage of the flashlamp could be kept relatively low. Thus a thyristor
could be employed as switching element instead of the spark gap which is
usually used in such circuits. In this way time and amplitude jitter of the current
discharge was reduced and the time and amplitude fluctuations of the laser
output were considerably diminished. Owing to the absence of line narrowing
intracavity elements, the output spectrum of the almost 10 μs long laser pulses
was broadband (600 to 610 nm). At a repetition rate of 100 Hz an average output
power of 6 W was achieved with an overall efficiency of 0.2%.

Although pulse energies of 12 J have been generated from a rhodamine 6G dye
solution excited by 8 linear flashlamps (Anliker et al. 1972), extremely high pulse
energies are best obtained with large coaxial lamps. Alekseev et al. (1972)
constructed a lamp of this type which was 30 cm long and had a 4 cm inner and
12 cm outer diameter. The lamp was filled with 1 Torr Xe. Discharging an
electrical energy of 17 kJ, the authors obtained pulse energies up to 32 J from an
alcoholic solution of rhodamine 6G. The efficiency of this system was about
0.2% and the specific output energy close to 1 J/cm^3. The output power was
reported as 10 MW, so that the half-width of the pulse should have been about
3.2 μs. The laser reported by Baltakov et al. (1973) was operated with the same

type of flashlamp and generated 110 J in 20 μs long pulses. In a further version of this laser, pulses of 400 J energy and 30 MW peak power could be generated at 50 kJ of electrical input (Baltakov et al. 1974).

Pulse energies of more than 1 J/cm^3 are higher than the energy stored in the dye solution and can be obtained only with multiple pumping of the dye molecules by the intense pump light pulse. At 10% inversion, 600 nm laser wavelength and a concentration of 10^{-3} M/ℓ, the stored energy is only of the order of 20 mJ/cm^3. Since this energy is typically stored for only a few nanoseconds, the molecules can be pumped many times during a pulse of several micro-seconds duration. Thus, even multikilojoule pulses appear obtainable from a few litres of excited dye solution if sufficiently intense light sources can be developed for the pumping of such large volumes.

For the generation of high average output power the flashlamps must be capable of handling energies of several hundred joules per flash at high repetition rates. Flashlamps of the commercial linear or coaxial design have a number of limitations which make them adequate for pumping dye lasers only in single shots or at low repetition rates. The fast discharge produces an acoustic shock wave which can explode the tube if the discharge energy is too high and the pulse width too small. Moreover, under high power operation, erosion of the wall and erosion of the electrodes with the subsequent deposition of opaque materials on the walls limit the lifetime of the lamps to a few hundred thousand pulses. In addition, the average input power is limited because nearly all the heat generated inside the tube must be transferred through the glass wall. Many of these difficulties can be overcome by use of a flashlamp with flowing gas as reported by Mack (1974). In this type of lamp a fast flow of Ar gas is injected into the tube near the walls and exits through holes in the electrodes. The vortex flow thus produced stabilizes the position of a low current glow discharge which runs continuously in the center of the lamp. When the flashlamp is fired, the discharge ignites in the path of the stabilized sustainer. The main discharge is inertially confined and cannot expand out to the wall of the tube during the pulse. In this manner erosion of the wall is eliminated. This increases the lifetime and the average power loading that can be tolerated. Morey and Glenn (1976a, b) used such a lamp in an elliptical pumping cavity with the lamp at one focus. The dye (2.5×10^{-4} M/ℓ of rhodamine 6G in ethanol) was circulated at a high flow rate of 4 m/s through a 3 mm thick transverse flow channel. The lamp with 10 cm arc length was fired at repetition rates up to 350 Hz. The energy per input pulse was about 200 J which gives an average power input up to 75 kW. The half-width of the laser output pulses was about 1.8 μs. The average power output increased nearly linearly with repetition rate to a maximum of 114 W at 255 Hz. The achievement of an even higher power output was limited by the replacement rate of dye in the laser channel and by a drop in flashlamp intensity at higher repetition rates. The authors speculate that a further optimization of such laser systems should yield operation at an average power level up to 500 W.

As pump source for very small dye lasers, a low energy unconfined spark can be used. As developed by Ferrar (1973), such lasers consist of a spark gap, dye cell and a spherical reflector which images the spark into the dye cell, and a pair of mirrors forming the optical cavity (fig. 10). In this design the spark gap is about 4 to 5 mm long. The dye cell of 10 mm length has an inner diameter of the order of 1 mm, Ferrar (1973) obtained lasing from a 10^{-3} M/ℓ solution of rhodamine 6G at electrical spark input energies as low as 5 mJ. With input energies of 0.1 J/pulse, the laser could be operated at pulse rates up to 2 kHz. The laser pulses exhibited a rise time of less than 10 ns and a pulse duration of about 130 ns. Using this scheme of excitation, Ferrar (1976) constructed a spark-pumped laser as a compact unit with kilowatt peak output pulses at subjoule spark input energy. Pulse durations of the order of 10^{-7} s and repetition rates up to 100 Hz were typical. The maintenance-free operation exceeded 10^6 pulses.

In flashlamp-pumped dye lasers with a nondispersive optical cavity, the spectral bandwidth of the laser output is of the order of several tenths of nanometers. For most applications it is necessary to narrow the bandwidth by suitable wavelength-selective cavity elements. Spectral narrowing down to 0.05 to 0.1 nm can be achieved by replacing one of the resonator mirrors by a diffraction grating (Soffer and McFarland, 1967). About the same amount of spectral narrowing is obtained by placing a set of prisms (Strome and Webb 1971, Schäfer and Müller 1971) or an interference filter into the cavity (Bonch-Bruyevich et al. 1968). A spectral width of about 1 pm was reported by Walther and Hall (1970) who inserted a two-stage polarization interference filter into the laser resonator. About the same bandwidth was obtained by Bradley et al. (1971d) with an intracavity Fabry–Perot interferometer. The spectral resolution of dispersive elements is limited by the divergence of the incident light beams. In flashlamp-pumped dye lasers the divergence of the laser light can be substantial (up to several mrad) owing to optical inhomogenities in the dye solution. These inhomogenities, produced by the intense pump light, generate thermal refraction effects which lead easily to the appearance of satellite lines in the output spectra if a grating is employed as wavelength selector. Anomalous spectral structure can also appear in multiple-prism tuned systems (Strome and Webb 1971), although Schäfer and Müller (1971) have reported a six-prism tuned ring laser, which produces a clean spectral line of 50 pm bandwidth.

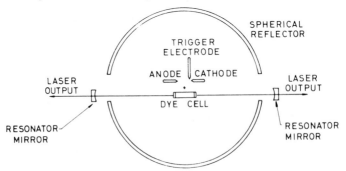

Fig. 10. Scheme of the the laser configuration in a spark-pumped dye laser (Ferrar 1973).

Since the wavelength selection of Fabry–Perot interferometers and (Fabry–Perot-type) interference filters is relatively insensitive to beam divergence, this type of wavelength selector appears superior for spectral narrowing. Bradley et al. (1971d), for example, narrowed the 250 mJ output of their dye laser (operating with an untuned bandwidth of 6.5 nm at 595 nm) to less than 0.3 nm with a single air-spaced Fabry–Perot of 27 nm free spectral range (7 μm spacing) and 75% reflectivity. With a second 100 μm thick Fabry–Perot interferometer the laser linewidth was reduced to 0.01 nm. While the reduction of the laser output energy by the first Fabry–Perot was negligible, the losses in the second tuning element reduced the output power to about half of the untuned energy output. With three intracavity Fabry–Perot interferometers, Gale (1973) was able to achieve single-mode operation of his flashlamp-pumped laser system. The first of the three Fabry–Perots with 5 μm spacing and 65% reflectivity (finesse $F = 7$) narrowed the untuned spectral width of about 8 nm down to a bandwidth of 0.2 nm. The linewidth was further reduced to 5 pm by a second Fabry–Perot of about the same finesse but 200 μm gap. Finally a third etalon with a finesse of 20 and 5 mm plate separation selected out a single longitudinal mode of the 0.5 m long laser cavity with 300 MHz longitudinal mode spacing. The resulting laser linewidth was of the order of 10 MHz. The insertion loss due to the etalons increased the laser threshold from 2.5 to 6 J. Single transverse and longitudinal mode operation could be maintained up to pump energies of 16 J. At this pump level the single-mode peak power density of the 400 ns long laser pulse was about 0.5 MW/cm^2 in the 1 mm diameter light beam. Efficient spectral narrowing with a combination of a low-loss narrowband interference filter and a solid quartz Fabry–Perot was reported by Kuhl et al. (1972). The interference filter of 6 Å passive bandwidth and 84% transmission narrowed the laser linewidth to about 0.1 nm, Adding a 0.25 mm thick solid quartz Fabry–Perot with dielectric coatings of 85% reflectivity reduced the linewidth further to less than 1 pm. With 8.5 J of pump energy, the 300 to 500 ns long laser pulses had a peak power of 17.5 kW. The losses in the tuning elements reduced the peak power to about 6 kW at narrowband operation.

Reliable single-mode operation of a flashlamp-pumped dye laser was achieved by Marowsky (1973). Besides an interference filter and two Fabry–Perot interferometers for spectral narrowing, the author used a Fox–Smith-type mode selector for isolation of a single longitudinal cavity mode. With this laser system, 300 to 400 ns long pulses of 1 kW peak power and 0.05 pm bandwidth were generated at 600 nm with 20 J of pump energy.

3.3 Tuning methods

A coarse selection of the wavelength of the dye laser emission is possible by an appropriate choice of dye, solvent, dye concentration and resonator losses. Spectral narrowing and tuning, however, are achieved with wavelength-selective resonators. Because the laser transition of an organic dye laser is homogeneously broadened, all of the energy stored in the upper laser level can be channeled into a spectrally narrow emission line. With suitable dispersive intracavity elements the spectral width of pulsed lasers can be narrowed down to almost its transform-limited bandwidth. The various methods of spectral narrowing and dye laser tuning are based on the dispersion of a prism, the diffraction of a grating, the wavelength-selective transmission of an interferometric filter or on the use of the wavelength-dependent rotation of the plane of polarization in birefringent materials. In the last-mentioned technique the dye laser is tuned by changing the angle of an optically active material with respect to the laser axis or by either electro-, magneto- or acousto-optical changes in the orientation of the plane of polarization. An analysis of these different methods for the spectral narrowing and tuning of dye lasers was given by Schäfer in Vol. 1, pp. 405–411.

In a laser-pumped dye laser of the design displayed in fig. 7, the wavelength is determined by the tilt angle of the grating and the optical distance between the reflecting surfaces of the intracavity Fabry–Perot. In order to tune the frequency of the laser output, the angle of the grating and the optical path length in the Fabry–Perot have to be changed simultaneously. Since the coherence length of the emitted laser light is of the order of the cavity length no discrete axial-mode structure is expected. Thus the wavelength can be tuned smoothly without complicated feedback control of the cavity length. The synchronization of the tuning elements, however, is difficult, and even with rather sophisticated electronic and mechanical devices, continuous tuning is usually limited to a small fraction of the operating range of a certain dye. Yamagishi and Szabo (1977), for example, reported synchronous tuning of a piezodriven echelle grating together with a piezotuned etalon of 15 GHz free spectral range over a frequency interval of 180 GHz. If an external confocal resonator is used for further spectral narrowing of the generated laser light, (fig. 7) the tuning of the optical length of this filter has to be synchronized to the wavelength tuning of the dispersive intracavity elements. The simultaneous tuning of grating, intracavity Fabry–Perot and external confocal filter over more than 10 to 20 pm is usually hampered by different nonlinear tuning characteristics of the wavelength-selective elements. The difficulties involved in the mechanical tuning can be avoided completely by simultaneous gas-pressure tuning of the wavelength-selective grating and the interferometers (Wallenstein and Hänsch, 1974). Inside a gas medium of variable pressure the reflecting grating and the Fabry–Perot select fixed wavelengths which depend only on the geometry of each tuning element. If grating and Fabry–Perot are tuned mechanically into resonance, a subsequent change of the pressure of the surrounding gas tunes the frequency of the wavelength selected by the grating at exactly the same rate as the frequency of the wavelength selected in the Fabry–Perot interferometer (Hirschberg and Kadesch 1958). Thus continuous smooth frequency tuning of the wavelength-selective laser elements is achieved without any mechanical change in their positions or spacings simply by changing the pressure of the gas which surrounds the grating and Fabry–Perot. With nitrogen as a scan gas, a pressure change of 760 Torr tunes the laser frequency over 150 GHz with a linearity better than 0.3%. The tuning range can be increased considerably by using a surrounding gas with higher refractive index or by changing the gas pressure over a range of several atmospheres. A pressure change of 760 Torr of propane, for example, gives a frequency tuning range of 524 GHz or 6 Å at 6000 Å.

As pointed out in § 2.2, Fabry–Perot interferometers and (low-order Fabry–Perot-type) interference filters are particularly suitable for spectral narrowing of flashlamp-pumped dye lasers. By inserting an interference filter into the cavity, the bandwidth of the laser output could be narrowed to less than 0.1 nm (Kuhl et al. 1972, Marowsky 1973). Frequency tuning of the emission is achieved over a range of several nanometers by tilting the filter relative to the axis of the resonator. If one or several intracavity Fabry–Perots are used for further narrowing of the spectral linewidth, the interferometers can be tuned in the same way. Although mechanical synchronization of the tilting of several intracavity Fabry–Perots is possible, a simultaneous pressure tuning (Green et al., 1973) seems to be more advantageous for several reasons. First of all, pressure tuning provides an easy and exact synchronization of the different Fabry–Perots. On the other hand, the spectral bandwidth $\Delta\lambda$ increases for a beam of divergence $\Delta\alpha$ with tilt angle α as $\Delta\lambda = \lambda\Delta\alpha \tan \alpha$ (Schäfer, Vol. 1, p. 409). Thus a large tuning range is possible only at the expense of a relatively large increase in bandwidth. In an interference filter, the influence of the beam divergence also increases with growing angle of incidence and it is advisable to choose a suitable filter for the desired wavelength so that it has to be tilted by less than 10° (Kuhl et al. 1972). In addition to an increasing bandwidth, the use of Fabry–Perot etalons at high tilt angles introduces walk-off losses, which become rather serious in systems with a large ratio of etalon thickness to beam diameter. Therefore, to realize angle tuning with a specific narrow spectral linewidth, the tuning range or beam divergence or both have to be reduced. If the laser emission is narrowed to a single cavity mode the length of the cavity has to be tuned simultaneously with the tuning of the dispersive intracavity elements.

As for Fabry–Perot interferometers, tuning elements based on the wavelength-dependent rotation of the plane of polarization are comparatively insensitive to angular variations in the optical path. This important property, and the particularly low optical losses, make polarization filters very useful devices for spectral narrowing and tuning of flashlamp-pumped dye lasers. A simple arrangement of this kind consists of a birefringent quartz plate cut parallel to the optical axis and a set of Brewster plate polarizers. This combination inserted into the laser cavity has transmission maxima for retardations in the quartz plate of multiple half-wavelengths. The wavelength can be tuned simply by changing the tilt angle of the birefringent plate (Soep 1970; details are presented also in Vol.1, p.410). Very satisfactory results can be obtained with a stacked plate birefringent tuning element (Bloom 1974). In a commercial flashlamp-pumped dye laser (Chromatix, CMX-4) such an element narrows the laser output to less than 0.1 nm spectral width and tunes the frequency continuously over a range of several tenths of nanometers.

If the quartz plate is replaced by a KDP crystal the laser wavelength can be tuned by electro-optical variation of the birefringence. This method was first applied by Walther and Hall (1970) who reported the electro-optic tuning of a dye laser with 1 pm bandwidth over a spectral range of 0.4 nm. Electronic tuning of a dye laser was achieved by Taylor et al. (1971) by inserting an acousto-optic filter into the dye laser cavity. In this way a rhodamine 6G dye laser, pumped by a Q-switched internally doubled Nd:YAG laser, was tunable over 78 nm. A filter linewidth of 0.68 nm resulted in a laser linewidth of 0.13 nm.

Both electro-optical and acousto-optical tuning have been used successfully for rapid tuning of a dye laser. This is of particular interest for applications in photolysis or plasma studies (Kopainsky 1975). Gerlach (1973) inserted an electro-optically tuned Lyot filter into the resonator of a flashlamp-pumped dye laser. With this device the laser emission was tuned at a speed of 35 nm/μs over a wavelength range of 20 nm. Rapid acousto-optic tuning of a N_2-laser-pumped dye laser was reported by Hutcheson and Hughes (1974a). The authors inserted an acousto-optic beam deflector between the dye cell and the diffraction grating. With the combination of deflector and grating as the frequency-selective tuning element, the authors reported the possibility of tuning rates up to 100 nm/μs over a wavelength range of up to 35 nm.

4. Dye laser amplifiers

In dye laser systems, as described in the previous sections, the amplifying medium is enclosed in an optical cavity for frequency-selective feedback. It is possible, however, to use the dye medium without enclosing resonator as a travelling light amplifier. The high gain and wide bandwidth make laser- or flashlamp-pumped dye solutions particularly attractive for such applications. From the analysis of strongly excited dye solutions of concentrations on the order of 10^{-2} M/ℓ the gain coefficient is expected to be as large as 300 dB/cm. In experimental investigations, Shank et al. (1970) and Hänsch et al. (1971) confirmed that the gain of a N_2-laser-pumped dye medium can be on the order of 20 to 200 dB/cm. As in laser-pumped amplifiers, the amplification in flashlamp-pumped dye solutions has been investigated extensively. Compared to laser-pumped amplifiers the single-pass gain of a flashlamp-pumped dye medium is in general considerably lower. A first gain measurement by Huth (1970) yielded a gain coefficient of 95 dB/m.

Unlike solid-state laser amplifiers, dye amplifiers cannot store pump energy for times much longer than a few nanoseconds. This is due to the short lifetime of the excited singlet state. Consequently, dye amplifiers are not appropriate to generate laser pulses of extremely high energy but are very useful for the amplification of weak signals to rather high intensities. The usefulness of dye amplifiers, however, can be limited severely by the background of amplified spontaneous emission (ASE). Owing to the high value of the cross section for stimulated emission (10^{-16} cm^2) – which is responsible for the high small-signal gain – the ASE intensity is usually rather high. Intense ASE, however, causes considerable losses of the stored energy and can reduce the single-pass gain by saturation. Therefore the ASE should be well below the saturation intensity, which is of the order of 0.1 to 1 MW/cm^2. In practice, this is possible at moderately high pumping rates if the gain of an amplifier of active radius r and length l is kept well below the value l^2/r^2. At a gain of this magnitude the ASE intensity at the output approaches the saturation intensity (Gassmann and Weber 1971). In practical systems, amplification factors up to one thousand have been realized without violating this condition. A considerably larger gain can be obtained with a multistage amplifier in which the ASE background is reduced by inserting suitable spectral and spatial filters in between the stages. In addition to an increase in the maximum useful gain, multistage amplifiers can provide superior intensity stabilization by gain saturation. Such stabilization is particularly desirable when the output of a short pulse dye laser is sent through an ultra-narrow external bandpass filter in order to reduce the bandwidth of the transmitted light to a few tens of megahertz. As discussed in § 2.1, spectral filtering reveals large spectral fluctuations in the laser output within the bandwidth of the laser. These spectral fluctuations are transformed into amplitude fluctuations of the filtered radiation. A high gain dye laser amplifier which boosts the intensity of the weaker laser pulses but exhibits lower gain for the more intense pulses owing to saturation, appears to be an attractive instrument for reducing these fluctuations. The advantages of such a scheme for the stabilization of a fluctuating laser source have been studied theoretically and experimentally by Curry et al. (1973).

Besides experimental investigations of dye amplifiers, theoretical studies have been reported in several publications. As mentioned already in § 1.2, Gassmann and Weber (1971) studied a theoretical model for a flashlamp-pumped high gain dye amplifier by solving a system of coupled rate equations. They took triplet–triplet absorption into account and calculated the gain in the limit of weak pumping as a function of dye concentration and amplifier geometry. From these calculations, conditions were derived for optimizing the dye concentration, and the length and radius of a dye laser amplifier in which a weak signal is to be amplified with a certain gain to the desired intensity. Flamant and Meyer (1973) analyzed a similar model for high pump intensities and substantial saturation where the total gain no longer increases exponentially with the length of the amplifier. A theoretical investigation of dye media transversely pumped by powerful pulsed lasers is reported by Ganiel et al. (1975). Their results describe the temporal and spatial dependence of the excited state population densities and ASE photon fluxes on pumping rate, spectral narrowing and gain saturation. The coupled rate equations for the excited state population densities and ASE photon fluxes were solved numerically for practical situations. It is found that under laboratory conditions the gain of such a system saturates rapidly. As a consequence, at high pumping rates the ASE output varies linearly with pump intensity. The performance of dye laser amplifiers was also analyzed as a function of input intensity and pumping rate. Fig. 11 displays, for example, the

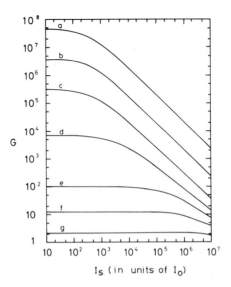

Fig. 11. Calculated gain (G) of a N_2-laser-pumped rhodamine 6G amplifier as function of the input signal I_S and pumping rates W. I_S is expressed in I_0 (see text). For input radiation, a bandwidth of 0.05 nm was assumed at $\lambda = 575$ nm. (a) $W = 5 \times 10^8$ (s^{-1}), (b) 5×10^7, (c) 10^7, (d) 5×10^6, (e) 3×10^6, (f) 2×10^6, (g) 10^6 (Ganiel et al. 1975).

gain G as a function of input signal I_S and pumping rate W as it was computed for a N_2-laser-pumped rhodamine 6G dye amplifier with an active volume of 1 cm length and with a radius of 0.2 mm. In this figure I_S is given in I_0, which is the input photon flux per unit wavelength for which the rate of stimulated emission equals the rate of spontaneous emission into the same spatial mode. For very small input signals the gain is constant, but, as the input signal becomes more intense, saturation effects reduce the gain. At high pump rates and high enough input signals, the gain becomes approximately proportional to the pump rate and inversely proportional to the input signal. Measurements of the gain in an N_2-laser-pumped rhodamine 6G amplifier were in rather good agreement with the prediction of fig. 11 (Ganiel et al. 1975).

5. Oscillator–amplifier systems

The application of dye laser amplifiers to boost the output of weak but highly monochromatic laser light to high intensities has been demonstrated in several reported experiments. In one of the first reported dye laser oscillator–amplifier systems, Itzkan and Cunningham (1972) pumped an oscillator and amplifier with two synchronized N_2 lasers. The cavity of the low power oscillator consisted of a grating and a semi-transparent mirror with an additional tilted Fabry–Perot interferometer inserted. An aperture between the oscillator and the amplifier reduced the beam divergence to about three times the diffraction-limited value. With this system, peak power of 5 to 50 kW in 5 ns long pulses with 2 pm spectral width could be generated throughout the visible from 360 to 670 nm with an efficiency of up to 25%. Extremely high light intensities are obtained with amplifiers pumped by powerful solid-state lasers. The good spatial beam quality of these pump sources permits longitudinal pumping of both dye laser oscillator and amplifier. A major benefit gained from a longitudinaly pumped system is the small beam divergence of the dye laser light and the good overlap of pump and dye laser beam in the active volume. Carlsten and McIlrath (1973) described an oscillator–amplifier system pumped by a Q-switched ruby laser firing every 15 s. The dye system generated 30 ns long pulses of 5 MW peak power in a linewidth of 50 pm. In the single-stage amplifier a gain of about 20 and a conversion efficiency up to 12% was obtained near 790 nm. An output energy of 1 J and 60 MW peak power in 15 ns long pulses was reported by Loth et al. (1976). Their ruby-laser-pumped dye laser oscillator followed by a four-stage amplifier generated pulses of 6.5 pm spectral width at a rate of 6 pulses/min at 770 nm. The scheme of a very powerful oscillator–amplifier system was developed by Moriarty et al. (1976). An illustration is given in fig. 12. The system is pumped longitudinally by the second-harmonic output of a Nd : YAG laser operating at 10 pps. The pump energy passes the beam height adjuster BHA 1 and is partitioned between oscillator and amplifier by mirrors M3 and M4. With about 20% of the the total pump radiation the dye oscillator (composed of M1, M2, three Fabry–Perots, the pressure box (PCC) and the dye cell DC1) generates a narrow line with less than 3 pm spectral width. This radiation is injected into the amplifier dye cell (DC2), which is pumped by about 70% of the initial pump light. With a pump energy of 185 mJ output pulses up to 55 mJ, were obtained in 9 ns long pulses from a rhodamine 6G solution at 564 nm. The output at 564 nm was frequency doubled to give a maximum power level of 600 kW at 282 nm.

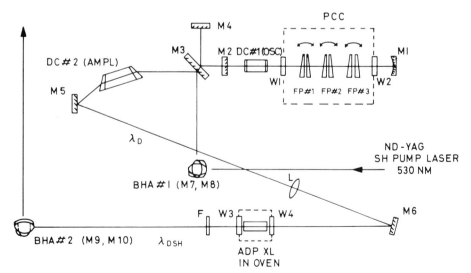

Fig. 12. Scheme of an oscillator–amplifier system pumped by 530 nm radiation of a frequency-doubled Nd : YAG laser (Moriarty et al. 1976).

An oscillator–amplifier system with an efficiency more than three orders of magnitude higher than that of a single narrowband oscillator was described by Wallenstein and Hänsch (1975). The scheme of this dye laser system is shown in fig. 13. The pressure-tuned oscillator is pumped by about 10% (100 kW) of the output of a 1 MW N₂ laser. Because rather high losses can be tolerated in the low power oscillator, a high finesse intracavity etalon ($F = 80$ to 90) with a large free spectral range (76 GHz) is used to reduce the linewidth to less than 1 pm and to

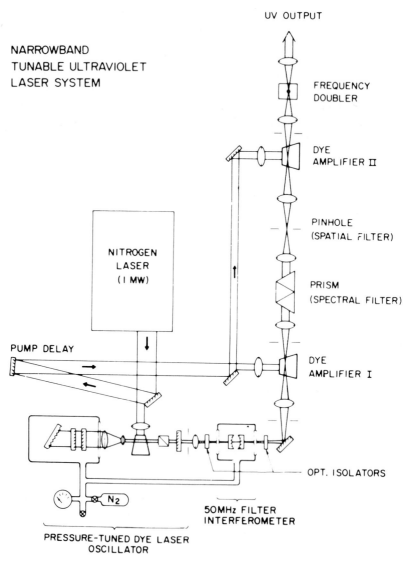

Fig. 13. Components of a narrowband oscillator–amplifier system pumped by a N₂ laser (Wallenstein and Hänsch 1975).

ensure single-line operation. Further reduction of the linewidth to 10 or 50 MHz is achieved with a confocal interferometer of 2 or 8 GHz free spectral range and a finesse of almost 200. The pump light for the two amplifier stages (100 to 200 kW for the first stage and 400 to 600 kW for the second) is obtained by geometrically dividing the N_2 laser beam. The pump light is delayed so that it arrives slightly later than the narrowband input signal from the oscillator in order to maximize the length of the output pulse. This is important because the linewidth of the output signal cannot be narrower than the Fourier transform limit of its short duration (50 MHz for a gaussian pulse of 10 ns length). The small-signal gain is about 30 dB per stage. With the 10 MHz confocal interferometer, an output power of the order of 50 kW has been obtained near 4600 Å, using a solution of 5×10^{-3} M/ℓ of 7-diethylamino–4-methylcoumarine in ethanol. Because both amplifier stages are decoupled for spontaneous emission by a spectral and spatial filter, the background of amplified spontaneous emission was less than a few percent of the total output. The bandwidth at the amplifier output is somewhat larger than the Fourier transform limit. Some additional line broadening apparently results from the amplifier gain saturation. Because this saturation is homogeneous, it does not affect the linewidth of any individual pulse, but it can broaden the width of the statistical average. Pulses from the oscillator which happen to be somewhat off resonance and hence are attenuated by the confocal interferometer experience relatively stronger amplification because of their reduced saturation. It is thus important to operate the confocal filter with its narrowest bandwidth (10 MHz) to minimize this effect. In this case, a total linewidth of about 80 MHz or 0.06 pm has been measured for the pulses of 50 kW peak power near 460 nm.

The high peak power and extremely narrow linewidth make such narrowband laser-pumped dye laser systems especially suitable for linear and nonlinear high resolution spectroscopy. The average power output, however, is rather low and does not exceed 1 W. Thus, for applications in which a narrow linewidth is of minor importance but which require high average power, flashlamp-pumped amplifier systems might be a more appropriate light source. Despite their comparatively low single-pass gain, flashlamp-pumped amplifiers are capable of high pulse energies and high average power if they are operated at high repetition rates.

Flamant and Meyer (1971) constructed a flashlamp-pumped dye laser oscillator followed by a six-stage amplifier. Each of the amplifier stages consisted of an 8 cm long cylindrical dye cell pumped by a linear flashlamp in an elliptical reflector. In the amplifier chain a gain of 700 in intensity was measured for rhodamine 6G near 590 nm. The energy gain was about 6 at an output of 60 mJ. With a later version of the flashlamp-pumped amplifier, Flamant and Meyer (1973) obtained a power amplification of 230 and output energies up to 1 J/pulse with 0.5 nm spectral width. About the same pulse energy was obtained by Loth and Megie (1974) in pulses of only 10 pm spectral width. In their system the 15 cm long dye cells of oscillator and amplifier were pumped by two linear flashlamps in a bielliptical reflector configuration at a repetition rate of 1 Hz. It should be noted that amplification of picosecond pulses to a peak power of more than 3 GW was reported by Adrain et al. (1974) in a three-stage flashlamp-pumped dye amplifier.

A very useful flashlamp-pumped amplifier configuration has been developed by Burlamacchi et al. (1975). For high gain and good efficiency it is desirable to operate a dye laser amplifier with dye solutions of high concentration (10^{-3} to 10^{-2} M/ℓ). At such concentrations, however, the laser medium strongly absorbs and attenuates the pumping radiation as it travels across the pumped solution. The resulting inhomogeneous distribution of absorbed pump intensity causes a similar distribution for inversion density and gain. In addition, the heat released into the medium by nonradiative transitions induces a nonuniform distribution of the index of refraction. Thus, for maximum output power and given cell thickness, an optimum concentration exists which results from a compromise of uniform inversion over the cross section of the active dye solution and maximum utilization of the pump intensity. The inhomogenieties of inversion and gain become particularly important if the transverse dimension of the cell is nearly equal or substantially larger than the absorption length of the pump light in the active dye solution. Because the absorption coefficients α of commonly used dyes are very large ($\alpha \approx 120\,\text{cm}^{-1}$ for a 10^{-3} M/ℓ solution of rhodamine 6G in ethanol), the corresponding absorption lengths are very small ($\alpha^{-1} \approx 8 \times 10^{-3}$ cm). Since the path of the pumping radiation will be no more than a few times the absorption length, the optimum transverse dimension of the cell is expected to be of this order. To take advantage of the high gain present in a concentrated dye solution near the wall of the dye cell, a cell is required which has small thickness in the direction of the incident pumping radiation, but provides an active volume sufficiently large for high power operation. These requirements are fulfilled best by planar wave guide cells. Burlamacchi et al. (1975) studied the spatial inhomogeniety of gain and refractive index across a concentrated dye solution as it is produced by the nonuniform distribution of the absorbed pump radiation. As result, the authors constructed a slab waveguide amplifier, with an active dye volume which was only 0.4 mm thick, but 8 mm wide and 150 mm long. The amplifier was pumped by two linear flashlamps in a bielliptical or a close coupled reflector configuration. In a double-pass amplifier of this type a low signal gain of up to 8.5×10^4 has been obtained. With an input energy of 20 μJ and 100 J of electrical energy discharged in the amplifier, an output energy of 100 mJ was obtained. This still corresponds to a partly saturated gain of 5000 and to an efficiency of 0.1%. The light propagation and amplification in such planar waveguide dye lasers have been analyzed in detail by Pratesi and Ronchi (1976). The thermal processes which affect the optical properties of the dye solution in planar waveguide cells were examined by Balucani and Tognetti (1976) by means of the appropriate hydrodynamic equations.

6. Injection-locking

Although pulse power and bandwidth of the radiation generated in pulsed laser systems are quite impressive, cw dye lasers are certainly superior in oscillation bandwidth and frequency stability. For free-running cw dye lasers, linewidths of the order of 1 MHz have been reported (Hartig and Walther 1973, Schröder et al. 1973). In systems which are stabilized by frequency-locking to an external frequency-stable resonator cavity, linewidths well below 1 MHz and long-term stabilities of a few hundreds of Hz could be achieved (Wu et al. 1974, Barger et al. 1975). In pulsed laser systems the bandwidths and frequency instabilities are orders of magnitude larger. Therefore it seems very desirable to combine the high peak power of pulsed lasers with the high monochromaticity and frequency stability of a cw dye laser. Owing to the homogeneous broadening of the dye laser medium this should be possible simply by amplifying the radiation of a cw dye laser in a pulsed travelling wave amplifier. When the output of the narrow-band oscillator is high enough to nearly saturate an amplifier of the travelling wave type, this is the most reasonable approach. In many cases, however, the narrowband signal which is to be amplified is of low power and thus a large fraction of the pump energy of the amplifier is converted into broadband fluorescence or into amplified spontaneous emission. In this case it is more advantageous to control the wavelength of a high power pulsed oscillator by the low power cw laser light, which is injected into the cavity of the pulsed power oscillator. The feasibility of such a laser scheme was demonstrated by Erickson and Szabo (1971) who injected the 514.5 nm light of a 50 mW Ar ion laser into the cavity of a broadband dye laser side-pumped by a N_2 laser. With the monochromatic injection, the dye laser output is spectrally narrowed from 40 nm down to a few axial modes of the dye laser cavity with 0.16 pm linewidth and 7 GHz spacing (fig. 14). Vrehen and Breimer (1972) obtained similar results by injecting the light of a 15 mW He–Ne laser operating at 632.8 nm into the cavity of a broadband cresyl violet dye laser longitudinally pumped by a frequency-doubled Nd:YAG laser. In these experiments, the dye laser amplifiers were operated with optical feedback. Compared to travelling wave dye laser amplifiers, amplifiers with optical feedback provide higher effective power gain. It may be rather difficult, however, to keep such devices below oscillation threshold, and in the absence of a sufficiently strong input signal, the amplifier may operate simply as a broadband laser oscillator. As shown by Vrehen and Breimer (1972), the spectral content of the output of such an amplifier is not only determined by the bandwidth of the injected radiation but also by transient characteristics of the pumping process. High amplifier gain together with a long build-up time of the cavity modes create a favorable condition at the beginning of the pumping pulse for amplification of the injected radiation, which competes with the growing cavity modes. Thus a substantial amplification of the injected radiation is possible even if its frequency is not matched to a particular mode of the amplifier cavity. An alternative means to achieve spectrally pure amplification in a feedback amplifier consists of the simultaneous but complicated

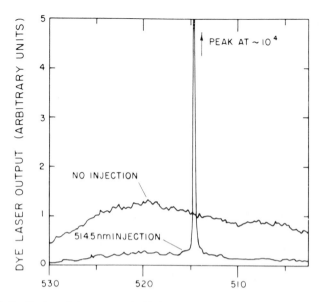

Fig. 14. Spectral distribution of the output of a N₂-laser-pumped coumarin dye laser without and with injection of monochromatic 514.5 nm light from an Ar-ion laser (Erickson and Szabo 1971).

frequency tuning of oscillator and amplifier cavity. This has been demonstrated by Turner et al. (1975) by injection-locking of an electro-optically tuned flash-lamp pumped dye laser to a tuned cw dye laser.

The problem of coupling the cavity of the cw dye laser and pulsed amplifier was avoided in a system reported by Pinard and Liberman (1977). The Ar-ion laser light which pumps a cw dye laser is superimposed with intense light pulses of a frequency-doubled Nd:YAG laser. Thus the single-mode cw output of the dye laser and its superimposed more intense pulsed output have essentially the same spectral properties. The intensity of the pulsed part of the dye laser emission was sufficient to be amplified in a two-stage travelling wave amplifier which was pumped by the major part of the frequency-doubled output of the Nd:YAG laser. With frequency-doubled output pulses of the Nd:YAG laser of about 3 kW peak power and 120 ns duration, a single-mode dye laser pulse power of 200 W was reported. A peak power output of 100 kW at a near fourier-transform-limited linewidth of 60 MHz has been generated by amplifying the output of a cw dye laser in a three-stage travelling wave dye amplifier, which was pumped by the 10 ns long pulses of a 1 MW N₂ laser (Salour 1977). Owing to the high low-signal gain, laser-pumped amplifiers are very suitable to amplify the weak output of a single-mode cw dye laser. The pulse duration in such amplifiers, however, is usually less than 10 ns. Since the Fourier transform limit of a 10 ns pulse is more than 50 MHz, these short-pulse amplifiers are not suitable for the amplification of very narrowband cw laser radiation. This would require a pulse length of the order of 1 μs. At the present time laser-pumped amplifiers with such long pulses are not available, but should be possible, for example, with newly developed pulsed lasers such as Xe-ion lasers or KrF and XeF excimer lasers.

Flashlamp-pumped amplifiers are also quite suitable for this purpose. The relatively low single-pass gain of such systems, however, necessitates the use of an amplifier configuration with optical feedback. Magyar and Schneider-Muntau (1972) injected pulses with an energy of a few millijoules generated with a small coaxial oscillator into a large amplifier cell of 9 mm diameter and 160 mm length. This cell, pumped by six linear flashlamps in a close coupled configuration, was placed in between two mirrors for optical feedback. Combining in this way the spectral purity of the small oscillator with the high energy of the amplifier, the authors obtained an effective power gain up to 400 and reported the generation of 600 mJ in pulses of 10 pm bandwidth. Feedback waveguide amplifiers can be very attractive provided that the resonator losses are sufficiently high to quench freely oscillating modes by gain competition. As demonstrated by Burlamacchi and Salimbeni (1976), the round-trip gain of a double-pass slab waveguide amplifier can be of the order of 10^5. By injection-locking a single amplifier stage with high loss resonator to a 10 mW narrowband cw dye laser, these authors reported an output energy of 10 mJ/pulse with a pulse width of 2μs. At high repetition rates such systems would be capable of generating narrowband radiation at substantial average power levels.

A very efficient system for injection-locking of a pulsed flashlamp-pumped coaxial dye laser to the radiation of a tunable cw dye laser was described by Blit et al. (1977). The cavity of the amplifying laser (fig. 15) was designed for single longitudinal and transverse mode operation, with a mode diameter large enough to fill the active medium. The ring design of the locked laser cavity has the advantage of avoiding optical feedback into the injecting light source. With an injection power of about 10 mW, almost all the laser power was channeled into

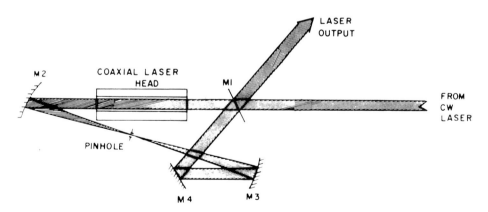

Fig. 15. Ring cavity of a flashlamp-pumped laser with injection system. The flat mirrors M1 and M2 are of 40 and 100 percent reflectivity. The total length of the cavity is 150 cm. The astigmatism introduced by the curved mirrors M2 and M3 (50 cm radius of curvature) is compensated by placing M3 out of the plane of M1, M2 and M4, and setting the planes defined by M1, M2, M3 and M2, M3, M4 perpendicular to each other (Blit et al. 1977).

the bandwidth of the injected radiation. At 100 J electrical input, $0.5\,\mu$s long pulses with an energy of 50 mJ and 30 MHz spectral width were obtained from a $10^{-4}\,M/\ell$ methanolic solution of rhodamine 6G. Continuous frequency tuning of this system requires, of course, feedback locking of the cavities of both lasers.

A rate equation analysis of the behavior of a pulsed dye laser with injected narrowband signal was presented by Ganiel et al. (1976). Their calculations consider the temporal evolution of the spectral components at all wavelengths, the tunability of the laser output as the injected wavelength is tuned across the gain curve of the dye laser and its dependence on the injected signal intensity.

In the future, injection-locking of powerful pulsed lasers seems to be one of the most promising methods of obtaining coherent radiation of high peak power, high average power and extremely high spectral stability and purity. This is of particular interest because efficient frequency doubling and mixing of such radiation in nonlinear optical crystals or frequency tripling in metal vapors will provide intense radiation with a bandwidth of only a few megahertz, tunable throughout the ultraviolet and even vacuum ultraviolet regions of the optical spectrum.

7. Two-wavelength operation

Many applications of dye lasers require tunable radiation at two or even three wavelengths simultaneously. This can be achieved, of course, by several independent dye laser systems, pumped by the same pulsed laser or by synchronized flashtubes. A number of systems have been developed, however, which produce two wavelengths from a single dye laser. Although the simple design makes such systems economically attractive, multi-wavelength operation of a single laser has several restrictions. Since the two wavelengths are produced from the same dye, the spacing of the wavelengths is limited by the width of the dye gain profile. Although the gain profile can be broadened by mixing different dyes together, this is limited to dyes with rather near-lying fluorescence bands. If this were not the case the increasing overlap of the fluorescence of the shorter wavelength dye with the absorption bands of the dye operating at longer wavelength would limit output and tuning of the laser radiation at the shorter wavelength. If a single dye is used, both wavelengths deplete the same excited state population. Consequently an increase in the intensity of one beam causes a decrease in the intensity of the other. This competition, analyzed by Flamant and Meyer (1975), not only affects the total intensity of the output pulses but may also cause a complicated time dependence of the intensity distribution between the laser pulses at both wavelengths. As was observed by the authors, the intensity of their multiline flashlamp-pumped laser switches from one wavelength to another several times during a laser pulse. In the analysis, this behavior was explained as due to thermal gradient effects in the dye medium.

Despite these fundamental difficulties several satisfactory working laser systems have been described in the literature. Stansfield et al. (1971) produced two beams in a dual cavity dye laser pumped by a Q-switched ruby laser. Both laser cavities, consisting of an outcoupling mirror and a grating, shared a common dye cell which was pumped nearly collinearly by the ruby laser beam. Hilborn and Brayman (1974) obtained simultaneous broadband two-wavelength emission from an untuned N_2-laser-pumped dye laser, which was operated with mixtures of rhodamine 6G and cresyl violet or rhodamine 6G and coumarin dyes. Zalewski and Keller (1971) reported dual-wavelength operation of a flashlamp-pumped dye laser by placing a grating behind each of the two mirrors of their untuned dye laser cavity. A very effective method for simultaneous two-wavelength selection in transversely laser-pumped dye lasers has been reported by Pilloff (1972) and Young et al. (1973). By placing a Glan Thompson polarizer between the dye cell and wavelength-selective grating (fig. 16) the laser beam can be split into two orthogonal polarized beams. With a second grating, feedback can be provided for a certain wavelength of the broadband fluorescence which is reflected by the polarizer out of the laser cavity. Instead of a beam splitting polarizer, Marx et al. (1976) and Wu and Lombardi (1973) used a partly transparent mirror to reflect part of the light beam out of the cavity. In the design of Marx et al. (1976), this light was reflected by a further mirror on to the same grating which is used for the laser light which passes the partly transparent mirror in the laser resonator. Since, for this second light beam the angle of incidence at the grating is different compared to the angle of the intracavity light, optical feedback is provided for two different wavelengths. By inserting Fabry–Perot interferometers into both beams between the beam splitting element and the grating, the output can be narrowed at both operating wavelengths (Marx et al. 1976). The same scheme for dual-wavelength selection has been applied by Inomata and Carswell (1977). In the system reported by these authors, simultaneous frequency doubling was achieved at both wavelengths with bidirectional illumination of a single KDP crystal. Using standard schemes of wavelength selection, Burlamachi et al. (1977) generated independently tunable two-wavelength output of their slab waveguide dye laser system.

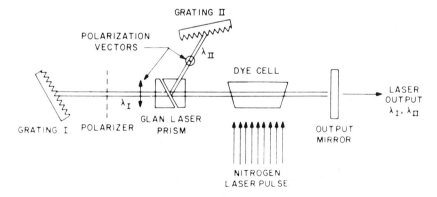

Fig. 16. Schematic diagram of a two-wavelength N_2-laser-pumped dye laser (Pilloff 1972).

A particularly simple method for generating dual-wavelength output in a
N_2-laser-pumped dye laser has been described by Schmidt (1975) and Lotem and
Lynch (1975). A wedge which is placed in front of the tuning grating intersects
only part of the cross section of the light beam (fig. 17). Feedback is therefore
provided simultaneously for two different wavelengths. In contrast to other
techniques, the beams at the two wavelengths pass through different regions of
the active medium. Thus there is little or no competition between the two
wavelengths and the power of both beams varies smoothly over the whole tuning
range.

The problem of gain competition was circumvented also with a system of two
coupled cavities as reported by Marowsky and Zaraga (1974). A powerful
flashlamp-pumped dye ring laser pumped a small intracavity dye laser. Both
emission wavelengths were independently tunable and covered the spectral
range 580 to 652 nm for a dye combination of rhodamine 6G and cresyl violet. A
different approach for multiple-wavelength operation of pulsed dye lasers utilizes two holographic gratings or a single composite holographic grating (Friesem
et al. 1973a, b). In the discussion of acousto-optic dye laser tuning, Taylor et al.
(1971) mentioned the possibility of multiple-wavelength operation through application of multiple rf signals to the acousto-optic tuning filter. The presence of
two or more acoustic frequencies in the filter creates a double or multiple peaked
passband characteristic with a possible transmission at each band of nearly one
hundred percent. With a method very similar to this proposed scheme, Hutcheson and Hughes (1974b) obtained multiple-wavelength output of their laser
which was tuned with a grating and an acousto-optical beam deflector.

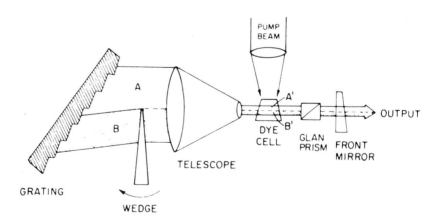

Fig. 17. Scheme of a double-wavelength dye laser. the upper part of the laser beam operates at the
original wavelength while the lower part is shifted towards the red because of the wedge (Lotem and
Lynch 1975).

8. Dye vapor lasers

At present, the overall efficiency of flashlamp-pumped dye laser systems is limited to about 1% and the efficiency of laser excited systems is even less. An improved efficiency would certainly increase the usefulness of dye lasers for many applications. In view of this, many research groups have investigated new schemes of excitation such as the direct excitation of dye vapors by an electrical discharge or fast electron beam.

In early studies, Ahmed and Pappalardo (1972a, b) observed fluorescence of several optically and discharge excited dye vapors. Compared to the optical bands of dye solutions the absorption and emission bands of vapors were somewhat broadened and shifted to shorter wavelengths, but had relatively the same shape and position. In experiments reported by Lempicki et al. (1972), several dye vapors were excited electrically by an Ar-buffered low current discharge or optically with a mode-locked frequency-doubled ruby laser as was reported by Pappalardo et al. (1974). Despite the strong fluorescence which was observed in all these experiments, no conclusive indications of population inversion or laser action could be detected. The first successful operation of a dye vapor laser was obtained with optical excitation (Borisevich et al. 1973). Vapor of the scintillator dye POPOP, buffered with high pressure pentane was excited with the second harmonic of a ruby laser. The peak pump intensity was of the order of 100 MW/cm^2. Independently, Steyer and Schäfer (1974) and Smith et al. (1974) observed laser action in the same dye which was pumped transversely by the output of a N$_2$ laser. In further studies of optically excited POPOP, Basov et al. (1975) used a longitudinal pumping geometry to pump the dye vapor with frequency-doubled ruby laser pulses of 2 MW peak power. The threshold intensity required for laser action was about 80 MW/cm^2. The observed amplified spontaneous emission peaked at the two wavelengths of 383.5 nm and 399.5 nm. In POPOP vapor longitudinally pumped by a N$_2$ laser, Smith et al. (1975) reported a power output of 30 kW with an efficiency of 8%. The laser emission could be tuned over a range of 25 nm and was spectrally narrowed to about 1 nm. These results were obtained in the absence of a buffer gas and at dye vapor pressures low enough that the average collision time was much longer than the fluorescence lifetime. It was thus concluded that the intraband relaxation time must be shorter than 1 ns. This was confirmed by Shapiro et al. (1974) from measured fluorescence rise times of a dye vapor after excitation with picosecond pulses. These results demonstrate that excited dye vapors can have a homogeneously broadenend gain profile, which, of course, is an important condition for an efficient spectral narrowing of the laser output by frequency-selective intracavity elements.

Many dyes have been investigated for application in dye vapor lasers by Stoilov and Trusov (1974) and Steyer and Schäfer (1975). In several dye vapors the measured quantum efficiency was between 50 and 100%, and laser action was reported for dyes like α-NPO, BBO or 1, 2-Di-(5-methyl-2-benzoxazolyl) ethylene. Pikulik et al. (1975) reported on laser action in vapors of pthalamide dyes with output wavelengths in the region 520 to 570 nm. It is interesting to note that the vapor of rhodamine 6G, which is one of the most efficient dyes in liquid lasers, has a quantum efficiency of 13% at low vapor pressures but decomposes before a vapor density appreciably higher than 10^{15} cm^{-3} is reached (Stoilov and Trusov 1974, Sakurai and de Winter 1975). The dye vapor fluorescence and laser output power usually increase if certain buffer gases are added (Borisevich et al. 1974, 1975a, b, c). This effect, initially thought to be due to more rapid thermalization of the laser levels, is caused primarily by a change in the triplet absorption spectrum (Borisevich 1975, Borisevich et al. 1975d).

Although investigations of optically excited dye vapors are very important for the determination of their spectroscopic properties, the efficiency of an optically excited dye vapor laser will be necessarily lower than that of a liquid dye laser. A higher overall efficiency may be obtainable only from a discharge- and electron-beam-pumped dye vapor laser medium. These excitation schemes, however, introduce serious problems. The most severe difficulties include decomposition of the dye and direct excitation of the metastable triplet state. From data on optically excited vapor fluorescence, it is possible to estimate the discharge conditions required for a certain gain. In optically excited POPOP, for example, a gain of 1 cm^{-1} requires a pumping power of about 2×10^7 W/cm^3 (Smith et al. 1975). Because the quantum efficiency in discharge-pumped dyes will be lower than in optical excitation, a value of the order of 2×10^8 W/cm^3 should be more realistic. Thus the excitation of a dye vapor in a transverse discharge scheme appears appropriate for the achievement of sufficiently high current densities. Smith et al. (1976) excited 20 dyes in a transverse discharge and studied the excitation as a function of buffer gas, of temperature and of discharge parameters. For the best dyes the measured dye concentration in the upper laser level corresponded to a calculated gain of 0.2 cm^{-1}. Saturation of the output with increasing vapor pressure occurred for all dyes studied. This saturation may be caused by concentration quenching and by the influence of the dye molecules on the electron energy distribution. The net gain is less than that computed from the observed fluorescence power. This is due to various loss mechanisms, such as ground state absorption, excitation of the triplet state and decomposition of the dye molecules. Thus no net gain was found for any of the molecules studied by Smith et al. (1976).

An alternative to discharge excitation is the pumping of dye vapors with an electron beam. As observed by Freund and Schiavone (1976), the fluorescence of POPOP vapor excited by direct electron collisions reaches an optimum for electrons of about 30 eV energy. The estimated excitation cross section for electrons of this energy (10^{-16} cm^2) is of the same order as the cross section for optical excitation. The spectrum of the POPOP fluorescence appeared to be the same for both types of excitation. More recently, Marowsky et al. (1976) studied the fluorescence of mixtures of POPOP dye vapor and various buffer gases after excitation with a 200 J electron beam. Xenon was found to be the most efficient buffer gas for energy deposition of the electron beam and energy transfer to the POPOP dye vapor, but no estimate was reported of the density of excited

POPOP vapor or of the corresponding gain. Edelstein et al. (1977) reported excitation by electron-beam pumping of rare gas–organic vapor mixtures for the dyes POPOP and α-NPO. Assuming complete population inversion between the excited singlet state and high vibronic levels of the ground state, the authors estimated from the observed fluorescence a maximum gain of 0.03 cm^{-1} and 0.01 cm^{-1} for POPOP and α-NPO respectively.

Although dye vapor lasers have not yet realized their potential for efficient scalable sources of tunable radiation, new excitation schemes may be found which provide suitable conditions for reaching this goal. As pointed out by Smith (1976) in his detailed review on dye vapor lasers, the selection of simpler molecules or the transfer of excitation from a long-lived excited state of one species to the desired excited state of the dye molecules (as studied by Sakurai et al. 1976 and Ryan et al. 1976) may be a useful scheme for excitation in dye vapor lasers.

9. Generation of tunable UV and IR

With presently available laser dyes, dye laser operation is limited to the wavelength range from about 340 to 1200 nm. The wavelength range can be extended towards shorter and longer wavelengths by frequency mixing of dye laser radiation in nonlinear optical materials. In this way, tunable ultraviolet radiation down to 196 nm has been obtained by second-harmonic and sum frequency generation in nonlinear crystals. Tunable radiation of even shorter wavelengths generated by third-harmonic generation and frequency mixing in metal vapors as nonlinear optical medium has been obtained. In the long wavelength range the generation of tunable near- and middle-infrared as the difference frequency of two laser oscillators has been very successful.

9.1 Harmonic generation and sum frequency mixing

Frequency doubling of visible laser light in nonlinear optical materials is a technique which was well established with fixed frequency lasers. The physics of this process has been discussed in several excellent reviews (Bloembergen 1965, Terhune and Maker, 1968). If the peak power of the laser radiation is sufficiently high (in the kW to MW range), the doubling crystal can be placed outside the laser cavity. At smaller power levels, as supplied by cw dye lasers, low-loss laser cavities with an intracavity doubling crystal provide the best efficiency for frequency conversion. Because the phase matching conditions require different crystal orientations for different wavelengths, doubling of the frequency-tuned output of a dye laser necessitates the rotation of the crystal, synchronously with dye laser tuning. At low laser intensities the energy conversion of the visible laser radiation into coherent ultraviolet light is proportional to the input intensity and the uv output increases as the square of the input. At sufficiently high intensities, the conversion efficiency approaches a constant value, which is usually of the order of 10 to 40%. To obtain an optimum in conversion, the dye laser light is focused in general into the doubling crystal. If the focusing is too tight the efficiency may be reduced, however, because the angular spread of the beam is too large to achieve simultaneous phase matching for the different angular directions. According to calculations by Kuhl and Spitschan (1972) the angular spread of the incident monochromatic laser radiation should not exceed 0.66 mrad for efficient second-harmonic generation in their 38 mm long crystal of KDP, which was cut under $\theta = 64.4°$.

In one of the first reported experiments on second-harmonic generation of dye laser light, Bradley et al. (1971a) used an ADP crystal to double the output of an interferometrically narrowed rhodamine 6G dye laser side-pumped by a frequency-doubled Nd:glass laser. The generated uv light pulses of 20 mJ and megawatt peak power were continuously tunable from 280 to 290 nm with a bandwidth of 0.2 nm. The conversion efficiency was about 10%. Ultraviolet pulses of similar energy and 5 pm bandwidth, tunable between 340 and 400 nm, were obtained by Hamadani and Magyar (1971) by doubling the output of a ruby-pumped dye laser in a $LiIO_3$ crystal. Dunning et al. (1972) generated intense ultraviolet radiation, which was continuously tunable in the range of 250 to 325 nm by frequency doubling the output of a N_2-laser-pumped dye laser in a 25 mm long ADP crystal (60° z-cut). The efficiency approached 11% at a laser peak power as low as 60 kW. In these experiments the linewidth of the dye laser was narrowed to less than 0.2 nm by a diffraction grating and beam-expanding prism as wavelength-selective intracavity elements. The beam convergence of the focused light inside the crystal was 3.2 mrad. Tunable uv light at wavelengths from 300 to 230 nm was obtained by Dunning et al. (1973) with a 10 mm long doubling crystal of lithium formate monohydrate. In this crystal, the conversion efficiency reached 2% at fundamental powers in excess of 50 kW. At present, the shortest second-harmonic output attained lies at 217 nm obtained by Dewey (1976), Dewey et al. (1975) and Zacharias et al. (1976) in a crystal of KB_5O_8. In this crystal the conversion efficiency is strongly wavelength dependent and increases for input pulses of about 10 to 20 kW from less than 0.1% at 300 nm to about 9% at 217 nm. As described in § 5, oscillator–amplifier systems are able to generate tunable coherent light of particularly high peak power, narrow bandwidth and almost diffraction-limited spatial beam quality. These qualities make the output of such systems well suited for efficient second-harmonic generation in nonlinear optical materials. The fundamental output of the oscillator–amplifier system reported by Moriarty et al. (1976) (fig. 12) was frequency doubled to give 7 ns long pulses of 1.5 pm spectral width and 600 kW peak power at 282 nm. Tunable uv light of only 0.03 pm spectral width with peak powers in the kW range was generated by frequency doubling the output of the N_2-laser-pumped oscillator–amplifier system (Wallenstein and Hänsch 1975). In the wavelength range 260 to 350 nm, conversion efficiencies of 10 to 20% were obtained with angle-tuned KDP or ADP crystals. Frequency doubling in lithium formate monohydrate yielded an efficiency of 2% near 240 nm at input powers as low as

2 kW, whereas in experiments with laser radiation of 0.2 nm bandwidth (Dunning et al. 1973) peak powers of 60 kW were required to achieve the same efficiency.

Although KDP and ADP crystals are transparent down to 200 nm, they do not permit phase matching for second-harmonic generation at uv wavelengths below 262 nm (ADP) and 259 nm (KDP), if the crystals are kept at room temperature. Temperature tuning of a 90° phase matched ADP crystal between -116°C and 52°C extends the generation of coherent radiation to the spectral range 246 to 265 nm (Jain and Gustafson 1976).

Ultraviolet radiation at shorter wavelengths can be obtained in these crystals as the sum frequency of two laser waves of different frequencies. This was demonstrated by Akhmanov et al. (1969) who generated 212 nm as the sum frequency of the infrared light (1.06 μm) of a Nd laser and its fourth harmonic in the ultraviolet. The same scheme was used by Massey (1974) for parametric up-conversion of tunable uv light in the range 250 to 325 nm into light of 200 to 235 nm. In a temperature-tuned 90° phase matched ADP crystal, the author mixed infrared Nd:YAG laser radiation with the frequency-doubled output of a dye laser. For the conversion of 266 nm to 213 nm, efficiencies of more then 50% with a peak and average power above 5 kW and 0.5 mW respectively were reported.

Kato (1977a) reported mixing in ADP the fourth harmonic of a Nd:YAG laser with the radiation of a near-infrared dye laser pumped by the second harmonic of the same Nd:YAG laser. Tunable radiation was generated at wavelengths as short as 196.6 nm with peak power as high as 40 kW. In a similar way Kato (1977b) generated ultraviolet light between 207.3 nm and 217 nm by mixing the fundamental and second harmonic of a high power visible dye laser in potassium pentaborate. A peak power as high as 250 kW with an average power of 15 mW was obtained at 207.3 nm.

It is interesting to note that Jantz and Koidl (1977) converted 10.6 μm laser radiation into the green spectral range by nonlinear mixing of the infrared radiation with the output of a dye laser in a 90° phase matched crystal of $AgGaS_2$. Quantum efficiencies were as high as 40% for this process.

Tunable radiation in the range from 216 to 234 nm was generated by Dinev et al. (1972) in a somewhat different way. Using the third-order nonlinear susceptibility of calcite for sum frequency generation, the 70 MW output of a Q-switched Nd:glass laser with a two-stage amplifier was mixed with its second harmonic (20 MW) and with the output of an untuned broadband rhodamine 6G or rhodamine B dye laser (2 to 4 MW) pumped by part of the second-harmonic Nd:glass laser light. The output peak power reached 15 kW. The wavelength could be tuned simply by rotating the crystal, since the strong dispersion of the calcite crystal restricted sum frequency generation to a 15 pm wide frequency interval of the incident broadband radiation.

The output of flashlamp-pumped dye lasers has been doubled successfully with output powers in the kilowatt range. With a 90° phase matched temperature-tuned ADP crystal outside the laser cavity, Jennings and Varga (1971) generated a peak power of the order of 10 W in 0.6 μs long pulses which were tunable from

250 to 290 nm. Kuhl and Spitschan (1972, 1975) doubled the output of a flashlamp-pumped rhodamine 6G laser in a 38 mm long angle-tuned KDP crystal. They achieved power conversion efficiencies of 10 to 18% for laser powers as low as 15 to 25 kW. The bandwidth of this laser could be varied between 8 nm and 1 pm by inserting an interference filter and Fabry–Perot interferometers into the cavity. The oscillation could be restricted even to a single TEM_{00} mode. In their report, Kuhl and Spitschan (1972) also described the detailed study of the influence of spectral and spatial characteristics of the laser output on the conversion into the ultraviolet. Hirth et al. (1977) reported tunable uv emission in the range 280 to 310 nm by doubling the output of a flashlamp-pumped rhodamine dye laser in KDP or ADP crystals. With conversion efficiencies of about 8%, the output energies reached almost 50 mJ in 0.75 μs long pulses of 5 pm bandwidth.

For dye lasers of low power and continuous or long pulse operation, intracavity doubling provides an optimum in efficiency. In a tunable uv laser pumped by a Nd:YAG laser, an angle-tuned ADP crystal was placed inside the dye laser resonator (Wallace 1971). The output of 200 W peak power, of 0.2 to 0.4 μs pulse length and of 2 cm^{-1} bandwidth could be tuned between 261 and 315 nm. Intracavity second-harmonic generation in continuous wave dye lasers was analyzed by Ferguson and Dunn (1977). It was used by Gabel and Hercher (1972) to double the radition of a cw dye laser with an intracavity crystal of lithium formate to generate continuous tunable ultraviolet in the region of 290 to 315 nm. The output powers reached 5 mW and 0.1 mW for multimode and single-mode operation respectively. Efficient frequency doubling of cw radiation of a rhodamine 6G dye laser is reported by Fröhlich et al. (1976). A 90° phase matched ADA crystal in a dye laser of particular low-loss cavity design, generated 50 mW multimode or 4 mW single-mode radiation tunable from 285 to 315 nm. From a cw ring dye laser, the same authors obtained from the same doubling material single-mode uv light exceeding 40 mW (Schröder et al. 1977). Although ADA is damaged at relatively low power densities, it has been used successfully in a flashlamp-pumped dye laser. Atkinson and Schuyler (1975) doubled the 55 ns long output of their flashlamp-pumped rhodamine 6G dye laser to generate tunable radiation in the 290 to 300 nm region at power levels greater than 1 kW. Finally it should be noted that single-frequency cw uv light of 0.1 mW was generated at 240 nm by summing the output of a 500 mW Kr^{+} laser at 4131 Å and a 200 mW single-mode rhodamine 6G dye laser in a 5 cm long ADP crystal cooled to 160 K (Hänsch 1976).

The generation of tunable coherent radiation at wavelengths shorter than 200 nm is possible by third-harmonic generation and sum frequency mixing in rare gases (Ward and New 1969) or in metal vapors as nonlinear medium. The use of metal vapors was first proposed by Harris and Miles (1971) and experimentally demonstrated by Young et al. (1971) who tripled the frequency of a nontunable picosecond Nd laser in Rb vapor. In further experiments, Harris and coworkers succeeded in generating coherent light as short as 118.2 nm (Kung et al. 1972) and 88.6 nm (Harris et al. 1973) by tripling in Cd vapor and in Ar respectively, the frequency-tripled (532 nm) or the frequency-quadrupled (266 nm) radiation of a Nd:YAG laser. Tunable vacuum ultraviolet in spectral regions between 118 and 199.5 nm was reported by Kung (1974). In this experiment the frequency-quadrupled radiation of a mode-locked Nd:YAG laser was mixed in Xe with visible tunable coherent light. This light was generated in an optical parametric ADP oscillator, pumped by the 266 nm light of the same Nd:YAG laser. Detailed calculations of the third-order susceptibility for light mixing in metal vapors (Miles and Harris 1973, Elgin and New 1976) show a resonant enhancement if the frequency of the incident radiation corresponds to one-, two- or three-photon allowed transitions of the metal vapors. Phase matching of incident and generated vuv can be achieved by compensating an anomalous dispersion with an appropriate buffer gas such as He, Ar or Xe. The resonant enhancement of the third-order susceptibility by a two-photon allowed transition was observed by mixing two visible photons (ω_1) and one infrared photon (ω_2) to yield light of $\omega_3 = 2\omega_1 + \omega_2$ by Harris and Bloom (1974) in Mg and by Bloom et al. (1974) in Na. The two-photon enhancement in third-harmonic generation was demonstrated also in frequency-tripling experiments of ruby laser light in Cs (Leung et al. 1974, Ward and Smith 1975, Georges et al. 1976).

The resonant enhancement of the third-order mixing process made possible the first generation of tunable vuv radiation by four-wave sum mixing of dye laser light in Sr vapor (Hodgson et al. 1974). The authors employed two tunable dye lasers, pumped simultaneously by a 1 MW N_2 laser (fig. 18). One of the dye lasers was tuned to a two-photon allowed transition and the other to a frequency ω_2, so that the sum $2\omega_1 + \omega_2$ corresponded to a transition from the ground state to an (autoionizing) state in the continuum above the ionization limit. In this way the authors succeeded in generating continuously tunable radiation from 177.8 to 181.7 nm and from 183.3 to 195.7 nm. At 179 nm, for example, a vuv signal of 4×10^9 photons was produced with a laser of 16 kW peak power ($0.1\,\text{cm}^{-1}$ bandwidth) at ω_1 and 18 kW ($1\,\text{cm}^{-1}$ bandwidth) at ω_2. The laser radiation was

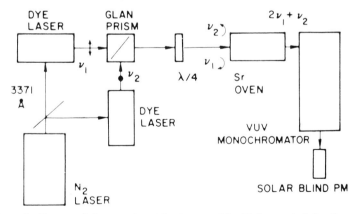

Fig. 18. Schematic diagram of the experimental set-up used by Hodgson et al. for the generation of tunable coherent vacuum-ultraviolet radiation by resonantly enhanced sum frequency mixing in Sr vapor (Hodgson et al. 1974).

mixed in 25 Torr of Sr and 460 Torr of Xe, added for phase matching. A similar method has been used by Wallace and Zdasiuk (1976) to generate tunable radiation in Mg vapor from 140 to 160 nm. With resonantly enhanced four-wave mixing of the fourth harmonic of a Nd:YAG laser and dye laser radiation of about 626 nm, Hsu et al. (1976) obtained tunable radiation at 120.3 nm in Hg vapor. In this mixing experiment a commercial dye laser was amplified in a parametric $NH_4H_2PO_4$ amplifier which was pumped by the fourth harmonic of a Nd:YAG laser. The authors reported 15 ps long vacuum ultraviolet light pulses of 300 W peak power.

Besides two-photon transitions, autoionizing states were found to strongly enhance the third-order susceptibility of metal vapors (Hodgson et al. 1974, Armstrong and Wynne 1974, Armstrong and Beers 1975, Sorokin et al. 1976, Wynne and Sorokin 1977). In an experiment similar to that described by Hodgson et al. (1974), Bjorklund et al. (1977) investigated four-wave sum mixing in Sr with three lasers with frequencies ω_1, ω_2 and ω_3. By adjusting the input frequencies so that they are related in a proper way to a single-photon, a two-photon and an autoionizing transition of the Sr atom, the observed mixing efficiency became exceptionally high. Output powers of $P = 1.6$ W (at $\lambda = 169.7$ nm) and $P = 3.2$ W (at $\lambda = 171.2$ nm) have been detected with laser input powers at ω_1, ω_2 and ω_3 between 1 and 100 kW. In addition to experiments mentioned so far, resonant third-harmonic generation has been achieved with tunable picosecond pulses in Na vapor by Drabovich et al. (1977) using a dye laser pumped by a mode-locked Nd:glass laser and by Taylor (1976) with a mode-locked flashlamp-pumped dye laser. Tunable picosecond pulses have also been generated in Ca (Sibbett et al. 1976) and Sr vapor (Royt et al. 1976). A detailed discussion of the physics and experimental achievements in the generation of tunable vacuum ultraviolet by optical mixing in atomic vapors has been presented recently by Wynne and Sorokin (1977). Four-wave sum mixing has also been achieved in molecular gases. In a first experiment, Wallace et al. (1976) used NO as nonlinear medium and converted tunable dye laser radiation into coherent light emission in broad regions, which were centered at 151, 143, 136 and 130 nm. Although the third-order nonlinear susceptibility of NO is only about 1% of that observed for the equivalent metal vapor systems like Ca or Sr, reasonable conversion efficiencies were realized by choosing higher number densities. Thus, with NO pressure of 200 Torr and appropriate phase matching, a power conversion efficiency of 3×10^{-7} or 10^7 photons/pulse was obtained at vuv wavelengths of 151 and 142 nm with dye laser input powers up to 20 kW.

9.2 Difference frequency generation

The operation of tunable dye lasers is limited to wavelengths shorter than about 1.2 μm. Tunable infrared light of longer wavelength can be generated with dye laser light by producing the difference frequency between light of a tunable dye laser and a fixed frequency laser or by generating the difference frequency of two tunable laser sources. For the mixing process a variety of crystals are found to provide suitable properties, such as adequate birefringence for phase matching and transparency at both the pump wavelength and at the generated infrared wavelength. Such crystals include, for example, $LiNbO_3$, Ag_3AsS_3(proustite) and $AgGaS_2$. In addition, metal vapors have been employed successfully as a medium for nonlinear frequency mixing.

In one of the first experiments, Dewey and Hocker (1971) produced in an angle-tuned $LiNbO_3$ crystal coherent infrared radiation as the difference frequency of a ruby laser and a ruby-pumped DTTC iodide dye laser. The ir light had peak powers of several kilowatts and 3 to $5\,cm^{-1}$ bandwidth, and was tunable between 3 and $4\,\mu m$. The crystal material, $LiNbO_3$, used in this experiment, restricts the generation of difference frequencies to wavelengths shorter than $4.5\,\mu m$. The generation of longer wavelengths is possible, for example, in proustite, which covers, in appropriate orientations, the full range from 2.5 to $13\,\mu m$. With proustite Hanna et al. (1971) have obtained tunable radiation in the middle infrared between 10.1 and $12.7\,\mu m$ as the difference of a $290\,kW$ ruby laser and a cryptocyanine dye laser which was tuned from 0.73 to $0.75\,\mu m$. The peak power of the infrared pulses was of the order of $100\,mW$. Kilowatt peak power in the wavelength range 3.2 to $5.6\,\mu m$ has been generated in the same material by Decker and Tittel (1973a) with a Q-switched ruby laser of $900\,kW$ and a ruby-pumped dye laser of 20 to $40\,kW$ peak power, operating between 790 and 890 nm. By mixing two dye lasers in this material, Decker and Tittel (1973b) observed mid-infrared radiation tunable from 5.82 to $7.25\,\mu m$ with pulse powers up to $100\,W$. The two dye lasers with output powers of 120 and $200\,kW$ (at 0.25 and 0.44 nm bandwidth respectively) were pumped by a ruby laser.

A $LiIO_3$ crystal was used by Meltzer and Goldberg (1972) to mix a DTTC iodide dye laser with part of the $4\,MW$ ruby pump source. The authors placed the crystal inside the dye laser cavity and obtained tunable radiation from 4.1 to $5.2\,\mu m$ with a peak power of about $100\,W$. Short pulse output, tunable from 1.13 to $5.6\,\mu m$, was reported by Moore and Goldberg (1976) by difference frequency mixing in $LiIO_3$ a dye laser (549 to 727 nm) with the 1.06 and $0.53\,\mu m$ pulse trains from its mode-locked Nd:YAG pump. $LiIO_3$ has also been used by Gerlach (1974) for generation of the difference frequency of two tunable flashlamp-pumped dye lasers in the ir range $1.5\,\mu m$ to $4.8\,\mu m$. With dye laser pulses of 14 and $30\,kW$ peak power and 200 to 300 ns duration, a conversion efficiency of about 10^{-4} was reported for this experiment.

Tunable infrared down-conversion in silver thiogallate ($AgGaS_2$) has been reported by Hanna et al. (1973). With a ruby laser and a ruby-pumped dye laser (0.74 to $0.81\,\mu m$) tunable infrared radiation of 4.6 to $12\,\mu m$ and $0.3\,W$ peak power could be produced. With two flashlamp-pumped dye lasers (tunable from 0.58 to 0.62 nm), Hanna et al. (1974) reported the generation of infrared light from 8.7 to $11.6\,\mu m$ and with about $10^{-4}\,W$ peak power.

As in the near infrared, lithium niobate can be utilized again for the generation of light in the far infrared. This was demonstrated by Faries et al. (1971) by mixing the outputs of two ruby lasers, operating at slightly different frequencies.

In this way the authors produced submillimeter waves between 20 and 38 cm^{-1}. Mixing the frequencies of two ruby-pumped dye lasers in this material, Auston et al. (1973) could generate far-infrared radiation with a few milliwatts peak power and tunable from 2 to 50 cm^{-1}.

CW dye lasers have also been used successfully for the generation of tunable infrared light. In LiNbO$_3$, Pine (1974, 1976) mixed the output of an Ar-ion laser at 514 nm or at 488 nm with the radiation of a single-mode cw rhodamine 6G dye laser. The infrared radiation of about 1μW was tunable in the range 2.2 to 4.2 μm. With radiation of the same laser sources, but using LiIO$_3$ as the nonlinear optical material, Wellegehausen et al. (1976) generated cw infrared of up to 4 μW in multimode and 0.5 μW in single-mode dye laser operation. The cw ir was tunable from 2.3 to 4.6 μm. CW infrared radiation tunable from 1.28 to 1.62 μm with power levels of 35 μW was obtained by Lahmann et al. (1976) by

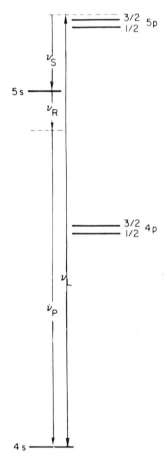

Fig. 19. Scheme of parametric four-wave mixing in K vapor. Tunable dye laser radiation of frequency ν_L excites stimulated electronic Raman scattering at frequency ν_S. Mixing with a second dye laser frequency ν_P results in generation of infrared light with frequency ν_R (Sorokin et al. 1973).

difference mixing in LiIO$_3$ the output of a rhodamine 6G dye laser with the intracavity radiation of a Nd:YAG laser.

A very interesting scheme for the generation of tunable infrared radiation was first demonstrated by Sorokin et al. (1973). In this scheme, potassium vapor was used as a nonlinear medium for a parametric four-wave mixing process. For mixing, two pulsed dye lasers provide two input beams at frequencies ν_L and ν_P. As indicated in fig. 19, the laser frequency ν_L is close to the 4s–4p resonance transition and excites stimulated Raman emission at a frequency ν_S to the state 5s. Mixing with the tunable frequency ν_P of the second laser results in the generation of an infrared beam of frequency ν_R. Phase matching can be achieved without any buffer gas by proper choice of the frequency ν_L. In the experiment by Sorokin et al., the two dye lasers were pumped by the same N$_2$ laser and provided peak powers of about 1 kW at a bandwidth of 0.1 cm^{-1}. With K contained in a heat pipe at pressures between 0.5 and 20 Torr, it was possible to generate wavelengths between 2 and 4 μm with peak powers of about 0.1 W. Rubidium permitted a continuous tuning over the range from 2.9 to 5.4 μm. Using a similar experiment, Kärkkäinen (1977) generated tunable infrared radiation between 1.8 and 20 μm by two-photon resonant four-wave mixing of two dye lasers beams in K vapor (1.8 to 4.8 μm) and Cs vapor (3.5 to 20 μm). The two dye lasers were pumped by the same N$_2$ laser and delivered radiation of 0.1 cm^{-1} spectral width and typically 5 to 20 kW peak power. At 2.4 μm, for example, the ir power generated in K was measured to be 3 W. The highest ir peak power in Cs was almost 400 mW, and was measured at 5.5 μm. The ir power at 20 μm was still of the order of 2 mW. It should be noted that the physics and experimental results of nonlinear infrared generation with tunable laser sources in crystals and vapors have been analyzed and reviewed in detail by Byer and Herbst (1977) for nonlinear mixing in crystals and by Wynne and Sorokin (1977) for four-wave mixing in metal vapors.

10. Short pulse generation

Many dye laser applications require frequency tunable light pulses of short duration. In this respect, dye lasers are extremely versatile. Flashlamp-pumped devices provide pulses ranging from a fraction of a microsecond to several milliseconds. Laser-pumped dye lasers operate in the region 1 to 200 ns. Pulses of considerably shorter duration are obtainable by controlled resonator transients or by mode-locking techniques.

10.1 Resonator transients

Relaxation oscillations and initial spiking at the beginning of laser pulses were studied in detail in pulsed solid-state lasers (Birnbaum 1964, Evtuhov and Neeland 1966, Johnson 1966). They are due to the interaction of the inversion population of the active medium in excess of steady state and of the energy of the photon field inside the resonator. The shape of relaxation oscillations depends primarily on the rate of the changes in the population inversion caused by pumping, spontaneous decay and stimulated emission, and on the rate of build-up and decay of photon density in the resonator due to stimulated emission and various loss mechanisms. A rate equation analysis of the transient effects in dye lasers (Miyazoe and Maeda 1971, Lin 1975a) makes evident that a regular, damped relaxation oscillation is to be expected for certain experimental conditions. The most favorable conditions for oscillations at the beginning of the dye laser output pulse are short photon decay times of the laser cavity and low, properly adjusted, pumping rates. This was demonstrated experimentally by Lin and Shank (1975) and Lin (1975a, b) with dye lasers pumped by 10 ns long pulses of a N_2 laser. The short cavity ($l = 1.3$ cm) had a photon decay time of only 50 to 60 ps. Relaxation oscillations were controlled by adjusting the pump intensity. The oscillations observed in various dyes throughout the visible exhibit pulses of 0.6 to 0.9 ns length (fig. 20). Such resonator transients have been used by Salzmann and Strohwald (1976) for pulse shortening on a time scale below 100 ps. A dye cell of only 40 μm length (corresponding to a photon-cavity decay time $\tau_p = 0.04$ ps) was filled with 10^{-2} M/ℓ solution of rhodamine 6G in methanol. It was pumped longitudinally by 100 ps long pulses, generated with a high pressure N_2 laser of 100 kW peak power. The dye laser emission consisted of ultrashort pulses of less than 10 ps duration.

Although a grating was used by Lin and Shank (1975), line narrowing is rather poor in the extremely short cavity and the output linewidth still exceeded 1.5 nm. Because a low pumping level (10 kW) has to be used to control the resonator transients, the power of the output pulses was typically 50 to 100 W. Since, however, only a small fraction of the N_2 laser is used for pumping the dye oscillator, the subnanosecond pulses can be amplified (after spectral narrowing by suitable filters) in amplifiers, pumped by the rest of the N_2 laser light. Powerful single, subnanosecond pulses, generated with this simple method may certainly be useful for the study of many relaxation processes occurring on a nanosecond time scale.

10.2 Mode-locking

The generation of ultrashort pulses by mode-locking techniques is well known from gas and solid-state lasers. Detailed summaries of the principles of these techniques have been presented by De Maria et al. (1969), Smith (1970), De Maria (1971) and Allen and Jones (1971). Mode-locking in dye lasers has been reviewed in detail in excellent articles by Shank and Ippen (1973) and by Bradley (1977).

By applying mode-locking techniques to a variety of dye laser systems, ultrashort pulses have been produced with durations down to less than 1 ps. Since the frequency bandwidth which is required to produce a pulse of picose-

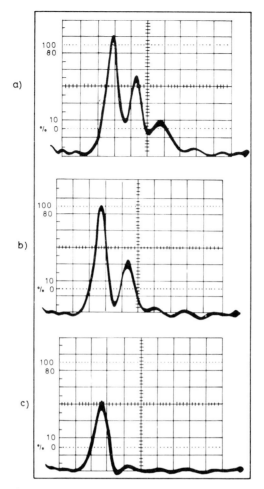

Fig. 20. Damped relaxation oscillations observed in N_2-laser-pumped dye laser with short cavity (13 mm), as the pump power is decreased from (a) to (c). Time scale: 1 ns/cm (Lin and Shank 1975).

cond duration is much smaller than the broad emission band of a laser dye, dye lasers can provide picosecond pulses which are wavelength tunable over several tenths of nanometers. In addition, dye lasers are the only sources which are capable of producing mode-locked picosecond pulses continuously at high repetition rates. In early attempts organic dye lasers were mode-locked by exciting the dye medium with the output of a mode-locked laser source. In this way, the dye medium is excited in a periodic manner, which provides the modulation necessary for mode-locking. In experiments reported by Glenn et al. (1968) and Soffer and Lin (1968), a mode-locked frequency-doubled Nd:glass laser was used to pump a rhodamine 6G or rhodamine B dye laser. The length of the dye laser resonator had to be adjusted to be equal to (or a multiple of) the length of the resonator of the pump laser. If the dye cell is placed at one end of the dye laser cavity, trains of pulses are observed with a period equal to (or a submultiple of) the pump pulse period. Dye lasers mode-locked in this way have been excited with mode-locked ruby lasers (Bass and Steinfeld 1968, Bradley et al. 1968b, Bradley and Durrant 1968), with a frequency-doubled mode-locked ruby source (Topp and Rentzepis 1971) and with frequency-doubled Nd:YAG lasers (e.g., Royt et al. 1974, Goldberg and Moore 1975). The wavelength of such mode-locked dye lasers has been tuned by using a grating as wavelength-selective end reflector (Soffer and Linn 1968). The pulse width of this Nd:glass-laser-pumped rhodamine 6G dye laser was measured to be 10 ps. From this pulse length Soffer and Linn estimated a peak power of 10^6 W at a pump power of about 10^8 W.

Fan and Gustafson (1976) and Cox et al. (1977) have demonstrated that picosecond pulses can also be generated with ultrashort cavity dye lasers pumped by picosecond pulses. The dye laser reported by Fan and Gustafson, for example, had a length of $50\,\mu$m with a corresponding cavity photon lifetime of only 0.1 ps. Filled with rhodamine B and pumped with the second-harmonic megawatt pulses of a mode-locked Nd:glass laser, this dye laser provided pulses of about 5 ps duration with an efficiency as high as 40%.

In a different laser scheme, a weak mode-locked signal obtained from a cw dye laser was injected into the cavity of a flashlamp-pumped dye laser (Moses et al. 1976). The injected radiation synchronously mode-locked the high power device, which produced pulses of about 20 ps duration and peak energies exceeding $20\,\mu$J.

Active mode-locking of a dye laser was reported by Ferrar (1969b). The resonator loss of a flashlamp-pumped coumarin laser was modulated with an acoustic modulator inside the resonator. Mode-locked pulses of several hundred picoseconds have been generated at $\lambda = 460$ nm in pulse trains of a few microseconds duration. Lill et al. (1977) demonstrated the possibility of active mode-locking of a flashlamp-pumped dye laser by bleaching the saturable absorber inside the laser resonator with picosecond pulses from a passively mode-locked flashlamp-pumped dye laser oscillator. As result, pulses tunable in the region 550 to 640 nm were produced with energies between 15 and $80\,\mu$J and with a duration of typically 5 ps.

A more interesting and successful method for short pulse generation in dye lasers is passive mode-locking, which was first observed by Schmidt and Schäfer (1968). A scheme of their laser is displayed on p. 415 in Vol. 1. Mode-locking a flashlamp-pumped rhodamine 6G dye laser with the organic dye DODCI as saturable absorber, the dye laser emission consisted of a train of equally spaced pulses with a $c/2L$ repetition frequency of 1 GHz. The saturable absorber DODCI used in this early experiment still remains one of the most effective mode-locking dyes for rhodamine 6G dye lasers. The results of Schmidt and Schäfer were essentially reproduced by Bradley and O'Neill (1969) in their rhodamine 6G and rhodamine B dye laser. From measurements of two-photon fluorescence, Bradley et al. (1969) determined that the pulselength of their mode-locked flashlamp-pumped dye laser should be as short as 6 ps. Time-resolved measurements with an ultrafast streak camera, developed by Bradley et al. (1971b, c), confirmed that well-isolated ultrashort pulses were produced in the passive mode-locked dye laser. Arthurs et al. (1972a) recorded pulse half-widths of 2.5 ps. Studying the time development of the mode-locked pulse, Arthurs et al. (1973) found that a picosecond pulse can be developed in about 15 round trips. At high power output, the mode-locked dye lasers produced pulses whose spectral content was not directly related to the pulse width. Laser pulses of less than 5 ps duration often exhibited a spectral bandwidth of 3 to 10 nm, which is up to 100 times larger than the Fourier transform limit given by the pulse durations (Bradley et al. 1969). To reduce the spectral bandwidth, Arthurs et al. (1971) inserted an interference filter into the resonator. Owing to the spectral narrowing the output corresponded to the Fourier transform limit set by the pulse width. The transform-limited pulses could be tuned over the wavelength range 602 to 625 nm. The pulse bandwidth was about 0.2 nm and the pulse duration was typically 2 to 4 ps.

The operating range of passively mode-locked lasers is limited only by the availability of laser dyes and suitable absorbers. With various combinations of three laser dyes and four polymethine absorbers, Arthurs et al. (1972b), for example, generated tunable picosecond pulses of less than 6 ps over the range 584 nm 704 nm. The pulse power in such short pulses can be as high as 20 MW (Bradley et al. 1971c). With a series of dye amplifiers, the pulse power can be increased up to several gigawatts (Adrian et al. 1974). In addition to providing power amplification, dye laser amplifiers can be useful for extraction of a single pulse out of a train of mode-locked pulses (Schmidt 1975, Urisu and Kajiyama 1977). Since an amplifier pumped by a N_2 laser provides amplification over a time period which can be shorter than the time difference between two pulses of the mode-locked train, only a single pulse is amplified to high peak powers.

As mentioned above, mode-locked cw dye lasers are capable of generating continuous trains of pulses with subpicosecond duration. Dienes et al. (1971) achieved mode-locking of a cw dye laser with an acousto-optic loss modulator in the dye laser resonator. The width of the generated pulses was determined to be 55 ps from autocorrelation measurements using second-harmonic generation in an ADP crystal. Active mode-locking was also attempted with a phase modulator (Kuizenga 1971). Stable mode-locking, however, appeared difficult to achieve and the output pulses were as long as 200 to 500 ps owing to incomplete mode-locking.

Mode-locking with a continuous mode-locked pump source was reported by Runge (1971, 1972). The author placed a laser dye in the resonator of a He–Ne laser, operating at 632.8 nm. The dye acted as a laser medium and simultaneously as a saturable absorber to mode-lock the He–Ne laser. The mode-locked dye laser pulses almost replicated the He–Ne pulses, which were 270 ps in duration. Mode-locking was observed with different dyes such as cresylviolet, nile blue, DODCI and DTDC. Shank et al. (1973) pumped a rhodamine 6G dye laser with an actively mode-locked Ar laser. The length of the dye laser cavity had to be adjusted for synchronization of the round-trip time in the Ar and dye laser cavity. The pulses of the dye laser were observed to be somewhat shorter than the pumping pulse, but were about 125 ps in duration. Further experiments of this kind have been reported by Mahr and Hirsch (1975), Harris et al. (1975), Chan et al. (1976) and by de Vries et al. (1976). Since this method does not require a spectral matching of gain medium and saturable absorber, the dye laser can be tuned anywhere within its wavelength range. It thus allows the generation of pulses at wavelengths where passively mode-locked systems are not available owing to the lack of an appropriate dye–absorber combination. A major disadvantage of this technique is the fact that typical pulse lengths are about 10 to 50 ps, which is much longer than those produced with passive mode-locking.

Passive mode-locking of cw dye lasers as reported by Ippen et al. (1972) and O'Neill (1972) produced continuous trains of pulses as short as 1.5 ps. The oscillation bandwidth was on the order of 0.5 to 0.65 nm and the peak power ranged from 50 to 100 W. Pulses of subpicosecond duration (0.5 to 1 ps) could be generated by Shank and Ippen (1974). In this experiment both the gain medium (rhodamine 6G) and the saturable absorber (DODCI) were contained in a single free-flowing dye stream located near the center of the resonator. With acousto-optical dumping of the mode-locked pulses from the optical cavity, peak pulse powers of several kilowatts were obtained at repetition rates up to 10^5 pps. Further development of this system resulted in a reliable, high repetition rate source of optical pulses as short as 0.3 ps (Ippen and Shank 1975). From a similar system Ruddock and Bradley (1976) also reported pulses of only 0.3 ps duration. In a somewhat different technique reported by Yasa (1975), Mahr (1976) and Yasa et al. (1977), the dye laser cavity contains two different dye jets of rhodamine 6G and cresyl violet. Since both the gain medium and the saturable absorber are allowed to lase, the double-mode-locked system produces pulses as short as 0.6 ps simultaneously at two wavelengths. The two-wavelength output may be of particular use in time-resolved experiments where two short synchronized pulses are required at two different wavelengths and are used for excitation and probing, for example.

Acknowledgement

The author is indebted to Dr. S.V. Filseth for a critical reading of the manuscript.

References

Adrain, R.S., E.G. Arthurs, D.J. Bradley, A.G. Roddie and J.R. Taylor (1974), Opt. Commun. **12**, 140.
Ahmed, S. and R. Pappalardo (1972a), IEEE J. Quantum Electronics **QE–8**, 525.
Ahmed, S. and R. Pappalardo (1972b), J. Chem. Phys. **56**, 5135.
Akhmanov, A.G., S.A. Akhmanov, B.V. Zhdanov, A.I. Kovrigin, N.K. Podsotskaya and R.V. Kholov (1969), JETP Letters **10**, 154.

Alekseev, V.A., I.V. Antonov, V.E. Korobov, S.A. Mikhnov, V.S. Prokudin and B.V. Skortsov (1972), Soviet. J. Quantum Electronics **1**, 643.

Allen, L. and D.G.C. Jones (1971), Progress in Optics (E. Wolf, ed., North-Holland; Amsterdam), Vol. IX, p. 179.

Anliker, P., M. Gassmann and H. Weber (1972), Opt. Commun. **5**, 137.

Armstrong, J.A. and J.J. Wynne (1974), Phys. Rev. Letters **33**, 1183.

Armstrong, L. and B.L. Beers (1975), Phys. Rev. Letters **34**, 1290.

Arthurs, E.G., D.J. Bradley and A.G. Roddie (1971), Appl. Phys. Letters **19**, 480.

Arthurs, E.G., D.J. Bradley, B. Liddy, F.O'Neill, A.G. Roddie, W. Sibbett and W.E. Sleat (1972a), Proc. Int. Conf. on High Speed Photography, Nice, France.

Arthurs, E.G., D.J. Bradley and A.G. Roddie (1972b), Appl. Phys. Letters **20**, 125.

Arthurs, E.G., D.J. Bradley and A.G. Roddie (1973), Appl. Phys. Letters **23**, 88.

Atkinson, G.H. and M.W. Schuyler (1975), Appl. Phys. Letters **27**, 285.

Atkinson, J.B. and F.P. Pace (1973), IEEE J. Quantum Electronics **QE–9**, 569.

Ault, E.R., R.S. Bradford and M.L. Bhaumik (1975), Appl. Phys. Letters **27**, 413.

Auston, D.H., A.M. Glass and P. Lefur (1973), Appl. Phys. Letters **23**, 47.

Baczynski, A., A. Kossakowski and T. Marszalek (1976), Z. Physik **B23**, 205.

Balucani, U. and V. Tognetti (1976), Optica Acta **23**, 923.

Baltakov, F.N., B.A. Barikhin, V.G. Kornilov, S.A. Mikhnov, A.N. Rubinov and L.V. Sukhanov (1973), Sov. Phys. Techn. Phys. **17**, 1161.

Baltakov, F.N., B.A. Barikhin and L.V. Sukhanov (1974), JETP Letters **19**, 174.

Barger, R.L., J.B. West and T.C. English (1975), Appl. Phys. Letters **27**, 31.

Basov, N.G., V.S. Zuev, Yu. Stoilov and K.K. Trusov (1975), Sov. J. Quantum Electronics **4**, 1178.

Bass, M. and J.I. Steinfeld (1968), IEEE J. Quantum Electronics **QE–4**, 53.

Birnbaum, G. (1964), Optical Masers (Academic Press; New York).

Bjorklund, G.C., J.E. Bjorkholm, R.R. Freeman and P.F. Liao (1977), Appl. Phys. Letters **31**, 330.

Blit, S., A. Fischer and U. Ganiel (1974), Appl. Optics **13**, 335.

Blit, S., U. Ganiel and D. Treves (1977), Appl. Phys. **12**, 69.

Bloembergen, N. (1965), Nonlinear Optics (Benjamin; New York).

Bloom, A.L. (1974), J. Opt. Soc. Amer **64**, 447.

Bloom, D.M., J.T. Yardley, J.F. Young and S.E. Harris (1974), Appl. Phys. Letters **24**, 427.

Bonch-Bruyevich, A.M., N.N. Kostin and V.A. Khodovoi (1968), Opt. Spectrosc. **24**, 547.

Borisevich, N.A. (1975), Spectrosc. Letters **8**, 607.

Borisevich, N.A., I.I. Kalosha and V.A. Tolkachev (1973), Zh. Prikl. Spectrosk. **19**, 1108.

Borisevich, N.A., I.I. Kalosha, V.A. Tolkachev and V.A. Tugbaev (1974), Zh. Prikl. Spectrosk. **21**, 92.

Borisevich, N.A., I.I. Kalosha and V.A. Tolkachev (1975a), Sov. Phys. – Dokl. **19**, 578.

Borisevich, N.A., L.M. Bolot'ko, I.I. Kalosha and V.A. Tolkachev, (1975b), Izv. Akad. Nauk. SSR **39**, 1812.

Borisevich, N.A., G.B. Tolstorozhev, V.A. Tugbaev and D.M. Khalimanovich (1975c), Zh. Prikl. Spectrosk. **23**, 1098.

Borisevich, N.A., L.M. Bolot'ko and V.A. Tolkachev (1975d), Dokl. Akad. Nauk. SSR **222**, 1361.

Bradley, D.J. (1977), Topics in Applied Physics (S.L. Shapiro, ed.; Springer–Verlag; Berlin), Vol. 18.

Bradley, D.J. and A.J.F. Durrant (1968), Phys. Letters **27A**, 73.

Bradley, D.J. and F. O'Neill (1969), J. Opto-Electronics **1**, 69.

Bradley, D.J., G.M. Gale, M. Moore and P.D. Smith (1968a), Phys. Letters **26A**, 378.

Bradley, D.J., A.J.F. Durrant, G.M. Gale, M. Moore and P.D. Smith (1968b), IEEE J. Quantum Electronics **QE–4**, 707.

Bradley, D.J., A.F.J. Durrant, F. O'Neill and B. Sutherland (1969), Phys. Letters **30A**, 535.

Bradley, D.J., J.V. Nicholas and J.R.D. Shaw (1971a), Appl. Phys. Letters **19**, 172.

Bradley, D.J., B. Liddy and W.E. Sleat (1971b), Opt. Commun. **2**, 391.

Bradley, D.J., B. Liddy, A.G. Roddie, W. Sibbett and W.E. Sleat (1971c), Opt Commun. **3**, 426.

Bradley, D.J., W.G.I. Caughey and J.I. Vukusic (1971d), Opt Commun. **4**, 150.

Burlamacchi, P. and R. Salimbeni (1976), Opt. Commun. **17**, 6.

Burlamacchi, P., R. Pratesi and R. Salimbeni (1975), Appl. Optics 14, 1311.

Burlamacchi, P., R. Coisson, R. Pratesi and D. Pucci (1977), Appl. Optics 16, 1553.

Byer, R.L. and R.L. Herbst (1977), Topics in Applied Physics (Y.R. Shen, ed., Springer-Verlag; New York), Vol. 16.

Carlsten, J.L. and T.J. McIlrath (1973), Opt. Commun. 8, 52.

Champagne, L.F., J.G. Eden, N.W. Harris, N. Djeu and S.K. Searles (1977), Appl. Phys. Letters 30, 160.

Chan, C.K., S.O. Sari and R.E. Foster (1976), J. Appl. Phys. 47, 1139.

Cox, A.J., G.W. Scott and L.D. Talley (1977), Appl. Phys. Letters 31, 389.

Curry, S.M., R. Cubeddu and T.W. Hänsch (1973), Appl. Phys. 1, 153.

Decker, C.D. and F.K. Tittel (1973a), Appl. Phys. Letters 22, 411.

Decker, C.D. and F.K. Tittel (1973b), Opt. Commun. 8, 244.

De Maria, A.J. (1971), Progress in Optics (E. Wolf, ed.; North-Holland; Amsterdam), Vol. IX, p. 31.

De Maria, A.J., W.H. Glenn Jr., M.J. Brienza and M.E. Mack (1969), Proc. IEEE 57, 2.

Dewey, C.F. and L.O. Hocker (1971), Appl. Phys. Letters 18, 58.

Dewey, C.F., W.R. Cook, R.T. Hodgson and J.J. Wynne (1975), Appl. Phys. Letters 26, 714.

Dewey, H.J. (1976), IEEE J. Quantum. Electronics QE–12, 303.

Dienes, A., E.P. Ippen and C.V. Shank (1971), Appl. Phys. Letters 19, 258.

Dinev, S.G., K.V. Stamenov and I.V. Tomov (1972), Opt. Commun. 5, 419.

Drabovich, K.N., D.I. Metchkov, V.M. Mitev, L.I. Pavlov and K.V. Stamenov (1977), Opt. Commun. 20, 350.

Drexhage, K.H. (1973), Topics in Applied Physics (F.P. Schäfer, ed.; Springer-Verlag; Berlin), Vol. 1.

Dunning, F.B., E.D. Stokes and R.F. Stebbing (1972), Opt. Commun. 6, 63.

Dunning, F.B., F.K. Tittel and R.F. Stebbing (1973), Opt. Commun. 7, 181.

Edelstein, S.A., H.H. Nakano, T.F. Gallagher and D.C. Lorents (1977), Opt. Commun. 21, 27.

Eesley, G.L. and M.D. Levenson (1976), IEEE J. Quantum Electronics QE–12, 440.

Elgin, J.N. and G.H.C. New (1976), Opt. Commun. 16, 242.

Erickson, L.E. and A. Szabo (1971), Appl. Phys. Letters 18, 433.

Evtuhov, V. and J.K. Neeland (1966), in: Lasers (A.K. Levine, ed.; Dekker; New York).

Ewanizky, T.F., R.H. Wright, Jr. and H.H. Theissing (1973), Appl. Phys. Letters 22, 520.

Fan, B. and T.K. Gustafson (1976), Appl. Phys. Letters 28, 202.

Faries, D.W., P.L. Richards, Y.R. Shen and K.H. Yang (1971), Phys. Rev. A3, 2148.

Farmer, G.I., B.G. Huth, L.M. Taylor and M.R. Kagan (1968), Appl. Phys. Letters 12, 136.

Ferguson, A.I. and M.H. Dunn (1977), IEEE J. Quantum Electronics QE–13, 751.

Ferrar, C.M. (1969a), IEEE J. Quantum Electronics QE–5, 621.

Ferrar, C.M. (1969b), IEEE J. Quantum Electronics QE–5, 550.

Ferrar, C.M. (1973), Appl. Phys. Letters 23, 548.

Ferrar, C.M. (1976), Optica Acta 23, 911.

Flamant, P. and Y.H. Meyer (1971), Appl. Phys. Letters 19, 491.

Flamant, P. and Y.H. Meyer (1973), Opt. Commun. 7, 146.

Flamant, P. and Y.H. Meyer (1975), Opt. Commun. 13, 13.

Fletcher, A.N. (1977a), Appl. Phys. 12, 327.

Fletcher, A.N. (1977b), Appl. Phys. 14, 295.

Fletcher, A.N., D.A. Fine and D.E. Bliss (1977), Appl. Phys. 12, 39.

Flusberg, A., T. Mossberg and S.R. Hartmann (1977), Phys. Rev. Letters 38, 59.

Freund, R.S. and J.A. Schiavone (1976), unpublished; for results refer to Smith, P.W. (1976), Optica Acta 23, 901.

Friesem, A.A., U. Ganiel and G. Neumann (1973a), Appl. Phys. Letters 23, 249.

Friesem, A.A., U. Ganiel, G. Neumann and D. Peri (1973b), Opt. Commun. 9, 149.

Fröhlich, D., L. Stein, H.W. Schröder and H. Welling (1976), Appl. Phys. 11, 97.

Furumoto, H. and H.L. Ceccon (1968), Appl. Phys. Letters 13, 335.

Furumoto, H.W. and H.L. Ceccon (1969), Appl. Optics 8, 1613.

Gabel, C. and M. Hercher (1972), IEEE J. Quantum Electronics QE–8, 850.

Gale, G.M. (1973), Opt. Commun. **7**, 86.

Ganiel, U., A. Hardy, G. Neumann and D. Treves (1975), IEEE J. Quantum Electronics **QE-11**, 881.

Ganiel, U., A. Hardy and D. Treves (1976), IEEE J. Quantum Electronics **QE-12**, 704.

Gassmann, M.H. and H. Weber (1971), Opto-Electronics **3**, 177.

Gavronskaya, E.A., A.V. Groznyi, D.I. Staselko and V.L. Strigun (1977), Opt. Spectrosc **42**, 213.

Georges, A.T., P. Lambropoulos and J.H. Marburger (1976), Opt. Commun. **18**, 509.

Gerlach, H. (1973), Opt. Commun. **8**, 41.

Gerlach, H. (1974), Opt. Commun. **12**, 405.

Gibbs, W.E.K. and H.A. Kellock (1968), IEEE J. Quantum Electronics **4**, 293.

Glenn, W.H., M.J. Brienza and A.J. De Maria (1968), Appl. Phys. Letters **12**, 54.

Goldberg, L.S. and C.A. Moore (1975), Appl. Phys. Letters **27**, 217.

Green, J.M., J.P. Hohimer and F.K. Tittel (1973), Opt. Commun. **9**, 407.

Hamadani, S.M. and G. Magyar (1971), Opt. Commun. **4**, 310.

Hammond, P.R. (1977), Appl. Phys. **14**, 199.

Hänsch, T.W. (1972), Appl. Optics **11**, 895.

Hänsch, T.W. (1976) Tunable Lasers and Applications, Proc. Loen Conf., Norway (A. Mooradian, T. Jaeger and P. Stokseth, eds.; Springer–Verlag; Berlin).

Hänsch, T.W., F. Varsanyi and A.L. Schawlow (1971), Appl. Phys. Letters **18**, 108.

Hänsch, T.W., A.L. Schawlow and P. Toschek (1973), IEEE J. Quantum Electronics **QE-9**, 553.

Hanna, D.C., R.C. Smith and C.R. Stanley (1971), Opt. Commun. **4**, 300.

Hanna, D.C. V.V. Rampel and R.C. Smith (1973), Opt. Commun. **8**, 151.

Hanna, D.C., V.V. Rampel and R.C. Smith (1974), IEEE J. Quantum Electronics **QE-10**, 461.

Harper, C.D. and M. Gundersen (1974), Rev. Sci. Instr. **45**, 400.

Harris, S.E. and D.M. Bloom (1974), Appl. Phys. Letters **24**, 229.

Harris, S.E. and R.E. Miles (1971), Appl. Phys. Letters **19**, 385.

Harris, S.E., J.F. Young, A.H. Kung, D.M. Bloom and G.C. Bjorklund (1973), Laser Spectroscopy (R.C. Brewer and A. Mooradian, eds.; Plenum; New York).

Harris, J.M., R.W. Chrisman and F.E. Lytle (1975), Appl. Phys. Letters **26**, 16.

Hartig, W. and H. Walther (1973), Appl. Phys. **1**, 171.

Herbst, R.L., H. Komine and R.L. Byer (1977), Opt. Commun. **21**, 5.

Hertel, I.V. and A.S. Stamatovic, IEEE J. Quantum Electronics (1975), **QE-11**, 210.

Hilborn, R.C. and H.C. Brayman (1974), J. Appl. Phys. **45**, 4912.

Hirschberg, J.G. and R.R. Kadesch (1958), J. Opt. Soc. Amer **48**, 177.

Hirth, A., K. Vollrath and J.Y. Allain (1977), Opt. Commun. **20**, 347.

Hodgson, R.T., P.P. Sorokin and J.J. Wynne (1974), Phys. Rev. Letters **32**, 343.

Hsu, K.S., A.H. Kung, L.J. Zych, J.F. Young and S.E. Harris (1976), IEEE J. Quantum Electronics **QE-12**, 60.

Hutcheson, L.D. and R.S. Hughes (1974a), Appl. Optics **13**, 1395.

Hutcheson, L.D. and R.S. Hughes (1974b), IEEE J. Quantum Electronics **QE-10**, 462.

Huth, B.G. (1970), Appl. Phys. Letters **16**, 185.

Inomata, H. and A.I. Carswell (1977), Opt. Commun. **22**, 278.

Ippen, E.P. and C.V. Shank (1975), Appl. Phys. Letters **27**, 488.

Ippen, E.P., C.V. Shank and A. Dienes (1972), Appl. Phys. Letters **21**, 348.

Itzkan, I. and F.W. Cunningham (1972), IEEE J. Quantum Electronics **QE-8**, 101.

Jacobs, R.R., H. Samelson and A. Lempicki (1973), J. Appl. Phys. **44**, 263.

Jain, R.K. and T.K. Gustafson (1976), IEEE J. Quantum Electronics **QE-12**, 555.

Jantz, W. and P. Koidl (1977), Appl. Phys. Letters **31**, 99.

Jennings, D.A. and A.V. Varga (1971), J. Appl. Phys. **42**, 5171.

Jethwa, J. and F.P. Schäfer (1974), Appl. Phys. **4**, 299.

Johnson, L.F. (1966), in: Lasers (A.K. Levine, ed.; Dekker; New York).

Juramy, P., P. Flamant and Y.H. Meyer (1977), IEEE J. Quantum Electronics **QE-13**, 855.

Kärkkäinen, P.A. (1977), Appl. Phys. **13**, 159.

Kato, K. (1975), IEEE J. Quantum Electronics **QE-11**, 373.

Kato, K. (1976), Opt. Commun. **18**, 447.

Kato, K. (1977a), Appl. Phys. Letters **30**, 583.

Kato, K. (1977b), IEEE J. Quantum Electronics **QE–13**, 544.

Keller, R.A. (1970), IEEE J. Quantum Electronics **QE–6**, 411.

Kopainsky, J. (1975), Appl. Phys. **8**, 229.

Kuhl, J. and H. Spitschan (1972), Opt. Commun. **5**, 382.

Kuhl, J. and H. Spitschan (1975), Opt. Commun. **13**, 6.

Kuhl, J., G. Marowsky, P. Kunstmann and W. Schmidt (1972), Z. Naturforsch. **27A**, 601.

Kung, A.H. (1974), Appl. Phys. Letters **25**, 653.

Kung, A.H., J.F. Young, G.C. Bjorklund and S.E. Harris (1972), Phys. Rev. Letters **29**, 985.

Kuizenga, D.J. (1971), Appl. Phys. Letters **19**, 260.

Lahmann, W., K. Tibulski and H. Welling (1976), Appl. Phys. **17**, 18.

Lawler, J.E., W.A. Fitzsimmons and L.W. Anderson (1976), Appl. Optics **15**, 1083.

Lempicki, A., S. Ahmed and R. Pappalardo (1972), Technical Report TR 72–841.1, Office of Naval Research, Alington Va. USA.

Leung, K.M., J.F. Ward and B.J. Orr (1974), Phys. Rev. **A9**, 2440.

Levenson, M.D. and G.L. Eesley (1976), IEEE J. Quantum Electronics **QE–12**, 259.

Lill, E., S. Schneider and F. Dörr (1977), Opt. Commun. **22**, 107.

Lin, C. (1975a), IEEE J. Quantum Electronics **QE–11**, 602.

Lin, C. (1975b), J. Appl. Phys. **46**, 4076.

Lin, C. and A. Dienes (1973), Opt. Commun. **9**, 21.

Lin, C. and C.V. Shank (1975), Appl. Phys. Letters **26**, 389.

Liu, R.S.H. and R.E. Kellog (1969), J. Amer. Chem. Soc. **91**, 250.

Lotem, H. (1973), Opt. Commun. **9**, 346.

Lotem, H. and R.T. Lynch, Jr. (1975), Appl. Phys. letters **27**, 344.

Loth, C. and G. Megie (1974), J. Phys. **E7**, 80.

Loth, C., Y.H. Meyer and F. Bos (1976), Opt. Commun. **16**, 310.

Mack, M.E. (1968), J. Appl. Phys. **39**, 2483.

Mack, M.E. (1974), Appl. Optics **13**, 46.

Magyar, G. and H.J. Schneider-Muntau (1972), Appl. Phys. Letters **20**, 406.

Mahr, H. (1976), IEEE J. Quantum Electronics **QE–12**, 554.

Mahr, H. and M.D. Hirsch (1975), Opt. Commun. **13**, 96.

Marling, J.B., D.W. Gregg and L.L. Wood (1970), Appl. Phys. Letters **17**, 527.

Marling, J.B., L. Wood and D.W. Gregg (1971), IEEE J. Quantum Electronics **QE–7**, 498.

Marowsky, G. (1973), Rev. Sci. Instr. **44**, 890.

Marowsky, G. and K. Kaufmann (1976), IEEE J. Quantum Electronics **QE–12**, 207.

Marowsky, G. and F. Zaraga (1974), Opt. Commun. **11**, 343.

Marowsky, G., F.P. Schäfer, J.W. Keto and F.K. Tittel (1976), Appl. Phys. **9**, 143.

Marx, B.R., G. Holloway and L. Allen (1976), Opt. Commun. **18**, 437.

Massey, G.A. (1974), Appl. Phys. Letters **24**, 371.

Meltzer, D.W. and L.S. Goldberg (1972), Opt. Commun. **5**, 209.

Meyer, Y.H. and P. Flamant (1976), Opt. Commun. **19**, 20.

Miles, R.B. and S.E. Harris (1973), IEEE J. Quantum Electronics **QE–9**, 470.

Miyazoe, Y. and M. Maeda (1971), IEEE J. Quantum Electronics **QE–7**, 36.

Moore, C.A. and L.S. Goldberg (1976), Opt. Commun. **16**, 21.

Morey, W.W. and W.H. Glenn (1976a), IEEE J. Quantum Electronics **QE–12**, 311.

Morey, W.W. and W.H. Glenn (1976b), Optica Acta **23**, 873.

Moriarty, A., W. Haeps and D.D. Davis (1976), Opt. Commun. **16**, 324.

Morrow, T. and M. Quinn (1973), J. Photochem. **2**, 343.

Moses, E.I., J.J. Turner and C.L. Tang (1976), Appl. Phys. Letters **28**, 258.

Myers, S.A. (1971), Opt. Commun. **4**, 187.

Neporent, B.S. and V.B. Shilov (1971), Opt. Spectrosc. **30**, 576.

O'Neill, F. (1972), Opt. Commun. **6**, 360.

Pappalardo, R., H. Samelson and A. Lempicki (1970a), Appl. Phys. Letters **16**, 267.

Pappalardo, R., H. Samelson and A. Lempicki (1970b), IEEE J. Quantum Electronics **QE–6**, 716.

Pappalardo, R., S.L. Shapiro, S. Ahmed and R. Alfano (1974), J. Chem. Phys. **60**, 3368.

Peterson, O.G., J.P. Webb, W.C. McColgin and J.H. Eberly (1971), J. Appl. Phys. **42**, 1917.

Pike, C.T. (1974), Opt. Commun. **10**, 14.

Pike, H.A. (1971), Ph. D. Thesis University of Rochester.

Pikulik, L.G., V.A. Yakovyenko and A.D. Dasko (1975), Zh. Prikl. Spektrosk. **23**, 493.

Pilloff, H.S. (1972), Appl. Phys. Letters **21**, 339.

Pinard, J. and S. Liberman (1977), Opt. Commun. **20**, 344.

Pine, A.S. (1974), J. Opt. Soc. Amer. **64**, 1683.

Pine, A.S. (1976), J. Opt. Soc. Amer. **66**, 97.

Pratesi, and L. Ronchi (1976), Optica Acta **23**, 933.

Richardson, J.H., L.L. Steinmetz and B.W. Wallin (1977), Appl. Optics **16**, 1133.

Royt, T.R., W.L. Faust, L.S. Goldberg and C.H. Lee (1974), Appl. Phys. Letters **25**, 514.

Royt, T.R., C.H. Lee and W.L. Faust (1976), Opt. Commun. **18**, 108.

Ruddock, I.S. and D.J. Bradley (1976), Appl. Phys. Letters **29**, 296.

Rulliere, C., J.P. Morand and O. de Witte (1977), Opt. Commun. **20**, 339.

Runge, P.K. (1971), Opt. Commun. **4**, 195.

Runge, P.K. (1972), Opt. Commun. **5**, 311.

Ryan, T., B. Davoodzadeh and T.K. Gustafson (1976), Opt. Commun. **18**, 183.

Sakurai, T. and H.G. de Winter (1975), J. Appl. Phys. **46**, 875.

Sakurai, T., I.M. Littlewood and C.E. Webb (1976), Appl. Phys. Letters **28**, 533.

Salour, M.M. (1977), Opt. Commun. **22**, 202.

Salzmann, H. and H. Strohwald (1976), Phys. Letters **57A**, 41.

Sam, C.L. and M.M. Choy (1977), Appl. Phys. Letters **30**, 199.

Schäfer, F.P. (1973), Topics in Applied Physics (F.P. Schäfer, ed., Springer-Verlag; Berlin), Vol. 1.

Schäfer, F.P. and H. Müller (1971), Opt. Commun. **2**, 407.

Schäfer, F.P. and L. Ringwelski (1973), Z. Naturforsch. **28a**, 792.

Schäfer, F.P., W. Schmidt and J. Volze (1966), Appl. Phys. Letters **9**, 306.

Schaerer, L.D. (1975), IEEE J. Quantum Electronics **QE-11**, 935.

Schmidt, A.J. (1975), Opt. Commun. **14**, 294.

Schmidt, W. (1970), Laser **2**, 47.

Schmidt, W. and F.P. Schäfer (1967), Z. Naturforsch. **22a**, 1563.

Schmidt, W. and F.P. Schäfer (1968), Phys. Letters **26A**, 558.

Schröder, H.W., H. Welling and B. Wellegehausen (1973), Appl. Phys. **1**, 343.

Schröder, H.W., L. Stein, D. Frölich, B. Fugger and H. Welling (1977), Appl. Phys. **14**, 377.

Shank, C.V. and E.P. Ippen (1973), Topics in Applied Physics (F.P. Schäfer, ed.; Springer-Verlag; Berlin), Vol. 1.

Shank, C.V. and E.P. Ippen (1974), Appl. Phys. Letters **24**, 373.

Shank, C.V., A. Dienes and W.T. Silfast (1970), Appl. Phys. Letters **17**, 307.

Shapiro, S.L., R.C. Hyer and A.J. Campillo (1974), Phys. Rev. Letters **33**, 513.

Sibbett, W., D.J. Bradley and S.F. Bryant (1976), Opt. Commun, **18**, 107.

Singer, L., Z. Singer and S. Kimel (1976), Appl. Opt. **15**, 2678.

Smith, P.W. (1970), Proc. IEEE **58**, 1342.

Smith, P.W. (1976), Optica Acta **23**, 901.

Smith, P.W., P.F. Liao, C.V. Shank, T.K. Gustafson, C. Lin and P.J. Maloney (1974), Appl. Phys. Letters **25**, 144.

Smith, P.W., P.F. Liao, C.V. Shank, C. Lin and P.J. Maloney (1975), IEEE J. Quantum Electronics **QE-11**, 84.

Smith, P.W., Liao, P.F., and P.J. Maloney (1976), IEEE J. Quantum Electronics **QE-12**, 539.

Smith, W.V., and P.P. Sorokin (1966), The Laser (McGraw-Hill; New York), p.74.

Snavely, B.B. (1969), Proc. IEEE **57**, 1374.

Snavely, B.B. (1973), Topics in Applied Physics (F.P. Schäfer, ed.; Springer-Verlag; Berlin), Vol. 1.

Snavely, B.B. and O.G. Peterson (1968), IEEE J. Quantum Electronics **QE-4**, 540.

Snavely, B.B. and F.P. Schäfer (1969), Phys. Letters **28A**, 728.

Soep, B. (1970), Opt. Commun. **1**, 433.

Soffer, B.H. and B.B. McFarland (1967), Appl. Phys. Letters **10**, 266.
Soffer, B.H. and Linn, J.W. (1968), J. Appl. Phys. **39**, 5859.
Sorokin, P.P. and Lankard, J.R. (1966), IBM J. Res. Develop. **10**, 162.
Sorokin, P.P. and J.R. Landard (1967), IBM J. Res. Develop. **11**, 148.
Sorokin, P.P., J.R. Lankard, E.C. Hammond and V.L. Moruzzi (1967), IBM J. Res. Develop. **11**, 130.
Sorokin, P.P., J.R. Lankard, V.L. Moruzzi and E.C. Hammond (1968), J. Chem. Phys. **48**, 4726.
Sorokin, P.P., J.J. Wynne and J.R. Lankard (1973), Appl. Phys. Letters **22**, 342.
Sorokin, P.P., J.J. Wynne, J.A. Armstrong and R.T. Hodgson (1976), Ann. N.Y. Acad. Sci. **267**, 30.
Stansfield, B.L., R. Nodwell and J. Meyer (1971), Phys. Rev. Letters **26**, 1219.
Steyer, B. and F.P. Schäfer (1974), Opt. Commun. **10**, 219.
Steyer, B. and F.P. Schäfer (1975), Appl. Phys. **7**, 113.
Stoilov, Yu, Yu, and K.K. Trusov (1974), Sov. J. Quantum Electronics **4**, 807.
Stokes, E.D., F.B. Dunning, R.F. Stebbings, G.K. Walters and R.D. Rundel (1972), Opt. Commun. **5**, 267.
Strome, F.C., Jr. and S.A. Tuccio (1973), IEEE J. Quantum Electronics **QE-9**, 230.
Strome, F.C., Jr. and J.P. Webb (1971), Appl. Optics **10**, 1348.
Taylor, J.R. (1976), Opt. Commun. **18**, 504.
Taylor, D.J., S.E. Harris, S.T.K. Nieh and T.W. Hänsch (1971), Appl. Phys. Letters **19**, 269.
Terhune, R.W., and P.D. Maker (1968), Lasers (A.K. Levine, ed.; Dekker; New York), Vol. 2.
Teschke, O., A. Dienes, and J.R. Whinnery (1976), IEEE J. Quantum Electronics **QE-12**, 383.
Tisone, G.C., A.K. Hays and J.M. Hoffman (1975), Opt. Commun. **15**, 188.
Topp, M.R. and P.M. Rentzepis (1971), Phys. Rev. **3A**, 358.
Tuccio, S.A. and F.C. Strome, Jr. (1972), Appl. Optics **11**, 64.
Turner, J.J., E.I. Moses and C.L. Tang (1975), Appl. Phys. Letters **27**, 441.
Urisu, T. and K. Kajiyama (1977), Opt. Commun. **20**, 34.
Vrehen, Q.H.F. and A.J. Breimer (1972), Opt. Commun. **4**, 416.
de Vries, J., D. Bebelaar and J. Langelaar (1976), Opt. Commun. **18**, 24.
Wallace, R.W. (1971), Opt. Commun. **4**, 316.
Wallace, S.C. and G. Zdasiuk (1976), Appl. Phys. Letters **28**, 449.
Wallace, S.C., K.K. Innes and B.P. Stoicheff (1976), Opt. Commun. **18**, 110.
Wallenstein, R. and T.W. Hänsch (1974), Appl. Optics **13**, 1625.
Wallenstein, R. and T.W. Hänsch (1975), Opt. Commun. **14**, 353.
Walther, H. and J.L. Hall (1970), Appl. Phys. Letters **17**, 239.
Ward, J.F. and G.H.C. New (1969), Phys. Rev. **185**, 57.
Ward, J.F. and A.V. Smith (1975), Phys. Rev. Letters **35**, 653.
Weber, M.J. and M. Bass (1969), IEEE J. Quantum Electronics **QE-5**, 175.
Wellegehausen, B., D. Friede, H. Vogt and S. Shahdin (1976), Appl. Phys. **11**, 363.
Wu, Ch.Y. and J.R. Lombardi (1973), Opt. Commun. **7**, 233.
Wu, F.Y., R.E. Grove and S. Ezekiel (1974), Appl. Phys. Letters **25**, 73.
Wynne, J.J. and P.P. Sorokin (1977), Topics in Applied Physics (Y.R. Shen, ed.; Springer-Verlag; Berlin.), Vol. 16.
Yamagishi, A. and A. Szabo (1977), Appl. Optics (in press).
Yariv, A. (1967), Quantum Electronics (Wiley; New York).
Yasa, Z.A. (1975), J. Appl. Phys. **46**, 4895.
Yasa, Z.A., A. Dienes and J.R. Whinnery (1977), Appl. Phys. Letters **30**, 24.
Young, J.F., G.C. Bjorklund, A.H. Kung, R.B. Miles and S.E. Harris (1971), Phys. Rev. Letters **27**, 1551.
Young, K.H., J.R. Morris, P.L. Richards and Y.R. Shen (1973), Appl. Phys. Letters **23**, 669.
Zacharias, H., A. Anders, J.B. Halpern and K.H. Welge (1976), Opt. Commun. **19**, 116.
Zalewski, E.F. and R.A. Keller (1971), Appl. Optics **10**, 2773.

Note added in proof

Since the preparation of the manuscript numerous important contributions have been published in the fast growing field of dye lasers. Publications which came to the attention of the author after the completion of this article are listed below.

Dye laser analysis

Computer simulation of an energy transfer dye laser, by S. Speiser and R. Katraro (1978), Opt. Commun. **27**, 287.

Photoquenching I. The dependence of the primary quantum yield of a monophotonic laser induced photochemical process on the intensity and duration of the exciting pulse, by S. Speiser, R. van der Werf and J. Kommandeur (1973), Chem. Phys. **1**, 297.

Photoquenching II. Pulsed-laser-pumped dye laser systems, by S. Speiser (1974), Chem. Phys. **6**, 479.

Photoquenching III. Analysis of the dependence of pulsed laser pumped dye laser performance on pumping conditions and on the dye molecular characteristics, by S. Speiser and A. Bromberg (1975), Chem. Phys. **9**, 191.

Threshold singularities of optically pumped dye lasers, by W. Heudorfer and G. Marowsky (1978), Appl. Phys. **17**, 181.

Second and first order phase transition analogy in the operation of an organic dye laser, G. Marowsky and W. Heudorfer (1978), Opt. Commun. **26**, 381.

Dye laser construction and performance

Power-scaling effects in dye lasers under high-power laser excitation, by C.A. Moore and C. David Decker (1978), J. Appl. Phys. **49**, 47.

A 1 MW p-terphenyl dye laser, by H. Bücher and W. Chow (1977), Appl. Phys. **13**, 267.

A high power dye-laser pumped by the second harmonic of a Nd–YAG laser, by W. Hartig (1978), Opt. Commun. **27**, 447.

Dye laser multipass amplifier system, by P. Ewart and J.M. Catherall (1978), Opt. Commun. **27**, 439.

Amplified spontaneous emission and spatial dependence of gain in dye amplifiers, by G. Dujardin and P. Flamant (1978), Opt. Commun. **24**, 243.

Nitrogen-laser-pumped single-mode dye laser, by S. Saikan (1978), Appl. Phys. **17**, 41.

Conversion d'énergie dans les amplificateurs à colorants en présence de superfluorescence, by G. Dujardin and P. Flamant (1978), Optica Acta **25**, 273.

Narrowband operation of a pulsed dye laser without intracavity beam expansion, by I. Shoshan, N.N. Danon, and U.P. Oppenheim (1977), J. Appl. Phys. **48**, 4495.

The use of a diffraction grating as a beam expander in a dye laser cavity, by I. Shoshan and U.P. Oppenheim (1978), Opt. Commun. **25**, 375.

Spectrally narrow pulsed dye laser without beam expander, by M.G. Littmann and H.J. Metcalf (1978), Appl. Opt. **17**, 2224.

Single mode operation of grazing-incidence pulsed dye laser, by M.G. Littmann (1978), Opt. Lett. **3**, 138.

On the dispersion of a prism used as a beam expander in a nitrogen laser pumped dye laser, by L.G. Nair (1977), Opt. Commun. **23**, 273.

A prism anamorphic system for gaussian beam expander, by T. Kasuya, T. Suzuki, and K. Shimoda (1978), Appl. Phys. **17**, 131.

Comment on the dispersion of a prism used as a beam expander in a nitrogen laser pumped dye laser, by R. Wyatt (1978), Opt. Commun. **26**, 9.

Wavelength tuning of nitrogen pumped dye laser, by R. Konjević and N. Konjević (1977), Opt. Commun. **23**, 187.

Fabry-Perot with short pulse lasers: spectral selection and spectral analysis in dye lasers, by H. Daussy, R. Dumanchin and O. de Witte (1978), Appl. Opt. **17**, 451.

Affinements spectral dans les lasers à colorants pulsés, by P. Flamant (1978), Appl. Opt. **17**, 955.

Birefringent filters for tuning flashlamp-pumped dye lasers: simplified theory and design, by I.J. Hodgkinson and J.I. Vukusic (1978), Appl. Opt. **17**, 1944.

A reliable high average power dye laser, by J. Jethwa, F.P. Schäfer and J. Jasny (1978), IEEE J. Quantum Electronics **QE-14**, 119.

High-energy dye laser pumped by Wall-ablation lamps, by T. Okada, K. Fujiwara, M. Maeda and Y. Miyazoe (1978), Appl. Phys. **15**, 191.

Optimization of a long pulse dye laser, by A. Hirth, H. Fagot and F. Wieder (1977), Opt. Commun. **23**, 315.

Efficient cavity dumped dye laser, by R.G. Morton, M.E. Mack, and I. Itzkan (1978), Appl. Opt. **17**, 3268.

Unstable resonator mode control in a transverse flow dye laser, by D.B. Northam, M.E. Mack and I. Itzkan (1978), Appl. Opt. **17**, 931.

Polarization of light from a pulsed dye laser, by C.H. Dugan, A. Lee and F.J. Morgan (1978), Appl. Opt. **17**, 1012.

Spatially inhomogeneous saturation of gain in organic solutions caused by amplified spontaneous emission, by I. Ketskemety, Zs. Bor, B. Racz and L. Kozma (1977), Opt. Commun. **21**, 25.

Intermolecular interactions and spectral shifts in energy transfer dye lasers, by M. Kleinerman and M. Dabrowski (1978), Opt. Commun. **26**, 81.

Coherence and beam geometry of a superradiant dye laser, by H.P. Grieneisen, R.E. Francke and A. Lago (1978), Appl. Phys. **15**, 281.

Concentration dependence of the gain spectrum in energy transfer dye mixtures, by T. Urisu and K. Kajiyama (1976), J. Appl. Phys. **47**, 3563.

Bleaching and diffusion of laser dyes in solution under high power UV irradiation, by E. Sahar and D. Treves (1977), Opt. Commun. **21**, 20.

Photodegradation relationships for bicyclic dyes in alcohol solutions, by A.N. Fletcher (1978), Appl. Phys. **16**, 93.

Effect of chemical substituents of bicyclic dyes upon photodegradation parameters, by A.N. Fletcher and D.E. Bliss (1978), Appl. Phys. **16**, 289.

Laser properties of triazinyl-stilbene compounds, by H. Telle, U. Brinkmann and R. Raue (1978), Opt. Commun. **24**, 33.

Dye laser action at 330 nm using benzoxazole: a new class of lasing dyes, by C. Rullière and J. Joussot-Dubien (1978), Opt. Commun. **24**, 38.

New efficient and stable laser dyes for cw operation in the blue and violet spectral region, by J. Kuhl, H. Telle, R. Schieder and U. Brinkmann (1978), Opt. Commun. **24**, 251.

Broadly tunable dye laser emission to 12850 Å, by K. Kato (1978), Appl. Phys. Lett. **33**, 509.

Flame locking of dye lasers to atom lines, by Y.H. Meyer, C. Loth and R. Astier (1978), Opt. Commun. **25**, 100.

Frequency locking of dye lasers by selective reflection on a glass/sodium vapor interface, by G. Dujardin and Y.H. Meyer (1978), Opt. Commun. **24**, 21.

On injection locking of homogeneously broadened lasers, by M.M. Ibrahim (1978), IEEE J. Quantum Electronics **QE-14**, 145.

Rate equations for dye lasers: Comments on the spiking phenomenon, by P. Flamant (1978), Opt. Commun. **25**, 247.

Two orthogonal polarized wavelengths generated in a N_2-laser pumped dye laser, by E. Winter, G. Veith and A.J. Schmidt (1978), Opt. Commun. **25**, 87.

Quantum efficiency and thermal stability of organic dye vapors, by T. Sakurai, A. Ogishima and M. Sugawara (1978), Opt. Commun. **25**, 75.

Electron energy loss spectra of dye vapors, by H.W. Hermann, I.V. Hertel and G. Marowsky (1978), Appl. Phys. **15**, 185.

Intense laser emission from electron-beam-pumped ternary mixtures of Ar, N_2 and POPOP vapor, by G. Marowsky, R. Cordray, F.K. Tittel, W.L. Wilson and C.B. Collins (1978), Appl. Phys. Lett. **33**, 59.

Some characteristics of efficient dye laser emission obtained by pumping at 248 nm with a high power KrF* discharge laser, by U.I. Tomin, A.J. Alcock, W.J. Sajeant and K.E. Leopold (1978), Opt. Commun. **26**, 396.

Narrow linewidth, short pulse operation of a nitrogen laser pumped dye laser, by R. Wyatt (1978), Opt. Commun. **26**, 429.

The 3-CN-4-MU dye laser in ethanol containing various amounts of water, acid or alkali, by M. Takakusa and U. Itah (1978), Opt. Commun. **26**, 401.

Laser a colorant pompe longitudinalement emettant a deux longueurs d'onde accordables, by R. Dorsinville (1978), Opt. Commun. **26**, 419.

Energy transfer in flashlamp pumped organic dye lasers, by P. Burlamacchi and D. Cutter (1977), Opt. Commun. **22**, 283.

Frequency conversion

Generation of tunable radiation below 2000 Å by phase matched sum-frequency mixing in $KB_5O_8 \cdot 4D_2O$, by J.A. Paisner, M.L. Spaeth, D.C. Gerstenberger and I.W. Ruderman (1978), Appl. Phys. Lett. **32**, 476.

Generation of tunable coherent radiation below 250 nm at MW power levels, by R.E. Stickel, Jr. and F.B. Dunning (1978), Appl. Opt. **17**, 1313.

Generation of tunable coherent vacuum uv radiation in KB5, by R.E. Stickel, Jr. and F.B. Dunning (1978), Appl. Opt. **17**, 981.

Generation of tunable 16 μm radiation by stimulated hyper-raman effect in strontium vapor, by J. Reif and H. Walther (1978), Appl. Phys. **15**, 361.

Continuous wave uv radiation tunable from 285 nm to 400 nm by harmonic and sum frequency generation, by S. Blit, E.G. Weaver, T.A. Rabson and F.K. Tittel (1978), Appl. Opt. **17**, 721.

Difference frequency conversion of $\lambda = 1.06 \mu$m Nd:YAG laser radiation in two-photon dye-laser resonantly pumped Rb vapor, by V.G. Arkhipkin, A.K. Popov and V.P. Timofeev (1978), Opt. Commun. **25**, 111.

Generation of coherent radiation tunable from 201 nm to 212 nm, by R.E. Stickel, Jr. and F.B. Dunning (1977), Appl. Opt. **16**, 2356.

A tunable, frequency doubled, continuous-wave dye laser using ADA, by A.I. Ferguson and M.H. Dunn (1977), Opt. Commun. **23**, 177.

Generation of tunable continuous-wave ultraviolet radiation from 257 to 320 nm, by S. Blit, E.G. Weaver, F.B. Dunning and F.K. Tittel (1977), Opt. Lett. **1**, 58.

Tunable UV-radiation by stimulated raman scattering in hydrogen, by V. Wilke and W. Schmidt (1978), Appl. Phys. **16**, 151.

Tunable coherent radiation source covering a spectral range from 185 to 880 nm, by V. Wilke and W. Schmidt (1979), Appl. Phys. **18**, 177.

Generation of cw VUV coherent radiation by four-wave sum frequency mixing in Sr vapor, by R.R. Freman, G.C. Bjorklund, N.P. Econo-mou, P.F. Liao and J.E. Bjorkholm (1978), Appl. Phys. Lett. **33**, 739.

Nonlinear generation of Lyman-alpha radiation, by Rita Mahon, T.J. McIlrath and D.W. Koopman (1978), Appl. Phys. Lett. **33**, 305.

Tunable, coherent radiation in the Lyman-α region (1210–1290 Å) using magnesium vapor, by T.J. McKee, B.P. Stoicheff and S.C. Wallace (1978), Opt. Lett. **3**, 207.

Upconversion of $\lambda = 3.39 \mu$m He–Ne laser radiation in two-photon resonantly pumped Na vapor, by V.G. Arkhipkin, A.K. Popov and V.P. Timofeev (1978), Appl. Phys. **16**, 209.

An improved automatically tunable second harmonic generation of dye laser, by J. Krasinski and A. Sieradzan (1978), Opt. Commun. **26**, 389.

Short pulse generation

Generation of synchronized cw trains of picosecond pulses at two independently tunable wavelengths, by R.K. Jain and J.P. Heritage (1978), Appl. Phys. Lett. **32**, 41.

Subpicosecond pulses from a tunable cw mode-locked dye laser, by J.P. Heritage and R.K. Jain (1978), Appl. Phys. Lett. **32**, 101.

Generation of near infrared picosecond pulses by mode locked synchronous pumping of a jet-stream dye laser, by J. Kuhl, R. Lambrich and D. von der Linde (1977), Appl. Phys. Lett. **31**, 657.

Characteristics of picosecond pulses generated from synchronously pumped cw dye laser system, by D.M. Kim, J. Kuhl, R. Lambrich and D. von der Linde (1978), Opt. Commun. **27**, 123.

Comparison of synchronous pumping and passive mode-locking of cw dye lasers for the generation of picosecond and subpicosecond light pulses, by J.P. Ryan, L.S. Goldberg and D.J. Bradley (1978), Opt. Commun. **27**, 127.

Ultrashort pulses from a cw dye laser using passive-active mode locking technique, by T. Kurobori, Y. Cho and Y. Matsuo (1978), Opt. Commun. **24**, 41.

Streak camera investigation of an actively mode-locked flashlamp-pumped dye laser, by E. Lill, S. Schneider, F. Dörr, S.F. Bryant and J.R. Taylor (1977), Opt. Commun. **23**, 318.

A high-efficiency tunable picosecond dye laser, by D. Huppert and P.M. Rentzepis (1978), J. Appl. Phys. **49**, 543.

Pulse-width dependence on intracavity bandwidth in synchronously mode-locked cw dye lasers, by C.P. Ausschnitt and R.K. Jain (1978), Appl. Phys. Lett. **32**, 727.

Subpicosecond pulse generation in a synchronously mode-locked cw rhodamine 6G dye laser, by R.K. Jain and C.P. Ausschnitt (1978), Opt. Lett. **2**, 117.

Subpicosecond light pulses from a synchronously mode-locked dye laser with composite gain and absorber medium, by G.W. Fehrenbach, K.J. Gruntz and R.G. Ulbrich (1978), Appl. Phys. Lett. **33**, 159.

Generation and measurement of 200 femto-second optical pulses, by J.C. Diels, E. van Stryland and G. Benedict (1978), Opt. Commun. **25**, 93.

Mode-locking of cw dye laser using externally modulated pumping laser, by T. Kobayashi, T. Hosokawa and T. Sueta (1978), Opt. Commun. **27**, 431.

Novel method for active mode-locking and tuning of dye lasers, by J. Jasny, J. Jethwa and F.P. Schäfer (1978), Opt. Commun. **27**, 426.

Production of deep blue tunable picosecond light pulses by synchronous pumping of a dye laser, by J.N. Eckstein, A.I. Ferguson, T.W. Hänsch, C.A. Minard and C.K. Chan (1978), Opt. Commun. **27**, 466.

A subpicosecond dye laser directly pumped by a mode-locked argon laser, by A.I. Ferguson, J.N. Eckstein and T.W. Hänsch (1978), J. Appl. Phys. **49**, 5389.

Nanosecond pulse generation from a self-injected laser-pumped dye laser using a novel cavity-flipping technique, by Y.S. Liu (1978), Opt. Lett. **3**, 167.

Generation of intense subnanosecond 0.58–0.80 μm pulses with a flashlamp-pumped dye laser, by T.J. Negran and A.M. Glass (1978), Appl. Opt. **17**, 2812.

PART B

Laser Applications

C. D. CANTRELL
S. M. FREUND
E. V. GEORGE
R. A. HAAS
D. HON
W. KOECHNER
W. F. KRUPKE
H. E. LESSING
W. H. LOWDERMILK
J. L. LYMAN
A. von JENA

B1 | Technology of Bandwidth-Limited Ultrashort Pulse Generation

W.H. LOWDERMILK

Lawrence Livermore Laboratory,
University of California,
Livermore, California 94550

Contents

© *North-Holland Publishing Company, 1979*
Laser Handbook, edited by M.L. Stitch

Abstract

Theoretical models are reviewed for bandwidth-limited ultra-short pulse generation by active and passive mode locking. Experiments on the generation techniques and pulse properties are summarized, and results compared with model predictions.

1 Introduction

This chapter deals with the physical processes and experimental techniques which are involved in generating ultrashort, bandwidth-limited light pulses. From the beginning of laser technology, the generation of shorter duration and more powerful pulses has been a subject of intense interest and development. Ultrashort light pulses are now applied in such diverse fields as photochemistry, communications and laser-induced fusion. By far the most successful technique for generating ultrashort pulses is that of mode locking. Mode-locked lasers are capable of producing pulses of duration less than 1 ps limited only by the gain bandwidth of the amplifying medium. The techniques of active and passive mode locking have been well known for a number of years and have been successfully applied to mode-locking a wide variety of gas lasers, solid-state lasers and liquid dye lasers. A detailed historical review of mode locking was given by Smith (1970).

Our attention in this chapter is confined to longitudinal mode-locking solid-state Nd lasers and dye lasers since these are the only widely used lasers which characteristically produce pulses with durations less than 100 ps. In § 2 the mathematical description of an optical pulse is summarized to clearly define the meaning of a 'bandwidth-limited' pulse. The formation of an ultrashort pulse by longitudinal mode locking is formally presented as well as a brief review of the techniques for ultrashort pulse measurements. Active mode locking is discussed in § 3. Modeling the widely used homogeneously broadened Nd: YAG laser is of specific interest, although the results presented also apply to medium- and high-pressure CO_2 lasers. The actively mode-locked dye laser is touched upon briefly to outline the difficulties involved in its operation.

Passive mode locking is discussed in considerable detail, primarily because the shortest available pulses are generated by this technique. Here, particular attention is paid to the details of the pulse-formation process. There is a basic difference between the formation processes in solid-state and dye lasers which requires separate models. This difference is associated with the excited-state lifetime of the active laser transition.

Passive mode locking of the solid-state laser is described in § 4. There the excited-state lifetime is very long compared to the transit time of a pulse through the laser. A single mode-locked pulse is selected from the fluctuating noise background by the saturable absorber and amplified by the gain medium. Although the gain medium assists in discriminating between two nearly equal fluctuation pulses, it plays no role in determining the final pulse shape. Advances in understanding of the Nd: glass laser are presented, as this laser is one of the most widely used laboratory sources of ultrashort pulses.

Passive mode locking of dye lasers is discussed in § 5. The excited-state lifetime of a dye molecule is short compared to the cavity transit time, a condition which allows the gain to saturate and recover quickly enough to play an important role in forming the mode-locked pulse. The pulses formed by active and passive mode-locking techniques have different characteristic properties. Attempts to combine the attractive features of both techniques by simultaneous active and passive mode locking are reviewed in § 6. Finally, in § 7 brief consideration is given various additional techniques which have recently been applied to generating and shaping ultrashort light pulses.

2 Bandwidth-limited pulses

This section summarizes the mathematical description of an optical pulse following the notation of Bradley and New (1974) and defines the terms which are important in discussing theoretical and experimental results presented in the following sections. Rigorous treatments have been given by Mandel and Wolf (1965) and Born and Wolf (1970). Throughout this section circular frequencies (Hz) are denoted by the symbol ν. The corresponding radian frequency (rad s^{-1}) is represented by ω and is related to ν by $\omega = 2\pi\nu$.

2.1 Structure of an optical pulse

The magnitude of the electric field at a fixed point in space is represented by $E(t)$ and is related to the frequency spectrum by the Fourier integral

$$E(t) = (2\pi)^{-1/2} \int_{-\infty}^{+\infty} e(\omega) \exp(-i\omega t)\, d\omega. \tag{1}$$

Since the field $E(t)$ is real,

$$e(-\omega) = e^*(\omega), \tag{2}$$

which means that the negative frequencies contribute no new information. It is convenient to define a function $V(t)$ such that $E(t) = \mathrm{Re}\, V(t)$. This function, called the complex analytic signal, is given by the expression

$$V(t) = (2\pi)^{-1/2} \int_{0}^{\infty} 2e(\omega) \exp(-i\omega t)\, d\omega$$

$$= (2\pi)^{-1/2} \int_{0}^{\infty} v(\omega) \exp(-i\omega t)\, d\omega, \tag{3}$$

where

$$v(\omega) = (2\pi)^{-1/2} \int\limits_{-\infty}^{+\infty} V(t) \exp(i\omega t) \, dt$$

$$= 2e(\omega), \qquad \omega > 0$$
$$= 0, \qquad \omega < 0. \tag{4}$$

The functions $V(t)$ and $v(\omega)$ define the optical pulse in the time and frequency domains respectively. The function $v(\omega)$ may be written as

$$v(\omega) = a(\omega) \exp[i\phi(\omega)], \tag{5}$$

where $a(\omega)$ and $\phi(\omega)$ are the spectral amplitude and spectral phase respectively. Laser pulses are quasi-monochromatic, which means that $a(\omega)$ is appreciably greater than zero only in a narrow frequency band $\Delta\omega_p$ about the carrier frequency ω_p. In this case, $V(t)$ may be separated into a slowly and a rapidly varying function of time as

$$V(t) = V'(t) \exp(i\omega_p t), \tag{6}$$

where the complex amplitude may be written as

$$V'(t) = A(t) \exp[i\Phi(t)]. \tag{7}$$

Here $A(t)$ and $\Phi(t)$ are the temporal amplitude and temporal phase of the carrier wave.

The quantity $V(t)$ has the useful property that the envelope of the pulse intensity is given by

$$I(t) = V^*(t)V(t) = A^2(t). \tag{8}$$

The spectral intensity $i(\omega)$, which is the quantity recorded by a spectrograph, is similarly defined as

$$i(\omega) = v^*(\omega)v(\omega) = a^2(\omega). \tag{9}$$

A complete description of the optical pulse requires both the amplitude and phase in either the time or frequency domain. Knowledge of the pulse's amplitude and phase in one domain can be transferred to the other by Fourier inversion. There is, of course, no unique relation between the temporal and spectral intensities because they do not contain the phase information. The only general relationship between the two intensity distributions is

$$\tau_p \frac{\Delta\omega_p}{2\pi} = \tau_p \Delta\nu_p \geq K, \tag{10}$$

where τ_p and $\Delta\omega_p$ are the full widths at half maximum (fwhm) of $I(t)$ *and* $i(\omega)$ respectively, and K is a constant which depends on the pulse shape. Table 1 gives values of K which have been calculated for several different pulse shapes. The shortest possible pulse obtainable for a given bandwidth is called bandwidth-limited or transform-limited and its duration is $\tau_p = 2\pi K(\Delta\omega_p)^{-1}$.

Table 1
Values of K calculated for several temporal amplitude functions $A(t)$.[a]

$A(t)$	K
$\exp[-\frac{1}{2}(t/t_0)^2]$	0.441
$\exp(-\frac{1}{2}|t/t_0|)$	0.142
$\exp(-\frac{1}{2}t/t_0)\theta(t)$	0.11
$1/\cosh(t/t_0)$	0.315
$\sin(t/t_0)/(t/t_0)$	0.892
$\sin^2(t/t_0)/(t/t_0)^2$	0.366
$\text{rect}(t/t_0)$	0.892
$[1+(t/t_0)^2]^{-1}$	0.142

[a] $\text{rect}(t/t_0)$ – rectangular pulse of width $t_p = t_0$. $\theta(t)$ – unit step function.

2.2 Longitudinal mode locking

Lasers usually oscillate simultaneously in a number of longitudinal modes, each separated by the cavity frequency $\nu_c = \Omega/2\pi = c/2L = T^{-1}$. Here L and T are the optical length and round-trip time of the cavity. Defining the amplitude, frequency and phase of the kth mode by A_k, ω_k and Φ_k respectively, the resultant analytic signal due to m oscillating modes is

$$V(t) = \sum_{k=1}^{m} A_k \exp i[\omega_k(t-z/c) + \Phi_k] \tag{11}$$

$$= \exp[i\omega_p(t-z/c)] \sum_{l=-m/2}^{m/2} A_l \exp i[l\Omega(t-z/c) + \Phi_l]. \tag{12}$$

Thus, the output is a modulated carrier of frequency $\omega_p = \omega_l - l\Omega$, whose envelope is periodic with the cavity period T and travels at the velocity c. The shape of the envelope depends on the amplitudes and phases of the modes. A laser is considered to be perfectly mode-locked when all the mode phases are equal.

When m modes oscillate with equal amplitude A and phase ϕ (which can be set to zero since the time origin is arbitrary), the laser's intensity is given by

$$I(t) = V^*(t)V(t) = A^2 \frac{\sin^2(m\Omega t/2)}{\sin^2(\Omega t/2)}. \tag{13}$$

The intensity maxima, occurring at $t = 0, T, 2T, \ldots$, are m times the average intensity, and the width of the intensity maxima (between consecutive zeros) is

$$\tau = 4\pi/m\Omega. \tag{14}$$

Thus, for ideal mode locking, the radiation in the laser cavity is a pulse whose duration is inversely proportional to the oscillating bandwidth.

2.3 Ultrashort pulse measurements

The development of mode-locked lasers has made feasible the generation of light pulses having durations less than 1 ps. Accurate measurement of such pulse widths is a significant problem which has been reviewed extensively by Bradley and New (1974).

The simplest and most direct method of recording the pulse's intensity $I(t)$ is provided by a photodiode and high-speed oscilloscope. The minimum risetime of this combination is ~ 100 ps. If the pulse is repetitive, sampling techniques can be employed which are capable of resolutions of ~ 25 to 50 ps.

A direct measurement of shorter pulses can be made by fast image-converter streak cameras. The streak camera's time resolution is limited theoretically to $\sim 10^{-14}$ s by the spread of photoelectron transit times resulting from the initial energy distribution of electrons leaving the photocathode. However, the extremely high streak velocity required for subpicosecond resolution reduces the image intensity. Attempts to compensate for the reduction by increasing the photocathode current density lead to photocathode saturation and space charge effects which limit the camera's dynamic range. Thus, for many applications the practical requirement of a dynamic range >10 results in limiting the camera's resolution to ~ 1 ps. These considerations are more completely discussed by Schelev et al. (1972). Bradley and Sibbett (1975) reported measurements with the Photochron II streak camera which indicated a temporal resolution of 0.7 ps.

The most popular method for studying the temporal structure of ultrashort pulses involves recording various correlation functions of $I(t)$. These techniques, which are capable of 0.1 ps resolution, include second-harmonic generation (Armstrong 1967), third-harmonic generation (Eckardt and Lee 1969), two-photon fluorescence (Giordmaine et al. 1967), three-photon fluorescence (Rentzepis et al. 1970) and four-photon parametric mixing (Auston 1971).

3 Active mode locking

The longitudinal modes of a laser oscillator can be locked in phase by the action of some active element in the cavity which modulates the gain or loss at the cavity frequency ν_c. The way in which the modulating element forms a short pulse can be viewed in either the time domain or the frequency domain. In the time-domain picture, the cavity radiation consists of some electromagnetic field distribution which passes back and forth through the cavity with the round-trip period T. The modulator reshapes the field distribution on successive passes until only a stable single pulse remains. In the frequency-domain picture, the cavity radiation consists of a superposition of discrete longitudinal modes which are separated in frequency by ν_c. The modulator creates weak sidebands on each longitudinal mode which overlap the adjoining modes. This coupling between the modes causes them eventually to oscillate in phase with each other.

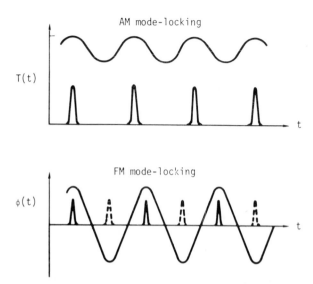

Fig. 1. Location of optical pulse relative to modulation cycle.

The cavity modes can be locked by amplitude modulation (AM) or frequency modulation (FM). Fig. 1 shows the location of the optical pulses relative to the modulation cycle for both cases. In the AM case, the modulator's transmission is varied and a short pulse is formed in the cavity such that it passes through the modulator at the time of minimum loss. Pulses are formed in the FM oscillator by a less direct process. The modulator's index of refraction is varied, which introduces an additional phase term $\phi(t)$ on the field passing through the modulator. The short pulse is formed in the cavity such that it passes through the modulator when $\dot{\phi} = d\phi/dt = 0$. At other times the radiation receives a frequency shift which is proportional to $\dot{\phi}$. The accumulation of frequency shifts on successive passes eventually shifts the radiation frequency outside the gain bandwidth of the active medium. Because $\dot{\phi} = 0$ twice in the modulation period, there are two equally probable positions of the optical pulse relative to the modulation signal. In addition, the quadratic variation of $\phi(t)$ around the time $\dot{\phi} = 0$ produces a frequency chirp on the pulse.

The nature of the mode-locking behavior is significantly different for laser transitions which are homogeneously or inhomogeneously broadened. The inhomogeneous laser oscillates freely and simultaneously on a large number of longitudinal modes. When no modulation signal is applied, the phases of the modes are randomly distributed. Mode locking over a large spectral bandwidth is easily obtained since the modulation need only produce weak sidebands which can capture the adjoining modes by injection locking. Early theoretical studies of active AM mode locking inhomogeneous lasers by DiDomenico (1964), Yariv (1965) and Crowell (1965) presented linearized solutions of the problem. More detailed nonlinear calculations by Harris and McDuff (1965) treated FM mode locking and later AM mode locking (McDuff and Harris 1967) numerically using the coupled mode equations, including effects of gain saturation and frequency pulling, with the assumption that the axial modes saturate independently. This inhomogeneous mode-locking model applies to the Doppler-broadened transitions of gas lasers such as the He–Ne laser studied by Harris and Targ (1964) and Ammann et al. (1965), and the argon laser discussed by Osterink and Targ (1967). Reviews have been given by Harris (1966) and McDuff (1972).

The homogeneous laser oscillates on only one or at most a small number of longitudinal modes near the center of the atomic transition when no modulation signal is applied. Formation of a stable optical pulse is the result of a delicate balance between the action of the modulator, which broadens the frequency bandwidth and narrows the pulse's duration and the gain medium, whose frequency-dependent amplification acts to narrow the pulse's bandwidth and increase its duration.

Mode-locking the homogeneous laser has been treated in a series of papers by Kuizenga and Siegman (1969, 1970a, b, 1974). The basic idea of their analysis is to begin with a single gaussian pulse circulating in the oscillator cavity and then determine how the pulse shape and amplitude is modified by each cavity element through which it passes. The steady-state pulse properties are obtained by equating the pulse waveform after a round trip of the cavity with the initial waveform. The required self-consistent solutions lead to expressions for the mode-locked pulse width which show its dependence on such important laser parameters as the atomic linewidth, the saturated gain, and the frequency and depth of the modulation. The effects of detuning the modulation frequency from the cavity frequency are also considered as well as the effects of intracavity etalons and dispersion.

This procedure relies on the assumption that the pulse bandwidth is much smaller than the atomic linewidth and the pulse width is short compared to the modulation frequency so that the gain line shape and modulator transmission may be approximated as gaussian. The analysis, of course, identifies only one steady-state pulse shape, the initial gaussian, while one expects the existence of an infinite set of modes constructed from the set of longitudinal cavity modes. Haken and Pauthier (1968) obtained approximate solutions of the coupled mode equations of homogeneous AM mode locking from which they identified an infinite set of 'supermodes'. However, they were unable to predict which mode should dominate. Solutions of the coupled mode equations given by Nelson (1972) also predicted a set of supermodes, and calculations of the supermode gains showed the simple gaussian to be the dominant mode.

More recently, Haus (1975a) presented an analysis of active mode locking in both the time and frequency domain which included methods for obtaining pulse shapes generated by other than sinusoidal modulation. The hermite–gaussian supermode solutions were obtained, and all solutions except the lowest-order one were shown to be unstable and therefore not attainable in the steady state.

The following review is confined to homogeneous mode locking, which applies to the widely used Nd:YAG laser and to medium- and high-pressure CO_2 lasers.

The analysis of Kuizenga and Siegman is followed, and typical results for Nd:YAG are presented.

3.1 Homogeneous model

The initial mode-locked pulse shape is assumed to be gaussian. Approximations are then made to the atomic gain line and the modulator transmission functions under the assumptions that $\Delta\omega_p \ll \Delta\omega_a$ and $\tau_p \ll T$, which maintain the gaussian shape. Properties of a gaussian pulse are reviewed first and then modifications of the pulse as it passes through the active medium and the modulator are considered.

3.1.1 Gaussian pulse

The electric field of a light pulse with a gaussian envelope can be written in the form

$$E(t) = \tfrac{1}{2}E_0 \exp[-at^2 + i\phi(t)] + \text{c.c.} \tag{15}$$

Here a determines the width of the envelope. The general phase function $\phi(t)$ may be expanded in a Taylor series about $t = 0$ and written in the form

$$\phi(t) = \omega_p t + bt^2 + \ldots, \tag{16}$$

where the initial phase is set equal to zero since it is arbitrary. The first term identifies the carrier frequency ω_p and the second term accounts for a linear frequency shift (chirp) $\Delta\omega = bt$ during the pulse. Higher-order terms in the expansion will not be considered. A complex gaussian envelope parameter w is defined by the relation

$$w = a - ib, \tag{17}$$

so that the pulse may be expressed as

$$E(t) = \tfrac{1}{2}E_0 \exp(-wt^2 + i\omega_p t) + \text{c.c.} \tag{18}$$

The pulse's spectrum, given by the expression

$$E(\omega) = \tfrac{1}{2}E_0(\pi/\omega)^{1/2} \exp[-(\omega - \omega_p)^2/4w] + \text{c.c.} \tag{19}$$

is also gaussian.

The full pulse width at the half-power points is

$$\tau_p = [(2\ln 2)/a]^{1/2}, \tag{20}$$

and the spectral bandwidth is

$$\Delta\nu_p = (1/\pi)[2\ln 2(a^2 + b^2)/a]^{1/2}. \tag{21}$$

Eq. (21) shows that a frequency chirp contributes to the bandwidth. For these gaussian pulses, the pulse width–bandwidth product is given by the expression

$$\tau_p \Delta\nu_p = (2\ln 2/\pi)[1 + (b/a)^2]^{1/2}. \tag{22}$$

3.1.2 Gain Medium

When the pulse passes through the active medium, its spectral amplitude $E_1(\omega)$ is multiplied by the complex frequency-dependent gain function $g_a(\omega)$ which is given by

$$g_a(\omega) = \exp g_0/[1 + 2\mathrm{i}(\omega - \omega_a)/\Delta\omega_a]. \tag{23}$$

Here g_0 is the saturated round-trip amplitude gain coefficient at the atomic line center ω_a, and $\Delta\omega_a$ is the full linewidth at half maximum of the fluorescence intensity. Near line center, the lorentzian line shape is well approximated as a gaussian which is obtained by expanding g_a about ω_a in the form

$$g_a(\omega) \simeq \exp\left\{g_0\left[1 - 4\left(\frac{\omega - \omega_a}{\Delta\omega_a}\right)^2 - 2\mathrm{i}\left(\frac{\omega - \omega_a}{\Delta\omega_a}\right)\right]\right\}. \tag{24}$$

The amplified pulse

$$E_2(\omega) = g_a(\omega)E_1(\omega) \tag{25}$$

remains gaussian with an envelope parameter w', where $(w')^{-1} = w^{-1} + 16g_0/\Delta\omega_a^2$. With the present approximation that the pulse bandwidth is small compared to the gain bandwidth ($\omega \ll \Delta\omega_a$), the envelope parameter is given by

$$w' \simeq w - 16g_0 w^2/\Delta\omega_a^2. \tag{26}$$

This result shows that the pulse bandwidth is reduced by passing through the active medium.

3.1.3 Modulator

When the spectrally narrowed pulse passes through the modulator, its amplitude $E_2(t)$ is multiplied by a time-dependent transmission function. Since this function takes a different form for amplitude and phase modulation, each case is considered separately.

Nd: YAG is often mode-locked by an acousto-optic amplitude modulator, which at $1.06\ \mu\mathrm{m}$ operates in the Bragg diffraction region. The single-pass amplitude transmission function is then given by

$$T_{\mathrm{AM}}(t) = \cos(\theta_A \sin \omega_m t). \tag{27}$$

Here ω_m is the modulation frequency and θ_A is the depth of modulation which varies as $(P_m)^{1/2}$, where P_m is the RF drive power into the modulator.

For ideal mode locking, the pulse passes through the modulator at a transmission maximum. Since there are two such times in a complete cycle, the modulation frequency is one-half the cavity frequency. When the pulse is short compared to the modulation period, $T_{\mathrm{AM}}(t)$ may be expanded about $t = 0$ and approximated by the gaussian form

$$T_{\mathrm{AM}}(t) \simeq \exp[-\tfrac{1}{2}(\theta_A\omega_m t)^2]. \tag{28}$$

After a round trip through the modulator, the pulse amplitude $E_3 = T^2 E_2$ has a new gaussian envelope parameter given by

$$w''_{AM} = w' + (\theta_A \omega_m)^2, \tag{29}$$

where the second term expresses the pulse width reduction produced by the modulator.

The sinusoidal phase variation introduced by a phase modulator results in a single-pass amplitude transmission of the form

$$T_{FM}(t) = \exp(-i\theta_F \cos \omega_m t). \tag{30}$$

Expanding T_{FM} about the phase extremum at $t = 0$ gives $T_{FM} \simeq \exp[\pm i\theta_F(\omega_m t)^2/2]$. After a round trip of the cavity, the FM gaussian pulse parameter is

$$w''_{FM} = w' + i\theta_F \omega_m^2, \tag{31}$$

where the second term is the chirp impressed on the pulse by the modulator.

3.1.4 Steady-state pulse

In the steady state, the pulse shape remains unchanged after a round trip through the cavity. This condition requires $w'' = w$ from which the steady-state value of w is determined. In the usual cases of interest, the AM solution is well approximated by

$$w_{AM} = a_{AM} \simeq \theta_A \omega_m \Delta \omega_a/(4\sqrt{g_0}), \tag{32}$$

and the steady-state pulse width is given by eq. (20) as

$$\tau_{PAM} = 2(g_0)^{1/4}[2\ln 2/(\theta_A \omega_m \Delta \omega_a)]^{1/2}. \tag{33}$$

Making use of the identity $(2i)^{1/2} = 1 + i$, the steady-state FM gaussian parameter is given in analogy with (32)

$$w_{FM} = a_{FM} - ib_{FM} = \frac{(1+i)}{4\sqrt{2}} \frac{\theta_F}{g_0} \omega_m \Delta \omega_a, \tag{34}$$

and the pulse width is given by the expression

$$\tau_{PFM} = 2(g_0/\theta_F)^{1/4}[2\sqrt{2}\ln 2/(\omega_m \Delta \omega_a)]^{1/2}. \tag{35}$$

The important feature to notice in comparing the AM and FM pulse widths given by eqs. (33) and (35) is that $\tau_{PAM} \propto (P_m)^{1/4}$ while $\tau_{PFM} \propto (P_m)^{1/8}$. Consequently, attempts to decrease the pulse duration by increasing the RF power are more effective in the AM case.

The more complete calculations of Kuizenga and Siegman (1970a) also presented expressions for the exact modulation frequency and for the saturated gain coefficient. However, for most applications the approximate values $\omega_m(FM) = 2\omega_m(AM) = 2\pi(c/2L)$ and $g_0 = 1/2 \ln[1/(1-l)]$, where l is the cavity loss, are sufficiently accurate.

Typical values for a Nd:YAG laser are: atomic linewidth, $\Delta \omega_a/2\pi = 120$ GHz; modulation frequency, $\omega_m/2\pi \sim 100$ MHz; saturated gain, $g_0 \sim 0.1$ and depth of modulation, $\theta_a \sim 0.5$. Using (33) and (22) we find for the AM case $\tau_p = 85$ ps and $\Delta \nu_p = 5.2$ GHz.

3.1.5 Summary

The difference in the physical pulse formation process for the AM and FM mode-locked laser is restated in summary. In the AM laser, the pulse is shaped directly by the time-dependent modulator transmission. The steady-state duration is determined by an equilibrium between the narrowing effected by the modulator and the broadening which results from the frequency-dependent gain. A pulse is formed in the FM laser only because the gain bandwidth is finite. The steady-state duration is reached when the increase of the pulse's bandwidth caused by the chirp it receives in passing through the modulator is balanced by the spectral narrowing of the gain medium. In this equilibrium condition the pulse is chirped.

3.2 Detuning

When the modulation frequency ω_m is detuned from the ideal mode-locking frequency ω_{m_0}, the pulse passes through the modulator at some phase angle ϕ away from the transmission maximum or extreme phase variation. The modulator transmission functions (28) and (30) must then be replaced by the following expressions:

$$T_{AM}(t) \simeq \exp\{-\theta_A^2/2[\sin^2 \phi + \omega_m t \sin 2\phi + (\omega_m t)^2 \cos 2\phi]\}, \tag{36}$$

$$T_{FM}(t) \simeq \exp\{\pm i\theta_F[\cos \phi - \omega_m t \sin \phi + \tfrac{1}{2}(\omega_m t)^2 \cos \phi]\}. \tag{37}$$

The behavior of detuned FM and AM lasers is considerably different. For the FM laser, the first term in the expansion (37) merely represents an overall phase shift of the pulse. The second term shows that the pulse carrier frequency ω_p is Doppler-shifted by the modulator away from the atomic line center ω_a. A new equilibrium is reached with $\omega_a = \omega_p + \delta$ when the shift on each pass is cancelled by an opposite shift which occurs when the active medium is traversed off line center. Expanding the frequency-dependent gain $g_a(\omega)$ given by eq. (23) about ω_p gives the same result as eq. (24) with ω_a replaced by ω_p, g_0 replaced by $g_0/(1 + 2i\eta)$, and $\Delta\omega_a$ replaced by $\Delta\omega_a(1 + 2i\eta)$ where $\eta = \delta/\Delta\omega_a$. The last term of the expansion shows that the pulse receives a chirp with the effective modulation index $\theta_F \cos\phi$.

Properties of the steady-state pulse are determined by the self-consistent analysis discussed above. These calculations have been given by Kuizenga and Siegman (1970a) and Siegman and Kuizenga (1970) along with numerical calculations which illustrate the laser's behavior. Fig. 2 and 3 show the variation of the pulse width and bandwidth as a function of detuning $\Delta\omega_m = \omega_{m_0} - \omega_m$. The pulse duration depends on detuning because the chirped pulse is compressed by the anomalous dispersion in the spectral region of the atomic resonance. The magnitude and sign of the chirp depend on detuning and, as fig. 2 shows, the maximum compression occurs for slight negative detuning. The power of the

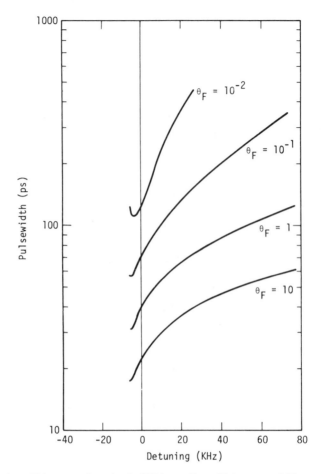

Fig. 2. Pulse width versus detuning in FM laser. From Kuizenga and Siegman (1970a).

optical pulse also depends on detuning, falling rapidly for negative detuning and more slowly for positive detuning, especially when far above threshold. However, the power does not depend very much on the depth of modulation θ_F.

Behavior of the detuned AM laser is much simpler than that of the FM laser. Because the amplitude modulator cannot change the pulse's frequency or introduce chirp, the effect of the active medium is the same as expressed by eq. (24) for $\Delta\omega_m = 0$. The pulse amplitude suffers a loss in the modulator given by the first term in the expansion (36). In addition, the modulator attenuates one edge of the pulse more than the other as the second term of the expansion shows. This effect introduces a temporal 'shift' of the pulse so that its repetition frequency remains $2\pi\omega_m$. The pulse width is determined by the effective depth of modulation $\theta_m \cos 2\phi$ given by the last term of eq. (36). Solution of the self-consistency equations provides expressions for the pulse width and

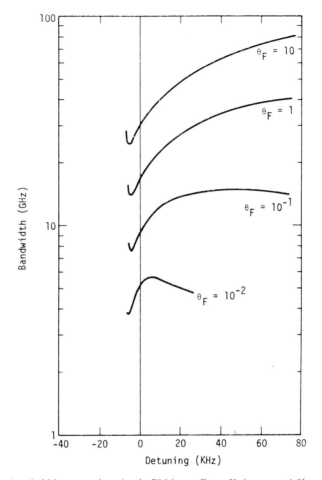

Fig. 3. Pulse bandwidth versus detuning in FM laser. From Kuizenga and Siegman (1970a).

bandwidth and for the exact modulation frequency. In general, the pulse width increases and the power decreases for both positive and negative detuning.

3.3 Etalon effects

The steady-state pulse width can be altered significantly by placing a Fabry–Perot etalon inside the laser cavity. Since the etalon's transmission depends on frequency, its main effect is to change the bandwidth of the laser. When the pulse width is much larger than the transit time of the etalon, the round-trip amplitude transmission T_e, expanded about a maximum at ω_e, is given by

$$T_e = \exp[i(8R)^{1/2}(\omega - \omega_e)/\Delta\omega_e - 4(\omega - \omega_e)^2/\Delta\omega_e^2]. \tag{38}$$

The bandwidth of the etalon $\Delta\omega_e$ is related to its reflectivity R and optical thickness d by the expression

$$\Delta\omega_e = (c/d)(1 - R)R^{-1/2}. \tag{39}$$

The effective laser gain is the product of T_e and $g_a(\omega)$. One may then calculate, for example, the pulse width of the FM laser with no detuning:

$$\tau_{\text{PFM}} = 2[2\ln2(2/\theta_m)^{1/2}/\omega_m]^{1/2}[g_0/\Delta\omega_m^2 + 1/\Delta\omega_e^2]^{1/4}. \tag{40}$$

Fig. 4 illustrates the effect of an uncoated quartz etalon in controlling the pulse width for the typical Nd: YAG laser whose parameters were given earlier. Etalons also generally improve the laser's stability by limiting its effective bandwidth.

3.4 Transient pulse formation

For many applications it would be useful for the energy of the optical pulse to be much greater than the typical 10^{-8} J obtained by cw mode locking. One approach

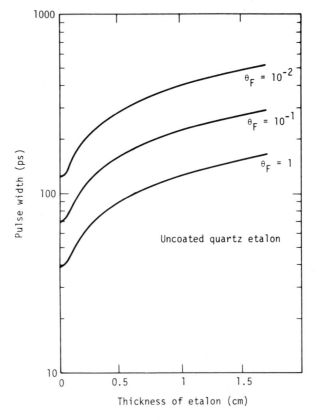

Fig. 4. Pulse width in FM mode-locked laser with uncoated quartz etalon. From Kuizenga and Siegman (1970a).

to obtain greater pulse energies is to simultaneously Q-switch and mode-lock the oscillator. A description of the transient formation of the mode-locked pulse has been developed by Kuizenga et al. (1973) and Krivoshchekov et al. (1973). The basic idea is that the spontaneous fluorescence emission is shaped into an approximately gaussian envelope by repeated passes through the modulator. While the width of the envelope is continually narrowed in time, the radiation bandwidth is narrowed by the frequency-dependent gain until the steady-state limits derived earlier are reached.

The AM laser is considered because it forms a pulse more rapidly by direct action of the modulator. Assuming that the change of the gaussian parameter $w_{AM} = a$ during the kth round trip is small, the envelope evolves according to the equation

$$da/dk = (\theta_A \omega_m)^2 - 16 g_0 a^2 / \Delta \omega_a^2. \tag{41}$$

The solution to this equation is $a(k) = a_0 \tanh (k/k_0)$, where a_0 is the steady-state value given by (32) and k_0 is given by

$$k_0 = \Delta \omega_a / 4 \theta_A \omega_m \sqrt{g_0}. \tag{42}$$

The transient narrowing of the pulse width is then given by

$$\tau_p(k) = \tau_{p_0} [\tanh (k/k_0)]^{-1/2}, \tag{43}$$

where τ_{p_0} is the steady-state pulse width. After a number of round trips $k \simeq 1.5 k_0$ (typically 10^3 to 10^4), τ_p varies from τ_{p_0} by less than 5%. Fig. 5 illustrates the transient narrowing of the pulse width for various values of the depth of modulation θ_A. The 10 to 100 μs required to reach the steady-state pulse width is long compared to the typical Q-switch build-up times of 0.5 to 1 μs reported by

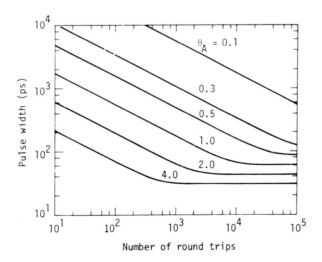

Fig. 5. Mode-locked pulse width versus number of round trips for Nd:YAG laser with $\Delta \omega_a / 2\pi = 1.2 \times 10^{11}$, $\omega_m / 2\pi = 2 \times 10^8$, $g_0 = 0.05$ and various values of θ_A. From Kuizenga and Siegman (1974).

Chesler et al. (1970). Consequently, steady-state conditions are not normally achieved in Q-switched, mode-locked operation.

Even after the pulse envelope reaches its steady-state value, it may still contain temporal substructure or frequency chirp. Since most of the transient mode locking occurs before the gain saturates, a large number of longitudinal modes can initially lase independently. The modulator generates a set of sidebands on each longitudinal mode, while the number of independent modes decreases owing to the natural mode selection of the frequency-dependent gain. The steady-state spectrum corresponds to the surviving longitudinal mode at line center with its sidebands. Kuizenga (1975) calculated an upper limit for the number of round trips required to reach the steady state of

$$k > \frac{\ln 2}{8g_0} \left(\frac{\Delta\omega_a}{\Omega} \right)^2, \tag{44}$$

where $\Omega = \pi c/L$ is the longitudinal mode spacing. For a Nd: YAG laser, about 10^5 round trips (~ 1 ms) would be needed for the pulse to reach its steady-state bandwidth. Clearly, then, under usual conditions a steady-state pulse is not generated by a simultaneously mode-locked and Q-switched oscillator.

However, another mode of Q-switch operation used by Ammann and Yarborough (1972) and later with mode locking by Kuizenga et al. (1973), Krivoshchekov et al. (1973) and Tomov et al. (1976) does allow the steady-state pulse parameters to be reached. The Q-switch loss is initially adjusted so that the laser slightly exceeds threshold. This 'pre-lase' condition is maintained until the pulse parameters reach nearly steady-state values. The Q-switch is then opened and the mode-locked pulse is rapidly amplified with little change in duration or bandwidth.

3.5 Nd:YAG laser

Mode-locking the Nd: YAG laser has been extensively investigated by Kuizenga, Siegman and co-workers. Their experimental measurements have verified many aspects of the theoretical model.

3.5.1 Steady state

Kuizenga and Siegman (1970b) mode-locked the cw Nd:YAG laser using a LiNbO$_3$ electro-optic phase modulator. The laser's bandwidth $\Delta\nu_p$ was monitored by a scanning Fabry–Perot interferometer and its pulse width τ_p was measured using a fast photodiode. These important parameters were then studied as functions of the depth of modulation θ_F. Fig. 6 illustrates the measured results compared with the $(\theta_F)^{1/4}$ dependence predicted by the theory. Modulation of the frequency spectrum produced by an etalon was clearly observed and the predicted increase in pulse width was verified.

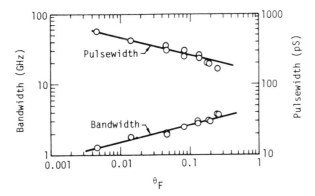

Fig. 6. Pulse width and bandwidth versus depth of modulation (θ_F) for FM Nd:YAG laser. From Kuizenga and Siegman (1970b).

Becker et al. (1972) presented results for mode-locking the Nd:YAG laser by intracavity phase modulation at the Nth harmonic ($N = 2, 3, 4, 5$) of the fundamental $c/2L$ cavity frequency. The resulting mode-locked spectra were shown to depend on the modulation frequency, depth of modulation and cavity losses as predicted by the homogeneous mode-locking theory.

3.5.2 Transient

The transient period of pulse formation in the Nd:YAG laser was studied experimentally by Kuizenga et al. (1973) using an optical correlator based on Type II second-harmonic generation (SHG). Fig. 7 shows the observed correlator signals. The top trace, obtained when the repetitively Q-switched laser was not mode-locked, contains a central 'coherence spike' characteristic of the gain-narrowed, but still noise-like, spectrum of the Q-switched pulse. The remainder of the trace is flat since the Q-switched pulse is much longer than the correlator's maximum delay. The middle trace was observed when weak amplitude modulation was applied. The laser pulse continues to exhibit rapid amplitude and phase fluctuations resulting in the same coherence spike. However, in addition the modulator has formed a 530 ps gaussian envelope. The bottom trace illustrates that increasing the depth of modulation brings the mode-locking processes nearer to completion during the Q-switch buildup period.

Fig. 8 shows the pulse width as a function of the build-up time τ_B with the depth of modulation held constant, compared with the $\tau_B^{-1/2}$ dependence predicted theoretically.

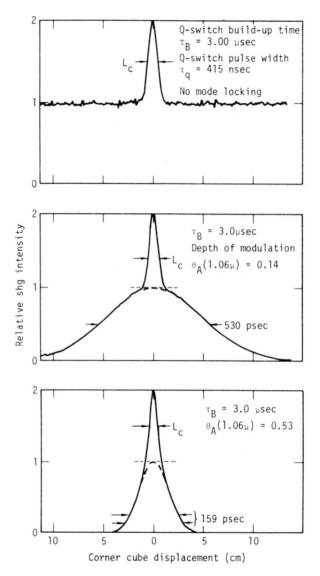

Fig. 7. Typical correlator traces for simultaneous Q-switching and mode locking. From Kuizenga et al. (1973).

3.5.3 Quasi-steady state

Kuizenga (1976) has recently improved the performance of the Q-switched actively mode-locked oscillator which makes use of a 'pre-lase' period to reach near-steady-state conditions. The laser was both mode-locked and Q-switched by acousto-optic amplitude modulators. This oscillator was specifically developed

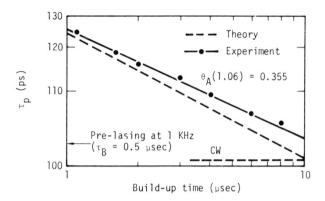

Fig. 8. Pulse width versus build-up time in *Q*-switched and mode-locked Nd:YAG laser. From Kuizenga et al. (1973).

to generate short pulses with complete reliability whose amplitude and duration would fluctuate less than ±5%. The laser's principle of operation is illustrated in fig. 9. During the 5 ms pre-lase period, near-steady-state pulses are generated. Their duration can be varied from 60 ps to over 1 ns by changes in the depth of modulation and use of intracavity etalons. At the maximum flashlamp current of 24 A, the single pulse energy switched out at the peak of the pulse train envelope was 400 μJ. The statistical distribution of the peak pulse amplitudes was carefully measured. In a sample of 500 consecutive shots, the amplitudes of all the pulses varied less than ±3% from the mean.

Two additional features of this oscillator which contribute to its stability are the use of all Brewster-angle components in the cavity (both acousto-optic modulators and the Nd:YAG rod) and water cooling of the modulators to hold their temperature-sensitive resonant frequencies constant. Similar techniques have been successfully applied to mode-locking ruby by Krivoshchekov et al. (1973) and Nd:glass by Tomov et al. (1976).

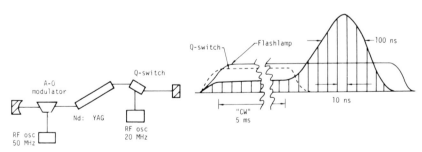

Fig. 9. Actively mode-locked and *Q*-switched oscillator and representation of operation.

3.6 Dye laser

The cw rhodamine 6G dye laser was actively mode-locked by Dienes et al. (1971) using an acousto-optic amplitude modulator, and by Kuizenga (1971) using an electro-optic phase modulator. Pulse widths were obtained in the range 50–100 ps. However, considerable difficulty was experienced in obtaining stable mode-locked behavior.

It was also difficult to prevent multiple pulses in the cavity round-trip time, especially when the pumping level was increased above threshold. A possible explanation for this difficulty is that the extremely wide effective atomic line-width $\Delta\omega_a/2\pi \sim 10^{13}$ Hz may allow separate pulses to build up in different spectral regions and the pulses then become separated in time by the frequency dependence of the group velocity.

Its broad linewidth makes the dye laser capable of generating pulses of duration ~ 1 ps (pulses of this duration have been achieved by the passively mode-locked cw rhodamine 6G dye laser). However, such short duration pulses are difficult to generate by active mode locking because the time required to reach steady-state conditions is very long. From eqs. (42) and (33), the time required to reach the pulse duration τ_p is

$$t \sim \frac{2L}{c} k_0 = 2\pi \ln 2/(\tau_p^2 \omega_m^3 \theta_A^2). \tag{45}$$

For typical values $\omega_m/2\pi = 100$ MHz and $\theta_A = 0.5$, the time needed to generate a 1 ps pulse is about 0.1 s. The circulating pulse and modulator transmission must remain in phase with an accuracy of $\sim 1:10^8$ for that length of time. Achievement of this condition is a severe, and as yet unsolved, technological problem.

4 Passive mode locking of solid state lasers

The use of an intracavity saturable absorber to simultaneously Q-switch and mode-lock a laser has been by far the most successful technique for producing ultrashort light pulses. The saturable absorber as a Q-switch was first suggested by Sorokin et al. (1964) and Masters et al. (1964). Passive mode locking with a saturable absorber was described in a patent application by Rigrod (1965); however, the first publication on the subject was by Mocker and Collins (1965) who observed that a ruby laser Q-switched with cryptocyanine frequently mode-locked. A short time later, DeMaria et al. (1966) mode-locked a Nd:glass laser.

The theory of passive mode locking has been described (DeMaria et al. 1969) in analogy with the electronic regenerative pulse generator of Cutler (1955). The periodic perturbation to the cavity radiation in passing through the saturable absorber is represented as gradually adding sidebands to the oscillating mode spectrum, finally producing in the steady state a periodic sequence of ultrashort pulses. Although this analysis is accurate for cw lasers, Letokhov (1968) pointed out that when a saturable absorber is used to Q-switch and mode-lock a laser, steady-state conditions are not reached during the Q-switched pulse and a transient analysis must be used.

These considerations led to proposal by Letokhov (1968, 1969a, b) of a 'fluctuation model' for pulse generation in a passively mode-locked oscillator. This model reveals the inherent limitations of passive mode locking and explains many difficulties which investigators have experienced, such as lack of reproducibility, background emission and satellite pulse generation. The basic idea of the model is that the final ultrashort pulse is formed by a combination of nonlinear processes which select and amplify noise fluctuations present in the initial fluorescence emission. Similar conclusions were reached independently by Fleck (1968, 1970) through detailed calculations of the amplification process beginning from the spontaneous noise emission. The model was further developed by Kuznetsova (1969), Zel'dovich and Kuznetsova (1972), Kryukov and Letokhov (1972) and Hausherr et al. (1973). The most recent work on this model has been presented by Zherikhin et al. (1974a) and Glenn (1975). The following description of the mode-locked pulse-formation process is based largely on these two papers.

4.1 Fluctuation model

The pulse-formation process may be separated into the four distinct regions illustrated in fig. 10. In region 1, the laser is below threshold and the intensity in the cavity is dominated by spontaneous emission. In region II, the gain exceeds cavity losses and the radiation experiences a period of linear amplification which lasts until it becomes sufficiently intense to begin bleaching the dye. Region III is the critical period for generating a mode-locked pulse. A noise fluctuation spike

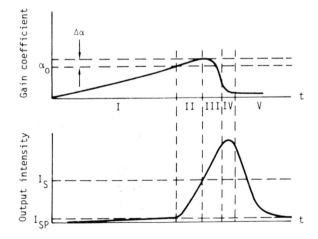

Fig. 10. Behavior of gain and average intensity in passively mode-locked laser.

is selected and amplified by the combined action of the dye nonlinearity and saturation of the gain. In region IV, the pulse intensity can finally become so large that it is affected by the nonlinear refractive index of the laser rod and the dye solvent. These nonlinearities cause self-phase modulation and sometimes destructive self-focusing. Although these effects are important in considerations of ultrashort pulse applications, they occur after the mode-locking process is complete. Consequently, further discussion of this region is deferred to § 4.4.

We now consider the first three regions in greater detail. The model is based on a simple laser resonator which contains the gain medium and a saturable absorber. In the following discussion, parameters of the gain medium are denoted by the subscript a and those of the saturable absorber by the subscript b. The round-trip power gain of the laser rod is given by e^{α} and the transmission of the dye cell is $e^{-\kappa}$. The linear losses of the cavity are represented by the round-trip transmission $e^{-\gamma}$.

4.1.1 Region I—Spontaneous emission

When the flashlamp pumping begins, ions are excited to the upper laser level and begin to fluoresce. The fluorescence bandwidth is the full characteristic linewidth of the laser transition and the emission's time dependence is noise-like. At the end of region 1, the gain coefficient has increased to the threshold value

$$\alpha = \alpha_0 = \kappa_0 + \gamma, \tag{46}$$

where κ_0 is the unbleached absorption coefficient of the dye.

4.1.2 Region II—Linear amplification

Under continued flashlamp pumping, the gain exceeds cavity losses and the radiation intensity begins to increase. Initially the population inversion and the dye transmission are not affected by the radiation, so the entire distribution of noise fluctuations is equally amplified. During this period of linear amplification, the radiation begins to acquire periodic structure with the cavity period T. The total radiation inside the cavity is obtained by summing over the longitudinal modes as discussed in § 2.2.

Because the fluorescence bandwidth $\Delta\omega_a$ is large, a great number of longitudinal modes of the cavity are initally excited. For example, the bandwidth of the inhomogeneously broadened transition of Nd:glass is $\sim 200\,\text{cm}^{-1}$ and the homogeneously broadened transition of Nd:YAG has a bandwidth of $\sim 4\,\text{cm}^{-1}$. A typical cavity length is 1.5 m, so initially the number of longitudinal modes m excited in a Nd:glass oscillator is $\sim 6 \times 10^4$, and in a Nd:YAG oscillator $m \sim 1.2 \times 10^3$.

At the beginning of region II, the amplitudes and phases of the modes are random functions of time which are determined by the statistical properties of the spontaneous emission. Interference of the cavity modes having random phase relations caused the radiation intensity to fluctuate. The average duration τ_a of the fluctuation peaks is determined by the spectral width of the fluorescence by the relation

$$\tau_a \simeq (1/\Delta\omega_a) = T/m. \tag{47}$$

For Nd:glass, $\tau_a \sim 0.1$ ps and for Nd:YAG, $\tau_a \sim 1$ ps.

In the classical limit, each mode contains a large number of photons and the field in each mode is gaussian noise. The random noise amplitude is described by the Rayleigh distribution, and its phase is uniformly distributed in the interval $(0, 2\pi)$. Although the field in each mode becomes coherent during amplification, the mode phases remain independent and randomly distributed. Under these conditions, the total field remains gaussian noise as shown by Mandel and Wolf (1965). The probability density of the fluctuation pulse intensities is then given by

$$W(I) = \langle I \rangle^{-1} \exp(-I/\langle I \rangle), \tag{48}$$

where the bracket indicates an average over the cavity period.

It is important that while the total number of fluctuations in the period T is initially large, a small number of these have intensities that significantly exceed the average. As shown by Letokhov (1969a, b), the average number of pulses whose intensity exceeds $\langle I \rangle$ by the factor β is

$$m \int_{\beta\langle I \rangle}^{\infty} W(I)\, dI = m \exp(-\beta). \tag{49}$$

Thus, the largest fluctuation pulse exceeds the average intensity by the factor $\beta = \ln m \sim 7$ to 9. This factor becomes important in the process of selecting one or a small number of peaks in region III. Kryukov and Letokhov (1972) also noted that the mean duration at half-maximum of the most intense fluctuations is

$$\tau_m \simeq \tau_a (\ln m)^{-1/2}. \tag{50}$$

The fact that τ_m is less than the average duration τ_a was interpreted as resulting from the contribution of the spectral wings to formation of the largest peaks.

During the linear amplification period, the fluctuation pulse duration increases as a result of gain narrowing by the amplifying medium. If the initial pulse is gaussian with pulse width τ_a, then after N round trips of the cavity with the gain coefficient α, the pulse duration at the end of the linear amplification period becomes

$$\tau_\ell = (\tau_a^2 + 16 \ln 2 N\alpha/\Delta\omega_a^2)^{1/2}. \tag{51}$$

A Nd:glass oscillator, for example, has a round-trip net gain at line center of a few percent in the linear region. The radiation is amplified from the spontaneous intensity to the dye's saturation intensity $\sim 10^7$ W/cm^2 in $\sim 10^3$ cavity transits, which corresponds to a buildup period of 10^{-5} s. At the end of this period, the fluctuation pulse durations have increased from ~0.1 to ~5 ps. This behavior has been used to change the duration of mode-locked pulses as discussed in § 7.

The end of region II is reached when the radiation intensity has grown to the point that nonlinearities of the dye become important. At this point, the gain has increased from its threshold value by the amount $\Delta\alpha$, a quantity which plays an important role in the mode-locking behavior.

4.1.3 Region III—nonlinear pulse selection and amplification

Region III is the critical region for mode locking. At the beginning of this region, the radiation inside the cavity is essentially a set of m pulses of average duration $\tau_\ell \sim 50 \times T/m$ whose intensities follow the probability density given by eq. (48). During this period the random collection of pulses must evolve to the condition that one pulse is much larger than all the others.

Interaction between the radiation and the amplifying medium, treated as a two-level system, is described by simple rate equations. Thus, the model is confined to treating pulse durations $\tau_p \gg 1/\Delta\omega_a$. The peak intensity of each pulse at the kth round trip of the cavity evolves according to the equation

$$dI_{k,i}/dk = I_{k,i}(\alpha_{k,i} - \gamma - \kappa_{k,i}) + (dI_{k,i}/dk)_{sp}. \tag{52}$$

The subscript i denotes the ith pulse of the set of m pulses circulating in the cavity. The last term, which is the contribution of spontaneous emission, dominates during region I but is negligible when region II is reached.

The gain of solid-state lasers saturates on the total energy extracted. Consequently, the gain coefficient at the kth pass is the same for all the pulses and follows the equation

$$d\alpha_k/dk = -\alpha_k(2\sigma_a/\hbar\omega_p) \sum_{i=1}^{m} I_{k,i}(T/m) + fT. \tag{53}$$

Here σ_a is the stimulated emission cross section and f is the pumping rate, which for solid-state lasers can be ignored during region III. Eq. (53) does not apply to all mode-locked lasers. The gain of a dye laser, for example, can be saturated by a single pulse. In that case, the gain medium plays a direct role in shaping the pulse as discussed in § 5.1.

The saturable absorber has two effects on the pulse formation. The first effect is to select one, or a small number, of fluctuation peaks. The most intense fluctuation peaks are least attenuated by the dye and grow rapidly in comparison with weaker pulses. This process, which is the main concern of the fluctuation model, is affected by two important properties of the dye: the recovery time τ_R of the ground-state population and the intensity I_s required to saturate or bleach the dye.

The dye is most effective in selecting a pulse when τ_R is short compared to the time between pulses ($\sim \tau_\ell$). Transmission through the dye is then given, following Hercher (1967), by

$$\exp(-\kappa) = \exp[-\kappa_0(1 + I/I_s)^{-1}].$$ (54)

The saturation intensity I_s may be expressed for a two-level system with absorption cross section σ_b as

$$I_s = \hbar\omega_p/(2\sigma_b\tau_R).$$ (55)

Fig. 11 illustrates the dye transmission function (54) for incident pulse intensities in the vicinity of I_s. The mode-locking behavior is considerably altered when τ_R is longer than the fluctuation pulse duration, as further discussed in § 4.3.

The second effect of the dye is to shorten and distort the pulse shape. Letokhov (1968) showed the amount by which the pulse duration changes on each pass through the dye cell is given by the expression

$$\frac{d\tau_p}{dk} = \frac{-\tau_p\kappa_0 I_k/I_s}{2\ln2(1 + I_k/I_s)(2 + I_k/I_s)}.$$ (56)

The rate of pulse shortening reaches its maximum at $I_k/I_s = \sqrt{2}$ and tends toward zero for both very large and very small intensities.

Letokhov (1969b) further showed the total amount by which the pulse is shortened during mode locking is greatest when the laser is operated near threshold. However, in general the pulse broadening in the linear amplification region exceeds its shortening in the nonlinear stage. Consequently, this effect is not retained in the model, and the calculations are based on 'square' pulses of constant duration.

4.1.4 Gain saturation

Early treatments of the fluctuation model considered only the nonlinearity of the dye to be responsible for selection of the single pulse. In addition, the selection

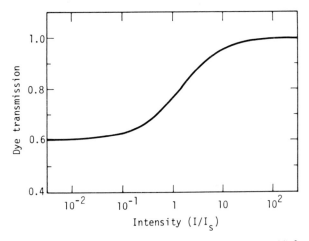

Fig. 11. Nonlinear transmission of saturable absorber given by eq. (54) for $\kappa_0 = 0.5$.

process was assumed to be complete before appreciable gain saturation occurred so that α was constant in region III. The pulses in the cavity then are statistically independent, and Glenn (1975) showed the probability of producing a 'single pulse' train is only 0.37 regardless of the initial distribution of pulse amplitudes. Since, experimentally, single pulse trains are produced a much higher percentage of the time, Glenn concluded that the pulses are not statistically independent; that there is interaction which allows one pulse to capture a large fraction of the stored energy at the expense of the others. Glenn proposed that gain saturation could play this important role in the following manner.

In the nonlinear region, the net excess gain experienced by a pulse depends on its intensity through the relation

$$\Delta\alpha(I_{k,i}) = \alpha_k - \gamma - \kappa_0(1 + I_{k,i}/I_s)^{-1}. \tag{57}$$

Initially, $\Delta\alpha$ is positive for all pulses. As the pulses are amplified, they extract energy from the laser rod and the gain coefficient α_k decreases. The low-intensity pulses can then experience $\Delta\alpha < 0$ and begin to decay while the higher-intensity pulses continue to grow.

This process may also be described by introducing a critical intensity I_c defined by the relation $\Delta\alpha(I_c) = 0$. Thus,

$$I_c/I_s = \kappa_0(\alpha_k - \gamma)^{-1} - 1 = \kappa_0(\kappa_0 + \Delta\alpha_k)^{-1} - 1. \tag{58}$$

Only those pulses of intensity $I_{k,i} > I_c$ are amplified. At the beginning of region III, $I_c < 0$ so all the pulses grow. As energy is extracted, I_c becomes positive and continues to increase. A single pulse train is produced when the intensity of only one pulse remains above I_c.

4.1.5 Initial pulses

To verify this picture by numerical simulation, it was necessary to choose an initial set of pulse amplitudes. Rather than starting with a set of random intensities distributed according to eq. (48), Glenn observed from the picture of pulses being eliminated by a rising critical intensity that the formation of a single pulse train depends on finally eliminating the second largest pulse. Thus, the most important property of a set of pulses which affects whether a well-mode-locked train will be produced is the ratio of the two largest pulses. The statistical distribution of that property was used to pick the starting set, and the distribution of amplified pulses was then completely determined by eqs. (52) and (53). The following numerical example and calculations will help to clarify this argument.

For a set of m starting pulses with the probability density given by eq. (48), the most probable intensity of the Nth largest pulse is

$$I_N = \log(m/N). \tag{59}$$

For example, if $m = 100$, the most probable intensities of the four largest pulses

Table 2
Representative values of the function $P(m, C)$.

C	1	1.01	1.02	1.04	1.10	1.20	1.50	2.0
P	1.0	0.96	0.93	0.86	0.69	0.48	0.15	0.02

are 4.6, 3.9, 3.5 and 3.2. The probability $P(m, C)$ that the largest pulse is C times or larger than the second largest pulse is

$$P(m, C) = \frac{\Gamma(C + 1)\Gamma(m - 1)}{\Gamma(m + C)} \approx \frac{C}{(m + 1)^{C-1}}, \tag{60}$$

where the approximation applies for $C \geq 1$. Table 2 shows the behavior of $P(m, C)$ for $m = 100$.

The mode-locking behavior was studied numerically for two interesting sets of 100 starting pulses. All the pulses of the first set had their most probable amplitudes. For this set, $C = 4.6/3.9 = 1.18$ and $P(100, 1.18) \approx 0.5$, which means the ratio of the two largest pulses exceeds 1.18 about 50% of the time. In the second set, all pulses except the second most intense had their most probable amplitude. The remaining amplitude was chosen so that $C = 1.02$ which, from table 2, would occur 93% of the time.

4.1.6 Numerical simulation

The input parameters used by Glenn in the numerical simulation were the linear loss coefficient γ, the unbleached dye absorption coefficient κ_0, the stimulated emission cross section for the gain medium σ_a (taken as $\sigma_a/\hbar\omega_a = 0.16 \text{ cm}^2/\text{J}$ characteristic of ED-2 glass), the saturation intensity of the dye I_s (taken as $5 \times 10^7 \text{ W/cm}^2$), the cavity transit time (taken as 5 ns) and the gain at the beginning of region III, $\alpha = \alpha_0 + \Delta\alpha$. The critical parameter was the excess gain, $\Delta\alpha = \alpha - \gamma - \kappa_0$, at the start of the calculation. The growth of the largest pulse is shown in fig. 12 for the case $\kappa_0 = 0.2$, $\gamma = 0.1$ and various values of $\Delta\alpha$. As $\Delta\alpha$ decreases, successively more passes are made before the maximum intensity is reached. Finally, when $\Delta\alpha = 0.01$ (1% excess gain), the pulse intensity remains small. In this case, the amplifier gain saturates before any pulse bleaches the dye. When the initial excess gain is slightly larger, there is a long period of slow pulse growth followed by a rapid rise as the dye bleaches.

A measure of the extent to which ideal mode locking is achieved is provided by the ratio $\langle I^2 \rangle / \langle I \rangle^2$. This quantity is often called the second harmonic enhancement ratio (SHER) because it can be measured experimentally by performing second-harmonic generation with the mode-locked pulse train and observing the enhancement of the signal when the dye is added. When m longitudinal modes are perfectly locked, the SHER takes its maximum value of m. In fig. 13, the value of this ratio calculated at the peak pulse amplitude as a function of $\Delta\alpha$ for several values of κ_0 shows that ideal mode locking occurs for

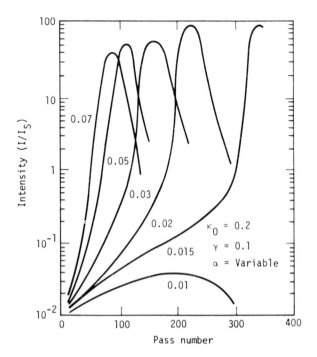

Fig. 12. Behavior of largest pulse intensity for $\kappa_0 = 0.2$, $\gamma = 0.1$ and various values of $\Delta\alpha$. From Glenn (1975).

only a narrow range of $\Delta\alpha$. When $\Delta\alpha$ is too small, the dye is not bleached; and when $\Delta\alpha$ is too large, the pulse intensities grow through the nonlinear region of the dye so rapidly that the amplitude discrimination is poor. In fig. 13 the initial pulse amplitudes were assigned their most probable value ($C = 1.18$). For the smaller values of C which occur more frequently, the range of $\Delta\alpha$ that produces good mode locking is more restricted.

Fig. 14 shows the amplitudes of the five largest pulses initially increasing and then one-by-one decreasing as the critical intensity rises. The calculation was performed for the second set of starting pulses ($C = 1.02$) and $\Delta\alpha = 0.015$. The ratio of the two largest pulses at the envelope peak was about 9000.

The importance of gain saturation is illustrated by fig. 15. The behavior of the five largest pulses under good mode-locking conditions is shown in fig. 15a. In fig. 15b, the gain was held constant so all pulse amplitudes grow without limit and little amplitude discrimination occurs.

4.1.7 Summary

The essential features of the fluctuation model may be summarized as follows:

(1) The pulse formation begins with a set of random noise fluctuation pulses;

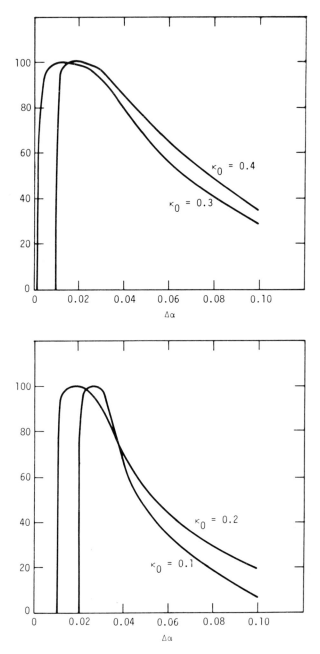

Fig. 13. SHER at pulse-train envelope maximum versus $\Delta\alpha$ for various values of κ_0. From Glenn (1975).

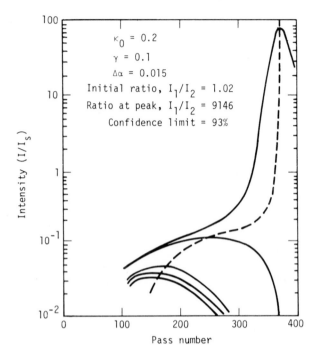

Fig. 14. Behavior of the five largest pulses with initial conditions given on figure. From Glenn (1975).

(2) Bleaching of the dye alone is not sufficient to form a single pulse train with high probability;

(3) The combination of gain saturation and dye bleaching acts strongly to amplify the most intense initial fluctuation relative to the weaker fluctuations;

(4) The most important parameter in the mode-locked pulse's formation is the initial gain above threshold $\Delta\alpha = \alpha - \gamma - \kappa_0$.

The predictions of the fluctuation model are in good qualitative agreement with experimental observations as discussed in the following section. Quantitative agreement is, however, not to be expected since many physically important aspects of the mode-locked oscillator have not been included, such as the finite recovery time of the dye which can significantly reduce its ability to select a single pulse and interference between pulses traveling in opposite directions which affects their statistical dependence.

4.2 Experimental studies

The original fluctuation model of Letokhov (1968) considered amplitude discrimination among the noise pulses due to the bleachable dye alone. Following this model, Hausherr et al. (1973) derived expressions for the probability of two

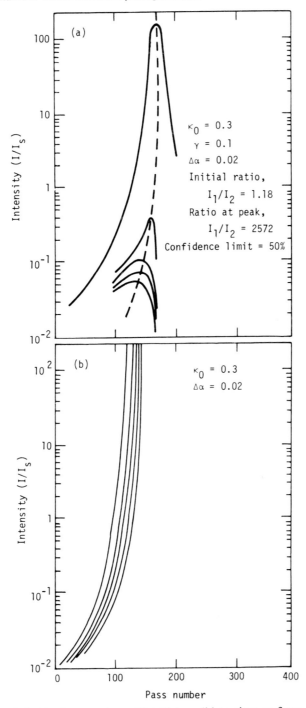

Fig. 15. Behavior of the five largest pulses with initial conditions given on figure (a) including gain saturation, (b) in absence of gain saturation. From Glenn (1975).

pulses being generated with a given intensity ratio. Using a Nd:glass oscillator, they found more reproducible mode-locking behavior occurred when stronger dye concentrations were used and when the laser gain was only slightly above threshold.

Direct evidence of the growth of short pulses from fluctuations in the amplified spontaneous emission was obtained by Zakharov et al. (1974) using an image-converter streak camera to photograph the emission from a Nd:glass oscillator which was in an unstable, ring-resonator configuration. The emission was recorded for 125 ns intervals with 250 ps resolution. The oscillator's emission was photographed at various times relative to the maximum of the pulse-train envelope. Fig. 16, which shows the intensities recorded at 1200, 900, 600 and 300 μs before the envelope maximum, demonstrates the transformation from irregular fluctuations to a periodic sequence of pulses.

Furthermore, two gain thresholds were observed in the pulse-formation process. At the first threshold, a free oscillation consisting of random intensity fluctuations occurs. Only for a definite excess gain above the first threshold does the single pulse train appear. The evolution of pulses with various amplitudes was obtained from the experimental data and is illustrated by fig. 17. Pulse amplitudes measured at the times recorded by figs. 16a, b, c, d are plotted in the

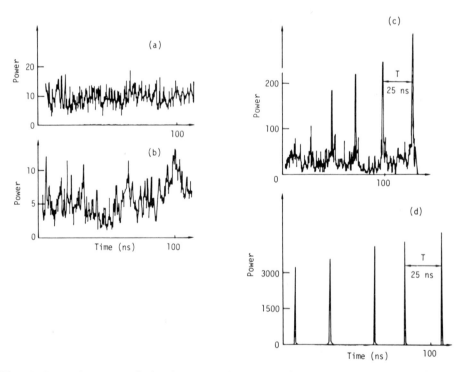

Fig. 16. Successive stages of ultrashort pulse development in Nd:glass ring laser. Densitometer traces of streak-camera photographs taken (a) 1 200 ns, (b) 900 ns, (c) 600 ns, (d) 300 ns, before pulse-train envelope maximum. Temporal resolution – 250 ps. From Zakharov et al. (1974).

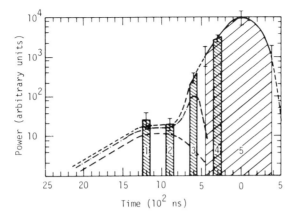

Fig. 17. Time dependence of pulse intensity for various initial amplitudes. The fluctuation amplitudes measured from figs. 16a, b, c, d are plotted in regions 1, 2, 3, 4 respectively. From Zakharov et al. (1974).

regions 1, 2, 3, 4 respectively in fig. 17. This figure shows that, while initially all the pulse amplitudes grow, a point is reached after which the most intense pulses continue to grow while the weaker pulses decay. This behavior was explained in § 4.1 as resulting from the fact that amplification of the pulses reduces the population inversion and the gain. Eventually only the most intense pulses, which strongly bleach the dye, experience a net positive gain. This behavior was initially surprising because the dye's absorption cross section is four orders of magnitude larger than the stimulated emission cross section of the active medium, so it was thought that nonlinear discrimination of the dye would occur much earlier than gain saturation. In fact, the dye population re-establishes itself in a time less than the cavity period while the gain does not. Consequently, both dye and gain can saturate at about the same time.

It was then clear that the time delay between bleaching the dye and the onset of gain saturation would affect the pulse formation. The effect was studied by Zherikhin et al. (1974a) who varied this delay time by changing the ratio of the pulse intensities in the dye cell and laser rod. The experiment was performed using a Nd:YAG oscillator in an unstable resonator configuration. A telescope inside the cavity allowed the ratio of the beam cross section in the dye cell to that in the laser rod to be set at 4:1, 2:1 and 1:1.

When the ratio was 4:1, only the free oscillation spikes appeared. Their number increased with pumping energy, but a single pulse train was not formed. In this case, the gain saturated before the most intense fluctuation pulses could bleach the dye. When the ratio was 1:1, a pulse train was formed but there were also many satellite pulses accompanying the main pulse. Fig. 18a illustrates the pulse-train envelope which was observed in this case. The dye bleached well before gain saturation occurred and, as a result, the amplitude discrimination due to the dye nonlinearity alone was poor. When the ratio was 2:1, a single pulse train was formed with the fewest observed satellites. Fig. 18b shows the pulse-

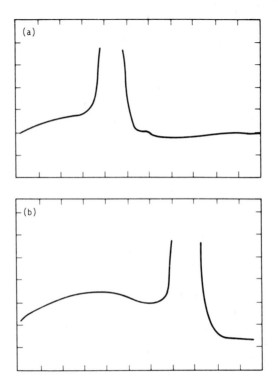

Fig. 18. Oscillograms showing mode-locked pulse development (0.1 μs/div) for beam-area ratio of (a) 1:1, (b) 2:1. From Zherikin et al. (1974a).

train envelope with a characteristic dip after the first maximum. The dip occurs because the photodiode measures the average intensity of all pulses in the cavity and at that point only a few of the most intense pulses continue to grow while the others are attenuated.

These experiments clearly show that formation of a single pulse train results not only from preferential growth of the strongest fluctuations but also from decay of the weaker pulses due to saturation of the gain. These and other experimental results concerned with ultrashort pulse-formation processes were reviewed by Chekalin et al. (1974). In addition, recent theoretical and experimental studies of a Nd:YAG ring oscillator by Wilbrandt and Weber (1975) also demonstrated the 'second threshold' for appearance of a mode-locked pulse train and that the mode-locking occurred with only a certain probability.

4.3 Absorber lifetime effects

In the preceding discussion of the fluctuation model, the recovery time τ_R of the saturable absorber was assumed to be instantaneous. Kryukov and Letokhov (1972) showed that if τ_R is longer than the duration of the fluctuation peaks τ_ℓ at the end of the linear amplification period, the shortest and most intense fluctuations are poorly discriminated against the background of surrounding fluctuations. The dye instead tends to select a fluctuation burst whose duration is approximately τ_R.

4.3.1 Ruby Laser

Effects of the saturable absorber's lifetime on the short pulse-formation process were studied by Kryukov et al. (1972a) using a streak camera with 20 ps resolution to photograph the emission from a ring, ruby laser, mode-locked with vanadium phtalocyanine in nitrobenzene ($\tau_R = 1.2 \pm 0.6 \, \mu s$). The fluctuation pulse durations τ_ℓ were adjusted by inserting intracavity etalons to control the oscillating bandwidth. With the ratio $\tau_R/\tau_\ell \simeq 10$, the pulse generation was unstable and consisted of a group of short pulses. The envelope of the group had a sharp leading edge and an exponential decay with a time constant of about τ_R. With $\tau_R/\tau_\ell \simeq 1$, the laser's stability improved. A 1 ns pulse envelope was generated with 0.1 ns substructure. In the case $\tau_R/\tau_\ell < 1$, the stability improved and a 2 ns pulse with no internal structure was generated.

Arthurs et al. (1974a) also studied the effect of absorber recovery time on pulse generation in a ruby laser. They used two different saturable absorbers, DDCI and DTDCI, with recovery times corresponding to the cases $\tau_R/\tau_\ell \simeq 1$ and $\tau_R/\tau_\ell \gg 1$ respectively. Their observations, made using a streak camera, showed that the absorber lifetime affects the pulse-generation process and sets a lower limit to the duration of the mode-locked pulse's envelope.

4.3.2 New infrared mode-locking dye

Studies of absorber lifetime effects in Nd lasers have been facilitated by a new class of Q-switch and mode-lock compounds for infrared lasers reported by Drexhage and Müller-Westerhoff (1972). A typical example is the transition-metal complex bis(4-dimethylaminodithiobenzil)nickel (BDN). These dyes have exceptional photochemical stability compared to the commonly used cyanine dyes Eastman Kodak A9860 and A9740, making them potentially reliable mode-locking saturable absorbers. However, compared with the later dyes whose recovery times of 7 ± 1 ps and 11 ± 1 ps were measured by von der Linde and Rodgers (1973a), BDN has a much longer lifetime which, in addition, depends strongly on the solvent. Magde et al. (1974) measured the lifetime to be 6.2 ± 1.6 ns in chloroform and 1.5 ± 0.2 ns in carbon disulfide, making the dye ineffective for mode locking in these solvents. However, the recovery time in two groups of electronically active solvents is sufficiently short that well-defined mode-locked pulse trains in Nd:glass were observed by Drexhage and Reynolds (1974) and Fan and Gustafson (1975). One group of solvents reduces the recovery time by a heavy-atom interaction and the other by a reversible electron transfer. Iodoethane ($\tau_R \sim 100$–200 ps) is an example of the first group, and the second group includes methysulfoxide ($\tau_R \simeq 65$ ps), pyridine ($\tau_R \simeq 60$ ps) and ethyl sulfide ($\tau_R \simeq 25$ ps).

4.3.3 Nd laser

The possibility of continuously varying τ_R in solvent mixtures was used by Al-Obaidi et al. (1975) to study the effect of τ_R on mode-locked pulse generation in Nd:YAG. With Kodak A9870 as the saturable absorber, the condition $\tau_R < \tau_\ell$ was satisfied and nearly transform-limited pulses of ~ 30 ps duration were observed using a streak camera. Using BDN in iodoethane ($\tau_R > \tau_\ell$), the measured pulse duration of ~ 55 ps was about twice the transform-limited duration. For further increases in τ_R obtained in mixtures of iodoethane and dichloromethane, the pulse duration did not increase significantly, but multiple pulses were generated.

Recent studies by Petukhov and Krymova (1976) of a ring Nd:glass oscillator mode-locked by two different dyes with $\tau_R \sim 8$ ps and 60 ± 15 ps agree with the general picture described in this section of absorber lifetime effects.

4.4 Nd:glass laser

The Nd:glass oscillator is one of the most widely used sources of high-power, ultrashort light pulses. Pulses with duration ~ 5 to 15 ps and energy ~ 1 mJ are usually achieved using the Eastman Kodak bleachable dyes A9740 or A9860 in the typical cavity arrangement illustrated in fig. 19. For many experiments it is convenient to select a single pulse from the mode-locked pulse train, which can be accomplished using an electro-optic shutter described by Kachen et al. (1968).

The body of experimental evidence accumulated on the Nd:glass oscillator also serves well to illustrate the effects which occur in the final stage (region IV) of the pulse-generation process.

4.4.1 Two difficulties

Since the first report by DeMaria et al. (1966) of saturable-absorber mode locking of Nd:glass, intensive studies have revealed two major problems: (1) the

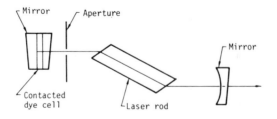

Fig. 19. Passively mode-locked laser.

lack of reproducibility of the mode-locked pulse train and (2) degradation of the optical and temporal properties of the pulse when its intensity exceeds $\sim 10^9 \, W/cm^2$.

The first problem results from the statistical nature of the events occurring in the first three regions of the passive mode-locking process which were discussed in § 4.1. The pulse amplitude fluctuates from shot to shot, and the oscillator frequently produces multiple pulses or no pulse at all.

The second problem arises from the nonlinear index of refraction of the glass which leads to self-focusing and self-phase modulation of the pulse. These effects may dominate the ultrashort pulse properties in the region IV. The action of these nonlinear mechanisms, and perhaps others such as two-photon absorption, is very sensitive to small perturbations of the pulse properties, resulting in large variations among successive pulses. This condition caused confusion and lead to contradictory observations—especially of effects which were cumulative over the pulse train.

4.4.2 Early studies

Many studies of mode-locked pulses made use of the intensity autocorrelation techniques described by Armstrong (1967), Weber (1967) and Giordmaine et al. (1967). Two difficulties were encountered with these measurements which indicated that the pulses from a mode-locked Nd:glass laser were more complicated than a perfectly mode-locked pulse.

The first difficulty, reported by many authors including Armstrong (1967), Giordmaine et al. (1967), Glenn and Brienza (1967), Klauder et al. (1968) and Kachen et al. (1968), was that the pulse duration appeared to be about ten times longer than would be inferred from the observed pulse bandwidth, which suggested that the pulse contained short-duration substructure. The pulse duration also varied during development of the pulse train.

The second difficulty was the low value of the contrast ratio measured in the two-photon fluorescence (TPF) experiments, which suggested imperfect mode locking. Several models were advanced for imperfect mode locking including random phase variations among the modes by Harrach (1969), Picard and Schweitzer (1969) and Grütter et al. (1969); systematic mode phase variations by Cubeddu and Svelto (1969), Smith (1969), Svelto (1970) and Treacy (1970); and a 'domain model' by Picard and Schweitzer (1970). These models could explain the broad pulse bandwidth but not the low TPF contrast ratio. Treacy (1968, 1969) showed the pulses were frequency-chirped by compressing them using a dispersive delay line. At the same time, subpicosecond structure was detected without the dispersive delay line by Bradley et al. (1969), Shapiro and Duguay (1969) and Eckardt and Lee (1969). Many other papers have addressed these two difficulties. Complete reviews are given by DeMaria et al. (1969), Duguay et al. (1970) and Greenhow and Schmidt (1974).

4.4.3 Nonlinear mechanisms

In attempts to resolve some of these issues, Duguay et al. (1970) found their pulse envelope, whose duration was 8 ps, contained internal structure of 0.4 to 0.8 ps duration. Furthermore, they found the pulse's spectral width expanded from $<20 \, \text{cm}^{-1}$ at the beginning of the pulse train to $\sim 80 \, \text{cm}^{-1}$ near the envelope maximum. The rapid spectral broadening was attributed to self-phase modulation due to the nonlinear refractive index n_2, which they further evaluated as $n_2 = (2 \pm 1) \times 10^{-22} m^2/V^2 (1.8 \times 10^{-13} \, \text{esu})$. The nonlinear index results in a frequency shift $\Delta \nu$ impressed on the wave after passing through the length l which is given by

$$\Delta \nu = -\frac{l n_2}{\lambda} \frac{\delta}{\delta t} E \cdot E. \tag{61}$$

Although the appropriate value of δt is uncertain since the pulse shape is changing, using the values $\delta t = 1$ ps, power density $= 1 \, \text{GW/cm}^2$, $l = 40$ cm, $\lambda = 1.06 \, \mu\text{m}$, eq. (61) gives $\Delta \nu = 10 \, \text{cm}^{-1}$. This result agrees well with the observed mean spectral broadening per round trip of $5 \, \text{cm}^{-1}$.

The important points were that the spectral broadening occurred after the dye had Q-switched and mode-locked the laser and that the nonlinear index of the glass n_2 is sufficiently large to account for the observed broadening. In addition, the group velocity dispersion of the laser glass is large enough to convert the FM structure to AM structure over typical propagation distances, which was also noted by Letokhov and Morozov (1967) and Armstrong and Courtens (1969). For example, a short pulse at $1.06 \, \mu\text{m}$ is delayed by 1 ps, after traversing 1 m of glass, compared to a pulse whose center frequency is $100 \, \text{cm}^{-1}$ lower. Combination of these effects with self-focusing and multiphoton absorption created a complicated picture of the mode-locked pulse development.

Finally, von der Linde et al. (1970) and von der Linde (1972) carefully investigated a single pulse switched out from various positions in the pulse train and then amplified. The temporal structure was studied by the TPF method and the pulse's spectrum was recorded using a 2 m spectrograph. The energy distribution was determined by photoelectric measurements, quantitative measurement of the TPF contrast ratio and three-photon fluorescence efficiency measurements. Fig. 20 shows an example of the pulse width and bandwidth of a 0.2 mJ pulse selected from the leading part of the train envelope. These pulses were found to be approximately bandwidth-limited, with the average measured pulse width τ_p of 5 ps and bandwidth $\Delta \nu_p$ of $4 \, \text{cm}^{-1}$. The product $p = \tau_p \Delta \nu_p = 0.6$, compared with $p = 0.441$ for gaussian pulses, allows for only a small amount of frequency modulation. In addition, the TPF contrast ratio of 3.0 gave no evidence of substructure. Fig. 21 shows the very different results obtained for pulses selected beyond the envelope maximum. The TPF trace contained a sharp subpicosecond component and the frequency spectrum, with a half-power width of $50 \, \text{cm}^{-1}$, exhibited strong modulation.

Bradley and Sibbett (1975) confirmed by direct linear measurement that pulses

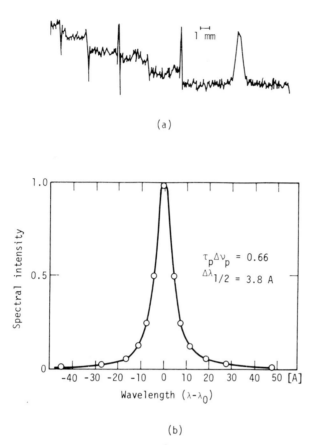

(a)

(b)

Fig. 20. Experimentally measured TPF pattern (a) and spectrum (b) of two different pulses switched from the front of the pulse train. The value of τ_p was calculated to be 4 ps. In (b) the open circles are measured points corresponding to $(1/2)^n$ of the maximum. The solid curve was determined using the measured characteristic curve of the photographic plate. From von der Linde (1972).

of transform-limited duration are produced at the beginning of the pulse train. They used an infrared sensitive streak camera of ~ 2.5 ps time resolution to measure the pulse duration and a 1 m spectrograph to record the spectrum. Measuring the fifth observable pulse in the train, they found the average values for a twenty-shot sample of $\tau_p = 3.8 \pm 0.8$ ps and $\Delta\nu_p = 7.3 \pm 1.7$ cm^{-1}, giving a value for the product $\tau_p\Delta\nu_p = 0.83 \pm 0.3$. These values were obtained for pulse energies up to 1 mJ. They also observed that later in the train, self-phase modulation increased the spectral width by ~ 10 cm^{-1} per round trip producing half-power spectral widths of 50 cm^{-1}.

More recent work on Nd:glass has confirmed and clarified the roles played by the major nonlinear mechanisms: self-focusing, self-phase modulation and also two-photon absorption on development of the mode-locked pulse. Eckardt et al. (1971, 1974) obtained time-resolved spectrograms of a Nd:glass ring oscillator

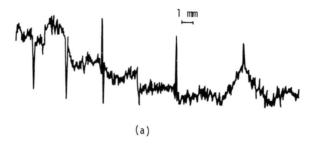

(a)

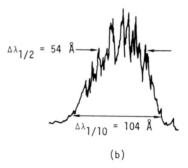

(b)

Fig. 21. TPF pattern (a) and spectrum (b) of a pulse switched from the trailing part of the pulse train. In (a) the widths of the base and the sharp components are 11 ps and 0.7 ps respectively. From von der Linde (1972).

which showed spectral broadening in agreement with earlier results. Pulse propagation including the effects of amplification, saturable absorption, self-phase modulation and dispersion was numerically simulated. Fig. 22 shows the intensity and phase of two pulses selected from different parts of the pulse train. The calculations made use of the parameters given in table 3.

Miroñov and Shatberashvili (1974) photographed the spectrum of a single pulse selected from different parts of the pulse train. Their results clearly showed the increasing width and complexity of the spectrum as the pulse train develops.

Self-focusing also leads to complicated temporal structure. Fig. 23 illustrates that stronger focusing of the more intense portions of the pulse allows them to escape from the resonator. Zherikhin et al. (1974b) calculated the pulse amplitude development including the effects of self-focusing and self-phase modulation in addition to dispersion and gain narrowing. Fig. 24 shows the pulse amplitude and second-harmonic enhancement ratio after each pass as well as the pulse shapes at several positions in the pulse train.

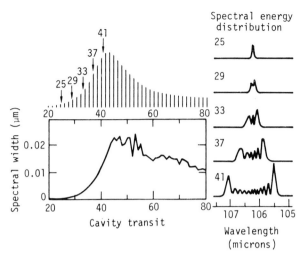

Fig. 22. Calculated spectral evolution of the Nd:glass laser mode-locked pulse. Upper left: schematic diagram of the calculated pulse intensity as a function of pass number. Lower left: spectral width. Right: spectral intensity for selected pulses. From Eckardt et al. (1974).

Table 3
Representative parameter values for Nd:glass laser.

Gain medium   ED-2 Glass
   Cross section $\sigma_a = 3.03 \times 10^{-20}$ cm^2
   Fluorescence linewidth $\Delta\nu$(fwhm) = 260 Å

Saturable absorber   Kodak A9860
   Cross section $\sigma_b = 3.7 \times 10^{-16}$ cm^2
   Saturation intensity $I_s = 5.6 \times 10^7$ W/cm^2
   Relaxation time $\tau_D = 6$ ps

Self-phase modulation
   Nonlinear index $n_2 = 2 \times 10^{-22}$ m^2 V^{-1}
   Relaxation – instantaneous

Dispersion
   Index $\quad n = \left(A + \dfrac{B}{\lambda^2 - C} - D\lambda^2 \right)^{1/2}$

     $A = 2.41870$
     $B = 0.01355$
     $C = 0.01520$
     $D = 0.00960$

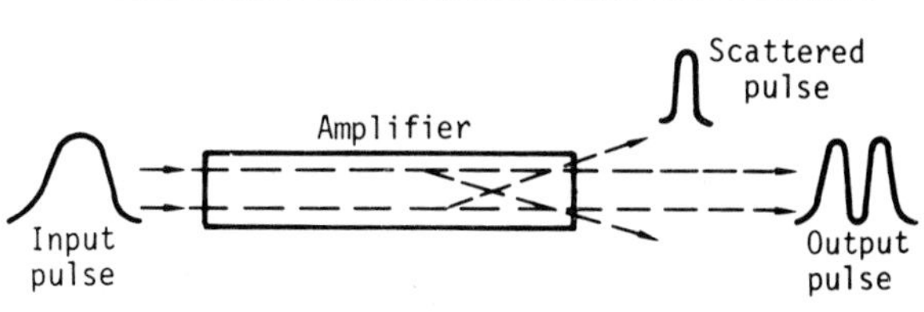

Fig. 23. Illustration of the role played by self-focusing in creating complicated time structure.

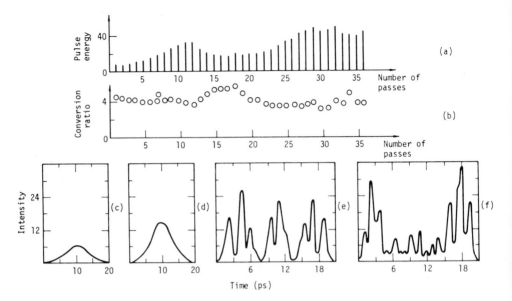

Fig. 24. Calculated dependence of (a) pulse energy and (b) SHER on number of passes through resonator. Initial pulse shape (c) and calculated shapes after 10 (d), 25 (e), and 35 (f) passes for initial pulse energy of 10^{-5} J. From Zherikhin et al. (1974b).

Eckardt et al. (1974) studied the effect of self-focusing alone using an iterative Fraunhofer diffraction calculation. Fig. 25 shows that the pulse intensity increases slowly at first, then more rapidly as the dye bleaches. At intensities of about 5×10^8 W/cm^2, self-focusing severely distorts the transverse field distribution and limits further increase in intensity. After self-focusing has occurred, the pulse rapidly deteriorates into a burst of noise. For higher-order transverse modes, the pulse deteriorated at an intensity of a few times 10^9 W/cm^2.

Two-photon absorption (TPA) in the laser rod may also be important at these intensities. The TPA loss is described by the equation $dI/dx = -\gamma I^2$, where $\gamma = N\sigma_2$ is the product of the neodymium ion concentration N and the TPA cross section σ_2. Penzkofer and Kaiser (1972) measured and calculated the value $\sigma_2 = 1.3 \times 10^{-32}$ cm^4/W for the Schott laser glass LG-630, which contains 3% by weight Nd$_2$O$_3$ ($N = 3.2 \times 10^{-20}$ cm^{-3}). As an example, the TPA loss in a 15 cm rod for a pulse of intensity 2×10^9 W/cm^2 is 12%.

TPA was introduced by Caruso et al. (1973) into the rate equations for mode-locked pulse amplification in a ring oscillator. It was shown to be effective in limiting the calculated peak intensity to $2-4 \times 10^9$ W/cm^2 in agreement with measurements. Nonlinear loss measurements by Speck and Bliss (1973) on Owens-Illinois ED-2 glass gave a value for the cross section σ_2 that was 0.1–0.2 times the value given above.

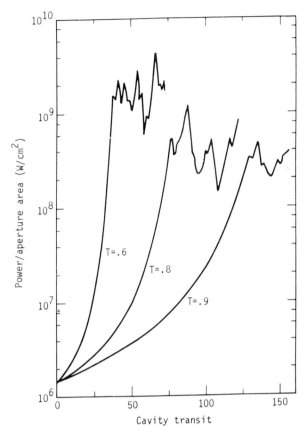

Fig. 25. Calculated power development of laser oscillation with self-focusing in plane-parallel resonator cavity (Fresnel number $N = 5$) for different values of initial dye transmission. From Eckardt et al. (1974).

4.5 New materials

While the nonlinear index n_2 affects the mode-locked oscillator through self-focusing and self-phase modulation, it also limits the performance of high-power, Nd:glass laser systems. Self-focusing, for example, reduces the fraction of a pulse's energy which can be focused to a small spot, and it can even result in damage to the laser. The n_2 coefficient and other important laser parameters such as the emission cross section and fluorescence lifetime depend on properties of the host material. To optimize these parameters, measurements of optical and spectroscopic properties of many different glasses have been made by Jacobs and Weber (1975, 1976), Milam and Weber (1976), Weber et al. (1976) and Weber (1976). Most laser glass used up to the present has been oxide glass, principally silicates and phosphates. These recent results show the low-index

Table 4

Comparison of the refractive indices and spectroscopic properties of the $^4F_{3/2} \rightarrow \,^4I_{11/2}$ transition of
Nd^{3+} in oxide and fluoride glasses.[a,b]

	Silicate	Phosphate	Fluoro-silicate	Fluoro-phosphate	Fluoro-beryllate
Refractive index n_D	1.567	1.507	1.480	1.492	1.347
Nonlinear index $n_2[10^{-13}$ esu]	1.4	0.91	0.9	0.8	~0.4 (est)
Cross section $\sigma[10^{-20}$ cm^2]	2.9	4.7	1.9	2.5	2.9
Linewidth – fwhm [nm]	29	19	27	26	20
Branching ratio	0.47	0.47	0.48	0.49	0.50
Fluorescence peak [nm]	1062	1054	1057	1054	1047
Radiative lifetime [μs]	340	330	510	460	600
Relative optical absorption efficiency	1.0	0.78	0.85	0.73	0.81

[a] Glass compositions (mol.%): Silicate – 60 SiO$_2$, 27.5 Li$_2$O, 10 CaO, 2.5 Al$_2$O$_3$
Phosphate – 50 P$_2$O$_5$, 33K$_2$O, 17 BaO
Fluorosilicate – SiO$_2$, K$_2$O, B$_2$O$_3$, F$_2$
Fluorophosphate – P$_2$O$_5$, Al$_2$O$_3$, MgF$_2$, CaF$_2$, SrF$_2$, BaF$_2$
Fluoroberyllate – 48 BeF$_2$, 27 KF, 15 caF$_2$, 10 AlF$_3$.
[b] From Weber et al. (1976).

fluoride glasses to be promising for high-power Nd lasers. Table 4 presents properties of several representative oxide and fluoride glasses.

These and other considerations have stimulated interest in generating stable mode-locked pulses at new wavelengths and using new materials. McMahon (1976) reported passively mode-locking a Nd:YAG oscillator on its 1.052 μm transition. Intracavity prisms provided sufficient frequency discrimination to allow complete suppression of the 1.064 μm and the 1.0614 μm transitions. No self-focusing effects were noted with total pulse-train energies of 10 to 15 mJ.

Two compositions of Nd:phosphate glass, LHG-5 from Hoya and EV-2 from Owens-Illinois, have been passively mode-locked by Fleming et al. (1977) using the Kodak A9860 dye. The reliability of pulse generation appeared about the same as with silicate glass. A single mode-locked pulse at 1.054 μm with energy of 1 to 2 mJ was switched out, amplified and photographed using a streak camera. Pulse durations of 6 ps were recorded with an average bandwidth $\Delta \nu_p = 46$ cm^{-1}. The product $\tau_p \Delta \nu_p = 8.3$ indicated that the pulse durations were not transform-limited. Veduta et al. (1974) reported reproducible passive mode locking a Nd:niobium phosphate glass to generate 20 ps pulses with bandwidths of 1.2 to 1.5 cm^{-1}.

Neodymium-doped calcium lanthanum silicate oxyapatite (Nd:CaLaSOAP) has an induced-emission cross section $\sigma_a \simeq 1.6 \times 10^{-19}$ cm^2 and fluorescence bandwidth of 45 cm^{-1} centered at 1.06 μm. These values are intermediate between those for Nd:silicate glass and Nd:YAG, which make it an interesting oscillator material. Passive mode locking with Kodak A9740 dye was reported by Eckardt et al. (1974). Pulses in the early part of the pulse train appeared bandwidth-limited with durations of 9 to 11 ps. Time-resolved spectra showed a broadening from 2 cm^{-1} to 25 cm^{-1} near the peak of the pulse-train envelope where the pulse energy was typically $8 \times 10^{-5} \text{ J}$.

Goldberg and Bradford (1976) passively mode-locked Nd:lanthanum beryllate (Nd:BEL), a biaxial, monoclinic crystal shown by Smith and Bechtel (1976) to have greater resistance to optical damage and lower n_2 than Nd:YAG. The material has two linearly polarized transitions for different crystal orientations: one at $1.070 \mu\text{m}$ with $\sigma_a = 2.1 \times 10^{-19} \text{ cm}^2$ and the other at $1.079 \mu\text{m}$ with $\sigma_a = 1.5 \times 10^{-19} \text{ cm}^2$. The fluorescence linewidth for both transitions is about 30 cm^{-1} at room temperature. More stable mode locking was observed for the higher-gain $1.070 \mu\text{m}$ transition. Spectral broadening from 2 cm^{-1} (fwhm) early in the pulse train to $\geq 20 \text{ cm}^{-1}$ in the later part was observed at representative peak pulse energies of $250 \mu\text{J}$ and durations of 15 ps.

5 Passive mode locking of dye lasers

The mode-locked pulse in a solid-state laser develops in several hundred transits of the cavity. There the basic role of the saturable absorber is to select a single intense fluctuation pulse. To effectively discriminate against secondary pulses, the recovery time of the dye cannot be much longer than the pulse duration, as discussed in § 4.3. Gain saturation in the solid-state laser acts to improve the discrimination between the largest and smaller fluctuation pulses but does not play a role in shaping the final pulse.

The mode-locked pulse in a dye laser is formed in a different manner. Studies by Arthurs et al. (1973. 1974b) showed the pulse develops very rapidly from a burst of spontaneous noise and the final pulse duration is much shorter than the recovery time of the saturable absorber used. New (1972) proposed that the pulse in a dye laser is generated by the combined action of the saturable absorber and saturation of the laser's gain. The basic principles of the model, called the 'quasi-continuous model', are reviewed in the following section.

5.1 Quasi-continuous model

Since the recovery time τ_R of the laser dye molecule is comparable to the cavity period, the pulsed dye laser operates in a near equilibrium, or quasi-continuous condition. The transition cross section of the amplifying medium is sufficiently large that significant gain saturation can occur on a single pass through the medium even at pulse intensities which are below damage thresholds. Using rate equations, New (1972), 1974) showed that an operating regime exists in which a pulse circulating in the cavity experiences a net loss on its leading edge owing to the saturable absorption and also on its trailing edge owing to saturation of the gain, but with no net loss of energy. This condition leads to rapid compression of the pulse even when its duration is much less than the absorber's recovery time.

The pulse compression requires a delicate balance between amplification and the two saturation mechanisms. The important parameters in determining the stability of compression are the unsaturated amplifier gain α and absorber loss β, the round-trip linear loss γ (the product $\alpha\beta\gamma$ is the net, small-signal, round-trip gain g_0), the ratio of gain to absorber cross section $s = \sigma_a/\sigma_b$ and the ratio $\xi = T/\tau_R$ of the cavity period to the gain recovery time. Fig. 26 illustrates the calculated regions of stability for the case $\beta = 0.2$, $\gamma = 0.4$. New found that a stable region does not exist for values of s much smaller than 2. This limiting case is indicated in fig. 26 by dotted lines. Qualitatively, the boundaries of the stable region are determined by two requirements. Initially, gain must exceed loss so the pulse can grow, but in the quasi-steady state, when significant single-pass gain saturation occurs, the leading edge of the pulse must experience net loss. Thus the amplifier gain must not completely recover in a cavity period. This requirement sets an upper limit on the length of the resonator. However, if the resonator is too short to allow sufficient gain recovery, the inverted population is rapidly extracted and the pulse does not make enough passes through the amplifier for the saturation mechanism to shorten the trailing edge of the pulse. The range of resonator lengths over which pulse compression occurs is determined by the gain and loss parameters of the laser. Outside these stable regions, long pulses or multiple pulses are generated.

Yasa et al. (1976) retained the relaxation times of the active and passive dyes in their rate-equation calculations of the net, time-dependent gain, and later Yasa et al. (1977) presented numerical simulations of the pulse development beginning

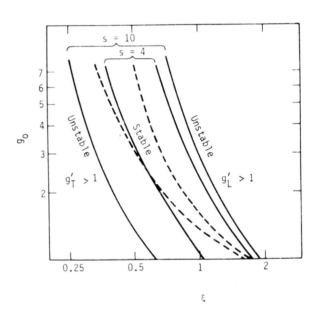

Fig. 26. Stable and unstable regions as function of g_0 and ξ for three values of s, for case where $\beta = 0.2$ and $\gamma = 0.4$. Dotted lines denote limiting case $s = 2$. From New (1972).

with an exponential distribution of random noise pulses. Their results show that with proper choice of the system parameters the combination of nonlinear gain and loss produces net gain only in a small region about one of the most intense noise spikes.

New and Rea (1976) later studied the stability of the steady-state solutions to small perturbations. Using similar perturbation techniques, Garside and Lim (1973a, b, 1974a, b) predicted stable mode-locking conditions for cw laser operation and presented numerical solutions for the steady-state pulses. In their calculations, coherent interactions between the active medium and the laser pulse were included using a density matrix formalism

Since the quasi-continuous model of New (1972) contained no bandwidth limitation, the pulse compression would continue indefinitely. In recent theoretical studies, Haus (1975b, c) introduced a bandwidth limitation on the gain into a rate-equation analysis of passive mode locking. The bandwidth-limiting element was shown to be equivalent to a diffusion operator in the time domain which acts to broaden the pulse. When this broadening is balanced against the pulse compression, a closed-form solution may be obtained for the pulse shape which was found to be a secant hyperbolic. The pulse intensity $I(t)$ is proportional to

$$I(t) \propto 1/\cosh^2(t/\tau_p), \tag{62}$$

where the pulse width τ_p is inversely proportional to the energy in the pulse, the gain bandwidth and the unsaturated absorber loss.

Experimental measurements were made by Haus et al. (1975) of the cw mode-locked dye laser's pulse shape using a background-free autocorrelation technique. When the laser was well above threshold, relatively long pulses were obtained with considerable substructure, but the pulse envelope was well described by the secant hyperbolic. When the laser was near threshold, a shorter pulse was produced. Although the pulse did exhibit exponential wings as predicted by the theory, its correlation function did not correspond to a secant hyperbolic for which several possible explanations were advanced.

New et al. (1976) derived two mathematical functions which characterize the stability of single- and multiple-pulse operation in a ring laser. These functions allow determination of the longest and shortest cavity transit times for which single and multiple pulses can be supported as well as qualitative estimates of the pulse durations.

5.2 Experimental studies

5.2.1 CW mode-locked laser

The cw mode-locked dye laser has been the subject of many recent dye laser experimental studies. Passive mode locking of the cw rhodamine 6G dye laser was first reported by Ippen et al. (1972), O'Neil (1972) and Arthurs et al. (1972).

Fig. 27 shows the optical resonator of Ippen et al., a five-mirror arrangement which is an extension of the resonator with astigmatic compensation described by Kogelink et al. (1972) and Dienes et al. (1972). The beam from a continuous argon laser at 514.5 nm was coupled into the resonator through a quartz prism and focused into the cell containing rhodamine 6G. Asimilar cell at the other end of the resonator contained the mode-locking dye DODCI. With this laser, continuous trains of pulses as short as 1.5 ps were observed.

The experimental and theoretical studies of pulsed and cw mode-locked dye lasers through 1973 have been reviewed by Shank and Ippen (1973), Bradley (1974), Dienes (1974) and Smith et al. (1974).

Recent experiments on mode-locked cw dye lasers have largely been devoted to obtaining the shortest possible pulse durations. Shank and Ippen (1974) mixed the active medium, rhodamine 6G and the saturable absorber, DODCI, together in a single free-flowing ethylene glycol stream. Pulse durations, measured by autocorrelation, were found to be 0.5 to 1 ps. Peak pulse powers of several kilowatts were obtained by using a pulsed acousto-optic output coupler to dump the resonator cavity at repetition rates up to 10^5 Hz.

Ippen and Shank (1975) reported mode-locking a cw dye laser using a combination of two saturable absorbers which improved the pulse-generation stability. Two free-flowing dye streams were used; one contained rhodamine 6G and the other contained a mixture of DODCI and malachite green. Operation with only DODCI, which has a long recovery time, occurred near threshold, and thus required careful adjustment and stabilization of the laser parameters, while malachite green, whose recovery time is in the picosecond range, did not produce stable mode locking in their laser. Using the dye mixture, the output pulse characteristics remained constant without readjustments for several hours.

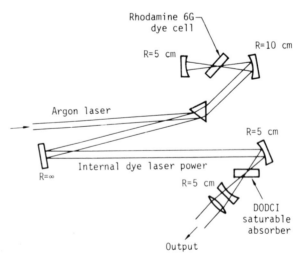

Fig. 27. Resonator configuration used to passively mode-lock a cw dye laser. From Ippen et al. (1972).

Measurements of the pulse's frequency spectrum showed an excess bandwidth and suggested the possibility of a frequency chirp. Direct temporal compression to pulse widths of about 0.3 ps was achieved using the grating-pair technique described by Treacy (1969).

A cw rhodamine 6G dye laser was mode-locked using a thin, optically contacted saturable absorber cell containing DODCI by Ruddock and Bradley (1976). The best fit to the SHG autocorrelation function was obtained for secant-hyperbolic-squared pulses of 0.3 ps duration. Simultaneous measurement of the oscillating bandwidth showed the pulse's duration was nearly bandwidth-limited. The product $\tau_p \Delta \nu_p = 0.45$ was obtained in comparison with the ideal value of 0.32 for $sech^2$ pulses. Pulses of duration 0.3 to 0.9 ps were obtained over the spectral region 598 to 615 nm.

5.2.2 Double mode locking

A new approach for generating ultrashort pulses was recently introduced by Yasa (1975). The technique, called double mode locking, is similar to work reported by Runge (1971) on a He–Ne laser mode-locked by cresyl violet. Two laser dyes are used with the absorption spectrum of one overlapping the fluorescence spectrum of the other. The basic idea is that the saturable absorber dye, also a laser dye with a long relaxation time, can become inverted by absorption of the main pulse. Subsequent stimulated emission from the inverted absorber dye then rapidly restores the absorber loss at the trailing edge of the main pulse. As a result, the main pulse experiences greater compression than in the conventional, single-mode-locked case which relies solely on gain saturation to sharpen the trailing edge of the main pulse. A by-product of double mode locking is the generation of a secondary short pulse at a longer wavelength than the main pulse. Fig. 28 illustrates the basic concepts of both single and double mode locking.

Double mode locking a rhodamine 6G and cresyl violet mixture pumped by a cw argon laser was demonstrated by Yasa and Teschke (1975a). Two short pulses were generated with wavelengths of 574 nm and 644 nm. The pulse shapes and spectra were measured by Yasa et al. (1976) who found the best fit to the SHG autocorrelation function was obtained by assuming pulses with $sech^2$ intensity distribution. The rhodamine 6G pulse duration decreased as the pump power increased. Pulse durations as short as 0.55 ps were recorded with a product $\tau_p \Delta \nu_p = 0.73$.

5.2.3 Synchronous pumping

Picosecond pulses can also be generated by pumping a laser dye with a pulse from another mode-locked laser. Even without a laser resonator, ultrashort pulses of superfluorescent emission can be obtained in this manner as demonstrated by Mack (1969) and more recently by Rubinov et al. (1975). The duration

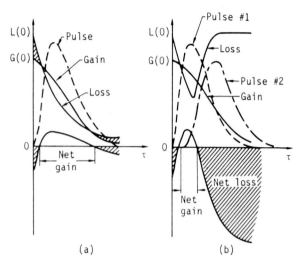

Fig. 28. Nonlinear gain and loss experienced by a pulse for (a) conventional mode locking, (b) double mode locking. The net gain duration is much shorter in (b) than in (a). From Yasa (1975).

of the emission may be several picoseconds with typically broad bandwidths of a few hundred cm^{-1}.

The dye laser can be mode-locked by the periodic gain modulation which results from pumping with a mode-locked pulse train. To produce mode locking, the dye laser cavity length must be equal to, or a submultiple of, the cavity length of the mode-locked pump laser. This technique of mode locking by synchronous pumping was first applied to rhodamine 6G and rhodamine B dye lasers pumped by a frequency-doubled, mode-locked Nd:glass laser by Glenn et al. (1968) and Soffer and Linn (1968). Other early work is reviewed by Shank and Ippen (1973). Numerical studies of mode locking by synchronous pumping using a rate-equation analysis have been presented by Yasa and Teschke (1975b).

Royt et al. (1974) used a streak camera to study the temporal properties of rhodamine 6G synchronously pumped by frequency-doubled Nd:glass. The pulse durations varied from 15 to 35 ps, with each pulse about 50% longer than the coincident pump pulse and delayed from it by 4 to 12 ps. They found severe variations in the pulse character for deviations from the optimum cavity length as small as 0.1 mm. Doubled Nd:YAG was used by Goldberg and Moore (1975) to pump rhodamine 6G, rhodamine B, cresyl violet perchlorate and carbazine 122. The dye laser incorporated Fabry–Perot tuning elements which allowed generation of ~30 ps pulses over the tuning range from 549 to 727 nm.

A continuous train of pulses from rhodamine 6G pumped by an actively mode-locked cw argon laser was referred to by Shank and Ippen (1973). Using this method, Chan and Sari (1974) generated pulses of 3 ps duration which were tunable from 560 to 610 nm with peak power up to 50 W. A similar system with the addition of an acousto-optic output coupler was discussed by Harris et al. (1975).

6 Combinations of active and passive mode locking

The discussions of the preceding section have shown that pulses generated by active or passive mode locking have different characteristic properties. Passively mode-locked pulses can have durations close to the limit set by the width of the gain profile. But since these pulses are developed from fluorescence noise under transient conditions, the pulse's duration and amplitude exhibit statistical fluctuations. Actively mode-locked pulses are highly reproducible, but their minimum duration is much greater than the gain-bandwidth limit. Attempts to generate stable minimum-duration pulses by adding a saturable absorber to an actively mode-locked laser have been made by several groups with similar success.

Krivoshchekov et al. (1975) placed a cell of cryptocyanine with small signal transmission of 40% in a ruby ring laser mode-locked by a LiNbO$_3$ electro-optic loss modulator. The cavity Q was actively controlled by a separate KDP modulator to produce a $\sim 1.7\,\mu s$ 'pre-lase' period before the Q-switch was completely opened. Without the saturable absorber, pulses of $\sim 100\,\text{ps}$ duration were generated for which $\tau_p \Delta \nu_p = 2.7$. With the saturable absorber, a 5 ps pulse duration was inferred from the expanded oscillating bandwidth.

Similar operation of a pulsed Nd:glass ring laser was reported by Tomov et al. (1977). With no saturable absorber in the cavity, the pulse duration was observed to decrease from 1.3 ns to 200 ps with increasing depth of modulation θ_A from 0.2 to 0.7, in approximate agreement with the transient buildup theory of Kuizenga et al. (1973). When a cell of Kodak A9740 with small-signal transmission of 80% was placed in the cavity, the pulse duration reduced to $\sim 15\,\text{ps}$ with $\Delta\nu_p \sim 2.4\,\text{cm}^{-1}$ ($\tau_p \Delta\nu_p = 0.96$) and energy $\sim 0.2\,\text{mJ}$.

Washio et al. (1977) mode-locked a cw SELFOC Nd:glass laser using the combination of a saturable absorber in a free-flowing dye stream and an acousto-optic modulator. To raise the pulse's peak power to near the saturation flux of the dye, the beam was strongly focused on the dye stream and relaxation oscillations were induced by pulsing the modulator. Under simultaneous spiking and active mode-locking conditions without the saturable absorber, the pulse envelope was 230 ps long and contained short-duration substructure. When dye was added, the pulse duration decreased to about 12 ps. The peak power of the mode-locked pulse was about 70 W.

7 Additional techniques for generating and shaping ultrashort pulses

This review is concluded by a brief discussion of some additional, linear and nonlinear optical techniques which have been used to generate or shape short-duration light pulses. More detailed information can be obtained from the original references cited in each case.

Reduction of spectral narrowing

During the linear amplification period of passive mode locking, the fluctuation pulse durations increase owing to spectral narrowing imposed by the frequency-dependent gain profile. An intracavity etalon with a dip in transmission at the center of the gain line can reduce the spectral narrowing, leading to generation of shorter pulses. Using an etalon in a ruby laser, McGeoch (1973) was able to reduce the pulse duration from 32 to 22 ps. Similar results were recorded by Varnavskii et al. (1975). Von der Linde and Rodgers (1973b) further analyzed the technique and generated bandwidth-limited pulses as short as 1.5 ps with a Nd:glass laser.

Pulse compression

A time-varying carrier frequency (chirp) can be induced on an optical pulse by passing it through a synchronously driven phase modulator or through a medium with a nonlinear refractive index. If the pulse is then passed through a dispersive delay line, such as a grating pair, it will be compressed in time if the carrier frequencies present at the leading edge of the pulse experience the greatest delay. The technique has been discussed and demonstrated by numerous authors. Duguay and Hansen (1969) compressed mode-locked He–Ne pulses from 500 to 270 ps using an interferometer proposed by Gires and Tournois (1964). Treacy (1968, 1969), using a grating pair, compressed mode-locked Nd:glass laser pulses from 5 to 0.4 ps. Laubereau (1969) and Laubereau and von der Linde (1970) demonstrated compression by a factor of 10 for 20 ps pulses from a Nd:glass laser which were chirped by self-phase modulation in carbon disulfide. The same technique was used recently by Lehmberg and McMahon (1976) to compress 100 ps pulses from a Nd:YAG laser by a factor ≥ 12. Chirped dye laser pulses were compressed by Grischkowsky (1974) taking advantage of the anomalous dispersion near an atomic resonance.

A cell containing a saturable absorber outside the laser resonator can act to shorten a light pulse by preferentially absorbing the leading and trailing edges. Use of this technique was studied by Penzkofer et al. (1972), Kryukov et al. (1972b) and Penzkofer (1974), who reported the duration of a mode-locked Nd:glass laser pulse was reduced from 8 to 2.6 ps by a single pass through a dye cell with small-signal transmission 10^{-7}. Pulses as short as 0.5 ps were generated by a series of saturable-absorber cells and amplifiers. Herrmann et al. (1975) proposed pulse shaping by a combination of two absorbers: the first, a common single-photon saturable absorber, mainly steepens the leading edge of the pulse, while the second absorber shortens the trailing edge of the pulse by single-photon absorption from an excited level which is populated by a two-photon absorption. Their calculations, assuming reasonable absorption cross sections, predict single-pass compression by a factor of 5 with peak intensity transmission of 77%.

A saturable absorber may also be used to suppress the background intensity between the pulses of a mode-locked pulse train. In this application, the intensity of the main pulse should be much greater than the saturation intensity so it experiences a high transmission while the low-intensity background is strongly attenuated. The performance of a sequence of dye cells and amplifiers as such a 'discrimination amplifier' was simulated numerically by Seka and Stüssi (1976). A relative background suppression of 2 000 was calculated with 84% transmission of the main pulse of intensity $I/I_s \sim 100$.

Background suppression and pulse shortening have also been accomplished by injecting an ultrashort pulse into a resonator which contains an amplifier and a saturable absorber. The behavior of this arrangement, which is called a stable two-component amplifier (STA) or a regenerative amplifier, has been discussed by Basov et al. (1969), Bykovskii et al. 1972) and Zherikhin et al. (1974c).

Stimulated Raman scattering

A transient stimulated Raman oscillator was used by Colles (1971) to generate Stokes pulses with durations as short as 0.3 ps. The Raman active medium, benzene, was pumped by a train of ultrashort pulses from a mode-locked Nd:glass oscillator. The Raman oscillator's length was adjusted so that a single Stokes pulse was amplified by successive pump pulses. This synchronous amplification arrangement allowed the group velocity mismatch to be compensated on each pass.

Bradley and Sibbett (1975) used transient stimulated Raman scattering in ethanol to generate a Stokes pulse whose duration they measured to be 0.57 ps. A streak camera was used to measure the duration of the Stokes pulse which was excited by the 1.5 ps pulse from a rhodamine 6G dye laser.

Pulses of ~ 30 ps duration were generated by backward Raman amplification of scattering initiated by self-focusing in CS_2 by Maier et al. (1969). The short-pulse generation in this case results from saturation of the forward-traveling pump pulse. This technique may prove to be useful for compressing the energy in a long-duration, low-intensity light pulse into a short-duration, high-intensity pulse.

Laser breakdown switch

There are a number of techniques for generating short optical pulses which rely on laser-induced gas breakdown. Reflection of laser light from the plasma created by gas breakdown can terminate the transmitted pulse with a fall time in the picosecond range, and several methods have been used to convert this relatively long pulse with a sharp trailing edge into a short optical pulse. Milam et al. (1974) made use of a Michelson interferometer and Yablonovitch (1974) used a grating monochrometer. Yablonovitch and Goldhar (1974) generated a

CO_2 laser pulse by optical free induction decay whose duration was later measured by Kowk and Yablonovitch (1977) to be 30 ps. Yablonovitch (1975) proposed that the problem of generating a short optical pulse of arbitrary shape and phase may be treated as that of designing an appropriate filter for the rich spectrum of the step-function pulse.

Picosecond opto-electronic switch

Auston (1975) demonstrated that the photoconductivity produced by absorption of picosecond optical pulses in a silicon transmission line structure could be used to perform elementary switching and gating functions. The techniques were extended by LeFur and Auston (1976) to switching an electrical signal as large as 1.5 kV. The signal was used to drive a traveling-wave $LiTaO_3$ Pockels cell with a measured risetime of 25 ps which can transmit a short-duration segment of a longer light pulse.

A final comment

The goal of this chapter was to review the theoretical models for the processes by which ultrashort pulses are generated and also to provide a comprehensive overview of recent experimental research on generation techniques and properties of ultrashort pulses—in particular, those which bear on the validity of these models. Although the treatment is certainly incomplete in several aspects, it has established that the basic dynamic interactions which lead to generation of ultrashort pulses are now reasonably well understood in the sense that the model predictions are in good agreement with experimental results. The task of further research will be to develop techniques for meeting specific experimental requirements such as those of stringent reliability over a wide range of operating parameters, subpicosecond pulse widths, particular or tunable wavelengths, ultrashort wavelengths and synchronizable sources. These and other goals continue to make studies of ultrashort pulse generation techniques interesting and fruitful.

Acknowledgment

The author is indebted to D.J. Kuizenga and J.E. Murray for numerous stimulating discussions on mode locking and for their careful review of the manuscript. He is also particularly grateful to K.J. Wenzinger for her generous and skillful effort in preparing the manuscript for publication.

References

Al-Obaidi, H., R.J. Dewhurst, D. Jacoby, G.A. Oldershaw and S.A. Ramsden (1975), Opt. Commun. **14**, 219.

Ammann, E.O. and J.M. Yarborough (1972), Appl. Phys. Letters **20**, 117.
Ammann, E.O., B.J. McMurtry and M.K. Oshman (1965), IEEE J. Quantum Electronics **QE-1**, 263.
Armstrong, J.A. (1967), Appl. Phys. Letters **10**, 16.
Armstrong, J.A. and E. Courtens (1969), IEEE J. Quantum Electronics **QE-5**, 249.
Arthurs, E.G., D.J. Bradley, B. Liddy, F. O'Neill, A.G. Roddie, W. Sibbett and W.E. Sleat (1972), in: Proc. 10th International Congress on High-Speed Photography (A.N.R.T.; Paris), 117.
Arthurs, E.G., D.J. Bradley and A.G. Roddie (1973), Appl. Phys. Letters **23**, 88.
Arthurs, E.G., D.J. Bradley and T.J. Glynn (1974a), Opt. Commun. **12**, 136.
Arthurs, E.G., D.J. Bradley, P.N. Puntambekar, I.S. Ruddock and T.J. Glynn (1974b), Opt. Commun. **12**, 360.
Auston, D.H. (1971), Appl. Phys. Letters **18**, 249.
Auston, D.H. (1975), Appl. Phys. Letters **26**, 101.
Basov, N.G., P.G. Kryukov, V.S. Letokhov and Yu.A. Matveetz (1969), Sov. Phys. JETP **29**, 830.
Becker, M.F., D.J. Kuizenga and A.E. Siegman (1972), IEEE J. Quantum Electronics **QE-8**, 687.
Born, M. and E. Wolf (1970), Principles of Optics, 4th ed. (Pergamon Press; Oxford), 494 et seq.
Bradley, D.J. (1974), Opto-Electron. **6**, 25.
Bradley, D.J. and G.H.C. New (1974), Proc. IEEE **62**, 313.
Bradley, D.J. and W. Sibbett (1975), Appl. Phys. Letters **27**, 382.
Bradley, D.J., G.H.C. New and S.J. Caughey (1969), Phys. Letters **30A**, 78.
Bykovskii, N.E., V. Kan, P.G. Kryukov, Yu. A. Matveetz, N.L. Ni, Yu.V. Senat-skii and S.V. Chekalin (1972), Sov. J. Quantum Electronics **2**, 56.
Caruso, A., R. Gratton and W. Seka (1973), IEEE J. Quantum Electron **QE-9**, 1039.
Chan, C.K. and S.O. Sari (1974), Appl. Phys. Letters **25**, 403.
Chekalin, S.V., P.G. Kryukov, Yu.A. Matveetz, and O.B. Shatberashvili (1974), Opto-Electron. **6**, 249.
Chesler, R.B., M.A. Karr and J.E. Geusic (1970), Proc. IEEE **12**, 1899.
Colles, M.J. (1971), Appl. Phys. Letters **19**, 23.
Crowell, M.H. (1965), IEEE J. Quantum Electronics **QE-1**, 12.
Cubeddu, R.R. and O. Svelto (1969), IEEE J. Quantum Electronics **QE-5**, 495.
Cutler, C.C. (1955), Proc. IRE **43**, 140.
DeMaria, A.J., D.A. Stetser and H. Heynau (1966), Appl. Phys. Letters **8**, 174.
DeMaria, A.J., W.H. Glenn, M.J. Brienza and M.E. Mack (1969), Proc. IEEE **57**, 2.
DiDomenico, M. (1964), J. Appl. Phys. **35**, 2870.
Dienes, A. (1974), Opto-Electron. **6**, 99.
Dienes, A., E.P. Ippen and C.V. Shank (1972), IEEE J. Quantum Electronics **QE-8**, 388.
Drexhage, K.H. and V.T. Müller-Westerhoff (1972), IEEE J. Quantum Electronics **QE-8**, 759.
Drexhage, K.H. and G.A. Reynolds (1974), Opt. Commun. **10**, 18.
Duguay, M.A. and J.W. Hansen (1969), Appl. Phys. Letters **14**, 14.
Duguay, M.A., J.W. Hansen and S.L. Shapiro (1970), IEEE J. Quantum Electronics **QE-6**, 725.
Eckardt, R.C. and C.H. Lee (1969), Appl. Phys. Letters **15**, 425.
Eckardt, R.C., J.N. Bradford and C.H. Lee (1971), Appl. Phys. Letters **19**, 420.
Eckardt, R.C., J.L. DeRosa and J.P. Lettelier (1974), IEEE J. Quantum Electronics **QE-10**, 620.
Fan, B. and T.K. Gustafson (1975), Opt. Commun. **15**, 32.
Fleck, J.A. (1968), Appl. Phys. Letters **13**, 365.
Fleck, J.A. (1970), Phys. Rev. **B1**, 84.
Fleming, G.R., I.R. Harrowfield, A.E.W. Knight, J.M. Morris, R.J. Robbins and G.W. Robinson (1977), Opt. Commun. **20**, 36.
Garside, B.K. and T.K. Lim (1973a), J. Appl. Phys. **44**, 2325.
Garside, B.K. and T.K. Lim (1973b), Opt. Commun. **8**, 297.
Garside, B.K. and T.K. Lim (1974a), Opt. Commun. **12**, 8.
Garside, B.K. and T.K. Lim (1974b), Opt. Commun. **12**, 240.
Giordmaine, J.A., P.M. Rentzepis, S.L. Shapiro and K.W. Wecht (1967), Appl. Phys. Letters **11**, 216.
Gires, F. and P. Tournois (1964), Compt. Rend. **258**, 6112.
Glenn, W.H. (1975), IEEE J. Quantum Electronics **QE-11**, 8.

Glenn, W.H. and M.J. Brienza (1967), Appl. Phys. Letters **10**, 221.
Glenn, W.H., M.J. Brienza and A.J. DeMaria (1968), Appl. Phys. Letters **12**, 54.
Goldberg, L.S. and J.N. Bradford (1976), Appl. Phys. Letters **29**, 585.
Goldberg, L.S. and C.A. Moore (1975), Appl. Phys. Letters **27**, 217.
Greenhow, R.C. and A.J. Schmidt (1974), Picosecond Light Pulses, in: Advances in Quantum Electronics (D.W. Goodwin, ed.; Academic Press; London), **2**, 158–295.
Grischkowsky, D. (1974), Appl. Phys. Letters **25**, 566.
Grütter, A.A., H.P. Weber and K. Dändliker (1969), Phys. Rev. **185**, 629.
Haken, H. and M. Pauthier (1968), IEEE J. Quantum Electronics **QE-4**, 454.
Harrach, R.J. (1969), Appl. Phys. Letters **14**, 148.
Harris, J.M., R.W. Chrisman and F.E. Lytle (1975), Appl. Phys. Letters **26**, 16.
Harris, S.E. (1966), Proc. IEEE **54**, 1401.
Harris, S.E. and O.P. McDuff (1965), IEEE J. Quantum Electronics **QE-1**, 245.
Harris, S.E. and R.Targ (1964), Appl. Phys. Letters **5**, 202.
Haus, H.A. (1975A), IEEE J. Quantum Electronics **QE-11**, 323.
Haus, H.A. (1975B), IEEE J. Quantum. Electronics **QE-11**, 736.
Haus, H.A. (1975C), J. Appl. Phys. **46**, 3049.
Haus, H.A. C.V. Shank and E.P. Ippen, (1975), Opt. Commun. **15**, 29.
Hausherr, B., E. Mathieu and H. Weber (1973), IEEE J. Quantum Electronics **QE-9**, 445.
Hercher, M. (1967), Appl. Optics **6**, 947.
Herrmann, J., J. Wienecke and B. Wilhelmi (1975), Opt. and Quantum Electronics **7**, 337.
Ippen, E.P. and C.V. Shank (1975), Appl. Phys. Letters **27**, 488.
Ippen, E.P., C.V. Shank and A. Dienes (1972), Appl. Phys. Letters **21**, 348.
Jacobs, R.R. and M.J. Weber (1975), IEEE J. Quantum Electronics **QE-11**, 846.
Jacobs, R.R. and M.J. Weber (1976), IEEE J. Quantum Electronics **QE-12**, 102.
Kachen, G., L. Steinmetz and J. Kysilka (1968), Appl. Phys. Letters **13**, 229.
Klauder, J.R., M.A. Duguay, J.A. Giordmaine and S.L. Shapiro (1968), Appl. Phys. Letters **13**, 174.
Kogelnik, H., E.P. Ippen, A. Dienes and C.V. Shank (1972), IEEE J. Quantum Electronics **QE-8**, 373.
Krivoshchekov, G.V., L.A. Kulevskii, N.G. Nikulin, V.M. Semibalamut, V.A. Smirnov and V.V. Smirnov (1973), Sov. Phys. JETP **37**, 1007.
Krivoshchekov, G.V., N.G. Nikulin and V.A. Smirnov (1975), Sov. J. Quantum Electronics **5**, 1096.
Kryukov, P.G. and V.S. Letokhov (1972), IEEE J. Quantum Electronics **QE-8**, 766.
Kryukov, P.G., Yu.A. Matveetz, S.A. Churilova and O.B. Shatberashvili (1972a), Sov. Phys. JETP **35**, 1062.
Kryukov, P.G., Yu.A. Matveetz, S.V. Chekalin and O.B. Shatberashvili (1972b), Sov. Phys. JETP Letters **16**, 81.
Kuizenga, D.J. (1971), Appl. Phys. Letters **19**, 260.
Kuizenga, D.J. (1975), Generation of Short Optical Pulses for Laser Fusion, Microwave Laboratory Report #2451, Stanford University.
Kuizenga, D.J. (1976), accepted for publication by Opt. Commun.
Kuizenga, D.J. and A.E. Siegman (1969), Appl. Phys. Letters **14**, 181.
Kuizenga, D.J. and A.E. Siegman (1970a), IEEE J. Quantum Electronics **QE-6**, 694.
Kuizenga, D.J. and A.E. Siegman (1970b), IEEE J. Quantum Electronics **QE-6**, 709.
Kuizenga, D.J. and A.E. Siegman (1974), Opto-Electron. **6**, 43.
Kuizenga, D.J., D.W. Phillion, T. Lund and A.E. Siegman (1973), Opt. Commun. **9**, 221.
Kuznetsova, T.I. (1969), Sov. Phys. JETP **30**, 904.
Kwok, H.S. and E. Yablonovitch (1977), Appl. Phys. Letters **30**, 158.
Laubereau, A. (1969), Phys. Letters **29A**, 539.
Laubereau, A. and D. von der Linde (1970), Z. Naturforsch. **25A**, 1626.
LeFur, P. and D.H. Auston (1976), Appl. Phys. Letters **28**, 21.
Lehmberg, R.A. and J.M. McMahon (1976), Appl. Phys. Letters **28**, 204.
Letokhov, V.S. (1968), Sov. Phys. JETP **27**, 746.
Letokhov, V.S. (1969a), Sov. Phys. JETP Letters **28**, 562.
Letokhov, V.S. (1969b), Sov. Phys. JETP **28**, 1026.

Letokhov, V.S. and V.N. Morozov (1967), Sov. Phys. JETP **25**, 862.
Mack, M.E. (1969), Appl. Phys. Letters **15**, 166.
Magde, D., B.A. Bushaw and M.W. Windsor (1974), IEEE J. Quantum Electronics **QE-10**, 394.
Maier, M., W. Kaiser and J.A. Giordmaine (1969), Phys. Rev. **177**, 580.
Mandel, L. and E. Wolf (1965), Rev. Mod. Phys. **37**, 231.
Masters, J.I., P. Kafalas and E.M.E. Murray (1964), Report No. T0136415, Office of Naval Research.
McDuff, O.P. (1972), in: Laser Handbook (F.T. Arecchi and E.O. Schulz-Dubois, eds.; North-Holland.; Amsterdam) 631.
McDuff, O.P. and S.E. Harris (1967), IEEE J. Quantum Electronics **QE-3**, 101.
McGeoch, M.W. (1973), Opt. Commun. **7**, 116.
McMahon, J.M. (1976), Opt. Commun. **18**, 170.
Milam, D. and M.J. Weber (1976), J. Appl. Phys. **47**, 2497.
Milam, D., R.A. Bradbury, A. Hordvik, H. Schlossberg and A. Szöke (1974), IEEE J. Quantum Electronics **QE-10**, 20.
Miroñov, A.B. and O.B. Shatberashvili (1974), Sov. J. Quantum Electronics **4**, 805.
Mocker, H.W. and R.J. Collins (1965), Appl. Phys. Letters **7**, 270.
Nelson, T.J. (1972), IEEE J. Quantum Electronics **QE-8**, 29.
New, G.H.C. (1972), Opt. Commun. **6**, 188.
New, G.H.C. (1974), IEEE J. Quantum Electronics **QE-10**, 115.
New, G.H.C. and D.H. Rea (1976), J. Appl. Phys. **47**, 3107.
New, G.H.C., K.R. Orkney and M.J.W. Nock (1976), Opt. and Quantum Electronics **8**, 425.
O'Neil, F. (1972), Opt. Commun. **6**, 360.
Osterink, L.M. and R. Targ (1967), Appl. Phys. Letters **10**, 115.
Penzkofer, A. (1974), Opto-Electron. **6**, 87.
Penzkofer, A. and W. Kaiser (1972), Appl. Phys. Letters **21**, 427.
Penzkofer, A., D. von der Linde, A. Laubereau and W. Kaiser (1972), Appl. Phys. Letters **20**, 351.
Petukhov, V.A. and A.I. Krymova (1976), Sov. J. Quantum Electronics **6**, 1025.
Picard, R.M. and P. Schweitzer (1969), Phys. Letters **29A**, 415.
Picard, R.M. and P. Schweitzer (1970), Phys. Rev. **1**, 1803.
Rentzepis, P.M., C.J. Mitschele and A.C. Saxman (1970), Appl. Phys. Letters **17**, 45.
Rigrod, W.W., (1965), U.S. patent 3,492,599.
Royt, T.R., W.L. Faust, L.S. Goldberg and C.H. Lee (1974), Appl. Phys. Letters **25**, 514.
Ruddock, I.S. and D.J. Bradley (1976), Appl. Phys. Letters **29**, 296.
Rubinov, A.N., M.C. Richardson, K. Sala and A.J. Alcock (1975), Appl. Phys. Letters **27**, 358.
Runge, P.K. (1971), Opt. Commun. **4**, 195.
Schelev, M.Ya., M.C. Richardson and A.J. Alcock (1972), Rev. Sci. Instr. **43**, 1819.
Seka, W. and E. Stüssi (1976), J. Appl. Phys. **47**, 3538.
Shank, C.V. and E.P. Ippen (1973), in: Dye Lasers (F.P. Schafer, ed.; Springer-Verlag; New York), 121–143.
Shank, C.V. and E.P. Ippen (1974), Appl. Phys. Letters **24**, 373.
Shapiro, S.L. and M.A. Duguay (1969), Phys. Letters **28A**, 698.
Siegman, A.E. and D.J. Kuizenga (1970), IEEE J. Quantum electronics **QE-6**, 803.
Smith, A.W. (1969), Appl. Phys. Letters **15**, 194.
Smith, P.W. (1970), Proc. IEEE **58**, 1342.
Smith, P.W., M.A. Duguay and E.P. Ippen (1974), Mode Locking of Lasers, in: Progress in Quantum Electronics (J.H. Sanders, and S. Stenholm, eds.; Pergamon Press; Oxford), **3**, Pt. 2, 107–229.
Smith, W.L. and J.H. Bechtel (1976), Appl. Phys. Letters **28**, 606.
Soffer, B.H. and J.W. Linn (1968), J. Appl. Phys. **39**, 5859.
Sorokin, P.P., J.J. Luzzi, J.R. Lankard and G.D. Petit (1964), IBM J. Res. Develop. **8**, 182.
Speck, D.R. and E.S. Bliss (1973), Laser Fusion Program Semiannual Report, January–June 1973 (UCRL-50021-73-1), Lawrence Livermore Laboratory, 20–29.
Svelto, O. (1970), Appl. Phys. Letters **17**, 83.
Tomov, I.V., R. Fedosejevs, M.C. Richardson and W.J. Orr (1976), Appl. Phys. Letters **29**, 193.
Tomov, I.V., R. Fedosejevs and M.C. Richardson (1977), Appl. Phys. Letters **30**, 164.

Treacy, E.B. (1968), Phys. Letters **28A**, 34.
Treacy, E.B. (1969), Appl. Phys. Letters **14**, 112.
Treacy, E.B. (1970), Appl. Phys. Letters **17**, 14.
Varnavskii, O.P., A.M. Leontovich and A.M. Mozharovskii, (1975), Sov. J. Quantum Electronics **5**, 1280.
Veduta, A.P., A.F. Solokha, N.P. Furzikov and G.V. Ellert (1974), Sov. J. Quantum Electronics **3**, 385.
von der Linde, D. (1972), IEEE J. Quantum Electronics **QE-8**, 328.
von der Linde, D. and K.F. Rodgers (1973a), IEEE J. Quantum Electronics **QE-9**, 960.
von der Linde, D. and K.F. Rodgers (1973b), Opt. Commun. **8**, 91.
von der Linde, D., O. Bernecker and W. Kaiser (1970), Opt. Commun. **2**, 149.
Washio, K., K. Kuizumi and Y. Ikeda (1977), IEEE J. Quantum Electronics **QE-13**, 47.
Weber, H.P. (1967), J. Appl. Phys. **38**, 2234.
Weber, M.J. (1976), Optical Materials for Neodymium Fusion Lasers, in: Critical Materials Problems in Energy Production (Academic Press; New York), 261–279.
Weber, M.J., C.B. Layne, R.A. Saroyan and D. Milam (1976), Opt. Commun. **18**, 171.
Wilbrandt, R. and H. Weber (1975), IEEE J. Quantum Electronics **QE-11**, 186.
Yablonovitch, E. (1974), Phys. Rev. **A10**, 1888.
Yablonovitch, E. (1975), IEEE J. Quantum Electronics **QE-11**, 789.
Yablonovitch, E. and J. Goldhar (1974), Appl. Phys. Letters **25**, 580.
Yariv, A. (1965), J. Appl. Phys. **36**, 388.
Yasa, Z.A. (1975), J. Appl. Phys. **46**, 4895.
Yasa, Z.A. and O. Teschke (1975a), Appl. Phys. Letters **27**, 446.
Yasa, Z.A. and O. Teschke (1975b), Opt. Commun. **15**, 169.
Yasa, Z.A., O. Teschke and L.W. Braverman (1976), J. Appl. Phys. **47**, 174.
Yasa, Z.A., A. Dienes and J.R. Whinnery (1977), Appl. Phys. Letters **30**, 24.
Zakharov, S.D., P.G. Kryukov, Yu.A. Matveetz, S.V. Chekalin, S.A. Churilova and O.B. Shatberashvili (1974), Sov. J. Quantum Electronics **3**, 395.
Zel'dovich, B.Ya and T.I. Kuznetsova (1972), Sov. Phys. Uspekhi **15**, 25.
Zherikhin, A.N., V.A. Kovalenko, P.G. Kryukov, Yu.A. Matveetz, S.V. Chekalin and O.B. Shatberashvili (1974a), Sov. J. Quantum Electronics **4**, 210.
Zherikhin, A.N., P.G. Kryukov, Yu. A. Matveetz and S.V. Chekalin (1974b), Sov. J. Quantum Electronics **4**, 525.
Zherikhin, A.N., P.G. Kryukov, E.V. Kurganova, Yu. A. Matveetz, S.V. Chekalin, S.A. Churilova and O.B. Shatberashvili (1974c), Sov. Phys. JETP **39**, 52.

B2 | High Average Power, Efficient Second Harmonic Generation

DAVID HON*

Laser systems Division,
Aerospace Groups,
Hughes Aircraft Company, Culver City, California

Contents

*Now at Hughes Research Labs, Malibu, California.

© *North-Holland Publishing Company, 1979*
Laser Handbook, edited by M.L. Stitch

Abstract

This chapter addresses the problems associated with high average power, high efficiency second harmonic generation (SHG), and how these problems may be circumvented by the application of recently developed experimental techniques. Thermal instability can be compensated by various feedback mechanisms using the electro-optic effect, the piezo-optic effect, or the angular orientation of the crystal. Thermal gradients can be eliminated by beam shaping techniques which distribute laser energy evenly across the face of a slab-shaped crystal. On the basis of a review of the principle of SHG, these techniques are developed theoretically. A comprehensive review of lasers and crystals used for SHG is given, emphasizing those suitable for high average power applications.

1 Introduction

Frequency conversion techniques expand the usefulness of existing lasers. Of these techniques, second harmonic generation (SHG), commonly called frequency doubling, has been particularly well developed. This is evidenced by the reliable SHG lasers with outputs of more than a watt that are commercially available.

When used in conjunction with other conversion processes, SHG offers the means for obtaining many output frequencies from a single, fixed frequency laser. A particularly noteworthy example is the set of wavelengths for which useful output power may be obtained from a Nd:YAG laser operating at 1.064 μm. Fig. 1 shows some of the combinations possible with dye lasers, optical parametric oscillators and stimulated Raman scattering systems, which may be employed to alter the Nd:YAG output frequency.

The Nd:YAG and CO_2 lasers, both good laser sources for SHG and which generate output powers in the range of hundreds of watts, have proven reliability in a host of industrial applications. Other lasers, desirable in certain respects as optical sources for SHG, are not yet suitable for applications that require stable outputs over extended periods of time. This problem is, of course, expected with newly developed laser sources, and considerable research is currently directed toward obtaining reliable operation. However, because of the maturity of high power Nd:YAG laser, most of the examples in this chapter draw on experience with it. The underlying principles are, of course, generally applicable.

Although the output power of the laser source used for SHG can limit the level of second harmonic power produced, it is usually the inability of the nonlinear material itself to withstand the effects of high incident laser power that limits the SHG output. Given that a laser with the necessary performance exists, high average power SHG depends primarily on two factors: (1) the availability of a sufficiently large and uniform SHG crystal with adequate nonlinear coefficients and a damage threshold, and low absorptivity at the frequencies involved and (2) the elimination of thermal instabilities and thermal gradients arising from optical absorption. Advances in high power SHG technology made in recent years have alleviated these problems considerably. The production of one joule near diffraction-limited second harmonic pulses at a repetition rate of 10 Hz (Kogan and Crow 1977) and 35 W of frequency doubled Nd:YAG output by the use of electro-optical tuning and beam shaping (Hon 1977, Hon et al. 1977) are

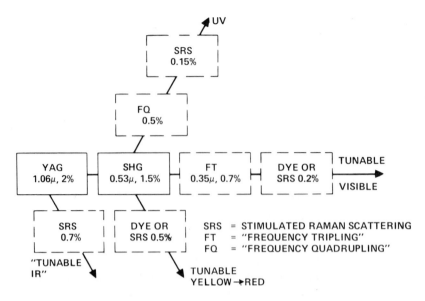

Fig. 1. Frequency conversion schemes based on the Nd:YAG laser. The near uv to near ir spectral range is covered by various combinations of techniques with nominal wall-plug efficiencies as indicated.

two examples in which high average powers have been obtained without degradation of the nonlinear medium.

This chapter specifically addresses the problems associated with high average power, high efficiency frequency doubling and how these problems may be circumvented by the application of recently developed experimental techniques. As such, the discussion is not intended to be a comprehensive survey of all aspects of SHG. However, the literature cited provides considerable information about many of the experimental and theoretical advances that had been made in the field prior to late 1977. In many cases of practical interest a semi-empirical approach is more useful than a full treatment based on first principles. Mathematically complete derivations will be deferred to the references. Instead, illustrative simpler models will be used throughout this chapter. Although the present discussion relates primarily to frequency doubling processes, the results are also applicable to the closely related processes of optical parametric amplification and frequency mixing in nonlinear media.

§ 2 of this chapter provides a brief review and discussion of the relevant aspects of the theory of SHG, including the nature of thermal effects in nonlinear media and how these effects can contribute to SHG inefficiencies. § 3 describes lasers which are used for SHG. Evaluation criteria are given and techniques for obtaining optimized laser output are presented. § 4 discusses the physical and optical properties of the nonlinear crystals used for high average power SHG. Selection criteria and crystal engineering as well as a detailed description of 23 nonlinear crystals in tabular form are included. Finally, § 5 discusses solutions to the optically produced thermal problems in the SHG crystals. In particular, electro-optical tuning, piezo-optical tuning, angle tuning and beam shaping are considered.

Recent review articles relevant to the subject of this chapter include: D.A. Kleinman's Optical Harmonic Generation in Nonlinear Media (1972), Y.R. Shen's Recent Advances in Nonlinear Optics (1976), K.F. Hulme's Nonlinear Optical Crystals and Their Applications (1973) and S.K. Kurtz's Nonlinear Optical Materials (1972). F. Zernike and J.E. Midwinter's Applied Nonlinear Optics (1973) and W. Koechner's Solid State Laser Engineering (1976) are tutorial books of particular value. V.D. Volosov and A.G. Kalintsev's Design of Optical Frequency Doublers (1974) and D.N. Nikogosyan's Nonlinear Optics Crystals (Review and Summary of Data) (1977) are both useful.

A list of symbols and their definitions is given at the end of the chapter.

2 Elements of second harmonic generation (SHG)

Second harmonic generation (SHG) (Franken et al. 1961, Dowley 1968, Hagen and Magnante 1969, Chester et al. 1970, Nath et al. 1971) results from the interaction of radiation with a medium possessing an electric susceptibility that is a nonlinear function of the intensity of the radiation. In principle, any solid substance without inversion symmetry can produce second harmonic radiation providing the peak power of the incident electromagnetic field is sufficiently large (nonaligned fluids are not useful as media for SHG because the existence of random molecular orientations necessarily implies the presence of bulk inversion symmetry).

As will be shown in ensuing sections of this chapter, efficient SHG depends not only upon the nonlinearity of the susceptibility but also upon matching the velocities of the fundamental and second harmonic electromagnetic waves that propagate through the nonlinear medium. This process, commonly referred to as phase-matching, will be briefly reviewed following an introduction to the elements that are particularly important for generation of second harmonic optical signals.

The second harmonic power $P(2\omega)$ generated by (a) monochromatic plane wave(s) of angular frequency ω and power $P(\omega)$ incident on a very 'thin' plane parallel slab of thickness L of a nonabsorbing nonlinear crystal is given by

$$P(2\omega) = \frac{8(\mu_0)^{3/2}(\epsilon_0)^{1/2}\omega^2 \cdot d_{ij}^2 \cdot P^2(\omega) \cdot L^2}{\pi h^2 \cdot n(2\omega) \cdot [n(\omega)]^2} \cdot \left[\frac{\sin(L\Delta k/2)}{(L\Delta k/2)}\right]^2, \quad (1)$$

where

h = spot diameter of the input at the fundamental wavelength,
$n(\omega)$ = refractive index of the crystal at the fundamental wavelength,
$n(2\omega)$ = refractive index of the crystal at the second harmonic wavelength,
d_{ij} .   = pertinent SHG coefficient of the crystal,
μ_0 = permeability of free space.
ϵ_0 = permittivity of free space.

The quantity Δk is the magnitude of the wave vector mismatch $\Delta \mathbf{k}$ between the fundamental wave(s) and the second harmonic wave, that is,

$$\Delta \mathbf{k} = \mathbf{k}_1(\omega) + \mathbf{k}_2(\omega) - \mathbf{k}_3(2\omega), \tag{2}$$

where $\mathbf{k}_1$ and $\mathbf{k}_2$ are the wave vectors of the fundamental wave(s), and $\mathbf{k}_3$ is the wave vector of the second harmonic. For the case of type I phase-matching (defined in § 2.2),

$$\mathbf{k}_1(\omega) = \mathbf{k}_2(\omega), \tag{3}$$

and eq. (2) then becomes

$$\Delta \mathbf{k} = 2\mathbf{k}(\omega) - \mathbf{k}(2\omega). \tag{4}$$

If the depletion of the fundamental beam during the SHG process is taken into account, it has been shown by Bloembergen (1965) that

$$P(2\omega) = P(\omega) \tan{(CL)}, \tag{5}$$

where

$$C = \sqrt{2} \frac{377^{3/2}}{n(\omega)} \omega d_{ij} \left[\frac{4P(\omega)}{\pi h^2} \right]^{1/2}.$$

Eq. (5) is plotted in fig. 2 as a function of power density $[4P(\omega)/\pi h^2]$ for the generation of $0.53 \, \mu$m radiation (second harmonic of $1.06 \, \mu$m) in a lithium niobate (LiNbO$_3$) crystal. Note the much lower experimental values.

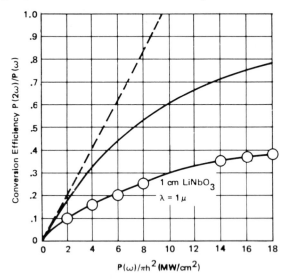

Fig. 2. Conversion efficiency versus power density of the fundamental [eq. (5)]. The dashed line is the approximation of eq. (1). Circles represent typical experimental data based on a high quality LiNbO$_3$ crystal and a multimode Nd:YAG Laser Operating Q-switched at 100 mJ per pulse. The lower efficiency than that predicted is due to absorption, crystal imperfections and multimode operation. Two upper curves by courtesy of Chromatic.

2.1 Phase-matching

As mentioned in the previous section, second harmonic generation occurs in a material when an electromagnetic field induces a polarization that is a nonlinear function of the field amplitude. The polarization wave produced by the interaction of the input radiation with the nonlinear material propagates through the medium and radiates an electromagnetic wave with a frequency that is different from that of the input fundamental wave. Since normal dispersive effects are present in the medium, the propagation velocity of the radiated (second harmonic) wave will differ from that of the polarization wave which produced it. As a result, destructive interference between frequency components will severely limit the SHG efficiency. A technique known as phase-matching (Maker et al. 1962, Giordmaine 1962) (also called velocity or phase synchronism) may be used to match the propagation velocities of the fundamental and second harmonic waves and thereby significantly reduce the contribution of dispersive effects to SHG inefficiency.

The remainder of this section is a review of the essential features of phase-matching in nonlinear crystals. The primary types of phase-matching are discussed and parameters which relate directly to phase-matching optimization are introduced in ensuing sections.

A wave vector $k(\omega)$ with magnitude $k(\omega)$ is related to the frequency (ω) and polarization ($\hat{e}$) dependent index of refraction [$n_{\hat{e}}(\omega)$] by

$$k(\omega) = |k(\omega)| = \omega n_{\hat{e}}(\omega)/c, \tag{6}$$

where c is the speed of light in a vacuum. $n_{\hat{e}}(\omega)$ is obtained from the index ellipsoid

$$\sum_{ij} B_{ij}(\omega)x_ix_j = 1; \qquad i, j = 1, 2, 3. \tag{7}$$

The x_i define a Cartesian coordinate system and the B_{ij} form a symmetric matrix. Although this form is not as familiar as the expression for the index ellipsoid in principle Cartesian coordinates [eq. (10)], it will be used in later sections of this chapter and is therefore introduced at this time. The index $n_e(\omega)$ for an arbitrary polarization $\hat{e} = (e_1, e_2, e_3)$ is given by

$$n_{\hat{e}}(\omega) = \left| \sum_{ij} B_{ij}(\omega)e_ie_j \right|^{-1/2}. \tag{8}$$

In a principal Cartesian coordinate system, cross terms of B_{ij} vanish, i.e.,

$$n_i(\omega) = B_{ii}(\omega)^{-1/2}; \qquad i = 1, 2, 3, \tag{9}$$

and eq. (7) may be written in the more familiar form

$$\frac{x^2}{n_x^2(\omega)} + \frac{y^2}{n_y^2(\omega)} + \frac{z^2}{n_z^2(\omega)} = 1. \tag{10}$$

For simplicity, the discussion in this chapter will be confined to uniaxial crystals (those possessing only one optic axis). Extension of the treatment to biaxial crystals is tedious but straightforward. For uniaxial crystals, eq. (10) becomes

$$\frac{x^2 + y^2}{n_o^2(\omega)} + \frac{z^2}{n_e^2(\omega)} = 1, \tag{11}$$

where $n_o(\omega)$ = ordinary refractive index, $n_e(\omega)$ = extraordinary refractive index.

Fig. 3 shows the index ellipsoid for a negative ($n_e < n_o$) uniaxial crystal. The wave normal is along S. Since the ellipsoid in this case is invariant to a rotation about the z axis, S is chosen to lie in the y-z plane.

The plane normal to S which contains the origin intersects the index ellipsoid along an ellipse (partially dotted) whose semiaxes define the 'ordinary' index n_o (which is θ-independent for uniaxial crystals) and 'extraordinary' index $n_e(\theta)$. Their directions define the directions of the D vectors of the two corresponding polarized waves. A projection on the y-z plane is illustrated in fig. 4. Note that the Poynting vector P, which gives the direction of power flow and is normal to the ellipsoid at its intersection with S (Born and Wolf 1975), is, in general, not collinear with S except in the special cases of $\theta = 0$ or 90°. As will be seen later, this noncollinearity gives rise to deleterious beam walk-off.

Substitution of $x = 0$, $y = n_e(\omega, \theta) \cos \theta$ and $z = n_e(\omega, \theta) \sin \theta$ in eq. (11) yields

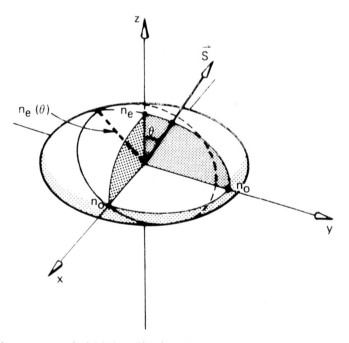

Fig. 3. Negative ($n_e < n_0$) uniaxial index ellipsoid with wave-normal S in the y-z plane making an angle θ with the z-axis. The plane normal to S intersects the index ellipsoid along an ellipse (partially dotted) whose semiaxes define the (θ-independent) 'ordinary' index n_0 and 'extraordinary' index $n_e(\theta)$.

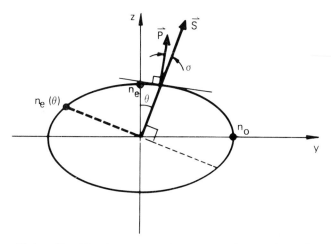

Fig. 4. Projection of index ellipsoid and wave-normal vector S on the S–z plane. The plane normal to S and which defines $n_e(\theta)$ is represented by a dotted line. The Poynting vector P is normal to ellipsoid surface at its intersection with S. Except when $\theta = 0$ or $90°$, P is not collinear with S, and this gives rise to beam walk-off.

an expression for $n_e(\omega, \theta)$ in polar coordinates:

$$n_e(\omega, \theta) = \left[\frac{\cos^2\theta}{n_o^2(\omega)} + \frac{\sin^2\theta}{n_e^2(\omega)}\right]^{-1/2}. \tag{12}$$

The three-dimensional surfaces in polar coordinates which give the indices $n_e(\theta, \phi)$ and $n_o(\theta, \phi)$ as a function of the wave normal S direction (θ, ϕ) are called the normal surfaces. For a uniaxial crystal, the azimuthal angle ϕ is redundant and the normal surfaces become ellipsoids of revolution about the z-axis. The intersections of these ellipsoids with the S–z plane are shown for a negative uniaxial crystal in fig. 5. The interior curve is a plot of eq. (12). The exterior curve is a circle of radius $n_o(\omega)$.

2.2 Angle phase-matching

Since the index of refraction is frequency dependent, phase-matching for waves of different frequencies does not occur naturally in normal dispersing media. Achievement of the phase-matching condition by utilization of crystal bire-fringence was obtained in 1962 (Maker et al. 1962, Giordmaine 1962) by orientat-ing the crystal such that the phase-mismatch term Δk given in eqs. (2) and (4) equals zero. This technique of angle phase-matching is illustrated in fig. 6 for the four possible cases of phase-matching:type I or type II each for a negative or positive uniaxial crystal. For type I-negative, the ordinary ray normal surface for the fundamental frequency ω is seen in fig. 6a to intersect the extraordinary ray normal surface of the second harmonic 2ω along the direction labeled on this projection on the y–z plane. In three dimensions, the intersection describes a cone (and its mirror image) of wave normal vectors $k(\omega)$ that will satisfy [using eq. (6)] the condition that eq. (4) equals zero. This is described by eq. (13) and

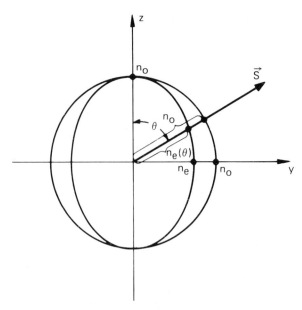

Fig. 5. The intersections of normal surfaces and the S–z plane for a negative uniaxial crystal. Intersections by wave-normal vector S provide n_0 and $n_e(\theta)$ for the ordinary ray and extraordinary rays respectively.

(14). Type II phase-matching is achieved if the fundamental wave can be coherently split between an ordinary and an extraordinary wave (by orientation of the crystal) such that the condition $\Delta k = 0$ in eq. (2) is satisfied. This is described by eq. (15) and (16). Type I and type II phase-matching conditions for a positive uniaxial crystal are depicted in fig. 6b and described by eqs. (17) through (20). The notation commonly used in the East European literature [(oo–e), etc.] to denote the four cases is also indicated below:

Type I – negative uniaxial (oo–e)

$$\left.\begin{array}{l} n_1(\omega, \theta) = n_2(\omega, \theta) = n_o(\omega), \\ n_3(2\omega, \theta) = n_e(2\omega, \theta); \end{array}\right\} \tag{13}$$

$$n_e(2\omega, \theta_m) = n_o(\omega) \tag{14}$$

where θ_m is the phase-match angle given by eqs. (21) through (24) or obtained experimentally.

Type II – negative uniaxial (eo–e)

$$\left.\begin{array}{l} n_1(\omega, \theta) = n_e(\omega, \theta), \ n_2(\omega, \theta) = n_o(\omega), \\ n_3(2\omega, \theta) = n_e(2\omega, \theta). \end{array}\right\} \tag{15}$$

$$n_e(2\omega, \theta_m) = \tfrac{1}{2}[n_e(\omega, \theta_m) + n_o(\omega)]. \tag{16}$$

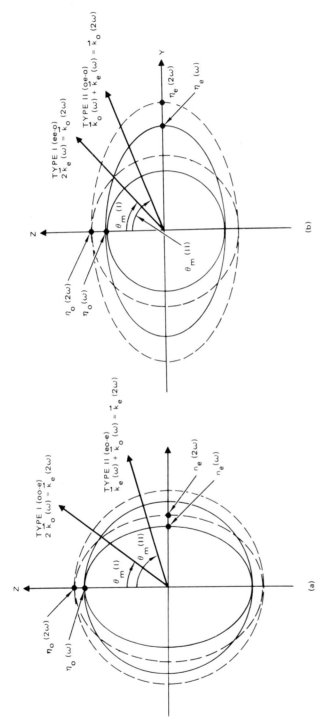

Fig. 6. (a) Type I and II phase-matching for SHG in a negative uniaxial crystal. The ordinary ray normal-surface of the fundamental (ω) wave intersects the extraordinary ray normal-surface of the second harmonic (2ω) wave, defining a cone (and its mirror image) of phase-matching wave normals [$\boldsymbol{k}(\omega)$] shown here in a y-z plane projection. θ_m (I) and (II) are the phase-match angles. When either θ_m(I) or θ_m(II) equals 90°, noncritical phase-matching is obtained. (b) The corresponding diagram for a positive uniaxial crystal.

Type I – positive uniaxial (ee–o)

$$n_1(\omega, \theta) = n_2(\omega, \theta) = n_e(\omega, \theta), \left.\right\}$$
$$n_3(2\omega, \theta) = n_o(2\omega). \qquad (17)$$

$$n_o(2\omega) = n_e(\omega, \theta_m). \qquad (18)$$

Type II – positive uniaxial (oe–o)

$$n_1(\omega, \theta) = n_o(\omega), \ n_2(\omega, \theta) = n_e(\omega, \theta), \left.\right\}$$
$$n_3(2\omega, \theta) = n_o(2\omega). \qquad (19)$$

$$n_o(2\omega) = \tfrac{1}{2}[n_e(\omega, \theta_m) + n_o(\omega)]. \qquad (20)$$

By combining eqs. (14) and (16) each with (12), the phase-matching angles $\theta_m(\text{I})$ and $\theta_m(\text{II})$ for a negative uniaxial crystal are given by (Yariv 1967, for example)

$$\sin^2 \theta_m(\text{I}) = \frac{[n_o(\omega)]^{-2} - [n_o(2\omega)]^{-2}}{[n_e(2\omega)]^{-2} - [n_o(2\omega)]^{-2}} \qquad (21)$$

and

$$\left[\frac{\cos^2\theta_m(\text{II})}{[n_o(2\omega)]^2} + \frac{\sin^2\theta_m(\text{II})}{[n_e(2\omega)]^2}\right]^{-1/2} = \frac{1}{2}\left\{n_o(\omega) + \left[\frac{\cos^2\theta_m(\text{II})}{[n_o(\omega)]^2} + \frac{\sin^2\theta_m(\text{II})}{[n_e(\omega)]^2}\right]^{-1/2}\right\}. \qquad (22)$$

The phase-matching angles for a positive uniaxial crystal can be similarly obtained by combining eqs. (18) and (20) each with (12):

$$\sin^2 \theta_m(\text{I}) = \frac{[n_o(2\omega)]^{-2} - [n_o(\omega)]^{-2}}{[n_e(\omega)]^{-2} - [n_o(\omega)]^{-2}}, \qquad (23)$$

$$\sin^2 \theta_m(\text{II}) = \frac{\{n_o(\omega)/[2n_o(2\omega) - n_o(\omega)]\}^2 - 1}{[n_o(\omega)/n_e(\omega)]^2 - 1}. \qquad (24)$$

For phase-matching in biaxial crystals, the reader is referred to Hobden (1967) and Singh et al. (1970a, b).

From fig. 6, the Poynting vector $P(\omega)$ is collinear with $k(\omega)$, but $P(2\omega)$ is not collinear with $k(2\omega)$, even though $k(\omega)$ and $k(2\omega)$ are collinear. This gives rise to beam 'walk-off' which can severely limit SHG efficiency. The walk-off angle σ between the two waves can easily be derived (Nelson 1964, Boyd et al. 1965) from the fact that P is normal to the index ellipsoid at its intersection with $k(\omega)$ or $k(2\omega)$. For type I-negative, for example, the walk-off angle is

$$\sigma \sim \tan \sigma = \tfrac{1}{2}n_o^2(\omega)\{[n_e(2\omega)]^{-2} - [n_o(2\omega)]^{-2}\} \sin 2\theta_m. \qquad (25)$$

In addition to beam walk-off, the usefulness of angle phase-matching (with θ_m other than 90°) is sometimes limited by the divergence of the fundamental wave. It can be seen from fig. 6, and is shown by eq. (31), that a change $\Delta\theta$ from θ_m leads linearly to a nonzero Δk.

To derive the variation of Δk with θ, expand Δk in a Taylor's series about θ_m:

$$\Delta k = \frac{d(\Delta k)}{d\theta}\Big|_{\theta_m} \Delta\theta + \frac{1}{2}\frac{d^2(\Delta k)}{d\theta^2}\Big|_{\theta_m} \Delta\theta^2 + \ldots, \tag{25a}$$

where, according to eq. (2) (assuming all vectors are collinear) and eq. (6),

$$\frac{d(\Delta k)}{d\theta}\Big|_{\theta_m} = \frac{\omega}{c}\frac{dn_1(\omega, \theta)}{d\theta}\Big|_{\theta_m} + \frac{\omega}{c}\frac{dn_2(\omega, \theta)}{d\theta}\Big|_{\theta_m} - \frac{2\omega}{c}\frac{dn_3(2\omega, \theta)}{d\theta}\Big|_{\theta_m},$$

or, using primes to denote differentiation with respect to θ,

$$= \frac{\omega}{c}[n_1'(\omega, \theta_m) + n_2'(\omega, \theta_m) - 2n_3'(2\omega, \theta_m)]. \tag{26}$$

Similarly,

$$\frac{d^2(\Delta k)}{d\theta^2}\Big|_{\theta_m} = \frac{\omega}{c}[n_1''(\omega, \theta_m) + n_2''(\omega, \theta_m) - 2n_3''(2\omega, \theta_m)]. \tag{27}$$

Since $n_o(\omega, \theta)$ is independent of θ,

$$n_o'(\omega, \theta_m) = n_o''(\omega, \theta_m) = 0 \quad \text{for all } \omega. \tag{28}$$

Meanwhile, on the basis of eq. (12),

$$n_e'(\omega, \theta_m) = -\tfrac{1}{2}n_e^3(\omega, \theta_m)[n_e^{-2}(\omega) - n_o^{-2}(\omega)] \sin 2\theta_m, \quad \text{for all } \omega \tag{29}$$

and

$$n_e''(\omega, \theta_m) = -n_e^3(\omega)[n_e^{-2}(\omega) - n_o^{-2}(\omega)], \text{ for all } \omega. \tag{30}$$

For angle phase-matching, the first (linear) term in eq. (25a) dominates and

$$\Delta k \sim \frac{d(\Delta k)}{d\theta}\Big|_{\theta_m} \Delta\theta. \tag{31}$$

Type I – negative uniaxial (oo–e)

Use of eqs. (13), (26), (28), (29) and (31) (in that order) yields

$$\Delta k = -\frac{\omega}{c} n_o^3(\omega)[n_e^{-2}(2\omega) - n_o^{-2}(2\omega)] \sin 2\theta_m \cdot \Delta\theta. \tag{32}$$

Type II – negative uniaxial (eo–e)

Use of eqs. (15), (26), (28), (29) and (31) yields

$$\Delta k = -\frac{\omega}{c}\{-\tfrac{1}{2}n_e^3(\omega, \theta_m)[n_e^{-2}(\omega) - n_o^{-2}(\omega)]$$

$$+ n_e^3(2\omega, \theta_m)[n_e^{-2}(2\omega) - n_o^{-2}(2\omega)]\} \sin 2\theta_m \cdot \Delta\theta \tag{33}$$

$$\sim -\frac{\omega}{2c} n_o^3(\omega)[n_e^{-2}(2\omega) - n_o^{-2}(2\omega)] \sin 2\theta_m \cdot \Delta\theta.$$

The approximation can best be understood by inspection of fig. 6(a).

Type I – positive uniaxial (ee–o)

Use of eqs. (17), (26), (28), (29) and (31) yields

$$\Delta k = -\frac{\omega}{c} n_e^3(\omega, \theta_m)[n_e^{-2}(\omega) - n_o^{-2}(\omega)] \sin 2\theta_m \cdot \Delta\theta, \left.\rule{0pt}{24pt}\right\} \tag{34}$$

where $n_e^3(\omega, \theta_m) = n_o^3(2\omega)$.

Type II – positive uniaxial (oe–o)

Use of eqs. (19), (26), (28), (29) and (31) yields

$$\Delta k = -\frac{\omega}{2c}[2n_o(2\omega) - n_o(\omega)][n_e^{-2}(\omega) - n_o^{-2}(\omega)] \sin 2\theta_m \cdot \Delta\theta. \tag{35}$$

The phase mismatch varies linearly with $\Delta\theta$ and leads to an inefficiency which can be calculated from eq. (1). Because of the angular sensitivity, angle phase-matching is also called critical phase-matching. The first minimum occurs in eq. (1) when $\Delta k = 2\pi/L$ [see also eq. (11)]. Eq. (32) through (35) can therefore be used to define the acceptance angles $\delta\theta$'s by setting $\Delta k = 2\pi/L$ in each of the four phase-matching conditions. As defined, $\delta\theta$ involves the interaction length L. The quantity $\delta\theta L$ is an intrinsic property of the particular material. When the second harmonic power is plotted as a function of polar angle θ for a uniaxial crystal, the acceptance angle is equal to the full width of the curve at 0.405 maximum. It is listed for many common nonlinear crystals in table 1. For angle phase-matched crystals $\delta\theta$ is typically of the order of one milliradian.

Uniaxial crystals have rotational symmetry around z, and the acceptance angle is therefore very large in the azimuthal direction. This fact may be utilized to increase SHG efficiency by cylindrically focusing a beam to a line thereby increasing the radiation power density at the crystal while not exceeding the acceptance angle (Volosov and Nilov 1966, Volosov and Rashchektseva 1970). Although the analytical situation is more complex for biaxial crystals, the normal surfaces for ω and 2ω still must intersect on lines (curved) which define directions of large acceptance angle for any phase-match condition.

2.3 Ninety degree phase-matching

Ninety degree phase-matching occurs if the phase-match angle, θ_m equals 90°. When θ_m is 90°, the k and E vectors of all participant beams are collinear with the crystal axes.

The first (linear) term in the Taylor's expansion eq. (22) of Δk vanishes, as can be verified by eqs. (30), (32), (34) and (36). The second (quadratic) term must be used. Thus,

$$\Delta k \sim \frac{1}{2} \frac{d^2(\Delta k)}{d\theta^2}\bigg|_{\theta_m} \cdot \Delta\theta^2, \tag{36}$$

which, in conjunction with eqs. (27), (28) and (30), provide the following expressions for Δk:

Type I – negative uniaxial (oo–e)

By using eq. (13),

$$\Delta k = -\frac{\omega}{c} n_e^3(2\omega)[n_e^{-2}(2\omega) - n_o^{-2}(2\omega)] \cdot \Delta\theta^2, \tag{37}$$

where $n_e(2\omega) = n_o(\omega)$.

Type II – negative uniaxial (eo–e)

By using eq. (15),

$$\Delta k = -\frac{\omega}{c}\{-\tfrac{1}{2}n_e^3(\omega)[n_e^{-2}(\omega) - n_o^{-2}(\omega)] + n_e^3(2\omega)[n_e^{-2}(2\omega) - n_o^{-2}(2\omega)]\} \cdot \Delta\theta^2$$

$$\sim \frac{\omega}{2c} n_o^3(\omega)[n_e^{-2}(2\omega) - n_o^{-2}(2\omega) \cdot \Delta\theta^2. \tag{38}$$

Type I – positive uniaxial (ee–o)

By using eq. (17),

$$\Delta k = -\frac{\omega}{c} n_e(\omega)[n_e^{-2}(\omega) - n_o^{-2}(\omega)] \cdot \Delta\theta^2. \tag{39}$$

Type II – positive uniaxial (oe–o)

By using eq. (19), it follows that

$$\Delta k = -\frac{\omega}{2c} [2n_o(2\omega) - n_o(\omega)][n_e^{-2}(\omega) - n_o^{-2}(\omega)] \cdot \Delta\theta^2. \tag{40}$$

Once again, the acceptance angles $\delta\theta$'s can be obtained by setting $\Delta k = 2\pi/L$ in eqs. (37) through (40). 90° phase-matching is also called noncritical phase-matching because $\delta\theta$ is typically relatively large.

Acceptance angles (in the polar direction) of tens of milliradians are common for SHG utilizing the 90° phase-matching condition. It is listed for some crystals in table I. As in the case of angle phase-matching, rotational invariance for uniaxial crystals implies that phase-matching is not sensitive to azimuthal changes. Also, since all Poynting vectors are collinear with their respective wave normals, beam walk-off does not occur when 90° phase-matching is used.

2.4 SHG acceptance temperature, δT

To maintain an achieved phase-matching condition it is important to keep the crystal at a precise temperature. In an angle phase-matched crystal, a departure from this temperature can usually be compensated by a reorientation of the crystal. In a 90° phase-matched crystal where the normal surfaces (fig. 6) are made to match up at the 'equator' by carefully bringing the crystal to the unique phase-matching temperature, reorientation works less well. In one temperature direction, reorientation can regain phase-matching only at the expense of progressively losing the advantages of noncritical phase-matching in terms of acceptance angle and beam walk-off. It does not work at all in the other temperature direction because the relevant normal surfaces will have drifted apart (this can best be visualized for a type I phase-matching).

The width of a SHG output versus temperature curve is called the acceptance temperature, δT, defined as the temperature distance between the maximum and the first minimum, such as shown in fig. 7. Although δT is theoretically approximately equal to the full width at 0.405 maximum, this is the case only for weak signal SHG in an optically uniform crystal. In practice δT is roughly equal to the fwhm with ± 5 to $\pm 15\%$ uncertainty.

As defined, δT can be obtained for the four phase-matching conditions by using eqs. (13), (15), (17) and (19), each in conjunction with eq. (2) and (6), and by using

$$2\pi/L = \Delta k \equiv \frac{\partial(\Delta k)}{\partial T}\,\delta T. \tag{41}$$

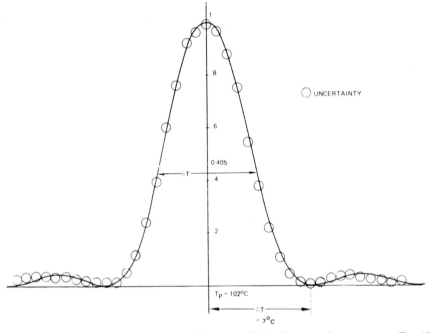

Fig. 7. Normalized weak-signal SHG power $P(2\omega)$ around the phase-match temperature $T_p \sim 102°C$ for a 1 cm $\sim$ 80% deuterated cesium dihydrogen arsenate (CD*A) crystal. Experimental data (O) are best fitted by the curve $[\sin^2\xi(T - T_P)/(\xi(T - T_P))^2]$. The distance between the maximum and the first minimum $- \delta T -$ is also the full width at 0.405 maximum and is called the SHG acceptance temperature.

The results are:

Type I – negative uniaxial (oo–e)

$$\delta T \sim (\pi c/L\omega)\left|\frac{\partial}{\partial T}[n_o(\omega) - n_e(2\omega, \theta_m)]\right|^{-1}. \tag{42}$$

Type II – negative uniaxial (eo–e)

$$\delta T \sim (2\pi c/L\omega)\left|\frac{\partial}{\partial T}[n_e(\omega, \theta_m) + n_o(\omega) - n_e(2\omega, \theta_m)]\right|^{-1}. \tag{43}$$

Type I – positive uniaxial (ee–o)

$$\delta T \sim (\pi c/L\omega)\left|\frac{\partial}{\partial T}[n_e(\omega, \theta_m) - n_o(2\omega)]\right|^{-1}. \tag{44}$$

Type II – positive uniaxial (oe–o)

$$\delta T \sim (2\pi c/L\omega)\left|\frac{\partial}{\partial T}[n_e(\omega, \theta_m) + n_o(\omega) - n_o(2\omega)]\right|^{-1}. \tag{45}$$

A typical plot of second harmonic output versus temperature is shown in fig. 7 for temperature-tuned cesium dideuterium arsenate (CD*A). The CD*A crystal used was about 80% deute rated and had a phase-match temperature of $\sim 100°C$. The data agrees closely with the curve obtained when $[\sin(L\Delta k/2)/(L\Delta k/2)]^2$ in eq. (1) is represented by $[\sin T\xi/T\xi]^2$, where ξ is a best fit parameter, indicating excellent optical quality CD*A. Table 1 gives δT for many crystals. A large acceptance temperature is desirable for high average power SHG, as will be shown later.

2.5 Optical nonuniformity in SHG crystals

The discussion so far has assumed perfect optical quality for the SHG crystals. While this is nearly the case with many crystals, notably isomorphs of potassium dihydrogen phosphate and good quality lithium niobate (see §§ 4.2.1 and 4.2.2), many other crystals exhibit optical nonuniformity resulting from such imperfections as microtwinning, dislocation and nonuniform impurity distributions. In the first approximation, such optical nonuniformity can be characterized by a spatially varying phase-match temperature. The SHG efficiency is limited because the usable crystal size is limited (along and/or perpendicular to the laser beam axis) and because an effective nonzero Δk is introduced into eq. (1), which leads to partial reconversion of energy from the second harmonic back to the fundamental when the phase-relationship between the two waves reverses as the waves progress through the crystal. Several authors have studied this problem, particularly relating to the effects on SHG by periodic optical nonuniformity such as striations in earlier lithium niobate and barium sodium niobate crystals

(Midwinter 1967, 1968, Bergman et al. 1968, Nash et al. 1970, Smith 1970a, b, Tsuya et al. 1970, Srivastava et al. 1976). Butyagin et al. (1974) studied the influence of linear inhomogeneity. Tagiev and Chirkin (1974) discussed the general case of inhomogeneity. Since crystal quality tends to vary from grower to grower and even from boule to boule, it is important to characterize each crystal in terms of its SHG efficiency as a function of power density and length, at least on a spot check basis, before integrating it into a SHG system. A nominal crystal length *d* for a crystal type by a given grower can often be chosen experimentally to designate the length that provides the most efficient SHG. This quantity will be used in § 5.5.

There are other effects, namely thermal effects (next section) and beam quality (§ 3.1), besides crystal inhomogeneity that make the naive application of eqi. (1) and (5) hazardous in practice. This is illustrated in fig. 2, in which the data based on a high-quality $LiNbO_3$ crystal and a multimode Nd:YAG laser operating Q-switched at 100 mJ per pulse is only half that predicted by eq. (5). Such experimental data using a crystal and a laser that are close facsimiles to the intended application is essential in the design of an efficient, high average power SHG system.

2.6 Self-heating in SHG crystals

Optical absorption of the traversing laser beam in the SHG crystal is expressed first as localized self-heating and then, after thermal gradients are established, as average power self-heating. The former is important in pulsed systems when the local rise in temperature produces a deterioration of the phase-matched condition during the pulse, thereby limiting the SHG efficiency during the trailing portion of the pulse.

Average power self-heating produces a steady-state spatial temperature distribution in the SHG crystal when either a continuous wave (cw) laser or a repetitively pulsed laser whose repetition rate is fast compared to the thermal diffusivity of the crystal is employed. This is the case most commonly encountered. Regardless of the energy distribution profile of the fundamental laser beam, any attempt to maintain temperature stability by extracting the generated heat creates a radial temperature gradient in the active medium. For a spatially gaussian beam of average power $\bar{P}$, spot diameter *h*, and a radial average power distribution given by (Okada and Ieiri 1971)

$$\bar{\rho}(r) = \left(\frac{4\bar{P}}{\pi h}\right) \exp\left(\frac{-4r^2}{h^2}\right), \tag{46}$$

the steady-state radial temperature distribution can be approximated by

$$T(r) = \left(\frac{\bar{P}\eta}{2K}\right) \exp\left(\frac{-2r^2}{h^2}\right) + T_0, \tag{47}$$

for $r < h/2$, where K = thermal conductivity and T_0 = a constant determined by boundary conditions.

The situation is shown in fig. 8. A normalized gaussian beam $\bar{\rho}(r)$ is represented by the solid line. The actual temperature distribution is shown by the dotted line (Woodbury 1977). This is closely approximated by eq. (47) (shown by the dashed line), the height of which at $r = 0$ is normalized to

$$4\Delta'T = \left(\frac{\bar{P}\eta}{2K}\right). \tag{48}$$

$\Delta'T$ is called the degradation temperature since it measures the severity of self-heating in the medium. As the magnitude of $\Delta'T$ approaches that of δT (see fig. 7), simultaneous phase-matching in all parts of the crystal becomes impossible, and the SHG efficiency is degraded.

An exact solution for the degraded efficiency can be obtained from eqs. (1) and (47) and fig. 7. The result is plotted in fig. 9, where the independent variable is $\Delta'T/\delta T$. It can be seen that when $\Delta'T/\delta T = 2$, the SHG output decreases by 20%. This occurs in a CD*A crystal when the input power is ~ 20 W. The curve shown in

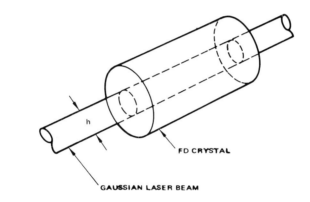

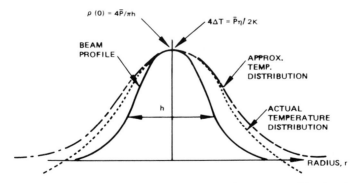

Fig. 8. Steady-state temperature profile (dotted) of a gaussian beam (solid) of radius h passing through a cylindrically symmetric crystal. (The actual solution is shown by the fine dotted line. Coarse broken line from eq. (29) is a good approximation for $r < h/2$.) All curves are normalized to equal heights at $r = 0$ for comparison.

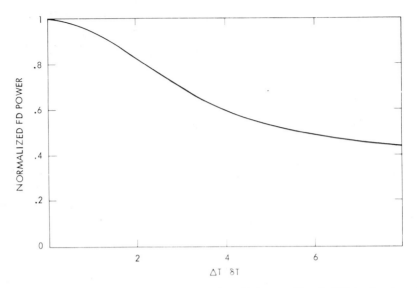

Fig. 9. Deleterious effect of thermal profile on FD efficiency. $\Delta T = P\eta/8K$ is the degradation
temperature due to self-heating; δT is the SHG tolerance temperature of the crystal.

fig. 9 is optimistic because it does not account for crystal deterioration and eventual
damage resulting from differential thermal stress.

Efficient SHG requires that the degradation temperature be no greater than the
acceptance temperature, i.e.,

$$\Delta'T = \frac{\bar{P}\eta}{8K} \leq \delta T. \tag{49}$$

It will be shown later that beam shaping can reduce $\Delta'T$ and thus increase the
power handling capability of a SHG crystal.

3 Lasers for high average power SHG

Eqs. (1), (4) and (5) are based on idealized assumptions about both the laser
source and the SHG crystal. That is, the laser output is considered to be a train of
plane waves, and the crystal is assumed to be of uniform optical quality. In addition,
the equations cited above do not consider the detrimental effects of self-heating in
high power applications. Nevertheless, these equations are remarkably accurate
when applied to weak signal SHG of a single-mode laser in a small uniform crystal.
The results obtained under these conditions therefore serve as an upper bound for
practical SHG performance, and the physical assumptions that underlie a
quantitative treatment of the results serve to guide the design of an optimal SHG
system. In this section, the qualities that define the ideal laser for frequency
doubling applications are described. Methods for optimizing laser brightness and
controlling divergence and linewidth are discussed, and a survey of the most
common lasers which are used in high power SHG applications is presented.

3.1 Properties of the ideal laser for high average power SHG

The important attributes for an ideal laser for high power SHG include: high peak power, narrow spectral linewidth and small beam divergence; i.e., the laser should be characterized by high brightness (spectral intensity per steradian). The requirement for high peak power is necessitated by the $P^2(\omega)$ term in eq. (1). Narrow spectral linewidth ensures that the phase-matching condition, described in eqs. (13) through (20), is satisfied for all frequencies that comprise the laser line-width. The desirability of a small beam divergence arises because many SHG crystals, particularly those that are angle phase-matched, possess a narrow acceptance angle $\delta\theta$ (see § 2.2). Furthermore, focusing is often required to increase the radiation power density at the crystal, and this in turn leads to a proportionate increase in divergence.

The laser that best approximates this brightness requirement is a pulsed laser operating in a single transverse (TEM$_{00}$) and longitudinal mode. Furthermore, a single-mode beam, whether pulsed or continuous, tends to be smooth in time and space. This lessens the risk of crystal damage due to the temporal and spatial fluctuations that are inherent in multimode laser beams. Unfortunately single-mode operation is usually obtained using techniques that severely limit both efficiency and average power. A laser with n randomly phased longitudinal modes of equal amplitude (but a single transverse mode) tends to frequency double more efficiently than a single-mode beam by a factor of $(2-1/n)$ (Bloembergen 1965, Francois 1966, Dowley 1968, Hagen and Magnante 1969). This is true only for low conversion efficiencies, where second harmonic generation follows a square-law dependence. For high conversion efficiencies the factor approaches one. However, in practice, a high average power laser has many transverse modes as well, giving rise to a beam many times diffraction-limited. That the resulting SHG efficiency is in fact then lower than that of the single-mode case can be seen from the following approximate argument. Assuming eq. (1) is applicable for each of the N mutually uncorrelated transverse modes in a multimode beam, the total average second harmonic power is

$$\overline{P_m(2\omega)} = b \sum_i^N \{\overline{P_i^2(\omega)}\}, \tag{50}$$

where $P_i(\omega)$ is the power of the ith mode and b is a constant of proportionality. However, if the same laser power is concentrated in a single mode, that is, if

$$P_s(\omega) = \sum_i^N P_i(\omega),$$

then the total average SHG power is

$$\overline{P_s(2\omega)} = b\overline{P_s^2(\omega)} = b \left\{\overline{\sum_i^N P_1(\omega)}\right\}^2. \tag{51}$$

This is greater than the multimode power described by eq. (50) since the square of the sum [eq. (51)] is greater than or equal to the sum of the squares [eq. (50)].

This fact, along with a larger power density safety margin to avoid crystal damage, explains why multimode lasers tend to have a lower SHG efficiency than do single-mode lasers. A 'good' single-pass conversion efficiency for second harmonic generation with a multimode laser is typically 40% as compared to an efficiency of approximately 65% for the case of a 'good' SHG with a single transverse mode laser.

3.2 Methods for increasing laser brightness

The brightness (spectral power density per steradian) of a given laser can be maximized by one or more of the techniques discussed in this section normally at the price of efficiency.

3.2.1 Q-Switching and cavity dumping

A substantial (as much as several orders of magnitude) increase in laser peak power is achieved by reduction of the laser pulse width using laser Q-switching or cavity dumping (less frequently used) – two techniques that are particularly effective for generating optical pulses of nanosecond duration. Since short pulse production by these methods has been discussed extensively elsewhere [e.g., Koechner (1976)], it will not be treated here. Although high peak powers are generated by these techniques, intracavity losses associated with the pulse production process often result in a moderate loss (10 to 60%) in overall average power.

3.2.2 Mode-locking

Laser mode-locking may be used to generate optical pulses of extremely short duration. Either active modulation (employing an externally driven modulation element) or passive modulation (employing a bleachable filter with intensity dependent transmission characteristics) of the intracavity flux produces a train of up to hundreds of picosecond duration pulses with an interpulse separation equal to the round-trip transit time of light in the resonator. Smith et al. (1974) give a comprehensive review of laser mode-locking and production of ultrashort pulses.

The enormous increase in peak power obtained with mode-locking is accompanied by a decrease of overall average power. However, the short temporal duration of the pulse results in an increased damage threshold of the nonlinear medium (Hurwitz 1966) The highest SHG conversion efficiency reported (70%) was achieved with a mode-locked beam.

Mode-locking broadens the spectral linewidth $\Delta\omega$ of the pulse in accordance with the uncertainty principle:

$$\Delta\omega\Delta t > 1.$$

This line broadening may interfere with phase-matching as noted above. it is possible to control $\Delta\omega$ by purposely preventing the temporal width Δt from becoming too narrow – a kind of deliberate 'sloppy' mode-locking. Mode-locking may be employed in conjunction with Q-switching to obtain short trains of intense pulses.

3.2.3 Linewidth narrowing

Linewidth narrowing of broadband lasers such as dye lasers and Nd : glass lasers is often necessary if optimal SHG is to be obtained. Linewidth compression is usually achieved by use of etalons and/or diffraction gratings. Depending on the laser, loss in overall average and peak power may result.

3.2.4 Divergence control

Beam divergence is a concern in almost all laser applications and therefore continues to be of central importance in laser engineering. Various techniques [e.g., unstable resonators (Siegman 1965), double rod rotators (Scott and DeWitt 1971) and improved medium uniformity] have proven successful. The challenge that confronts the laser designer is optimization of beam divergence without significantly sacrificing efficiency.

3.2.5 Intracavity SHG

SHG efficiency can be substantially improved for a low power laser by placing the SHG crystal inside the laser resonator. Not only are the photons forced to pass through the crystal repeatedly, but the one-way peak power is usually larger than the output peak power by at least a factor of $[(1 - \Omega)/\Omega]$, where Ω, the transmission of the output mirror, is small for low power lasers. By carefully adjusting the various parameters, Geusic et al. (1968) converted almost all of the 1.06 μm energy from a single-mode Nd : YAG laser into the second harmonic using an intracavity $Ba_2Nb_5O_{15}$ crystal. Smith (1970a, ba) explored the theory in greater detail. The problem was pursued further by Barry and Kennedy (1975) and Hanamitsu et al. (1976) Volosov et al. (1973, 1975) Dmitriev et al. (1974, 1975) and Barry (1976) deal primarily with the simpler case of a single-mode laser. However, the general case of a multimode (or single-mode) laser with insertion loss by the SHG crystal may be usefully treated in the following semi-empirical manner. Assume that the laser has been adequately characterized (experimentally) in terms of the pulse energy output E_{out} as a function of the pulse energy input E_{in} and the transmission Ω of the output mirror. (For a cw laser, E_{out} and E_{in} should be replaced by P_{out} and P_{in}.) If L is the two-way insertion loss of the SHG crystal, and if the extracted second harmonic is the sole output coupling of the cavity, the the optimal two way SHG coupling should be

$$\Omega_{SHG} = \Omega_0 - L, \tag{52}$$

where Ω_0 is the experimentally optimized value of Ω for the given E_{in}. To achieve the optimal Ω'_{SHG}, a number of parameters must be carefully chosen, including: spot size at the crystal, length of crystal and the specific crystal. Unlike the case of external-cavity SHG, maximum single-pass efficiency is not necessarily optimal. Instead, the single-pass SHG efficiency versus power density must be used to find Ω'_{SHG}. Only experimental values should be used because they depend on the characteristics of the laser (§ 3.1), uniformity (§ 2.5) and thermal profile (§ 2.6) of the crystal. It also should be determined with single-pass experiments as close to the actual operating condition as possible because absorption is a nonlinear function of second harmonic intensity. It is often possible to achieve (with some final tuning) a maximum SHG pulse energy E_{SHG} given by

$$E_{SHG} = E_{out}[(\Omega_0 - L)/\Omega_0], \tag{53}$$

where E_{out} is the optimal output of the laser before the SHG crystal is introduced. This corresponds to a 100% SHG efficiency less the insertion loss.

Since photons are traveling in both directions inside the laser resonator, SHG outputs in both directions are also generated. The two SHG outputs can be 'folded' inside the cavity to form one linearly polarized output if the laser is single-transverse-mode. The distance between the back mirror and the SHG crystal must be carefully optimized so that the dispersiveness of air cancels the effects of the relative phase retardations caused by intermedia interfaces and non-phase-matching media, such that phase-matching occurs for the ω and 2ω waves in the SHG crystal for both directions of propagation. It has been found, however, that in the case of a multimode laser, the two SHG outputs must be extracted separately if optimal SHG efficiency is to be achieved; otherwise, the second harmonic coverts part of its energy back to the fundamental frequency as it traverses the crystal the second time. This down conversion is a parametric process with the idler power being supplied by the laser.

3.3 Common lasers in use for high power SHG

Any laser with an output frequency that is compatible with the SHG media can be used. The most commonly used are those which most closely resemble the 'ideal' laser described in § 3.1. Nd:YAG, ruby, CO_2 and dye lasers (Huth et al. 1968, Bokut et al. 1972) together with the frequency shifted outputs from these sources are among those most often encountered in SHG applications. High peak power, multiwatt copper vapor lasers (Hargrove and Kan 1977) appear promising. SHG experiments involving various laser wavelengths (as well as various crystals) are included in table 1 along with a large number of references. Only Nd:YAG and CO_2 lasers are discussed in greater detail below.

3.3.1 Neodymium: YAG laser

Nd:YAG (Nd^{3+} doped into a yttrium aluminum garnet crystal), which has a primary laser transition in the near infrared (1.06 μm) spectral region, has been widely used as a source for SHG for many years. The maturity and versatility of this laser have been primarily responsible for its popularity.

Since this lasing medium possesses an energy storage time of about 200 μs, the flashlamp pumped/Q-switched mode of operation has been the most effective in high peak power SHG applications. Depending upon the specific design, 5 to 20 ns pulses at average powers of ~3 W multimode or ~1 W single mode at repetition rates as high as 200 Hz may be routinely generated with a $4 \times 1/4$ inch ND:YAG rod. The output, typically less then 150 mJ, can be amplified by successively larger rods to achieve as much as 80 W average power with a 2% (nominal) electrical to optical conversion efficiency. Two joule, single mode output at 10 Hz has recently been reported (Kogan and Crow 1977) for an Nd:YAG (unstable) oscillator and several amplifiers.

The beam divergence θ_0 of a 1/4 inch multimode beam is typically 2 to 3 mrad. From experience, a brightness of 8×10^{11} W/cm^2 rad^2 (for example, from a 20 mJ, 10 ns 3 mrad laser) can be considered the minimum for efficient (>20%) SHG in most 90° phase-matched crystals (assuming the availability of beam focusing optics). Since brightness is inversely proportional to the square of the beam divergence, a single-mode beam with its higher brightness is preferred for angle phase-matching in crystals with small acceptance angles.

There is a class of flashlamp-pumped Nd:YAG lasers that produces average powers in the hundreds of watts but does not employ any Q-switch elements (since none exist that can sustain the high average power). However, since the output pulses are temporally broad, these lasers generally do not possess the necessary brightness for efficient SHG.

Nd:YAG lasers can be pumped with cw krypton lamps to produce powers as high as 200 W with electrical to optical conversion efficiencies as large as 3% (again assuming a $4 \times 1/4$ inch laser rod). A multi-rod resonator of this type has been tested and shown to produce cw powers approaching 1 kW. These peak powers are again too low for efficient SHG. However, the peak power can be increased significantly by either Q-switching (Chesler et al. 1970a, b), cavity-dumping (Maydan and Chesler 1971), or mode-locking (Kuizenga and Siegman 1970). For example, a continuously pumped, acousto-optically Q-switched Nd:YAG laser operating at repetition rates of 5 to 30 kHz produces only slightly reduced average powers when compared with the output obtained in cw operation. Beam divergence depends on average power and is typically 8 to 16 mrad.

Pulsewidths of 80 to 350 ns (dependent on repetition rate and average power) are obtained and peak powers in excess of 100 kW may be achieved with commercial multi-rod systems.

Cavity dumping has produced pulses of < 100 ns duration at repetition rates of 10^5 to 10^7 Hz with no reduction in laser efficiency. However, the active modulation technology required to fully exploit this high power technique is still at a relatively early stage of development. Higher pulse rates may be achieved by mode-locking the Nd:YAG laser. Mode-locking has been used to produce 30 ps to 1 ns duration pulses at a repetition frequency of 10^8 Hz (Kuizenga and Siegman 1970). Such lasers, when frequency doubled to produce 0.53 μm radiation so as to exploit detector sensitivity, offer great potential for high data rate optical communications.

The Nd:YAG laser can lase in 18 lines in the spectral range 1.05 to 1.44 micron, some of them with efficiency only slightly below the 1.064 micron primary laser line (Smith 1968, Marling 1977). Tuning was achieved at room temperature with an etalon.

For a comprehensive survey of the design variations and performance specifications of the Nd:YAG laser see Kiss and Pressley (1966), Findlay and Goodwin (1970) and Weber (1971). Koechner's Solid-State Laser Engineering (1976) is an excellent source book for Nd:YAG laser technology.

3.3.2 CO₂ laser

Use of a CO_2 laser as the input to a frequency doubling unit offers an attractive means for obtaining spectral flexibility in the mid-infrared. With over 400 CO_2 lasers lines available from 8.7 to 11.8 μm, at maximum efficiencies in the range of 25%, frequency doubling may be used to provide effective coverage of the 4.3 to 5.8 μm band. Owing to the vigorous research and development efforts directed toward CO_2 laser technology, a wide variety of designs must be considered as candidates for use in second harmonic generators.

Devices using axial flow are often characterized by low peak powers and continuous wave operation. While many do fit this description, significant exceptions exist in short duration, pulsed devices operating with very high voltage power supplies to produce output pulses of up to 1 MW with 20% efficiency at 30 Hz rates (Hill 1970). Other techniques, such as Q-switching (Tyte 1970), and mode-locking (Bridges and Cheo 1969), have been utilized to boost the peak powers attainable in cw discharge, pulsed output configurations. However, kinetic and thermal limitations restrict these techniques to power levels on the order of 10 kW per meter of discharge.

More suitable sources for frequency doubling applications are lasers in which the gas flow is transverse to the beam axis. Although cw and pulsed gas dynamic lasers operating at peak powers in the 60 to 450 kW range have been reported (Konyukof et al. 1970, Christianson and Herzberg 1973), transversely excited atmospheric pressure (TEA) lasers appear most compatible with the short duration, high energy pulsed mode required for most nonlinear processes. Although specific devices vary widely in physical configuration, discharge geometry and pre-ionization scheme, peak power-levels in the 100 MW to 3 GW range (Seguin et al. 1972, Richardson et al. 1973), are attainable for pulse durations in the 50 to 200 ns range. Mode locking, occurring spontaneously as the result of insertion of active (e.g., acousto-optic or electro-optic modulators) and passive (e.g., SF_6 absorption cell) intracavity elements, has been effective in generating trains of pulses with temporal widths of 1 to 5 ns (Wood et al. 1970, Nurmikko et al. 1971) and efficiences as high as 80% of those achieved in normal operation. With significant capability for adapting the basic laser design to meet required beam size, shape and quality specifications, the TEA configuration is an attractive choice for frequency doubling applications.

In addition to specifying the discharge geometry, additional factors must be considered in selecting a CO_2 laser for frequency doubling. Although stable resonator designs may be employed successfully, theoretical and experimental results suggest that unstable resonators provide several significant advantages (Siegman 1965, 1974, Krupke and Tory 1969). These include single transverse mode operation at the high Fresnel numbers typical of large volume discharges, efficient energy extraction from the large gain region and elimination of easily damaged partially reflecting cavity output mirrors. Choice of the appropriate design permits achievement of the high brightness levels required for efficient frequency doubling.

An additional constraint on CO_2 laser configuration is imposed by the phase-matching requirements of the nonlinear material chosen for use. Most high gain CO_2 lasers oscillate simultaneously on several spectral lines. Since typical nonlinear media do not exhibit dispersive properties that allow simultaneous phase matching over the entire wavelength range, line selection is required. This is generally accomplished by the addition of a grating, prism or Fabry–Perot etalon of appropriate design within the laser resonator, with some loss of efficiency.

A final consideration relating to frequency doubling of CO_2 laser output relates to the optical media available for beam control. Although active laser medium may be scaled almost arbitrarily in length to increase output power density, the finite damage thresholds of available SHG crystals significantly limit the achievable powers. With maximum allowable energy densities in the range of 1 to 75 J/cm^2 for typical TEA laser outputs of 75 to 200 ns pulse duration, maximum accessible peak power densities are in the range of 4 to 400 MW/cm^2 [see, for example, Braunstein et al. (1973) Hanna et al. (1972), Davit (1973)].

The highest SHG conversion efficiency to date is 21% (average power 30% peak power) when a 1.3 cm long $CdGeAs_2$ crystal was used by Menynk et al. (1976). Using a 0.9 cm $CdGeAs_2$ crystal, Kildal and Mikkelsen (1974) obtained 15% conversion with a 1.4×10^6 W/cm^2 power density. 15% was also reported by Berezovskii et al. (1975) in SHG in 0.9 cm of Te using a 5×10^6 W/cm^2, 30 ns CO_2 laser. Small size and high absorption of infrared nonlinear crystals are currently the major difficulties in attempts to obtain high average power SHG.

4 Crystals for high average power SHG

Just as it was useful for defining the ideal laser, the general discussion of SHG presented in § 2 is likewise helpful for identifying the characteristics of the ideal crystal for frequency doubling.

4.1 Properties of the ideal crystal

An ideal crystal for high average power SHG is one in which generation of second harmonic radiation occurs with high efficiency even when the average power level of the incident fundamental radiation is very high. The ideal crystal must have a:
(1) High nonlinear coefficient.
(2) High damage threshold – so that power density can be maximized.
(3) Large size with optical uniformity. Spatial variations in the refractive index in any direction imply variations in Δk for a given temperature.
(4) Large acceptance angle. This is required because most efficient high average power lasers are characterized by outputs that are not diffraction limited. Furthermore, to achieve a desirable power density, the beam must often be focused, and this results in a proportionately larger divergence.
(5) Low optical absorption η at both ω and $2\,\omega$.
(6) High thermal conductivity K.
(7) Large acceptance temperature δT.
(8) Low cost and be easy to optically fabricate.
Crystal uniformity was discussed in § 2.5. Items (1) through (3) came from eq. (1); item (4) has been discussed in §§ 2.2 and 2.3; and, items (5) through (7) are implied by eq. (49) of § 2.6.

4.2 High average power SHG crystals

Intracavity SHG conversion efficiencies approaching 100% (Geusic et al. 1968) and extracavity single-pass efficiencies as high as 70% (Machewirth et al. 1976) have been reported. However, in practice, the 'ideal' laser and 'ideal' SHG crystal are seldom simultaneously available for a given application, and single-pass conversion efficiencies of ~40% in the visible spectral region and ~10% in the ir and uv are usually considered good. At visible and near-visible wavelengths, there are a variety of SHG crystals with low optical absorptions (0.5 to 5% cm^{-1}). Restrahl (fundamental lattice vibrations) or multiphonon absorptions in the ir and intrinsic electron band absorptions in the uv limit the availability of suitable crystals for SHG in these spectral regions.

Table 1 lists some of the commonly used SHG crystals and their pertinent characteristics, i.e., transparency range, refractive indices, nonlinear coefficients, damage thresholds, absorption coefficients and nominal sizes. The table also lists representative SHG experiments that have been conducted with each crystal and includes the fundamental frequencies, critical or noncritical, and type I or type II phase-matching conditions, temperature, acceptance temperature, length of crystal used and (where available) the efficiency obtained. Phase-match angles θ_m's and acceptance angles $\delta\theta$'s not included may be calculated from the indices by application of eqs. (21) through (24), eqs. (32) through (35) and eqs. (37) through (40). Although some data are presented here for the first time most of the information was obtained from the literature. Although the table attempts to review the current status of available SHG crystals as completely as is feasible, it should not be considered an exhaustive survey. For additional information, the reader is referred to an excellent review article by Nikogosyan (1977). Earlier, less exhaustive reviews (Harris 1969, Pressley (ed.) 1971, Warner 1971, Milton 1972, Colles and Pidgeon 1975, Koechner 1976), are also helpful. In addition, the book Applied Nonlinear Optics by Zernike and Midwinter (1973) offers an excellent treatment of this general subject.

Several crystals and crystal classes shown in table 1 have been of particular importance in SHG applications. They are therefore discussed in more detail below.

4.2.1 Isomorphs of potassium dihydrogen phosphate (KDP)

These materials comprise a versatile group of SHG crystals which span the broad spectral range from ~0.2 to 1.5 μm. In particular, 90° phase-matching is possible with many of 'hese crystals and may be used to obtain SHG in many bands within the above spectral region [see fig. 10, courtesy of R.S. Adhav (1974)]. The nonlinear coefficients of these materials are generally quite low (related to their low indices) (Miller 1964). However, their damage thresholds are large and high peak power density is therefore necessary and acceptable for SHG.

Large, uniform crystals (4 to 30 cm cubes) can be grown from aqueous solutions, and coherence lengths of 2 to 4 cm can be obtained. The large variety of crystals available permits efficient SHG with a lare number of sources, including ruby, neodymium, dye and argon lasers. For example, rubidium dihydrogen phosphate (RDP) is an efficient material for room temperature frequency doubling of the ruby laser output (0.6943 μm), while rubidium dihydrogen arsenate (RDA) is suitable for 90° phase-matching of the ruby fundamental frequency.

For Nd lasers, cesium dihydrogen arsenate (CDA) and its deuterated form (d-CDA or CD*A) have exhibited SHG conversion efficiencies in excess of 50%. CDA phase-matches at ~50·C, while ~80% deuterated CD*A phase matches at ~100°C; the acceptance temperature δT for both is ~3°C for a crystal length of 2 cm. Deuterated potassium dihydrogen phosphate (d-KDP or KD*P) crystals have been used in a type II angle phase-matching configuration to frequency

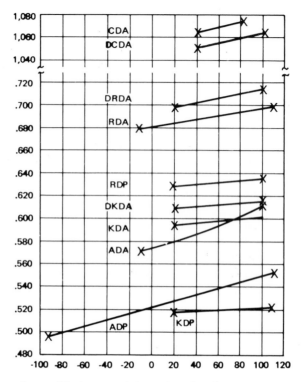

Fig. 10. KDP isomorphs can 90° phase-match in many regions of the visible spectrum. Shown here are crystal selections based on fundamental wavelength versus temperature curves. (Courtesy Adhav).

double the 1.064 μm fundamental with conversion efficiencies of up to 70%. KD*P can also be used to frequency double the resulting 0.523 μm light to 0.266 μm light to 0.266 μm. Ninety degree, type I phase-matching is used to obtain this fourth harmonic wavelength with high efficiency.

Ammonium dihydrogen arsenate (ADA) phase-matches to dye and argon laser lines at temperatures between 18 and 100°C to generate uv radiation. When ammonium dihydrogen phosphate (ADP) is employed as the frequency doubling medium, phase-matching of argon lines as short as 0.4965 μm can be obtained.

These crystals are all hygroscopic and, because of their softness, often present some polishing and coating difficulties. Nevertheless, excellent optical polish and anti-reflection coatings have been achieved on all these crystals. Deuteration (substitution of hydrogen by its heavier isotrope deuterium) has often been used successfully in this family of crystals to lower the linear absorption in the visible (which is largely attributed to multiphonon absorption tails from the high frequency portion of the Restrahl involving the light atom hydrogen) and to shift the phase-matching conditions in terms of temperature, angle of orientations and/or fundamental frequency by a slight modification of the index ellipsoid. The absorption coefficient of KD*P is low for SHG involving 1.06 μm. However, due to nonlinear effects involving 0.53 μm radiation, the corresponding coefficient for CD*A is high and has only recently been reduced from 0.07 to 0.015 cm^{-1} by purification of the raw materials (Hon et al. 1979). Because the dominant absorption during SHG is nonlinear in origin, the level of deuteration in CDA (which affects the smaller linear absorption only) tends to have relatively little effect on the overall SHG performance beyond increasing the phase-matching temperature. Presently available CD*A cannot be stored at room temperature for more than a few years owing to slow chemical decomposition. The rate of decomposition is accelerated by elevated storage temperatures. The crystal disintegrates at about 125°C. The most complete compilation of data on KDP isomorphs can be found in Adhav (1968) and Milek and Welles (1970).

4.2.2 Lithium niobate (LiNbO$_3$)

Lithium niobate crystals can be grown with high optical quality and in relatively large sizes (4 cm^3). This material is 90° phase-matchable for 1.06 μm radiation at approximately 0°C (Boyd et al. 1964, Miller et al. 1965). However, LiNbO$_3$ is particularly susceptible to optical index damage resulting from propagation of high average power visible or ultraviolet radiation. This situation is reversible if the ambient temperature of the crystal is increased to ~170°C or if the crystal is left alone for several days (Ashkin et al. 1966, Chen 1969, Peterson et al. 1971). Nassau et al. (1966) describe growth, domain structure and crystal structure studies. Other papers report studies of some of the practical problems that arise in the use of lithium niobate (Boyd et al. 1964, Miller et al. 1965, Ashkin et al. 1966, Bjorkholm 1968). So-called 'hot' LiNbO$_3$ which is prepared by increasing the proportion of the Li$_2$O or MgO in the melt can be used for 90° phase-matching for 1.06 μm at ~170°C, and is capable of rather efficient frequency conversion in high power applications. Unfortunately, even though boules with diameter larger than 5 cm can be grown, the nonstoichiometrically grown 'hot' crystals suffer from optical inhomogeneities in the form of striations normal to the growth direction. Growth along the x or y-axes instead of the z-axis, as well as improvement in growth techniques would enhance the usefulness of this crystal in high power applications. The advantages of this crystal include a relatively high nonlinear coefficient, large acceptance angle, high thermal conductivity and structural features that allow ease of handling and polishing. An additional detriment to those mentioned above is a relatively low damage threshold (~80 MW/cm^2, multimode).

4.2.3 Barium sodium niobate (Ba$_2$NaNb$_5$O$_{15}$)

This crystal is physically similar to LiNbO$_3$. However, it possesses higher nonlinear coefficients and is not readily susceptible to (reversible) optical index damage. Unfortunately 5 mm cubes are the largest sizes commercially available.

Voronov et al (1975) reported that partial substitution of sodium by potassium (0 to 80%) can both improve the optical quality and lower the 90° phase-match temperature for 1.06 μm from ~100 to ~23°C when the percentage of substitution is ~35%. However, the size of the crystal attained was not reported.

4.2.4 Potassium titanium phosphate, KTiOPO₄ (KTP)

This recently developed crystal shows promise of being the ideal crystal for SHG in the spectral region 0.35 to 4.5 μm (Zumsteg et al. 1976, Zumsteg 1978). The two most striking features about this slightly biaxial, angle phase-matched crystal are its large acceptance angle ($\delta\theta \sim 25$ mrad for a 4 mm crystal used for frequency doubling 1.06 μm), and large tolerance temperature (>50°C). Beam walk-off is minimal. It has a damage threshold of ~ 150 MW/cm² and an absorption coefficient $\eta \ll 1\%$. It has high optical indices of ~ 1.8, and thus, high nonlinear coefficients. It is a hard material, chemically stable, optically uniform and is easily polished and coated. A 4.5 mm crystal has exhibited 42% SHG conversion efficiency. In an intracavity experiment, 50 W of output was briefly reached. (Laser Focus 6, 1978). However, the maximum size attained to date is only ~ 5 mm for this newly developed crystal. This crystal was grown by DuPont using a hydrothermal growth procedure that relied on spontaneous nucleation.

4.2.5 Organic SHG crystals with charge-transfer resonance enhancement

Efficient harmonic frequency conversion has been reported in numerous organic crystals which exhibit pronounced charge-transfer bands in their electronic spectra (Davydov et al. 1970, Jerphagnon 1971, Southgate and Hall 1972, Bolognesi et al. 1973). The highly efficient conversion results directly from the resonance enhancement of the nonlinear susceptibility by a charge-transfer process.

Inter- and intramolecular charge transfer in a conjugated system containing donor and acceptor substituents will result in an appreciable change in the dipole moment of the molecule. Davydov et al. (1970) have proposed that, since the nonlinear susceptibility depends on the magnitude of the dipole moment change as well as on the strength of the allowed electronic transition nearest to the frequency 2ω, significant frequency conversion should be observed in these systems. In addition to the high nonlinear susceptibility, these biaxial substances generally exhibit a strong birefringence, high optical damage threshold (Bass et al. 1969, Southgate and Hall 1971) and a relatively weak dependence of the (angle) phase-matching conditions on temperature (Kotovschchikov et al. 1975). These properties have made charge-transfer organic crystals attractive candidates for frequency doubling at optical and infrared frequencies.

A large number of organic crystalline media have been surveyed and suggested for use in nonlinear applications. These materials are listed and discussed by Kurtz (1972) and Bolognesi et al. (1973); an example, meta-nitroanaline (mNA), is described in table 1. A 1.5 mm thick crystal gave a SHG efficiency of 15%. Research on these and other organic compounds is continuing and should provide valuable information about the nature of the nonlinear response in these materials and how the response can be optimized by modifications of the molecular structure. New, practical SHG crystals suitable for high average power applications may emerge as a result.

4.2.6 Cadmium germanium arsenide (CdGeAs₂)

4.2.6 Cadmium germanium arsenide (CdGeAs$_2$)

Cadmium germanium arsenide was used by Kildal et al. (1974) to frequency double the 10.6 μm output from a CO_2 laser. A conversion efficiency of 15% was obtained. The phase-matching angle was 48.4°, and a radiation power density of 1.4×10^7 W/cm^2 was employed. Menyuk et al. (1976) found that the absorption at 10 μm and 0.5 μm decreased to as low as 0.1 cm^{-1} and 0.4 cm^{-1} respectively when the crystal is cooled to 77°K. Using a 1.3 cm crystal at this temperature they obtained average and peak conversion efficiencies of 21 and 30% respectively in a type I SHG where the phase-match angle was 32°. The crystal is transparent in the spectral region from 2.4 to 18 μm. Although the absorption coefficient is large ($\sim$0.23 cm^{-1}) when compared with crystals operating in the visible, it is small in comparison with the absorption coefficients of other ir crystals. Thermal problems associated with this coefficient are currently the biggest difficulties facing high average power SHG in the ir. Solutions discussed in § 5 should be helpful.

4.3 Crystal engineering

4.3 Crystal engineering

The search for crystals for high power frequency doubling applications is continuing in many laboratories with results that have produced both short-term benefits and long-term potential. The search has primarily been directed along two paths: improvement of existing materials and identification and synthesis of new materials.

Impurities are known to be a major reason for absorption loss and heating as well as for lowering damage thresholds. It is further recognized that nonlinear processes, such as multiphoton absorptions or generation of higher harmonics (Bliss 1969, Ng and Woodbury 1971) often increase the overall losses during SHG when the peak power density is very high. Many of these mechanisms have been traced to impurities and stoichiometric imperfections. Considerable progress toward alleviating these problems has been made in recent years. However, significant research is still necessary in these areas as well as toward eventual development of large, optically uniform crystals for high power applications.

To obtain phase-matching for a desired frequency or to obtain a noncritical phase-matched condition, the refractive indices of a SHG crystal may be manipulated by crystal engineering. Techniques such as isotopic substitution, impurity doping (both stoichiometric and nonstoichiometric) and growth of mixed crystals have been successfully employed. Deuteration of KDP isomorphs and lithium-rich 'hot' lithium niobate are good examples.

Perhaps the most straightforward approach for selecting new candidates for SHG is to search for promising combinations among all the available data on optical crystalline materials for refractive index, transmission and crystal class. Such information is available from extensive tables compiled by Larsen and Berman (1934), Dana and Ford (1951) and Winchell and Winchell (1964). Proustite (Ag_3AsS_3) and cinnabar (HgS) were found in this manner (Hulme et al. 1967, Boyd et al. 1968). Miller (1964) suggested that a close relationship exists between the refractive indices and the nonlinear coefficient, i.e., between the linear and nonlinear polarizabilities. This rule has been confirmed in numerous cases and can be used to direct investigators toward high index materials.

Attempts to understand the microscopic origin of nonlinear susceptiability (in terms of bond charge anharmonicity) have been fruitful (Levine 1969, 1970, Jerphagnon 1970, 1971, Shigorin and Shipulo 1973). Such scientific insight has led to the successful discovery and design of a whole class of organic and inorganic SHG compounds where resonant charge-transfer is utilized to provide a large nonlinear susceptibility (see § 4.2). Future work in this area will concentrate on improving the transmission, fine tuning the absorption edge and growth of larger crystals.

Experimentally, the powder method proposed by Kurtz and Perry (1968) continues to be a powerful and cost-effective way to evaluate the nonlinear coefficients and phase-matching possibilities of candidate crystals.

For an in-depth treatment on the subject of crystal engineering, the reader is referred to S.K. Kurtz's treatise Nonlinear Optical Materials' which is chapter D1 of vol. 1 of this handbook.

5 Thermal problems in high average power SHG

The ideal laser and SHG crystal have been described. It was shown that, using appropriate laser power and proper phase-matching conditions, high SHG efficiency can be obtained for low average powers. However, in order to scale the performance to high average powers, thermal effects in the nonlinear crystal must be considered. Absorption of radiation by the crystal (see § 2.3) creates two average power problems. One is thermal instability, which prevents the establishment of the phase-matching condition for other than brief moments. The other is the generation of thermal gradients, which prevents simultaneous phase-matching in all parts of the crystal (Okada and Ieiri 1971, Gorokhov et al. 1974).

Thermal instabilities can be eliminated by servo-controlled electro-optical tuning, piezo-optical tuning or angle tuning. Thermal gradients can be reduced by laser beam shaping. Thirty-five watts of 0.53 μm radiation have been generated by using a combination of some of these techniques when frequency doubling a

Nd:YAG laser (Hon 1977, Hon et al. 1977). The techniques are scalable, and the potential for much higher second harmonic output power is promising.

5.1 Electro-optical (EOT) and piezo-optical (POT) tuning

Phase-matching for SHG involves establishing the proper relation between the dispersion, birefringence and temperature of the refractive index of the SHG crystal. Thus, the most common techniques involve (1) choice of material, (2) angular tuning and (3) temperature tuning. Electro-optical tuning (EOT) (van der Ziel 1964, Soffer and Winder 1967, Hon 1976) and, more recently, piezo-optical tuning (POT) (Hon, 1978) are more sophisticated techniques that further extend the possible range of high power SHG. Only one tuning technique is required to stabilize SHG output, but both tuning concepts are treated together in this discussion.

The second harmonic power realized relative to the maximum second harmonic power achievable may be written as a function of the phase mismatch Δk by rearrangement of eq. (1):

$$P(2\omega)/P(2\omega)_{\max} = \sin^2(\Delta kL/2)/(\Delta kL/2)^2. \tag{54}$$

Eq. (54), plotted in fig. 11, is maximized when $\Delta k = 0$. In high average power applications, the temperature fluctuations ΔT cause Δk to depart from zero. However, under certain conditions, this phase mismatch can be compensated by either an externally applied electric field E or stress field σ (without rotation of the crystal). We assume that at the outset the temperature and orientation of the crystal are adjusted so that a phase-matched condition exists. This provides the proper relationship between the wave vectors k and polarizations of the participant waves. Δk may be re-expressed as:

$$\Delta k = \Delta k(2\omega) - 2\Delta k(\omega), \tag{55}$$

where $\Delta k(\omega)$ and $\Delta k(2\omega)$ are the changes in $k(\omega)$ and $k(2\omega)$ that have been produced by ΔT, E and σ. For simplicity type I phase-matching has been assumed in eq. (55).

Disruptions caused by crystal heating ΔT can in some cases be offset by judicious application of E and/or σ, so that eq. (55) remains equal to zero. That is:

$$\Delta k = \Delta k(2\omega) - 2\Delta k(\omega) = 0. \tag{56}$$

It is assumed that ω and 2ω, as well as polarizations and directions of propagation, are held unchanged from the initial phase-match condition.

Under the influence of ΔT, E and σ, the index of refraction at any frequency is still represented by an ellipsoid, although somewhat changed, and eq. (7)

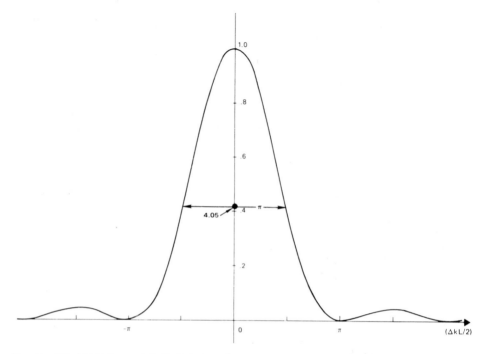

Fig. 11. $P(2\omega)/P(2\omega)_{max} = \sin^2(\Delta kL/2)/(\Delta kL/2)^2$, second harmonic power realized relative to the maximum obtainable as a function of $(\Delta kL/2)$, eq. (54).

becomes

$$[B_{ij}(\omega) + \Delta B_{ij}(\omega)]x_i x_j = 1, \tag{57}$$

where

$$\Delta B_{ij}(\omega) = \theta_{ij}(\omega)\Delta T + z_{ijk}(\omega)E_k + \pi_{ijkl}\sigma_{kl}, \quad i, j, k, l = 1, 2, 3, \tag{58}$$

and θ_{ij}, z_{ijk} and π_{ijkl} are components of the pyro-optic, electro-optic and piezo-optic constants, which are second-, third- and fourth-rank tensors respectively.

By using eqs. (6) and (58) and differentiating eq. (8), eq. (56) can be rewritten as

$$[e_i(2\omega)e_j(2\omega)][\theta_{ij}(2\omega)\Delta T + z_{ijk}(2\omega)E_k + \pi_{ijlm}(2\omega)\sigma_{lm}]$$
$$+ [e_i(\omega)e_j(\omega)][\theta_{ij}(\omega)\Delta T + z_{ijk}(\omega)E_k + \pi_{ijlm}(\omega)\sigma_{lm}] = 0, \tag{59}$$

where m also ranges from 1 to 3.

This is the general equation for electro-optical and piezo-optical tuning of SHG for the compensation of temperature fluctuations. It usually reduces to a simpler form in specific applications.

5.2 Example: second harmonic generation in CD*A

As an example, consider the case of EOT or POT applied to SHG in temperature tuned, 90° phase-matched CD*A. The optimal SHG orientation is shown in fig. 12.

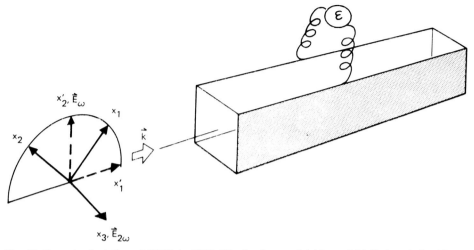

Fig. 12. Crystal orientation of CD°A in SHG. The fundamental $1.06\,\mu$m field E_ω is polarized in the degenerate $x_1 x_2$ (or xy) plane at 45° with either axis. The green light $E_{2\omega}$, produced by SHG, is polarized along the x_3-(or z or c) axis. Voltage applied along x_3 results in a 45° rotation of x_1 and x_2 to x_1' and x_2'.

In temperature tuned, 90° phase-matching, the normally degenerate $x_1(x)$ and $x_2(y)$ axes are at 45° to the propagation direction k of the $1.06\,\mu$m fundamental beam which is polarized in the $x_1 x_2$ plane. The $0.53\,\mu$m SHG beam is polarized along the $x_3(z)$ axis. The polarization unit vectors are therefore

$$\hat{e}(2\omega) = \frac{1}{\sqrt{2}}(\hat{x}_1 + \hat{x}_2), \qquad \hat{e}(\omega) = \hat{x}_3. \tag{60}$$

Substituting eq. (60) into the general equation eq. (59), and noting that $n_{\hat{e}}(2\omega) = n_{\hat{e}}(\omega) = n$ by definition of phase-matching, yield

$$\{\tfrac{1}{2}[\theta_{11}(2\omega) + \theta_{22}(2\omega) + \theta_{12}(2\omega) + \theta_{21}(2\omega)] - \theta_{33}(\omega)\}\Delta T$$
$$+ \{\tfrac{1}{2}[z_{11k}(2\omega) + z_{22k}(2\omega) + z_{12k}(2\omega) + z_{21k}(2\omega)] - z_{33k}(\omega)\}E_k$$
$$+ \{\tfrac{1}{2}[\pi_{11lm}(2\omega) + \pi_{22lm}(2\omega) + \pi_{12lm}(2\omega) + \pi_{21lm}(2\omega)] - \pi_{33lm}(\omega)\}\sigma_{lm} = 0. \tag{61}$$

Further simplification arises from crystal symmetry which dictates the forms of the θ, z and π tensors. In the reduced notation (see Nye (1957) for definition) for CD*A, eq. (61) simplifies to

$$[\theta_1(2\omega) - \theta_3(\omega)]\Delta T + z_{6k}(2\omega)E_k + [\pi'_{1l}(2\omega) - \pi'_{6l}(2\omega) - \pi'_{3l}(\omega)]\sigma_l = 0,$$
$$k = 1, 2, 3, \quad l = 1, \ldots, 6, \tag{62}$$

where, from the definition of the reduced notation for the π tensor,

$$\pi'_{il} = \pi_{il} \text{ if } l = 1, 2, 3, \quad \text{but} = \tfrac{1}{2}\pi_{il} \text{ if } l = 4, 5, 6.$$

5.2.1 Electro-optical tuning

In practice, either EOT or POT, but not both, are used for thermal compensation. For EOT, σ_l equals zero in eq. (62). From fig. 12 the applied field, E, might conveniently take the form of either $E = E\hat{x}_3$ or $E = (E/\sqrt{2})(\hat{x}_1 + \hat{x}_2)$.

EOT Case I
$E = E\hat{x}_3$. Eq. (62) becomes,

$$[\theta_1(2\omega) - \theta_3(\omega)]\Delta T + z_{63}(2\omega)E = 0.$$

It follows that

$$(\Delta T/E)_1 = z_{63}(2\omega)/(\theta_1(2\omega) - \theta_3(\omega)). \tag{63}$$

This gives the shift in the phase-match temperature due to an applied electric field. Equivalently,

$$(\Delta T/E)_1 = n^3 z_{63}(2\omega)/2 \frac{d}{dT}[n_1(2\omega) - n_3(\omega)],$$

the last step being a simple application of eqs. (10) and (58). However, a more useful expression in terms of experimental observables is written as (Hon 1976)

$$(\Delta T/E)_1 = Ln^3 \delta T z_{63}(2\omega)/\lambda(\omega), \tag{64}$$

where $\lambda(\omega)$ is the vacuum wavelength of the fundamental beam and δT is the SHG acceptance temperature defined in § 2.2.

EOT Case II
In Case II, where $E = (E/\sqrt{2})(\hat{x}_1 + \hat{x}_2)$, eq. (41) becomes

$$[\theta_1(2\omega) - \theta_3(\omega)]\Delta T + (E/\sqrt{2})[z_{61}(2\omega) + z_{62}(2\omega)] = 0.$$

Since both z_{61} and z_{62} vanish (Nye 1957), there is no EOT.

5.2.2 Stabilization of SHG by EOT

In fig. 13 it is evident that the applied electric field shifts the curve of SHG output vs. temperature. Depending on the duration τ of the applied E-field pulse, either a 'clamped' ($\tau \sim 200$ ns) or 'unclamped' ($\tau \sim 200$ μs) (Nye 1957) effect may be achieved. The EOT coefficients ($\Delta T/\Delta E$) are found to be 0.32 and 0.55°C cm/kV respectively.

By modulating the amplitude of the E-field by a small increment and by comparing the respective SHG outputs, it can be established whether more or less field would increase the efficiency by zeroing Δk. A closed loop such as that shown in fig. 14 can quickly compensate for any thermal fluctuations within the crystal. In addition to the limited EOT compensation range, the oven temperature can also be controlled by the same electronics to give an almost arbitrarily large total range. In a reported experiment (Hon 1976) a 0.52 J, 15 ns, 20 Hz Nd:YAG laser was used with an uncoated $1 \times 1 \times 2$ cm CD*A crystal housed in a dry oven with maximum thermal conductivity and minimum thermal mass. The output from the uncompensated crystal vanished after less than one

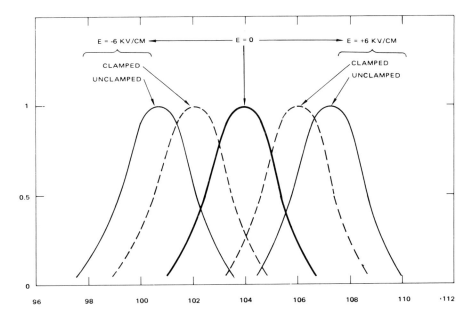

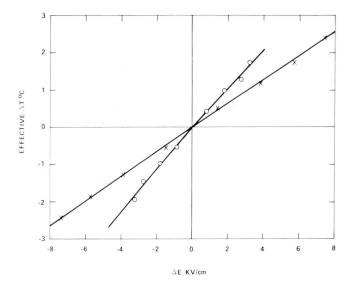

Fig. 13. Electro-optical effects in SHG in CD*A. (a) Normalized SHG output versus temperature for $E = 0$ and $E = \pm$KV/cm, clamped and unclamped. (b) Shift of phase-match temperature ΔT versus applied field ΔE. The steeper (shallower) straight line correspond to the unclamped (clamped) case. ($\Delta T/\Delta E$), the EOT coefficients, are determined to be 0.55 and 0.32°C cm kV^{-1}.

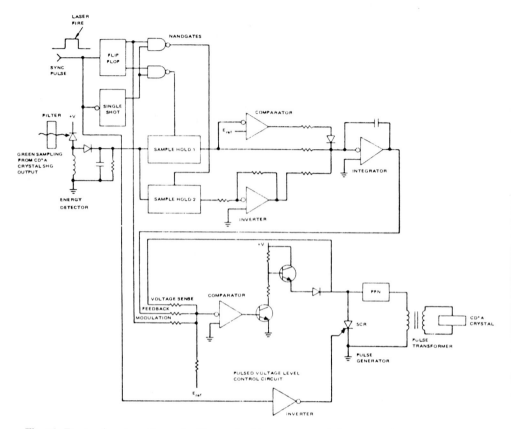

Fig. 14. Electronic schematic used with a pulsed laser system of the automatic electro-optic SHG fine-tuning device.

minute. However, when EO compensation was employed, a conversion efficiency of 57% (Fresnel reflection considered) was obtained and the output remained constant indefinitely. This is shown in fig. 15. The EO compensation response time of the loop was four times faster than the maximum self-heating rate of 0.5°C/s. Oven response was slow but adequate for the long-term stabilization. Ambient fluctuations of ±60° C did not affect the performance.

This technique is useful in other SHG materials and also in parametric oscillators. The power limit is determined by temperature gradients that ultimately cause unacceptable loss of efficiency. This will be considered in § 5.5.

5.2.3 Piezo-optical tuning (POT)

For POT, $E_k = 0$ in eq. (59). Two directions of stress can be conveniently applied, namely along $\hat{x}_3$ or along $(\hat{x}_1 + \hat{x}_2)$.

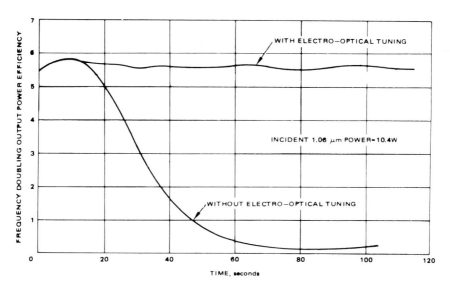

Fig. 15. Frequency doubling conversion efficiency versus time for an uncoated $1 \times 1 \times 2$ cm CD*A crystal with and without EO tuning. The $1.06\,\mu$m source was a multimode 0.52 J per pulse 20 Hz Q-switched Nd:YAG laser.

POT Case I

For this case, $\boldsymbol{\sigma} = (0, 0, 1, 0, 0, 0)\sigma/\sqrt{3}$. On using the symmetry rules for the tensor $\boldsymbol{\pi}$ (Nye 1957), eq. (62) becomes

$$[\theta_1(2\omega) - \theta_3(\omega)]\Delta T + [\pi_{13}(2\omega) - \pi_{33}(\omega)]\sigma = 0.$$

The POT coefficient $(\Delta T/\sigma)_1$, which describes the shift in the phase-match temperature due to an applied stress σ, is

$$(\Delta T/\sigma)_1 = [\pi_{33}(\omega) - \pi_{13}(2\omega)]/[\theta_1(2\omega) - \theta_3(\omega)]$$

and

$$(\Delta T/\sigma)_1 = [\pi_{33}(\omega) - \pi_{13}(2\omega)]Ln^3\delta T/\lambda(\omega), \tag{65}$$

by reasoning identical to that leading to eq. (64).

POT Case II

In Case II, where the applied stress is parallel to $(\hat{\boldsymbol{x}}_1 + \hat{\boldsymbol{x}}_2)$,

$$\boldsymbol{\sigma} = (\sigma/\sqrt{2})(1, 1, 0, 0, 0, 2)$$

in the reduced notation. On application of the symmetry rules for the CD*A $\boldsymbol{\pi}$ tensor, eq. (62) becomes

$$[\theta_1(2\omega) - \theta_3(\omega)]\Delta T + [\pi_{11}(2\omega) - 2\pi_{31}(\omega) + \pi_{12}(2\omega) + \pi_{66}(2\omega)/2](\sigma/\sqrt{2}) = 0.$$

It follows that

$$(\Delta T/\sigma)_{11} = [2\pi_{31}(\omega) - \pi_{11}(2\omega) - \pi_{12}(2\omega) - \pi_{66}(2\omega)/2]/[\theta_1(2\omega) - \theta_3(\omega)].$$

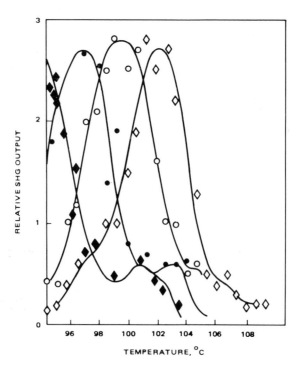

Fig. 16. Piezo-optic tuning (POT) in CD*A. Relative SHG output versus temperature for various pressures, σ is along $(\hat{x}_1 + \hat{x}_2)$.

And, as before,

$$(\Delta T/\sigma)_{11} = [2\pi_{31}(\omega) - \pi_{11}(2\omega) - \pi_{12}(2\omega) - \pi_{66}(2\omega)/2]Ln^3\delta T/\lambda(\omega), \qquad (66)$$

for the POT coefficient in Case II. Experimental results are shown in fig. 16.

5.2.4 POT by hydrostatic pressure

Hydrostatic pressure can be used for POT by submerging the crystal in a fluid. For some applications this technique may offer practical advantages over the unidirectional POT discussed above. Eq. (58) now reads

$$\begin{bmatrix} \Delta B_1 \\ \Delta B_2 \\ \Delta B_3 \\ \Delta B_4 \\ \Delta B_5 \\ \Delta B_6 \end{bmatrix} = \sigma \begin{bmatrix} \times & \bullet & \cdot & \cdot & \cdot & \cdot \\ \bullet & \bullet & \cdot & \cdot & \cdot & \cdot \\ \bullet & \bullet & \bullet & \cdot & \cdot & \cdot \\ \cdot & \cdot & \cdot & \bullet & \cdot & \cdot \\ \cdot & \cdot & \cdot & \cdot & \bullet & \cdot \\ \cdot & \cdot & \cdot & \cdot & \cdot & \bullet \end{bmatrix} \begin{bmatrix} 1 \\ 1 \\ 1 \\ \cdot \\ \cdot \\ \cdot \end{bmatrix} = \sigma \begin{bmatrix} \pi_{11} + \pi_{12} + \pi_{13} \\ \pi_{11} + \pi_{12} + \pi_{13} \\ 2\pi_{31} + \pi_{33} \\ \cdot \\ \cdot \\ \cdot \end{bmatrix}$$

and the equivalent of eqs (65) and (66) for hydrostatic POT is

$$(\Delta T/\sigma)_H = [2\pi_{31}(\omega) + \pi_{33}(\omega) - \pi_{11}(2\omega) - \pi_{12}(2\omega) - \pi_{13}(2\omega)]Ln^3\delta T/\lambda(\omega). \qquad (67)$$

Table 2

POT and EOT coefficients for three 90° phase-matching crystals. SHG orientation for CD*A is given in § 5.2. For LiNbO$_3$ and Ba$_2$NaNb$_5$O$_{15}$, $\boldsymbol{k}$ is along $\hat{y}$, $\boldsymbol{E}(\omega)$ is along $\hat{x}$ and $\boldsymbol{E}(2\omega)$ along $\hat{z}$. Definitions of $(\Delta T/\sigma)_{\mathrm{I}}$ $(\Delta T/\sigma)_{\mathrm{II}}$ and $(\Delta T/\sigma)_{\mathrm{H}}$ for LiNbO$_3$ and Ba$_2$NaNb$_5$O$_{15}$ are similar to those for CD*A, for which the relevant equations are eqs. (65), (66) and (67).

Coefficient	Crystal	LiNbO$_3$	Ba$_2$NaNb$_5$O$_{15}$	CD*A
POT coefficients[a] $\left(\dfrac{°C\ cm^2}{kg}\right)$	$(\Delta T/\sigma)_{\mathrm{I}}$	− 0.006	0.009	0
	$(\Delta T/\sigma)_{\mathrm{II}}$	0.0043	0.019	0.44
	$(\Delta T/\sigma)_{\mathrm{H}}$	− 0.009	− 0.045	0.063
EOT coefficients[b] $\left(\dfrac{°C\ cm}{KV}\right)$	$(\Delta T/E)$ E along $\hat{z}$	0.058	0.66	0.55
$\delta T \times$ Crystal Length (°C cm)		0.75	2	5

[a]Uncertainty ~ 15%.
[b]Uncertainty ~ 7%.

5.3 Other applications of EOT and POT

The EOT and POT coefficients of three 90° phase-matchable crystals – CD*A, LiNbO$_3$ and Ba$_2$NaNb$_5$O$_{15}$ – are presented in table 2.

The electro-optical effect has also been used to compensate for optical inhomogeneities and thereby increase the overall performance of SHG crystals. Regardless of the direction of the inhomogeneity relative to the lasing axis, proper application of the EO compensation field increases the total useful size of a single crystal. Furthermore, this can usually be achieved with just one or two pairs of electrodes. In one experiment involving a 2.5 cm long 'hot' LiNbO$_3$ crystal (typically crystal inhomogeneity limits the optimal length to about 1.5 cm) the application of a 600 V signal to a pair of electrodes spanning half the crystal length increased the SHG efficiency by greater than a factor of two (Hughes Aircraft Co. IR & D report 1975). Similarly, local application of the piezo-optical effect can be used to maximize effective SHG crystal sizes.

5.4 Thermal compensation by angle tuning

In angle phase-matched crystals, the acceptance angle is typically very small (see § 2.1). Any change in temperature caused by either environment or absorption will require a continuous adjustment of the crystal orientation. Two closed-loop servo schemes, both similar in principle to the EOT described above can be used. One involves dithering the crystal orientation; the second involves monitoring the SHG profile in the far field. In the latter, more practical scheme, a telescope that has an effective focal length of 10^2 cm focuses a 0.5 mrad beam into a 0.5 mm spot at a quadrant detector (according to the formula: waist = focal length × divergence). The relative signals indicate and control the necessary angular adjustment of the crystal to re-optimize the phase-matching condition (Kuhl and Spitschan 1975).

5.5 Control of thermal gradients by beam shaping

We will now discuss methods of handling very high average powers in SHG – in principle limited only by the size of the system.

As shown in § 2.3, regardless of the energy distribution profile of the fundamental laser beam, the absorption of radiant energy by the SHG crystal produces a temperature gradient across the crystal which for sufficiently high power makes it impossible to obtain uniform phase-matching over the entire interaction volume. For 1.06 μm radiation, gradients have been found to significantly lower the conversion efficiency of SHG crystals for average input power levels greater than approximately 20 W. The levels are lower for the longer ir due to the higher absorptivity of the SHG crystal. The temperature profile for a gaussian beam is represented by eq. (47) and shown in fig. 8 for a gradient across the crystal of $\Delta'T = (\bar{P}\eta/8K)$. § 4.1 pointed out that the ideal SHG crystal must have low absorptivity η and high thermal conductivity K. But, given a crystal, and an average power $\bar{P}$, $\Delta'T$ can still be drastically reduced by distributing the laser energy over an elongated cross section instead of focusing it on one spot (Hon 1977, Hon et al. 1977, Moses et al. 1978, Hon and Bruesselbach 1979). This is easily seen by considering the simplified case of a uniform laser beam incident over the entire face of a thin slab-shaped SHG crystal of thickness h, width W and length L, as illustrated in fig. 17. The steady-state solution of the heat flow equation

$$\frac{dQ}{dt} = -KA\frac{dT}{dx},$$

is

$$T(x) = T(0) - \frac{\bar{P}\eta x^2}{8KWh}.$$

It follows that

$$\Delta'T = T(0) - T(x = h/2) = \frac{\bar{P}\eta h}{8KW}, \tag{68}$$

where $\Delta'T$ is the temperature between the center and the edge of the beam. $\Delta'T$ is proportional to the slab geometry factor h/W, which is small for the thin slab configuration. Eq. (68) reduces to eq. (48) when h equals W (although the $\Delta'T$ defined in the two equations are only roughly equivalent).

In the case of a laser with a high peak as well as a high average power, a cylindrical lens system can shape the beam entering the appropriately shaped

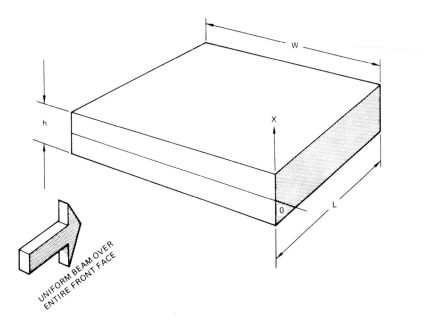

Fig. 17. A uniform beam incident on a crystal slab of aperture $h \times W$ and length L. The steady-state temperature profile is given by eq. (47).

crystal face to an elliptical cross section. This is called *beam fanning*. As an example, a reduction of h and an increase of W, each by a factor of 4, which would keep the illuminated area constant, would reduce ΔT by a factor of 16. In the case of a laser with high average power but low peak power, such as a cw or a high-repetition-rate laser, *beam scanning* should be used. In beam scanning, the focused beam spot is scanned repetitively and uniformly across the surface of a (set of) crystal slab(s) to achieve the desired distribution of thermal heating. Analogous to the technique of transverse flow in high-power gas lasers in that the thermally distorted medium is replaced by a fresh volume in every pulse, the power-handling capability here is limited only by the size of the scanning system. Both beam fanning and scanning are discussed below, the objective being to develop a set of general design guidelines for any adequate laser system. A common SHG crystal, cesium dideuterium arsenate (CD*S), will be used as an example though its actual application is still limited.

For simplicity, two assumptions are made:
(1) The laser beam profile is always "square" in near and far fields and, if it is pulsed, in time.
(2) The SHG system, which includes the laser and the crystal, is one that is nominally efficient for a single pulse. We concentrate on problems and solutions associated with high average powers only. Optimization of an inherently inefficient system is not discussed*.

*Boyd and Kleinman (1968) explained the procedure of optimization of a single-pass SHG system based on a number of idealistic assumptions about the quality of both the beam and the crystal.

In designing a high power SHG unit, a convenient criterion is to require that $\Delta'T \le \delta T$ [eq. (49)]. That is, from eq. (68),

$$W \ge \frac{\bar{P}\eta h}{8K\delta T}. \tag{69}$$

Since practical considerations usually demand that W, the crystal width, be minimized, we may write

$$W = \frac{\bar{P}\eta h}{8K\delta T}. \tag{70}$$

Eq. (70) is the basic design equation. However, several other conditions involving the characteristics of the laser and the SHG crystal must be met. For the laser, the pertinent characteristics are: beam divergence θ_0, beam spot diameter D_0, peak power P, average power $\bar{P}$, pulse repetition rate f and pulsewidth t. For the crystal, the pertinent characteristics are: absorption coefficient η, acceptance angle(s) $\delta\theta$, thermal conductivity K, acceptance temperature δT, desirable power density for safe, efficient SHG ρ_d and nominal crystal length d. These are listed in table 3 and discussed below.

A list of values of the characteristics for three common 90° phase-matching crystals (LiNbO$_3$, CD*A, and Ba$_2$NaNb$_5$O$_{15}$) and for KTP are given in table 4.

Table 3
Laser beam and FD crystal characteristics.

Fundamental beam	FD crystal
θ_0 = beam divergence	η = absorption coefficient
D_0 = beam spot diameter	$\delta\theta$ = acceptance angle
P = peak power	K = thermal conductivity
$\bar{P}$ = average power	δT = acceptance temperature
f = pulse repetition rate	ρ_d = desirable power density for safe, efficient SHG
t = pulsewidth	d = nominal SHG crystal length

Table 4.
SHG (1.06 μm) crystal parameters for three temperature-tuned 90° phase-matched crystals

Materials	$\delta\theta$ Acceptance angle[a] [mr]	δT Acceptance temperature [°C]	η Absorp. coeff.[c] [cm^{-1}]	K Conductivity [W/°CCcm]	ρ_d Desirable power density [MW/cm^2]	d Nominal crystal length [cm]	Presently available size [cm^2]
CD*A	50	3.5	0.04	0.01	100	2.5	4 × 4
Hot LiNbO$_3$	55	0.6	0.025	0.034	20	1.5	0.3 × 4
Ba$_2$NaNb$_5$O$_{15}$	100	1.2	0.02	0.02	15	0.5	0.5 × 0.5
KTP	16	>50	<0.01	~0.05	40	0.5	0.4 × 0.4

[a]Based on the nominal crystal length listed and refers to polar direction.
  Acceptance angle in azimuth is theoretically infinite (see §§ 2.1 and 2.2).
[b]During SHG.
[c]Based on a round spot containing 90% energy of a gaussian-like multimode beam.

5.5.1 Beam divergence and acceptance angles

In this section the limitation imposed on the power density obtainable through focusing by the beam divergence of the laser output and the acceptance angle of the SHG crystal is derived for the case where diffraction is not important, that is, for a multimode beam whose nonspherical beam divergence is at least several times the diffraction limit. In fact, with minor redefinition the results are generally applicable for any laser beam (Goubau and Schwering 1961, Pierce 1961, Kogelnik and Li 1966). If a multimode (collimated) laser beam is brought to a focus of diameter h by a lens (or a mirror) of focal length F, then the (nonspherical) beam divergence θ_0 is defined as

$$\theta_0 \equiv \frac{h}{F}, \tag{71}$$

h may be measured at any convenient intensity or energy level providing it is done consistently.

Generally, the beam divergence in any direction varies inversely with the magnification in that direction. If a beam of diameter D_0 and inherent divergence θ_0 is recollimated to an oval shape of width W and height h then the angles between the limiting rays at the focus are

$$\theta_h = (\theta_0 D_0 / h), \qquad \theta_W = (\theta_0 D_0 / W). \tag{72a, b}$$

If SHG is to be efficient, θ_h and θ_W must not exceed the acceptance angles $\delta\theta_h$ and $\delta\theta_W$ in those two directions (see §§ 2.1 and 2.2). That is:

$$\theta_h = (\theta_0 D_0 / h) \le \delta\theta_h, \qquad \theta_W = (\theta_0 D_0 / W) \le \delta\theta_W$$

or

$$h \ge (\theta_0 D_0) / \delta\theta_h, \qquad W \ge (\theta_0 D_0) / \delta\theta_W. \tag{73a, b}$$

Consequently, 90° phase-match crystals, which usually have large acceptance angles (often in the tens of milliradians), will accept the focused output of a high power multimode laser whose beam divergence may be several milliradians. On the other hand, angle phase-matched crystals (often with sub-millradian acceptance angles) are better adapted to doubling the frequency of single mode laser beams, which for solid-state lasers are invariably lower in power and less efficient. Since by the definitions of h and W, more demagnification occurs in the h-direction, the attendant increase in divergence is greater. The beam sometimes can be magnified in the W-direction from the beam diameter D_0. Proper orientation of the oblong beam shape with respect to the crystal is important if the larger acceptance angle in the azimuthal direction is to be utilized, as suggested by Volosov et al. (1969) and discussed in § 2.2. Orientation is also important if any of the tuning techniques is needed.

5.5.2 Nominal SHG crystal length, d

In general, the length of a crystal used in SHG must be optimized as part of the overall system. Theoretically d is coupled to all the other laser and crystal parameters listed in table 3. However, in applications involving high average power, high peak power, multimode lasers, the crystal length is usually determined or limited by the size or uniformity of the crystal, which varies somewhat from sample to sample as well as from crystal to crystal. For many practical purposes, it is sufficient to assume, as here, fixed, experimental values of d for each type of crystal, such as those listed in table 4.

The nominal length imposes a condition on how tightly the beam can be focused. Fig. 18 shows a beam of diameter D_0 focused to a waist h inside a crystal of length d. From eq. (71),

$$h = F_h \theta_0,$$

where F_h is the focal length of the lens in the vertical plane. To maintain a reasonably high power density throughout the entire interaction volume, a condition such as that which follows, must be imposed on F_h,

$$\frac{D_0}{F_h} \frac{d}{2} \leqq h. \tag{74}$$

This condition will ensure proper utilization of the SHG crystal by requiring the vertical beam diameter inside the crystal to fall between h and $\sqrt{2}h$.

The combination of eqs. (71) and (74) gives,

$$h^2 \geq \tfrac{1}{2} d \, D_0 \theta_0. \tag{75}$$

Inability to satisfy eq. (75), for whatever reason, leads to inefficiency. If the reason is insufficient brightness of the laser, Boyd and Kleinman's (1968) approach may be used to optimize the focusing and crystal length.

5.5.3 Desired power density ρ_d

SHG efficiency generally improves with peak power density until the crystal is damaged. The quantity ρ_d is defined as the power density which gives acceptable efficiency without the risk of crystal damage. Since the efficiency curve often

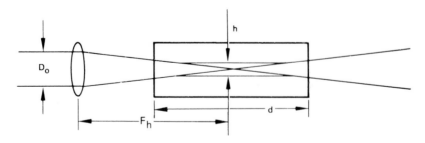

Fig. 18. A beam of diameter D_0 focused to a spot diameter h by a lens of focal length F_h. To properly utilize the crystal of length d, eq. (53) must be satisfied.

(though not always) levels off at power density levels lower than the damage threshold, ρ_d is experimentally selected. The listing in table 4 are based on the presently attainable bulk and surface qualities for the respective crystals. For efficient SHG by beam fanning,

$$\rho_d = P/Wh. \tag{76a}$$

For efficient SHG by beam scanning, W in eq. (76) should be replaced by h:

$$\rho_d = P/h^2. \tag{76b}$$

5.5.4 *Minimum peak power,* min P

The minimum peak power min P required for efficient SHG is determined by the desired power density ρ_d, beam divergence θ_0, acceptance angle $\delta\theta$, and the nominal crystal length d. The following simultaneous inequalities can easily be derived for beam fanning:

$$\min P \geq \begin{cases} [\rho_d \theta_0^2 D_0^2 / \delta\theta_h \delta\theta_w] \text{ from eqs. (73) and (76)} & (77a) \\ [\rho_d d D_0 \theta_0 / 2] \text{ from eqs. (75) and (76).} & (77b) \end{cases}$$

To obtain eq. (77b), W in eq. (76) is set equal to h in the limit of minimum peak power. For beam scanning, the denominator $(\delta\theta_h \delta\theta_w)$ in eq. (77a) should be replaced by $(\delta\theta)^2$, where $\delta\theta$ is the smallest acceptance angle. Depending on the values of the parameters involved, either eq. (77a) or eq. (77b) will dominate. For cesium dideuterium arsenate, min P is calculated according to table 4 to be 3×10^4 W, and is shown on fig. 19.

Intracavity SHG may be used to increase the peak power at the crystal while simultaneously decreasing the single-pass efficiency required for optimal SHG, but this will complicate the design of the laser (see § 3.2.5). Wave-guides and light-pipes have been used to significantly enhance the SHG efficiency of a laser beam whose peak power would otherwise be insufficient (Sohler and Suche 1978). However, these devices can handle very low average powers.

5.5.5. *Beam fanning conditions*

Generally speaking, beam fanning, which is passive, is preferred over beam scanning. If min P is available from a given laser and if we assume eq. (73b) is satisfied (as is usually the case), then a condition for beam fanning can be obtained. Equations (70) and (76) are combined separately with Eqs. (73a) and (75) to obtain

$$ft = \bar{P}/P \leq \begin{cases} (8K\delta T\delta\theta_h^2)/(\eta\rho_d\theta_0^2 D_0^2) & (78a) \\ (16K\delta T)/(\eta d D_0 \theta_0 \rho_d), & (78b) \end{cases}$$

where all terms shown are defined in table 3. Depending on parameter values, either eq. (78a) or eq. (78b) will dominate. The condition for successful beam fanning depends on the average power $\bar{P}$ divided by the peak power P of the laser. $\bar{P}/P$ is the duty cycle of the system (ft). For uniaxial crystals (as are most

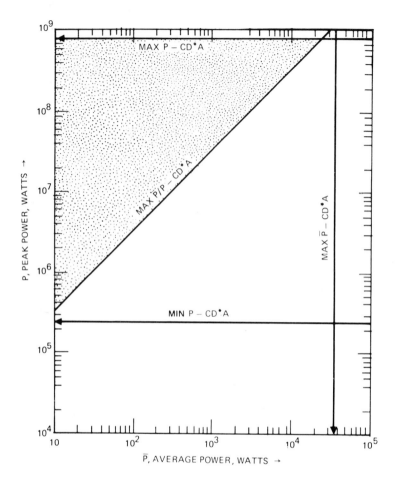

Fig. 19. Laser average powers ($\bar{P}$) and peak powers (P) that can utilize beam fanning in CD*A (90° phase-matched for 1.06 μm). The shaded area is bounded by maximum $\bar{P}/P$ as defined by eq. (78b) and maximum P as given by eq. (80). Typical YAG parameters are used: $D_0 = 0.6$ cm and $\theta_0 = 3$ mr.

of the common SHG crystals), the large azimuthal acceptance angle is usually taken as $\delta\theta_h$. In particular, for 90° phase-matching, $\delta\theta_h$ is essentially infinite and inequality (78a) becomes trivial. Two examples are (hot) lithium niobate and cesium dideuterium arsenate when used in SHG of 1.06 μm. The application of eq. (78b) is displayed in fig. 19, where P is plotted against $\bar{P}$ and in fig. 20, where f is plotted against t. Values in table 4 are used.

Points that lie in the area above the curve in fig. 19 and below the curve in fig. 20 satisfy the inequalities (78), and beam fanning is applicable provided other conditions to be developed below are also satisfied. Otherwise, beam scanning must be considered.

It is obvious from table 4 that KTP, with its large acceptance temperature δT,

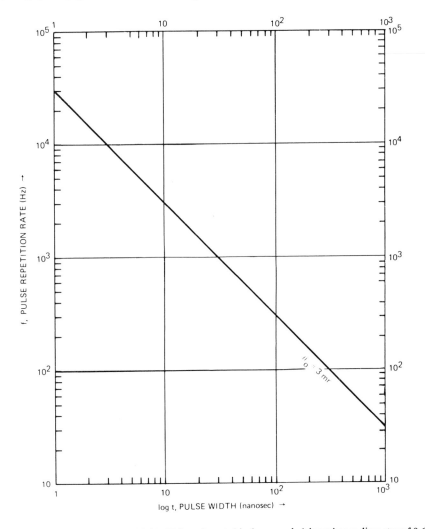

Fig. 20. Maximum ft [eq. (57)] for CD*A. Values from table 4 are used. A laser beam diameter of 0.6 cm (D_0) is used. Points under and to the left of the curves permit beam fanning.

large conductivity K, and small absorptivity η, is particularly suited for beam fanning in spite of its small available sizes at present.

5.5.6 Size and shape in beam fanning

To obtain the beam shape necessary to control thermal gradients, eq. (70) may be used in conjunction with eq. (76). We obtain

$$h = (8KP\delta T/\eta\rho_d\bar{P})^{1/2} = (8K\delta T/\eta\rho_d ft)^{1/2}$$
$$W = (\eta\bar{P}P/8K\delta T\rho_d)^{1/2} = \bar{P}(\eta/8ftK\delta T\rho_d)^{1/2}. \tag{79a, b}$$

5.5.7 *Maximum powers in beam fanning*

If Eq. (79b) stipulates a W value larger than available crystal size, an array
formed by carefully butting together two or more slabs would be necessary.
These crystal slabs, however, must be cut from the adjacent parts of the same
boule to achieve uniformity of phase-match temperature. At present, the maxi-
mum crystal size of sufficient optical uniformity is about $4 \times 4\ cm^2$ and $0.3 \times 4\ cm^2$,
for CD*A and $LiNbO_3$, respectively. ($BaNaNbO_3$ and KTP, on the other hand,
can only be grown in sizes up to 5 mm and 4 mm cubes.) If, in practice, the beam
has to underfill the crystal by a factor of two to avoid beam damage to the
crystal oven, then the maximum peak power (max P) that can be frequency
doubled without damage is obtained from eq. (76). Using values of ρ_d from table
4

$$\text{max } P = 8 \times 10^8\ W \text{ for CD*A}. \tag{80}$$

The maximum average power (max $\bar{P}$) can be derived from eq. (70):

$$\text{max } \bar{P} = \frac{8 \times (\text{area}/2)k\delta T}{\eta h^2}. \tag{81}$$

Since h must simultaneously satisfy eqs. (73a) and (75), max $\bar{P}$ is the smaller of
the two possible values in eq. (81). For uniaxial CD*A, only eq. (75) is applicable
and

$$\text{max } \bar{P} = \frac{8 \times \text{area } k\delta T}{\eta dD_0\theta_0}. \tag{82}$$

Taking $D_0 = 0.6$ cm and $\theta_0 = 3$ mrad (typical of a YAG laser) and using the values
from table 4,

$$\text{max } \bar{P} = 3.6 \times 10^4\ W \text{ for CD*A}. \tag{83}$$

Fig. 19 shows max P and max $\bar{P}$ for the crystal. The shaded areas correspond
to SHG systems in which beam fanning can be used.

5.5.8 *Beam scanning*

If a proposed system satisfies eq. (77) but not eq. (78), then beam fanning, the
preferred method, is not applicable, and beam scanning should be considered. In
beam scanning, the beam is focused to a spot of diameter h and *repetitively*
scanned across an array of crystal slabs cut from the same boule and carefully
butted together to provide a scan width of W. h is already given by eq. (76b). To
rewrite:

$$h = (P/\rho_d)^{1/2}. \tag{84}$$

W can be obtained by substituting Eq. (84) into Eq. (70):

$$W = \eta\bar{P}\sqrt{P}/8K\delta T\sqrt{\rho_d}. \tag{85}$$

5.5.9 Maximum powers in beam scanning

If we again assume that $16 \, \text{cm}^2$ is the respective maximum available sizes of good quality CD*A and $LiNbO_3$ crystals and that slices of these crystals can be arranged in an appropriate array to accept an illumination area half the size, then the maximum powers that these crystals can handle can be calculated by using eq. (81), which shows that h should be minimized. However, unlike the case of beam fanning with a uniaxial crystal, both inequalities (73a) and (75) must be used to determine the smallest allowable h. Using values in table 4 and $D_0 = 0.6$ and $\theta_0 = 9$ mr (typical of highly pumped YAG lasers), max $\bar{P}$ is calculated to be 5.6×10^3 W for CD*A. This value is shown in fig. 21.

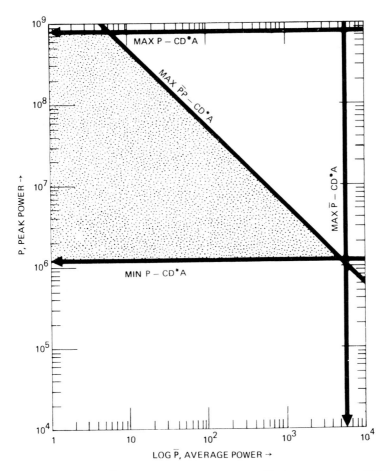

Fig. 21. Laser average powers ($\bar{P}$) and peak powers (P) that can utilize beam scanning in CD*A. Shaded areas are bounded by (a) max P, which is identical to those obtained for fanning in eq. (80) (fig. 19), (b) min P, which is calculated according to 5.5.4 for $\theta_0 = 9$ mr, typical of high average power YAG laser and (c) $P\bar{P}$ as given by eq. (86).

Furthermore, using eqs. (84) and (85),

$$h \times W = \text{area}/2 = \frac{[P/\rho_d]^{1/2}(\bar{P}\eta\sqrt{P})}{8K\delta T\sqrt{\rho_d}},$$

we have

$$\bar{P}P = (4K\delta T\rho_d \times \text{area})/\eta = 5.6 \times 10^9 \, \text{W}^2 \, \text{for CD*A}. \tag{86}$$

Fig. 21 shows the regions on a P versus $\bar{P}$ plot in which scanning is expected to eliminate temperature profile degradations. For this particular crystal, a shaded area is bound by (a) max P, which is identical to that obtained for fanning in eq. (80); (b) min P, which is calculated according to § 5.5.4 for $\theta_0 = 9$ mr; and (c) $\bar{P}P$, as given in eq. (86). Unfortunately, YAG lasers with large $\bar{P}$ tend to have small P, as discussed in § 3.3.1.

5.6 Design of high average power SHG experiments

5.6.1 General guidelines of beam shaping

A complete set of guidelines for designing a high-power SHG experiment has been presented. It is based on fanning or scanning the output from a fundamental laser having any given characteristics. Limits of performance imposed by existing crystal growth technology have been defined. Fig. 22 presents a flow chart summarizing a step-by-step approach starting with a set of specifications of the output of a fundamental laser. In this chart, if the proposed fundamental passes a test step, in the form of an indicated equation and/or graph, the next test is applied. In general, beam fanning, which is easier to implement, should be given preference over beam scanning. If a laser does not pass *all* the test steps from beginning to end, it must be discarded and an alternate laser must be proposed. It must be emphasized, however, that, although these guidelines are very useful for a very large range of systems, they are only approximate because of the two assumptions stated before eq. (69). A theoretical or experimental fine tuning of a design must be performed subsequently.

5.6.2 Some experimental techniques

In beam fanning, it is advantageous to place the crystal at the far field of the beam where the power profile in time and space is generally smoother compared to the near field. This can be vital to avoiding damage from hot spots. The beam should slightly overfill the crystal slab at the narrow sides to better approximate the case of thermal flow in an infinite slab. This will also avoid thermal stress near the beam edges, hence avoiding undesirable birefringence and breakage. The beam should underfill by as little as possible the broad sides of the slab so that the heat reservoir in good thermal contact with the crystal can maintain a

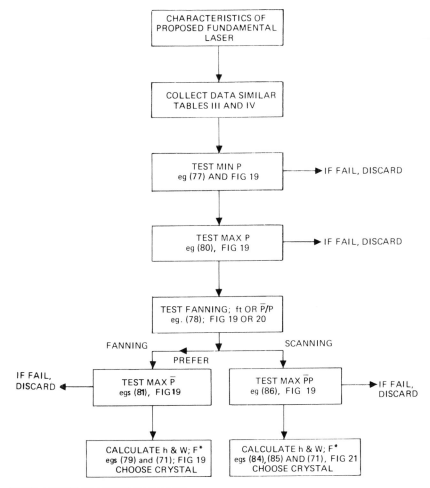

Fig. 22. Design flow chart of high power SHG unit using beam shaping techniques.

nearly uniform temperature in spite of a slight but inevitable intensity profile across the width of the slab. The heat reservoir should not be in contact with the narrow sides of the slab. EO tuning has been used compatibly with beam fanning in SHG of Nd:YAG to provide a stability that would otherwise be unachievable (Hon 1977, Hon et al. 1977). In beam scanning using crystals with a slight optical nonuniformity, either due to the quality of the crystal itself or to the inability of the oven to keep a perfectly uniform temperature across a large width W, EO tuning is especially essential to maintain optimal SHG output on a shot-to-shot basis. For nonuniformities too large to be handled by EO tuning alone, sectional heating of each wide slab can be used. In a recent experiment (Moses et al. 1978, Hon and Bruesselbach 1979), these techniques were used on an array of five (hot) lithium niobate crystals, each 1.25 in long and each heated with three small sectional heaters. One common EO tuning system was used for all five crystals, and the error function, which indicates whether the crystal at any given time is slightly below or above the phase-match temperature, was used to control both the small heaters and the compensating E-field itself. The specially built oven rode on a rail and bounced back and forth at 4 Hz between two short, powerful springs. This scanning system was used with a cw-pumped, acousto-optically q-switched, double-rod YAG laser. The nominal output of this system at 1.06 μm was 200 W in average power, 120 nsec in pulse width, 10 KHz in repetition rate, and 0.6 cm in beam diameter; it had a divergence of 9 mr. Thirty-five watts of 0.53 μm (green) was produced in an intracavity configuration where the 1.06 μm flux was regularly 500 W and could be as high as 900 W. The theory of beam shaping was verified in that little if any degradation of the SHG efficiency was detected that was attributable to thermal heating. The efficiency was not higher because of insufficient 1.06 μm brightness. Hot lithium niobate, which phase-matches at 170°C, was used to avoid optical index damage at lower temperatures. But because of the lithium-rich nonstoichiometric growth, a substantial amount of nonuniformity in the form of striation is still unavoidable in the crystal. This necessitated the sectional heating.

Other scanning schemes are possible. One is a confocal scanning, shown in fig. 23, where the beam is scanned across a stationary array of slabs. Another scheme is to arrange the slabs into a circular array in a drum-like oven that rotates at 3 Hz or faster. To avoid the fabrication and performance problems at slab-to-slab interfaces, an annular ring of a uniaxial crystal may be cut and polished so that the beam travels radially across the crystal in its x–y plane (fig. 24). Again, the crystal annular ring rotates at 3 Hz or faster.

For laser systems that lack the peak power for efficient SHG, a possible solution is to use a series of waveguides formed into a continuous array ring (fig. 25) that rotates synchronously with the high-repetition-rate laser. By confining the light intensity over a long distance in a waveguide or a light pipe, orders of magnitude of increase in efficiency have been achieved (Sohler and Suche 1978).

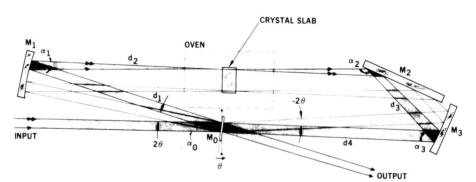

Fig. 23. Confocal scanning scheme. The saw-tooth scanning mirror M_0 scans the focused beam uniformly across the crystal slab placed in the confocal plane of M_1 and M_3. The back side of M_0 descans the output.

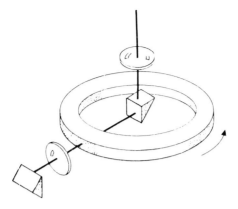

Fig. 24. Scanning by using an annular ring of a uniaxial crystal.

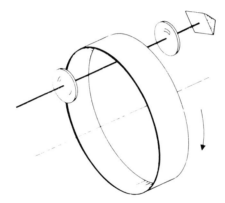

Fig. 25. Synchronized scanning by using an array of waveguides or light pipes.

6 Conclusion and acknowledgment

A comprehensive discussion of efficient, high-average-power second harmonic generation has been presented. A survey of the state of the art has been made and is summarized in Table 1, which includes extensive references to the literature. Design criteria for both the laser and the SHG crystal have been posted along with techniques to circumvent problems and optimize outputs. The basic problems in scaling an efficient SHG system to high-average-power applications are thermal instability and thermal profile. These have been defined and various up-to-date solutions discussed: electro-optical, piezo-optical, and angle tuning techniques for controlling the phase-match condition and stabilizing the output. The techniques of beam shaping by fanning and scanning were discussed in some detail.

The author wishes to thank Drs. Eric Woodbury and Gregory Olson for careful proof reading and useful suggestions. Dr. Olson helped compile literature data on SHG materials, and Mr. Steven Guch contributed to the CO_2 laser discussion. I particularly wish to acknowledge Mr. Hans Bruesselbach who assisted and collaborated with me in many of the high average power SHG experiments. Mr. Dean Stovall and, more recently, Drs. Edward Moses and Fred Wu also contributed to the experimental effort. Dr. Won Ng supplied earlier SHG data on $LiNbO_3$ and $Ba_2NaNb_5O_{15}$. Dr. Richard Zumsteg supplied test samples of KTP.

I offer my apology for any omission due to oversight in a chapter of this scope.

Finally, I thank all the fellow physicists and engineers who contributed to the field of nonlinear physics and whose results are quoted or utilized in this chapter.

List of symbols

c	speed of light
d	nominal length of crystal
d_{ij}	pertinent nonlinear coefficient
D_0	beam spot diameter
$\hat{e}(\omega)$	unit vector of polarization of light
$e_i(\omega)$	components of unit vector of polarization
E	magnitude of E
E	electronic field vector
$\mathscr{E}$	energy
f	pulse repetition rate
F	focal length of a lens
h	height of beam shape at crystal or spot diameter
$k(\omega)$	wave vector of laser beam with angular frequency ω
K	thermal conductivity
L	length of crystal
$\mathscr{L}$	round-trip insertion loss of a SHG crystal
$n_o(\omega, \theta), n_e(\omega, \theta)$	ordinary and extraordinary indices of refraction at angular frequency ω and at polarization making an angle θ from the uniaxial z-axis
$n_o^{(\omega)}, n_e^{(\omega)}$	ordinary and extraordinary indices
P	peak power
$\bar{P}$	average power
S	wave normal
t	pulse width
T	temperature
W	width of a beam shape at the crystal
(x_1, x_2, x_3)	(x, y, z) Cartesian coordinate
z_{ijk}, z_{ij}	components of electro-optic tensor
$\delta\theta$	acceptance angle

$\delta\theta_h$ acceptance angle in the h direction

$\delta\theta_W$ acceptance angle in the W direction

δT acceptance temperature

Δk $|\Delta k|$

Δk phase mismatch quantity

ΔT change in temperature

η optical absorption coefficient

θ_h beam divergence in the h direction

$\theta_{i,j}, \theta_i$ components of pyro-optic tensor

θ_m phase-match angle

θ_0 beam divergence

θ_W beam divergence in the W direction

λ wavelength

π_{ijkl}, π_{ij} components of the piezo-optic tensor

$\bar{\rho}$ average power density

ρ_d desirable peak power density

δ pressure

σ walk-off angle

σ stress tensor

ω angular frequency, often referring to the fundamental laser source

Ω output coupling

Ω_0 optimal output coupling of a laser without intracavity SHG

Ω_{SHG} output coupling through optimal SHG.

References

Abdullaev, G.B., L.A. Kulevskii, A.M. Prokhorov, A.D. Savel'ev, E.Yu. Salaev and V.V. Smirnov (1972), Pis'ma Zh. Eksp. Teor. Fiz. **16**, 130 (English transl.: 1972, JETP Letters **16**, 90).

Abdullaev, G.B., L.A. Kulevskii, P.B. Nikles, A.M. Prokhorov, A.D. Savel'ev, E. Yu. Salaev and V.V. Smirnov (1976), Kvantovaya Elektron. (Moscow) **3**, 163 (English transl.: 1976, Sov. J. Quantum Electronics **6**, 88).

Adhav, R.S. (1968), Research on Electro-optic Crystals for Display Systems, Technical Rpt. No. RADC–TR–68–307, Rome Air Development Center, Air Force Systems Command, Griffis Air Force Base, New York.

Adhav, R.S. and A.D. Vlassoponlos (1974), Guide to Efficient Doubling, Laser Focus, May.

Adhav, R.S. and R.W. Wallace (1973), IEEE J. Quantum Electronics **QE–9**, 855.

Akhmanov, S.A., A.I. Kovrigin, A.S. Piskarskas and R.V. Khokhlov (1965), Pis'ma Zh. Eksp. Teor. Fiz. **2**, 223 (English transl.: 1965, JETP Letters **2**, 141).

Akhmanov, S.A., A.I. Kovrigin, V.A. Kolosov, A.S. Piskarskas, V.V. Fadeev and R.V. Khokhlov (1966), Pis'ma Zh. Eksp. Teor. Fiz. 372 (English transl.: 1966, JETP Letters **3**, 241).

Ammann, E.O., C.D. Decker and J. Falk (1974), IEEE J. Quantum Electronics **QE–10**, 463.

Andreev, R.B., V.D. Volosov and A.G. Kalintsev (1974), Optika; Spektrosk. **37**, 294 (English transl.: 1974, Opt. Spectrosc. (USSR) **37**, 169).

Antonov, E.N., V.G. Koloshnikov and D.N. Nikogosyan (1974a), Optika; Spektrosk. **36**, 768 (English transl.: 1976, Opt. Spectrosc., (USSR) **36**, 446).

Antonov, E.N., V.R. Mironenko, D.N. Nikogosyan and M.I. Golovei (1974b), Kvantovaya Elektron. (Moscow) **1**, 1742 (English transl.: 1975, Sov. J. Quantum Electronic. **4**, 963).

Ashkin, A., G.D. Boyd, J.M. Dziedzic, R.G. Smith, A.A. Ballman, J.J. Levinstein and K. Nassau (1966), Appl. Phys. Letters **9**, 72.

Atanesyan, V.G., K.V. Karmenyan and S.A. Sarkisyan (1974), Pis'ma Zh. Eksp. Teor. Fiz. **20**, 537 (English transl.: 1974, JETP Letters **20**, 244).

Attwood, D.T., E.L. Pierce and L.W. Coleman (1975), Opt. Commun. **15**, 10.

Avdeenko, K.I., I.I. Zubrinov, B.I. Kidyarov and D.V. Sheloput (1976), Abstracts of Papers presented at Eighth All-Union Conf. on Coherent and Nonlinear Optics, Tbilisi [in Russian].

Badikov, V.V., O.N. Pivovarov, Yu, V. Skokov, O.V. Skrebneva and N.K. Trotsenko (1975), Kvantovaya Elektron. (Moscow) **2**, 618 (English transl.: 1975, Sov. J. Quantum Electronics **5**, 350).

Barry, J.D. (1976), IEEE J. Quantum Electronics 254.

Barry, J.D. and C.J. Kennedy (1975), IEEE J. Quantum Electronics **QE-11**, 575.

Bass, M., D. Bua, R. Mozzi and R.R. Monchamp (1969), Appl. Phys. Letters **15**, 393.

Berezovskii, V.V., Yu.A. Bykovskii, M.I. Goncharov and I.S. Rez (1972), Kvantovaya Elektron. (Moscow) No. 2(8), 105 (English transl.: 1972, Sov. J. Quantum Electronics **2**, 180).

Berezovskii, V.V., Yu.A. Bykovskii, S.N. Potanin and I.S. Rez (1973), Kvantovaya Elektron (Moscow) **2** (14), 74 (English transl.: 1973, Sov. J. Quantum Electronics **3**, 134).

Berezovskii, V.V., Yu.A. Bykovskii, A.V. Lebedev, A.I. Maimistov, E.A. Manykin and K.N. Kirsov (1975), Abstracts of Papers presented at Seventh All-Union Conf. on Coherent and Nonlinear Optics, Tashkent [in Russian].

Bergman, J.G., A. Ashkin, A.A. Ballman, J.M. Dziedzic, H.J. Levenstein and R.G. Smith (1968), Appl. Phys. Letters **12**, 92.

Bhar, G.C. and R.C. Smith (1974), IEEE J. Quantum Electronics **QE-10**, 546.

Bjorkholm, J.E. (1968), Appl. Phys. Letters **13**, 36, 53 and 399.

Bliss, E.S. (1969), EASTM Special Technical Publication 469: Damage in Laser Glass, ed. by A.J. Glass, A.H. Guenther, C.M. Stickley and J.D. Myers, p. 9.

Bloembergen, N. (1965), Nonlinear Optics (Benjamin; New York), 89.

Boggett, D.M. and A.F. Gibson (1968), Phys. Letters **28A**, 33.

Bokut', B.V., N.S. Kazak, A.G. Mashchenko, V.A. Mostovnikov and A.N. Rubinov (1972), Pis'ma Zh. Eksp. Teor. Fiz **15**, 26 (English transl.: 1972, JETP Letters **15**, 18.

Bolognesi, G.P., S. Mezzetti and F. Pandarese (1973), Opt. Commun. **8**, 267.

Chemla, D.S., P.J. Kupecek, D.S. Robertson and R.C. Smith (1971), Opt. Commun. **3**, 29.

Chemla, D.S., P.K. Kupecek and C.A. Schwartz (1973), Opt. Commun. **7**, 225.

Chen, F.S. (1969), J. Appl. Phys. **40**, 3389.

Chesler, R.B., M.A. Karr and J.E. Geusic (1970a), J. Appl. Phys. **41**, 4125.

Chesler, R.B., M.A. Karr and J.E. Geusic (1970b), Proc. IEEE **58**, 1899.

Christiansen, W.H. and A. Herzberg, (1973), Proc. IEEE **61**, 1060.

Colles, M.J. and C.R. Pidgeon (1975), Rep. Prog. Phys. **38**, 329.

Dana, E.S. and W.E. Ford (1951), A Textbook of Minerology, 4th ed. (Wiley; New York).

David, J. (1973), in; Laser Induced Damage in Optical Materials (A.J. Glass and A.H. Guenther, eds.) NBS Spec. Pub. 387.

Davydov, B.L., L.D. Derkacheva, V.V. Dunina, M.E. Zhabotinskii, V.F. Zolin, L.G. Koreneva and M.A. Samokina (1970), Sov. Phys. JETP Letters **12**, 16.

Davydov, B.L., M.E. Zhabotinskii, V.F. Zolin, L.G. Koreneva and M.A. Samikhina (1971), Sov. Phys. JETP Letters **13**, 238.

Davydov, B.L., L.A. Kulevskii, A.M. Prokhorov, A.D. Sal'ev and V.V. Smirnov (1972), Pis'ma Zh. Eksp. Teor. Fiz. **15**, 725 (English transl.: 1972, JETP Letters **15**, 513).

Davydov, B.L., L.G. Koreneva and E.A. Lavrovskiy (1974), Radio Eng. and Electron. Phys. (USA) **19**, 130.

Davydov, B.L., S.G. Kotovshikov and V.A. Nefedov (1977), Sov. J. Quantum Electronics **7**, 129.

Day, G.W. (1971), Appl. Phys. Letters **18**, 347.

Deserno, U. and G. Nath (1969), Phys. Letters **30A**, 483.

Dmitriev, V.G. and I. Ya. Itskhoki (1975), Sov. J. Quantum Electronics **5**, 735.

Dmitriev, V.G. A.G. Ershov, A.I. Kovrigin, V.R. Kushnir, S.R. Rustamov and N.V. Shkunov (1971a), Kvantovaya Elektron. (Moscow) No. **5**, 133 (English transl.: 1972, Sov. J. Quantum Electronics **1**, 550).

Dmitriev, V.G., A.G. Ershov, P.I. Zudkov, G.A. Sharif and E.M. Shvom (1971b) Kvantovaya Elektron. (Moscow) No. 1, 116 (English transl.: 1971, Sov. J. Quantum Electronics **1**, 84).

Dmitriev, V.G., E.A. Shalaev and E.M Shoom (1974), Sov. J. Quantum Electronics **3**, 450.

Dmitriev, V.G., V.A. Konovalov and E.A. Shalaev (1975), Sov. J. Quantum Electronics **5**, 282.

Dowley, M.W. (1968), Appl. Phys. Letters **13**, 395.

Dowley, M.W. and E.B. Hodges (1968), IEEE J. Quantum Electronics **QE–4**, 552.

Ernst, G.J. and W.J. Witteman (1972), IEEE J. Quantum Electronics **QE–8**, 382.

Fay, H., W.J. Alford and H.M. Dess (1968), Appl. Phys. Letters **12**, 86.

Feichtner, J.D. and G.W. Roland (1972), Appl. Optics **11**, 993.

Feichtner, J.D., R. Johannes and G.W. Roland (1970), Appl. Optics **9**, 1716.

Findlay, D. and D. Goodwin (1970), Advances in Quantum Electronics **1**, 77.

Francois, G.E. (1966), Phys. Rev. **143**, 597.

Franken, P.A., A.E. Hill, C.W. Peters and G.Weinreich (1961), Phys. Rev. Letters **7**, 118.

Geusic, J.E., H.J. Levinstein, S. Siugh, R.G. Smith and L.G. Van Vitert (1968), Appl. Phys. Letters **12**, 306.

Giordmaine, J.A. (1962), Phys. Rev. Letters **8**, 19.

Golyaev, Yu. D., V.G. Dmitriev, I.Ya. Itskhoki, V.N. Krasnyanskaya, I.S. Rez and E.A. Shaleev (1973), Sov. J. Quantum Electronics **3**, 72.

Gorokhov, Yu.A., D.P. Krindach, V.S. Marorov and V.S. Shevera (1974), Opt. Spectrosc. **36**, 587.

Goubau, G. and F. Schwering (1961), Proc. IRE **AP–9**, 3, 248.

Hagen, W.F. (1968), IEEE J. Quantum Electronics **QE–4**, 36.

Hagen, W.F. and P.G. Magnante (1969), J. Appl. Phys. **40**, 219.

Hanamitsu, K., Y. Cho and Y. Matsuo (1976), Jap. J. Appl. Phys. **15**, 503.

Hanna, D.C., B. Luther-Davies, H.N. Rutt, R.C. Smith and C.R. Stanley (1972) IEEE J. Quantum Electronics **QE–8**, 317.

Hanna, D.C., V.V. Rampal and R.C. Smith (1974), IEEE J. Quantum Electronics **QE–10**, 461.

Hargrove, S. and T. Kan (1977), IEEE J. Quantum Electronics **QE–13**, 280.

Harris, S.E. (1969), Proc. IEEE **57** 2096.

Henningsen, T., J.D. Feichtner and N.T. Melamed (1971), IEEE J. Quantum Electronics **QE–7**, 248.

Herbst, R.L. and R.L. Byer (1972), Appl. Phys. Letters **21**, 189.

Hill, A.E. (1970), Appl. ᵖhys. Letters **16**, 423.

Hobden, M.V. (1967), J. Appl. Phys. **38**, 4365.

Hon, D.T. (1976), IEEE J. Quantum Electronics **QE–12**, 148.

Hon, D.T. (1977), IEEE J. Quantum Electronics **QE–13**, 9, 990.

Hon, D.T. (1978), J. Appl. Phys. **49**, 369.

Hon, D.T. and H. Bruesselbach (1979), IEEE Conf. on Laser Engineering and Applications, Wash. D.C. (May 1979).

Hon, D.T., H. Bruesselbach and E. Woodbury (1977), High Average Power Frequency Doubling of Nd:YAG for Pumping Dye Lasers, SPIE Conference, San Diego, August 1977.

Hon, D.T., H. Bruesselbach and R.S. Adhav (1979), Opt. Commun. (to be published).

Huber, P. (1975), Opt. Commun. **15**, 196.

Hulme, K.E. (1973), Rep. Prog. Phys. **36**, 497.

Hulme, K.E., P. Jones, P.H. Davies and M.V. Hobden (1967), Appl. Phys. Letters **10**, 133.

Hurwitz, C.E. (1966), Appl. Phys. Letters **8**, 121.

Huth, B.G. and Y.C. Kiang (1969), J. Appl. Phys. **40**, 4976.

Huth, B.G., G.I. Farmer, L.M. Taylor and M.R. Kagan (1968), Spectrosc. Letters **1**, 425.

Izrailenko, A.I., A.I. Kovrigin and P.V. Nikles (1970), Pis'ma Zh. Eksp. Teor. Fiz. **12**, 475 (English transl.: 1970, JETP Letters **12**, 331).

Jain, R.K. and T.K. Gustafson (1973), IEEE J. Quantum Electronics **QE–9**, 859.

Jennings, D.A. and A.J. Varga (1971), J. Appl. Phys. **42**, 5171.

Jerphagnon, J. (1970), Phys. Rev. **B2**, 1091.

Jerphagnon, J. (1971), IEEE J. Quantum Electronics **QE–7**, 42.

Jerphagnon, J. and M. Bernard (1968), IEEE J. Quantum Electronics **QE–4**, 395.

Jerphagnon, J. and H.W. Newkirk (1971), Appl. Phys. Letters **18**, 245.

Jerphagnon, J., E. Batifol and M. Sourbe (1967), C.R. Acad. Sci. Ser. **B265**, 400.

Johnson, F.M. and J.A. Duardo (1967), Laser Focus **3**, No. 6, 31.

Juyal, D.P. and G.C. Thomas (1975), Opt. Commun. **15**, 26.

Kato, K. (1974), Appl. Phys. Letters **25**, 342.

Kato, K. (1974a), IEEE J. Quantum Electronics **QE–10**, 616.

Kato, K. (1974b), IEEE J. Quantum Electronics **QE–10**, 622.

Kato, K. (1975a), J. Appl. Phys. **46**, 2721.

Kato, K. (1975b), Opt. Commun. **13**, 93.

Kato, K. (1975c), Opt. Commun. **13**, 361.

Kato, K. (1975d), IEEE J. Quantum Electronics **QE–11**, 373.

Kato, K. (1975e), IEEE J. Quantum Electronics **QE–11**, 939.

Kato, K. and S. Nakao (1974), Jap. J. Appl. Phys. **13**, 1681.

Kato, K., A.J. Alcock and M.C. Richardson (1974), Opt. Commun. **11**, 5.

Kielich, S. (1970), Opto-electronics **2**, 125.

Kildal, H. and J.E. Mikkelsen (1974), Opt. Commun. **10**, 306.

Kiss, Z.J. and R.J. Pressley (1966), Appl. Optics **5**, 10, 1474.

Kleinman, D.A. (1972), in: Laser Handbook (North-Holland; Amsterdam), **2**, chapter E4.

Kleinman, D.A., A. Ashkin and G.D. Boyd (1966), Phys. Rev. **145**, 338.

Koechner, W. (1976), Solid-State Laser Engineering (Springer–Verlag; New York).

Kogan, R.M. and T.G. Crow (1977), IEEE J. Quantum Electronics, No. 9.

Kogelnik, H. and T. Li (1966), Proc. IEEE, **54**, 1312.

Konyukof, V.K. et al. (1970), JETP Letters **12**, 321.

Kotovschikov, S.G., E.A. Lavrovskii and B.L. Davydov (1975), Zh. Prikl. Spektrosk. **12**, 161.

Krupke, W.F. and W.R. Tory, (1969), IEEE J. Quantum Electronics **QE–S**, 575.

Kryukov, P.G., Yu.A. Matvetts, D.N. Nikogosyan, A.V. Sharkov, E.M. Gordeev and S.D. Franchenko (1977), Kvant Elektron. **4**, 211 (English transl.: 1977, Sov. J. Quantum Electronics **6**, 127).

Kuhl, J. and H. Spitschan (1975), Opt. Commun. **13**, 6.

Kuizenga, D.J. and A.E. Siegman (1970), IEEE J. Quantum Electronics **QE–6**, 694.

Kupecek, P., E. Batifol and A. Kuhn (1974a), Opt. Commun. **11**, 291.

Kupecek, P.J., C.A. Schwartz and D.S. Chemla (1974b), IEEE J. Quantum Electronics **QE–10**, 540.

Kurtz, S.K. (1972), in: Laser Handbook (North Holland; Amsterdam), **1**, chapter D1.

Kurtz, S.K. and T.T. Perry (1968), J. Appl. Phys. **39**, 3798.

Kushida, T., Y. Tanaka, M. Ojima, and Y. Nakazaki (1975), Jap. J. Appl. Phys. **14**, 1097.

Labuda, E.F. and A.M. Johnson (1967), IEEE J. Quantum Electronics **QE–3**, 164.

Lacina, W.B., R.G. Eguchi, M.M. Mann and M.L. Bhaumik (1971), Appl. Opt. **10**, 221.

Larsen, E.S., and H. Berman (1934), The Microscopic Determination of Non-Opaque Minerals, 2nd ed., Washington D.C., Government Printing Office, Geological Survey Bulletin, p. 848.

Lauberaeu, A., L. Greiter and W. Kaiser (1974), Appl. Phys. Letters **25**, 87.

Levine, B.F. (1969), Phys. Rev. Letters **22**, 787.

Levine, B.F. (1970), Phys. Rev. Letters **25**, 440.

Levine, B.F. and C.G. Bethea (1972), Appl. Phys. Letters **20**, 272.

Machewirth, J.P., R. Webb. and D. Anafi (1976), Laser Focus **12**, No. 5, 104.

Maker, P.D., R.W. Terhune, M. Nisenoff and C.M. Savage (1962), Phys. Rev. Letters **8**, 21.

Marling, J. (1977), IEEE J. Quantum Electronics **QE–13**, 94D.

Massey, G.A. and R.A. Elliott (1974), IEEE J. Quantum Electronics **QE–10**, 899.

Massey, G.A., J.C. Johnson and R.A. Elliott (1976), IEEE J. Quantum Electronics **QE–12**, 143.

Mayden, D. and R.B. Chesler (1971), J. Appl. Phys. **12**, 1031.

McCarthy, D.E. (1968), Appl. Optics **7**, 1997.

Menyuk, G.W. Iseler and A. Mooradian (1976), Appl. Phys. Letters **29**, 422.

Midwinter, J.E. (1967), Appl. Phys. Letters **11**, 128.

Midwinter, J.E. (1968), J. Appl. Phys. **39**, 3033.

Milek, J.T. and S.J. Welles (1970), Linear Electro-Optic Modulator Materials Report AD704556, Hughes Aircraft Corp., Culver City, CA.

Miller, R.C. (1964), Appl. Phys. Letters **5**, 17.

Miller, R.C., G.D. Boyd and A. Savage (1965), Appl. Phys. Letters **6**, 77.

Milton, A.F. (1972), Appl. Optics **11**, 231.

Moses, E., H. Bruesselbach, D. Stovall and D.T. Hon (1978) Conf. on Lasers and Their Applications, Soc. Opt. Quant. Electron. (Orlando, Florida, Dec. 1978).

Nash, F.R., J.G. Bergman, G.D. Boyd and E.H. Turner (1969), J. Appl. Phys. **40**, 5201.

Nash, F.R., G.D. Boyd, M. Sargent III and P.M. Bridenbaugh (1970), J. Appl. Phys. **41**, 2564.

Nassau, K., J.J. Levinstein, G.M. Laiacano, S.C. Abrahams, J.M. Reddy, J.L. Bernstein and J. Hamilton (1966), 5 papers, J. Phys. Chem. Solids **27**, 983.

Nath, G. and S. Haussuhl (1969), Appl. Phys. Letters **14**, 154.

Nath, G., H. Mehmanesch and M. Gsanger (1970), Appl. Phys. Letters **17**, 286.

Nelson, D.F. and E.H. Turner (1968), J. Appl. Phys. **39**, 3337.

Nelson, T.J. (1964), Bell System Tech. J. **43**, 821.

Ng, W.K. (1976), Hughes Aircraft Company, Culver City, CA private communication.

Ng, W.K. and E.J. Woodbury (1971), Appl. Phys. Letters **18**, 12, 550.

Nikogosyan, D.N. (1977), Sov. J. Quantum Electronics **7**, No. 1.

Nikogosyan, D.N., A.P. Sukhorukov and M.I. Golovie (1975), Kvantovaya Elektron. **2**, 609 (English transl.: 1975, Sov. J. Quantum Electonics **5**, 344).

Nurmikko, A.V., T.A. DeTempo and S.E. Schwarz (1971), Appl. Phys. Letters **18**, 130.

Nye, J.F. (1972), Physical Properties of Crystals (Oxford University Press; London).

Okada, M. and S. Ieiri (1971), IEEE J. Quantum Electronics **QE-7**, 12.

Patel, C.K.N. (1965), Phys. Rev. Letters **15**, 1027.

Patel, C.K.N. (1966), Phys. Rev. Letters **16**, 613.

Pearson, J.E., G.A. Evans and Y. Yariv (1972), Opt. Commun. **4**, 366.

Peterson, G.E., A.M. Glass and T.J. Negran (1971), Appl. Phys. Letters **19**, 130.

Pierce, J.R. (1961), Proc. Nat'l. Acad. Sci. **47**, 1808.

Pressley, R.J., ed. (1971), Handbook of Lasers with Selected Data on Optical Technology, The Chemical Rubber Co., Cleveland, OH.

Rabson, T.A., H.J. Ruiz, P.L. Shah and F.K. Tittel (1972), Appl. Phys. Letters **20**, 282.

Richardson, M.C., A.J. Álcock, K. Leopold and P. Burtyn (1973), IEEE J. Quantum Electronics **QE-9**, 236.

Schinke, D.P. (1972), IEEE J. Quantum Electronics **QE-8**, 86.

Scott and DeWitt (1971), Appl. Phys. Letters **1**, 3.

Seguin, H.J., K. Manes and J. Tulip (1972), Rev. Sci. Instrum. **43**, 1134.

Shen, Y.R. (1976), Rev. Mod. Phys. **48**, No. 1.

Sherman, G.H. and P.D. Coleman (1973a), J. Appl. Phys. **44**, 238.

Sherman, G.H. and P.D. Coleman (1973b), IEEE J. Quantum Electronics **QE-9**, 403

Shigorin, V.D. and G.P. Shipulo (1973), Optika Spectrosk **34**, 83.

Siegman, A.E. (1965), Proc. IEEE **53**, 277.

Siegman, A.E. (1974), Appl. Opt. **13**, 353.

Singh, S., W.A. Bonner, J.R. Potopowicz and L.G. Van Uitert (1970a), Appl. Phys. Letters **17**, 292.

Singh, S., D.A. Draegert and J.E. Geusic (1970b), Phys. Rev. **B2**, 2709, 2724.

Smith, P.W., M.A. Duguay and E.P. Ippen (1974), Mode-Locking of Lasers, in: Progress in Quantum Electronics (Pergamon Press; New York), **3**.

Smith, R.G. (1968), IEEE J. Quantum Electronics **QE-4**, 505.

Smith, R.G. (1970a), IEEE J. Quantum Electronics **QE-6**, 215.

Smith, R.G. (1970b), J. Appl. Phys. **41**, 3014.

Soffer, B.H. and I.M. Winder (1967), Appl. Phys. Letters **24A**, 5.

Sohler, W. and H. Suche (1978), Appl. Phys. Lett. **33**, 518.

Sonin, A.S. and A.S. Vasilevskaya (1971), Electro-Optic Crystals [in Russian] (Atomizdat; Moscow).

Southgate, P D. and D.S. Hall (1971), Appl. Phys. Letters **18**, 456.

Southgate, P.D. and D.S. Hall (1972), J. Appl. Phys. **43**, 2765.

Srivastava, G.P., M.L. Goyal, S. Mohan and N.M. Nahar (1976), Optica Acta, **23**, No. 4, 331.

Sukhorukov, A.P. and I.V. Tomov (1970), Optika i Spektrosk. **28**, 1211 (English transl.: 1970, Opt. Spectrosc. (USSR) **28**, 651).

Sullivan, S. and E.L. Thomas (1975), Opt. Commun. **14**, 418.

Suvorov, V.S. and A.A. Filimonov (1967), Fiz. Tverd. Tela (Leningrad) **9**, 2131. (English trans.: 1968, Sov. Phys. Solid State **9**, 1674.)

Suvorov, V.S. and I.S. Rez (1969), Optika i Spektrosk. **27**, 181 (English transl.: 1969), Opt. Spectrosc. USSR) **27**, 94).

Suvorov, V.S. and A.S. Sonin (1966), Kristallografiya **11**, 832 (English transl.: 1966, Sov. Phys. Crystallog. **11**, 711).

Suvorov, V.S., A.S. Sonin and I.S. Rez (1967), Zh. Eksp. Teor. Fiz. **53**, 49 (English transl.: 1968, Sov. Phys. JETP **26**, 33).

Tagiev, Z.A. and A.S. Chirkin (1974), Sov. J. Quantum Electronics **4**, 1452.

Taynai, J.D., R. Targ and W.B. Tiffany (1971), IEEE J. Quantum Electronics **QE-7**, 412.

Teich, M.C., R.L. Abrams and W.B. Gandrud (1969), Opt. Commun. **2**, 206.

Tsuya, H., Y. Fujino and K. Sugilbuchi (1970), J. Appl. Phys. **41**, 2557.

Tyte, D.C. (1970), in: Advances in Quantum Electronics (D.W. Goodwin, ed., Academic Press New York), **1**, 29.

Umegaki, S., S. Yabumoto and S. Tanaka (1972), Appl. Phys. Letters **21**, 400.

van der Ziel, J.P. (1964), Appl. Phys. Letters **5**, 27.

Volkova, E.N. and V.V. Fadeev (1968), in Nonlinear Optics [in Russian], Nauka, Novosibirsk, p. 185.

Volosov, V.D. and A.G. Kalinstev (1974), Sov. J. Quantum Electronics **4**, 451.

Volosov, V.D. and E.V. Nilov (1966), Optika i Spektrosk. **21**, 715 (English transl.: 1966, Opt. Spectrosc. (USSR) **21**, 392).

Volosov, V.D. and M.I. Rashchektseva (1970), Optika Spektrosk. **28**, 105.

Volosov, V.D., Yu.E. Kamach, E.N. Kozlovskii and V.M. Ovchinnikov (1969), Opt. Mekh. Promst. No. **10**, 3 (English transl.: 1969, Sov. J. Opt. Technol. **36**, 656).

Volosov, V.D., V.N. Krylov, V.A. Serebryakov and D.V. Sokolov (1973), Pis'ma Zh. Eksp. Teor. Fiz **19**, 38 (English transl.: 1974, JETP Lett. **19**, 29).

Volosov, V.D., S.G. Karperko, N.G. Korneinko, V.N. Krylov, A.A. Manko and V.L. Stuzhevskii (1975), Sov. J. Quantum Electronics **5**, 500.

Voronov, V.V. Yu.S. Kuz'minov, V.V. Osiko and A.M. Prokhorov (1975), Sov. J. Quantum Electronics **5**, 297.

Wallace, R.W. (1971), Opt. Commun. **4**, 316.

Wallace, R.W. and S.E. Harris (1969), Appl. Phys. Letters **15**, 111.

Warner, J. (1971), Opto-electronics **3**, 37.

Weber, H.P., E. Mathieu and K.P. Meyer (1966), J. Appl. Phys. **37**, 3584.

Weber, M.J. (1971), Insulating Crystal Lasers, in: Handbook of Lasers, (R.J. Pressley, ed.) 371–420. The Chemical Rubber Co., Cleveland, OH.

Weiss, J.A. and L.S. Goldberg (1974), Appl. Phys. Letters **24**, 389.

Wenzel, R.G. and G.P. Arnold (1976), Appl. Optics **15**, 1322.

Winchell, A.N. (1974), Optical Properties of Organic Compounds (New York).

Winchell, A.N. and H. Winchell (1964), Microscopical Characters of Artificial Inorganic Solid Substances (Academic Press; New York).

Wood, O.R., R.L. Abrams and T.J. Bridges (1970), Appl. Phys. Letters **17**, 376.

Woodbury, E.J. (197x), private communication.

Yariv, A. (1967), Quantum Electronics (Wiley; New York).

Zernike, F. Jr. (1965), J. Opt. Soc. Amer. **55**, 210.

Zernike, F. and J. Midwinter (1973), Applied Nonlinear Optics, (Wiley; New York).

Zumsteg, F.C., J.D. Bierlein and T.E. Gier (1976), J. Appl. Phys. **47**, 11, 4980.

Zumsteg, F.C. (1978), Conf. on Laser and Electro-optical Systems, San Diego, CA.

Zverev, G.M., E.A. Levchuk, V.A. Pashkov and Yu. D. Poryadin (1972), Kvantovaya Elektron. (Moscow) No. **2**(8), 94 (English transl.: 1972, Sov. J. Quantum Electronics, 167).

B3 | Laser-Induced Chemical Reactions and Isotope Separation*

C.D. CANTRELL,
S.M. FREUND,
and J.L. LYMAN

University of California,
Los Alamos Scientific Laboratory,
P.O. Box 1663, Los Alamos, New Mexico 87545

Contents

* Work performed under the auspices of the United States Department of Energy.

© *North-Holland Publishing Company, 1979*
Laser Handbook, edited by M.L. Stitch

Abstract

In this chapter we review the rapidly expanding field of infrared laser chemistry concentrating on those facets which are close to our individual interests. No attempt has been made to achieve completeness of coverage, and only papers available to us by December 1977 are referenced.

The discussion is divided into three sections. The first presents the phenomenological problem with a broad overview of reported experiments. The second deals with some quantitative aspects of reactions induced with intense laser radiation, and the third establishes a formalism for a unified theoretical treatment of the absorption process.

1 Introduction

Laser-induced or laser-enhanced chemistry has blossomed into a full-scale discipline in just a few short years. So much so, that in this chapter we must restrict our attention to infrared laser chemistry, leaving out entirely any excited electronic state photochemistry. The breadth of just this area is truly amazing. Within this first restriction, we have addressed only laser-induced chemistry in gaseous homogeneous systems. Experiments range from simple rate enhancement of bimolecular chemical reactions with laser excitation, to the stimulation of previously unknown chemical reactions. Theoretical interpretations have explained many of the results, but at the time of the writing of this chapter it is fair to say that there are large gaps in our understanding of these phenomena.

We have divided our discussion into three sections. The first states the phenomenological problem and gives a broad overview of reported experiments. The second deals with the quantitative aspects of reactions induced with intense laser radiation. From the chemical reactions investigated some speculation is possible as to the distribution of the absorbed energy among vibrational modes, which in turn leads to some fascinating conclusions. The third establishes a formalism for a unified theoretical treatment of the absorption process. No attempt has been made to achieve completeness in coverage of this field. Rather we have concentrated on those facets which are closest to our individual interests, in the hope that this represents a fair and tutorially useful treatment. Only papers that were available to us in December 1977 are referenced in this chapter.

In §2, the subject of laser-enhanced chemical reaction, from bimolecular reactions of vibrationally excited molecules to photodissociation as a result of weak or intense infrared laser excitation, is discussed. The occurrence of increased rate of reaction when one vibrational quantum of energy is added to one of the reaction partners in a gas mixture is well documented. Details of the interpretation of such rate enhancements are treated. Dissociation of molecules under the influence of intense infrared laser radiation is also well known. However, experimental results have not always led to unambiguous or even correct interpretations. Some of the pitfalls in the analysis of such experiments are discussed and several of the more definitive papers are outlined. In keeping with the tutorial nature of our review, we have questioned the interpretation of some experiments in the light of more recent work. We would like to express our apologies to those authors whose papers have been thus questioned, since it is often true in a new and initially largely empirical field that the early experiments contribute heavily to the initial development of the field, but are subsequently found to require different or more complex interpretations than were originally given. Special mention is made of the possibility of thermally heating gas mixtures with lasers and of inducing some novel reactions. This latter class of reactions is termed 'laser alchemy'. The entire discussion runs in a loose chronological order, with the more modern experiments providing a few of the missing pieces to a more complete understanding of such processes.

In §3, a more detailed analysis of the available experimental results on the absorption of intense laser radiation by SF_6 and the subsequent dissociation will be given. The absorption and dissociation data are compared with predictions of a computer model of the processes. The model includes collisional energy transfer processes, dissociation formulated using RRKM unimolecular reaction rate theory and a simple formulation of absorption of laser radiation. These comparisons are used to draw conclusions about the absorption mechanism, the rate of V–V transfer process (collisionless) and the role of collisions in the process.

In §4 a formalism is constructed for treating the absorption of intense infrared laser radiation by a polyatomic molecule. In contrast to §3, where absorption is treated as the least well-defined part of a somewhat phenomenological model of the laser-induced reaction, the approach here is to establish the groundwork for a complete and correct theory of absorption. A reduced-density-matrix formalism is employed for this problem and is applied to the absorption of intense CO_2 laser radiation by SF_6. Coupling of the laser field to the ν_3 normal mode of SF_6 and the damping of the ν_3-mode oscillations by interaction with the other normal modes of the molecule are both included in the formalism. A detailed treatment of collisional effects is not given in this section, since no definitive theoretical treatment is available. Much of the recent theoretical work on this phenomenon will be reviewed in the process of describing the formalism, and some recent calculations on the role of anharmonic mode mixing that were obtained using the reduced-density-matrix formalism will be presented.

2 Infrared laser chemistry

In this section we address the subject of infrared laser-induced or laser-enhanced chemical reactions from a phenomenological, qualitative chemical viewpoint. The quantitative aspects concerning the absorption of enough infrared radiation to produce a chemical reaction, be it thermal, or truly laser stimulated, as well as the details of the actual molecular excitation process will be treated in §§3 and 4. Here we are interested in determining where the laser energy is absorbed and its role in promoting the reaction, and what species are produced after multiple-photon absorption and their ultimate fate. The ramifications of much milder excitations in the increase of chemical reaction rates are discussed. New synthetic routes (laser alchemy) are also pursued in this section.

As a discipline, laser-induced chemistry has four distinct manifestations. The first, laser enhancement of chemical reactions, has shown that reaction rates can be accelerated by orders of magnitude with just one or two quanta of excitation in the molecules. The second, laser isotope separation, has intrigued many because of its great economic potential with regard to the nuclear fuel problem and to other industrial applications of isotopes. The third relates to the possibility of bond-specific chemistry induced with a laser, and has as its attraction the hope of changing the normal thermal chemical routes. That is, chemical reactions might be coerced to provide very specific products and become much more efficient in the process. Novel syntheses can also be envisioned. The final and most common, laser pyrolysis, has a stigma associated with it. The identification of the laser as a heating element which can provide high gas temperatures while allowing the container walls to remain cool, has with few exceptions (Shaub and Bauer 1975) been avoided in the literature. It is quite likely that much of the reported 'bond-specific' laser-induced chemistry is simply homogeneous catalysis at high temperatures. Indeed our approach will be to extract from the mass of experimental data those reactions which lead to isotopic selectivity or products not normally obtained from heterogeneous catalysis.

Different products as well as altered product distributions may well indicate that the laser-induced reaction route has deviated from the high-temperature reaction route. However, the identification of such a process as laser or bond specific is difficult unless isotopic enrichment – which precludes the domination of collisional processes – has simultaneously occurred, or the products are monitored in real time and found to follow the laser pulse on a time scale short compared to general scrambling processes.

We shall begin with some unexplored details of laser-enhanced bimolecular reactions, go on to a description of the possibly misinterpreted laser chemistry experiments, then to the difficult-to-understand papers, and finish with several somewhat more clearcut experiments and suggestions for improvement of this genre of experiment. There are many reviews of laser effects on chemical reactions (Basov et al. 1974, 1977, Karlov 1974, Knudtson and Eyring 1974, Bergamann et al. 1975, Klein et al. 1975, Ronn 1975, Lin 1976, Letokhov and Moore 1976a, b, Aldridge et al. 1976, Jensen et al. 1976, Birely et al. 1976, Karlov and Prokhorov 1976, Wolfrum 1977). Therein are presented detailed descriptions of the lasers used.

2.1 Laser-enhanced bimolecular reactions

There are a number of experiments that have been performed that have the effect of vibrational energy on the rate of bimolecular reactions is measured. Since lasers have been employed in many of these to produce vibrationally exicted species, the subject may properly be treated in this paper. However, an extensive review of this type of reaction will not be given, since the subject is somewhat orthogonal to our main theme and since several recent reviews have treated laser-enhanced bimolecular reactions (Birely and Lyman 1975, Wolfrum 1977).

The question addressed in these experiments is what is the effect of vibrational energy on the rate of bimolecular reactions. In order to illuminate some of the problems that may arise when analyzing the results of a vibrationally enhanced bimolecular chemical reaction, we will consider in detail the analysis of the most studied of these, the reaction of vibrationally excited ozone with nitric oxide. The effect of reactant vibrational energy on the rates of these reactions has been studied by exciting either ozone (Gordon and Lin 1973, 1976, Kurylo et al. 1974/75, 1975, Braun et al. 1974, Freund and Stephenson 1976) or nitric oxide (Stephenson and Freund 1976) with infrared lasers. The effect of the NO electronic state on the rates of these reactions has also been studied (Redpath and Menzinger 1975). The majority of this work has been performed by measuring the enhanced rate when ozone is vibrationally excited. In these experiments the ozone was excited to the (001) vibrational level by absorbing radiation from a CO_2 laser (Q-switched, TEA or square-wave chopped cw). The effect of vibrational excitation on the reaction rate was observed by monitoring the visible luminescence from NO_2^*. The contribution to the rate from vibrationally excited ozone was determined from either the relaxation time of the luminescence after the laser excitation or from the depth of modulation of the luminescence when a chopped cw laser was used.

The rate of reaction of ozone with nitric oxide has also been measured under thermal conditions between 216 and 322 K (Clough and Thrush 1967). The reaction produces nitrogen dioxide in both the electronic ground state and an electronically excited state by

$$NO + O_3 \rightarrow NO_2 + O_2, \tag{1}$$

$$NO + O_3 \rightarrow NO_2^* + O_2, \tag{2}$$

where the asterisk (*) indicates electronic excitation. They found that

$$k(1) = (4.3 \pm 1.0) \times 10^{11} \exp(-2330 \pm 150/RT) \, cm^3 \, mole^{-1} \, s^{-1}, \tag{3}$$

$$k(2) = (7.6 \pm 1.5) \times 10^{11} \exp(-4180 \pm 300/RT) \, cm^3 \, mole^{-1} \, s^{-1}. \tag{4}$$

The value of $k(1)$ was obtained by subtracting $k(2)$ from the overall rate coefficient (Clyne et al. 1964).

The experiments to date have established that the rate constant for vibrationally excited ozone ($k^\dagger$) is greater than the thermal rate constant (k) in each case [(1) and (2)]. The rate constant ratios ($k^\dagger/k$) range from 6 to 17 for reaction (1) and from 4 to 7 for reaction (2) (Kurylo et al. 1974/75, Gordon and Lin 1976 and earlier papers). There has been some discussion about which vibrational states contribute to $k^\dagger$. The laser energy is absorbed by the ν_3 vibrational mode. This state is nearly resonant with ν_1. There is now agreement that excitation of the 100 and 001 states enhances the reaction rate but that the low-lying ν_2 vibrational state (010) contributes only slightly, if at all (Freund and Stephenson 1976). There appears to be general agreement that vibrational energy is less effective than translational energy in promoting reactions (1) and (2) since $k^\dagger/k$ is less than $\exp(h\nu/RT)$, where ν is the laser frequency. It has been pointed out, however, that from the temperature dependence of reaction (1), the vibrational enhancement is about what one would expect if all forms of energy were equally effective in promoting reaction (Gordon and Lin 1976).

In the following discussion an assessment will be made of what one would expect from this type of experiment if vibrational and translational energy were equally effective in promoting reaction. Some of the conclusions reached by others will also be evaluated.

Ozone is a nonlinear triatomic molecule and therefore has three normal vibrational modes with fundamental frequencies $\nu_1 = 1110 \text{ cm}^{-1}$, $\nu_2 = 705 \text{ cm}^{-1}$ and $\nu_3 = 1043 \text{ cm}^{-1}$. For the following discussion we will number the vibrational states of the molecule in increasing order of energy to simplify notation. The energy of the ith vibrational state will be designated as E_i, where $i = 1$ for the ground state (000) and $i = 2$ for the state (010), and so on. The specific rate constant for the reaction of ozone in vibrational state i by reaction (1) above will be written as $k_i(1)$. The overall rate constant for reaction (1) or (2) is related to the specific rate constants by

$$k = \sum_{i=1}^{n} k_i X_i,$$ (5)

where the sum is over all n vibrational states of the molecule and X_i is the fraction of the ozone molecules in energy state i. It is a major purpose of the laser-enhanced reaction experiments to gain information about the specific rate constants $k_i(1)$ and $k_i(2)$.

A bimolecular rate constant can be written in terms of the relative translational energy of the collision partners as (LeRoy 1969)

$$k(T) = (2/RT)^{3/2}(1/\pi\mu^{1/2}) \int_0^\infty E_t \sigma(E_t) \exp(-E_t/RT) \, dE_t,$$ (6)

where E_t is the relative translational energy, μ is the reduced mass of the collision partners and $\sigma(E_t)$ is the excitation function (LeRoy 1969) for the reaction. If it is assumed that only relative translational energy along the line of centers is effective in promoting reaction, and that reaction does occur when that energy exceeds some threshold E_0, then

$$\sigma(E_t) = C(1 - E_0/E_t), \quad E_t \geq E_0,$$
$$= 0, \quad\quad\quad\quad E_t < E_0,$$

where C is a constant.

With an excitation function of this form, eq. (6) becomes

$$k(T) = C(8RT/\pi\mu)^{1/2} \exp(-E_0/RT),$$

which reduces to

$$k(T) = A_0 T^{1/2} \exp(-E_0/RT).$$

In order to calculate the specific rate constants we initially postulate that the vibrational energy and the line of centers translational energy are equally effective in promoting reaction. We then calculate the consequences of this assumption and make comparisons with the experimental data. With this postulate the specific rate constants become

$$k_i = AT^{1/2} \exp[-(E_0 - E_i)/RT], \quad E_i \leq E_0,$$
$$= AT^{1/2}, \qquad\qquad\qquad\qquad E_i > E_0. \tag{7}$$

The pre-exponential constant A is independent of i, but it is different in reaction (1) and (2). If one assumes that an equilibrium distribution over vibrational states was maintained for the conditions of the thermal rate measurement, then the mole fraction of molecules in state i is

$$X_i = \exp(-E_i/RT)/Q_v, \tag{8}$$

where

$$Q_v = \sum_{i=1}^{n} \exp(-E_i/RT)$$

is the vibrational partition function. It follows then, from eqs. (5), (7) and (8), that

$$k = \sum_{i=1}^{n} k_i X_i = \sum_{i=1}^{i'} \frac{AT^{1/2}}{Q_v} \exp(-E_0/RT) + \sum_{i=i'+1}^{n} \frac{AT^{1/2}}{Q_v} \exp(-E_i/RT)$$
$$= k_1/Q_v \left[i' + \sum_{i=i'+1}^{n} \exp(-(E_i - E_0)/RT) \right] \tag{9}$$
$$= v'k_1,$$

where i' is the index of the highest energy vibrational level less than E_0, n is the number of bound vibrational states and v' is the effective number of vibrational levels that contribute to the overall reaction rate

$$v' = i'/Q_v + \sum_{i=i'+1}^{n} \exp[-(E_i - E_0)/RT]/Q_v \approx i'. \tag{10}$$

We note that under the assumptions stated above that all vibrational states with $E_i < E_0$ contribute equally to the total rate constant at all temperatures. For example, even though the population of state $i = 4$ (100) is only 0.005 times that of the ground state (000), both states make equal contribution to the rate constant of reaction (2). Also to be noted from eq. (9), is that the thermal rate constant is significantly greater than k_1, the ground state rate constant ($v' \approx 5$ for reaction (2), see below).

If v' is assumed to be independent of temperature (which it very nearly is) this rate constant in Arrhenius form is

$$k = A_{Arr} \exp(-E_{act}/RT), \tag{11}$$

where $E_{act} = E_0 + \frac{1}{2}RT$ and $A_{Arr} = v'AT^{1/2} e^{1/2}$.

From the data of Clough and Thrush (eqs. 3 and 4) we find by the use of eqs.

(7) and (10) at 300 K,

$$E_0(1) = 2.0 \text{ kcal/mole}, \qquad E_0(2) = 3.9 \text{ kcal/mole},$$

$$v'(1) = 2.3, \qquad v'(2) = 4.9, \tag{12}$$

$$A(1) = 6.6 \times 10^9, \qquad A(2) = 5.2 \times 10^9.$$

If this formulation is correct one may then calculate the rates of these two reactions for any temperature or vibrational energy distribution. This we now do for two distributions that differ in the way vibrational energy transfer processes are treated.

Let us first consider the effect of absorbed laser radiation on the reaction rate in a dilute gas mixture where the translational temperature remains constant and V–V energy transfer processes are negligible. We assume that V–T energy transfer among the *excited* vibrational states is immediate and that the excess population in excited states, ΔX, relaxes to the ground state at a finite rate. After a laser pulse the ground state will be depleted by ΔX and the populations of the excited states are related by Boltzmann factors at the translational temperature, or

$$\begin{aligned} X_1 &= X_1(\text{eq}) - \Delta X, \\ X_i &= X_i(\text{eq}) + C \exp(-E_i/RT), \quad i \neq 1, \end{aligned} \tag{13}$$

where C is a constant to be determined from the condition that the X_i's must sum to unity, or

$$C = \frac{\Delta X}{Q_v - 1}. \tag{14}$$

The rate constant [(eqs. (7), (10), (13) and (14)] for the laser-excited sample is then

$$\begin{aligned} k &= \sum_{i=1}^{n} X_i k_i \\ &= [X_1(\text{eq}) - \Delta X] k_1 + \sum_{i=2}^{n} \left[X_i(\text{eq}) + \frac{\Delta X}{Q_v - 1} \exp(-E_i/RT) \right] k_i \\ &= k(\text{eq}) + \Delta X \left[\frac{1}{Q_v - 1} \sum_{i=2}^{n} \exp(-E_i/RT) k_i - k_1 \right] \\ &= k(\text{eq}) + \Delta X k_1 \frac{v' Q_v - Q_v}{Q_v - 1}, \end{aligned} \tag{15}$$

where $k(\text{eq})$ is the rate constant at equilibrium ($\Delta X = 0$). This may be written as

$$k = k(\text{eq}) + \Delta X k^{\dagger}, \tag{16}$$

where

$$k^{\dagger} \equiv \frac{dk}{d\Delta X} = k_1 \frac{v' Q_v - Q_v}{Q_v - 1} \tag{17}$$

is the rate constant for the 'excited' species. We see that this quantity is independent of time.

From this expression, along with eq. (9), the ratio $k^\dagger/k$ can be obtained under the stated assumptions of equal effectiveness of translational and vibrational energy with the relative populations of the vibrationally excited states determined by fast V–T processes. At 300 K this calculation gives

$$k^\dagger(1)/k(1) = 11.5 \quad \text{and} \quad k^\dagger(2)/K(2) = 16.7.$$

The ratio for reaction (1) is within the range of the experimental data while that for reaction (2) is high by about a factor of three. Note that this is not the ratio of the rate constant for the laser excited state ($i = 3$) to that of the ground state ($i = 1$) [$\exp(E_3/RT)$]. For example, $k_3(2)/K_1(2) = 150$, which is almost an order of magnitude greater than $k^\dagger(2)/k(2)$.

As an example of how rate constants of vibrationally excited species are obtained, we now discuss how the quantity $k^\dagger(1) + k^\dagger(2)$ is obtained from a typical experiment (Kurylo et al. 1974/75 or Gordon and Lin 1973). A commonly observed quantity is the time rate of change of the chemiluminescent light intensity from NO_2^* produced by reaction (2). The observed emission intensity is proportional to the rate of the reaction

$$NO_2^* \rightarrow NO_2 + h\nu,$$

which, if a steady-state concentration of NO_2^* is assumed, is proportional to the rate of reaction (2) or

$$I = Dk(2)[NO][O_3], \tag{18}$$

where I is the observed chemiluminescent intensity and D is a constant. If it is further assumed that changes in NO and O_3 concentration are slow compared to changes in I due to laser excitation, and if we consider only the time after the laser is off, we get from eq. (18)

$$dI/dt = D[NO][O_3](dk(2)/dt). \tag{19}$$

But

$$dk(2)/dt = (dk(2)/d\Delta X)(d\Delta X/dt). \tag{20}$$

From eq. (17),

$$dk(2)/d\Delta X = k^\dagger(2),$$

which is independent of ΔX (and time).

For a small displacement ΔX from equilibrium

$$d\Delta X/dt = -\lambda\Delta X, \tag{21}$$

where λ is the total rate constant for the approach of ΔX to zero by reaction, vibrational relaxation and pumping the reactants out of the viewing zone. We have

$$\lambda = k^\dagger(1)[NO] + k^\dagger(2)[NO] + k_M[M] + k_p, \tag{22}$$

where k_M is the vibrational deactivation rate constant and k_p is a pumping term.

Eq. (19), with eqs. (20), (21) and (22), becomes

$$dI/dt = -D[NO][O_3]k^\dagger(2)\Delta X\lambda = -I'\lambda,$$

where I' is just the intensity in excess of the equilibrium intensity [eq. (18)]. Then

$$dI/dt = dI'/dt \qquad \text{and} \qquad dI'/dt = I'\lambda, \tag{23}$$

which is just a simple exponential rate expression.

Experimentally $k^\dagger(1) + k^\dagger(2)$ is determined by measuring λ as a function of [NO] by monitoring the visible light emitted from NO_2^* and thus determining [from eq. (22)]

$$d\lambda/d[NO] = k^\dagger(1) + k^\dagger(2).$$

In deriving eq. (23) it was assumed that V–T processes determined the distribution over vibrational states. If instead we assume that V–V processes are rapid and the distribution over *all* vibrational states is described by a vibrational temperature T_V, which in turn relaxes to the translational temperature by V–T processes and reaction, one obtains the following:

$$X_i = (1/Q_v(T_v)) \exp(-E_i/RT_v),$$

where $Q_v(T_v)$ is the vibrational partition function at temperature T_v. The fraction of molecules excited by the laser from the ground state will then be [from eq. (8)] $\Delta X = (1/Q_v(T)) - (1/Q_v(T_v))$ and the total rate constant is

$$\begin{aligned}
k = \sum_{i=1}^{\infty} k_i X_i &= \sum_{i=1}^{i'} \frac{AT^{1/2}}{Q_v(T_v)} \exp\{[-(E_0 - E_i)/RT] - (E_i/RT_v)\} \\
&\quad + \sum_{i=i'+1}^{n} \frac{AT^{1/2}}{Q_v(T_v)} \exp(-E_i/RT_v) \\
&= k_1/Q_v(T_v)\left\{ \sum_{i=1}^{i'} \exp\left[\frac{-E_i}{R}\left(\frac{1}{T_v} - \frac{1}{T}\right)\right] \right. \\
&\quad \left. + \sum_{i=i'+1}^{n} \exp[(-E_i/RT_v) + (E_0/RT)]\right\}.
\end{aligned} \tag{25}$$

This expression is considerably more complex than the expression obtained assuming V–T processes dominate [eq. (15)]. We see that

$$k^\dagger = dk/d\Delta X \tag{26}$$

is not independent of ΔX (and therefore, time). This means that the simple relation of chemiluminescent intensity to the rate constants $k^\dagger(1)$ and $k^\dagger(2)$ is not strictly correct [see eqs. (23) and (24)]. One can, however, calculate $k(1)$ and $k(2)$ using eqs. (25) and (26) and obtain $k^\dagger(1)$ and $k^\dagger(2)$ from the slopes of $k(1)$ and $k(2)$ versus ΔX (small ΔX). This was done over a range of temperatures from 150 to 400 K. A similar calculation assuming only V–T processes was made [eq. (17)] and the results are plotted along with the experimental data of Kurylo et al. (1975) in fig. 1. The V–V curve does reproduce the experimental data to

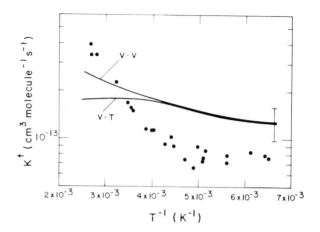

Fig. 1. Arrhenius plot for $k' = k'(1) + k'(2)$. The individual data points are from Kurylo et al. (1975). The line marked V–T was calculated assuming only vibrational–translational energy transfer process [no V–V, see eq. (17)]. The line marked V–V was calculated assuming fast vibrational–vibrational energy transfer processes [see eqs. (25) and (26)]. The error bar is a measure of the uncertainty in the thermal rate measurements (Clough and Thrush 1967) on which the calculations were based.

within a factor of two, even though the thermal data on which the calculations are based has large error bars [eqs. (3) and (4)]. We see that the two methods of calculating $k^\dagger = k^\dagger(1) + k^\dagger(2)$ give similar results at low temperature, but at high temperature the V–V energy transfer significantly alters the population of the more highly excited states and a large rate constant results. Such a large change in $k^\dagger$ from such a small change in the distribution over vibrational states shows how sensitive $k^\dagger$ is to the distribution. Other phenomena that could change the vibrational state distribution are laser absorption by excited states, depletion of population in excited states by reaction and finite V–T rates among excited states. These phenomena could certainly affect $k^\dagger$, but since little is known about them no attempt has been made to estimate their effect.

It has been argued (Gordon and Lin 1976) that since the NO_2^* luminescence [and therefore $k(2)$] varies linearly with laser intensity (and therefore ΔX), the doubly excited states such as (020) do not contribute significantly to the reaction rate since they would give quadratic dependence on intensity. However, under the assumptions employed in deriving eq. (17) $k(2)$ is exactly proportional to ΔX (i.e., no quadratic dependence), and the state (020), $(i = 5)$ does contribute substantially to $k^\dagger(2)$.

We hoped to make several points with this analysis. First, reaction from vibrationally excited states may contribute significantly to the total reaction rate even though their populations may be low. Second, the fact that the rate constant ratio $k^\dagger/k$ is significantly less than $\exp(h\nu/kT)$ does not necessarily indicate that vibrational energy is less effective than translational energy in promoting reaction. Third, the rate constant $k^\dagger$ for reaction of vibrationally excited species is quite sensitive to the distribution over vibrational states produced by laser excitation, reaction and collisional processes. This is especially true if one of the reactive species is a polyatomic molecule and if the reaction involves a significant activation energy. Fourth, because of the third item just mentioned it is difficult to assess whether vibrational or translational energy is more important in promoting the reaction of ozone with nitric oxide [reactions (1) and (2)]. It does appear, however, that the two types of energy are nearly equally effective in promoting reaction (1), but that vibrational energy is less effective in promoting reaction (2).

2.2 Possibly misinterpreted experiments

There are several cw laser experiments (Mayer et al. 1970, Yogev et al. 1972, Robinson et al. 1974, Freeman et al. 1974, Zitter et al. 1975, Tardieu de Maleissye et al. 1976, Zitter and Koster 1976) and two (Dever and Grunwald 1976, Grunwald and Olszyna 1976) pulsed laser photolyses which are almost certainly thermal reactions where the results can be explained by high temperature thermochemistry. Two additional works (Orr and Keentok 1976, Lesiecki and Guillory 1977a) describe nonthermal pathways to their observations. However, thermal chemistry may also be an important factor.

Mayer et al. (1970) essentially started the pursuit of infrared laser-induced chemistry with their now well-known reaction of a gas-phase mixture of CH_3OH, CD_3OD and Br_2. They found from infrared spectroscopy of the products that a reaction of the form

$$H_3COH + Br_2 \rightarrow 2HBr + H_2CO \tag{27}$$

occurred almost to completion (5% H_3COH remaining) when an equimolar mixture of D_3COD, H_3COH and Br_2 at a total pressure of 39 Torr was irradiated for 60 s by a 90 W cw HF laser, while the D_3COD remained basically untouched (15% reacted). Recent unsuccessful attempts (Willis et al. 1976) to repeat this separation using a pulsed HF laser suggest that reaction (27), the known thermal reaction, occurred by heating the gas mixture while the D_3COD reacted more slowly, as is reasonable kinetically at the temperatures produced. As the H_3COH was consumed the temperature returned to ambient leaving the unreacted D_3COD in such a condition that further reaction was unfavorable. The reacting system then shuts itself off before all reactants are depleted since the remaining reactants do not absorb. This certainly represents a very novel use of lasers. However, infrared analysis of complex reaction mixtures is not entirely unambiguous, and therefore the reported results should be treated with some suspicion.

Karlov in his review (Karlov 1974) mentions his study of the laser-induced reaction

$$BCl_3 + C_2H_2 \rightarrow HCl + HC \equiv CBCl_2 \tag{28}$$

which follows a different route than the known thermal reactions where direct addition occurs. Attempts to induce reaction (28) at somewhat lower pressure with a CO_2 TEA laser capable of similar performance (Freund and Ritter 1973) resulted in extremely small yields of the boron compound ($\sim 2\%$). Subsequent communication with Karlov uncovered that he was monitoring the HCl production only and inferring the HC_2BCl_2 yield from the supposed stoichiometry. Therefore, although it is clear from the fact that HCl is produced that the normal thermal route is not the one followed, one cannot understand the results unless further product identification is undertaken.

Yogev et al. (1972), Robinson et al. (1974) and Tardieu de Maleissye et al. (1976) made the almost certainly incorrect assumption that the gas temperature in the region of the laser was measured by thermocouple readings taken nearby. Clearly, the relatively high heat capacity of the temperature measuring device and the fact that the cell walls were at ambient temperature contributed to the low temperatures measured, thereby masking the thermal effects occurring at the elevated temperatures produced by the cw CO_2 laser in these experiments. Yogev et al. (1972) additionally make the argument that the decomposition of isoprene under the influence of 5 W of cw CO_2 laser light at $944 \, \text{cm}^{-1}$ is not a thermal process because N_2 added as a diluent quenches the reaction more efficiently than does He, although their respective thermal conductivities might suggest the opposite behavior. Although the V–T quenching rate for N_2 and He on isoprene is not measured in this paper, and this rate may be larger for He than N_2 in some cases (Weitz and Flynn 1973), it is possible that N_2 could deactivate the isoprene more rapidly than He. In this case then, we suggest that the higher specific heat for N_2 could account for a lower gas temperature and consequently less of the pyrolysis products reported in this work as compared to the He case. Higher pressure results, where it is demonstrated that dimerization occurs at the same rate regardless of whether the isoprene absorbs the laser radiation or whether it is absorbed by a nonreactive bath molecule, show that in this regime as well, thermal reaction is the dominant path.

Freeman et al. (1974) in a cw CO_2 laser photolysis of CF_2Cl_2 have neglected thermal conductivity effects in ascribing the lowering of the decomposition rate of the Freon with increasing SF_6 concentration in a flowing system to other than heating effects. By increasing the size of the laser beam they reduced the laser intensity in order to keep the total energy absorbed constant as the partial pressure of SF_6 (much larger absorption coefficient) was increased. This substantially reduced the final temperature of the mixture which is governed by its thermal conductivity and heat capacity, thus reducing the reaction yield. Their observation that the absorption of laser energy by CF_2Cl_2 is independent of intensity while that of SF_6 is quite intensity dependent is not very surprising. Nowak and Lyman (1975) have shown that drastic changes in absorption coefficient for SF_6 can occur with changes in temperature. For example, the absorption coefficient at the SF_6 band center (P(16) line at $947.75 \, \text{cm}^{-1}$) decreases by a factor of ten between 300 and 750 K, while at a slightly lower frequency (P(32) line at $933 \, \text{cm}^{-1}$) it increased by a factor of ten over the same temperature range. One last comment on the work by Freeman et al. (1974) is that they observed a fifth power dependence of the reaction rate on laser intensity. They suggested that this may be due to either a thermal process or to multiphoton absorption. At these power densities multiphoton processes are insignificant. As will be shown in the next section, such dependence on the power is expected for thermal reactions.

Bailey et al. (1974) arrive at a reaction temperature of 900 K by measuring the pressure increases in the static cell, by examining the ir band contours of the C_2H_4 produced in the cw-laser-promoted decomposition of C_2H_5Cl and by using the Arrhenius parameters and rate equation for the decomposition of this compound. All three methods agree roughly, from which the authors conclude that the laser provides a selective nonthermal reaction pathway, since the small quantity of butadienes produced from the product C_2H_4 implies temperature in excess of 1100 K. However, these measurements reflect the average state of the contents of the cell rather than the very high temperatures occurring in the laser path and severe gradients occurring away from this region so that their conclusions as to the mechanism are not justified by the experimental measurements. Further, butadiene absorbs in the same region ($972\,cm^{-1}$) where the C_2H_5Cl irradiation was performed, which could perhaps hasten its removal. Some methane is also produced, whereas the thermal unimolecular process at 900 K yields only C_2H_4 and HCl. Again, at the very much higher temperatures in the beam path, other thermal reactions can easily occur.

The problem of using low-power absorption coefficients to determine the energy absorbed when there can be a temperature change is brought up again in two papers by Zitter and co-workers (Zitter et al. 1975, Zitter and Koster 1976). In the first of these, pressures of SF_6–CF_2Cl_2 mixtures are adjusted such that the low-power absorbances are equal at the frequency of absorption for SF_6 in one case, and for CF_2Cl_2 in the other. It is then assumed that the laser power absorbed under intense laser radiation will be the same for the two mixtures. This is almost certain to be incorrect, which makes the conclusion of nonthermal chemistry suspect. A similar problem exists in the second paper where the conclusion is drawn that mode-specific reaction is occurring when irradiation on the weaker of two absorption bands induces a reaction at a faster rate than for the stronger absorption. The authors have recognized errors in interpretation and in a more recent paper (Zitter and Koster 1977) measured the absorption at higher intensity and found it to be quite different from low intensity. Their conclusion of mode-specific reaction was dropped, but the fact that their reaction rates vary sharply with wavelength while absorbed energy is roughly constant was presented as evidence of a nonthermal component of the reaction.

Perhaps the most extensively publicized body of data that we feel to be misinterpreted is that by Grunwald and co-workers (Dever and Grunwald 1976, Grunwald and Olszyna 1976, Chem. Eng. News, 1976). In this work pulsed-CO_2-laser-induced reactions of several halocarbons (CCl_3F, $CClF_3$ and $HCClF_2$) were studied at fairly high pressure (60 Torr and above). Product distributions and the effect of absorbed laser energy on the reaction probabilities were studied.

The most significant conclusion drawn from this work was that the primary process of dissociation takes place while the absorbed laser energy remains in one vibrational mode. If true, this conclusion certainly alters the currently accepted concepts of collisional intramolecular energy transfer processes. While it may be true for some species under some conditions that energy absorbed from an infrared laser may remain in the absorbing mode long enough to directly promote reaction, we believe for several reasons that this has not been demonstrated for this set of data. First, redistribution within the molecule is too fast for the conditions of these experiments. While there is some question as to the amount of vibrational energy necessary to promote rapid, collisionless intramolecular vibrational energy transfer, it is generally conceded that collisional intermode transfer of vibrational energy in an excited polyatomic molecule is very rapid. There is a very large body of ultrasonic absorption and dispersion data that has been interpreted employing this assumption (see, for example, Cottrell and McCoubrey 1961). In fact a new theoretical approach was necessary to 'explain' the conclusion of reaction from a single normal mode (Chem. Eng. News 1976). Second, the suggested primary photolysis reaction for at least one of the species studied has since been shown to be incorrect. The primary dissociation reaction of CCl_3F was suggested to be

$$CCl_3F \rightarrow CClF + Cl_2.$$

Lee et al. (Coggiola et al. 1977, Grant et al. 1977) have reported that under collisionless conditions similar to their work on the ir laser photolysis of SF_6, Cl, not Cl_2, is formed from photolysis of CCl_3F. This certainly weakens the description of the mechanism for product formation. Third, even though single-mode oscillation results in some energy in the carbon–chlorine bond (Bauer and Chien 1977), some intramolecular energy transfer must occur since, for all three molecules studied, the laser energy was absorbed into what is predominantly a carbon–fluorine stretching motion and the carbon–fluorine bonds remain intact in each case. Fourth, others (Mukamel and Ross 1977, Shamah and Flynn 1977) have suggested that even if the conclusion of non-statistical behavior is correct, an explanation based on fast rather than slow intermode energy transfer may equally well account for the data. Fifth, in order for the authors to explain the reaction products, it was necessary to postulate that a large fraction of the reacting reagent molecules is consumed by free radical reactions and not by primary laser photolysis reactions. These subsequent reactions certainly contribute to the overall activation energy of the reaction.

Finally, we consider the argument that only a model based on reaction from a single mode gives the proper activation energy. The authors argue that a purely thermal reaction could not account for the data since the final temperatures from absorbed laser energy are too small (up to 1200 K) and that the activation energies assuming only thermal reactiona are too low. For example, the activation energy assuming complete thermalization of the absorbed energy is 22 ± 2 kcal/mole for the $CClF_3$ reaction (Dever and Grunwald 1976) but the C–Cl bond strength is 86 kcal/mole, which was assumed to be the pyrolysis activation energy. However, it has been demonstrated that thermal pyrolysis of these species does indeed occur at the temperatures achieved by the laser (Trenwith and Watson 1957, Bidinosti and Porter 1961). Trenwith and Watson have measured the conversion of $CClF_3$, CCl_2F_2 and CCl_3F in an alumina vessel over a range of temperatures from 700 to 1200 K. From these data a first-order rate constant for the thermal pyrolysis of $CClF_3$ can be extracted. A least-squares fit to an Arrhenius plot of their data that results in more than 1% reaction gives

$$K(T) = 2.8 \times 10^6 \exp[-(48/RT)],$$

where the activation energy is in kcal/mole. We see that the thermal activation energy is somewhat lower than the reaction enthalpy and that the laser therefore deposits sufficient energy for thermal pyrolysis. It is highly probable that both types of experiments are influenced by heterogeneous reactions.

The thermal pyrolysis of CCl_3F (Trenwith and Watson 1957) gives an activation energy of 23 kcal/mole compared with 17 kcal/mole for the laser reaction and 77 kcal/mole for the reaction enthalpy (Dever and Grunwald 1976). Again, there is no need to postulate reaction from a single vibrational mode.

The differences in products between the laser experiments and the thermal pyrolysis could be explained by differences in wall conditions, temperature, etc., and by the fact that many of the potential products are further photolyzed by the laser.

Orr and Keentok (1976) mixed SF_6 with various hydrocarbons and observed the luminescence from the CH or C_2 produced upon irradiation at wavelengths absorbed by the SF_6. The delay of onset was found to be inversely proportional to the total pressure with the hydrocarbon concentration held constant and independent of the choice of hydrocarbon. They concluded from the facts that at a total pressure of 1 Torr the peak of the luminescence signal is reached within $3.5 \mu s$ after the laser pulse, and that the vibration–translation relaxation time for SF_6 is $150 \mu s$ Torr, that the hydrocarbon dissociation cannot result from a thermal mechanism involving a translationally hot gas. The relaxation rate for highly vibrationally excited SF_6 is almost certainly very much higher than the number quoted, which is measured for $v_3 = 1$, making this conclusion suspect. However, the pressure may be sufficiently low that the author's mention of a fluorine atom chemistry route to the luminescence via dissociation of the SF_6 could also explain the results. The observation of a sixth-power dependence of the intensity of luminescence on the incident laser power again does not necessarily indicate a deviation from thermal chemistry.

Finally, Lesiecki and Guillory (1977a, b) investigated the CO_2 TEA laser photolysis of cyclopropane at moderate pressures (1 to 20 Torr) in static cells. They observed that the relative abundance of the products from the ir experiments is slightly different from that found in a thermal decomposition at ~ 1900 K, and concluded that a high energy non-Boltzmann reaction must be occurring. In an attempt to account for the further photolysis of ethylene and propylene products at the same laser wavelength, static cells of the pure gases were filled to 10 to 20 Torr and irradiated with a similar number of pulses as used in the cyclopropane experiments, with the result that only 10% decomposition for either gas occurs. Since the high power absorption coefficients for these gases were not measured it is unreasonable to expect that the final gas temperature achieved during their irradiation will be the same as that calculated for cyclopropane.

At these high temperatures, achievable in the focused beam, ethylene will surely decompose to acetylene and hydrogen, making acetylene the primary product of the photolysis, as is observed, whereas at ~ 1900 K, ethylene is found to be the major product of cyclopropane pyrolysis. The laser photolysis does indeed yield almost exclusively acetylene and presumably hydrogen. This is again another instance where careful experimentation must be undertaken to decide whether other than thermal routes are accessible to a reacting system.

2.3 Difficult to interpret experiments

Unlike the experiments in §2.2, where it is very difficult to ascribe the results to laser-specific chemistry, there are two ir photolysis experiments (Bachmann et al. 1974, Rockwood and Hudson 1975) which yield products very much different from those obtainable thermally, yet under conditions where many gas-phase collisions occur. In these reactions, the key to understanding the mechanism may be that the laser induces high temperature chemistry away from the vessel walls, or homogeneous chemistry (Shaub and Bauer 1975), whereas conventional high temperature chemistry often has a dominant heterogeneous component.

Bachmann et al. (1977) have selectively converted diborane to icosaborane $B_{20}H_{16}$, in high yields using a cw CO_2 laser to initiate the reaction. The conventional pyrolysis of B_2H_6 results in a mixture of products which lacks any evidence of $B_{20}H_{16}$ formation. It is suggested that a chain mechanism can explain the high product yield with low laser input at the high pressures used. In an attempt to reproduce this result, Shatas et al. (1977) found that the cw-laser-induced reaction of diborane gave *no* $B_{20}H_{16}$; however, smaller boranes ($B_{10}H_{14}$ and B_5H_9) were observed. This discrepancy remains unresolved.

Another such reaction is described by Rockwood and Hudson (1975). In this paper they investigate the CO_2-TEA-laser-stimulated high pressure (~ 100 Torr) reaction

$$BCl_3 + H_2 \rightarrow BHCl_2 + HCl.$$

The known thermal route consists of heating BCl_3 and H_2 in the presence of a catalyst, but is found to produce other products such as B_2H_6 in significant quantities in addition to the desired $BHCl_2$. It is believed by the authors that the reaction begins by a multiple-photon dissociation of BCl_3 to BCl_2 and Cl. However, recent high pressure cw-laser-initiated experiments (Shatas 1976) have shown similar selectivity in the yields of $BHCl_2$ from the same reaction. This casts some doubt as to the nature of the primary step. It is possible that at the very high temperatures produced in the laser path, H_2 may react directly with BCl_3 or either molecule may dissociate and then react. On the basis of this small set of experiments one can envisage that homogeneous, laser-heated chemistry may provide a selective efficient alternative to the well-known heterogeneous catalysis. However, real-time analysis of the products will be necessary to uncover the true mechanism, distinguishing among thermal, radical and wall reactions.

Shaub and Bauer (1975) have used SF_6 to provide the vehicle for absorption of cw infrared laser radiation in those instances where the reactants are transparent, thereby generating controlled temperatures from 500 to 1500 K. This technique, later used by Orr and Keentok (1976) provides generality to the homogeneous laser-power pyrolyses just described (Bachmann et al. 1974, Rockwood and Hudson 1975) since SF_6 is quite inert to temperatures in excess of 1500 K and has a reasonably fast vibration–translation relaxation time. The absorbed energy is therefore rapidly (within a millisecond) transferred to the ambient gas. Many gases were decomposed using SF_6 to effect heat transfer, among them CH_4, C_2F_6, C_2H_2, C_2H_6, cis- and trans-2-butene and t-BuCl, to name a few. All gave products characteristic of normal pyrolysis, but application of the methodology to synthesize materials which are catalytically decomposed on hot walls should be rewarding.

One final experiment which is difficult to interpret from the data given was performed by Ambartzumian et al. (1976a). Here the authors irradiate mixtures of trans- and cis-$C_2H_2Cl_2$, with a CO_2 TEA laser and observe that as the process continues, the quantity of trans isomer, which absorbs the laser light, decreases in favor of the cis form. This is what one would expect from thermodynamics since the cis form is more stable. Further, the sum of both the cis- and trans-$C_2H_2Cl_2$ decreases as the photolysis proceeds, indicating that the molecules are being torn apart rather than simply isomerized, and that whatever recombination of the fragments takes place to reform the dichloroethylene does so favoring the cis form. Luminescence from C_2 and CH fragments is observed under the conditions of the experiments. A simple set of experiments can be performed to demonstrate if indeed decomposition results in the effective isomerization. That is, the mixed ^{12}C-^{13}C trans-$C_2H_2Cl_2$ can be prepared and, assuming of course that we can still have absorption of the radiation by this species, the ^{13}C-^{13}C cis isomer can be searched for. If it is produced in any significant quantities, then dissociation is certainly the reaction path. Another experiment might be to use SF_6 as the laser absorber and investigate the product distribution. If the same product ratios are found, then the isomerization has to be purely thermal in origin. Again, as in the reactions of Bachmann et al. (1974) and Rockwood and Hudson (1975), real-time product monitoring would help dramatically to discover what is actually happening.

2.4 More clearcut laser-induced chemistry experiments

This subsection divides logically into three categories comprising (1) laser breakdown experiments, (2) laser-induced dissociations and (3) laser alchemy.

2.4.1 Laser breakdown experiments

Recently, laser-induced breakdown experiments have again become popular (Isenor and Richardson 1971, Chin 1976, Ronn 1976, Ronn and Earl 1977, Freund 1977). In this genre of experiment, high gas pressures (tens of Torr) and laser intensities ($> 10^7$ W/cm^2) are utilized to induce breakdown. The manifestations of the process are intense white light and very rapid destruction of the irradiated gas, generally proceeding to the smallest stable fragments under the rather rigorous conditions. Ethylene is found to yield carbon and some polymers (Chin 1976), for example, while COS decomposes to CO and S (Ronn 1976). It is not immediately obvious what advantages the method has over conventional high temperature pyrolysis except perhaps that, in some cases, more complicated fragmentation paths can be avoided. In one instance (Freund 1977), an efficient separation of B_2H_6 impurity in SiH_4 has been achieved using laser breakdown of the gas to selectively destroy the more unstable species, leaving the stable SiH_4 essentially untouched.

2.4.2 Laser-induced chemistry

An attempt will be made to effect a somewhat artificial separation between laser-induced dissociation and laser-induced chemistry (laser alchemy) experiments by observing that most of the early experiments in this field were performed by physicists who had access to state-of-the-art laser devices and concentrated most of their attention on the photophysics of the laser-induced dissociation or fragmentation process. Little attention was paid to the nature of the primary product or to subsequent chemical reactions. Rather, for many of the isotopically selective schemes to be mentioned, scavengers were introduced in a trial and error fashion to simply remove the fragmented species. Analyses of extent of reaction and isotopic enrichment were generally performed by mass spectroscopy, infrared spectrophotometry or gas chromatography on the un-reacted starting material rather than on the final products whose identification remained a mystery deemed of little or no interest to solve. There is no doubt that in these experiments laser-induced phenomena were being observed. That is, isotopic selectivity cannot occur to the extent observed in these reactions by any known thermal process. Put another way, the kinetic isotope effect is generally quite small, and only works in one direction – the lighter isotope generally reacts more rapidly. More recently, however, increasing involvement of chemists in the research and a natural attempt on the part of investigators to study all phases of a subject of much expanded interest has led to more detailed investigations of chemical applications of lasers. Very careful analysis of the results is necessary to avoid incorrect interpretation (Dever and Grunwald 1976) even though all of the products are identified.

Probably the most interesting laser-induced chemical reactions investigated to date are the multiple-photon isotope separation reactions. This phenomenon has been the subject of many experimental papers (Ambartzumian et al. 1974b, 1975b–g, Lyman et al. 1975, 1977a, b, Lyman and Rockwood 1975, 1976, Yogev and Benmair 1975, Campbell et al. 1976a–c, Hancock et al. 1976, Kompa 1976, Ritter and Freund 1976, Fuss and Cotter 1977, Gower and Billman 1977, Kolodner et al. 1977), including several with a molecular beam irradiated with the laser pulse (Brunner et al. 1977, Coggiola et al. 1977, Diebold et al. 1977), and several recent reviews (Aldridge et al. 1976, Letokhov and Moore 1976a, b, Ambartzumian and Letokhov 1977). In this class of reactions a low pressure gas sample consisting of some polyatomic molecule (such as SF_6) with or without diluent is irradiated with intense infrared laser radiation. If conditions are favorable, the polyatomic molecule absorbs sufficient energy from the laser field to dissociate. Since the rate of absorption of laser radiation is quite sensitive to frequency, the reaction can have a very high isotopic selectivity when the infrared absorption band has an isotope shift on the order of a few wavenumbers or more. The isotopes of H, B, C, Si, S, Cl, Mo and Os have been enriched in this manner (Aldridge et al. 1976). Most of the available experimental data is for SF_6. We will now discuss problems associated with interpreting the available experimental data for this class of reactions.

Experimental results reported for these reactions have generally consisted of some measure of the extent of reaction, isotopic enrichment, chemiluminescence, or amount of energy absorbed for a wide range of experimental parameters. These parameters include the number of laser pulses, the pulse energy or fluence (energy/area), peak power or intensity (power/area), gas pressure, partial pressure of constituents, laser frequency (of two lasers in some cases), irradiation geometry, cell material, etc. In order to obtain some understanding of the fundamental processes involved these data must be reduced to some form that can be simply represented on a molecular scale. For example, the data for enrichment and dissociation could be represented as the probability that a molecule reacts when it is irradiated with some given fluence for each isotopic form of the molecule (see §3).

A judicious choice of experimental conditions is essential to avoid undesirable complications. One such complication is the reaction of the sample in the absence of laser radiation. Many species will react with or are catalyzed by windows, wall materials, stopcock grease, etc. For example, the laser-induced reaction of BCl_3 is severely complicated by stopcock grease (Lyman and Rockwood 1976). It has also been reported that the products obtained from the laser-induced reaction of some hydrocarbons are influenced by the size of the reaction vessel, which indicates wall reactions are important (Karny and Zare 1977).

Another complication that may cloud the interpretation of an experiment is the extent of thermal or radical chain reactions after pulsed irradiation. Radical scavengers and other diluents have been widely used to prevent such unwanted reactions. We insert a caution at this point that a diluent may also change the rate of energy absorption from the laser field by pressure broadening of absorption lines, collisional deactivation of excited species or changing the gas temperatures as a result of the heat capacity difference.

A complication that is perhaps more subtle is the possibility of photolysis of reaction products or intermediates. This is especially significant when experiments are performed by irradiating a single sample repeatedly in a static cell. Several examples have been observed.

Freund and Ritter (1975) have presented evidence that a significant fraction of the laser-induced reaction of BCl_3 with H_2S proceeds by a laser-enhanced reaction of a thermal reaction product (BCl_2SH).

A detailed study of the laser-induced reaction of SF_5Cl has been made (Leary et al. 1978). This reaction is isotopically selective (for sulfur). The major stable products are SF_4, SF_6, S_2F_{10}, Cl_2 and possibly ClF. It has been demonstrated that SF_6, S_2F_{10} and possibly SF_5 (an unstable intermediate) are further photolyzed. In fact, since the photolysis of products and intermediates affects reactions that reform the reactant, SF_5Cl, the isotopic composition may, under appropriate conditions, be dominated by the secondary photolysis reactions. Isotope ratios opposite from that expected from the primary photolysis of SF_5Cl have been observed.

2.4.3 *Laser alchemy* (laser-induced chemistry)

Although the title 'laser alchemy' is being used somewhat facetiously to describe some of the 'real chemistry' experiments performed with infrared lasers, researchers in this area are just beginning to recognize its potential. Many of the chemical systems to be described show truly amazing and sometimes bizarre behavior under the influence of laser radiation, and often the details will have to be further investigated in order to fully appreciate exactly what is occurring. Until such complete characterization is performed, however, some of the results will maintain a certain 'alchemy' appearance.

Two interesting but peripheral applications of lasers to chemistry which we will mention, but not dwell on, are the initiation of explosions and laser action. Bellows and Fong (1975) irradiated ethyl iodide with a cw CO_2 laser and observed and explosion which could be explained by assuming the laser energy absorbed was converted directly into heat. It turns out that as the reaction proceeds, one of the products, ethylene, which has a stronger absorption coefficient than ethyl iodide, absorbs the laser radiation more efficiently so that an unstable condition develops with an ensuing rapid temperature rise.

HF laser action has been produced using CO_2 laser radiation in two chemical systems. Lyman and Jensen (1973) have reported HF gain in CO_2-irradiated mixtures of both SF_6 and N_2F_4 with H_2, while Akinfiev et al. (1974) and Balykin et al. (1975) irradiated mixtures of SF_6 and H_2, D_2 or C_2H_6, and Belotserkovets et al. (1976) irradiated mixtures of N_2F_4 and H_2 or D_2. Each of these latter three groups observed chemical reactions with subsequent laser output. SF_6 and N_2F_4 each absorb the CO_2 laser light. The potential of the method for frequency up-conversion has not been fully investigated, but recent results by Rockwood (1975) suggest that visible chemical lasers may be possible with CO_2 laser-initiated chemistry. He has observed intense visible luminescence from BCl_3–H_2S mixtures when the BCl_3 was pumped by a CO_2 laser.

There have been many workers who have investigated the luminescence of radicals produced by ir laser photodissociation of small molecules both immediately following the light pulse (collisionless regime), and some hundreds of collisions thereafter (Lyman and Jensen 1972, Isenor et al. 1973, Hallsworth and Isenor 1973, Ambartzumian et al. 1974a, b, 1975a, b, d, Bagratashvili et al. 1975, Campbell et al. 1976a–c, Bourimov et al. 1976, Lesiecki and Guillory 1977a, b). In one of the first works of its kind, Lyman and Jensen (1972) dissociated N_2F_4 with a CO_2 TEA laser and monitored the NF_2 produced by its 260 nm uv absorption. They concluded that the dissociation was not of thermal origin from the fact that, even assuming very fast V–T relaxation of the N_2F_4, the NF_2 was produced on a time scale short relative to that possible by heating. It is, however, quite possible that at the pressures employed (>2 Torr) a rapid collisionally induced dissociation could occur if the excited vibrational levels had much faster V–T rates than the lower ones, although Lyman (1973) has calculated a series of thermal bimolecular rate coefficients which bear no resemblance to the measured rates, indicating that this is not the case. Isenor et al. (Isenor et al. 1973, Hallsworth and Isenor 1973) in experiments initiated in 1971 and Ambartzumian et al. (1974a) in later work observed fluorescence from laser-induced dissociation of SiF_4 and BCl_3 respectively. In both molecules the researchers found two fluorescence peaks depending on the gas pressure. The first peak followed the laser peak directly with little or no delay. The delay of the second peak was pressure dependent, indicating that the peaks resulted from collisionless and thermal dissociation respectively. Ambartzumian et al. (1974a) followed this work with an observation of the $(0, 2)^2\Pi_{1/2} - {}^2\Sigma$ transition in emission of the BO radical produced when mixtures of BCl_3 and O_2 are irradiated. By viewing only the very first 150 ns of the BO fluorescence, the product corresponding to the collisionless dissociation of BCl_3 was observed. The authors found by using a spectrometer that they could resolve the emission of the ^{11}BO from that of ^{10}BO, and when ^{11}BCl$_3$ is excited by the CO_2 laser, ^{11}BO predominates in the emission spectrum, while when ^{10}BCl$_3$ is pumped, the ^{10}BO predominates.

Subsequently, Ambartzumian et al. (1975a) demonstrated the generality of the methodology by photolyzing molecules such as C_2H_4, C_2F_3Cl, CH_3OH, CH_3NO_2, CH_3CN and CF_2Cl_2, and observing the luminescence from the fragments produced. In the case of CH_3CN, for example, the same products were observed under collisionless conditions as are from the 'delayed' luminescence which results from many collisions among vibrationally excited molecules. The authors conclude from this that it is impossible to selectively break particular chemical bonds using ir photolysis, although differentiation between excited and unexcited molecules (different isotopic species) is certainly possible. That is, it is only the energy accumulated in the molecule which, when above a threshold, determines the mode of dissociation, not how it was deposited there. There will be more on this question in §3. Radicals formed in the ground electronic state were not detected, but it was suggested that laser-induced fluoresence techniques be used to further investigate these systems.

Campbell et al. (1976a–c) have more recently observed the collisionless dissociation of NH_3 into ground electronic state NH_2 fragments when the CO_2 laser is both resonant and off-resonant with the ground vibrational state absorption features using laser-induced fluorescence. More complete investigation of the BCl_3 laser-induced dissociation has extended the previous results to pressures as low as 0.03 Torr with the observation that the zero-delay luminescence is still present (Ambartzumian et al. 1975b). Addition of O_2 as a scavenger for the BCl_3 photofragments was found to yield enrichment factors of the order of 10 for ^{10}B (Ambartzumian et al. 1975b), and yet further research (Bourimov et al. 1976) has allowed the kinetics of the BO and BCl radicals to be followed, yielding a rate constant of $(8 \pm 4) \times 10^5 \, s^{-1} \, Torr^{-1}$ for the reaction of BCl_3 fragments with O_2. The C_2H_4 luminescence has been followed as a function of exciting laser frequency with the use of a high pressure CO_2 TEA laser (Bagratashvili et al. 1975) with the result that no direct correlation between the detailed rotational structure of this molecule and the intensity of luminescence was observed. Further, direct observation of the excitation of high vibrational levels ($\nu \geq 10$) in OsO_4 by changes in the uv spectrum was performed by Ambartzumian et al. (1975d). The authors also obtained a 15% enrichment of osmium isotopes, where the isotopic shift per unit mass is approximately $0.26 \, cm^{-1}$ for the ν_3 vibration which was the one excited by the laser.

From further work by the same group (Ambartzumian et al. 1977) the time dependence of the visible luminescence suggested that the small time delay of this luminescence might be due either to collisional dissociation or to chemical reaction between excited OsO_4 molecules giving electronically excited products. They additionally observed that exciting the center of the R-branch transition of ν_3 produced no luminescence, pointing again (Ambartzumian et al. 1976c) to a PQR excitation scheme to compensate for the anharmonicity for the pumping process. Finally, Lesiecki and Guillory (1977b) have observed the decomposition product for the CO_2 laser-induced photofragmentation of CH_3CN to be almost totally ground electronic and vibrational state CN with a rotational temperature of ~ 664 K. This work very nicely supplements the pioneering work of Ambartzumian et al. (1975a) on this system.

As a minor digression, we report that Ambartzumian et al. (1975g, 1976d) have studied the selective dissociation of SF_6 and CCl_4 on weak combination bands and found that the dissociation rate is proportional to the dipole moment of the transition and not to its square, which is a surprising result. References (Ambartzumian et al. 1975d, 1976d) are supported by recent work on the CO_2 TEA laser photodissociation of MoF_6 (Hudson et al. 1976). Here MoF_6, which has a $1 \, cm^{-1}$ per mass unit isotope shift for the $\nu_3 + \nu_5$ combination band, shows a significant enrichment (which can be driven in either direction) when irradiated neat or with CO as a scavenger. In the latter case COF_2 is found as one of the products. As yet the details of the dissociation process with regard to the primary products are very vague and further experiments will be necessary.

The final part of this section is devoted to research aimed at elucidating the mechanistic chemistry occurring during some laser-induced chemistry experiments. We begin with a paper by Freund and Ritter (1975) which points up several of the difficulties involved in analyzing even seemingly trivial chemical systems. In the CO_2 laser photolysis of a BCl_3–H_2S mixture the authors found that an isotopically selective reaction occurs when very modest laser energy (~ 0.1 J/cm^2 pulse) is focused by means of a 25 cm focal length BaF_2 lens into a static cell containing the gases. At this fluence, no reaction was found to occur with either H_2 or O_2 as a scavenger under conditions that produce a reaction at higher fluence (Lyman and Rockwood 1976). One might conclude from this that a reaction was taking place from an excited vibrational level of the ground electronic state of BCl_3. Although a rather extensive effort was made to extract the mechanism from the reaction products and disappearance of reactants, the products could not be identified other than as white solids containing B, Cl, H and S. Anomalies in the wavelength dependence of the reaction yields and enrichments suggested that a second species, $HSBCl_2$ ($DSBCl_2$), formed thermally in small quantities from the equilibrium reaction of BCl_3 and $H_2S(D_2S)$, was most likely also being photolyzed at the same wavelengths as BCl_3. Further experiments (Freund and Ritter 1975) on neat $HSBCl_2$ showed that similar enrichments in the unreacted $HSBCl_2$ could be obtained by its photolysis as for the BCl_3 irradiations. Unfortunately, BCl_3 was produced in these irradiations, further clouding the results.

Two recent papers (Bachmann et al. 1975, 1977) describe other laser-induced chemical reactions involving boron compounds. In the first (Bachman et al. 1975) the authors used an in-situ probe to define an upper limit to the temperature which enabled them to define the reaction of interest,

$$B(CH_3)Br_2 + HBr \rightarrow BBr_3 + CH_4,$$

as a cw-laser-induced chemical process, where the $B(CH_3)Br_2$ absorbs the laser radiation. It is known that the reaction

$$B(CH_3)_2Br + HBr \rightarrow B(CH_3)Br_2 + CH_4$$

goes thermally above 250°C, while the former reaction does not occur below 450°C. Since neither the $B(CH_3)_2Br$ nor the HBr absorbs the laser light, a mixture of $B(CH_3)Br_2$, HBr and $B(CH_3)_2Br$ which is stimulated by the laser to react away the first two compounds leaving the third untouched must have maintained a translational temperature below 250°C and undergone a nonthermal reaction. This is a surprising result in view of the high pressures (>300 Torr) used and the fact that a cw laser was the source of light. Perhaps we have another case of homogeneous versus heterogeneous reactions here. In the second paper (Bachmann et al. 1977) the trimerization of tetrachloroethylene sensitized by BCl_3 is reported. Large yields (88%) of the product C_6Cl_6 were obtained when mixtures of BCl_3 and C_2Cl_4 were irradiated by a cw CO_2 laser with a minimal loss of BCl_3. In order to rule out simple heating effects, mixtures of SF_6 and C_2Cl_4 were irradiated with no product formation. Again high gas pressures were employed and the results quite surprising. It is possible that the same temperatures were not achieved in both systems or that the BCl_3 can be dissociated to chlorine atoms which promote the reaction.

Very low pressure (several millitorr) cw irradiations of boron compounds have been investigated (Lory et al. 1975, Chien and Bauer 1976). In the CO_2-laser-induced decomposition of $H_3B \cdot PF_3$ (Lory et al. 1975) and D_3BPF_3 (Chien and Bauer 1976) to yield B_2H_6 and PF_3 or B_2D_6 and PF_3 respectively the authors used H_3BCO as a temperature probe, finding no more decomposition of this compound than was allowed at the temperature of the walls in the absence of laser radiation, while the trifluorophosphine boranes reacted at approximately 1%/min under the influence of the laser. The laser power and frequency dependence of the dissociation rate has led them to conclude further that the reaction proceeds with the absorption of more than one photon rather than by collisional up-pumping. Two further observations were (1) that the most efficient irradiation frequency did not correspond to the maximum of the low-power laser absorption profile but was displaced to the red, as is observed for many experiments using TEA lasers for photodissociation of molecules, and (2) that although the values of the actual absorption coefficients on P(24) (10.6 μm band) for D_3BPF_3 and P(34) for H_3BPF_3 are very nearly equal, the efficiency for decomposing the deuterated compound on P(24) is considerably greater than for the hydrogenated species, indicating some specific vibrational effects. The authors propose, on the basis of a normal mode analysis, that the latter observation can be explained by the theory that when the adduct absorbs infrared radiation in the ν_3 mode (as both do) a larger fraction of the energy is deposited into the B–P stretching vibration for D_3BPF_3 than for H_3BPF_3. It is also possible that we have another example where the increased temperature effects the absorption coefficient for the two compounds quite differently.

Three papers by Yogev and co-workers (Yogev and Lowenstein–Benmair 1973, Yogev and Benmair 1975, Glott and Yogev 1976) show similar effects using pulsed instead of cw CO_2 lasers. The first of these describes how low pressure (<30 Torr) trans-2-butene is decomposed by focused laser radiation. In a 1:1 mixture of the cis and trans isomers, 15% enrichment of the cis isomer was observed as a result of selective decomposition of the trans isomer. The authors make this conclusion on the basis that decomposition of 2-butene occurs at much higher temperature than that where significant cis–trans isomerization takes place. Methylene chloride photolysis is shown to be isotopically selective when a CO_2 laser is used (Yogev and Benmair 1975). Mixtures of CD_2Cl_2 and CH_2Cl_2 at pressures in the region of 2 Torr show preferential decomposition of the deuterium compound upon irradiation. Heating a similar mixture at higher pressures (41 Torr) with a cw CO_2 laser produced the opposite enrichment, as would be expected for a thermal reaction. That is, the hydrogenic species should react faster in a thermal decomposition process. The third work (Glott and Yogev 1976) describes a rather novel laser-induced isomerization where decomposition of the reactants and products is kept to a minimum. The authors irradiated several deuterated biallyls (1, 5 hexadiene) at low pressures (less than

5 Torr) and found in one case in particular that the Cope rearrangement

goes essentially to completion with only a small amount of decomposition occurring. Peak powers of about 0.3 MW were focused into the static cell and products analyzed by gas chromatography, NMR spectroscopy and ir spectrophotometry. This reaction can be restored to an equilibrium mixture of the two species by heating the product to 300°C for 3 h in a sealed tube.

Two final papers (Braun and Tsang 1976, Danen et al. 1977) have also used chemical reactions for temperature probes when investigating laser-induced reactions of large organic molecules by pulsed CO_2 lasers. In the first (Braun and Tsang 1976), the authors photolyze various alkyl halides dilute in helium. The question of whether these reactions proceed via specific excitation by the laser pulse or by a rapid thermal pulse is answered by experiments using propylene and isobutyl chloride alone and in a mixture. Use is made of the fact that propylene is much more stable than isobutyl chloride at all temperatures, and that it is often added as a stable inhibitor of chain decompositions. Now, most of the propylene is destroyed in the reaction zone with very little accompanying isobutyl chloride decomposition, which fact provides evidence for the specific, nonthermal nature of the laser excitation. Danen et al. (1977) have very recently investigated the decomposition of ethyl acetate using unfocused CO_2 laser light according to the reaction

$$CH_3-\overset{\displaystyle O}{\overset{\|}{C}}-OCH_2-CH_3 \rightarrow CH_3CO_2H + CH_2 = CH_2.$$

When mixtures of ethyl acetate and isopropyl bromide (~ 20 Torr) were irradiated in a region where the former compound absorbs but the latter does not, the ratio of propene produced by HBr elimination of the isopropyl bromide to the ethene produced from the ethyl acetate decomposition was what would be expected from the thermal rate for two compounds with very similar activation energy (the ratio of the thermal rate constants is essentially temperature independent). Adding 10 Torr of SF_6 to 10 Torr of a similar mixture of isopropyl bromide and ethyl acetate (1:3 ratio) and irradiating at 945 cm^{-1}, where only the SF_6 absorbs, produced a similar ratio of propene to ethene, showing that the majority of the ethene produced from irradiation of ethyl acetate alone under nonfocused conditions arose from thermal processes rather than from any selective, nonequilibrium laser initiation. The authors further found, however, that at much higher energy densities obtained by focusing the laser light, laser-induced chemistry prior to collisional redistribution of energy could be observed. That is, the ratio of propene to ethene produced in the reaction was reduced from the thermally calculated ratio of about 5 to approximately 0.6 at the higher energy density. Further, the ratio was monitored as a function of pressure at constant laser intensity, with the result that as the number of collisions was reduced, the rate at which thermal equilibrium was reached was decreased, allowing the nonequilibrium laser-induced pathway to become increasingly important. Thus the need to conduct experiments at low pressure and high laser intensity to minimize V–V and V–T,R intermolecular relaxation may be a general requirement for the laser-induced chemistry of large polyatomic molecules. One final observation is that since the laser-induced pathway for decomposition is identical to the thermal route, some intramolecular energy transfer must occur. More on this subject below.

Attempts at understanding the reaction chemistry involved in a laser-induced chemistry experiment have been made by two groups (Ritter and Freund 1976, Leary et al. 1978). In their study of the laser-induced reaction of CCl_2F_2 with O_2, Ritter and Freund found the COF_2 formed to be enriched in the opposite carbon isotope from the residual CCl_2F_2, which positively identified the reaction as a nonthermal one. A second laser-induced reaction between CCl_2F_2 and HCl to yield $HCClF_2$, again with the appropriate enrichment, led the authors to conclude that the reaction intermediate was $:CF_2$. However, no C_2F_4 was observed. Further experiments with Cl_2 scavengers such as NO and $Me_2C{=}CH_2$ demonstrated significant C_2F_4 product formation, again with the requisite enrichment, thus further implicating the difluorocarbene radical as being formed in the photochemical step. Ritter (1978) has extended the work of Ritter and Freund (1976) to reactions of CF_2Br_2 with olefins, further implicating $:CF_2$ as one of the primary products in the photolysis. He also has suggested the use of laser-produced, isotopically specific reactive intermediates such as $:CF_2$ for the direct synthesis of labeled compounds. King et al. (1977) used laser-induced fluorescence techniques to detect $:CF_2$ from CO_2-laser-induced dissociation of both CF_2Cl_2 and CF_2Br_2. Not only was $:CF_2$ observed directly in the laser-induced fluorescence experiments, but the vibrational state distribution in $:CF_2$ was also determined. Since the distribution was Boltzmann with a vibrational temperature that was independent of the absorbing vibrational mode, but different for each molecule, the authors concluded that the vibrational energy was randomly distributed within the excited parent molecule. The question of whether or not $:CF_2$ is a primary product of CF_2Cl_2 photolysis remains unsettled. Time-dependent studies should be most revealing.

In experiments with a molecule of similar structure Bittenson and Houston (1977) have demonstrated highly selective carbon isotope enrichment by CO_2 laser photolysis of CF_3I. In this case the reaction products are C_2F_6 and I_2, which indicates C–I bond cleavage is the primary photolytic reaction. This is another example of reaction proceeding by the lowest energy dissociation pathway available.

Leary et al. (1978) investigated the laser-induced decomposition of SF_5Cl. Many experiments involving the laser-dissociated fragments of this molecule and various scavengers have been performed with the result that at least part of the reaction most certainly occurs following the route

$$SF_5Cl \xrightarrow{h\nu_{laser}} SF_5 + Cl.$$

The circumstantial evidence supporting this reaction has been given a somewhat stronger base by the recent work of Karl and Lyman (1978). The authors found that chlorine atoms are formed in the laser-induced dissociation of SF_5Cl by using chlorine atom resonance detection techniques. Implications of these results are that since the square–planar sulfur–fluorine stretching motion (ν_8) is excited by the CO_2 laser and the molecules fall apart along the sulfur–chlorine bond, intramolecular energy transfer must occur. Also pointed out by Leary et al. (1978) is that the laser photolyzes some of the products as well, complicating the results. It is necessary that reactions where further laser photolysis of the products can occur be performed in a flowing system.

Additional investigations of the mechanism for laser-induced chemical reactions have been made by Ronn and co-workers (Earl and Ronn 1976, Lin and Ronn 1977). Chemical reactions of vibrationally excited methyl halides with chlorine (except for CH_3I) principally follow the route

$$CH_3X + Cl_2 \rightarrow CH_2XCl + HCl,$$

where X = F, Cl or Br. Reactions between CH_3F and Br_2, H_2 and NO_2 did not occur under similar conditions of laser power and pressure. The primary process is speculated to be dissociation of Cl_2 by collisions with excited CH_3X molecules (no dimers of any kind were observed). Chlorine atoms can then readily react with other CH_3X molecules producing CH_2XCl and hydrogen chloride. Of additional interest is that the product formation in this reaction is essentially independent of the pumping laser line to within $\pm 50\,cm^{-1}$ of the closest coincidences with the laser, a fact attributed to the large permanent electric dipole moment of these molecules contributing to the dynamic Stark effect produced by the intense laser electric field. In experiments with CH_2F_2 (Lin and Ronn 1977) a reaction mechanism similar to that proposed for the CH_3X work explains the results for low pressures ($<1\,Torr$), whereas at higher pressure ($>2\,Torr$), dimerization occurs and a unimolecular decomposition yielding $\cdot CHF_2$ radicals is used to explain the results. It is, however, conceivable that at the higher pressures, a truly thermal process may be occurring as a result of gas heating. Photolysis of various of the products by laser radiation further complicates matters. The mechanisms can be checked by looking for chlorine atom formation using resonant fluorescence detection techniques as was done by Karl and Lyman (1978).

3 Multiple-photon reaction model

In this section a more detailed analysis is given of some of the available
experimental results for the multiple-photon reaction of SF_6. A model based on
RRKM unimolecular reaction rate theory is employed for this purpose. Nomenclature and definitions, which follow Lyman et al. (1977b), along with a
discussion of geometrical effects are also included in the section. Only reactions
of SF_6 are considered; however, the concepts may be generally applied.

3.1 Nomenclature and definitions

In a typical isotope enrichment experiment a sample is irradiated i times with
pulse energy E. One may then determine the fraction f of reactant (SF_6)
remaining and the concentration of products (SF_4, etc.). The isotope ratios in the
residual reactant (and possibly some product) can also be measured. By convention, the isotopes are compared to the most abundant isotopic form. For SF_6,
$R_m = [^m SF_6]/[^{32} SF_6]$, where m labels the isotopic mass. The enrichment factor β_m
is then defined as $\beta_m = R_m/R_m^0$, where R_m^0 is the ratio before irradiation.

Many workers have reported results in terms of β_m of the reactant or some
variation of it. This quantity by itself has limited quantitative significance since
its value is influenced by so many nonfundamental parameters such as the
number of laser pulses and the size of the reaction vessel. On the other hand, the
enrichment factor in a product species does have quantitative significance for
small extent of reaction since it is a direct measure of the reaction yield ratio for
two isotopic forms of the reactant. (This is the selectivity to be defined below.)

If the number of pulses is treated as a continuous variable, then the equation
describing the reaction of $^m SF_6$ is

$$d[^m SF_6]/di = -(V_m/V_0)[^m SF_6], \tag{29}$$

where V_0 is the total cell volume and V_m is the equivalent volume of $^m SF_6$
reacting per pulse. The latter will be referred to as the reaction yield. The total
yield for all isotopes is designated as V with no subscript. Since ^{32}S is the most
abundant (95%), V is nearly equal to V_{32} for most experiments. We will see
below that V_m is a good measure of the reaction yield since it is simply related to
the probability of reaction.

If V_m is assumed to be independent of the extent of reaction (i.e., of product
or scavenger concentration) eq. (29) can be integrated to give

$$V_m = -\frac{V_0}{i} \ln \frac{[^m SF_6]}{[^m SF_6]_0} = -\frac{V_0 \ln(\beta_m f)}{i}, \tag{30}$$

where $f = [^{32} SF_6]/[^{32} SF_6]_0$ and $[^m SF_6]_0$ is the concentration before irradiation
begins.

One may divide eq. (29) by a similar equation for $m = 32$ to obtain

$$d[^m SF_6]/d[^{32} SF_6] = \alpha_m([^m SF_6]/[^{32} SF_6]),$$

where α_m is the selectivity factor which is the yield ratio of two isotopic forms or

$$\alpha_m = V_m/V_{32} = 1 + (\ln \beta_m (\text{reactant})/\ln f), \tag{31}$$

and in terms of a product enrichment factor

$$\alpha_m = \ln[1 - (1 - f)\beta_m (\text{product})]/\ln f \approx \beta_m (\text{product}), \tag{32}$$

where the approximation is good for a small extent of reaction, i.e., $f \approx 1$.

The selectivity factor is not only a quantity of fundamental interest but also may serve as a useful diagnostic for the occurrence of nonselective reactions. These include thermal reactions occurring after all isotopically selective vibrational excitation is randomized, and long-chain-free radical reactions. The most commonly observed example is the loss of selectivity with increasing pressure that has been observed in many systems.

3.2 Geometrical effects

Many experiments have been performed using focused laser radiation and therefore the reaction yield consists of contributions from a wide fluence range. The probability of reaction $P(\Phi)$ for some given fluence Φ is related to the yield by

$$V = \int_0^\infty g(\Phi)P(\Phi)\,d\Phi, \tag{33}$$

where $g(\Phi)$ is the fluence distribution function or the volume at fluence Φ per unit fluence in the irradiation cell. It may be obtained from $G(\Phi)$, the volume with fluence greater than Φ, by $g(\Phi) = -dG(\Phi)/d\Phi$, and $G(\Phi)$ may easily be obtained from a consideration of the irradiation geometry.

We give three examples here. Further details and a more complicated example are given elsewhere (Lyman et al. 1977b). The simple double cone geometry of a pulse focused into the center of a reaction cell gives

$$G(\Phi) = G_0(\Phi_w/\Phi)^{3/2}, \quad \Phi > \Phi_w,$$
$$= G_0, \quad \Phi \leq \Phi_w,$$

where G_0 is the total irradiated volume and Φ_w is the fluence at the cell window. Therefore,

$$g(\Phi) = \tfrac{3}{2}G_0\Phi_w^{3/2}/\Phi^{5/2}, \quad \Phi > \Phi_w,$$
$$= 0, \quad \Phi \leq \Phi_w. \tag{34}$$

We see then, as a number of authors have indicated (Isenor et al. 1973, Fettweis and Mevergnies 1977, Speiser and Jortner 1976, Lyman et al. 1977a, b, Fuss and Cotter 1977, Keefer et al. 1976), if the yield data are presented in the form

$$V = CE^n, \tag{35}$$

where C and n are constants and E is the pulse energy ($E \propto \Phi_w$), that from eq. (33) with eq. (34) the exponent n is not simply related to the dependence of $P(\Phi)$ on Φ. If $P(\Phi)$ increases sharply with fluence, that is, if $P(\Phi)$ is proportional to some high power of Φ, then n in eq. (35) will be $\frac{3}{2}$ when the experiment is performed with the focal point inside the reaction cell. This upper limit of $n = \frac{3}{2}$ is determined only by geometry.

As a second example we consider unfocused laser radiation of uniform fluence Φ_0 passing through a reaction cell. This simple arrangement gives

$$G(\Phi) = G_0, \quad \Phi \geq \Phi_0,$$
$$= 0, \quad \Phi < \Phi_0,$$

where G_0 is the volume irradiated. Therefore $g(\Phi) = G_0 \delta(\Phi_0)$ and

$$V = G_0 P(\Phi_0). \tag{36}$$

The use of unfocused laser radiation is certainly the best way to determine $P(\Phi)$.

Since it is nearly impossible to obtain a laser beam that has uniform fluence across the entire beam area, one may ask how minor fluctuations across the beam influence the accuracy of a measurement of reaction probability when eq. (36) is assumed to be correct. Let us consider a laser beam that has circular symmetry with a gaussian radial distribution truncated at the inflection point. That is

$$\Phi = \phi_c \exp(-r^2/2a^2), \quad r \leq a,$$
$$= 0, \quad r > a, \tag{37}$$

where ϕ_c is the fluence at the beam center, a is the beam radius and r is the radial distance from the center. The mean fluence (pulse energy/beam area) will then be from eq. (37)

$$\langle \phi \rangle = 2(1 - e^{-0.5})\phi_c = 0.787\phi_c \tag{38}$$

and the fluence distribution function becomes

$$g(\phi) = 2G_0/\phi, \quad e^{-0.5}\phi_c \leq \phi \leq \phi_c,$$
$$= 0, \quad \text{otherwise.} \tag{39}$$

If the actual value of $P(\phi)$ is increasing with the nth power of ϕ in the region around $\langle \phi \rangle$, then by use of eqs. (33), (38) and (39) we find that the observed reaction probability, assuming eq. (36) to be correct, is related to the actual reaction probability by

$$P_{obs.}(\langle \phi \rangle)/P_{act.}(\phi) = (2/n)(1.27^n - 0.77^n).$$

So the error introduced by a nonuniform fluence of this sort is small. For example, a 10% error occurs only for $n > 3.6$. A 3% error in determining $\langle \phi \rangle$ results in a 10% error in $P(\phi)$. We conclude then that fluence nonuniformities of this magnitude are insignificant compared to the uncertainty in the pulse energy (and area) measurement.

3.3 Unimolecular reaction model

A computer model of the laser-induced reaction of SF_6 has been developed that will be useful in analyzing some of the available data. The model will help to give an interpretation of these experiments as well as indicate directions for further experimental investigations. At this point we will concentrate on absorption and reaction measurements with fluence and pressure as experimental parameters. The experimental results concerned with the effect of laser frequency require a more detailed development of the absorption mechanism and will therefore be reserved for a later section. This model, which has been described in greater detail elsewhere (Lyman 1977), will be more concerned with the chemistry and will also help define what the data do or do not say about the absorption mechanism. In the past much of the data have been misinterpreted because of failure to consider all aspects of the laser-induced reaction.

A rate equation formalism is employed with three processes being included: absorption and emission of laser radiation, collisional energy transfer (V–T process) and dissociation (to SF_5 and F). Unless otherwise specified intramolecular energy transfer will be assumed to occur infinitely fast, and the probability of finding a molecule in a given vibrational state within some small energy range is therefore equally likely for all states within the range. This is the basic assumption of RRKM unimolecular reaction rate theory (Robinson and Holbrook 1972) which is employed to calculate the dissociation rate. It will be assumed that other processes such as spontaneous light emission, bimolecular reaction of SF_6, further reaction of SF_5 and F, intermolecular V–V energy transfer, diffusion and expansion do not significantly affect the reaction, and these types of processes are not specifically included in the model. However, attempts will be made to estimate the importance of some of these. We recognize that the rate equation formalism neglects coherent effects, etc. (see §4) and that several of the approximations employed reduce the accuracy of the calculations. We feel, however, that much can be learned from this approach, particularly about the energy distribution and dynamics among the higher vibrational energy levels.

Consider a system of molecules distributed among a series of vibrational energy levels with the density of molecules in level i designated as n_i. This density may change by the three processes mentioned above: collisional energy transfer, radiant energy transfer and reaction. The rate of collisional energy transfer from level j to level i is $\omega P_{ij} n_j$, where ω is the rate constant for gas kinetic collision and P_{ij} is the probability that a collision changes the molecule from state j to state i. The normalization condition for P_{ij} is

$$\sum_i P_{ij} = 1. \tag{40}$$

The rate of radiative energy transfer from level j to level i is $\rho f_{ij} n_j$, where ρ is the laser intensity divided by the energy spacing between levels j and i, and f_{ij} is the cross section for the transition induced between level j and level i.

The rate of dissociation from level i is $k_i n_i$, where k_i is obtained from RRKM unimolecular reaction rate theory.

The rate of change of density in level i is then

$$\frac{dn_i}{dt} = \rho \sum_j f_{ij} n_j - \rho n_i \sum_j f_{ji} + \omega \sum_j P_{ij} n_j - \omega n_i - k_i n_i. \tag{41}$$

An additional differential equation for the translational (and rotational) temperature was added, and the full set of equations was solved numerically.

Since the density of SF_6 vibrational states increases sharply with energy it is not practical to treat each state as a separate level. The vibrational states will be partitioned into levels of energy spacing $\Delta E = 944 \text{ cm}^{-1}$, or the energy of a CO_2 laser photon. The number of states in each level g_i is obtained by using the Whitten–Rabinovitch method (Robinson and Holbrook 1972) for approximating the density of vibrational states by treating the molecule as a set of harmonic oscillators.

Collisional energy transfer is assumed to occur only between adjacent levels, or $P_{ij} = 0$, $|i - j| > 1$, and the collisional excitation and de-excitation probabilities are related by

$$P_{i,i-1} = P_{i-1,i}(g_i/g_{i-1}) \, e^{-\Delta E/kT}. \tag{42}$$

An expression for $P_{i-1,i}$ was derived by assuming that the energy transfer occurs by unit changes in the quantum number of the lowest frequency vibrational mode, ν_6. And as in a harmonic oscillator, the transition probability is proportional to ν_6. We obtain then

$$P_{i-1,i} = C_6 \langle \nu_6 \rangle, \tag{43}$$

where $\langle \nu_6 \rangle$ is the average value of ν_6 at vibrational energy $E_v = i\Delta E$, and C_6 is the probability of a transition between $\nu_6 = 1$ and $\nu_6 = 0$. We note that for highly excited molecules the vibrational relaxation rate is large (Tardy and Rabinovitch 1977). This is in turn related to the measured vibrational relaxation time of SF_6, $P\tau$, by

$$C_6 = P/(P\tau)\omega f[1 - \exp(-h\nu_6/kT)], \tag{44}$$

where f is the fraction of the vibrational energy in the ν_6 mode ($f \approx 0.2$). The normalization condition [eq. (40)] then gives P_{ii}. The values of $P\tau$ for SF_6 and Ar were obtained from the work of Breshears and Blair (1973):

$$(P\tau)_{SF_6} = 2.22 \times 10^{-6} \exp(38.0/T^{1/3}) \text{ Torr s},$$

$$(P\tau)_{Ar} = 6.17 \times 10^{-6} \exp(34.9/T^{1/3}) \text{ Torr s}. \tag{45}$$

From the work of Steinfeld et al. (1970)

$$(P\tau)_{H_2} = 0.02(P\tau)_{Ar}. \tag{46}$$

The dissociation rate constants k_i were obtained by the use of RRKM unimolecular reaction rate theory (Robinson and Holbrook 1972). The rate constant $K_a(E_v^*)$ for dissociation of a molecule with energy E_v^* is obtained by calculating from statistical considerations the rate that energized molecules (designated by *) pass through an activated complex (designated by $^+$) to products. The internal energy of the activated complex is

$$E_v^+ = E_v^* - E_0,$$

where E_0 is the energy barrier for reaction. The value of E_0 which, along with other features of the model, gave the best reproduction of the thermal rate data of Bott and Jacobs (1969) was

$$E_0 = 34.2 \Delta E = 91.3 \, \text{kcal/mole}.$$

This value is consistent with the SF_5–F bond strength of $92.6 \pm 3 \, \text{kcal/mole}$ obtained from a recent re-evaluation of heats of formation of sulfur-containing compounds (Benson 1978).

The RRKM formulation of the dissociation rate constant is

$$k_a(E_v^*) = L^\ddagger \frac{Q_1^+}{Q_1} \frac{1}{hN^*(E_v^*)} \sum_{E_v^+=0}^{E_v^+} N^+(E_v^+), \tag{47}$$

where h is Planck's constant and $L^\ddagger$ is the statistical factor, or the number of equivalent reaction paths ($L^\ddagger = 6$ for SF_6).

The quantities Q_1 and Q_1^+ are rotational partition functions for the undistorted molecule and the activated complex respectively. If it is assumed that the activated complex has the length of one S–F bond doubled, then

$$Q_1^+/Q_1 = 1.71,$$

which was used in the model and is about what one would expect from a 'loose' activated complex (Robinson and Holbrook 1972). The density of vibrational states $N(E_v^*)$ was calculated using the Whitten–Rabinovitch approximation mentioned earlier with the fundamental vibrational frequencies listed by McDowell et al. (1976a, b). The summation is over all vibration levels of the activated complex except for one normal mode that becomes the reaction coordinate. The vibrational frequencies of the activated complex are not known and they were therefore chosen such that the model properly reproduced the thermal dissociation rate data of Bott and Jacobs (1969). Some of the oscillation frequencies of SF_6 are more influenced by extending one S–F bond than others. The frequencies of SF_6^+ were chosen such that those most influenced by bond extension were 75% of the SF_6 value and those least influenced by bond extension were 95% of the SF_6 value. One ν_4 oscillation became the reaction coordinate. The summation over the vibrational states of SF_6^+ was again obtained using the Whitten–Rabinovitch approximation. With all of these features the model properly reproduced the thermal rate data of Bott and Jacobs, which was for SF_6 diluted with Ar over a broad temperature (1600 to 2100 K) and pressure (0.1 to 32 atm) range. The calculated high pressure limit of the unimolecular rate constant was

$$k_\infty = 1.20 \times 10^{15} \exp(-92.0/RT) \, \text{s}^{-1},$$

where the activation energy is in kcal/mole. Further details are given in the original reference (Lyman 1977).

For laser-induced dissociation it is necessary to know ρ and f_{ij}. For all calculations presented here only linear, incoherent absorption will be assumed. That is, $\rho = I/\Delta E$, where I is the laser intensity and $f_{ij} = 0$, $|i - j| \neq 1$, with f_{ij} independent of the laser intensity. The values of the remaining cross sections $f_{i-1,i}$ and $f_{i,i-1}$ will depend on frequency, the degree of excitation (i), the rate of intramolecular energy transfer, etc. Several formulations of these constants will be tried and the results compared with experimental data.

For a harmonic oscillator one would expect that the absorption cross section would increase linearly with i. However, the effect of anharmonicity and broadening of the absorption band with increasing energy would tend to make the cross section decrease with increasing i. Therefore, as an initial approximation we assume the $f_{i,i-1}$ is independent of i, or

$$f_{i,i-1} = f_{10}. \tag{48}$$

The rate of the reverse process, stimulated emission, will depend on the rate of intramolecular energy transfer. That is, the cross section ratio $f_{i-1,i}/f_{i,i-1}$ will be determined entirely by the relative number of states at levels i and $i - 1$ that are involved in the laser absorption process. Three separate cases are considered. If intermode coupling is weak (slow intramolecular energy transfer) and the laser energy is absorbed by a single nondegenerate mode, then only the vibrational states in that mode will contribute to the statistics of absorption and emission, and

$$f_{i-1,i}/f_{i,i-1} = 1. \tag{49}$$

For absorption by all three of the triply degenerate ν_3 modes and weak intermode coupling we get

$$f_{i-1,i}/f_{i,i-1} = i/(i + 2). \tag{50}$$

For rapid intermode coupling (collisional, laser-induced, or spontaneous) it is as if the entire molecule is absorbing the radiation and

$$f_{i-1,i}/f_{i,i-1} = g_{i-1}/g_i, \tag{51}$$

where g_i is the total number of vibrational states of level i. This ratio is what one would expect if absorption occurred in what has been referred to as the quasi-continuum.

Calculations have been performed for all three formulations of $f_{i-1,i}/f_{i,i-1}$. This type of calculation is only appropriate for intense short pulses at low pressure since a high collision rate will certainly induce rapid intramolecular energy transfer and eq. (51) will be the correct form.

Figs. 2, 3 and 4 show sample calculations for absorption of laser radiation by 1 mode [eq. (49)], 3 modes [eq. (50)] and 15 modes [eq. (51)]. In each case $f_{i,i-1}$ was given by eq. (48) and f_{10} was chosen to give $P(\Phi) = 0.1$ at $\Phi = 5$ J/cm^2. The laser pulse was of constant intensity for 400 ns. All three calculations are for 0.5 Torr

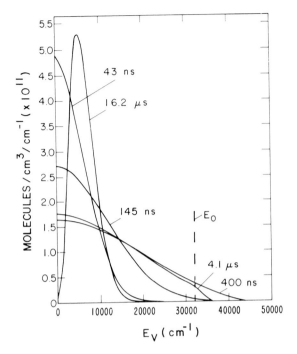

Fig. 2. Calculated distribution of SF$_6$ molecules over vibrational energy levels at several times during and after laser excitation of a single normal mode [eq. (49)]. Excitation was with a square pulse 400 ns long with fluence of $\Phi = 4.32$ J/cm^2. The cross sections for each level were given by eqs. (48) and (49) with $f_{10} = 9.24 \times 10^{-19}$ cm^2. For this calculation the mean number of laser photons absorbed per molecule was $\langle n \rangle = 15.8$ and the reaction probability was $P(\Phi) = 0.074$.

of 25% SF$_6$ in H$_2$ initially at 300 K. The graphs show the distribution of SF$_6$ molecules over vibration energy per molecule at successive times.

The most striking difference among these figures is the differences in the distribution produced by the laser. For these calculations the pressure is sufficiently low that collision effects are minimal during the pulse and the 400 ns curves are almost entirely determined by laser absorption and emission. Since the dissociation rate is a very rapidly increasing function of energy in excess of E_0, for E_v greater than about 38 000 cm^{-1} the dissociation rate will also influence the distribution at 400 ns. As an example of how rapidly the dissociation rate increases, the lifetime for dissociation is about 5 ms for 1 photon (944 cm^{-1}) in excess of E_0, 1.3 μs for 5 photons and 20 ns for 10 photons. Because of the different distributions produced in each case, which will of course be reflected in the reaction probability, comparisons of calculations with experimental data should give some indication of how many normal modes of SF$_6$ are actually involved in the absorption process.

As time proceeds beyond 400 ns the distribution is changed in all three cases by two processes: dissociation of molecules with $E_v > E_0$ and collisional deactivation. The collisional energy transfer is mainly by H$_2$ since it has a smaller

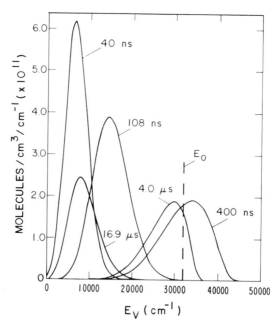

Fig. 3. Calculated distributions for excitation of three normal modes [eq. (50)]. Conditions were the same as for fig. 2 except $\Phi = 8.28$ J/cm^2 and $f_{10} = 4.81 \times 10^{-19}$ cm^2. This calculation gave $\langle n \rangle = 29.8$ and $P(\Phi) = 0.319$.

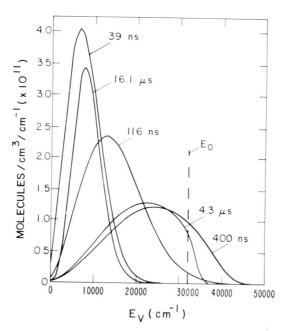

Fig. 4. Calculated distributions for excitation of all fifteen normal modes [eq. (51)]. Conditions were the same as for fig. 2 except $\Phi = 8.00$ J/cm^2 and $f_{10} = 1.99 \times 10^{-19}$ cm^2. This calculation gave $\langle n \rangle = 27.1$ and $P(\Phi) = 0.509$.

relaxation time than SF_6 [eqs. (45) and (46)]. An equilibrium distribution at a temperature that is too low to maintain reaction is achieved in about 16 μs. There are several processes that could further contribute to either the distribution over vibrational energy or to the temperature. The first is intermolecular V–V energy transfer, which will be discussed in greater detail later. Second, the exoergic reaction

$$F + H_2 \rightarrow HF + H$$

will tend to heat the gas on a time scale of about 10 μs. Third, the dissociation of SF_5, an endoergic reaction, will tend to keep the temperature below 1400 to 1600 K. Fourth, as the temperature of the irradiated gas rises it will begin to expand and therefore cool on the time scale of about ten microseconds.

Several laboratories (Yablonovitch et al. 1977, Kolodner et al. 1977, Ambart-zumian et al. 1975e, Stafast et al. 1977) have measured the reaction probability as a function of the laser fluence over a fairly wide range of conditions. In fig. 5 calculations of the type shown in figs. 2, 3, and 4 have been performed and the reaction probabilities compared with the experimental data. As mentioned above

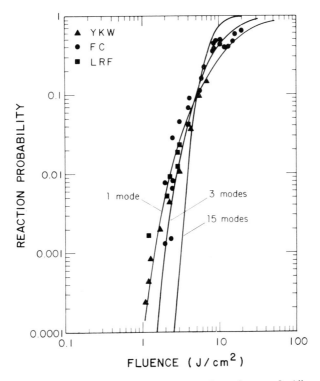

Fig. 5. A log–log plot of reaction probability $P(\Phi)$ versus laser fluence Φ. All experiments were performed with the $P(20)$ CO_2 laser line at 944 cm^{-1}. The labels YKW, FC and LRF refer to Kolodner et al. (1977), Fuss and Cotter (1977) and Lyman et al. (1977b) respectively. The methods used to obtain the calculated curves are given in the text.

the absorption cross section f_{10} was chosen to give $P(\Phi) = 0.1$ at $\Phi = 5\,\mathrm{J/cm^2}$ (values of f_{10} are given in the captions of figs. 2, 3 and 4). We note that the cross section (f_{10}) necessary to give the same reaction probability must increase as the number of modes involved in the absorption process decreases. This follows because a small number of modes is much more easily 'saturated' by the intense laser pulse.

We note that the agreement is reasonably good among the three different sources of experimental data. Only data from experiments performed at low pressure (<0.5 Torr) where the reaction probability was measured directly [eq. (36)] are included in the figure. However, other measurements (Lyman et al. 1977b) where a deconvolution of the reaction probability from a focused geometry was performed are also in good agreement with that shown in fig. 5 (for $P > 0.1$). Data at higher pressure are shifted only slightly to higher fluence. The calculations were performed for 0.5 Torr of 25% SF_6 in H_2, but the pressure is sufficiently low that changes in $P(\Phi)$ to lower pressures are minor.

We see that the function labeled 3 modes best reproduces the experimental data. The function for absorption by the full molecule (15 modes) is definitely too steep, and the function for absorption by one mode is not steep enough.

For confirmation of this conclusion we make a comparison with the experimental data of energy absorbed per molecule in fig. 6. The experimental data are from two sources using two measurement techniques, but the agreement is excellent (see also Deutsch 1977). The Russian data (AGLM) were obtained for SF_6 at 0.5 Torr by measuring incident and transmitted laser energy. Some of the points were obtained with focused laser radiation. These results were actually reported as a function of laser intensity. The intensity was converted to fluence assuming a square pulse of 90 ns duration. The second set of data (YKW) (Yablonovitch et al. 1977, Black et al. 1977) was obtained for 0.2 Torr of SF_6 by using an acoustical technique. The energy absorbed is reported as the average number of photons absorbed per molecule $\langle n \rangle$. The solid curves are from the same calculations used in fig. 5.

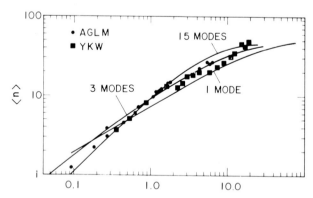

Fig. 6. A log–log plot of the mean number of photons absorbed per pulse $\langle n \rangle$ versus laser fluence Φ. The labels AGLM and YKW for the experimental data refer to Ambartzumian et al. (1975e) and Yablonovitch et al. (1977) respectively.

If harmonic oscillator selection rules ($\Delta v_3 = \pm 1$) and transition probabilities (proportional to v_3) were strictly obeyed (even for three or more absorbing modes), and if no reaction occurred during the pulse, then the slope of $\log \langle n \rangle$ versus $\log \Phi$ would be 1.0. The fact that the experimental data give a slope less than 1.0 (0.3 to 0.85) over the entire range indicates that $f_{i,i-1}$ increases less rapidly with i than one would expect from harmonic oscillator transition probabilities. Some of the turn-down above $2 \, \mathrm{J/cm^2}$ is probably due to dissociation of SF_6. Dissociation produces another species (SF_5) that may also absorb laser radiation (Grant et al. 1977). Here again we see that even though the fit is good for absorption by 15 modes (fast V–V) at low fluence ($\Phi < 2 \, \mathrm{J/cm^2}$), the curve that gives the best fit in the region where reaction is occurring ($\Phi > 2 \, \mathrm{J/cm^2}$) is the one for absorption by three modes. Absorption by one mode gives low values of $\langle n \rangle$ for $\Phi > 1.0 \, \mathrm{J/cm^2}$.

From the data of Yablonovitch and co-workers (1977) in figs. 5 and 6 a plot of $P(\Phi$ versus $\langle n \rangle$ can be be constructed (fig. 7). Comparison of these data with the calculated curves more convincingly demonstrates that absorption by the three ν_3 modes is correct. Note that these curves are independent of the assumed value of the cross section, f_{10}.

Several calculations have been made for $P(\Phi)$ and $\langle n \rangle$ versus Φ using difference forms of the absorption cross section. Expressions of the type

$$f_{i,i-1} = f_{10} i X^{i-1}$$

have been tried with absorption into one mode, where $X = 1.0, 0.95$, and 0.90. None of these gives any better fit to both the absorption and dissociation data than the assumption used for the calculations of figs. 5, 6 and 7 [eq. (48)]. An

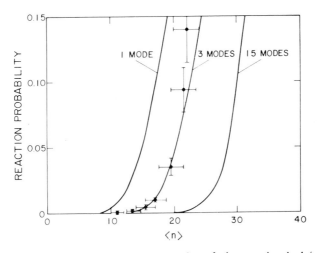

Fig. 7. Reaction probability $P(\Phi)$ versus the mean number of photons absorbed $\langle n \rangle$. The experimental points are from the data of Yablonovitch et al. (1977) and the solid curves were obtained from the same calculations as used in figs. 5 and 6.

additional calculation for absorption by all 15 modes was made with

$$f_{i,i-1} = f_{10} \, 0.95^{i-1}.$$

This gave only a slightly better fit to the dissociation data but gave a value for $\langle n \rangle$ that was large by 1.5 to 3.0 times over the range of the data of fig. 6. The number of modes involved in the absorption process actually influences the distribution to a greater extent than fairly extensive changes of $f_{i,i-1}$ with i.

These calculations give convincing evidence that energy is absorbed by all three ν_3 modes of SF_6 and that collisionless coupling to other modes is weak. It is almost certain that at or near E_0 rapid internal energy redistribution will ensue. From the velocity distribution of SF_6 dissociation fragments under molecular beam conditions (Coggiola et al. 1977, Grant et al. 1977) it has been concluded that the absorbed laser radiation is randomly distributed when the molecule has sufficient energy to dissociate. There certainly must be some energy redistribution, since the dissociation reaction involves feeding enough vibrational energy into a single S–F bond to break it. All we can say from the above analysis is that the absorbed energy is not rapidly randomized until the vibrational energy is a molecule exceeds about 60 to 80% of E_0. There is nothing in this analysis that suggests that the basic RRKM assumption of random energy distribution for $E_v > E_0$ is violated. Additional experiments that are more sensitive to the intermode coupling near E_0 must be performed to further investigate the energy randomization rate and the possibility of mode- or bond-selective chemical reactions. When this work was initiated it was expected that the models proposed by others of linear absorption by the 'quasi-continuum' where all modes are strongly mixed would be verified (Letokhov and Makarov 1976, Larsen and Bloembergen 1976). We have certainly not exhausted all possible formulations of the absorption cross section, some of which may bring the model into better agreement with the 'quasi-continuum' picture. The absorption cross section may have a different dependence of the amount of vibrational energy in the molecule than assumed, and the effect of initial rotational state was not addressed. How these and other properties of the molecule affect the absorption rate will be discussed in greater detail in §4. It does appear from this analysis, however, that the view of polyatomic molecules presented by Rice and co-workers (Oxtoby and Rice 1976, and other papers in series) is consistent with the data. In that series of papers the authors state that rapid intermode coupling occurs only when there is a resonance or when E_v is some large fraction (0.5 to 0.8) of E_0.

We would like to emphasize the point that in no case have we assumed anything other than linear absorption. That is, each f_{ij} is independent of intensity (I or ρ) and

$$f_{ij} = 0, \quad |i - j| > 1.$$

One may be tempted to suggest nonlinear absorption as an explanation for the 'cubic' dependence of $P(\Phi)$ on Φ, which was observed over a narrow range of Φ in early experiments (Ambartzumian et al. 1976c). We see, however, that the entire set of data can be reproduced assuming only linear absorption. We note that even a completely thermal reaction would give a reaction probability curve similar to that shown in fig. 5. For example, from the thermal dissociation rate data (Bott and Jacobs 1969) and the SF_6 heat capacity it can be calculated that the probability of thermal dissociation increases with about the eleventh power of the deposited thermal energy when the reaction probability is near 0.001. That is not to say that nonlinear absorption processes do not occur, but their role in determining the rate of absorption and reaction is minor. This has been independently demonstrated by several sets of experiments (Lyman et al. 1977a, b, Yablonovitch et al. 1977, Kolodner et al. 1977) where the laser pulse shape was varied at constant energy, which gave an independent measure of the relative roles of intensity (power) and fluence (energy). In all cases fluence was found to be more important than intensity, which verifies that linear absorption processes dominate.

The pressure dependence of the reaction yield should be useful in illuminating the role played by collisions. The only collisional processes included in the model were V–T energy transfer by SF_6, H_2 and Ar. Other collisional processes that may influence the laser-induced reaction of SF_6 are intramolecular V–V energy transfer, intermolecular V–V energy transfer, rotational energy transfer and collisional dephasing of the molecular oscillations. We postulate for reasons described below that the first two of these are most important.

Experiments have been performed by observing how the reaction yield or probability changes with diluent pressure (H_2), SF_6 pressure (with no diluent) or the total pressure of an SF_6–diluent mixture.

Fig. 8 shows how the reaction yield changes with partial pressure of the hydrogen diluent. The experimental points are from two different sources but are in reasonably good agreement with each other. Experiments of this sort, where the SF_6 partial pressure is held at some low constant value while the H_2 pressure is varied, show the effect of SF_6–H_2 collisions (which we assume is mainly V–T or V–R–T energy transfer). The H_2 deactivation pressure, or the H_2 pressure necessary to reduce the reaction probability to $1/e$ of its low pressure limit, can be seen from fig. 8 to be about 3.5 Torr. We see that for the H_2 vibrational relaxation time used [eq. (46)] the calculation for absorption by three modes again gives the best fit to the experimental results. The large differences in the slopes of the calculated curves are due to the differences in the distribution over vibrational energy produced by the laser absorption. (See figs. 2, 3 and 4.)

In fig. 9 we have the same type of plot for pure SF_6. The differences between the two curves could possibly be due to differences in fluence or some other experimental parameter. The curve of Gower and Billman (1977) was obtained at the P(22) CO_2 laser line (942.4 cm^{-1}) from enrichment data assuming $\alpha_{34} = 0.0$ over the entire pressure range. The model described above and used for the calculations to this point predicts no change in reaction probability for SF_6 over this pressure range. It is clear that V–T energy transfer processes by SF_6 are much too slow to account for the observed decrease of reaction probability with

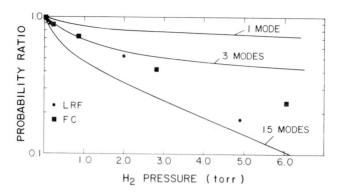

Fig. 8. Probability ratio $[P(\Phi)$ with hydrogen divided by $P(\Phi)$ with no hydrogen]. For both sets of experimental data, as well as the calculations, 0.1 Torr of SF_6 was used. For the data labeled LRF (Lyman et al. 1977b) focused laser radiation was used. For the data labeled FC (Fuss and Cotter 1977), the reaction probability was 0.27 with no hydrogen. For the calculated curves the reaction probabilities with no hydrogen were each within the range 0.02 to 0.10.

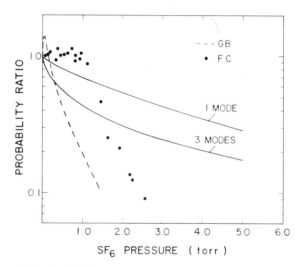

Fig. 9. Probability ratio $[P(\Phi)$ divided by $P(\Phi)$ in the low pressure limit] versus SF_6 pressure. The data labeled FC (Fuss and Cotter 1977) were taken at high fluence $[P(\Phi) = 0.58$ at low pressure] with the $P(20)$ CO_2 laser line. The data labeled GB (Gower and Billmann 1977) were taken with $\Phi \approx 15$ J/cm^2 with the $P(22)$ CO_2 laser line. The calculated curves were obtained by using the fast V–V relaxation time [eq. (52)] in place of the V–T relaxation time to calculate collisional transition probabilities.

SF_6 pressure. We propose that collisional V–V energy transfer process (inter- and intramolecular) are responsible for these experimental observations.

Intramolecular energy transfer will tend to redistribute the absorbed energy into all normal modes of the molecule without affecting the total vibrational energy of the molecule. On the other hand, intermolecular V–V energy transfer by processes such as

$$SF_6(v_3) + SF_6(v_3') \rightleftharpoons SF_6(v_3 - 1) + SF_6(v_3' + 1)$$

tends to give a Boltzmann distribution within each mode. The net effect of both processes is that the vibrational energy of the system approaches a distribution that is describable by a single vibrational temperature with the total vibrational energy of the system remaining approximately constant (i.e., changed only by V–T processes).

To answer the question of how this type of change of vibrational energy distribution affects the rate of reaction we refer again to the calculation results in figs. 2 to 4. For absorption by all 15 modes of the molecule we note from fig. 4 that the distribution produced is very nearly Boltzmann and that the collisional V–V transfer process will do little to change the rate of reaction. If anything, they will tend to increase the reaction rate. However, if the energy is absorbed by fewer than the full 15 modes (for example, three or one), the distribution change by the collisional V–V transfer processes will sharply decrease the reaction rate by decreasing the number of molecules with $E_v > E_0$. This may be seen more readily by noting in fig. 7 that energy absorbed into one mode that is sufficient to react 10% of the SF_6 produces no reaction when the energy is distributed over 15 modes. We see then that if the conclusion reached earlier is correct that energy is absorbed by the three v_3 modes and that collisionless intermode energy transfer is slow, the collisional V–V transfer processes will reduce the reaction rate.

Since intermolecular V–V energy transfer tends to destroy isotopic selectivity, the rate of this process may be estimated by using the selectivity (α_m) as a diagnostic. This has been done in an experiment performed with two lasers at different frequencies, one pulse at low fluence and a later one at high fluence that dissociates the SF_6 (Ambartzumian 1976). It was found that by increasing the time delay between the pulses the selectivity was decreased. A relaxation time of

$$P\tau_{V-V} = 1 \times 10^{-6} \text{ Torr s} \tag{52}$$

was obtained from this experiment. This result is consistent with the observation by others (Fuss and Cotter 1977, Lyman et al. 1977b) that isotopic selectivity degrades for SF_6 pressure greater than about 0.5 Torr.

Since the model as employed to this point designates the SF_6 molecules only by vibrational energy and not by a set of vibrational quantum numbers, a proper formulation of collisional V–V transfer processes cannot be included. However, in order to obtain a qualitative picture of the role of these processes, we artificially replace the V–T relaxation time for SF_6 [eq. (45)] by the V–V relaxation time listed above [eq. (52)] in eq. (44). From the discussion above this will clearly give an improper result for absorption by 15 modes (since the energy distribution is already nearly random), but for three or one this change should give a reasonably accurate reaction probability. These calculations are shown in fig. 9, where we see that absorption by three modes again gives the best result. The differences in the shapes of the experimental and calculated curves may reflect the inappropriateness of incorporating a nonlinear collisional process (V–V transfer) in a linear set of equations [eq. (41)]. The V–V transfer processes will also have a different dependence on temperature and vibrational excitation than the V–T processes.

In fig. 10 we see that for a mixture of SF_6 and H_2, where V–T energy transfer by H_2 and V–V energy transfer occur at roughly the same rate, the model reproduces the experimental results reasonably well when absorption into three modes is assumed.

We conclude this discussion of the multiple-photon processes with a brief summary of the points discussed.

In performing and interpreting an experiment of this type, attention must be given to a number of the experimental conditions such as irradiation geometry, spatial profile of the beam, photolysis of reaction products, heterogeneous reaction of reactants as well as products, etc. From the rate-equation model of the reaction of SF_6, which included laser energy absorption and emission, collisional energy transfer and reaction formulated from RRKM rate theory, we conclude that: first, linear absorption processes dominate the absorption kinetics. Nonlinear absorption processes such as two- and three-photon absorption, power broadening, etc. play a very minor role in the kinetics. Second, laser energy is absorbed by the three near-resonant ν_3 modes and it is likely that collisionless intermode coupling is weak, for $E_v \ll E_0$. This conclusion represents an alternative explanation to the 'quasi-continuum' explanation of the absorption and dissociation data. We stress that the interpretation based on this incomplete model is not unique and any absorption model that gives a vibrational energy distribution similar to that produced by the '3 mode' assumption would be equally acceptable. There is also no reason to question that the vibrational energy is randomized for high levels of excitation. Third, the pressure effects are best explained by fast V–T relaxation processes by species such as H_2 or perhaps even by radicals (H, F) and by intra–intermolecular V–V transfer processes. The isotopic selectivity was found to be a useful diagnostic, especially for the collisional processes.

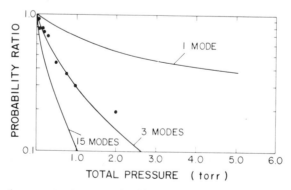

Fig. 10. Probability ratio versus total pressure for 25% SF_6 with 75% H_2. The experimental data were from Lyman et al. (1977b). The calculations were performed with $(P\tau)_{SF_6}$ given by eq. (52).

4 Theory of collisionless multiple-photon laser excitation of polyatomic molecules

Although theoretical treatments of vibrationally heated molecules and of the redistribution of vibrational energy through collisions and intramolecular transfer are common in the chemical literature, we know of no published theory which includes all of the major aspects of the vibrational excitation of polyatomic molecules by intense, pulsed infrared laser light. A correct theory (as opposed to a phenomenological model) should include at least the following: (1) the correct vibrational (and perhaps rotational) energy levels for the vibrational normal modes, including degeneracies, and a correct treatment of mode–mode coupling; (2) the dynamics by which the molecular state evolves in time, taking into account the interaction of some normal modes directly with the electromagnetic field of the incident laser light, and with the other normal modes of the molecule; (3) collisional effects. In order to be considered a theory, a calculation should, if at all possible, make use only of parameters which can be determined by experiment or by *ab initio* molecular calculations. Since no theory which meets these requirements has been published, we shall review the existing theories and phenomenological models by constructing a formalism which is sufficiently general to meet the above objectives, and which may be used to derive the starting points of a number of published calculations as special, approximate cases (Cantrell 1977b). As in the previous section of this review, we shall confine our attention to one of the most thoroughly studied molecules in the context of collisionless multiple-photon laser excitation: SF_6.

4.1 General physical description

Before introducing any formalism, we shall give a general description of our physical ideas. The vibrational modes of a molecule provide a basis in terms of which it is possible to expand an arbitrary displacement of the nuclei in a molecule, apart from rotations and translations. Only two of the normal modes of SF_6 or other octahedral XF_6 molecules are infrared-active (fig. 11), namely, ν_3 and ν_4 (McDowell et al. 1976a, b). Since this conclusion depends only on the symmetry of the molecular potential-energy surface (Louck and Galbraith 1976) and the irreducible representation of the symmetry group to which the dipole operator (which interacts with the laser field) belongs, it is true independently of the vibrational amplitude or the specific form of the dipole operator as a function of vibrational nuclear displacements. The frequencies of the two infrared-active normal modes in SF_6 are sufficiently different ($\nu_3 = 948$ cm^{-1}, $\nu_4 = 615$ cm^{-1}) that only ν_3 interacts strongly with the 10.6 μm CO_2 laser lines. Therefore, it is convenient to reduce the number of degrees of freedom which must be considered in describing the pumping of SF_6 by a CO_2 laser by retaining the description of molecular vibrations in terms of normal modes, while allowing for the fact that the modes are coupled together by interactions which are negligibly small for small nuclear displacements. Classically these couplings lead to

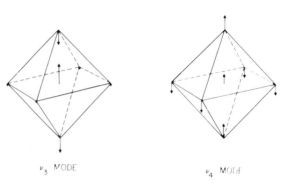

CARTESIAN DISPLACEMENT COORDINATES
FOR THE INFRARED-ACTIVE
FUNDAMENTAL VIBRATIONS OF SF_6

ν_3 MODE ν_4 MODE

Fig. 11. The infrared-active vibrational modes of octahedral XF_6 molecules, after McDowell et al. (1976a, b). The relative nuclear displacements shown are correct for SF_6. (Figure courtesy of R.S. McDowell).

motions which involve sizable amplitudes of other modes when only one mode is initially strongly excited, as shown pictorially by Herzberg (1960, p. 430). Quantum-mechanically the normal-mode product basis functions are not even approximate eigenstates of the vibrational hamiltonian for large rms nuclear displacements. Hence an interaction with an external field which results in strong excitation of a subset of the normal modes will also result in 'leakage' of the energy into other modes of the molecule. Such 'leakage' may have serious consequences for bond-specific chemistry in less symmetrical molecules than SF_6. Although it is tempting to regard this 'leakage' as an irreversible process which can lead to simple vibrational heating of the molecule (Black et al. 1977), it is important to recognize that the time evolution of the molecular state during and after excitation with a pulsed laser is in principle a coherent process in which the amplitudes of different molecular vibrational states have well-defined phase relations. It is possible that the coherence of the molecular state may play a role in some processes (Goodman et al. 1973), although it is not possible to make definite predictions at this time. It is also difficult at the present time to design unambiguous experiments to observe coherent effects in the state of a complex molecule excited by a short laser pulse. Collisions with other gas molecules will alter the phase relations among the quantum-mechanical amplitudes in a random manner, and will therefore lead to a true approach to vibrational thermal equilibrium. It may, however, happen that during laser excitation the temperatures of the vibrational modes are unequal, as has been discussed in §3. If this is true in a given case, the implications are significant for increasing the number density of selectively vibrationally excited molecules.

   The description of a phenomenon in which coherent excitation, complex energy-level structure, and interaction with both an external probe and unobserved internal degrees of freedom (such as all the modes except ν_3 in SF_6) must all be taken into account, is a well-studied problem in quantum statistical mechanics. To focus attention on the modes which interact strongly with the laser field it is natural to use a reduced density matrix $\rho^{(3)}$ for the ν_3 mode alone (Fano 1956, Ter Haar 1961, Lax 1964, 1966), which is obtained from the density matrix ρ for the entire molecule by averaging over the modes which do not interact directly with the laser field. We shall refer to the ν_3 mode as the *system*, and the remaining modes as the *reservoir*. If we label a set of basis states $|s\rangle$ of the system alone with an index s and a set of basis states of the reservoir alone with an index r, then the elements of the density matrix ρ for all the vibrational degrees of freedom carry the double indices $sr; s'r'$. Thus $\rho_{sr;s'r'}$ is the matrix element between the full-molecule basis states $|s\rangle|r\rangle$ and $|s'\rangle|r'\rangle$. A physical observable, such as the dipole-moment operator, which depends only on ν_3 normal coordinates, does not act at all on the reservoir states $|r\rangle$. (Here and elsewhere we put the ν_4 mode in the reservoir, although it could easily be included as part of the system if necessary.) When one is concerned with a pure system observable A with matrix elements $(A)_{sr,s'r'} = A_{ss'}\delta_{rr'}$, as in the interaction between the ν_3 mode and the electromagnetic field, the expectation value

$$\langle A \rangle = \text{Tr}(A) = \sum_{ss'} \sum_{rr'} \rho_{sr;s'r'}(A_{ss'}\delta_{rr'}) \tag{53a}$$

and all other, more complex averages involving A alone can be calculated by first summing $\rho_{sr;s'r'}$ over the unobserved reservoir states $|r\rangle$, and then summing over the system states $|s\rangle, |s'\rangle$. The new density matrix calculated in this way acts only on states of the system, and is called the *reduced density matrix* for the system (i.e. the ν_3 mode):

$$\rho^{(3)}_{ss'} = \sum_r \rho_{sr;s'r}, \tag{53b}$$

i.e.,

$$\rho^{(3)} = \text{Tr}_R \, \rho, \tag{53c}$$

where Tr_R denotes the trace over reservoir states, as in eq. (53b).

Since the diagonal density matrix element $\rho_{sr;sr}$ is the fraction of the molecular population found in the state $|s\rangle|r\rangle$, we see from (53b) that the 'population' $\rho^{(3)}_{ss}$ is in fact not just the population in the state $|s\rangle|0\rangle$ (where $|0\rangle$ denotes the ground state of the reservoir), but is the sum of the populations in all the vibrational basis states $|s\rangle|r\rangle$ which include $|s\rangle$. A ladder of states with increasing ν_3 excitation begins at every vibrational state $|0\rangle|r\rangle$ which lacks ν_3 excitation, i.e., at every state of the reservoir alone. The reduced density matrix $\rho^{(3)}$ describes completely the interaction of this huge set of ν_3 ladders with an external laser electromagnetic field, in a manner which we shall devote this section to explaining.

It can already be seen from the definition of $\rho^{(3)}$ that the reduced density matrix depends upon the state of the reservoir. We shall see below that as the laser field pumps the ν_3 mode, energy is fed into the reservoir and causes substantial changes in the interaction of the ν_3 mode with the laser. This is a consequence of the fact that the details of the absorption of laser energy by highly excited molecular states differ from the details of absorption by low-lying states. The reduced density matrix $\rho^{(3)}$ can be used to discuss these phenomena in a quantitative manner, which we shall review in some detail.

It might at first seem that the system has been singled out for special attention in eqs. (53b) and (53c). The reservoir can, however, be treated in an entirely symmetrical manner by introducing the corresponding reduced density matrix:

$$\rho^{(R)} = \mathrm{Tr}_s\, \rho. \tag{53d}$$

We shall see below that equations of motion may be derived for $\rho^{(3)}$, starting from one for ρ; the same is true of $\rho^{(R)}$. The coupled equations of motion for $\rho^{(3)}$ and $\rho^{(R)}$ provide a natural framework for discussing the flow of energy within the SF_6 molecule. We have seen in §3 that there is reason to believe that energy stays in the ν_3 mode of SF_6 during a large fraction of the time between the beginning of the laser pulse and dissociation. A formalism which singles out the ν_3 mode for attention and treats the reservoir more simply (e.g., in terms of a vibrational temperature) is a natural candidate for describing a situation in which the rate of excitation by the laser is initially large compared to intramolecular 'leakage'. For other molecules, or for very highly excited states of SF_6, a more detailed treatment of the reservoir may be desirable.

The concept of a reduced density matrix originated at the time the foundations of quantum mechanics were laid (von Neumann 1932). Since then a substantial body of techniques has been developed and successfully applied to the description of other highly nonequilibrium systems, such as the nuclear spins in nuclear magnetic resonance experiments (Redfield 1957; reviewed by Slichter 1963) and lasers (Lax 1966, Scully and Lamb 1967, Lax and Louisell 1967, Haken and Weidlich 1967, Fleck 1966, Gordon 1968, Lax and Zwanziger 1973, Cantrell et al. 1973). These techniques have not, to our knowledge, enjoyed a wide application in theoretical chemistry, although certain studies (Goodman and Thiele 1972, Jortner and Mukamel 1976, Hodgkinson and Briggs 1976, Hougen 1976, Goodman et al. 1976, Stone et al. 1976) fit naturally into the framework we shall describe. Other techniques which may be used to describe the rapid vibrational heating of a molecule, such as the RRKM statistical techniques employed in §3 or the Pauli master equation (Montroll and Shuler 1957, Artamonova et al. 1970, Gelbart et al. 1972, Basov et al. 1973, Karlov et al. 1973, Heller and Rice 1974, Kay 1974) and diffusion-equation approximations to the master equation (Afanas'ev et al. 1971, 1973) are special, approximate cases of the general reduced-density-matrix equation of motion, which we shall discuss after first reviewing the energy levels of the ν_3 mode of SF_6.

4.2 Vibrational energy levels

4.2.1 Introduction

In accordance with our earlier comments on the necessity for a physically correct description of the number, symmetry type and degeneracy of the excited vibrational states of the modes which are pumped directly by an infrared laser, we first propose a technique for approximating the potential-energy surface in the molecular electronic ground state. Such an approximation cannot usefully be limited to small nuclear displacements; this rules out power-series expansions in the normal coordinates such as are usually employed in molecular spectroscopy (Herzberg 1945). Restricting ourselves for the sake of clarity to octahedral molecules such as SF_6, we note that functions of (for example) the normal coordinates $q_1^{(3)}$, $q_2^{(3)}$, $q_3^{(3)}$ of the triply degenerate ν_3 normal mode, of the form

$$V_1(\{q^{(3)}\}) = \sum_{i=1}^{3} v(q_i^{(3)}), \tag{54}$$

where $v(-q) = v(q)$ but the function v is otherwise arbitrary, are invariant under the operations of the octahedral symmetry group of SF_6.

This may be proved by expressing the elements of the octahedral group in permutation-inversion form (Galbraith and Cantrell 1977). The same techniques show that functions of the following forms are also invariant under the operation of the octahedral group, and are acceptable contributions to the vibrational potential energy:

$$V_2(\{q^{(3)}\}) = \sum_{i<j} u(q_i^{(3)})u(q_j^{(3)}), \tag{55a}$$

$$V_3(\{q^{(3)}\}) = w(q_1^{(3)}q_2^{(3)}q_3^{(3)}), \tag{55b}$$

where $u(-q) = u(q)$ and $w(-q) = w(q)$, and so forth. The approximation of using only eq. (54) for the vibrational potential energy of the ν_3 mode may be adequate, for example, for an approximate description of the ν_3 mode of SF_6.

If the eigenfunctions of a one-dimensional Schrödinger equation with the potential energy $v(q)$ are $\psi_n(q)$, then the vibrational eigenfunctions for the ν_3 mode are the products

$$\psi_{n_1 n_2 n_3}(\{q\}) = \prod_{i=1}^{3} \psi_{n_i}(q_i), \tag{56}$$

and the approximate energy eigenvalues are

$$E_{n_1 n_2 n_3} = \sum_{i=1}^{3} E_i. \tag{57}$$

Those eigenvalues $E_{n_1 n_2 n_3}$ with the same but permuted values of the integers n_1, n_2, n_3 are degenerate. If all the n_i are different, the energy level is six-fold degenerate; if two are equal, three-fold degenerate; and if all are equal, non-degenerate. Energy levels with the same total number of ν_3 vibrational quanta, $\Sigma_1^3 n_i$, which are not required to be degenerate by this permutation symmetry, will be different. The model of eq. (54) therefore includes the phenomenon of anharmonic splitting of the vibrational overtone levels of a degenerate mode (Shaffer et al. 1939, Hecht 1960), which is known experimentally to be important in the $2\nu_3$ and $2\nu_4$ overtone levels of CH_4 and CD_4 (Fox 1962), and is suspected of being important in many other spherical-top molecules (Cantrell and Galbraith 1976, Akulin et al. 1977). Anharmonic splitting has been suggested as a general explanation for the ease with which many spherical-top molecules can be dissociated by a CO_2 laser (Cantrell and Galbraith 1976, Akulin et al. 1977, Cantrell and Galbraith 1977). The basis functions given in eq. (56) will be coupled together by potential-energy terms of the form of eqs. (55) making the actual vibrational eigenstates of the ν_3 mode linear combinations of the functions in eq. (56). It seems quite likely that functions of the form of eqs. (54) and (55) can be found which give a good representation of the section of the ground-state potential-energy hypersurface in a subspace spanned by the ν_3 normal coordinates and passing through the origin.

The dependence of the vibrational potential energy upon other modes must now be considered briefly. By way of example we consider the coupling of the ν_3 mode (of F_{1u} symmetry under the octahedral group) to another triply degenerate mode (but of F_{2g} symmetry) such as the ν_5 mode of SF_6. This coupling can be conveniently described in a manner which separates the system (ν_3) from one of the reservoir modes (ν_5) by functions of the forms

$$V_{35}(\{q^{(3)}\}, \{q^{(5)}\}) = \sum_{i=1}^{3} Q(q_i^{(3)})F(q_i^{(5)}) \tag{58}$$

or

$$U_{35}(\{q^{(3)}\}, \{q^{(5)}\}) = \sum_{i<j} Q(q_i^{(3)})F(q_j^{(5)}), \tag{59}$$

where $F(-q) = F(q)$ and $Q(-q) = Q(q)$, but F and Q are otherwise arbitrary functions. Many other functional forms for $\nu_3 - \nu_5$ coupling which are octahedrally invariant can be written down using the permutation-inversion representation of the operators of the octahedral group (Galbraith and Cantrell 1977) and the semidirect product form of the transformation operators for vibrational normal coordinates (Louck and Galbraith 1976). However, we shall make the assumption [which is plausible by virtue of the enormous freedom allowed in choosing the functions in eqs. (58) and (59)] that the interaction between the ν_3 mode and the reservoir in the SF_6 molecular can be represented well enough using a potential of the form of eqs. (58) or (59) to fit any experimental or theoretical results which are likely to become available.

4.2.2 Anharmonic energy levels for SF_6: details

In fig. 12 we show the results of a numerical calculation (Cantrell and Galbraith 1977) of the anharmonic splitting of the excited vibrational states of the ν_3 mode

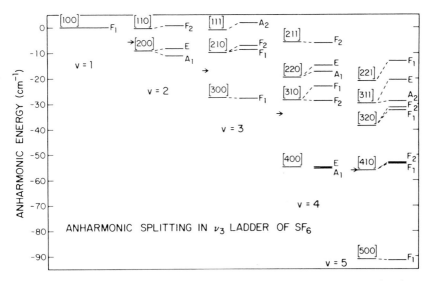

Fig. 12. The overtone energy levels of the ν_3 vibrational mode of SF$_6$. Small arrows give the energy levels without anharmonic splitting. The left-hand column for each value of v shows the levels predicted analytically by eq. (60) when $G_{33} = -2T_{33}$; the right-hand column shows the levels predicted by exact numerical diagonalization of the vibrational hamiltonian [eq. (60)].

of the SF$_6$ molecule, using a hamiltonian of the form (Galbraith and Cantrell 1977)

$$H^{(3)} = (\omega_3 + 3X_{33} - 8T_{33})N + (X_{33} - 6T_{33})N^2$$
$$+ (G_{33} + 2T_{33})l_3^2 + 10T_{33}M, \tag{60}$$

where N is the operator sum of the numbers of quanta in the three ν_3 modes,

$$N = n_1 + n_2 + n_3, \tag{61}$$

M is the operator sum of the squares of the numbers of quanta,

$$M = n_1^2 + n_2^2 + n_3^2 \tag{62}$$

and l_3 is the vibrational-angular-momentum operator for the ν_3 mode. The hamiltonian in eq. (60) is a re-ordered version of the anharmonic hamiltonian derived by Hecht (1960), and the parameters X_{33}, G_{33} and T_{33} have the same meaning as in his paper. For SF$_6$ these parameters are believed to be $X_{33} = -2.8 \text{ cm}^{-1}$, $G_{33} = 1.34 \text{ cm}^{-1}$ and $T_{33} = -0.44 \text{ cm}^{-1}$ (Jensen et al. 1977). The hamiltonian of eq. (61) may be derived from a three-dimensional harmonic-oscillator hamiltonian plus an anharmonic potential energy of fourth order in the ν_3 normal coordinates $\{q_i\}$ by neglecting operators which are off-diagonal in N, a procedure which may be justified by a simple perturbation argument. The three contributions to the anharmonic potential energy in the original hamiltonian from which eq. (61) is derived are the following: the spherically symmetric operator

$$(q_1^2 + q_2^2 + q_3^2)^2 \tag{63}$$

and the two octahedrally symmetric operators (Cantrell et al. 1978)

$$q_1^4 + q_2^4 + q_3^4 \tag{64}$$

and

$$q_1^2 q_2^2 + q_1^2 q_3^2 + q_2^2 q_3^2. \tag{65}$$

We note that eqs. (64) and (65) are of the forms of eqs. (54) and (55a) respectively.

The parameters X_{33}, G_{33} and T_{33} used by Cantrell and Galbraith (1977) to obtain the energy levels of fig. 12 were derived by Jensen et al. (1976) using a sum-of-Morse-functions model of the force field of SF_6. Since these parameters have not been confirmed by detailed spectroscopy of the excited states of SF_6, the energy levels shown in fig. 12 must be considered preliminary. Estimates of vibrational ahnarmonic parameters for SiF_4 have been published by Akulin et al. (1977); the anharmonic parameters of the $2\nu_3$ and $2\nu_4$ states of CH_4 and CD_4 have been determined through a detailed rotational assignment by Fox (1962). Explicit formulas for the energy levels determined by a hamiltonian such as eq. (61) have been given by Hecht through $v = 4$. For $v = 5$ or 6, the energy levels are awkward to express analytically, since the occurrence of four levels of the same octahedral symmetry type in each of the manifolds $v = 5$ and 6 gives rise to an eigenvalue equation of fourth order. For $v \geq 7$, there are five or more levels of at least one octahedral symmetry type, and the problem must be solved numerically (Cantrell and Galbraith, 1976, 1977). However, the parameter set of Jensen et al. (1976) is very close to a parameter set derived by assuming an anharmonic potential energy of the form of eq. (64) (Cantrell and Galbraith 1976). For such a potential, the parameters obey the equalities

$$X_{33} = 6 T_{33}, \tag{66}$$

$$G_{33} = -2 T_{33}, \tag{67}$$

and the energy levels may be expressed exactly in the form (Galbraith and Cantrell 1977)

$$E_{v,m} = v\nu_3 + 10 T_{33}(m - v), \tag{68}$$

where v and m are the eigenvalues of N and M respectively [eqs. (61) and (62)]. The eigenvalues determined in this way are also shown in fig. 12. Since the remaining l_3^2 term in eq. (60) will in general remove the 'accidental' degeneracy of the levels predicted by eq. (68), it is convenient to denote the excited states of the ν_3 mode (or other degenerate modes) as $|v\alpha\rangle$, where v is the total number of quanta of that mode and α is an index (such as $F_1^{(2)}$) which specifies the state uniquely within that manifold.

We summarize our conclusions about the features of the energy levels of a symmetrical polyatomic molecule which are important for understanding the phenomenon of collisionless multiple-photon dissociation by noting the theoretical calculations which remain to be done, and which we expect to have a major impact on the quantitative understanding of this phenomenon. First, very few calculations have been done of the rotational structure of vibrationally excited states of degenerate modes. A meaningful calculation must, at a minimum, include the Teller Coriolis interaction $(-2B\zeta_3 \mathbf{J} \cdot \mathbf{l}_3)$, since this has a major effect upon the spacing of the rotational energy levels (Herzberg 1945). Preliminary dynamical calculations using a rotational energy BJ^2 (Larsen 1976) indicate that rotational energy can indeed compensate for vibrational detuning of the molecular transition frequency from the incident laser frequency, as suggested by Ambartzumian et al. (1976b) and Larsen and Bloembergen (1976). Such compensation can strongly affect the order of magnitude of the excitation achieved, particularly with low laser intensities (Larsen 1976). Second, a first-principles estimate of the forms and strengths of the couplings between the active mode(s) and the remaining modes of the molecule is essential for a quantitative understanding of the damping and shifting of the excited active-mode vibrational states, which we shall discuss qualitatively below. Such information could be derived from force-field models such as Jensen et al. (1976) employed to calculate the anharmonic parameters of the ν_3 mode of SF_6, or from *ab initio* calculations of the electronic-ground-state potential-energy surface.

4.3 Reduced-density-matrix equation of motion

We shall now state an equation of motion for the reduced-density matrix $\rho^{(3)}$ defined in eqs. (53d) and (53c), drawing heavily on the derivations presented by Lax (1964, 1966) and Louisell (1973). We shall include the laser electromagnetic field in the hamiltonian H along with the molecular vibrational normal coordinates, because it is convenient to do so in one particular method of solving a portion of the equation of motion for $\rho^{(3)}$ which will be discussed below, and because including the field also allows for the possibility of averaging over a laser field which is only partially coherent. [A technique for laser field averaging appropriate for a partially coherent laser which does not require the laser field degrees of freedom to be included in the hamiltonian has been described by Eberly (1976).] We shall assume in this review, however, that the laser field is fully coherent. The hamiltonian H is the sum of the following hamiltonians: $H^{(F)}$ of the laser field; $H^{(3)}$ of the ν_3 mode; $H^{(R)}$ of the reservoir; $H^{(3F)}$ of the interaction between the laser field and the ν_3 mode; and $H^{(3R)}$ of the interaction between the ν_3 mode and the reservoir. Thus

$$H = H^{(F)} + H^{(3)} + H^{(R)} + H^{(3F)} + H^{(3R)}. \tag{69}$$

The exact equation of motion for the full density matrix ρ for the (molecule + field) is

$$i\hbar(\partial\rho/\partial t) = [H, \rho]. \tag{70}$$

The derivation of an equation for $\rho^{(3)}$ alone is described in detail by Louisell (1973), based on the work of Lax (1964, 1966). Certain details which must be changed in Louisell's derivation have been described by Cantrell (1977b). The derivation involves writing an equation for $\rho^{(3)}$ correct to second order in $H^{(3R)}$ and correct to all orders in $H^{(3F)}$,

$$
\begin{aligned}
\rho^{(3)}(t + \Delta t) - \rho^{(3)}(t) = (i\hbar)^{-1} & \int_t^{t+\Delta t} [H^{(3F)}(t'), \rho^{(3)}(t')]\, dt' \\
& + (i\hbar)^{-1} \int_t^{t+\Delta t} \mathrm{Tr}_R[H^{(3R)}(t'), \rho(t)]\, dt' \\
& + (i\hbar)^{-2} \int_t^{t+\Delta t} \int_t^{t'} \mathrm{Tr}_R[H^{(3R)}(t'), [H^{(3R)}(t''), \rho(t)]]\, dt''\, dt'.
\end{aligned}
$$

(71)

In eq. (71), the time dependence of the various operators in sight is to be computed using the hamiltonian $H_0 = H^{(F)} + H^{(3)} + H^{(R)}$. It is then assumed that $\rho(t')$ can be considered to be the product $\rho^{(3)}(t')\rho^{(R)}(t')$ over the limited time interval $[t, t + \Delta t]$. For an analysis of this approximation the reader should consult Lax (1964) and Louisell (1973). The separable form assumed for $H^{(3R)}$, as in the sample terms in eqs. (58) and (59), then permits one to write the terms in eq. (71) as the products of averages of reservoir operators and operators on the ν_3 mode, integrated over time. The assumptions that the coarse-graining time interval compared to the correlation time of certain reservoir averages, and short compared to the time required for significant damping of the ν_3 vibrational amplitude, then reduce eq. (71) to the following form expressed in terms of matrix elements taken in a basis of eigenstates of $H^{(3)}$:

$$
\partial \rho_{nn'}^{(3)} / \partial t = (\partial \rho_{nn'}^{(3)} / \partial t)_{\mathrm{laser}} + (\partial \rho_{nn'}^{(3)} / \partial t)_{\mathrm{res}},
$$

(72)

where the coherent driving of the ν_3 mode by the laser field is described by

$$
\left(\frac{\partial \rho_{nn'}^{(3)}}{\partial t} \right)_{\mathrm{laser}} = -\frac{i}{\hbar} \sum_{n''} \{ H_{nn''}^{(3F)}(t)\rho_{n''n'}^{(3)}(t) - \rho_{nn''}^{(3)}(t) H_{n''n'}^{(3F)}(t) \}
$$

(73)

and the interaction with the reservoir is described by

$$
\begin{aligned}
\left(\frac{\partial \rho_{nn'}^{(3)}}{\partial t} \right)_{\mathrm{res}} = -\sum_{n''} \sum_i & \{ [|Q_i(0)_{nn''}|^2 w_i(\omega_{nn''}) + |Q_i^\dagger(0)_{nn''}|^2 w_i^-(\omega_{n''n})^* \\
& + |Q_i^\dagger(0)_{n'n''}|^2 w_i^-(\omega_{n''n'}) + |Q_i(0)_{n'n''}|^2 w_i^+(\omega_{n'n''})^*] \rho_{nn'}^{(3)} \\
& - \delta_{nn'}[|Q_i(0)_{nn''}|^2 (w_i^-(\omega_{nn''}) + w_i^-(\omega_{nn''})^*) \\
& + |Q_i^\dagger(0)_{nn''}|^2 (w_i^+(\omega_{n''n}) + w_i^+(\omega_{n''n})^*)] \rho_{n''n''}^{(3)} \}.
\end{aligned}
$$

(74)

In eq. (74), the reservoir and the ν_3 mode are separated into two quite different types of terms in a manner which we shall now discuss. The time dependence of the interaction hamiltonian between the laser field and the ν_3 mode, $H^{(3F)}(t)$, and of the reduced density matrix $\rho^{(3)}(t)$, is given by the standard transformation to an interaction picture based on $H^{(3)}$:

$$
H^{(3F)}(t) = \left\{ \exp\left(\frac{i}{\hbar} H^{(3)} t \right) \right\} H^{(3F)} \left\{ \exp\left(-\frac{i}{\hbar} H^{(3)} t \right) \right\}
$$

(75)

and similarly for $\rho^{(3)}(t)$. The interaction between the active (ν_3) mode and the reservoir has been assumed to be of the separable form (Fano 1956, Lax 1964, 1966)

$$H^{(3R)} = \hbar \sum_i (Q_i F_i + Q_i^\dagger F_i^\dagger), \tag{76}$$

where the notation is intended to suggest a generalized dipole coordinate (Q_i) times field strength (F_i) interaction. In fact, where Q_i is an operator *function* of the active-mode normal coordinates, and F_i is an operator *function* of the remaining normal coordinates (i.e., F_i acts only on the reservoir). While the matrix elements of the active-mode operators Q_i appear explicitly in eq. (74), the reservoir operators F_i have been subsumed in the quantities

$$w_i^+(\omega) = \sum_{a,b} p_a |\langle a| F_i |b\rangle|^2 \int_0^\infty \exp(i(\omega + \omega_{ab})\tau)\, d\tau, \tag{77}$$

$$w_i^-(\omega) = \sum_{a,b} p_a |\langle a| F_i^\dagger |b\rangle|^2 \int_0^\infty \exp(i(\omega - \omega_{ab})\tau)\, d\tau. \tag{78}$$

The frequency ω which appears as an argument in $w_i^\pm$ is always one of the transition frequencies between eigenstates n, n' of the ν_3 mode:

$$\omega_{nn'} = \hbar^{-1}(E_n^{(3)} - E_{n'}^{(3)}). \tag{79}$$

The quantities $w_i^\pm(\omega)$ are proportional to the Fourier transform of the product of the ith term in the ν_3 mode–reservoir interaction $H^{(3R)}(t)$ with itself at a later time $t + \tau$. The 'constant' of proportionality is, of course, the modulus squared of a matrix element of the ν_3-mode operator Q_i, as is displayed in eq. (74). By the Wiener–Khinchin theorem (Kittel 1958) the quantities $w_i^\pm(\omega)$ are therefore to be regarded as *spectral densities* of the reservoir operator F_i in eq. (76). This terminology is derived from electrical engineering, where the power spectrum of a time-varying voltage $e(t)$ is proportional to the Fourier transform with respect to time of the correlation function $\langle e(t)e(t + \tau)\rangle$. The spectral density of a reservoir 'field' coordinate F_i, although a less concrete concept, is defined in a similar way. The concept of the spectral density of reservoir coordinate has proven to be useful in, for example, the theory of magnetic resonance (Slichter 1963, §§5.5 and 5.6), where the interaction between nuclear spins (the system) and a heat bath (the reservoir) lends itself well to a treatment in terms of the reduced density matrix (Redfield 1957). In what follows we shall refer to $w_i^\pm(\omega)$ as the *reservoir spectral densities* (Louisell 1973).

In the derivation of eqs. (72) to (74) (Cantrell 1977b) a number of simplifying assumptions have been employed. One of these amounts to a definition, albeit an important one: we have assumed that the reservoir expectation values of all terms in the interaction between the ν_3 mode and the reservoir which are diagonal in both ν_3 and reservoir coordinates have been included in $H^{(3)}$ (Lax

1964). Symbolically, we define

$$H^{(3)} = \mathrm{Tr}_R[\rho^{(R)}(H - H^{(3F)} - H^{(F)})].$$ (80)

This implies that $H^{(3R)}$ is fully off-diagonal in both ν_3 and reservoir coordinates. Even more important, eq. (80) implies that as the state of the reservoir is altered by the transfer of energy from the laser-pumped ν_3 mode, the energy levels of the ν_3 mode (eigenvalues of $H^{(3)}$) will be altered (Cantrell 1977a), owing to anharmonic interactions of the form $X_{3i}[(n_3 + (3/2))(n_1 + (d_i/2)]$ (Herzberg 1945) (where n_3, n_i are numbers of ν_3- and other-mode quanta, and d_i is the degeneracy of the mode $i \neq 3$), and of more complex forms as well. Second, we have assumed that the reservoir density matrix $\rho^{(R)}$ is diagonal in a basis of energy eigenstates $|a\rangle$ [eqs. (77) and (78)], with probability p_a that the reservoir will be in state $|a\rangle$. Although this assumption is merely convenient, and can be removed if necessary, it is in fact true for a thermal-equilibrium reservoir density matrix, which is the first in a hierarchy of approximations to $\rho^{(R)}$ (Lax 1964). In the explicit evaluation we give below of the reservoir spectral densities, we assume that the reservoir is simply heated by interaction with the ν_3 mode, and neglect all coherent reservoir effects. Third, we have assumed that there are no 'accidental' coincidences of transition frequencies $\omega_{nn'}$, such that $\omega_{nn'} + \omega_{n''n'''}$ vanishes for n, n', n'' and n''' all distinct. This is not the case for a harmonic oscillator (Lax 1964, 1966), but is correct for anharmonic oscillators. Finally, in the derivation of eqs. (72) to (74) a markovian approximation was made (Louisell 1973), which amounts physically to assuming that the interaction between the ν_3 mode and the reservoir is sufficiently chaotic that the time evolution of the system (ν_3 mode) density matrix at time t depends only on the state of the system and the reservoir at time t. This assumption will be correct provided the reservoir spectral bandwidth (the width in ω of the functions $(|w_i^\pm(\omega)|)$ is large compared to the reciprocal $(\Delta t)^{-1}$ of the finite interval of time Δt over which the derivative $\partial \rho^{(3)}/\partial t$ is computed. In turn $(\Delta t)^{-1}$ must be large compared to the rate at which energy decays from the ν_3 mode to the reservoir [see eq. (119)]. It should be noted that these assumptions are *not* equivalent to a rate-equation approximation for either the driving of the ν_3 mode by the laser field, or the evolution of $\rho^{(3)}$ owing to the interaction with the reservoir (see eq. (113) and the associated discussion). For a more detailed discussion of these approximations, the reader is urged to consult the excellent book of Louisell (1973).

The reduced-density-matrix equation of motion [eqs. (72) to (74)] may be derived straightforwardly (Cantrell 1977b) using the methods of Lax (1964, 1966) and Louisell (1973). Equations very similar in form to eqs. (72) to (74) have been written down on phenomenological grounds by Redfield (1957), Goodman and Thiele (1972) and Hougen (1976). Hodgkinson and Briggs (1976) have derived an equation of the same form as these using a formally different method, the projection technique of Zwanzig (1964). The advantage of a derivation of eqs. (72) to (74), as opposed to assuming phenomenological equations of the same form, is simply that as a result of a derivation one is in possession of expressions for the myriad constants which appear in the phenomenological treatments. These expressions can, in principle, be evaluated for any problem of interest. In practice, it will probably be difficult to obtain numerical values for many of the quantities in eqs. (72) to (74) for a complex polyatomic molecule which are correct in more than simply their order of magnitude. Despite these anticipated difficulties we believe it is preferable to have expressions for the quantities appearing in eqs. (72) to (74), since the form of these expressions can suggest much physical insight which is obscured in a purely phenomenological treatment. We hope that this section will provide a useful framework for future theoretical research into collisionless laser-driven molecular excitation, and that perhaps by pointing out difficulties we may stimulate further research.

4.4 Coherent driving. Rabi oscillations. Power broadening

The coherent driving described by eqs. (73) and (75) is profoundly affected by the anharmonic shifts and splittings implied, for example, by eqs. (57) or (60); by the alteration of the energy levels of the active mode(s) due to heating of the reservoir; and by rotations, which we exclude from the discussion for the sake of simplicity. For instance, if we let the hamiltonian for the ν_3 mode take the form [derived from eq. (53)]

$$H^{(3)} = \sum_{i=1}^{3} \left\{ \hbar \omega_3 a_i^\dagger a_i + \hbar \sum_{l=2}^{\infty} v_l (a_i^\dagger)^l a_i^l \right\} \tag{81}$$

then the creation ($a_i^\dagger$) and annihilation (a_i) operators in an interaction picture based on $H^{(3)}$ are

$$a_i(t) = \exp[-i(\omega_3 + \Omega(n_i))t] a_i(0) \tag{82}$$

$$a_i^\dagger(t) = [a_i(t)]^\dagger. \tag{83}$$

The anharmonic correction to the harmonic transition frequency ω_3 is

$$\Omega(n_i) = \sum_{l=1}^{\infty} \alpha_l [n_i]^l, \tag{84}$$

$$\alpha_l = \sum_{m=l+1}^{\infty} m v_m S_{m-1}^{(l)}. \tag{85}$$

We have expressed eqs. (82) to (85) in terms of the number operator $n_i = a_i^\dagger a_i$ and the Stirling numbers of the first kind, $S_m^{(l)}$. If we assume a coupling of the ν_3 mode of the form

$$H^{(3F)} = \tfrac{1}{2} \mu_0 E_0 \sum_{i=1}^{3} (a_i e^{i\omega t} + a_i^\dagger e^{-i\omega t}) \tag{86}$$

to a classical electric field

$$E(t) = E_0 \cos \omega t \tag{87}$$

in the rotating-wave approximation, then the driving term, eq. (73), becomes

$$\left(\frac{\partial \rho^{(3)}}{\partial t}\right)_{\text{laser}} = -\frac{i}{2\hbar} \mu_0 E_0 \sum_{i=1}^{3} \{[\exp(-i(\omega_3 - \omega + \Omega(n_i))t]a_i(0), \rho^{(3)}]$$

$$+ [a_i^\dagger(0) \exp[i(\omega_3 - \omega + \Omega(n_i))t], \rho^{(3)}]\}. \tag{88}$$

This operator differential equation will in general imply that the largest contributions to $(\partial \rho^{(3)}/\partial t)_{\text{laser}}$ come from pairs of states for which the oscillating factors $\exp[\pm i(\omega_3 - \omega + \Omega(n_i))t]$ are most nearly constant in time. The degeneracy of the ν_3 mode thus enters eq. 88 in two important respects: (i) the anharmonic splitting of the energy levels

$$E_{n_1 n_2 n_3} = \sum_{i=1}^{3} \left\{ \hbar \omega_3 n_i + \hbar \sum_{m=2}^{\infty} v_m \sum_{l=1}^{m} S_m^{(l)}(n_i)^l \right\} \tag{89}$$

with the resulting correction to the transition frequencies [eqs. (84) and (85)] reducing the destructive oscillation of the right-hand side of eq. (88); and (ii) the threefold degeneracy of the ν_3 mode provides three terms instead of one which contribute to the change with time of $\rho^{(3)}$. Other analytic approaches to the reduced-density-matrix description of a coherently driven nondegenerate system have been described by Oraevsky and Savva (1970).

For a pair of levels (n, n') which are resonantly coupled, the coherent-driving equation of motion, eq. (73), can be cast in the form

$$\frac{\partial}{\partial t}(\rho_{n'n'}^{(3)} - \rho_{nn}^{(3)}) = -\frac{2i}{\hbar} H_{nn'}^{(3F)}(\rho_{nn'}^{(3)} - \rho_{n'n}^{(3)}) + \left(\frac{\partial}{\partial t}(\rho_{n'n'}^{(3)} - \rho_{nn}^{(3)})\right)_{\text{NR}}, \tag{90}$$

$$\frac{\partial}{\partial t}(\rho_{nn'}^{(3)} - \rho_{n'n}^{(3)}) = -\frac{2i}{\hbar} H_{nn'}^{(3F)}(\rho_{n'n'}^{(3)} - \rho_{nn}^{(3)}) + \left(\frac{\partial}{\partial t}(\rho_{nn'}^{(3)} - \rho_{n'n}^{(3)})\right)_{\text{NR}}, \tag{91}$$

where n and n' stand for all the quantum numbers required to specify the state. The notation 'NR' in eqs. (90) and (91) is intended to denote all terms arising from eq. (73) for which $\omega_{nn''}$, $\omega_{n''n'}$, etc. are not equal to the laser frequency ω. In the case of exact resonance, which we have assumed, the (exponential) time dependences of $\rho_{nn'}^{(3)}$ and $H_{n'n}^{(3F)}$ exactly cancel (in the rotating-wave approximation.) Hence the operators in the first term on the right-hand side of eqs. (90) and (91) are to be evaluated in the Schrödinger picture, where they are independent of time. Eqs. (90) and (91) are identical to the form for the density-matrix equations of motion for a two-level system (Lamb 1964), apart from the nonresonant terms. In writing eqs. (90) and (91), we have assumed, without loss of generality, that the matrix element $H_{n'n}^{(3F)}$ is real, and hence equal to $H_{nn'}^{(3F)}$. If the nonresonant terms in eqs. (90) and (91) could be neglected, the population difference $(\rho_{n'n'}^{(3)} - \rho_{nn}^{(3)})$ and polarization $i(\rho_{nn'}^{(3)} - \rho_{n'n}^{(3)})$ would oscillate with the frequency $(2|H_{nn'}^{(3F)}|/\hbar)$, which we shall call the *Rabi frequency* for the transition between n and n'. The resonating levels (n, n') in eqs. (90) and (91) may also be coupled by a multiphoton (rather than a single-photon) transition. In that case the hamiltonian $H^{(3F)}$ may be replaced by an appropriate effective multiphoton hamiltonian $H^{(\text{eff})}$ for the two levels involved. The use of perturbation theory to evaluate the matrix element $H_{nn'}^{(\text{eff})}$ leads to an explicit formula for the multiphoton Rabi frequency (Larsen and Bloembergen 1976), which is valid as long as the amplitudes of all states other than n and n' are small compared to the amplitudes of n and n'. Since the latter situation does not often arise in molecular systems with the complex level structure characteristic of degenerate vibrational modes, it is frequently necessary to resort to a numerical solution of the equations of motion (eq. 73) by methods such as those described following eq. (92), which automatically include such phenomena as multiphoton transitions, power broadening and optical Stark shifts.

The transition between n and n' described by eqs. (90) and (91) will be close to resonance when the Rabi frequency $2|H_{nn'}^{(3\text{F})}|/\hbar$ is comparable to, or larger than, the detuning $|\omega - \omega_{nn'}|$, for in that case eqs. (90) and (91) predict significant growth in the population difference and polarization in the time $|\omega - \omega_{nn'}'|^{-1}$ in which destructive oscillation begins to occur. Thus the detuning range within which one may expect significant transfer of population between n and n' is proportional to the Rabi frequency. This phenomenon is often called *power broadening* of the resonant transition between n and n'. For an N-photon transition the power broadening is proportional to the Nth power of the laser electric field if the perturbative expression of Larsen and Bloembergen (1976) for the multiphoton Rabi frequency is valid.

The shift and splitting of molecular energy levels through interaction with a strong laser electromagnetic field is frequency referred to as the *optical Stark effect*. General perturbation–theoretical expressions for the shifts are well known (see, e.g., Sakurai 1967), and specific calculations have been published for diatomic and some polyatomic molecules, including rotational as well as vibrational contributions to the molecular energy levels (Pert 1973, Braun and Petelin 1974, Golger et al. 1976). Such approximate expressions are correct when only two molecular levels are near resonance with the laser frequency, i.e., when all other allowed transitions are detuned by an amount large compared to their Rabi frequencies, as in the recent experiment of Bischel et al. (1976). In complicated energy-level patterns such as is shown in fig. 12, where more than one transition can be near resonance, it appears to be essentially meaningless to attempt to calculate optical Stark shifts separately from multiphoton and power-broadening effects. All of these effects can, of course, be included in numerical calculations, although not in a manner which allows a clear separation of any one effect from the other two.

4.5 Multiphoton theories of collisionless laser excitation of SF₆ and other molecules

We have used the term *multiple-photon* to refer to any process involving the absorption of more than one photon. We use the term *multiphoton* to refer to calculations of quantum-mechanical amplitudes or rates of excitation, such as those of Larsen and Bloembergen (1976), which (i) involve the absorption of more than one quantum of laser light per molecule, and (ii) result in amplitudes or rates containing those energy denominators which are characteristic of the lowest nonvanishing order of perturbation theory in which the transition of interest can occur (Göppert-Mayer 1931, Bebb and Gold 1966). In response to the experiments of Isenor and Richardson (1971), Pert (1973) calculated the multiphoton rate of dissociation of polyatomic molecules which absorb 20 to 40 photons before dissociation. Earlier calculations of multiphoton molecular dissociation by Bunkin et al. (1964), Askar'yan (1965) and Bunkin and Tugov (1970) assumed that only a small number of quanta were required to dissociate the molecule. Recently Faisal (1976) has calculated the multiphoton dissociation rate using approximations which will be discussed below.

An N-photon process in which none of the intermediate steps are resonant would proceed at a rate proportional to I^N, where I is the laser intensity. Such an extraordinarily strong dependence of the collisionless dissociation rate upon the laser intensity is observed in atoms for N as large as 15 or 20, but is not at all in agreement with experiments on molecules (see §3). If one of the intermediate states n' in the multiphoton process is resonant with the laser frequency ω in the sense that ω_{0n} is equal to an integer multiple of ω, then the perturbation techniques break down, yielding very large (actually infinite) rates which are inconsistent with the conservation of probability. Faisal (1976) has suggested an approximate technique for evaluating multiphoton amplitudes in which resonances occur (because of rotational effects or anharmonic splitting, for example): replacement of the infinite factor by the maximum value allowed by conservation of probability, namely, 1. This leads, as one might expect, to a dependence of the dissociation rate upon the laser intensity I of the form I^m, where m is the number of nonresonant steps.

Pert (1973) has addressed the more general problem of calculating the dominant terms in the sums over intermediate states which are characteristic of multiphoton amplitudes (Göppert-Mayer 1931, Bebb and Gold 1966), using harmonic-oscillator vibrational energies and approximate anharmonic-oscillator matrix elements. In a transition in which N laser photons are exactly resonant with $(N - m)$ or $(N + m)$ vibrational quanta, the set of intermediate states for which the individual energy denominators are as close to resonance as possible yields a vanishing transition amplitude (Pert 1973). Explicit formulas for the next-most-resonant set of intermediate states were given by Pert, who also derived approximate expressions for multiphoton amplitudes including the effects of rotation.

Quantum-mechanical numerical calculations of multiple-photon excitation (Larsen and Bloembergen 1976, Jortner and Mukamel 1976, Hodgkinson and Briggs 1976, Goodman et al. 1976, Stone et al. 1976, Cantrell and Galbraith 1977, Walker and Preston 1977, Shore and Eberly 1977) have verified the occurrence of multiphoton resonances, even though the approximations made in perturbation-theoretical calculations of the amplitudes of these resonances are not valid. The sharp spikes in fig. 13 are multiphoton resonances in which v laser photons are resonant with a transition $|v''\alpha''\rangle \rightarrow |(v'' + v), \alpha'\rangle$. The selection rules

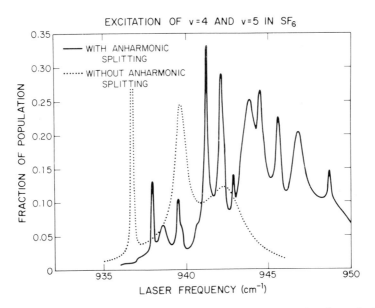

Fig. 13. The time-averaged fraction of the molecular population in $v_3 = 4$ and $v_3 = 5$ of SF_6 as a function of laser frequency, at a laser intensity of $3\,GW/cm^2$. Solid line: with anharmonic splitting. Dotted line: without anharmonic splitting.

for such transitions may be derived trivially. The perturbation-theoretical derivation of the multiphoton Rabi frequency (Larsen and Bloembergen 1976) predicts correctly that the multiphoton resonances take longer to build up, and hence give narrower spikes as functions of the laser frequency, the larger v is. The occurrence of such multiphoton resonances appears to be a characteristic feature which distinguishes quantum from classical behavior in the one-dimensional calculations of Walker and Preston (1977).

4.6 Solution of coherent multilevel equations of motion

The exact numerical and analytical study of resonant and nonresonant transitions in multilevel systems has reached a sophisticated level (Letokhov and Makarov 1976a, Sargent and Horowitz 1976, Einwohner et al. 1976, Makarov 1977, Eberly et al. 1977, Bialynicka-Birula et al. 1977). In general the Rabi oscillations among a system of N levels are nonperiodic (Eberly et al. 1977), even in a situation in which the successive transitions are all resonant, and driven with the same Rabi frequency.

For a system which is coherently driven and lacks significant damping (i.e., for which the coupling to the reservoir is zero), a wave function may be used to describe the time evolution of the system, rather than a density matrix. In that case the solution of the Schrödinger equation or the equivalent eq. (73) results in a unitary time-development operator $U^{(\text{eff})}(t, t_0)$ satisfying the equation

$$i\hbar(\partial/\partial t)U^{(\text{eff})}(t, t_0) = (H^{(3\text{F})} + H^{(3)} + H^{(\text{F})})U^{(\text{eff})}(t, t_0) \tag{92}$$

and describing the (coherent) evolution in time of the coupled molecule–laser field system. For a monochromatic laser field of frequency ω and for a molecular system in which radiative couplings between states which are not coupled in the harmonic approximation can be neglected, and in the rotating-wave approximation, the coupled basis $|\text{molecule}\rangle \otimes |\text{field}\rangle$ required to solve eq. (92) contains the same number of states as the original molecular basis (Mukamel and Jortner 1976). The basis states are conveniently represented as $|n - v; v\alpha\rangle$, where n is the initial number of photons (assumed to be larger than the largest number of vibrational quanta excited) and v, α index the vibrational state. For a laser pulse which is of zero intensity for times $t < t_0$ and constant intensity for a finite time thereafter, the solution of eq. (92) may be expeditiously accomplished by a single matrix diagonalization of the hamiltonian $H^{(\text{F})} + H^{(3)} + H^{(3\text{F})}$ (Mukamel and Jortner 1976, Mower 1966). The diagonal matrix elements of

$$H^{(\text{eff})} = H^{(\text{F})} + H^{(3)} + H^{(3\text{F})} \tag{93}$$

are

$$\begin{aligned} H^{(\text{eff})}_{v\alpha;v\alpha} &= n\hbar\omega + (E^{(3)}_{v\alpha} - v\hbar\omega) \\ &= (\text{constant}) + (E^{(3)}_{v\alpha} - v\hbar\omega), \end{aligned} \tag{94}$$

where the constant is independent of v and α and may therefore be neglected. The off-diagonal matrix elements of $H^{(\text{eff})}$ are the same as of $H^{(3\text{F})}$, and for large n are well approximated by the matrix elements for a classical laser field. The matrix elements of $U^{(\text{eff})}(t, 0)$ in the basis $|n - v; v\alpha\rangle$ are

$$U^{(\text{eff})}(t)_{v\alpha;v'\alpha'} = \sum_\lambda c(v\alpha; \lambda)c(v'\alpha'; \lambda)^* \exp[-i(E_\lambda/\hbar + v\omega)t], \tag{95}$$

where $c(v\alpha; \lambda)$ is an expansion coefficient of the eigenvector $|\lambda\rangle$ of $H^{(\text{eff})}$ in terms of the basis $|n - v; v\alpha\rangle$,

$$|\lambda\rangle = \sum_{v\alpha} c(v\alpha; \lambda)|n - v; v\alpha\rangle, \tag{96}$$

$$H^{(\text{eff})}|\lambda\rangle = E_\lambda|\lambda\rangle. \tag{97}$$

Eq. (95) may be used to find the probability for the molecular system plus field to be in the state $|n - v; v\alpha\rangle$ at time t, given that it was in the state $|n - v'; v'\alpha'\rangle$ at time $t = 0$:

$$p(v\alpha, t|v'\alpha', 0) = |U^{(\text{eff})}(t)_{v\alpha;v'\alpha'}|^2. \tag{98}$$

The time-averaged probability is

$$\overline{p(v\alpha|v'\alpha')} = \sum_\lambda |c(v\alpha; \lambda)|^2 |c(v'\alpha'; \lambda)|^2 \tag{99}$$

provided the time averaging is carried out over an interval which is long compared to $\max(\hbar/E_\lambda)$.

For energy levels of the active mode which are not strongly shifted and broadened by the reservoir interactions described by eqs. (74) and (76), the dependence of the fraction of the molecular population excited to a specific vibrational state upon the frequency ω of the incident laser light should be well described (as far as is possible using vibrational states alone) by solving eq. (88), or, equivalently, by diagonalizing $H^{(\text{eff})}$ [eq. (93)]. Previous numerical calculations of the laser excitation of a non-rotating one-dimensional anharmonic oscillator (Mukamel and Jortner 1976, Larsen and Bloembergen 1976, Bloembergen et al. 1976) do not produce a dependence of vibrational excitation upon laser frequency which agrees well with the experimental dependence of the dissociation rate of SF_6 upon the frequency of a low-intensity, nearly resonant CO_2 laser, as reported by Ambartzumian et al. (1976b). Fig. 13 shows the dependence of the fraction of the population in $v = 4$ and $v = 5$,

$$p(\omega) = \sum_\alpha \left[\overline{p(4, \alpha|0)} + \overline{p(5, \alpha|0)} \right], \tag{100}$$

upon the laser frequency ω, calculated for the triply degenerate ν_3 mode of SF_6 by diagonalizing $H^{(\text{eff})}$ [eq. (93)] (Cantrell and Galbraith 1977). The energy levels of the ν_3 mode were determined by numerical diagonalization of the vibrational anharmonic hamiltonian of Hecht (1960), using the anharmonic parameters X_{33}, G_{33} and T_{33} calculated by Jensen et al. (1976). The approximate dipole matrix elements are given by Galbraith and Cantrell (1977) and Overend et al. (1977). The results shown in fig. 13 thus include anharmonic splitting, but not rotations. The calculated frequency dependence, but not the order of magnitude of excitation, are in good agreement with the experimental results of Ambartzumian et al. (1976b) in which the excitation of low-lying vibrational states of cooled SF_6 by a low-intensity, pulsed CO_2 laser was probed by dissociating the molecule using a more intense, nonresonant pulsed CO_2 laser. The agreement between the predicted frequency dependence and that observed at a temperature of 190 K by Ambartzumian et al. (1976b) must be interpreted with caution. The vibrational states up to which the molecule is pumped by the weak, nearly resonant laser in the latter experiment are not known, but are merely estimated to be $v = 4$ and $v = 5$ (Bloembergen et al. 1976).

An alternative to exact diagonalization, which may be useful for large numbers of levels, has been employed by Akulin et al. (1977) to discuss the coherent driving of a one-dimensional, nondegenerate oscillator by an electromagnetic field. By expanding finite differences such as $f(n) - f(n - 1)$ in a formal Taylor series in n and retaining up to second derivatives, they reduced the Schrödinger equation to the form of a diffusion equation in one dimension. Additional quantum numbers such as appear in eqs. (56) and (57) would require the use of additional dimensions in the diffusion equation, increasing the difficulty of either numerical or analytical calculations.

4.7 Effects of rotation

The effect of rotational energy levels upon the coherent driving of a spherical-top molecule such as SF_6 by a laser field is still the subject of much debate. Certainly a change of rotational energy in a vibration–rotation transition can partially cancel a change of vibrational energy, and thereby help bring the frequencies of successive transitions closer to the same (laser) frequency. This effect has been suggested as the explanation for the absorption of up to four successive CO_2 laser photons of the same energy in SF_6 (Ambartzumian et al. 1976c, Larsen and Bloembergen 1976). However, this explanation suggests a strong dependence of the excitation rate upon the total angular momentum J, since such essentially fortuitous cancellations of vibrational anharmonic shifts by rotational energy occur for very specific values of J. At the present time, no such strong dependence has been experimentally observed in, for example, OsO_4 (Ambartzumian 1976), in which collisionless multiple-photon dissociation also occurs. As was mentioned above, the hamiltonian for a spherical-top molecule includes not only the rigid-rotator energy BJ^2 but also the Teller (1935) Coriolis energy $-2B\zeta_3 J \cdot l_3$, where ζ_3 is the magnitude of the vibrational angular momentum in units of $\hbar$. In the $v = 0$ to $v = 1$ transition frequencies these terms combine to produce a contribution $2B(1 - \zeta_3)(J + 1)$ to the R(J) line, and a contribution $-2B(1 - \zeta_3)J$ to the P(J) line. The Coriolis levels for low J for SF_6 are shown in fig. 14. For transitions other than $v = 0$ to $v = 1$, one is faced with the fact that the rotational manifolds of (for example) $v = 2$ are sufficiently close in energy that the expected small differences in the value of $\zeta_{v\alpha}$ for the A_1, E and F_2 vibrational sublevels shown in fig. 12 (Fox 1962) may lead to strong rotational perturbations (Rhodes and Cantrell 1977), making any simple model meaningless in fundamental terms. In addition, the rotational levels of SF_6 and other spherical-top molecules are split by perturbations involving quadratic or higher powers of both the total angular momentum and the vibrational normal coordinates into as many levels as the octahedral or tetrahedral symmetry of the molecule allows (Jahn 1938, Hecht 1960, Aldridge et al. 1975, McDowell et al. 1976b). This splitting is hardly an insignificant detail; it causes the P(60) rotational 'line' of the $v = 0$ to $v = 1$ transition in SF_6 to extend over 0.434 cm^{-1} (McDowell et al. 1976b), and must therefore be considered (after anharmonic splitting and scalar rotational energy) as a contributor to the resonant enhancement of laser excitation, along the lines discussed in connection with eqs. (90) and (91). An additional effect of including rotational energy levels is that the number and degeneracy of the levels which are significantly coupled in eq. (73) are both greatly increased over the comparable values when vibrational levels alone are included. These facts alone can lead to an increase of orders of magnitude in the time-averaged probability [eq. (99)], at low laser intensity. Quite possibly these comments may help to explain calculations (Larsen 1976, Ackerhalt and Galbraith 1978) showing that including just the rigid-rotator energy BJ^2 increases significantly the excitation and dissociation rates for a nondegenerate or degenerate anharmonic oscillator. However, it appears to be

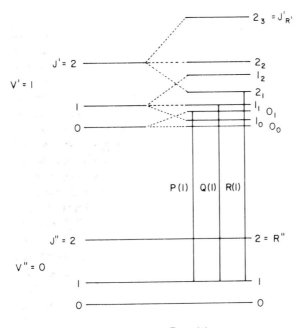

Coriolis   Sublevels   And   Transitions

Fig. 14. Teller Coriolis levels for low J values of SF_6. Dipole-allowed transitions from the vibrational ground state ($v = 0$) to the state in which one quantum of the ν_3 mode is excited ($v = 1$) preserve the value of the rotational angular momentum R.

certain (Cantrell and Fox 1978) that the lowest-order vibration–rotation interaction $-2B\zeta_3 \mathbf{J} \cdot \mathbf{l}_3$ will affect both the low-intensity spectrum and the dynamics of excitation, as has been shown to be the case in a classical calculation (Gorchakov and Sazonov 1976) for a rotating diatomic molecule. Perturbations of vibration–rotation levels of ($n\nu_3$) by other very close, weakly coupled levels may also lead to large isotope effects, as discussed by Rhodes and Cantrell (1977).

4.8 Intermode vibrational relaxation

4.8.1 Physical discussion

At a fundamental quantum-statistical-mechanical level, the problem of describing the evolution in time of a polyatomic molecule in a laser field is conceptually extremely similar to problems which have been previously considered in the quantum theory of the laser (Lax 1966, Scully and Lamb 1967, Lax and Louisell 1967, Haken and Weidlich 1967, Fleck 1966, Gordon 1968). A correct description of the laser involves a study of the time evolution of the energy and statistics of a subsystem (the laser field) in a situation in which the interaction between the subsystem and reservoir consists of the absorption and stimulated emission of light. Exactly the same ingredients are involved in the problem of collisionless laser-driven molecular photodissociation. Exactly the same general reduced-density-matrix equation of motion which applies to the laser (Lax 1966) also applies to the molecular problem, although this fact is not widely recognized in papers which apply the reduced-density-matrix equation of motion to molecular excitation (Oraevsky and Savva 1970, Goodman and Thiele 1972, Stone et al. 1973, Goodman et al. 1976, Hodgkinson and Briggs 1976). In the case of the laser, the reduced-density-matrix equation of motion is not simply the same as the Pauli master equation, because the off-diagonal elements of the density matrix must be considered in order to describe the time evolution of coherence. In the laser, the off-diagonal element $\rho_{n,n+k}$ of the reduced (laser field) density matrix is driven by the interaction $(\partial\rho/\partial t)_{\text{laser}}$ [eq. (73)], but because of the incoherence of the atomic system an element $\rho_{n',n'+k'}$ with a different degree of off-diagonality $(k' \neq k)$, is not coupled to $\rho_{n,n+k}$ (Scully and Lamb 1967). This simplification is, of course, inapplicable to the problem of molecular excitation by a coherent laser field. There are other, more fundamental differences of basic physics between the two problems. In the laser, the electromagnetic field acquires coherence primarily because the atomic driving becomes weaker as the laser field grows stronger, so that the large fluctuations of intensity which are characteristic of the initial (spontaneous-emission) state of the laser field are not amplified. In the molecule, the driving (laser) system is coherent, and since the excitation of the molecule does not significantly deplete the laser field, the interaction between the molecule and the laser does not automatically lead to an amplitude-stabilized motion (if it did, the molecules might not dissociate!). Further, in the molecule the coupling of reduced-density-matrix elements of different degrees of off-diagonality implies that it is absolutely essential to consider off-diagonal elements of the reduced density matrix simply to describe the evolution of state populations, which are given by the diagonal elements. This point has been repeatedly emphasized by Goodman (Goodman and Thiele 1972, Stone et al. 1973, Goodman et al. 1976, Stone and Goodman 1978a, b). The criterion for the inapplicability of population rate equations, derived for a more restricted case by Ackerhalt and Eberly (1976) is: population rate equations employing a laser-stimulated transition rate between molecular levels are inapplicable when the Rabi frequency is larger than the damping rate or the power-broadened width of the levels in question. Similar questions were considered previously by Wilcox and Lamb (1960) and Lamb and Sanders (1960). Finally, the fundamental similarity of the reduced-density-matrix equations of motion in the quantum theory of the laser and the quantum theory of collisionless multiple-photon laser excitation of molecules suggests that quantum-mechanicals equations of motion can be written down for the amplitude of molecular vibrational excitation, which are derived similarly to the quantum Langevin equations for the laser field (Lax 1966, Haken and Weidlich 1967). The fundamental connection between the reduced-density-matrix equation of motion and Langevin equations (Lax 1966) thus permits one to pass from a description using vibrational-state populations and density-matrix elements to a description using the physical amplitude of vibration, plus appropriate noise and damping terms. Although there appears to be little practical advantage in the latter type of description in a quantum-mechanical calculation, it is of interest to note that a detailed conceptual bridge can be built between the density-matrix equation and Langevin equations of motion for physical observables. The latter methods are closely related to calculations which employ classical vibrational coordinates (Lamb 1976, Cotter et al. 1976, Walker and Preston 1977). Specifically, a classical calculation of molecular excitation using only the active mode(s) of the molecule should employ a classical Langevin equation, including both damping and noise terms (Lamb 1976). The calculation of Cotter et al. (1976) included the coherent driving and damping portions of such an equation of motion for a single, nondegenerate vibrational degree of freedom.

A convenient, but highly idealized, approach to the description of the excitation of SF_6 and other molecules is to consider only the energy levels of the infrared-active normal mode [i.e., the eigenvalues of $H^{(3)}$, eq. (89)]. However, in an actual molecule a large number of levels involving the excitation of other modes lie near each level of the active mode (e.g., the ν_3 mode of SF_6). Owing to the anharmonic nature of the electronic-ground-state potential energy as a function of the vibrational normal coordinates, each level $v\nu_3$ will mix with the other nearby levels in a manner which can be calculated by perturbation theory if the anharmonic interaction is weak. If the interaction is strong, then the anharmonic hamiltonian must be diagonalized for the set of levels which are in Fermi resonance (Herzberg 1945). As a result of anharmonic interactions, a large number of energy levels (eigenstates of the full molecular hamiltonian) will have a nonzero component of the states comprising $v\nu_3$, and will therefore be accessible by allowed optical transitions from some of the states comprising $(v-1)\nu_3$. In the time domain, the optical excitation of a real level $v\nu_3$ will lead to a finite probability amplitude in all the nearby coupled levels, and will give rise to oscillations of the probability for being in any one of these levels. After some recurrence time, the level populations may almost repeat (Bixon and Jortner 1968). Only the initial phase of this fully coherent, quantum-mechanical evolution in time is similar to damping, i.e., to an irreversible flow of probability from $v\nu_3$ to the other, coupled levels (Freed 1976). Hence if any quantum-mechanical recurrence time is long compared to the time occupied by an experiment (e.g., a laser pulse length), it may be possible to view the states of the molecule which interact strongly with the laser field (i.e., $v\nu_3$ in SF_6) as isolated, discrete levels which are damped (i.e., undergo an irreversible decay of amplitude) owing to interaction with the reservoir (i.e., the optically inactive modes). In the frequency domain, such damping will cause the energy levels of the optically active (ν_3) mode to spread out, thus facilitating successive resonant absorptions of laser photons. In the treatment adopted here and previously by others (Lax 1964, 1966, Goodman and Thiele 1972, Goodman et al. 1976, Hodgkinson and Briggs 1976) it is possible to show, using the reduced-density-matrix equation of motion, that such damping phenomena are present and to derive equations from which the damping constants may be calculated.

One expects the damping rate of the level $v\nu_3$ to increase as v increases, for at least two reasons. The total molecular density of vibrational states increases almost exponentially with v (Haarhoff 1963); and the number of terms in a power-series expansion of the coupling hamiltonian $H^{(3R)}$ which can couple $v\nu_3$ to nearby states also increases with v. We consider the quasi-continuum (Isenor et al. 1973, Bloembergen 1975, Letokhov and Makarov 1976a, b, Jortner and Mukamel 1976, Bloembergen et al. 1976) to begin when resonant transitions are possible from $(v-1)\nu_3$ to $v\nu_3$ with a strength nearly equal to that of the strongest such transition. This is equivalent to requiring that the width due to damping be much larger than the anharmonic defect. Owing to the steady increase of the damping rate with v, the onset of the quasi-continuum is expected to be gradual, not sudden. Models in which the quasi-continuum is assumed to start at a specific value of v thus involve an idealization and simplification of the actual situation (Letokhov and Makarov 1976b, Larsen and Bloembergen 1976, Jortner and Mukamel 1976). In such models one may expect the uppermost discrete level to have an anomalously high population (Eberly et al. 1977), and one may also expect to see population cycling down the ladder (i.e., towards decreasing v) as well as upwards. The dynamics of a system with a steadily increasing damping as a function of v may, or may not, be similar to the dynamics of a system with a finite number of discrete levels, with the uppermost level coupled to a quasi-continuum. It appears that further theoretical work is needed on the calculation of damping rates as functions of ν_3 excitation using realistic models of the vibrational force field, and on the dynamics of a multilevel system with excitation-dependent damping constants.

There has been considerable and inconclusive discussion in the literature of the question of the distribution of energy among the normal modes of a laser-excited molecule at the time of dissociation (Black et al. 1977, Grant et al. 1977). Although the current experimental evidence for the velocity distribution of the dissociation fragments of SF_6 (Grant et al. 1977) and the number of photons absorbed in excess of those required to meet one of the published values of the thermochemical dissociation energy (Black et al. 1977) are both consistent with complete thermal equipartition of energy among the molecular vibrational modes, it is not clear that it has been proved that the energy is thermalized, i.e., that there is (in classical language) no coherent relationship among the phrases and amplitudes of the various normal modes. It is even less clear (see §3) that a statistical theory applies to the excitation process by which the dissociation energy is reached (Freed 1976). We are not able to resolve these questions at the present time. If an experimental observation of Rabi sidebands could be made in light scattered by molecules undergoing laser excitation, as suggested by Goodman and Thiele (1972), it would establish the importance of coherent excitation. A lack of observation of sharp sidebands would not settle the question, however, since the superposition of sidebands resulting from many Rabi frequencies (due to many different interlevel transition moments) might be expected to smear out any sharp structure in the spectrum of the scattered light. In the calculations summarized in the remainder of this review, we have assumed purely for the sake of convenience that the energy in the modes other than the active mode (which interacts directly with the laser field) is thermalized. In SF_6 we take the 'thermal' modes to be all those other than the ν_3 mode. We regard this at the moment as simply a useful approximation which enables an explicit calculation of the reservoir spectral densities [eqs. (78) and (79)] by the use of conservation of energy. Alternatively, it would be possible to avoid the use of a thermal hypothesis by solving a reduced-density-matrix equation for the reservoir (Lax 1966).

We find that the damping rate of a specific bond or complex of levels such as $\nu\nu_3$ should increase rapidly as the energy of excitation of the remaining modes (the reservoir) increases. This supports a suggestion by Hodgkinson and Briggs (1976) that the excitation of the reservoir initially proceeds slowly, then increasingly rapidly as the energy content of the reservoir builds up. According to their calculations the initial 'slow heating' phase occupies most of the time required to deposit the energy needed for disssociation. As Hodgkinson and Briggs (1976) point out, the final 'explosive' phase proceeds very rapidly, possibly so rapidly that the assumptions made in deriving the reduced-density-matrix equation of motion are invalid.

Radiationless processes in molecules, of which intermode vibrational relaxation is one example, have been discussed using a different formalism by Bixon and Jortner (1968) and Jortner (1970). They consider states ϕ_s (in the Born–Oppenheimer basis) which are radiatively coupled to the ground state (and possibly to one another), and which are also coupled weakly by nonradiative interactions to a multitude of 'background' states. The Born–Oppenheimer states ϕ_s are analogous to 'doorway' states in nuclear theory (Feshbach et al. 1967). We also use the Born–Oppenheimer basis, in which a general basis state is factored into electronic, vibrational and rotational states. In our development, all the states $|s\rangle|r\rangle = |\nu\nu_3\rangle|r\rangle$ (see §4.1) which involve excitation of the ν_3 vibrational ladder, play the role of the radiatively accessible Born–Oppenheimer states ϕ_s in the theories of Bixon and Jortner (1968) and Jortner (1970). Their background states collectively comprise our reservoir. Our description of the molecular state is more general than theirs to the extent that they employ a wave function to describe the state of the whole molecule, while we employ a density matrix ρ (§4.1) of which a wave function is a limiting case (von Neumann 1932). Apart from this rather minor difference, our description of intramolecular vibrational relaxation is physically equivalent to theirs, although we have not made in §4.10 some of the approximations used by Jortner (1970) for the form of the coupling between the optically accessible states and the remaining molecular degrees of freedom. Their formalism includes, in principle, both leakage of excitation out of the optically accessible states (Letokhov and Makarov 1976a, b), and dephasing due to interaction with the background states. Both of these effects are, of course, present in a general form in our treatment. As it happens, Bixon and Jortner (1968) and Jortner (1970) did not attempt to discuss the case in which the reservoir is significantly excited either owing to an initial thermal distribution of energy or owing to leakage from the laser-pumped states, although such a discussion could be given in their framework. There are problems, particularly where collisions are concerned, in which a density-matrix treatment is essential (Goodman and Thiele 1972; Goodman et al. 1973) and in which a wave function treatment of the molecule is inappropriate. For this more general set of problems, not considered in our review or by Bixon and Jortner (1968) and Jortner (1970), the formalism developed here is still applicable.

4.8.2 Damping, population rate equations and frequency shifts

Damping may be included in the Schrödinger equation by adding terms representing exponential decay:

$$i\hbar \frac{dc_n}{dt} = \sum_k H_{nk}(t)c_k - \frac{1}{2}i\hbar \sum_k \gamma_{nk}c_k \tag{101}$$

(Lamb 1964, 1976), where the c_n are the expansion coefficients of the wave function $|\psi\rangle$ in terms of a complete set $|n\rangle$, and the γ_{nk} (the elements of a hermitian damping matrix Γ) represent the irreversible decay of level $|n\rangle$ for $k = n$, or for $k \neq n$ represent an irreversible transfer of population from level $|k\rangle$ to level $|n\rangle$ (due, for example, to collisions). In terms of the density matrix ρ, eq. (101) may be transcribed to read (Lamb 1964, 1976)

$$i\hbar(\partial\rho/\partial t) = [H, \rho] - \tfrac{1}{2}i\hbar(\Gamma\rho + \rho\Gamma), \tag{102}$$

or equivalently

$$i\hbar \frac{\partial\rho_{nn'}}{\partial t} = \sum_{n''} (H_{nn''}\rho_{n''n'} - \rho_{nn''}H_{n''n'})$$

$$- \frac{1}{2}i\hbar \sum_{n''} (\gamma_{nn''}\rho_{n''n'} + \rho_{nn''}\gamma_{n''n'}). \tag{103}$$

The content of eqs. (102) or (103) may be summarized by the equation

$$i\hbar(\partial\rho/\partial t) = H^{(\tau)}\rho - \rho H^{(\tau)\dagger} \tag{104}$$

after the formation of the total hamiltonian

$$H^{(\tau)} = H - \tfrac{1}{2}i\hbar\Gamma \tag{105}$$

from the two hermitian matrices H and Γ. Jortner and Mukamel (1976) have used an effective Hamiltonian (Mower 1966) of the same form as $H^{(\tau)}$ to describe the ν_3 mode of SF_6, in a theoretical treatment of collisionless multiple-photon excitation and dissociation.

We turn now to the role of damping in the equation of motion for the reduced density matrix. If we resolve the reservoir spectral densities [eqs. (78) and (79)] into real and imaginary parts

$$w_i^{\pm}(\omega) = \tfrac{1}{2}\gamma_i(\omega) + i\Delta_i^{\pm}(\omega),\tag{106}$$

then we can cast eq. (74) for $(\partial\rho_{nn'}^{(3)}/\partial t)_{\text{res}}$ into the form

$$
\begin{aligned}
\left(\frac{\partial\rho_{nn'}^{(3)}}{\partial t}\right)_{\text{res}} = &-\sum_{n''}\sum_i \tfrac{1}{2}[|Q_i(0)_{nn''}|^2\gamma_i^+(\omega_{nn''}) + |Q_i^{\dagger}(0)_{nn''}|^2\gamma_i^-(\omega_{n''n}) \\
&+ |Q_i(0)_{n'n''}|^2\gamma_i^+(\omega_{n'n''}) + |Q_i^{\dagger}(0)_{n'n''}|^2\gamma_i^-(\omega_{n''n'})]\rho_{nn'}^{(3)} \\
&+ \delta_{nn'}\sum_{n''}\sum_i [|Q_i(0)_{nn''}|^2\gamma_i^-(\omega_{nn''}) + |Q_i^{\dagger}(0)_{nn''}|^2\gamma_i^+(\omega_{n''n})]\rho_{n''n''}^{(3)} \\
&- i\sum_{n''}\sum_i [|Q_i(0)_{nn''}|^2\Delta_i^+(\omega_{nn''}) - |Q_i^{\dagger}(0)_{nn''}|^2\Delta_i^-(\omega_{n''n}) \\
&+ |Q_i^{\dagger}(0)_{n'n''}|^2\Delta_i^-(\omega_{n''n'}) - |Q_i(0)_{n'n''}|^2\Delta_i^+(\omega_{n'n''})]\rho_{nn'}^{(3)}.\tag{107}
\end{aligned}
$$

Each of the three groups of terms on the right-hand side of eq. (107) has its own physical interpretation, which we shall discuss briefly.

(i) The first group of terms multiplying $\rho_{nn'}^{(3)}$ is of the same form as the terms arising from $\tfrac{1}{2}i\hbar(\Gamma\rho + \rho\Gamma)$ in eq. (103). Therefore for *off*-diagonal elements of the reduced density matrix $\rho_{nn'}^{(3)}$ ($n \neq n'$), the time evolution is of the same form as is described by eq. (101), with the elements of the damping matrix being given by

$$\gamma_{nn'} = \delta_{nn'}\sum_{n''}\sum_i [|Q_i(0)_{nn''}|^2\gamma_i^+(\omega_{nn''}) + |Q_i^{\dagger}(0)_{nn''}|^2\gamma_i^-(\omega_{n''n})].\tag{108}$$

We recall that it is the off-diagonal elements of $\rho^{(3)}$ which describe the quantum-mechanical coherence properties of the ν_3 mode; for a completely incoherent system $\rho_{nn'}^{(3)}$ would vanish for $n \neq n'$ (as in the random phase approximation).

(ii) For the *diagonal* elements of $\rho^{(3)}$, eq. (107) implies

$$
\begin{aligned}
\left(\frac{\partial\rho_{nn}^{(3)}}{\partial t}\right)_{\text{res}} = &-\sum_{n''}\sum_i [|Q_i(0)_{nn''}|^2\{\gamma_i^+(\omega_{nn''}) - \gamma_i^-(\omega_{nn''})\} \\
&+ |Q_i^{\dagger}(0)_{nn''}|^2\{\gamma_i^-(\omega_{n''n}) - \gamma_i^+(\omega_{n''n})\}]\rho_{n''n''}^{(3)}\tag{109}
\end{aligned}
$$

if we temporarily disregard the third group of terms in eq. (107). Eq. (109) is a familiar rate equation for the fraction of the population in level n, ρ_{nn}, in terms of the rate per molecule

$$w(n \to n'') = \sum_i \{|Q_i(0)_{nn''}|^2\gamma_i^+(\omega_{nn''}) + |Q_i^{\dagger}(0)_{nn''}|^2\gamma_i^-(\omega_{n''n})\}\tag{110}$$

at which population is transferred from n to n''. The population rate equations describe the relaxation of the ν_3 mode towards an equilibrium distribution of populations, with rates which depend upon the level of ν_3 excitation. For a one-dimensional *harmonic* oscillator in which only adjacent levels are coupled and all transition rates are proportional to the square of harmonic-oscillator dipole matrix elements, the population rate equations have been solved exactly by Montroll and Shuler (1957).

(iii) The third group of terms in eq. (107) represents a correction to the ν_3 transition frequencies $\omega_{nn''}$ arising from the interaction between the ν_3 mode and the reservoir. To see this, we note that this group of terms contributes to $ih(\partial \rho^{(3)}/\partial t)_{res}$ the quantity $[\Delta H^{(3)}, \rho^{(3)}]$, where the energy-shift operator

$$\Delta H_{nn'}^{(3)} = \delta_{nn'} \hbar \sum_{n''} \sum_i [|Q_i(0)_{nn''}|^2 \Delta_i^+(\omega_{nn''}) - |Q_i^\dagger(0)_{nn''}|^2 \Delta_i^-(\omega_{n''n})] \tag{111}$$

is diagonal in the ν_3 basis. The correction to the transition frequency implied by the third group of terms in eq. (107) is simply equal to

$$\Delta \omega_{nn'} = \hbar^{-1}[\Delta H_{nn}^{(3)} - \Delta H_{n'n'}^{(3)}]. \tag{112}$$

Apparently such reservoir-induced frequency shifts have not previously been discussed in the context of the dynamics of laser-driven molecules. The frequency shift expressed in eqs. (111) and (112) is not the same as that implied in eq. (77). In eq. (77) there is a contribution to $H^{(3)}$ from operators which are diagonal in a basis in which the reservoir density matrix is diagonal. However, as will become apparent when we given an example which shows how we intend $\gamma_i^\pm$ and $\Delta_i^\pm$ to be evaluated, the reservoir operators F_i are all fully off-diagonal, and so are the ν_3 operators Q_i.

We have shown that the reservoir-induced change of the reduced density matrix $\rho^{(3)}$ may be expressed in the form

$$\left(\frac{\partial \rho_{nn'}^{(3)}}{\partial t} \right)_{res} = -\tfrac{1}{2}(\gamma_{nn} + \gamma_{n'n'})(1 - \delta_{nn'})\rho_{nn'}^{(3)}$$

$$- \delta_{nn'} \sum_{n''} [\rho_{nn}^{(3)} w(n \rightarrow n'') - \rho_{n''n''}^{(3)} w(n'' \rightarrow n)]$$

$$- i\Delta\omega_{nn'}\rho_{nn'}^{(3)}, \tag{113}$$

where the new symbols are defined in eqs. (108), (110) and (112), and where each term in eq. (113) has a simple physical meaning: decay of coherence; relaxation to equilibrium populations; or frequency shift. We stress that all of the terms in eq. (113) are, in principle, of the same order of magnitude. It is not, in general, correct to discard one term, such as the frequency shift, while retaining the other terms. Equations of the same form as eq. (113) have been used by Goodman and Thiele (1972), Goodman et al. (1973, 1975, 1976), Stone et al. (1973, 1975, 1976), and Hodgkinson and Briggs (1976) to discuss the excitation of molecules by a laser field.

The first two sets of terms in eq. (113), representing the decay of coherence and relaxation towards an equilibrium distribution of populations, are frequently referred to as the optical Bloch equations (Sargent et al. 1974, p. 91, Ackerhalt and Eberly 1976). This terminology is borrowed from the field of magnetic resonance (Slichter 1963), where the (exponential) decay times for the diagonal and off-diagonal elements of the density matrix are called, respectively, T_1 and T_2 for a two-level system. For a multilevel system it is generally the case that the relaxation of coherence proceeds at a different rate from the relaxation of the populations. Relaxation phenomena in a multilevel system also can, and generally do, lead to a time dependence which cannot be represented by a single exponential.

Several well-known treatments of molecular vibrational excitation using lasers have employed only a subset of the terms in eq. (113). Jortner and Mukamel (1976) used an effective hamiltonian (Mower 1966) of the same form as $H^{(\tau)}$ [eq. (105)]. Thus in the density-matrix equation of motion which results from their hamiltonian, the term $\delta_{nn''}\Sigma_{n''}\rho_{n''n''}w(n'' \to n)$ of eq. (113) is absent, as are all the frequency-shift terms of eq. (113). The missing term $\delta_{nn''}\Sigma_{n''}\rho_{n''n''}w(n'' \to n)$ represents a rate-process transfer of population from other levels (n'') to level n, as a result of interaction with the reservoir. This omission is inconsistent with the conservation of probability for the reduced density matrix $\rho^{(3)}$. From the definition of $\rho^{(3)}$, eq. (70), it is evident that $\mathrm{Tr}_3\, \rho^{(3)}$ is equal to $\mathrm{Tr}\, \rho$, where ρ is the density matrix for the entire molecule (ν_3 mode plus reservoir) and the latter trace runs over all the molecular vibrational states of the electronic ground state. In our discussion we have been concerned with the process of excitation, not with the process of dissociation; hence the molecule is expected to stay in the electronic ground state, and $\mathrm{Tr}\, \rho$ will remain equal to 1. Therefore $\mathrm{Tr}_3\, \rho^{(3)}$ is also equal to 1. From eq. (113) it follows that in the Jortner–Mukamel theory, $\mathrm{Tr}_3\, \rho^{(3)}$ obeys the equation

$$\frac{\partial}{\partial t}(\mathrm{Tr}_3\, \rho^{(3)}) = -\sum_{n,\, n''} w(n \to n'')\rho_{nn}, \tag{114}$$

which predicts that $\mathrm{Tr}_3\, \rho^{(3)}$ will decay to zero in time. Replacement of the term $\delta_{nn'}\Sigma_{n''}\rho_{n''n''}w(n'' \to n)$, which is present in eq. (113) but missing in the corresponding equation for $\rho^{(3)}$ implied by the hamiltonian of Jortner and Mukamel (1976), causes the right-hand side of eq. (114) to vanish, as it should. Even if $\mathrm{Tr}\, \rho$ departs from the value 1 owing to dissociation, it is doubtful whether it is correct to neglect the transfer of population from other levels in the ν_3 ladder into the level n in calculating the de-excitation of the ν_3 mode due to interaction with the reservoir.

Karplus and Schwinger (1948) have derived a solution of equations of the form of eq. (74) and eq. (113) under the assumption that one may neglect those terms in eq. (74) which involve the product of $H^{(3F)}$, the interaction hamiltonian for ν_3 and the laser field, with the difference between the actual reduced density matrix $\rho^{(3)}$ and the equilibrium density matrix (which describes the equilibrium state determined by the population rate equations alone). This amounts to assuming that the actual $\rho^{(3)}$ deviates little from the equilibrium density matrix. This may be true in the presence of 'hard' collisions, but is probably not appropriate for the problem of collisionless laser excitation of molecules. In the latter problem it appears to be possible for the actual $\rho^{(3)}$ to deviate substantially from the equilibrium density matrix (which is not to be confused with the steady-state $\rho^{(3)}$ discussed below). Hougen (1976) has discussed the collisionless laser excitation of SF_6 using the Karplus–Schwinger techniques, supplemented by a Fourier expansion of the density-matrix elements.

4.8.3 The steady-state approximation; 'leakage' to the quasi-continuum

It is possible for a steady state to exist, in which the change of the density matrix due to one cause is balanced by that due to another. In such a dynamical equilibrium, the density matrix for the ν_3 mode would satisfy the equation

$$\partial\rho^{(3)}/\partial t = (\partial\rho^{(3)}/\partial t)_{\text{laser}} + (\partial\rho^{(3)}/\partial t)_{\text{res}} = 0. \tag{115}$$

This would imply that the mean energy of the ν_3 mode,

$$\overline{E^{(3)}} = \text{Tr}\,(H^{(3)}\rho^{(3)}), \tag{116}$$

would be constant in time:

$$\frac{\overline{\partial E^{(3)}}}{\partial t} = \text{Tr}\left(H^{(3)}\frac{\partial\rho^{(3)}}{\partial t}\right) = 0. \tag{117}$$

In writing down eq. (117) we have neglected the rate of change in time of $H^{(3)}$ [eq. (77)] due to a change with time of the average of reservoir operators coupled to ν_3 operators which are diagonal (or have large diagonal matrix elements) in a basis in which $H^{(3)}$ is diagonal. It is correct to neglect $\text{Tr}(\rho^{(3)}\partial H^{(3)}/\partial t)$ as long as it is much smaller than $\text{Tr}(H^{(3)}\partial\rho^{(3)}/\partial t)$. We expect this to be the case as long as the growth of excitation of the reservoir is sufficiently slow that the coupling between the ν_3 mode and the reservoir is not a rapidly increasing function of time. This will be true, for example, in the 'slow heating' phase of excitation described by Hodgkinson and Briggs (1976), the existence of which is plausible in view of the forms derived below for the reservoir spectral densities and for $\gamma_i^{\pm}, \Delta_i^{\pm}$.

If the mean energy in the ν_3 mode is constant in time, as implied by eq. (117), then it is reasonable to regard the two cancelling terms

$$\text{Tr}[H^{(3)}(\partial\rho^{(3)}/\partial t)_{\text{laser}}] + \text{Tr}[H^{(3)}(\partial\rho^{(3)}/\partial t)_{\text{res}}] = 0 \tag{118}$$

as representing the rate of change of $\overline{E^{(3)}}$ due to interaction with the laser and with the reservoir respectively. In this case the energy acquired by the reservoir in the dynamical steady state implied by eq. (115) is

$$(\overline{\partial E^{(\text{res})}}/\partial t)_{\text{SS}} = -\text{Tr}[H^{(3)}(\partial\rho^{(3)}/\partial t)_{\text{res}}]$$

$$= \sum_{n,\,n''} E_n^{(3)}[\rho_{nn}^{(3)}w(n \to n'') - \rho_{n''n''}^{(3)}w(n'' \to n)]. \tag{119}$$

This result has been tested numerically for a nondegenerate oscillator by Hodgkinson and Briggs (1976), who find that throughout most of the time prior to dissociation eq. (119) is a reasonably good approximation to the rate of growth of the reservoir energy, thus justifying their identification of the 'slow heating' phase of excitation. We are thus justified in regarding eq. (119) as expressing the 'leakage' of energy to the quasi-continuum (Letokhov and Makarov 1976a).

We note that the steady-state approximation of eq. (115) is not equivalent to a rate-equation approximation [using only the second term in eq. (113)], since eq.

(115) does not require that the off-diagonal elements of the density matrix vanish. All that is implied in eq. (115) is that the 'leakage' to the quasi-continuum is balanced by excitation due to the laser. It should also be remembered that if none of the reservoir states were unbound (i.e., able to decay in a truly irreversible manner) then the transfer of energy from the ν_3 mode to the reservoir, which is a coherent quantum-mechanical process, might at some later time lead to the reappearance of energy in the ν_3 mode (Bixon and Jortner 1968). Only over a restricted interval of time does the interaction between the ν_3 mode and the reservoir lead to something resembling decay or 'leakage' (Freed 1976). If dissociation occurs within this time interval, then it is legitimate to speak of 'leakage' without specifying all the restrictions on the applicability of this concept. These conditions imply a limit on the laser pulse length t_p:

$$t_p \ll \rho(\nu)/c, \tag{120}$$

where $\rho(\nu)$ is the total density of states (Bixon and Jortner 1968).

4.9 Master equations and approximations to master equations

A set of equations for $(\partial \rho^{(3)}_{nn}/\partial t)_{res}$ which include only the population rate equations [the second set of terms on the right-hand side of eq. (113)] are known as *master equations*. Master equations have been studied for many years in connection with vibrational excitation by collisions (Montroll and Shuler 1957) and by lasers in the collision-dominated regime (Artamonova et al. 1970, Afanas'ev et al. 1971, 1973, Basov et al. 1973, Karlov et al. 1973), and in connection with intramolecular vibrational energy transfer (Gelbart et al. 1972, Heller and Rice 1974, Kay 1974). The master equation for a nondegenerate harmonic oscillator interacting with a thermal reservoir can be solved exactly (Montroll and Shuler 1957); this fact undoubtedly has much to do with the popularity of this model. Unfortunately such one-dimensional models appear to have little applicability to the degenerate vibrational modes with which one is faced in practice. Approximate master equations which employ a continuum approximation in the excitation energy of the ν_3 mode or in the total ν_3 vibrational quantum number v may conceivably be more useful than the exact one-dimensional harmonic-oscillator methods, and for that reason they will be briefly summarized here. It is by now reasonably clear that the dynamics of excitation of the low-lying vibrational levels of SF_6 must be treated by a coherent equation of motion, and we have seen above that it is possible to describe the dynamics of absorption of laser energy in higher-lying states by equations which are the same in form as those required for the low-lying states. Under these circumstances one might ask why one need ever consider the use of a master equation, in which the absorption of laser energy must also be described by population rate equations instead of eq. (73). One possible reason for using a master equation is computational convenience, either in numerical calculations or in analytical studies designed to elicit qualitative information.

Another possible reason is that the laser field may be incoherent. An incoherent laser field which is not too strong for the validity of the same second-order perturbation theory used to obtain eq. (81) may be treated using a reduced-density-matrix equation of motion [see also Eberly (1976)]. For the diagonal elements of $\rho^{(3)}$ ('populations') this will give results of the same form shown in eq. (113): population rate equations. Finally, even if the driving laser field is coherent, Ackerhalt and Eberly (1976) have argued that appropriate population rate equations are valid if the damping rates or power-broadened level widths are large compared to the rate of laser excitation. A rigorous justification for the use of a master equation (as in §3) rather than the more general methods discussed in this section is difficult at present since definitive studies of the applicability of a master equation to coherently driven multilevel systems have just begun to appear (Ackerhalt and Eberly 1976, Stone and Goodman 1978b). We have based our use of a master equation (i.e., population rate equations) to describe optical excitation of SF_6 in §3 on the expectation that the use of an appropriate master equation will lead to approximately the correct time dependence of the vibrational populations and the absorbed energy (for a careful discussion see Stone and Goodman 1978b). Despite the fact that the precise limits of validity of the master equation are now known for laser-pumped multilevel systems, we hope that it will be tutorially useful for this review to contain both master-equation and more general approaches.

We conclude this discussion of the master equation with an outline of the techniques used to reduce $(\partial \rho_{nn'}^{(3)}/\partial t)$ laser [eq. (73)] to a master-equation form involving only diagonal elements of the density matrix $\rho_{nn}^{(3)}$, and the approximations implicit in these techniques. Since the ν_3–laser interaction $H_{nn'}^{(3f)}$ is fully off-diagonal, $\partial \rho_{nn}^{(3)}/\partial t$ for a diagonal element $\rho_{nn}^{(3)}$ will necessarily involve the off-diagonal elements $\rho_{nn'}^{(3)}$, with $n \neq n'$. For a two-level system, Lamb (1964, §16) has shown that formally solving the equation of motion for $\rho_{nn'}^{(3)}$ (with $n \neq n'$) in terms of the diagonal elements $\rho_{mm}^{(3)}$, and substituting this formal solution for $\rho_{nn'}^{(3)}(t)$ into the equation of motion for $\rho_{mm}^{(3)}$, leads to population rate equations for the diagonal elements $\rho_{mm}^{(3)}$ *provided* that rapid variations of $\rho_{mm}^{(3)}$ in time can be neglected. This derivation breaks down when, for example, coherent pumping is so strong that the populations oscillate at the Rabi frequency, as discussed in §4.4 (Ackerhalt and Eberly 1976). A careful discussion of the approximations inherent in eliminating the off-diagonal density matrix elements as sketched above has been given by Stone and Goodman (1978a). The population rate equations (master equation) we have employed in §3 may be regarded as a special, approximate case of the reduced-density-matrix equation of motion discussed in this section, subject to the considerations detailed by Stone and Goodman (1978a). A precise specification of the accuracy of the methods of §3 as compared with the methods of this section is a research problem at the present time, and is thus beyond the scope of this review.

4.10 Evaluation of reservoir spectral densities and vibrational couplings

In order to complete the detailed physical discussion outlined above, it is necessary to assume a specific form for Q_i and F_i in eq. (76) and evaluate the corresponding matrix elements and reservoir spectral densities. This will verify the statements already made concerning the growth of the damping matrix elements as functions of ν_3-mode excitation and of reservoir excitation, and will provide specific expressions for both the damping constants and the shifts of the ν_3 energy levels, both of which scale similarly with respect to the excitation of the ν_3 mode and of the reservoir.

As was noted after eqs. (77) and (78), the quantities $w_i^\pm(\omega)$ are the spectral densities of a particular portion (indicated by the index i) of the interaction energy between the ν_3 mode and the reservoir. In general the reservoir spectral densities are complex-valued functions of ω; the real part of $w_i^\pm(\omega)$ has the physical significance of a damping rate, while the imaginary part is a frequency shift [eqs. (107) to (114)]. The time integration in eqs. (77) and (78) may be performed with the help of the identity (Heitler 1954)

$$\int_0^\infty e^{\pm i\Omega\tau}\, d\tau = \pi\delta(\Omega) \pm i P \frac{1}{\Omega}, \tag{121}$$

where $\delta(\Omega)$ is the Dirac delta function and P denotes the principal value (when Ω is a variable of integration). The real and imaginary parts of $w_i^+(\omega)$ [eq. (106)] may then be expressed as

$$\gamma_i^+(\omega) = 2\pi \sum_{a,b} p_a |\langle a|F_i|b\rangle|^2 \delta(\omega + \omega_{ab}), \tag{122}$$

$$\Delta_i^+(\omega) = P \sum_{a,b} p_a |\langle a|F_i|b\rangle|^2 \frac{1}{\omega + \omega_{ab}}, \tag{123}$$

and similarly for $w_i^-(\omega)$. Although these equations are well adapted to problems in which one can replace the sums over $|a\rangle$ and $|b\rangle$ by integrals of continuous functions over ω_{ab} (Louisell 1973), the delta function and principal value have no well-defined meaning in a discrete sum such as occurs in eqs. (122) and (123). In fact, the upper limit of the integrals in the definition of $w_i^\pm(\omega)$ should be the 'coarse-graining' time interval Δt, which is employed in calculating a finite-difference approximation to $\partial\rho^{(3)}/\partial t$ [eq. (71)] that eventually results in eqs. (72) to (74) (Louisell 1973). In that case eq. (131) should be replaced by

$$\int_0^{\Delta t} e^{\pm i\Omega\tau}\, d\tau = \frac{\sin(\Omega\Delta t)}{\Omega} \pm i \frac{1 - \cos(\Omega\Delta t)}{\Omega} \tag{124}$$

so that γ_i^+ and Δ_i^+ become (Hodgkinson and Briggs 1976)

$$\gamma_i^+(\omega) = 2 \sum_{a,b} p_a |\langle a|F_i|b\rangle|^2 \frac{\sin((\omega + \omega_{ab})\Delta t)}{\omega + \omega_{ab}}, \tag{125}$$

$$\Delta_i^+(\omega) = \sum_{a,b} p_a |\langle a|F_i|b\rangle|^2 \frac{[1 - \cos((\omega + \omega_{ab})\Delta t)]}{\omega + \omega_{ab}}. \tag{126}$$

Although these equations have the apparently unattractive feature of depending on the rather arbitrary time interval Δt, in reality the dependence upon Δt is weak provided the bandwidth of the functions $|w_i^\pm(\omega)|$ is large compared to $(\Delta t)^{-1}$ (Louisell 1973). This slight disadvantage is outweighed by the fact that eqs. (125) and (126) are well suited for numerical evaluation. As we have already noted, the value of Δt is restricted not only by the reservoir bandwidth, but also by the requirement that Δt be small in comparison with the time in which the elements of $\rho^{(3)}$ are significantly damped (Louisell 1973).

For the sake of obtaining analytic expressions for the reservoir matrix elements appearing in $w_i^\pm(\omega)$, we shall use special forms for Q_i and F_i, expressed in terms of creation and annihilation operators for the ν_3 mode ($a_i^\dagger, a_i$; $i = 1, 2, 3$) and for the reservoir modes ($b_r^\dagger, b_r$; $r = 1, \ldots, 12$). We shall replace the single index i by the triplet (i, j_i, k_i), taking Q_i to be replaced by

$$Q_{i,j_i,k_i} = (a_i^\dagger)^{j_i}(a_i)^{k_i} \quad (i = 1, 2, 3) \tag{127}$$

and F_i to be replaced by

$$F_{i,j_i,k_i} = \sum_{\{m\},\{p\}} \kappa(i, j_i, k_i; \{m\}, \{p\}) \prod_r (b_r^\dagger)^{m_r}(b_r)^{p_r}. \tag{128}$$

We note that any expansion of Q_i and F_i in power series in normal coordinates may be re-expressed as an expansion in these normally ordered operators (Louisell 1973). These expressions for the operators which enter $H^{(3R)}$ are of a standard multiphonon form (Van Hove et al. 1961, Lax 1964, Englman and Jortner 1970). Essentially eqs. (127) and (128) display the operator Q_iF_i expanded in terms of its matrix elements between harmonic basis functions for the ν_3 and reservoir modes. It may be shown that with the choice for $H^{(3)}$ expressed in eq. (80), both F_i and Q_i are fully off-diagonal (Cantrell 1977b). We shall make the choices

$$j_i > k_i, \quad m_r < p_r, \tag{129}$$

in eqs. (127) to (129). If we specify the states of the ν_3 mode by the harmonic basis functions

$$|\{n\}\rangle = \psi_{n_1 n_2 n_3} = \prod_{i=1}^{3} \frac{(a_i^\dagger)^{n_i}}{\sqrt{n_i!}} |0, 0, 0\rangle, \tag{130}$$

then the matrix elements of the ν_3 operator Q_{i,j_i,k_i} are

$$\langle\{n\}|[Q_{i,j_i,k_i}|\{n'\}\rangle] = \delta_{n_i', n_i + k_i - j_i} \left[\frac{(n_i + k_i - j_i)!(n_i)!}{\{(n_i - j_i)!\}^2} \right]^{1/2}. \tag{131}$$

If we also use harmonic-oscillator basis functions for the reservoir, denoted $|\{v\}\rangle$, then the matrix elements of the reservoir operator F_{i,j_i,k_i} are

$$\langle\{v\}|[F_{i,j_i,k_i}|\{v'\}\rangle] = \sum_{\{m\},\{p\}} \kappa(i, j_i, k_i; \{m\}, \{p\}) \prod_r \left[\frac{(v_r + p_r - m_r)!(v_r)!}{\{(v_r - m_r)!\}^2} \right]^{1/2} \delta_{v_r', v_r + p_r - m_r} \tag{132}$$

The reservoir transition frequency in eq. (77) is

$$\omega_{ab} = -\sum_r \omega_r(p_r - m_r). \tag{133}$$

The matrix elements of $Q^\dagger_{i,j_i,k_i}$ and $F^\dagger_{i,j_i,k_i}$ may be evaluated similarly.

The explicit expression for the matrix element $(Q_{i,j_i,k_i})_{\{n\},\{n'\}}$ in eq. (131) shows that the damping and frequency-shift constants $\gamma_{\{n\},\{n'\}}$ and $\Delta\omega_{\{n\},\{n'\}}$ [eqs. (108) and (113)] depend strongly upon the level of ν_3 excitation, which is designated by the integers n_i. The operator Q_{i,j_i,k_i} creates j_i quanta and destroys k_i quanta in the ith ν_3 normal mode ($i = 1, 2, 3$); as is evident from eq. (131), when both n_i and n'_i are very large compared with k_i or j_i, the matrix element $|(Q_{i,j_i,k_i})_{\{n\},\{n'\}}|^2$ is approximately proportional to $(n_i)^{j_i+k_i}$; i.e., proportional to n_i raised to a power equal to the total number of vibrational quanta which are created or destroyed.

As we have mentioned previously, treating the reservoir density matrix $\rho^{(R)}$ as the density matrix for a system in thermal equilibrium is the first in a hierarchy of approximations to $\rho^{(R)}$ (Lax 1964). We shall treat the reservoir modes as harmonic oscillators, so that the reservoir density matrix $\rho^{(R)}$ factors into a product of thermal harmonic-oscillator density matrices (Messiah 1960, chapter 12) referring to single reservoir modes. This approximation of factorizability of $\rho^{(R)}$ has also been discussed in detail by Lax (1964). We anticipate neglecting the anharmonic nature of the reservoir to have mild consequences by comparison with our assumption of thermal equilibrium for $\rho^{(R)}$. In fact it is unnecessary in our calculation to assume that all the reservoir normal modes have the same temperature, and we shall not do so.

The thermal average over v_r of the square of any one of the quantities in square brackets in eq. (132) may be shown by a generating-function technique to be (Cantrell 1977b)

$$((v_r + p_r - m_r)!(v_r)!/\{(v_r - m_r)!\}^2)_r = m_r!p_r!\langle v_r\rangle_R^{m_r} \sum_{q_r=0}^{p_r} \binom{p_r}{q_r}\binom{m_r + p_r - q_r}{m_r}\langle v_r\rangle_R^{p_r-q_r}, \tag{134}$$

where the mean number of quanta in the reservoir mode r (with temperature T_r) is

$$\langle v_r\rangle_R = 1/(e^{\hbar\omega_r/kT_r} - 1). \tag{135}$$

As one might have anticipated, this quantity also contains numerical factors which grow rapidly with the number of reservoir quanta created (m_r) or destroyed (p_r). The dependence upon $\langle v_r\rangle_R$ is also important. If the thermal energy kT_r of the mode is low compared with the mode's quantum energy $\hbar\omega_r$, then $\langle v_r\rangle_R$ is $\ll 1$, and only the term $q_r = p_r$ in eq. (134) is significant. In that case the thermal average in eq. (134) reduces to

$$((v_r + p_r - m_r)!(v_r)!/\{(v_r - m_r)!\}^2)_r \cong m_r!p_r!\langle v_r\rangle_R^{m_r}\{1 + (m_r + 1)p_r\langle v_r\rangle_R\} \tag{136}$$

where the second term has been included in case $m_r = 0$. When $m_r = 0$, the quantities $m_r!$ and $\langle v_r\rangle_R^{m_r}$ are to be set equal to 1, even when $\langle v_r\rangle_R = 0$.

We conclude our derivation of the thermal average by substituting our results for the thermal average of the matrix elements of F_i [eq. (134)] and the matrix elements of Q_i [eq. (131)] into eqs. (125) and (108) to obtain the decay constant $\gamma_{\{n\}}^+$ for the emission of reservoir phonons from the ν_3 state in eq. (130):

$$
\begin{aligned}
\gamma_{\{n\}}^+ &= \sum_{\{n''\}} \sum_{i,j_i,k_i} \gamma_{i,j_i,k_i}^+(\omega_3(j_i - k_i)) |Q_{i,j_i,k_i}(0)_{\{n\},\{n''\}}|^2 \\
&= 2 \sum_{i,j_i,k_i} \sum_{\{m\}} \sum_{\{p\}} |\kappa(i, j_i, k_i; \{m\}, \{p\})|^2 \\
&\quad \times \frac{\sin\left[\omega_3(j_i - k_i) - \sum_r r(p_r - m_r)\right]\Delta t}{\omega_3(j_i - k_i) - \sum_r \omega_r(p_r - m_r)} \\
&\quad \times \frac{(n_i + k_i - j_i)! n_i!}{\{(n_i - j_i)!\}^2} \prod_r m_r! p_r! \langle v_r \rangle_R^{m_r} \\
&\quad \times \sum_{q_r=0}^{p_r} \binom{p_r}{q_r}\binom{m_r + p_r - q_r}{m_r}\langle v_r \rangle_R^{p_r - q_r}.
\end{aligned}
\tag{137}
$$

The above formalism appears to be complex, but its physical meaning is actually simple. First, let us discuss an aspect of the physical significance of the reservoir spectral density, $w_i^+(\omega)$. From the definition in eq. (77), it is clear from the general quantum-mechanical injunction to average over initial states and sum over final states (when calculating transition rates) that $|a\rangle$ is to be considered the initial state of the reservoir, and $|b\rangle$ the final state. The operator which works on the initial state $|a\rangle$ in $w_i^+(\omega)$ is clearly $F_i^\dagger$. According to the conventions in eq. (129), $F_i^\dagger$ causes a net increase in the number of reservoir vibrational quanta (phonons) in state $|b\rangle$, as compared with state $|a\rangle$. The physical interpretation of $w_i^+(\omega)$ [with the conventions of eq. (129)] is that it is the *reservoir spectral density for the emission of reservoir phonons*. The superscript + is intended to serve as a mnemonic for this physical interpretation. According to this interpretation, we should expect that $w_i^+(\omega)$ will be nonzero (for $\omega > 0$) even when the mode temperature is zero (so that $\langle v_r \rangle_R = 0$), because phonon emission can either be stimulated (proportional to a power of $\langle v_r \rangle_R$ when $m_r = 0$) or spontaneous. Inspection of eq. (134) shows that one term, representing the spontaneous emission of reservoir phonons, does survive when $m_r = 0$ (i.e. when F_i is purely a lowering operator, $(b_r)^{p_r}$) and when $T_r = 0$. At zero reservoir temperature (all $T_r = 0$) the terms in eq. (134) in which all m_r vanish give the contribution to the damping rate $\gamma_{\{n\}}^+$ of the anharmonic mixing of $|\{n\}\rangle$ with other states. Inspection of eq. (134) also reveals that the only operators F_i which give rise to nonzero averages [eq. (134)] when the mode temperature is zero are precisely those which have $m_r = 0$, i.e., those which contribute only reservoir phonon emission processes to $w_i^+(\omega)$. This is a very substantial limitation on the number of terms in the interaction energy $H^{(3R)}$ between the ν_3 mode and the reservoir which can actually contribute to the damping and shift of ν_3 energy levels when the reservoir temperature is low. It is tempting to speculate that this effect may be responsible for delaying the onset of the quasi-continuum until relatively high ν_3 excitation is reached when the reservoir is initially cold. As the reservoir heats up, the increase of the thermal occupation numbers per reservoir mode, $\langle v_r \rangle_R$, will cause the value of $w_i^{\mp}(\omega)$ to increase precipitously, possibly until states with only a few ν_3 quanta lie in the quasi-continuum. This represents an extension and quantification of the ideas of Hodgkinson and Briggs (1976) concerning the initial 'slow heating' and final 'explosive' phases of molecular excitation.

Another way to express these ideas is to consider the validity of attempting to determine when the quasi-continuum begins by calculating the total density of states near a given ν_3 level $|\{n\}\rangle$ and using a single estimated interaction strength between $|\{n\}\rangle$ and the other nearby molecular levels. Some of the states near $|\{n\}\rangle$ will have a different parity than $|\{n\}\rangle$, and hence cannot be coupled to $|\{n\}\rangle$ in an octahedral molecule. The remainder of the molecular states near $|\{n\}\rangle$ can each be coupled to $|\{n\}\rangle$ by an interaction term which destroys the ν_3 quanta which are not present in the non-ν_3 state, and creates the quanta of other modes which are present in the state in question. Thus the density of relevant states does not change as the molecule is heated. However, as the reservoir temperature rises, more and more interaction terms become significant, and the magnitude of each term increases as a high power of $\langle v_r \rangle_R$. Therefore the effective interaction strength, and with it the damping and level shift of $|\{n\}\rangle$, rise extremely rapidly as the molecule heats up. This means that the very approximate $|\{n_q\}\rangle$ at which the quasi-continuum begins is a function of the temperature of the reservoir. When the reservoir is cold, $|\{n_q\}\rangle$ will be a rather high-lying state. When the molecule is well heated, $|\{n_q\}\rangle$ may lie at a very low excitation energy. This behavior is an elementary consequence of the fact that the 'population' $\rho^{(3)}_{\{n_q\},\{n_q\}}$ is the sum of the actual populations of all the molecular vibrational basis states which contain $|\{n_q\}\rangle$, as in eq. (53b). When the molecule is not highly excited, $|\{n_q\}\rangle$ will lie at a high level of ν_3 excitation. When the molecule is highly excited, most absorptions of laser quanta begin at states with many vibrational quanta, only some of which will be ν_3 quanta. Further discussions of these points have been given by Cantrell et al. (1977) and Cantrell (1977b).

Finally, we note that the use of a single ν_3 state in eq. (131) is a matter of pedagogical convenience, and does not represent a fundamental limitation of our version of the individual terms of eqs. (108) and (110) in order to clarify the underlying physics. Clearly, computations should take into account the full variety of terms encountered in eqs. (108) and (110).

4.11 Computational studies of the reduced-density-matrix equation of motion

Reasonably extensive numerical studies of the reduced-density-matrix equation of motion [eqs. (72), (73) and (113)] have been conducted by Goodman, Stone et al. (Goodman et al. 1973, 1976, Stone et al. 1973, 1975, 1976) and by Hodgkinson and Briggs (1976). In this review we discuss the computational methods used by these authors, and refer the reader to the original articles for an exposition of the numerical results.

The reduced-density-matrix equation of motion [eqs. (72), (73) and (113)] is a set of coupled linear first-order differential equations in the matrix elements $\rho_{nn'}^{(3)}$ and, as such, can be solved numerically by step-by-step integration. [Analytical results have been presented for a two-level system by Scully and Lamb (1967) and Goodman and Thiele (1972). Analytical results for a three-level system are given by Brewer and Hahn (1975).] For systems with a large number N of levels or for long times, a direct numerical integration to obtain the N^2 functions $\rho_{nn'}^{(3)}(t)$ (where $n = 1, \ldots, N$) will be prohibitively expensive. An alternative method of numerical solution depends on an eigenvector expansion of the reduced density matrix $\rho^{(3)}$ (Scully and Lamb 1967, Goodman and Thiele 1972, Stone et al. 1973). We shall now motivate the formalism of Goodman and Thiele (1972) and Stone et al. (1973) by showing how the reduced-density-matrix formalism includes as a special case the coherent driving discussed in §4.6. In the absence of ν_3–reservoir coupling (i.e., when $(\partial \rho_{nn'}^{(3)}/\partial t)_{res} = 0$) and when the ν_3 density matrix represents a pure case (von Neumann 1932), we have in the Schrödinger picture

$$\rho^{(3)}(t) = |\psi(t)\rangle\langle\psi(t)|, \tag{138}$$

where

$$|\psi(t)\rangle = \sum_n c_n|n\rangle, \tag{139}$$

and n denotes the quantum numbers v, α (when a purely vibrational problem is considered). The introduction of dressed states $\tilde{c}_n$:

$$c_n = \tilde{c}_n \, e^{-in\omega t}, \tag{140}$$

when the ν_3 mode is driven by the semiclassical laser field $E(t)$ [eq. (87)], and when the ν_3–laser interaction is

$$H_{nn'}^{(3F)} = \mu_{nn'}E(t) \tag{141}$$

(with nonvanishing dipole matrix elements $\mu_{nn'}$ when $n' = n \pm 1$) leads to the time-independent Schrödinger equation (Letokhov and Makarov 1976b, Bialynicka-Birula et al. 1977)

$$i\hbar(\partial\tilde{c}_n/\partial t) = (E_n - n\hbar\omega)\tilde{c}_n + \sum_{n'} \tfrac{1}{2}\mu_{nn'}E_0\tilde{c}_{n'} \tag{142}$$

in the rotating-wave approximation.

The density matrix (138) therefore has the matrix elements

$$\rho_{nn'}^{(3)}(t) = c_n(t)c_{n'}(t)^* \tag{143}$$

$$= e^{i(n'-n)\omega t}\tilde{c}_n(t)\tilde{c}_{n'}(t)^*. \tag{144}$$

When the $\tilde{c}_n$ are chosen to be the components of an eigenstate (dressed state) of the time-independent effective hamiltonian evident in eq. (142), then the product $\tilde{c}_n(t)\tilde{c}_{n'}(t)^*$ is independent of time.

Stone et al. (1973) attacked the problem of re-stating the reduced-density-matrix equation of motion in a computationally more convenient form by expressing the elements of $\rho^{(3)}$ in the Schrödinger picture as follows:

$$\rho_{nn'}^{(3)} = C_{nn'} \, e^{i(n'-n)\omega t} \, e^{-\lambda t}, \tag{145}$$

where λ is an eigenvalue and the $C_{nn'}$ are the components of a time-independent eigenvector of dimension N^2. Comparison of eq. (145) with eq. (144) shows that for a coherently driven system, $\lambda = 0$ and $C_{nn'} = \bar{c}_n(0)\bar{c}_{n'}(0)^*$. Substitution of eqs. (145) and (141) into (72), (73) and (113) leads to the following eigenvalue problem after application of the rotating-wave approximation:

$$[i(n'-n)\omega - \lambda]C_{nn'} = -\frac{i}{\hbar}(E_{n'} - E_n)C_{nn'} - \frac{iE_0}{2\hbar}(\mu_{n,n+1}C_{n+1,n'} + \mu_{n,n-1}C_{n-1,n'}$$

$$- \mu_{n'+1,n'}C_{n,n'+1} - \mu_{n'-1,n'}C_{n,n'-1}) - \tfrac{1}{2}(\gamma_{nn} + \gamma_{n'n'})(1 - \delta_{nn'})C_{nn'}$$

$$- \delta_{nn'} \sum_{n''} [C_{nn}w(n \to n'') - C_{n''n''}w(n'' \to n)] - i\Delta\omega C_{nn'}. \tag{146}$$

In deriving eq. (146) we have used the fact that eq. (72) applies to a density matrix $\rho^{(3)}$ in an interaction picture (Lax 1964), while eq. (145) applies to a density matrix in the Schrödinger picture; hence the additional term $(i/\hbar)(E_{n'} - E_n)C_{nn'}$ on the right-hand side of eq. (146). In their original development, Stone et al. (1973) assumed equally spaced energy levels; we have removed that restriction in eq. (146).

To solve eq. (146) in general it is necessary to solve an $N^2 \times N^2$ matrix eigenvalue problem for the (possibly complex) eigenvalues λ_l and the corresponding eigenvectors $C_{nn'}^{(l)}$ where $l = 1, \ldots, N^2$. If this has been accomplished, then a general solution of the reduced-density-matrix equation of motion may be reconstructed by superposition:

$$\rho_{nn'}^{(3)}(t) = \sum_l a_l C_{nn'}^{(l)} \exp \lambda_l t \exp i(n'-n)\omega t, \tag{147}$$

where the coefficients a_l are to be determined by the initial conditions. The time development may therefore be calculated for all times for which the laser field strength and phase remain constant [as implied by eq. (87)] by solving a single eigenvalue problem, as in the solution of the Schrödinger equation discussed in §4.6. If we suppose for the moment that the eigenvalues λ_l are known, then it is possible to calculate the components $C_{nn'}^{(l)}$ of the eigenvectors by a straightforward iterative technique described by Stone et al. (1973), the largest step of which involves the inversion of an $(N-1) \times (N-1)$ matrix. The central numerical problem of solving the reduced-density-matrix equation of motion is thus to find the complex eigenvalues of the $N^2 \times N^2$ matrix of coefficients of $C_{nn'}$ in eq. (146). Procedures for doing this without brute-force diagonalization of the $N^2 \times N^2$ matrix are discussed by Stone and Goodman (1978a).

Acknowledgement

The authors appreciate the efforts of our colleagues who have critically read the manuscript prior to publication, particularly Dr. R.V. Ambartzumian, Prof. S.H. Bauer, Dr. H.S. Bennett, Prof. S.W. Benson, Dr. J.H. Birely, Prof. N. Bloembergen, Prof. J.H. Eberly, Prof. M.F. Goodman, Prof. W.A. Guillory, Prof. J. Jortner, Dr. S.R. Leone, Prof. V.S. Letokhov, Dr. L.A. Levin, Prof. A.N. Oraevsky, Prof. J.B. Orr, Dr. M.C. Richardson, Prof. J.W. Robinson, Prof. J.I. Steinfeld, Dr. J. Stone, Prof. J.M. Thorne, Prof. K.H. Welge, Dr. V.N. Sazonov, Prof. E. Yablonovitch, Prof. A. Yogev and Prof. R.N. Zitter.

References

Ackerhalt, J.R. and J.H. Eberly (1976), Phys. Rev. A **14**, 1705.

Ackerhalt, J.R. and H.W. Galbraith (1978), J. Chem. Phys. **69**, 1200.

Afanas'ev, Yu. V., E.M. Belenov, E.P. Markin and I.A. Poluektov (1971), JETP Letters **13**, 331.

Afanas'ev, Yu. V., E.M. Belenov and I.A. Poluektov (1973), Kvantovoya Elektron. **2**, 46.

Akinfiev, N.N., N.G. Basov, V.N. Galochkin, S.I. Zavorotnyi, E.P. Markin, A.N. Oraevsky and A.V. Pankratov (1974), JEPT Letters **19**, 383.

Akulin, V.M., S.S. Alimpiev, N.V. Karlov and B.G. Sartakov (1977), Zh. Eksp. Teor. Fiz. **72**, 88.

Aldridge, J.P., R.F. Holland, H. Flicker, K.W. Nil and T.C. Harman (1975), J. Molec. Spectrosc. **54**, 328.

Aldridge, J.P., J.H. Birely, C.D. Cantrell and D.C. Cartwright (1976), Physics of Quantum Electronics (S.F. Jacobs, M. Sargent, III, M.O. Scully and C.T. Walker, eds.; Addison-Wesley; New York), Vol. IV, p. 57.

Ambartzumian, R.V. (1976), Tunable Lasers and Applications, Proceedings of the Loen Conference, Norway, p. 150.

Ambartzumian, R.V., N.V. Chekalin, V.S. Doljikov, V.S. Letokhov and E.A. Ryabov (1974a), Chem. Phys. Letters **25**, 515.

Ambartzumian, R.V., V.S. Letokhov, E.A. Ryabov and N.V. Chekalin (1974b), JEPT Letters **20**, 273.

Ambartzumian, R.V., N.V. Chekalin, V.S. Letokhov and E.A. Ryabov (1975a), Chem. Phys. Letters **36**, 301.

Ambartzumian, R.V., V.S. Dolzhikov, V.S. Letokhov, E.A. Ryabov and N.V. Chekalin (1975b), Sov. Phys. – JETP **42**, 36.

Ambartzumian, R.V., Yu. A. Gorokhov, V.S. Letokhov and G.N. Makarov (1975c), JEPT Letters **21**, 171.

Ambartzumian, R.V., Yu. A. Gorokhov, V.S. Letokhov and G.N. Makarov (1975d), JEPT Letters **22**, 43.

Ambartzumian, R.V., Yu. A. Gorokhov, V.S. Letokhov and G.N. Makarov (1975e), Sov. Phys. – JETP **42**, 993.

Ambartzumian, R.V., Yu. A. Gorokhov, V.S. Letokhov, G.N. Makarov, E.A. Royabov and N.V. Chekalin (1975f), Sov. J. Quantum Electronics **5**, 1196.

Ambartzumian, R.V., Yu. A. Gorokhov, V.S. Letokhov and A.A. Puretskii (1975g), JEPT Letters **22**, 177.

Ambartzumian, R.V., N.V. Chekalin, V.S. Doljikov, V.S. Letokhov and V.N. Lokhman (1976a), Opt. Commun. **18**, 400.

Ambartzumian, R.V., N.P. Furzikov, Yu. A. Gorokhov, V.S. Letokhov, G.N. Makarov and A.A. Puretzky (1976b), Opt. Commun. **18**, 517.

Ambartzumian, R.V., Yu. A. Gorokhov, V.S. Letokhov, G.N. Makarov and A.A. Puretskii (1976c), JEPT Letters **23**, 22.

Ambartzumian, R.V., Yu. A. Gorokhov, V.S Letokhov, G.N. Makarov and A.A. Puretzki (1976d), Phys. Letters **56A**, 183.

Ambartzumian, R.V., Yu. A. Gorokhov, V.S. Letokhov, G.N. Makarov and A.A. Puretsky (1976e), ZhETF **71**, 440.

Ambartzumian, R.V., Yu. A. Gorokov, V.S. Letokhov, G.N. Makarov, A.A. Puretskii and N.P. Furzikov (1976f), JEPT Letters **23**, 194.

Ambartzumian, R.V., Yu. A. Gorokhov, V.S. Letokhov, G.N. Makarov, E.A. Ryabov and N.V. Chekalin (1976g), Sov. J. Quantum Electronics **6**, 437.

Ambartzumian, R.V. and V.S. Letokhov (1977), Acc. Chem. Res. **10**, 61.

Ambartzumian, R.V., Yu. A. Gorokhov, G.N. Makarov, A.A. Puretzki and N.P. Furzikov (1977), Chem. Phys. Letters **45**, 231.

Artamonova, N.D., V.T. Platonenko and R.V. Khokhlov (1970), Sov. Phys. – JETP **31**, 1185.

Askar'yan, G.A. (1965), Sov. Phys. – JETP **21**, 439.

Bachmann, H.R., H. Noth, R. Rinck and K.L. Kompa (1974), Chem. Phys. Letters **29**, 627.

Bachmann, H.R., H. Noth, R. Rinck and K.L. Kompa (1975), Chem. Phys. Letters **33**, 261.

Bachmann, H.R., R. Rinck, H. Noth and K.L. Kompa (1977), Chem. Phys. Letters **45**, 169.

Bagratashvili, V.N., I.N. Knyazev, V.S. Letokhov and V.V. Lobko (1975) Opt. Commun. **14**, 426.

Bailey, R.T., F.R. Bruickshank, J. Farrell, D.S. Horne, A.M. North, P.B. Wilmot and Tin Win (1974), J. Chem. Phys. **60**, 1699.

Balykin, V.I., Yu. R. Kolomiiskii and O.A. Tumanov (1975), Sov. J. Quantum Electronics **5**, 454.

Basov, N.G., A.N. Oraevskii, A.A. Stepanov and V.A. Shcheglov (1973), Sov. Phys. – JETP **38**, 918.

Basov, N.G., A.N. Oraevskii and A.V. Pankratov (1974), Chemical and Biochemical Applications of Lasers (C.B. Moore, ed.; Wiley-Interscience; New York), Vol. I, p. 203.

Basov, N.G., E.M. Belanov, V.A. Isakov, E.P. Markin, An. N. Oraevskii and V.I. Romanenko (1977), Sov. Phys. – Vsp. **20**, 209.

Bauer, S.H. and K-R Chien (1977), Chem. Phys. Letters **45**, 529.

Bebb, H.B. and A. Gold (1966), Phys. Rev. **143**, 1.

Bellows, J.C. and K. Fong (1978), J. Chem. Phys. **63**, 3035.

Belotserkovets, A.V., G.A. Kirillov, S.B. Kormer, G.G. Kochemasov, Yu. V. Kuratov, V.I. Mashendzhniov, Yu. V. Savin, E.A. Stankeev and V.D. Urlin (1976), Sov. J. Quantum Electronics **5**, 1313.

Benson, S.W. (1978), Chem. Rev. **78**, 23.

Bergmann, K., S.R. Leone, R.G. MacDonald and C.B. Moore (1975), Israel J. Chem. **14**, 105.

Bialynicka-Birula, Z., I. Bialynicki-Birula, J.H. Eberly and B.W. Shore (1977), Phys. Rev. A. **16**, 2048.

Bidinosti, D.R. and R.F. Porter (1961), J. Amer. Chem. Soc. **83**, 3737.

Birely, J.H. and J.L. Lyman (1975), J. Photochem. **4**, 269.

Birely, J.H., D.C. Cartwright and J.G. Marinuzzi (1976), SPIE **76**, 124.

Bischel, W.K., P.J. Kelly and C.K. Rhodes (1976), Phys. Rev. A **13**, 1817.

Bittenson, S. and P.L. Houston (1977), International Conference on Multiphoton Processes, Rochester, New York, June 6–9, 1977, Paper W2B.2.

Bixon, J. and J. Jortner (1968), J. Chem. Phys. **48**, 715.

Black, J.G., E. Yablonovitch, N. Bloembergen and S. Mukamel (1977), Phys. Rev. Letters **38**, 1131.

Bloembergen, N. (1975), Opt. Commun. **15**, 416.

Bloembergen, N., C.D. Cantrell and D.M. Larsen (1976), Tunable Lasers and Applications (A. Mooradian, T. Jaeger, P. Stokseth, eds.; Springer-Verlag; Berlin), pp. 162–176.

Bott, J.F. and T.A. Jacobs (1969), J. Chem. Phys. **50**, 3850.

Bourimov, V.N., V.S. Letokhov and E.A. Ryabov (1976), J. Photochem. **5**, 49.

Braun, P.A. and A.N. Petelin (1974), Sov. Phys. – JETP **39**, 775.

Braun, W. and W. Tsang (1976), Chem. Phys. Letters **44**, 354.

Braun, W., M.J. Kurylo, A. Kaldor and R.P. Wayne (1974), J. Chem. Phys. **61**, 461.

Breshears, W.D. and L.S. Blair (1973), J. Chem. Phys. **59**, 5824.

Brewer, R.G. and E.L. Hahn (1975), Phys. Rev. A **11**, 1641.

Brunner, F., T.P. Cotter, K.L. Kompa and D. Proch (1977), J. Chem. Phys. **67**, 1547.

Bunkin, F.V. and I.I. Tugov (1970), Sov. Phys. – Dokl. **14**, 678.

Bunkin, F.V., R.V. Karapetyan and A.M. Prokhorov (1964), Sov. Phys. – JETP **20**, 145.

Campbell, J.D., G. Hancock and K.H. Welge (1976a), Chem. Phys. Letters **43**, 581.

Campbell, J.D., G. Hancock, J.B. Halpern and K.H. Welge (1976b), Chem. Phys. Letters **44**, 404.

Campbell, J.D., G. Hancock, J.B. Halpern and K.H. Welge (1976c), Opt. Commun. **17**, 38.

Cantrell, C.D. (1977a), Laser Spectroscopy, (J.L. Hall and J.L. Carlson, eds.; Springer Series in Optical Sciences; Springer-Verlag; Berlin), Vol. 7, pp. 109–115.

Cantrell, C.D. (1977b), unpublished.

Cantrell, C.D. and K. Fox (1978), Opt. Letters **2**, 151.

Cantrell, C.D. and H.W. Galbraith (1976), Opt. Commun. **18**, 513.

Cantrell, C.D. and H.W. Galbraith (1977), Opt. Commun. **21**, 374.

Cantrell, C.D., M. Lax and W.A. Smith (1973), Phys. Rev. A **7**, 175.

Cantrell, C.D., H.W. Galbraith and J.R. Ackerhalt (1978), Multiphoton Processes (J.H. Eberly and P. Lambropoulos, eds.; Wiley; New York), pp. 307–330.

Chem. Eng. News (1976), Nov. 22, 18.

Chien, K-R. and S.H. Bauer (1976), J. Phys. Chem. **80**, 1405.

Chin, S.L. (1976), Can. J. Chem. **54**, 2341.

Clough, P.N. and B.A. Thrush (1967), Trans. Faraday Soc. **63**, 915.

Clyne, MA.A., B.A. Thrush and R.P. Wayne (1964), Trans. Faraday Soc. **60**, 359.

Coggiola, M.J., P.A. Schulz, Y.T. Lee and Y.R. Shen (1977), Phys. Rev. Letters **38**, 17.

Cotter, T.P., W. Fuss, K.L. Kompa and H. Stafast (1976), Opt. Commun. **18**, 220.

Cottrell, T.L. and J.C. McCoubrey (1961), Molecular Energy Transfer in Gases (Butterworths; London).

Danen, W.C., W.D. Munslow and D.W. Setser (1977), J. Amer. Chem. Soc. **99**, 6961.

Dever, D.F. and E. Grunwald (1976), J. Amer. Chem. Soc. **98**, 5055.

Deutsch, T.F. (1977), Opt. Letters **1**, 25.

Diebold, G.J., F. Engelke, D.M. Lubman, J.C. Whitehead and R.N. Zare (1977), J. Chem. Phys. **67**, 5407.

Earl, B.L. and A.M. Ronn (1976), Chem. Phys. Letters **41**, 29.

Eberly, J.H. (1976), Phys. Rev. Letters **37**, 1387.

Eberly, J.H., B.W. Shore, I. Bialynicki-Birula and Z. Bialynicka-Birula (1977), Phys. Rev. A **16**, 2038.

Einwohner, T.H., J. Wong and J.C. Garrison (1976), Phys. Rev. A **14**, 1452.

Englman, R. and J. Jortner (1970), Molec. Phys. **18**, 145.

Faisal, F.H.M. (1976), Opt. Commun. **17**, 247.

Fano, U. (1956), Rev. Mod. Phys. **29**, 74.

Feshbach, H., A.K. Kerman and R.H. Lemmer (1967), Ann. Phys. (NY) **41**, 230.

Fettweis, P. and M. Nève de Mévergnies (1977), Appl. Phys. **12**, 219.

Fleck, J.A. (1966), Phys. Rev. **149**, 322.

Fox, K. (1962), J. Molec. Spectrosc. **9**, 381.

Freed, K.F. (1976), Chem. Phys. Letters **42**, 600.

Freeman, M.P., D.N. Travis and M.F. Goodman (1974), J. Chem. Phys. **60**, 231.

Freund, S.M. (1977), unpublished.

Freund, S.M. and J.J. Ritter (1973), unpublished.

Freund, S.M. and J.J. Ritter (1975), Chem. Phys. Letters **32**, 255.

Freund, S.M. and J.C. Stephenson (1976), Chem. Phys. Letters **41**, 157.

Fuss, W. and T.P. Cotter (1977), Appl. Phys. **12**, 265.

Galbraith, H.W. and C.D. Cantrell (1977), The Significance of Nonlinearity in the Natural Sciences (B. Kursunoglu, A. Perlmutter and L.F. Scott, eds.; Plenum; New York), pp. 227–264.

Gelbart, W.M., S.A. Rice and K.F. Freed (1972), J. Chem. Phys. **57**, 4699.

Glott, I. and A. Yogev (1976), J. Amer. Chem. Soc. **98**, 7088.

Golger, A.L., V.S. Letokhov and S.P. Fedoseev (1976), Krantovaya Elektron. **3**, 1457.

Goodman, M.F. and E. Thiele (1972), Phys. Rev. A **5**, 1355.

Goodman, M.F., J. Stone and E. Thiele (1973), J. Chem. Phys. **59**, 2919.

Goodman, M.F., J. Stone and E. Thiele (1975), J. Chem. Phys. **63**, 2929.

Goodman, M.F., J. Stone and D.A. Dows (1976), J. Chem. Phys. **65**, 5052.

Göppert-Mayer, M. (1931), Ann. Physik **9**, 273.

Gorchakov, V.I. and V.Ṅ. Sazonov (1976), Sov. Phys. – JETP **43**, 241.

Gordon, R.G. (1968), J. Math. Phys. **9**, 1087.

Gordon, R.J. and M.C. Lin (1973), Chem. Phys. Letters **22**, 262.

Gordon, R.J. and M.C. Lin (1976), J. Chem. Phys. **64**, 1058.

Gower, M.E. and K.W. Billman (1977), Opt. Commun. **20**, 123.

Grant, E., M. Coggiola, Y. Lee, P. Shultz and Y. Shen (1977), Paper Presented at the 2nd Winter Colloquim on Laser Induced Chemistry, Park City, Utah, February 13–16, 1977. See also Multiphoton Processes (J.H. Eberly and P. Lambropoulos, eds.; Wiley; New York, 1978). pp. 359–370.

Grunwald, E. and K.J. Olszyna (1976), Laser Focus, June, 41.

Haken, H. and W. Weidlich (1967), Z. Phys. **205**, 96.

Hallsworth, R.S. and N.R. Isenor (1973), Chem. Phys. Letters **22**, 283.

Hancock, G., J.D. Campbell and K.H. Welge (1976), Opt. Commun. **16**, 177.

Hecht, K.T. (1960), J. Molec. Spectrosc. **5**, 355.

Heitler, W. (1954), The Quantum Theory of Radiation, 3rd edn. (Oxford University Press; New York).

Heller, E.J. and S.A. Rice (1974), J. Chem. Phys. **61**, 936.

Herzberg, G. (1945), Molecular Spectra and Molecular Structure. II, Infrared and Raman Spectra of Polyatomic Molecules (Van Nostrand; Princeton, New Jersey).

Herzberg, G. (1960), Molecular Spectra and Molecular Structure. III, Electronic Spectra and Electronic Structure of Polyatomic Molecules (Van Nostrand; Princeton, New Jersey).

Hodgkinson, D.P. and J.S. Briggs (1976), Chem. Phys. Letters **43**, 451.

Hougen, J.T. (1976), J. Chem. Phys. **65**, 1035.

Hudson, J.W., J.L. Lyman and S.M. Freund (1976), Proceedings of the Electro-Optics Systems Design Conference 1976 International Laser Exposition, New York, September 14–16, 1976, p. 309.

Isenor, N.R. and M.C. Richardson (1971), Appl. Phys. Letters **18**, 224.

Isenor, N.R., V. Merchant, R.S. Hallsworth and M.C. Richardson (1973), Can. J. Phys. **51**, 1281.

Jahn, H.A. (1938), Proc. Roy. Soc. (London) **A168**, 469.

Jensen, R.J., J.G. Marinuzzi, C.P. Robinson and S.D. Rockwood (1976), Laser Focus, May, 51.

Jensen, C.C., W.B. Person, B.J. Krohn and J. Overend (1977), Optics Commun. **20**, 275.

Jortner, J. (1970), Pure and Appl. Chem. **24**, 165.

Jortner, J. and S. Mukamel (1976), J. Chem. Phys. **65**, 5204.

Karl, R.R. and J.L. Lyman (1978), J. Chem. Phys. **69**, 1196.

Karlov, N.K. (1974), Appl. Optics **13**, 301.

Karlov, N.V. and A.M. Prokhorov (1976) Sov. Phys. – Vsp. **19**, 285.

Karlov, N.V., N.A. Karpov, Yu. N. Petrov and O.M. Stelmakh (1973), Sov. Phys. – JETP **37**, 1012.

Karny, Z. and R. Zare (1977), Chem. Phys, **23**, 321.

Karplus, R. and J. Schwinger (1948), Phys. Rev. **73**, 102C.

Kay, K.G. (1974), J. Chem. Phys. **60**, 2370.

Keefer, D.R., J.E. Allen and W.B. Person (1976), Chem. Phys. Letters **43**, 394 (1976).

King, D.S., P.K. Shenck and J.S. Stephenson (1977), International Conference on Multiphoton Processes, Rochester, New York, June 6–9, 1977, Paper W2B.8. See also Lasers in Chemistry (M.A. West, ed.; Elsevier Scientific Publishing Co.; Amsterdam), p. 340.

Kittel, C. (1958), Elementary Statistical Physics (Wiley; New York).

Klein, F., F.M. Lussier and J.I. Steinfeld (1975), Spectrosc. Letters **8**, 247.

Knudtson, J.T. and E.M. Eyring (1974), Annu. Rev. Phys. Chem. **25**, 255.

Kolodner, P., C. Winterfeld and E. Yablonovitch (1977), Opt. Commun. **20**, 119.

Kompa, K.L. (1976), Tunable Lasers and Applications (A. Mooradian, T. Jaeger, P. Stokseth, eds.; Springer; Berlin), pp. 177–189.

Kurylo, M.J., W. Braun, A. Kaldor, S.M. Freund and R.P. Wayne (1974/75), J. Photochem. **3**, 71.

Kurylo, M.J., W. Braun, C.N. Xuan and A. Kaldor (1975), J. Chem. Phys. **62**, 2065.

Lamb, W.E. (1964), Phys. Rev. A **134**, 1429.

Lamb, W.E. (1976), Paper Presented at First Winter Colloquium on Laser Induced Chemistry, Steamboat Springs, CO, February 1976.

Lamb, W.E. and T.M. Sanders (1960), Phys. Rev. **119**, 1901.

Larsen, D.M. (1976), Opt. Commun. **19**, 404.

Larsen, D.M. and N. Bloembergen (1976), Opt. Commun. **17**, 254.

Lax, M. (1964), J. Phys. Chem. Solids **25**, 487.

Lax, M. (1966), Phys. Rev. **145**, 110.

Lax, M. and W.H. Louisell (1967), IEEE J. Quantum Electronics **3**, 47.

Lax, M. and M. Zwanziger (1973), Phys. Rev. A **7**, 750.

Leary, K.M., J.L. Lyman, L.P. Asprey and S.M. Freund (1978), J. Chem. Phys. **68**, 1671.

LeRoy, R.L. (1969), J. Phys. Chem. **73**, 4338.

Lesiecki, M.L. and W.A. Guillory (1977a), J. Chem. Phys. **66**, 4317.

Lesiecki, M.L. and W.A. Guillory (1977b), J. Chem. Phys. **66**, 4239.

Letokhov, V.S. and A.A. Makarov (1976a), Opt. Commun. **17**, 250.

Letokhov, V.S. and A.A. Makarov (1976b), Coherent Excitation of Multilevel Molecular Systems in Intense Quasi-Resonant Laser ir Field, Institute of Spectroscopy, Akademgorodok, Podol'skii Rayon, Moscow, USSR.

Letokhov, V.S. and C.B. Moore (1976a), Sov. J. Quantum Electronics **6**, 129.

Letokhov, V.S. and C.B. Moore (1976b), Sov. J. Quantum Electronics **6**, 259.

Lin, C.T. (1976), Spectrosc. Letters **9**, 615.

Lin, S.T. and A.M. Ronn (1977), Chem. Phys. Letters **49**, 255.

Lory, E.R., S.H. Bauer and T. Manuccia (1975), J. Phys. Chem. **79**, 545.

Louck, J.D. and H.W. Galbraith (1976), Rev. Mod. Phys. **48**, 69.

Louisell, W.H. (1973), Quantum Statistical Properties of Radiation (Wiley; New York).

Lyman, J.L. (1973), Ph.D. Dissertation, Brigham Young University, Provo, Utah, U.S.A.

Lyman, J.L. (1977), J. Chem. Phys. **67**, 1868.

Lyman, J.L. and R.J. Jensen (1972), Chem. Phys. Letters **13**, 421.

Lyman, J.L. and R.J. Jensen (1973), J. Phys. Chem. **77**, 883.

Lyman, J.L. and S.D. Rockwood (1975), Proceedings of the Electro-Optics Systems Design Conference – 1975 International Laser Exposition, Anaheim, CA, November 11–13, 1975, p. 179.

Lyman, J.L. and S.D. Rockwood (1976), J. Appl. Phys. **47**, 595.

Lyman, J.L., R.J. Jensen, J.P. Rink, C.P. Robinson and S.D. Rockwood (1975), Appl. Phys. Letters **27**, 87.

Lyman, J.L., J.W. Hudson and S.M. Freund (1977a), Opt. Commun. **21**, 112.

Lyman, J.L., S.D. Rockwood and S.M. Freund (1977b), J. Chem. Phys. **67**, 4545.

Makarov, A.A. (1977), Zh. Eksp. Teor. Fiz. **72**, 1749.

Mayer, S.M., M.A. Kwok, R.W.F. Gross and D.J. Spencer (1970), Appl. Phys. Letters **17**, 516.

McDowell, R.S., J.P. Aldridge and R.F. Holland (1976a), J. Phys. Chem. **80**, 1203.

McDowell, R.S., H.W. Galbraith, B.J. Krohn, C.D. Cantrell and E.D. Hinkley (1976b), Opt. Commun. **17**, 178.

McDowell, R.S., H.W. Galbraith, C.D. Cantrell, N.G. Nereson and E.D. Hinkley (1977), J. Molec. Spectrosc. **68**, 288.

Messiah, A. (1960), Quantum Mechanics (North-Holland: Amsterdam), Vol. I.

Montroll, E.W. and K.E. Shuler (1957), J. Chem. Phys. **26**, 454.

Mower, L. (1966), Phys. Rev. **142**, 799.

Mukamel, S. and J. Jortner (1976), Chem. Phys. Letters **40**, 150.

Mukamel, S. and J. Ross (1977), J. Chem. Phys. **66**, 5235.

Nowak, A.V. and J.L. Lyman (1975), J. Quant. Spectrosc. Radiative Transfer **15**, 945.

Oraevsky, A.N. and V.A. Savva (1970), Short Communications in Physics, FIAN, **7**, 50.

Orr, B.J. and M.V. Keentok (1976), Chem. Phys. Letters **41**, 68.

Overend, J., H.W. Galbraith, C.D. Cantrell and W.B. Person (1977), J. Chem. Phys. (submitted for publication).

Oxtoby, D.W. and S.A. Rice (1976), J. Chem. Phys. **65**, 1676.

Pert, G.J. (1973), IEEE J. Quantum Electronics **QE-9**, 435.

Redfield, A.G. (1957), IBM J. Res. Develop. **1**, 19.

Redpath, A.E. and M. Menziger (1975), J. Chem. Phys. **62**, 1987.

Rhodes, C.K. and C.D. Cantrell (1977), pp. 293–325 in The Significance of Nonlinearity in the Natural Sciences (B. Kursunoglu, A. Perlmutter and L.F. Scott eds.; Plenum; New York), pp. 293–325.

Ritter, J.J. (1978), J. Amer. Chem. Soc. **100**, 2441.

Ritter, J.J. and M. Freund (1976), J.C.S. Chem. Commun. **1976**, 811.

Robinson, J.W., P.J. Moses and P.M. Boyd (1974), Spectrosc. Letters **7**, 395.

Robinson, P.J. and K.A. Holbrook (1972), Unimolecular Reactions (Wiley; London).

Rockwood, S.D. (1975), Chem. Phys. **10**, 453.

Rockwood, S.D. and J.W. Hudson (1975), Chem. Phys. Lett. **34**, 542.

Ronn, A.M. (1975), Spectrosc. Letters **8**, 303.

Ronn, A.M. (1976), Chem. Phys. Letters **42**, 202.

Ronn, A.M. and B.L. Earl (1977), Chem. Phys. Letters **45**, 556.

Sakurai, J.J. (1967), Advanced Quantum Mechanics (Addison-Wesley Series in Advanced Physics; Addison-Wesley; Reading, Mass.).

Sargent, M. and P. Horowitz (1976), Phys. Rev. A **13**, 1962.

Sargent, M., M.O. Scully and W.E. Lamb (1974), Laser Physics (Addison-Wesley; Reading, Mass.).

Scully, M.O. and W.E. Lamb (1967), Phys. Rev. **159**, 208.

Shaffer, W.H., H.H. Nielsen and L.M. Thomas (1939), Phys. Rev. **56**, 895 and 1051.

Shamah, I and G. Flynn (1977), J. Amer. Chem. Soc. **99**, 3192.

Shatas, R.A. (1976), Red Stone Arsenal, Alabama, private communication.

Shatas, S., D. Gregory, R. Shatas, and C. Riley (1977), Inorg. Chem. **17**, 163.

Shaub, W.M. and S.H. Bauer (1975), Int. J. Chem. Kin. **7**, 509 (1975).

Slichter, C.P. (1963), Principles of Magnetic Resonance with Examples from Solid State Physics (Harper and Row; New York).

Speiser, S. and J. Jortner (1976), Chem. Phys. Letters **44**, 399.

Stafast, H., W.E. Schmid and K.L. Kompa (1977), Opt. Commun. **21**, 121.

Steinfeld, J.I., I. Burak, D.G. Sutton and A.V. Nowak (1970), J. Chem. Phys. **52**, 5421.

Stone, J. and M.F. Goodman (1978a and b), to be published.

Stephenson, J.C. and S.M. Freund (1976), J. Chem. Phys. **65**, 4303.

Stone, J., E. Thiele and M.F. Goodman (1973), J. Chem. Phys. **59**, 2909.

Stone, J., E. Thiele and M.F. Goodman (1975), J. Chem. Phys. **63**, 2936.

Stone, J., M.F. Goodman and D.A. Dows (1976), J. Chem. Phys. **65**, 5062.

Tardieu de Maleissye, J., F. Lempereur, C. Marsal and R.I. Ben-Aim (1976), Chem. Phys. Letters **42**, 46.

Tardy, D.C. and B.S. Rabinovitch (1977), Chem. Rev. **77**, 369.

Teller, E. (1935), Z. Phys. Chem. **28**, 371.

Ter Haar, D. (1961), Pure Appl. Phys. **10**, 323.

Trenwith, A.B. and R.H. Watson (1957), J. Chem. Soc. **1957**, 2368.

Van Hove, L., N.M. Hugenholtz and L.P. Howland (1961), Quantum Theory of Many-Particle Systems (W.A. Benjamin; New York).

von Neumann, J. (1932), Mathematische Grundlagen der Quantenmechanik (Springer-Verlag; Berlin); The Mathematical Foundations of Quantum Mechanics (translated by R.T. Beyer) (Princeton University Press, 1955), Chapters IV, VI.

Walker, R.B. and R.K. Preston (1977), J. Chem. Phys. **67**, 2017.

Weitz, E. and G. Flynn (1973), J. Chem. Phys. **58**, 2679.

Wilcox, L.R. and W.E. Lamb (1960), Phys. Rev. **119**, 1915.

Wolfrum, J. (1977), Ber. Buns. Phys. Chem. **81**, 114.

Willis, C., R.A. Back, R. Conkum, D. McAlpine and F.K. McClusky (1976), Chem. Phys. Letters **38**, 336.

Yablonovitch, E., P. Kolodner, and C. Winterfeld (1977) (private communications).

Yogev, A. and R.M.J. Benmair (1975), J. Amer. Chem. Soc. **97**, 4430.

Yogev, A. and R.M.J. Lowenstein-Benmair (1973), J. Amer. Chem. Soc. **95**, 8487.

Yogev, A., R.M.J. Loewenstein and D. Amar (1972), J. Amer. Chem. Soc. **94**, 1091.
Zitter, R.N. and D.F. Koster (1977), J. Amer. Chem. Soc. **99**, 5491.
Zitter, R.N. and D.F. Koster (1976), J. Amer. Chem. Soc. **98**, 1613.
Zitter, R.N., R.A. Lau and K.S. Wills (1975), J. Amer. Chem. Soc. **97**, 2578.
Zwanig, R. (1964), Physica **30**, 1109.

B4 | Pulsed Holography

Walter KOECHNER

Science Applications Inc., McLean, Virginia 22102

Contents

Abstract

This chapter reviews the progress in pulsed holography which resulted from the development of long coherence length solid-state lasers with precise trigger capability. Available devices and equipment, such as laser light sources, optical components, and recording materials are described. The applications of pulsed holography in nondestructive testing, flow visualization and particle studies are reviewed.

© *North-Holland Publishing Company, 1979*
Laser Handbook, edited by M.L. Stitch

1 Introduction

In chapter F2 of volume 2 the principles of holography, the most important types of holograms and their properties, as well as a summary of general applications were discussed.

This chapter will be concerned with the practical aspects of pulsed holography. In particular, the operating characteristics of pulsed lasers and the equipment required to take pulsed holograms will be reviewed. Also, some of the precautions necessary in working with high energy lasers will be outlined and experimental arrangements will be discussed. The broad range of applications of pulsed holography will be illustrated by giving typical examples from the field of nondestructive testing, flow visualization and particle studies.

The most practical application of holography is holographic interferometry which permits one to measure very small displacements or deformations of an object. The displacements obtained from the holographic technique are known in terms of the wavelength of the laser light used to construct the hologram, and highly accurate displacements can thus be obtained from the resultant hologram fringe systems. As is the case with all holographic methods, it works without contact; consequently the object is unaffected.

In recording a hologram it is imperative that optical components of the holographic apparatus, as well as the object, must not move more than $\frac{1}{8}$ wavelength during the exposure of the holographic plate. When a hologram is made with a continuous-wave (cw) laser exposure times of seconds or minutes are required owing to the low power output of these light sources. This imposes strict requirements upon the holographic support structures.

The general practice has been to support all the components, including the object, upon a single massive optical table which can be suitably decoupled from the surrounding environment, thereby minimizing the components' relative motions with respect to each other. Thus, holographic interferometry conducted with cw lasers as the constructing source is generally limited to the examination of small components which are readily supportable by an existing holographic apparatus located in a relatively stable environment.

The high degree of stability required for recording cw holograms precludes the use of large components as a hologram subject. Furthermore, no holograms can be made of moving or rotating objects, or objects subjected to a transient vibration or located in a vibrating environment.

In early 1965, Q-switched pulsed ruby laser hologram interferometry,

579

pioneered by Brooks et al. (1965, 1966), was applied to make double-exposure holograms of fast moving objects. This was the starting point for a large research effort to explore the applicability of pulsed holography to a large variety of tasks.

Since the 20 to 50 ns long pulse from a Q-switched ruby laser eliminates the need for extreme mechanical stability of the object while making the hologram, the massive tables and clamps so vital for cw laser holography may be dispensed with, and laser holograms can be made not only of static scenes, but also of moving or dynamic scenes. Another advantage of pulsed holography is the fact that ambient illumination, present in an industrial environment, is not a problem. Fogging of the holographic plate due to ambient illumination during the exposure time can be alleviated in pulsed holography by employing a large aperture shutter in front of the photographic plate in conjunction with a spectral bandpass filter. In fact, many of the holographic methods would not be applicable in many environments which contain excessive amounts of ambient illumination or vibration were it not for the extremely short duration pulses that can be obtained from pulsed lasers.

As mentioned above, during the exposure time the object should not move more than $\frac{1}{8}$ wavelength. In the case of a 20 ns pulse of a ruby laser, this means that the tolerable speed of the object in the direction of the beam is about 3 m/s. If the object is illuminated from the side, the speed can be several orders of magnitude higher. With a pulsed laser single holograms can be made of transient events, such as shock waves. The range of objects extends all the way to holograms of bullets in flight and objects flying at Mach 10 velocity. Holographic portraits of people or whole groups of people can be taken.

Table 1 shows a comparison of the applicability of pulsed and c.w. holography to the investigation of objects in various dynamic situations. The table is not meant to indicate that pulsed lasers will replace cw lasers for holographic interferometry, but rather that they complement cw methods.

Table 1.
Comparison of pulsed and cw holography.

Test object	Pulsed holography	cw holography
Stationary	Single exposure	Time exposure
Stationary and distorting	Double exposure	Real time
Moving	Single exposure	Not applicable
Moving and distorting	Double exposure	Not applicable
Harmonic oscillating	Double exposure	Time averaging
Transient event	Double exposure	Not applicable

2 Coherent light sources

The ruby laser, by virtue of its high energy output capabilities and its wavelength, for which photographic emulsions of high sensitivity and resolution are available, is used almost exclusively as the coherent light source in pulsed holography. To a very limited extent frequency doubled Nd:YAG lasers are occasionally employed in pulsed holography. The performance and salient features of these lasers will be discussed in the following section.

Solid-state lasers employed for holography are characterized by a high degree of spatial and temporal coherence and by their ability to emit two Q-switched pulses with short interpulse separation. Spatial coherence of the reference beam is desirable to obtain high resolution holograms. The temporal coherence of the laser determines the size of the object from which a hologram can be made.

Typical holographic lasers have a gaussian beam profile and a coherence length of 1 to 2 m. Special ruby systems have been built with a coherence length as long as 10 m. The double-pulse capability is essential for applications in nondestructive testing. Usually two holograms, with time intervals between 1 and 1000 μs, are superimposed on the same photographic plate. Any perturbation of the test object during this time interval will show up as interference fringes on the double-pulsed hologram. The double-pulse technique makes it possible to apply holography to stress and vibration analysis, shock propagation and vibration studies, and to flow visualization of projectiles passing through air.

The major components of an optically pumped solid-state laser oscillator are a cylindrical laser rod, a helical or linear flashlamp, a pump cavity which provides good optical coupling between the flashlamp and the laser rod, and an optical resonator comprised of a totally and a partially reflective mirror. Auxiliary equipment of a laser oscillator includes a high-voltage power supply, energy-storage capacitor, flashlamp trigger unit and a water-cooling system.

For holographic applications it is necessary to modify the temporal, spectral and spatial output characteristics of a conventional oscillator. This is achieved by inserting additional optical components into the resonator, such as a Q-switch which reduces the output pulse length to tens of nanoseconds, an etalon or resonant reflector which reduces the number of axial modes, and an intracavity aperture which forces the laser radiation into a single transverse mode.

Essentially, all solid-state lasers employed for holography are Q-switched. The pulse length obtained from conventional operation of the laser is too long for forming holograms of many objects; furthermore, it is impossible to maintain a very narrow linewidth over the duration of the flashlamp pulse owing to heating effects in the laser host material.

The mode-selecting components inserted in the resonator drastically reduce the output energy from the laser oscillator. For many holographic applications it is necessary to increase the available energy from a holographic system by adding one or several amplifiers. In an oscillator–amplifier system, pulse width, spectral and spatial coherence are primarily determined by the oscillator whereas pulse energy and power are determined by the amplifier. Fig. 1 shows

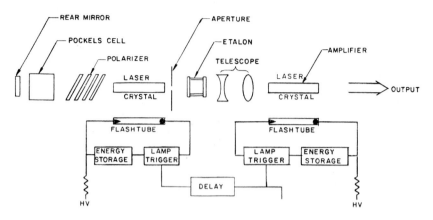

Fig. 1. Typical ruby oscillator–amplifier system employed in pulsed holography.

an optical schematic of a typical ruby oscillator–amplifier system employed in holography.

In the following subsections we will briefly describe some of the major properties and design features of solid-state lasers which are relevant to pulsed holography. A general introduction to lasers is given by Lengyel (1971) and a detailed account of the principles, design and operation of solid-state lasers can be found in the book by Koechner (1976).

2.1 Ruby and Nd:YAG laser

2.1.1 The ruby laser

The rod of a ruby laser is made of synthetic sapphire, Al_2O_3, which is doped with 0.05% by weight of Cr_2O_3. The substitution of a small percentage of the Al^{3+} with Cr^{3+} produces a pink-colored material. Laser action results from stimulation of Cr^{3+} ions by the pump light. The ruby laser emits red light at a wavelength of 6943 Å.

Typical dimensions of a ruby rod in an oscillator employed for holographic applications are 5 to 10 mm in diameter and 75 to 100 mm in length. Both end faces are polished parallel and are antireflection coated. The most important ingredient in a laser oscillator designed for good spatial and temporal coherence is a carefully selected ruby laser rod. For the laser crystal to cause minimum wavefront distortion it should be of the highest optical quality and free of inhomogeneities. The oscillator is Q-switched either with a Pockels cell or with a saturable absorber.

The advantage of passive dye Q-switches include low cost, simplicity of operation and the emission of the output pulse in a narrow linewidth. However, there are a number of distinct disadvantages associated with the saturable

Q-switch: the time between the triggering of the flashlamp and the emission of a *Q*-switched pulse is associated with a jitter which is typically of the order of 10 to 100 μs; furthermore, with a dye *Q*-switch it is not possible to obtain two *Q*-switched pulses.

In holographic applications very often precise timing between an event and the output pulse is required. In addition, holographic interferometry requires the generation of two *Q*-switched pulses. The principle advantage of the Pockels cell is that it is capable of triggering repeatably with respect to an external event. It is also possible to generate two or more laser pulses within one flashlamp pump cycle. For these reasons the Pockels cell *Q*-switch is employed on most commercial holographic lasers.

In applications requiring greater *Q*-switch energy than obtainable from an oscillator it is possible to build an oscillator–amplifier system. Ruby rods for amplifiers have diameters from 1 to 2 cm and lengths up to 20 cm. The spatial and temporal coherence is essentially preserved in the process of amplification. The diameter of the ruby crystal is the ultimate limit to the energy attainable without exceeding the damage threshold of the crystal. The highest operating energy density level commonly used is 6 J/cm^2 or 300×10^6 W/cm^2.

Table 2 shows typical performance data for holographic ruby laser systems. As can be seen from this table, an increase of coherence length requires operation of the oscillator closer to threshold, which results in a lower output energy. It should be noted that a long coherence length is useful only for taking holograms of large objects. For proper illumination of large objects the pulse energy has to be high, too.

The amount of energy required from the laser for a given object size is shown in fig. 2. This empirical data is based upon actual holograms of diffusely

Table 2.
Typical performance data of holographic ruby laser systems.[a]

Type of system	Coherence length [m]	Single-pulse operation energy [J]	Double-pulse operation[b] energy per pulse [J]
Oscillator	0.5–2 5–10	0.040 0.010	0.025 0.007
Oscillator and one amplifier	0.5–2 5–10	1 0.2	0.4 0.2
Oscillator and two amplifiers	0.5–2 5–10	10 4	4 2

[a] Output wavelength: 6943 Å; Pulse width: 15–40 ns; Pulse rate (typical): 2 ppm; Transverse mode: TEM$_{00}$.
[b] Double-pulse separation 1–1000 μs.

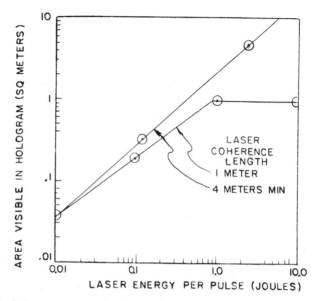

Fig. 2. Laser energy and coherence length requirement for reflecting objects.

illuminated objects, using Agfa 10E75 film. These curves are useful only as a guide, since the reflectance of the object is, of course, an important parameter, as is the degree of uniformity of illumination desired. Note that fig. 2 shows the amount of energy required in a single pulse. For holographic interferometry applications the laser must produce these amounts of energy in each of the two pulses.

Further discussions of ruby lasers employed in holography are given by Wuerker and Heflinger (1971), Gregor (1971) and Young and Hicks (1974).

2.1.2 The Nd:YAG laser

Neodymium-doped yttrium aluminum garnet (Nd:YAG) possesses a combination of properties uniquely favorable for laser operation. In particular the cubic structure of YAG favors a narrow fluorescent linewidth which results in high gains and low threshold for laser operation. The laser transition has a wavelength of 1.064 μm.

For holographic applications the advantage of Nd:YAG as compared to ruby, namely a more efficient operation, smaller size and weight, and a higher pulse repetition rate capability, is offset by two major disadvantages: a Nd:YAG laser is not capable of generating as much Q-switch energy as a ruby laser; furthermore, the output is in the infrared. In order to utilize a Nd:YAG laser the output wavelength has to be reduced to 5300 Å by employing a harmonic generator at the output. Frequency doubling of Nd:YAG is accomplished by means of a temperature controlled cesium dideuterium arsenate crystal (CD*A). Typical conversion efficiencies are in the order of 20 to 40%.

The maximum output energy obtainable from the largest frequency doubled Nd:YAG laser is about two orders of magnitude lower than the energies from large ruby lasers of comparable spatial and spectral quality. On the other hand, ruby is limited to a maximum pulse repetition rate of 1 pps, whereas a Nd:YAG laser is capable of up to 50 pps. The lower output capabilities of Nd:YAG combined with the added complexity of a harmonic generator have made Nd:YAG not a very successful contender in the field of holography. Table 3 shows typical performance data of frequency doubled Nd:YAG lasers useful for holographic applications. Even for very modest output energies Nd:YAG lasers require at least two or three amplifiers. Holographic Nd:YAG lasers are discussed by Way (1975) and Bates (1973).

2.2 Spatial coherence

Laser operation in several transverse modes introduces a degradation in holographic image quality because of the nonuniform phase and frequency distribution across the reference beam.

A laser rod of 6 to 10 mm in diameter located in a typical laser resonator 50 to 100 cm long will cause a large number of transverse modes to oscillate simultaneously across the rod diameter. Since the oscillation frequencies of transverse modes are unrelated, the spatial coherence of the output light is very poor. The most common technique to produce TEM_{00} mode output in holographic

Table 3.
Typical performance data of holographic Nd:YAG lasers.[a]

Type of system	Coherence length [m]	Single-pulse operation energy (J)	Double-pulse operation[b] energy per pulse (J)
Oscillator and two amplifiers and frequency doubler	1	0.040	0.020
Oscillator and three amplifiers and frequency doubler	1	0.150	0.070

[a] Output wavelength: 5300 Å; Pulse width: 15–25 ns; Pulse rate (typical): 5 pps; Transverse mode: TEM_{00}.
[b] Double-pulse separation 1–1000 μs.

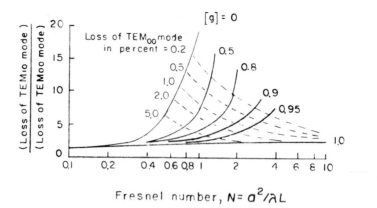

Fig. 3. Ratio of the losses per transit of the two lowest-order modes for a symmetrical resonator. The dotted curves are contours of constant loss for the TEM_{00} mode. After Li (1965).

systems is the use of a nearly plane-parallel resonator with an internal aperture for mode selection.

The single-mode operation of the laser results in a gaussian beam intensity profile and a uniphase wavefront. Since higher-order transverse modes have a larger spatial extent than the fundamental mode, a given size aperture will preferentially discriminate against higher-order modes in a laser resonator. Whether the laser will operate only in the lowest-order mode depends on the size of this mode and the diameter of the smallest aperture in the resonator.

The diffraction losses caused by a given aperture and the transverse mode selectivity achievable with an aperture of radius a are illustrated in fig. 3. In this figure the ratio of the loss of the TEM_{10} mode to the loss of the TEM_{00} mode is plotted as a function of the Fresnel number for a symmetrical resonator. Note that the mode selectivity is strongly dependent on the resonator geometry, and is greatest for a confocal resonator and smallest for the plane-parallel resonator. From fig. 3 it follows that the resonators of lasers operating in the TEM_{00} mode will have Fresnel numbers on the order of approximately 0.5 to 2.0. For Fresnel numbers much smaller than these, the diffraction losses will become prohibitively high, and for much larger values mode discrimination will be insufficient.

Without an aperture, a 50 cm long resonator with a 6 mm diameter ruby rod as the limiting aperture will have a Fresnel number of 19. For resonators 50 cm to 100 cm in length apertures of 1.5 to 2 mm are employed. The rear mirror typically has a radius of about 5 to 10 m, whereas the front mirror is flat and is comprised of one or several etalons for axial mode selection.

2.3 Temporal coherence

If a laser is operated without any axial-mode-selecting elements in the cavity, then the spectral output will be comprised of a large number of discrete

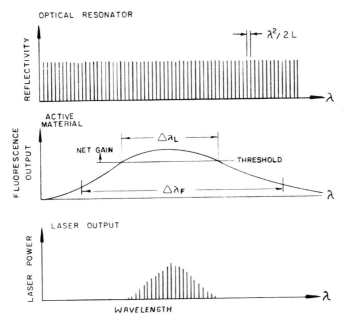

Fig. 4. Schematic diagram of the spectral output of a laser without mode selection. $\Delta\lambda_F$ – fluorescence linewidth, $\Delta\lambda_L$ – laser linewidth.

frequencies determined by the longitudinal modes. The linewidth of the laser transition limits the number of modes that have sufficient gain to oscillate. The situation is diagrammed schematically in fig. 4, which shows the resonance frequencies of an optical resonator and the fluorescence line of the active material. Laser emission occurs at those wavelengths at which the product of the gain of the laser transition and the reflectivity of the mirrors exceeds unity. In the idealized example shown, the laser would oscillate at 23 axial modes.

The wavelength separation of two adjacent longitudinal modes is given by

$$\Delta\lambda = \lambda_0^2/2L, \tag{1}$$

where L is the optical length of the resonator. With $L = 75$ cm and $\lambda_0 = 6943$ Å, one obtains $\Delta\lambda = 0.003$ Å. Depending on the pumping level for ruby and Nd:YAG, one finds a linewidth of approximately 0.3 to 0.5 Å for the laser emission in the absence of mode selection. Therefore, these lasers typically oscillate in about 100 to 150 longitudinal modes.

The spectral width of a single longitudinal mode can be expressed by (Koechner 1976)

$$\delta\lambda = \lambda^2[1 - (R_1)^{1/2}]/2L\pi(R_1)^{1/2}, \tag{2}$$

where R_1 is the reflectivity of the output mirror (we assume that the rear mirror has a reflectivity of $R_2 = 1$). A typical value for a laser cavity is $R_1 = 0.5$.

Introducing this value into eq. (2) and again assuming $L = 75$ cm, we obtain $\delta\lambda = 0.0004$ Å.

If the laser emits K axial modes, the bandwidth between the two extreme modes is

$$\Delta\nu = (K - 1)c/2L \text{ or } \Delta Z = (K - 1)Z^2/2L, \quad K \geq 2. \tag{3}$$

The wave emitted from a laser which emits a discrete number of integrally related wavelengths is strongly modulated. The situation can be illustrated by the simplest case of two superimposed traveling waves whose wavelengths are specified by adjacent axial modes.

This situation is shown schematically in fig. 5. The two waves interfere with one another, and produce traveling nodes which are found to be separated from one another in time by twice the cavity separation. That is, the output of such a laser is modulated at a frequency of twice the end-mirror separation ($\nu_m = c/2L$).

When three integrally related frequencies are emitted, the output becomes more modulated; however, the maxima are still separated from one another by a distance of twice the mirror separation. As the number of integrally related modes increases, the region of constructive interference – which is inversely proportional to the number of oscillating modes – becomes narrower.

The spectral characteristics of solid-state lasers employed in holography are commonly specified in terms of coherence length. The temporal coherence of any spectral source is defined as the path-length difference over which the

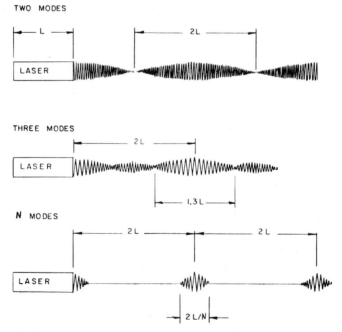

Fig. 5. Schematic of the output of a laser operating at two, three and N longitudinal modes. After Wuerker (1971).

radiation can still interfere with itself. Coherence length is defined as the path-length difference for which fringe visibility in a Michelson interferometer is reduced to $1/2^{1/2}$.

The fringe visibility of an interferometer is defined as

$$V = (I_{max} - I_{min})/(I_{max} + I_{min}). \tag{4}$$

In our subsequent discussion we will assume that the intensities of the axial modes in the laser output are equal. In this case coherence length is directly related to the observed visibility of the interference fringes, and a simple relation between a set of longitudinal modes and temporal coherence can be obtained. The general case of axial modes of unequal amplitudes is treated by Collier et al. (1971).

2.3.1 Laser emission at a single axial mode

The reflectivity versus wavelength around the resonance points of an optical resonator can be expressed by a gaussian distribution. When a pulse with a gaussian spectral distribution interacts with another identical pulse delayed by a time τ, the fringe visibility can be expressed as (Siebert 1971, Collier et al. 1971)

$$V = \exp[-(\pi\tau\delta\nu)^2/4\ln 2]. \tag{5}$$

If we express the transit time difference τ in terms of an optical path length difference $\Delta l = c\tau$, where c is the speed of light, we can write $\tau\delta\nu = \Delta l\delta\lambda/\lambda^2$. Using this substitution, eq. (5) is plotted in fig. 6. The fringe visibility V is reduced to $1(2)^{1/2}$ from the peak value for $\Delta l\delta\lambda/\lambda^2 = 2(\ln 2)/\pi(2)^{1/2} = 0.32$. Therefore, we obtain for the total path-length difference for which the fringe visibility is greater than $1/2^{1/2}$,

$$l_c = 0.64c/\delta\nu = 0.64\lambda^2/\delta\lambda. \tag{6}$$

The parameter l_c is referred to as the coherence length of the laser. Combining

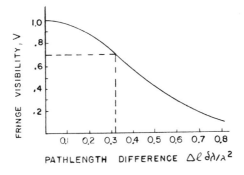

Fig. 6. Fringe visibility in a Michelson interferometer of a beam with a gaussian spectral profile. The parameters λ and $\delta\lambda$ are the wavelength and the linewidth of the beam, and Δl is the path length difference.

eqs. (2) and (6), we obtain for the coherence length l_c of a single-mode laser

$$l_c \approx 4L(R_1)^{1/2}/(1 - (R_1)^{1/2}), \tag{7}$$

where L is the resonator length and R_1 is the reflectivity of the front mirror (it is assumed that $R_2 = 1$).

The single-axial-mode output pulse from a ruby oscillator having a cavity length of 75 cm and a front-mirror reflectivity of $R_1 = 0.4$ will have a coherence length of $l_c = 5.2$ m. The spectral width of the single line, according to eq. (6), will be $\delta\nu = 37$ MHz. The bandwidth-limited pulse length is 17 ns.

2.3.2 Laser emission at N axial modes

The linewidth of a single line is assumed to be very narrow compared to the mode separation. Therefore, the power spectrum of the laser is represented by δ functions. In this case the fringe visibility can be expressed by

$$V = \left| \frac{\sin(N\pi\Delta l/2L)}{N \sin(\pi\Delta l/2L)} \right|. \tag{8}$$

This function, which has a periodicity of $2L$, is plotted in fig. 7 for the case of a laser oscillating in two, three and four modes. The coherence length, defined the same way as for the single-axial-mode case, follows from eq. (9),

$$l_c = 2L/N, \quad N \geq 2. \tag{9}$$

2.3.3 Linewidth control

It is possible to discriminate against most of the axial modes by adding additional reflecting surfaces to the basic resonator. If a Fabry–Perot type reflector is inserted between the two mirrors of the resonator, it will cause a strong amplitude modulation of the closely spaced reflectivity peaks of the basic laser resonator. This will prevent most modes from reaching threshold.

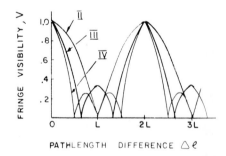

Fig. 7. Fringe visibility as a function of path-length difference for a laser operating in two (II), three (III) and four (IV) longitudinal modes. L is the length of the laser resonator.

The role of the resonant devices employed in interferometric mode selection is to provide high feedback for a single wavelength near the center of the fluorescence line while at the same time discriminating against nearby wavelengths. For example, by replacing the standard dielectrically coated front mirror with a single-plate resonant reflector, the number of oscillating modes can be greatly reduced.

Resonant reflectors featuring two, three or four plates have reflectivity peaks which are much narrower as compared to a single sapphire etalon; this makes such a unit a better mode selector. The etalons fabricated from quartz or sapphire have a thickness which is typically 2 to 3 mm. This assures a sufficiently large spectral separation of the reflectivity maxima within the fluorescence curve so that lasing can occur on only one peak. In multiple-plate resonators, the spacing between the etalons is 20 to 25 mm in order to achieve a narrow width of the main peak.

In an ideal resonant reflector the reflectivity as a function of wavelength shows very narrow peaks which are widely separated. The maximum reflectivity of a resonant reflector is

$$R_{\max} = (n^N - 1/n^N + 1)^2, \tag{10}$$

where n is the refractive index of the plates and N is the number of reflecting surfaces.

Fig. 8 shows as an example the spectral properties of a 2.5 mm thick single-element sapphire reflector and a three-plate quartz resonant reflector. The etalons are usually made from quartz or sapphire because both materials have high damage thresholds. The advantage of sapphire over quartz is that higher peak reflectivity can be achieved for the same number of surfaces. The peak reflectivities for single-, double-, and triple-plate resonant reflectors are 0.13, 0.40, and 0.66 if quartz is used, and 0.25, 0.66, and 0.87 in the case of sapphire.

Optimum mode selection from a resonant reflector is achieved only when the

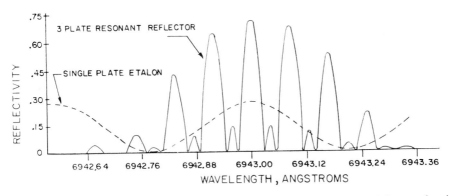

Fig. 8. Reflectivity versus wavelength of a single sapphire etalon (plate thickness 2.5 mm, refractive index 1.76), and a three-plate quartz resonant reflector (plate thickness 2.5 mm, spacing 25 mm, refractive index 1.45).

reflection maximum of the device is centered at the peak of the fluorescence curve of the active material. Ideally, the gain at the adjacent reflection maxima should be insufficient to produce oscillation. Temperature tuning is the normal means of shifting the reflectivity peaks of the reflectors relative to the laser linewidth.

2.3.4 Enhancement of longitudinal mode selection

Mode selection is considerably enhanced for operation close to threshold, reduction of the laser linewidth, shortening of the optical resonator and lengthening of the Q-switch pulse build-up time. The linewidth of solid-state lasers, in particular ruby, decreases for decreasing temperature. This reduces the number of axial modes which have sufficient gain to oscillate. Cooling of a ruby crystal as a means of obtaining single-axial-mode operation in combination with other mode-selecting techniques is discussed by McClung and Weiner (1965).

The spectral separation of adjacent axial modes is inversely proportional to the length of the resonator. Thus in a short resonator it is easier to discriminate against unwanted modes. For example, reliable single-mode output with energies per pulse of 2 to 4 mJ has been obtained from a ruby laser with a cavity only 28 cm long (Carman 1972). The laser consisted of a helical flashlamp, a pumped 7.5 cm by 0.6 cm rod, a 1.5 mm aperture, a two-plate resonant reflector and a dye cell for Q-switching. The 1.7 mm thick dye cell filled with cryptocyanine and methanol produced a 8 ns long pulse.

It was observed very early that passive dye Q-switches tend to act as mode selectors. The mode-selection property of the Q-switching dyes was explained by Sooy (1965) as a result of natural selection. Longitudinal mode selection in the laser takes place while the pulse is building-up from noise. During this build-up time, modes which have a higher gain or a lower loss will increase in amplitude more rapidly than the other modes. The difference in amplitude between two modes becomes larger if the number of round trips is increased. Therefore, for a given loss difference between the modes it is important for good mode selection to allow as many round trips as possible.

In fig. 9 the output power ratio of two modes as a function of number of round trips is plotted. The parameter is the difference in reflectivity which these modes experience at the mode selector.

The development of a pulse in a dye Q-switched laser takes longer than, for example, in the case of a Pockels cell Q-switched system (Daneu et al. 1966). As a result, in dye Q-switched systems, mode selectors such as etalons or resonant reflectors are more effective in discriminating against unwanted modes. In Sooy's paper specific reference is made to saturable Q-switching, but the analysis is equally applicable to the behavior of any Q-switch that ensures a large build-up time.

A Pockels cell Q-switch can be operated in a manner that ensures a large build-up time by increasing the rise time or by opening the switch in two steps.

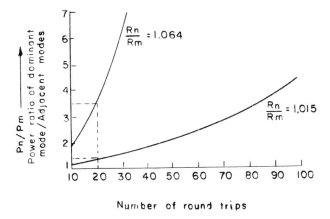

Fig. 9. Mode selection during the build-up of the pulse. The parameter R_n/R_m expresses the differences of the reflectivity which two modes experience at the mode selector during one round trip.

In this technique, initially the Pockels cell is only partially open and, therefore, presents a loss which is analogous to the low-level absorption loss of a saturable absorber. As a result of pumping, the net gain increases until it exceeds unity, and build-up of the giant pulse then starts. This build-up is monitored by a photodiode and its signal is used to trigger the Pockels cell to open completely when a preset signal level is reached. This is analogous to the bleaching of a saturable absorber at a particular intensity. Thus a long build-up time is achieved by an active Q-switch used in a way which is very closely analogous to a 'passive' saturable absorber Q-switch.

Combining several of the mode-control methods discussed in this section will provide essentially single-mode behavior from a Q-switched ruby laser. Commercial ruby laser systems designed for single-longitudinal-mode operation usually employ a Pockels cell Q-switch. As was mentioned before, this Q-switch provides a very precise external trigger capability and permits the generation of two Q-switch pulses during one pump cycle.

In a typical system, the rear reflector is a four-plate resonant reflector of the type shown in fig. 8. The fourth plate is added to increase the peak reflectivity of the device. The front reflector is a 2.5 mm thick single sapphire etalon. The main function of this device is to provide an optimum output coupling of 25% and to prevent oscillation at satellite peaks (spaced 0.086 Å from the main peak) of the multiplate resonator. Both mode-selecting elements are independently temperature controlled to within ±0.2°C. The laser head contains a 10 cm long by 0.96 cm diameter ruby rod pumped by a helical flashlamp. Operation at the TEM$_{00}$ mode is accomplished by inserting a 1.5 mm aperture into the 75 cm long resonator.

Single-axial-mode operation is achieved by drastically increasing the rise time of the Pockels cell voltage. In a standard Pockels cell Q-switch, the rise time of the voltage pulse on the crystal is typically 20 ns, and the laser output appears normally after 50 to 100 ns. In a 75 cm long resonator this time delay amounts to about 10 to 20 round trips for the energy to build up. In this particular case the rise time is reduced to 1 μs, and the Q-switch pulse is emitted at the end of this time period; this suggests that about 200 round trips occurred before the pulse was emitted.

The systems are all operated only slightly above threshold. Single-transverse-mode operation is obtained at output energies below 50 mJ from a single oscillator. Single-transverse- and longitudinal-mode operation is achieved at output levels between 10 and 15 mJ. In one system containing several amplifiers, single-mode output was 10 J, sufficient to obtain holograms with a scene depth of 5 m.

2.4 Multiple pulse operation

Techniques to extract multiple pulses from a ruby oscillator depend on the time separation between the pulses:

2.4.1 Pulse separation 1 μs to 1 ms

This time interval is the one most commonly used in double-pulse holography. The output is obtained from a standard single-pulse system by Q-switching the laser twice during its pump cycle. The longest pulse separation which can be achieved is determined by the length of the flashlamp pulse. The shortest time interval is determined by the switching electronics of the Pockels cell and the build-up time of the Q-switch pulse. The application of ruby lasers in double-pulsed holography requires that the energy in the two pulses is equal. This can be achieved by adjusting the delay between the flashlamp trigger and the first Q-switch pulse, by adjusting the voltage of the Pockels cell and by selecting the lamp input energy. By changing one or all of these parameters one can obtain equal output energies in both pulses over the time interval indicated above.

2.4.2 Pulse separation 20 ns to 1 ms

A technique which allows one to reduce the time interval between pulses essentially down to zero involves the utilization of a dual Q-switch oscillator. In these systems two giant pulses are extracted from different areas of the ruby rod. The system utilizes two separate Pockels cells with a double aperture in the cavity to select two separate TEM_{00} outputs from the rod, each one Q-switched with its own Pockels cell. The output of the laser consists of two beams separated by several millimeters and separable in time from essentially zero to 1 ms. The two output beams may be recombined by using a beam splitter and several mirrors or prisms. Multiple-cavity laser techniques are also discussed by Landry (1971, 1973) and Armstrong and Forman (1977).

2.4.3 Pulse separation 1 ms to 1 s

In order to extend the pulse interval time beyond the 1 ms range, the flashlamp has to be fired twice. Since the time interval is too short for a recharge of the pulse-forming network after the first pulse is issued, one usually resorts to the technique of charging two capacitor banks simultaneously. The two PFNs, decoupled from each other by means of diodes or ignitrons, are discharged at a time interval which is determined by the desired pulse separation.

2.4.4 Pulse separation larger than 1 s

In these cases the system is operated in the single Q-switched mode at a repetition rate according to the required interpulse separation.

2.5 Holographic cameras

In contrast to scientific applications usually performed in a laboratory, holographic lasers are often used as diagnostic instruments in an industrial environment. Therefore, protecting the optical components of the laser from dust and dirt can be important. Furthermore, a certain amount of mobility and transportability is also often required from a holographic system.

In response to these requirements several companies have developed so-called holographic cameras (Wuerker and Heflinger 1971, Koechner 1973, Close 1973, Rundle and Higgins 1975). In these units all components required to take a hologram are packaged in one enclosure. Fig. 10 shows a photograph of a holographic camera mounted on a tripod. The enclosure, which can be hermetically sealed and purged with nitrogen, protects the optical components from the environment. The unit contains a ruby oscillator and one amplifier stage, with appropriate beam splitters, mirrors, and lenses to obtain an off-axis reference beam. The output from the amplifier is projected through a beam splitter and expanded on to a diffusing screen by a negative lens. One of the two beams reflected from the wedge beam splitter is reflected within the holocamera by a series of mirrors to form an optical delay and is then expanded, collimated and reflected directly on to the hologram plate to form the reference beam. Since the oscillator beam is shifted in frequency as it propagates through the amplifier rod, it is standard procedure to derive the reference beam at the output of the amplifier. This assures that no frequency shift exists between the subject and reference beam (Siebert 1971).

The second beam reflected from the beam splitter is directed on to a photo detector to monitor the laser output. The object under investigation is placed some 1.5 m in front of the holocamera and illuminated from the diffuser output. The diffuser can be removed to use a diverging beam to illuminate the object. A selection of lenses is used to vary the angular divergence. The reference beam is

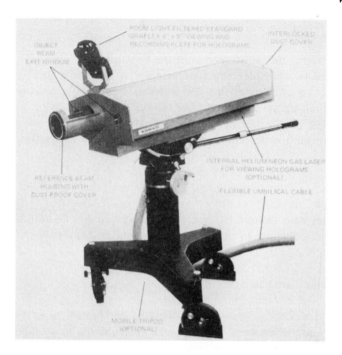

Fig. 10. Photograph of a holographic camera mounted on a tripod: (a) operational system; (b) with covers removed (Korad Div., Hadron Inc., Model KHC).

'folded' to allow object placement up to 2 m in front of the camera and still maintain equal reference and object beam path lengths.

The holocamera to object distance can be varied by modifying the path length of the reference beam. Within the reference beam path there is a provision to insert attenuators in order to adjust the intensity to the right level. Erf et al.

(1972) have found that while the reference-to-object beam ratio should be greater than unity, there is considerable latitude in this number, with ratios between 2:1 and 30:1 producing comparable results. The object beam emerges through a window on the front of the enclosure, whereas the reference beam is reflected onto the photographic plate mounted on top of the unit by means of a mirror housed in the tubular structure in front of the camera.

The oscillator consists of a high quality ruby rod which is optically pumped by a helical flash tube. The resonator is formed by a multi-element Fabry–Perot etalon rear reflector and a single-plate etalon output reflector. The etalons are used to select a single output mode. The temperature of the etalons is controlled to better than 0.05°C. A 1.7 mm diameter aperture is used to select a single transverse laser mode. The two laser pulses are produced by controlling the laser action by a Pockels cell. The laser oscillator produces pulses with an energy of 15 mJ/pulse.

The main technology of the laser is the design of the mode-selecting etalons and their temperature control to enable a single laser mode to be selected and to ensure that the same single mode is used for each of the two pulses. The quality of the laser mode structure defines the coherence length of the laser and hence the uninterrupted depth of field of the hologram.

The holographic camera shown in fig. 10 contains also a detector to monitor pulse energy and a He–Ne gas laser for alignment and pointing. Not shown in this photo is the electronics cabinet which is required to power the laser and electronically double-pulse the Pockels cell, and a small cooling unit required to maintain the laser's temperature. With two amplifiers containing 15 by 1 cm ruby rods pumped at 6 kJ, the system shown in fig. 10 is capable of producing 4 J of output in a single transverse and longitudinal mode. In the double-pulsed mode two pulses with 2 J of energy and time intervals from 1 to 1000 μs are generated.

At the high power levels which are achieved with pulsed lasers, the interaction of the laser radiation with the various optical components of the holographic apparatus must be considered in order to avoid damaging these components. Typically, a single-mode laser operated in the Q-switched mode has an energy output of approximately 40 mJ contained in a beam of 1.5 mm diameter. Assuming a 20 ns pulse width, the power density is 113 MW/cm^2.

Another point for consideration is that of multiple-element components (lenses, polarizers, etc.) which are cemented together. The cement, generally Canada balsam, has a low damage threshold – several orders of magnitude lower than glass – and cannot be used for laser applications. If multi-element components are required they must be air spaced.

At the higher power levels of Q-switched ruby lasers, care must be taken that reflections from curved surfaces are not transmitted back into the laser. The negative lens should preferably be a plano-concave element with the concave surface facing away from the laser and the flat surface slightly tilted such that the back reflection misses the output port of the laser.

The high pulse power also causes problems whenever the beam is focused to a point, because the focused high power ionizes air and damages any other material in its path. This means that spatial filters cannot be used to eliminate diffraction effects. Also, microscope objective lenses commonly used to expand the laser beam in cw holography must be replaced by diverging lenses instead.

Mirrors are essential to holographic systems. Because Al mirrors absorb 10% of ruby radiation, only the diverged beam should be reflected from Al-coated mirrors. The raw, nondiverged beam should be reflected only from mirrors capable of withstanding energies similar to those in the ruby oscillator. Beam splitters also need to be reflection- and antireflection-coated with dielectric materials. Neutral density filters, that are commonly used with cw lasers, will be destroyed at high pulsed laser energies. A set of glass filters is needed in which the absorbing material is diffused through the glass substrate.

A practical pulsed laser holographic system requires also diagnostic equipment to provide a measure of the laser operating characteristics on a shot-to-shot basis. The axial mode purity can be examined by photographically recording the fringes of a Fabry–Perot interferometer with a fixed spacing between 10 and 40 mm. A burn pattern taken by firing the laser beam onto a piece of unexposed Polaroid film is usually sufficient to obtain information about the transverse mode structure. Operation at the TEM_{00} mode will reveal itself as a smooth burn pattern with maximum intensity at the center.

Pulse shape and beam energy can be measured utilizing a fast-response detector and a calorimeter, such as a ballistic thermopile. With the fast-response detector one can determine, beside the pulse shape, the pulse separation and energy distribution in double-pulse operation. Also, ripple on the envelope of the pulse indicates mode-beating due to the presence of several axial modes. For alignment of a holographic system an autocollimator or a He–Ne laser can be used.

3 Recording materials

If one wishes to record a hologram, the recording medium employed must have a resolution capability which is at least equal to the fringe spacing d given by

$$d = \lambda/2 \sin \Theta, \tag{11}$$

where Θ is the angle between the normal of the photographic plate and the reference and object beam respectively, and λ is the wavelength of the laser. For a typical case, with $\Theta = 15°$ and $\lambda = 0.6943 \,\mu m$, the film must be able to resolve approximately $1/d = 750$ lines/per mm.

From eq. (11) it follows that practical off-axis holography requires high-resolution recording materials. Silver halide photographic emulsions are by far the most widely used recording materials. However, with the development of high-power coherent light sources it has become feasible to record holograms in materials other than photographic emulsions.

An illuminated hologram can reconstruct the wavefront it has recorded by spatially modulating either the amplitude or the phase of the illuminating light.

Photographic films which modulate the amplitude are, of course, suitable for forming photographs as well as holograms. On the other hand, phase-modulating materials, while of little use in photography, are ideal for holography.

Materials such as photopolymens have the potential for significant improvement over common photographic materials. At present, many of these materials have high resolution, but all are relatively insensitive.

In this subsection we consider materials especially made for the ruby or frequency doubled Nd:YAG laser or materials which show promise of being useful. Hologram recording materials are discussed in detail by Nassenstein et al. (1969, 1970), Collier et al. (1971) and Pennington (1971).

3.1 Silver halide photographic emulsions

Photographic emulsion primarily consists of extremely fine grains of silver halide compounds dispersed in gelatin. Also present in the gelatin are certain sensitizing agents. The emulsion is coated over a transparent substrate which is either a glass plate or flexible acetate film. When a two-beam interference pattern exposes a photographic plate and the plate is developed, an absorption hologram is formed. During the development process the exposed silver halide grains are converted to metallic silver. A bleach process may be used after development to convert the silver into a transparent compound; this in turn converts the absorption hologram to a phase hologram. Both transmission and reflection holograms can be recorded in photographic emulsions. For absorption transmission holograms the optimum average optical density to be achieved is about 0.6, while for absorption reflection holograms the density should be approximately 2.

In holography the choice of emulsion is much more critical for success than in photography. Choosing a film with too low of resolution for a particular problem in photography will lead to a substandard image. The same error in holography may lead to a recording with no capability to produce an image at all. The primary factors to be considered are resolution and required exposure. Other factors are concerned with noise properties, and special absorption and scattering properties. Table 4 provides properties of the most common photographic emulsions employed in pulsed holography.

Compared to standard photographic emulsions the materials listed in table 4 are fairly insensitive. This is the result of the requirement for high resolution. It is well known that high speed and high resolution are not obtained simultaneously in a silver halide emulsion. In fact, photographic speed decreases very rapidly with increasing resolution.

Once a laser has been chosen and the wavelength of the light to be used in recording a hologram has been determined, it is desirable to select the emulsion most sensitive to that wavelength. Fig. 11 shows spectral sensitivity curves for several emulsions. The curves illustrate that it is possible to obtain emulsions which are uniformly sensitive over the visible spectral range (panchromatic),

Table 4.
Silver halide emulsions suitable for pulsed holography (the emulsions are
typically 6 μm thick).

Recording material	Exposure[a] [erg/cm^2][b]	Resolution [lines/mm]	Laser source
Agfa 8E75	200	3000	ruby
Agfa 10E75	50	2800	ruby
Agfa 10E56	50	2800	YAG
Kodak 649F	1000	2000	ruby, YAG
Kodak SO-343	1000	2000	YAG
Kodak HRP	1000	2000	YAG

[a] Exposure required for density 1.
[b] 1 erg $\cong$ 0.1 μJ.

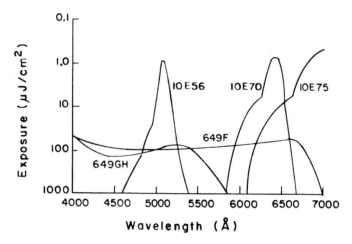

Fig. 11. Spectral sensitivity curves for photographic emulsions. From Collier et al. (1971).

sensitive only in the blue-green end of the spectrum (orthochromatic) or selectively sensitive at common laser wavelengths.

When exposed for the very brief period of a few tens of nanoseconds by a single pulse from a ruby laser, silver halide photographic emulsions exhibit a so-called reciprocity failure. That is, for a given energy density on the film and a given development process, an exposure time in the nanosecond range produces less optical density in the developed emulsion than longer exposure times. For example, for the Agfa emulsions listed in table 4 reciprocity failure associated with exposure times in the range of 10 to 50 ns requires an increase in energy density by a factor of 2 to 4 (Hercher and Ruff 1967, Collier et al. 1971).

The diffraction efficiency obtained in Agfa 10E75 as a function of optical density is shown in fig. 12. The maximum diffraction efficiency of 1.88%, which occurs at a density of 0.7, was observed to increase with development time (Landry and Phipps 1975). A hologram should be made with its normal bisecting

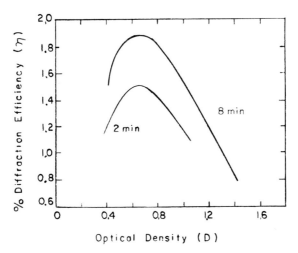

Fig. 12. Diffraction efficiency of single exposure plates as a function of optical densities for two development times. Intensity ratio of reference to object beam was one. Landry and Phipps (1975).

the angle between object and reference radiation to reduce the effects of emulsion shrinkage. Emulsion shrinkage is largely normal to the plane of the emulsion. Therefore, if the above criterion is met, fringe planes will also lie normal to the emulsion and shrinkage will not change the fringe spacing. This effect becomes more severe for thicker emulsions and for volume holograms.

3.2 Thermoplastic films

Several desirable properties make the thermoplastic film potentially an important recording material. It is highly photosensitive to all visible light, thus including the ruby wavelength, and it requires no wet development. A hologram recorded in this material has moderately high diffraction efficiency and is stable at room temperature until intentionally erased. The material can be reused a number of times. In addition, the recorded hologram behaves nearly ideally as a plane phase hologram (Collier et al. 1971, Friesem and Tompkins 1973).

Holograms can be recorded in a thermoplastic film by causing its surface to deform in accordance with the light intensity variations of holographic interference patterns. Thermoplastics are usually not photosensitive and must be combined with a photoconductor in a film structure which can respond to light.

The sequence of a complete hologram recording–erasure cycle is as follows. Initially a uniform electrostatic potential is applied to the surface of the thermoplastic film. This is accomplished by depositing charges from a corona discharge on the film surface. During the exposure the photoconductor causes the charges to neutralize at the areas where light is received. The incident light causes a spatially varying electrical field pattern in the thermoplastic film. During the development stage heat is applied momentarily to the thermoplastic, raising its temperature to near the softening or melting point, typically between 60 and 100°C. At this temperature the thermoplastic film deforms under the force of the local electric field and becomes thinner at high-field areas and thicker elsewhere. The surface relief variations or thickness deformations are a true image of the electric charge distributions on the film surface which in turn is caused by incident light distribution. Cooling quickly to room temperature, the deformation is frozen in and a hologram is thus recorded as a thickness variation of the thermoplastic. The recording is stable at room temperature.

If it is desired to erase the hologram, heat again is applied to the thermoplastic, raising its temperature higher than that for the development. At this elevated temperature the surface tension of the softened or molten thermoplastic evens out the thickness variation and hence erases the hologram. The increased electrical conductivities of both the photoconductor and the thermoplastic during the erasure usually cause a complete neutralization of the electrostatic charges.

Thermoplastic film is commercially available from Fuji Photo Optical Co., Ltd., Japan, and Kalle-Hoechst AG., Germany. Table 5 lists typical characteristics of thermoplastic film suitable for pulsed holography. As follows from this table, thermoplastic films have a lower resolution and require a higher exposure energy compared with the best photographic emulsions.

3.3 Photopolymers

Photopolymer materials usually consist of a vinyl or acrylic monomer, a photosensitive initiator system and a polymeric binder. The formation of phase volume holograms in these materials has been explained by Colburn and Haines (1971).

During exposure the monomer in areas of higher intensity illumination is polymerized to a greater extent than that in lower intensity areas. The binder reduces the shrinkage that would normally occur, so initially the effect of the exposure is to lower the refractive index because of the slightly lower molecular polarizability of the polymer. The exposure creates concentration gradients in the remaining monomer, and additional monomer diffuses into the region of higher exposure. The hologram is then fixed by a uniform over-all exposure that

Table 5.

Characteristics of thermoplastic film suitable for pulse holography[a].

Spectral sensitivity	0.35–0.7 μm
Exposure (ruby laser)	80 μJ/cm^2
Resolution	1000 lines/mm
Diffraction efficiency	10%

[a] Supplier: Fuji Photo Optical Co. Ltd., Japan; Kalle-Hoechst Ag., Germany.

polymerizes the remaining monomer. Because of the diffusion of the monomer, the resulting polymer concentration is greater in the regions that initially received the more intense exposure, and thus they have a higher refractive index.

Desirable properties of these materials are high diffraction efficiency, rapid in-place development, no wet processing and good resolution. At present, photopolymers have the disadvantage that they are not very sensitive and that they can be used only in the blue-green spectrum. No red-sensitive material for ruby has been developed yet.

As an example, table 6 lists the characteristics of an experimental holographic photopolymer made by DuPont (Booth 1975). The material is sensitized in the blue-green spectral region with a dye sensitizer. The photopolymer can be coated on any glass or polyester film base, starting with the material in liquid form. Exposures can be made with a frequency doubled Nd:YAG laser. Once exposed the material is stable even under extreme environmental conditions.

Recently a multi-component high-resolution photopolymer system was described in which a modulation of the refractive index is not caused by a density modulation but is due to a composition modulation caused by the interaction of several monomers with different rates of polymerization (Tomlinson et al. 1976).

4 Experimental techniques and applications

4.1 Pulsed holographic interferometry

Conventional interferometry can be used to measure small path-length differences of optically polished and specularly reflecting flat surfaces. Holographic interferometry extends this range by allowing measurements to be made on three-dimensional surfaces of arbitrary shape and surface condition. There are three variations to the basic holographic interferometric technique: real time, time average and double-exposure interferometric holography. In pulsed holography only the double-exposure technique is of interest (see, for example, Wuerker and Heflinger 1970, Collier et al. 1971, Erf 1974).

Table 6.
Properties of DuPont's holographic photopolymer
(Booth 1975).

Exposure wavelength	0.35–$0.53\,\mu$m
Exposure energy	20–$30\,\text{mJ/cm}^2$ in air
	2–$3\,\text{mJ/cm}^2$ in oxygen-free
	atmosphere, such as nitrogen
Spatial frequency	up to 3000 lines/mm
Diffraction efficiency	up to 100%
Thickness	5–$150\,\mu$m

The double-exposure method consists of recording two successive holograms of the object on the same photographic plate. We assume that the surface of the object is deformed or displaced between the two exposures. Upon reconstruction of the hologram, two three-dimensional images of the object will be formed. Since both reconstructed images are formed in coherent light and exist at approximately the same location in space, they will interfere with each other and produce a set of bright and dark interference fringes (in the reconstructed image). The fringes represent contours of equal displacement along the viewing axis.

In order to relate the fringe pattern to the relative change in object configurations we turn to fig. 13 which shows schematically two different positions of a vibrating object surface. From two holograms of these positions z_1 and z_2, successively recorded on the same plate, the object waves S_1 and S_2 can be reconstructed simultaneously so that they interfere.

The first object wave S_1 shall be described by an amplitude S_1 and a phase φ_1:

$$S_1 = |S_1| \exp i\varphi_1. \tag{12}$$

The second object wave S_2 shall have a phase difference $\Delta\varphi$ as compared to S_1:

$$S_2 = |S_1| \exp i(\varphi_1 + \Delta\varphi) = S_1 \exp i\Delta\varphi. \tag{13}$$

After reconstruction the intensity is given by

$$|S_1 + S_2|^2 = |S_1 + S_1 \exp i\Delta\varphi|^2 = 4|S_1|^2 \cos^2 \tfrac{1}{2}\Delta\varphi. \tag{14}$$

A minimum of the intensity is obtained if the following condition is satisfied:

$$\Delta\varphi = (2n + 1)\pi, \quad n = 0, 1, 2, 3, \ldots, \tag{15}$$

where n is an integer.

The phase difference $\Delta\varphi$ between two object points separated by d follows from fig. 13. The optical path of the object wave S_2 is longer than the optical path of the object wave S_1 by the amount of $d(\cos \alpha + \cos \beta)$. α and β are the angles of illumination and recording with respect to the surface normal. Relating

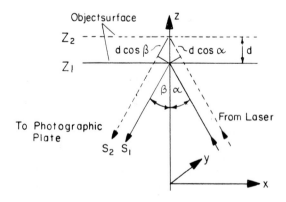

Fig. 13. Phase difference $\Delta\varphi$ between two object points.

the phase difference to the laser wavelength λ, we obtain:

$$\Delta\varphi = d(\cos\alpha + \cos\beta)2\pi/\lambda. \tag{16}$$

From eqs. (15) and (16) one obtains

$$d = (2n + 1)\lambda/2(\cos\alpha + \cos\beta). \tag{17}$$

The minima (dark fringes) are at those points where the displacement d satisfies eq. (17). For any two points on the object the relative displacement Δd can be calculated by counting the n interference fringes between them.

The distance between n interference fringes indicates a displacement normal to the surface of

$$\Delta d = n\lambda/(\cos\alpha + \cos\beta). \tag{18}$$

The division of Δd by the pulse separation time Δt gives the relative vibration velocity. It is proportional to the number of interference fringes per unit length. This number is called fringe density.

In case of several vibration centers with nodal lines between them on the object, the holographic measurement of amplitudes is possible if one knows at what vibration phases the first and second hologram was taken. In this case an accelerometer attached to the area of interest provides the trigger signal for the laser pulse. Employing electronic delay generators, any part of the oscillation can be examined.

In fig. 14 the lower curve is a graph of the measured acceleration signal of a point of the holographed object surface versus time. The two peaks at the times t_1 and t_2 are the laser trigger pulses. Because the vibration normally is of harmonic-type, double integration of the acceleration signal leads to the displacement versus time which corresponds to the upper curve. The surface

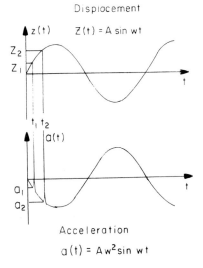

Fig. 14. Harmonic oscillations of an object point.

displacement between the first and second pulse is given by

$$z(t_2) - z(t_1) = A(\sin \omega t_2 - \sin \omega t_1). \tag{19}$$

Since $z(t_2) - z(t_1) = \Delta d$ we obtain from eqs. (18) and (19) the relationship between displacement and oscillation amplitude as follows:

$$A = n\lambda/(\sin \omega t_2 - \sin \omega t_1)(\cos \alpha + \cos \beta). \tag{20}$$

The number n can be determined by counting the fringes in the interference picture beginning at a nodal line and finishing at a vibration center. Using this technique, amplitudes of any magnitude can be determined. However, the times t_1 and t_2 have to be chosen such that well-defined interference patterns result. Eq. (20) represents the fringe pattern and the amplitude distribution for one instant in time.

The principles of fringe formation and localization in holographic interferometry are discussed in the works of Haines and Hildebrand (1966), Sollid (1969), Shibayama and Uchiyama (1971) and Collier et al. (1971).

4.2 Environmental considerations

Pulsed lasers, owing to their short exposure times, possess the qualities of overcoming the adverse effects of a practical industrial environment. The particular hostile environmental characteristics which prohibit the use of long exposure methods include vibration, object movement and ambient illumination.

4.2.1 Vibration and maximum allowable object movement

During the exposure the maximum subject motion must not change the optical path of the subject light reflected to the hologram by as much as $\frac{1}{4}\lambda$. Otherwise the interference fringe pattern will be completely blurred and not recorded. In single-pulse holography, vibrational effects upon the holographic process are essentially nonexistent since the pulse duration, on the order of 20 ns, is short enough to remove these effects entirely.

Let us consider the maximum speed of a subject point allowed in the hologram-forming geometry of fig. 13. Object motion along the ellipse, in which one of the foci represents the source and the other foci the photographic plate, is not restricted, since the relative distances remain constant. Maximum change in the relative path lengths occurs when the motion is perpendicular to the elliptical surface (along the z-axis). Suppose we define the maximum allowed displacement d in that direction during an exposure Δt to be that producing a change $\Delta l = \frac{1}{4}\lambda$ in the object path. If we assume $\alpha = \beta$ it is seen from fig. 13 that

$$\Delta l = 2d \cos \alpha = \frac{1}{4}\lambda \tag{21}$$

or

$$d = \lambda/8 \cos \alpha. \tag{22}$$

The maximum allowed speed of the object is then $v = d/\Delta t$ or

$$v = \lambda/8\Delta t \cos \alpha. \tag{23}$$

Assuming $\alpha = 30°$, $\lambda = 0.6943 \,\mu$m and $\Delta t = 30$ ns, we find $v = 3.3$ m/s.

4.2.2 Ambient illumination

Fogging of the holographic plate due to ambient illumination during the exposure will degrade reconstructed object intensity and, if the extraneous exposure is sufficient, obscure the image completely. The photographic plate has to be protected from fogging by a multilayer dielectric film bandpass filter or an absorption-type cut-off filter (for example, Schott glass # RG-665). If, in addition, a focal plane shutter is mounted in front of the filter–photographic plate assembly, holograms can be made under daylight or normal room light conditions. The opening of the mechanical shutter has to be synchronized with the laser. Solenoid-operated mechanical shutters with free apertures of up to 6 in are commercially available with minimum opening times of 0.4 to 0.6 s.

4.3 Nondestructive testing

From eq. (18) it follows that adjacent fringes correspond to relative changes in object configuration of $\frac{1}{4}\lambda$ (assuming $\alpha = \beta = 0$). With a ruby laser as coherent light source, the measurement of displacments in the order of $0.2 \,\mu$m is therefore feasible.

Because of the extreme sensitivity to surface deformations, holographic interferometry can be used to gain information with regard to the structural characteristics of a component by observing the surface movement produced when the component is subjected to a small perturbation. Ideally, one pulse is made with the subject in an unperturbed condition and the second pulse is made with the subject in a perturbed condition. Anomalies in the observed fringe pattern can then reflect the nature of a particular defect.

If the interpulse separation is long, the particular advantages of pulsed laser illumination are lost; components and subject must be free from any extraneous motion, otherwise a secondary set of fringes related to the motion will be formed which may obscure the stress-induced fringes. It is, therefore, preferable that both laser pulses be generated during the same flashlamp pump pulse so that pulse separation in the tens of microseconds regime is achieved; this eliminates the need for vibration-isolated mounting configurations. However, it also unfortunately precludes the use of static stressing of the component under examination, requiring instead the application of a ramp- or impulse-type of stressing force.

Typically, a nondestructive testing equipment comprises a holographic camera which contains the coherent light source and associated optical components necessary to take holograms, an appropriate test object perturber which modifies the surface of the test object between the two exposures and a hologram reconstruction set-up.

The surface of the test object can be modified in several ways. Common techniques are acoustic, thermal, pressure and mechanical stressing. By combining holography with any of these excitation techniques, it is possible to interferometrically record the physical properties of a large variety of objects ranging from laminated panels, composite materials, tires, engines, turbine blades to automobiles.

Pulsed laser hologram interferometry does not replace measurements made by other methods, such as strain gauges, but rather complements and enhances the accuracy of the older techniques. The hologram can show the deformation of the complete surface of a test object at one instant of time, while the strain gauge shows surface strain at one point for the entire test period. Hence, holography may yield a more complete 'picture' of an unknown object deformation. Also, the hologram proves to be useful for indicating the optimum location for placing the strain gauge. Another advantage of holography is that it does not require any loading of the object nor the installation of sensors. On the other hand, the holographic fringe patterns, which can be quite complicated, must be interpreted to obtain an understanding of the pattern. Interferometric holography, when taken in combination with various stress mechanisms, offers considerable promise in the area of nondestructive testing.

We will illustrate the viability and versatility of pulsed holographic interferometry by briefly describing some of the applications. It is, of course, impossible to review all the excellent work carried out in this area. For a detailed discussion of holographic nondestructive testing the reader is referred to Heflinger and Brooks (1970), Wuerker and Heflinger (1970), Collier et al. (1971), Chu et al. (1972), Erf et al. (1972) and Erf (1974).

4.3.1 Vibration analysis

One area in which holographic nondestructive testing (HNDT) has proved to be quite successful is that of surface vibration analysis, especially useful to industries like the automotive, aircraft and heavy equipment industries, or wherever deformations or motion must be measured with extreme accuracy. Such vibration studies can result in the elimination of a particular vibration, and therefore the elimination of a source of noise or a potential part failure.

In the automotive industry vibration studies have been made on the frame, engine, gearbox, tires, chassis, disc brakes and various other components of automobiles (Felske and Happe 1973, 1974, 1975). Figs. 15 and 16 illustrate some typical examples of the application of double-pulsed holography in the field of sound and vibration analysis. In order to reduce the transfer of solid body

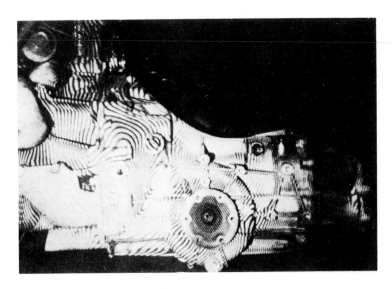

Fig. 15a. Vibration pattern of a gearbox at 6200 rev/min. The right side shows a nearly parallel and equidistant fringe system from which it can be concluded that this part of the gearbox had the required stiffness.

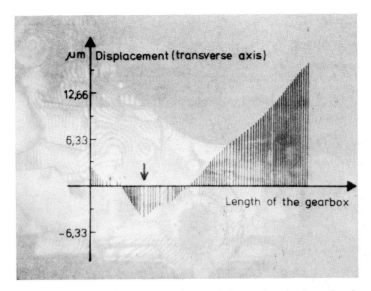

Fig. 15b. The evaluation of the interference picture of the gearbox is shown by the graph of the displacement in transverse axis versus the length of the gearbox. The displacement is evaluated along the horizontal axis of the drawn coordinate system. The displacement curve shows a bending within the area of the clutch-housing. Felske and Happe (1975).

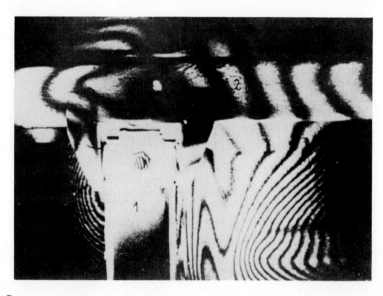

Fig. 16a. ① is a gearbox-carrier at 4020 rev/min. The parallel and equidistant fringe system refers to a rotational movement around an axis parallel to its own longitudinal axis. This movement is not transferred to the longitudinal member ②. The vibration pattern on ② is originated by the frontside engine carriers.

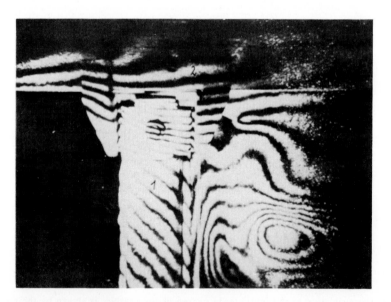

Fig. 16b. At 4680 rev/min the gearbox-carrier of fig. 16a shows quite another vibration pattern which is transferred to the longitudinal member ② and from this to the bottom of the car. Felske and Happe (1975).

vibrations from the engine–gearbox combination into the passenger cell of a car, it is very important to find out the points of relatively small vibration in the engine or gearbox shell so that one can use these points for mounting the engine. The normal way of mounting the drive unit into a car is to support it on both sides of the crankcase and at one point of the gearbox. Therefore, it is important to have a gearbox with a high degree of dynamic stiffness in order to keep solid body vibrations at a minimum. Fig. 15a shows the interference picture of a gearbox driven by an engine running at 6200 rev/min.

Fig. 15b shows a graph of the elongation in transverse axis versus the length of the gearbox as it is evaluated from fig. 15a. We see that there is a bend of the elongation curve where the clutch housing begins. That means that within this area the housing does not have the required degree of stiffness. The interference pattern on the clutch housing shows its surface deformation and gives information where stiffening ribs must be made.

Characteristic variations of the noise level of automobiles as a function of the speed of rotation are often caused by resonances of the engine and gearbox supporting structures. Fig. 16a shows an interference picture of the gearbox support at 4020 rev/min taken from underneath the car. The support member is covered by a parallel equidistant fringe system which stops at the end of the member. The movement of the support unit is a rotation around an axis parallel to its fringe system, but this movement is not transferred to the longitudinal member of the car. The fringe system covering this is originated by the frontside carriers of the engine. It is transferred to the bottom of the car body and from this the sound is transmitted into the passenger cell. Fig. 16b shows the interference picture of the same gearbox support holographed at 4680 rev/min. The fringe system on the support structure is now continued on the longitudinal member and from here transferred to the bottom of the car, but the fringe density on the bottom is now smaller than at 4020 rev/min.

4.3.2 Shock propagation

The double-pulsed holographic technique can be used to study transient vibrations, such as those induced by impact loads. It is possible to observe the movement of mechanical shock waves across an object (Evenson et al. 1972). With the extensive use of composite materials in the aircraft industry, shock and stress wave propagation data in these materials has been of interest.

The optical arrangement employed to study the stress wave propagation in a graphite composite is shown in fig. 17. Neutral density filters are used to control the intensity ratio of reference and object beams. A wedge beam splitter is used to separate the reference beam and the object illuminating beam. Two single-element diverging lenses expand the beams to the necessary diameter. A diffuser in the object illuminating beam helps to provide a uniform intensity distribution at the object (Gregor 1970).

The graphite composite plate was rigidly mounted to a frame structure and

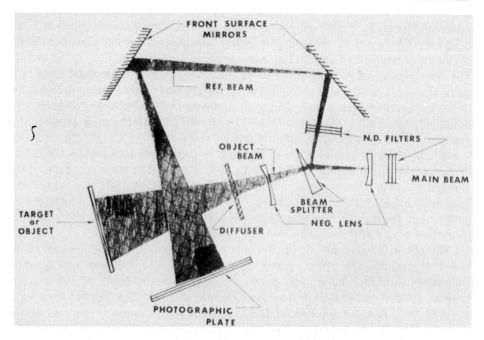

Fig. 17. Holographic camera set-up.

held in place with large clamps. For control experiments, an Al plate was similarly mounted. The samples were excited from behind by the impact of a ballistic pendulum at the center of the clamped plates. The pendulum comprises a steel ball attached to a thin wire. The triggering of the laser was accomplished by contact closure between the steel ball and the specimen plate. An adjustable accurate electronic delay provided a trigger signal to the laser at a specified time period after the impact of the steel ball. The steel ball was released by an electromagnet from a fixed height to give a consistently repeatable impulse to the specimen plate.

One sample was a 25 cm by 25 cm four-ply graphite composite material of 0.89 mm thickness. The second sample was a homogeneous Al plate with a thickness of 2.3 mm. Two holograms were recorded, one before and one after the impact of the steel ball. As shown in fig. 18a the stress wave velocity depends on the direction in the graphite composite material. However, for the Al plate, the shock wave propagates radially symmetrical from the point of impact as expected (fig. 18b).

The experimental results reported here demonstrate several important facts about the composite material under study. Since the stress wave velocity of propagation is proportional to the Young's modulus of the material, it can be readily observed that this modulus is a function of direction in the composite sample (fig. 18a). Thus effects of fiber orientation, fiber matrix designs and bonding techniques can be evaluated by using the data from holographic interferograms such as fig. 18a.

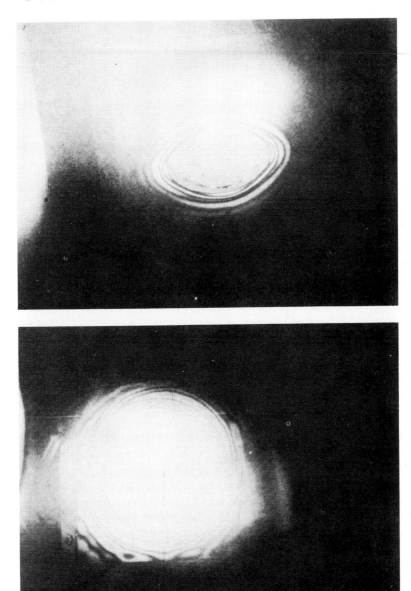

Fig. 18. Shock wave in a composite plate (a), and Al plate (b) 60 μs after impact of steel ball.

Double-pulsed holography has been shown to be a valuable tool in investigating the basic nature of shock transmission and bump absorption in tires. Extensive impact wave propagation tests have been conducted to examine the reaction of a tire when it strikes a sharp bump. In one experimental set-up (Potts 1973, Potts and Scora 1974) the holocamera is aimed at a tire mounted to a rigid

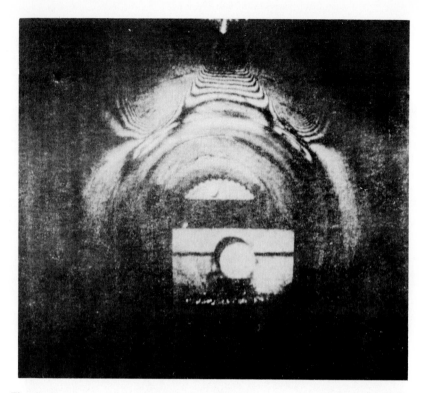

Fig. 19. Shock propagation signature in an H78-15 bias-ply tire. Potts and Scora (1974).

test stand. An impulsive force is imparted to the tire by dropping a weighted pendulum onto the tread of the tire. Dropping the pendulum from 45° above horizontal produces a simulated rolling speed of 20 mile/h. Timing of the laser pulses is accomplished by a switching circuit tripped by the pendulum just prior to striking the tire. The pulse timing was varied by adjusting the height of this switch via a micrometer screw. This adjustment was necessary in order to capture shock waves, in a series of exposures, progressing from the point of impact through the tire–wheel assembly.

The tests carried out on bias-ply, belted bias and radial-ply tires revealed that the shock waves follow the cord path through the body plies and into the wheel. Fig. 19 shows an example of the shock propagation in a bias-ply tire.

Double-pulsed holographic interferometry performed on transient vibrations indicates that extremely good quality results are obtainable and that there is a large potential of useful information relating to the understanding of how components react to impact loads. The observation of shocks travelling within a component is also a potential NDT technique, as faults can be detected by abnormalities in the progress of a shock wave.

4.3.3 Stress and deformation analysis

There is considerable interest in the observation of rotating components, especially for flutter studies. The tip speeds that can be observed by double-pulsed holography are limited by the movement due to rotation that occurs between the two pulses. The principle interest in applying double-pulsed holography to the investigation of rotating turbine blades is the measurement of vibration and aerodynamic effects with the blade-tip speeds in the region of 500 m/s.

Under certain conditions, which include special arrangements of optical elements and short interpulse separation, it is possible to record holograms of a spinning turbine blade (Overoye et al. 1973). It was shown that the holographic recording process was least sensitive to motion of the blade if the angles of the illuminating beam on the blade and the viewing direction (α and β respectively) through the hologram were selected so that $\cos \alpha + \cos \beta$ was approximately equal to 0.

The use of ultra-short pulse separations is not the solution to the problem since under the rotating conditions the blade vibrational frequencies, which are of interest, are below 1 kHz. To measure these vibrational deflections pulse separations of 50 to 500 μs are required. To permit such pulse separations Hockley (1976) proposed to use a rotating-prism device to compensate for the rotation of the disc so that image stability can be obtained over a 500 μs period.

Another interesting HNDT application involves a deformation study of a printer type-piece. In this study, holographic interferometry with a pulsed ruby laser was used to determine printer type-piece deformation caused by impact of the print hammer. A variety of type-piece designs was examined. Effects of impact velocity and hammer–type-piece alignments were measured. The investigators found a considerable amount of type-piece twisting caused by the impact of the print hammer, a condition that was unexpected and unknown prior to the study (Wilson and Strope 1972).

4.3.4 Structural analysis of laminated and composite materials

Holographic NDT methods have proven to be a preferred means for inspecting sandwich or laminated structures or panels used in the aircraft and aerospace industries. With a holographic system, defects such as the following can be located and identified: cell wall separation, separations between laminates, separation between skin and bonded structure members, debonds between laminate layers and adhesive porosity (Erf 1974).

Fig. 20 shows a double-pulse hologram of a laminated panel. The panel, comprised of a honeycomb core epoxied to thin Al surface plates, is a Fan Reverser panel from a jet engine. The panel was excited by a transducer which was clamped directly to the lower rear surface of the panel. The fringes indicate a small debonded area.

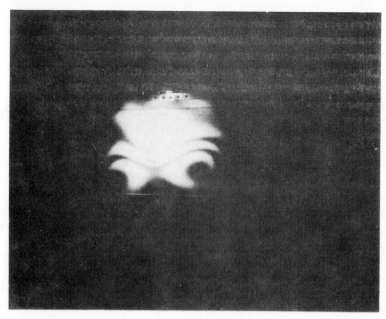

Fig. 20. Double-pulse hologram of honeycomb panel showing small debond near center of panel.

4.4 Flow visualization

An important application of pulsed laser holography is the study of gas dynamics in wind tunnels. Some of the information desired includes airflow patterns, formation of shock waves, density and velocity profiles, and the like. Holograms may be taken of a model or a full-size test object in a wind tunnel to aid in flow visualization. Three-dimensional density distribution as well as three-dimensional shock distribution are easily identified by holographic interferometry.

Shadowgraph, schlieren and interferometric techniques have long served the aerodynamicist as an invaluable tool in gas dynamic studies, providing a means for visualizing flow patterns and determining density gradients. It is not surprising, therefore, that considerable effort has been expended on integrating these highly sensitive flow visualization systems with the high-speed, three-dimensional recording and storage capabilities of pulsed holography (Chow and Mullaney 1967, Havener 1970, Reinheimer et al. 1970, Erf et al. 1973, Trolinger 1973, 1974).

Producing holograms of gas flows using transmission holography is less demanding on the laser coherence properties than normal reflection holography. The main requirement is to observe an event that will produce a sufficiently large refractive index change to form interferometric fringes.

Fig. 21 illustrates a typical off-axis trans-illumination holocamera. In this case a reference beam is split from the original laser beam and is usually routed around the test region. At the recording plane it is mixed at some angle with the

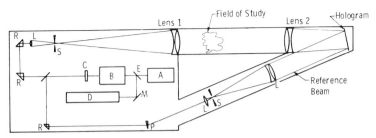

Fig. 21. Holographic flow visualization set-up. Trolinger (1974). A – Q-switch; B – Ruby laser head; C – Front laser etalon; D – Alignment laser; E – Removable mirror; R – Right-angle prisms; M – Mirror; L – Lens; S – Spatial filter.

object wave. Providing the reference wave in this manner allows more versatility in illuminating the scene. For example, the scene can now be illuminated with diffuse light. Interferometry is accomplished by superimposing two images of the same field as it existed at separate times. Slight changes in the optical path through the field (such as that caused by refractive index changes) cause phase differences in light emanating from the separate images and are manifested as interference fringes in the superimposed images.

Fig. 22 is a reconstruction of a spherical cone model in a Mach 8 flow. The field diameter was 50 cm. The double-pulse hologram was made with and without airflow.

Present-day transonic fan and compressor rotor blade rows operate with

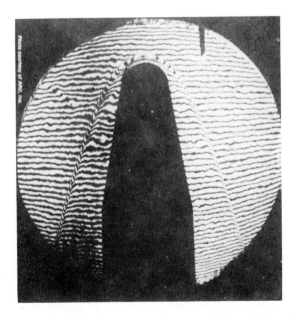

Fig. 22. Visualization of bow shock wave of a stationary model in a windtunnel at Mach 8 airflow. Trolinger (1973).

supersonic velocities relative to the blades over a large part of the blade span. To aid in the development of high performance designs, it is desirable to know the actual shock patterns and flow conditions within the rotor blade rows. Double-pulsed holography interferometry has been employed for flow visualization within rotating compressor blades (Wuerker et al. 1973, Benser et al. 1974, 1975).

In another application, holographic interferometry was used to measure the three-dimensional static density field in the transonic flow from large nozzles (Clark et al. 1976).

The development of pulsed lasers enables plasma physicists to take 'stop action' space-resolved interferograms of highly self-luminous, transient plasma discharges (Jahoda and Siemon 1972). Employing double-pulsed holographic interferometry, Bernard et al. (1973) measured the electron density in a plasma as a function of time and space through a noncylindrical glass vessel with very poor optical quality.

4.5 Particle studies

Microscopic objects can be photographed and studied in their three-dimensional environment using holography. Holographic techniques leave the flow undisturbed and permit one to determine the exact size, shape and distribution of particles suspended in liquids or in air. The short light pulse effectively 'freezes' the movement of small particles in dynamic aerosols. The hologram reconstructs the small objects in space. These reconstructed images form an exact three-dimensional model of the original distribution (Matthews et al. 1968, 1972, Trolinger 1974).

Pulsed holography is ideal for the high-resolution analysis of a dynamic, nearly transparent, three-dimensional field of light-scattering elements. Typical areas of application include the measurement of droplet size and distribution of fuel mixtures in carburetors, in the spray cone of fuel injector nozzles, in the spray emitted from spray cans, or in the spray of nozzles distributing insecticides.

Fig. 23 shows a set-up which permits one to study the distribution of aerosols emerging from a nozzle (for example, a spray of insecticides). The nozzle is mounted on one side of a chamber, the pressure of which can be changed, therefore allowing one to study the aerosol distribution as a function of atmospheric pressure. The object beam is directed through windows into the chamber. This beam which intersects the spray cone illuminates the droplets emerging from the nozzle. The forward-scattered light of the object beam exposes the photographic plate and causes interference with the reference beam which is directed around the spray cone.

If the particle field is not too dense, the more standard in-line holography technique can be applied. The set-up shown in fig. 23 has been successfully used to obtain qualitative and quantitative information about the spray produced by

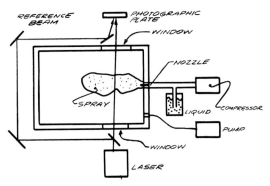

Fig. 23. Set-up for holographic measurement of aerosols emerging from a nozzle.

two prototype turbine fuel injector nozzles. Single- and double-exposure off-axis reference holograms were taken of the spray which was illuminated with a thin (5 mm thick) sheet of coherent light from a ruby laser. The partial illumination of the spray cone removed the scattering of the unilluminated part of the spray, and thus reduced the optical background noise sufficiently to reconstruct particles down to 15 μm with good edge contrast (Koechner 1973).

Fig. 24 shows the technique employed to reconstruct the hologram. The particle field is reconstructed by illuminating the hologram with the collimated beam from a He–Ne laser. A combination of lenses magnifies the reconstructed image, which is then focused onto the face of a closed-circuit television camera vidicon tube and observed on the monitor screen. A wire grid is imaged on the vidicon tube for calibration purposes. The TV camera and lenses remain fixed after the desired magnification is set. The reconstructed volume is scanned by moving the hologram, and consequently, the reconstructed images in three

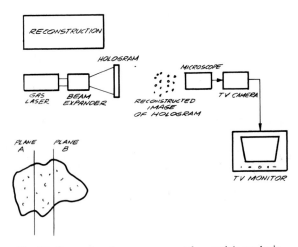

Fig. 24. Reconstruction arrangement for particle analysis.

orthogonal directions. As the reconstructed sample volume is scanned, the various particles come into sharp focus at their correct position in space.

Reconstructed images can initially be scanned under low magnification to provide a quick examination of the volume. Portions can then be enlarged by increasing the magnification of the optical system. Noise from irregularities on glass surfaces in the reconstruction system usually limits useful lens magnification to about 10. The television system electronically magnifies the images approximately 25 times, giving an overall maximum magnification of 250 times. The system could resolve particles down to 15 μm in size. Fig. 25 shows a photograph of a reconstructed hologram.

A single hologram of a particle field can reconstruct thousands of images. Analysis of each for size, velocity and position becomes a tedious and time-consuming process. To augment data acquisition, computerized image analyzers are sometimes applicable. For example, one such device analyzes the video

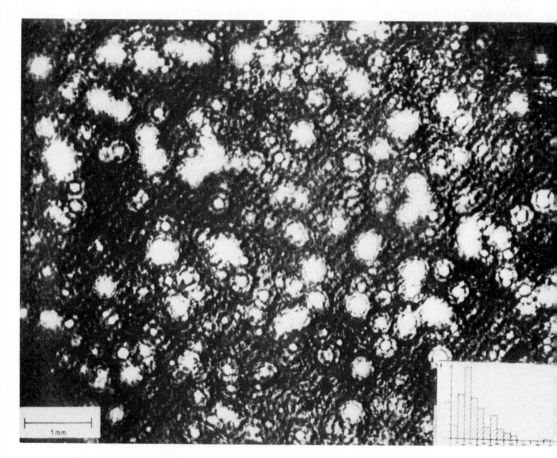

Fig. 25. Photograph of a hologram showing fuel droplets ejected by a nozzle. A histogram of the droplet size is shown to the right. Felske and Happe (1975).

signal arising from a television camera, counting number of images in focus and measuring the size distribution.

4.6 *Portraiture*

The development of multi-stage, long-coherence-length ruby lasers has made it possible to record human subjects (Siebert 1968, Ansley 1970). Owing to the short pulse duration of a Q-switched ruby laser, mechanical instabilities and object movements are eliminated. The distinctive features of the laser systems employed in holographic portraiture are a combination of high energy output and long coherence length. A hologram of a single person requires a minimum energy of 250 mJ and a coherence length of 1 m. Holographic group portraits are usually made with 4 to 10 J of energy and a coherence length of 5 to 10 m (Koechner 1977).

Fig. 26 shows the arrangement for forming a transmission hologram of a human subject. The most important consideration in taking a hologram of a person is to avoid eye injuries of the subject from the laser radiation. In fig. 26 the object beam is expanded by a negative lens and passed through a diffusing screen. If the beam incident on the diffusing screen is expanded enough so that the energy density is below

$$I_{\text{Dmax}} = 0.07 \text{ J/cm}^2, \tag{24}$$

the scattered light from the diffuser does not present a radiation hazard to the model (ANSI 1973, Sliney and Freasier 1973). Equally important is the optical layout of the reference beam. It is imperative that the portion (about 10%) of the reference beam which is reflected by the photographic plate is directed away from the human subject, as shown in fig. 26. The subject is typically 1 to 2 m

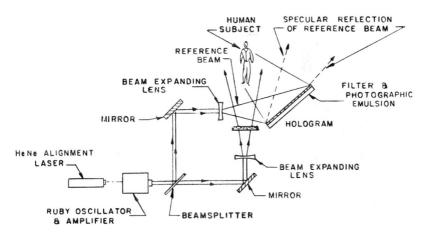

Fig. 26. Arrangement for forming a transmission hologram of a human subject employing a Q-switched ruby laser. After Ansley (1970).

away from the photographic plate. The optical paths of the object and reference beam should be matched at the position of the human subject.

5 Hologram reconstruction

Once the hologram has been recorded and properly processed it can be reconstructed with any laser. Holograms constructed with ruby lasers are usually reconstructed with the expanded beam from a He–Ne or Ar laser. Ideally, the reconstruction wavelength and geometry should be identical to the reference beam in the formation of the hologram.

When a thin emulsion hologram is illuminated by a reference source not identically matching the original reference source used to make the hologram, the following effects are observed:

Change in image position. The hologram-formation geometry is governed by Bragg's law which relates the fringe spacing to the diffraction angle and the wavelength according to eq. (11).

Image aberrations. Hologram aberrations are analogous to lens aberrations. Changing the radius of curvature of the reference wave causes spherical aberration. Changing its angle of illumination of the hologram causes distortion, coma and astigmatism. Changing its wavelength causes chromatic aberration. The effects of aberration become more serious as fringe spacing is reduced. Aberrations are examined by Champagne (1967) and Heflinger and Brooks (1970).

Image magnification. If λ_0 and λ are the wavelength used for recording and reconstruction respectively, the magnification of the virtual and real image becomes (Vienot 1972)

$$M = \lambda/\lambda_0. \tag{25}$$

The difference in wavelength between a ruby (6943 Å) and a He–Ne laser (6328 Å) is small enough not to cause any problems in most situations.

The holographic image can be recorded with conventional photographic equipment. Photographs of the holographic reconstructions are usually taken with single reflex cameras employing 35 or 50 mm focal length lenses and $f/2$ to $f/5.6$ lens settings. The aperture stop is a compromise between maximum depth of focus and minimum speckle. Increasing the f-number increases both the depth of focus and the speckle pattern on the holographic image. The trade-off between aperture size or depth of field and speckle noise is the most critical factor in obtaining good photographic records.

6 Laser safety

Pulsed holography involves the use of powerful ruby lasers. In this section we will briefly point out some of the hazards associated with solid-state lasers.

Hazards associated with solid-state lasers can be separated into two broad categories – those hazards related to the laser beam itself and those hazards related to the electronic equipment.

6.1 Radiation hazards

Two major safety problems resulting from the characteristics of laser radiation are the following: first, the laser beam can traverse great distances with little change in the radiation intensity; and second, extremely high intensities can be obtained. The collimated beam from a laser will be focused to a small spot at the retina with subsequent high flux densities. By comparison, conventional sources of illumination are extended, they are considerably less bright and they emit light in all directions. These light sources produce a sizable image on the retina with a corresponding lower power density. Suitable precautions must necessarily be taken to control spurious reflections of the holographic illuminating beams in order to prevent retinal damage.

For a Q-switched ruby laser the maximum safe energy level on the retina is (ACGIH 1972, ANSI 1973)

$$I_{Rmax} = 0.07 \text{ J/cm}^2. \tag{26}$$

This peak value of energy density that can be tolerated by the retina has to be related to the energy density on the cornea of the eye. Assume a large parallel beam incident on the cornea. The minimum spot size on the retina due to aberrations is 10μm. If the eye is adapted to night condition the pupil has a diameter of about 7 mm. In this worst-case situation, the focusing power of the eye increases the energy density of a parallel beam striking the cornea by a factor $(7 \text{ mm})^2/(10 \mu\text{m})^2 \approx 5 \times 10^5$. Dividing the retinal maximum safe energy level by this factor and assuming a safety factor of 10, one obtains a maximum permissible exposure level at the cornea for direct illumination or specular reflection from a Q-switched ruby laser of

$$I_{Cmax} = 1 \times 10^{-8} \text{ J/cm}^2. \tag{27}$$

For a daylight-adapted pupil (3 mm diameter) the safe energy density can be increased by a factor 5. For a laser beam reflected off a diffusing screen, the energy density must be kept below

$$I_{Dmax} = 0.07 \text{ J/cm}^2. \tag{28}$$

At that energy density the light diffusely reflected back into the eye is at a safe level (ANSI 1973, Sliney and Freasier 1973).

6.2 Electrical hazards

Although the hazards of laser radiation are receiving deserved attention from government agencies, users and manufacturers, the chief hazard around solid-state lasers is electrical rather than optical. Most solid-state lasers require high-voltage power supplies and the use of energy-storage capacitors charged to lethal voltages. Furthermore, associated equipment such as Q-switches, optical gates, modulators, etc., is operated at high voltages. The power supply and associated electrical equipment of a laser can produce serious shock and burns and, in extreme cases, can lead to electrocution (Franks and Sliney 1975).

6.3 Safety precautions applicable to solid-state lasers

Enclosure of the beam and target is the safest way of operating a laser. Interlocked doors, warning signs and lights, key-locked power switches, an emergency circuit breaker and like precautions should be taken to protect operators and passers-by from electrical and radiation hazards of the laser equipment.

In the laboratory it is often not possible to enclose fully a high-power laser. In these situations the following safety precautions should be observed: (1) Do not look into the beam or at specular reflections of the beam. (2) Wear adequate eye protection when the laser is operating. Laser safety goggles should be shatter-resistant and designed to filter out the specific-wavelength generated by the laser. (3) Employ a countdown or other audible warning before the laser is fired. (4) Control access to the laser area and have a flashing red light on the door when laser is in operation. (5) Provide protection against accidentally contacting charged-up capacitors in energy-storage banks, high-voltage power supplies, etc. These components should be installed in cabinets having interlocked doors. Furthermore, capacitor banks should be equipped with gravity-operated dump solenoids.

The key to a successful safety program is the training and familiarization of the personnel involved with laser hazards and subsequent control measures. A number of organizations and government agencies have developed laser safety standards for users and manufacturers of laser equipment (ANSI 1973, DHEW 1975).

References

ACGIH, American Conference of Governmental Industrial Hygienists (1972), P.O. Box 1937, Cincinnati, Ohio 45201.
ANSI, American National Standard for the Safe Use of Lasers (1973), Z136.1, American National Standards Institute, New York.
Ansley, D.A. (1970), Appl. Optics **9**, 815.
Armstrong, W.T. and P.R. Forman (1977), Appl. Optics **16**, 229.
Bates, H.E. (1973), Appl. Optics **12**, 1172.
Benser, W.A., E.E. Bailey and T.F. Gelder (1974), Application of Holography to Flow Visualization Within Rotating Compressor Blade Row, Final Report NASA Cr-121264, National Aeronautics and Space Administration, Washington, D.C.

Benser, W.A., E.E. Bailey and T.F. Gelder (1975), Trans. ASME, January, 75.

Bernard, A., A. Jolas, J. Launspach and J.P. Watteau (1973), Plasma Phys. 15, 1019.

Booth, B.L. (1975), Appl. Optics 14, 593.

Brooks, R.E., L.O. Heflinger, R.F. Wuerker and R.A. Briones (1965), Appl. Phys. Letters 7, 92.

Brooks, R.E., L.O. Heflinger and R.F. Wuerker (1966), IEEE J. Quantum Electronics QE-6, 275.

Carman, R. (1972), Semi-annual Report, 27, Lawrence Livermore Laboratory, Livermore, Calif.

Champagne, E. (1967), J. Opt. Soc. Amer. 57.

Chow, H.H.M. and G.J. Mullaney (1967), Appl. Optics 6, 981.

Chu, W.P., D.M. Robinson and J.H. Goad (1972), Appl. Optics 11, 1644.

Clark, L.T., D.C. Koepp and J.J. Thykkuttathil (1976), ASME J. Fluids Engng., July.

Close, D.H. (1973), A Pulsed Ruby Laser Source for Portable Holography in Holography and Optical Filtering, NASA report SP-299, National Aeronautics and Space Administration, Washington, D.C.

Colburn, W.S. and K.A. Haines (1971), Appl. Optics 10, 1636.

Collier, R.J., C.B. Burckhardt and L.H. Lin (1971), Optical Holography (Academic Press; New York).

Daneu, V., C.A. Sacchi and O. Svelto (1966), IEEE J. Quantum Electronics QE-2, 290.

DHEW, Department of Health, Education and Welfare (1975), Laser Products, Performance Standards, Federal Register 40, No. 148, 32252-32266.

Erf, R.K. (1974), Holographic Nondestructive Testing (Academic Press; New York).

Erf, R.K., J.P. Waters, R.M. Gagosz, F. Michael and G. Whitney (1972), Nondestructive Holographic Techniques for Structures Inspection, AFML Contract No. F33615-71-C-1874.

Erf, R.K., R.M. Gagosz and J.P. Waters (1973), Nondestructive Testing and Flow Visualization – Two Aerospace Applications of Holography, Report NASA SP-299, National Aeronautics and Space Administration, Washington, D.C.

Evenson, D.A., R. Aprahamian and J.L. Jacoby (1972), Holographic Measurement of Wave Propagation in Axisymmetric Shells, Report NASA CR-2063, National Aeronautics and Space Administration, Washington, D.C.

Felske, A. and A. Happe (1973), ATZ Automobiltechnische Zeitschrift 3, 96.

Felske, A. and A. Happe (1974), A Special Interferences Hologram Camera for Quick Vibration Analysis of Drive Units and Car Bodies, Paper Presented at 11th Int. Congress on High Speed Photography, London.

Felske, A. and A. Happe (1975), Double Pulsed Laser Holography as Diagnostic Method in the Automotive Industry, Paper Presented at the Conf. on Engineering Uses of Coherent Optics, University of Strathclyde, Glasgow.

Franks, J.K. and D.H. Sliney (1975), Electro-Optics Syst. Design 7, 20.

Friesem, A.A. and E.N. Tomkins (1973), Photoplastic Recording Materials in Holographic Memories in Holography and Optical Filtering, Report NASA SP-299, National Aeronautics and Space Administration, Washington, D.C.

Gregor, E. (1970), Holographic Interferometry in Shock Propagation Studies, Holographic Newsletter No. 2, Hadron Inc., Korad Division, Santa Monica, Calif.

Gregor, E. (1971), Proc. SPIE, April, 93.

Haines, K.A. and B.P. Hildebrand (1966), Appl. Optics 5, 595.

Havener, A.G. (1970), Holographic Applications in Shadowgraph, Schlieren and Interferometry Analyses of Heat Transfer and Fluid Flow Test Subjects, Report 70-0270, Aerospace Research Laboratories.

Heflinger, L.O. and R.E. Brooks (1970), Holographic Instrumentation Studies, NASA Contract or Report CR 114274, National Aeronautics and Space Administration, Washington, D.C.

Hercher, M. and B. Ruff (1967), J. Opt. Soc. Amer. 57, 103.

Hockley, B.S. (1976), Rolls-Royce Ltd., Derby Engine Division, England, private communication.

Jahoda, F.C. and R.E. Siemon (1972), Holographic Interferometry Cookbook, Report LA-5058-MS, Los Alamos, Available from NTIS, Springfield, Va. 22151.

Koechner, W. (1973), Ind. Res. 15, 44.

Koechner, W. (1976), Solid-State Laser Engineering, (Springer-Verlag; New York).

Koechner, W. (1977), Holographic Portraiture, in: Handbook of Optical Holography (H.J. Caulfield, ed.; Academic Press; New York).

Landry, M.J. (1971), Appl. Phys. Letters **18**, 494.

Landry, M.J. (1973), IEEE J. Quantum Electronics **QE-9**, 604.

Landry, M.J. and G.S. Phipps (1975), Appl. Optics **14**, 2260.

Lengyel, B.A. (1971), Lasers (Wiley-Interscience; New York).

Li, T. (1965), Bell Syst. Tech. J. **44**, 917.

Matthews, B.J., R.F. Wuerker and D.T. Harrje (1968), Small Droplet Measuring Technique, Report AD 828 005, Air Force Rocket Propulsion Lab., Research & Technology Div., Edwards, Calif.

Matthews, B.J., R.F. Wuerker, H.F. Chambers and J. Hojnacki (1972), Holography of JP-4 Droplets and Combusting Boron Particles, Paper Presented at Symposium on Instrumentation for Air-breathing Propulsion, Naval Postgraduate School, Monterey, Calif.

McClung, F.J. and D. Weiner (1965), IEEE J. Quantum Electronics **QE-1**, 94.

Nassenstein, H., H. Dedden, H.J. Metz, H.E. Rieck and D. Schultze (1969), Photogr. Sci. Eng. **13**, 194.

Nassenstein, H., H.T. Buschmann, H. Dedden, E. Klein, E. Moisar and H. Rieck (1970), An Investigation of the Properties of Photographic Materials for Holography, in: The Engineering Uses of Holography (E.R. Robertson and J.M. Harvey, eds.; Cambridge University Press, London).

Overoye, K.R. and R. Aprahamian (1973), Holographic Instrumentation of Turbine Blades, Report NASA SP-299, National Aeronautics and Space Administration, Washington, D.C.

Pennington, K.S. (1971), Holographic Parameters and Recording Materials, in: Handbook of Lasers (R.J. Pressley, ed.; The Chemical Rubber Co.; Cleveland, Ohio).

Potts, G.R. (1973), Tire Sci. Technol. **1**, 255.

Potts, G.R. and T.T. Scora (1974), Tire Vibration Studies – The State-Of-The-Art, Paper Presented at Akron Rubber Group, Winter meeting.

Reinheimer, C.J., C.E. Wiswall, R.A. Schmiege, R.J. Harris and J.E. Decker (1970), Appl. Optics **9**, 2059.

Rundle, W.J. and T.V. Higgins (1975), Present State-Of-The-Art of Ruby Laser Holocameras, 117th Technical Conference, Society of Motion Picture and Television Engineers (SMPTE), Los Angeles, Calif.

Shibayama, K. and H. Uchiyama (1971), Appl. Optics **10**, 2150.

Siebert, L.D. (1968), Proc. IEEE **56**, 1242.

Siebert, L.D. (1971), Appl. Optics **10**, 632.

Sliney, D.H. and B.C. Freasier (1973), Appl. Optics **12**, 1.

Sollid, J.E. (1969), Appl. Optics **8**, 1587.

Sooy, W.R. (1965), Appl. Phys. Letters **7**, 36.

Tomlinson, W.J., E.A. Chandross, H.P. Weber and G.D. Aumiller (1976), Appl. Optics **15**, 534.

Trolinger, J.D. (1973), Review of Holographic Instrumentation at Arnold Engineering Development Center, in: Holography and Optical Filtering, Report NASA SP-299, National Aeronautics and Space Administration, Washington, D.C.

Trolinger, J.D. (1974), Laser Instrumentation for Flow Field Diagnostics, Report AGARD-AG-186, Arnold Research Organization Inc., Tenn.

Vienot, J.C. (1972), Holography, in: Laser Handbook (F.T. Arecchi and E.O. Schulz-Dubois, eds.; North Holland; Amsterdam), Vol. 2.

Way, F.C. (1975), The Pulsed Nd:YAG Holographic Laser – Present Status and Applications, Proc. Electro-Optic System Design Conf., Anaheim, Calif.

Wilson, A.D. and D.H. Strope (1972), Holographic Interferometry Deformation Study of a Printer Type-Piece, IBM J. Res. Develop. **16**, 258.

Wuerker, R.F. and L.O. Heflinger (1970), Pulsed Laser Holography, in: The Engineering Uses of Holography, (E.R. Robertson and J.M. Harvey, eds.; Cambridge University Press; London), p. 99.

Wuerker, R.F. and L.O. Heflinger (1971), SPIE J. **9**, 122.

Wuerker, R.F., R.J. Kobayashi and L.O. Heflinger (1973), Application of Holography to Flow Visualization Within Rotating Compressor Blade Row, Report NASA CR-121264, AiResearch 73-9489.

Young, M. and A. Hicks (1974), Appl. Optics **13**, 2486.

B5 | Advanced Lasers for Fusion

William F. KRUPKE

Advanced Quantum Electronics,
Lawrence Livermore Laboratory,
University of California,
Livermore, California 94550

E. Victor GEORGE and Roger A. HAAS

Advanced Lasers Group,
Lawrence Livermore Laboratory,
University of California,
Livermore, California 94550

Contents

© *North-Holland Publishing Company, 1979*
Laser Handbook, edited by M.L. Stitch

Contents chapter B5 (cont.)

Abstract

Laser driver systems' performance requirements for fusion reactors are developed following a review of the principles of inertial confinement fusion and of the technical status of fusion research lasers (Nd:glass; CO_2, iodine). These requirements are analyzed in the context of energy-storing laser media with respect to laser systems design issues: optical damage and breakdown, medium excitation, parasitics and superfluorescence depumping, energy extraction physics, medium optical quality, and gas flow. Three types of energy-storing laser media of potential utility are identified and singled out for detailed review in this chapter: (1) Group VI atomic lasers, (2) rare earth solid state hybrid lasers, and (3) rare earth molecular vapor lasers. The use of highly-radiative laser media, particularly the rare-gas monohalide excimers, are discussed in the context of short pulse fusion applications. The concept of backward wave Raman pulse compression is considered as an attractive technique for this purpose. The basic physics and device parameters of these four laser systems are reviewed and conceptual designs for high energy laser systems are presented. Preliminary estimates for systems efficiencies are given.

1 Overview

1.1 Historical perspective

Since the detonation of the first thermonuclear explosive about a quarter century ago, mankind has held the dream of obtaining a virtually limitless supply of energy from the controlled burning of thermonuclear fuels. Turning this dream into reality has been perhaps the most difficult scientific and technological challenge ever faced by man. The central problem in producing thermonuclear power is the generation and confinement of a sufficiently dense ultra-high-temperature plasma of heavy hydrogen. In a thermonuclear explosive, the deuterium and tritium plasma is rapidly heated and is confined by its own inertia. That is, an adequate number of thermonuclear reactions are produced in the hot plasma during the time it takes the plasma mass to disassemble due to hydrodynamic motion. The approach to achieve *controlled* generation of thermonuclear energy, initiated in the 1950's, relies on some form of magnetic bottle to confine a hot plasma at relatively low density for a relatively long time. Great strides towards achievement of the required plasma conditions have been made, particularly in the last few years, and the demonstration of scientific feasibility of magnetic confinement fusion is now projected to occur in the mid 1980's.

With the invention of the laser, the notion of using inertial confinement for the production of *controlled* thermonuclear energy became a plausible concept. Shortly after the laser was demonstrated by T.H. Maiman in 1960, calculations were performed by Stirling Colgate, Ray Kidder, John Nuckolls, Ron Zabawski and Edward Teller to describe the implosion of tiny deuterium–tritium pellets to thermonuclear conditions by irradiation with intense laser pulses. Some of these calculations showed that efficient generation of fusion energy would not result from a simple heating of the fuel, but that the fuel would have to be imploded to 10000 times liquid density (Nuckolls et al. 1972).

The technical suitability of lasers for fusion studies was markedly improved with the invention of Q-switching and mode-locking techniques which permitted the generation of intense laser pulses of short duration. Experimental laser fusion studies began in 1963 and by the mid-1960's, R. Kidder and S. W. Mead had constructed a twelve-beam laser system for fusion studies at the Lawrence Livermore Laboratory (Mead 1970). Subsequently in 1972, Nikolai Basov and his colleagues at the Lebedev Physical Institute reported the production of

3×10^6 neutrons upon irradiation of a 100 μm diameter, solid CD_2 microsphere with a few hundred joule/few nanosecond nine-beam laser pulse (Basov et al. 1972). Evidence of neutron production (10^5 to 10^6 neutrons) by implosion was produced (Johnson 1974) at KMS Fusion in May 1974 using glass-shell microspheres irradiated with Nd:glass laser pulses. Similar results were also obtained using DT filled glass-shell microspheres at the Lawrence Livermore Laboratory in Dec. 1974 (Holzrichter 1975). During the past five years, laser fusion studies world-wide have grown enormously with significant programs underway in a half-dozen countries. With the successful physics and engineering development of multi-terwatt Nd:glass laser systems such as the JANUS, CYCLOPS, ARGUS and SHIVA SYSTEMS at the Lawrence Livermore Laboratory (Holzrichter 1976), the thermonuclear (TN) pellet gain of fusion pellets has been increased by many decades, as shown in fig. 1. The thermal nature of the fusion plasma was first confirmed with a pellet gain of 10^{-6} (10^6 neutrons) using α-particle spectra (Slivinsky et al. 1975). Achievement of this plasma condition was reconfirmed at a higher pellet gain of 10^{-4} (2×10^9 neutrons) using neutron time of flight spectra (Lerche et al. 1977). With the successful operation of the SHIVA laser in November 1977, experiments to establish conditions of significant TN burn (pellet gain of 10^{-2}) will commence early in 1978. Based on the projected technical capabilities of even larger lasers such as the SHIVA–

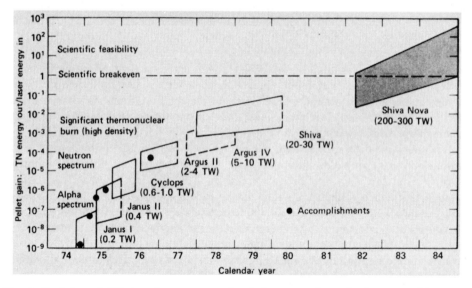

Fig. 1. Evolution of Nd:glass laser systems for fusion research at the Lawrence Livermore Laboratory, in relation to key physics milestones of the DOE Inertial Confinement Fusion (ICF) Program. The vertical extent of the boxes associated with each laser indicates the range of target gains expected with the corresponding laser power. The horizontal extent of the boxes indicates the duration of operation of the system. The representative data points, expressed as TN energy output divided by laser energy input to the target, show how well these projections have been attained. The JANUS II and ARGUS II Systems continue in operation. The ARGUS IV system was planned but never built. The SHIVA system became operational in Nov. 1977.

NOVA system (fig. 1), the demonstration of scientific feasibility of laser-driven fusion is now anticipated in the mid-1980's, on a time scale comparable to that of magnetic confinement fusion.

Looking beyond achievement of these scientific milestones toward power applications, it is clear that efficient, repetitively-pulsed laser systems will have to be developed which can provide not only high peak power pulses (as do present single-pulse Nd:glass, iodine, and CO_2 fusion research lasers) but high average power as well. To initiate the process of identifying and developing such systems, the performance requirements for a power plant laser system have been projected by combining the results of on-going implosion experiments (Manes et al. 1977) with LASNEX-computer simulations of high density implosions and with preliminary analyses of reactor systems (Maniscalco, 1976, Maniscalco et al. 1977). Although requirements derived in this way undoubtedly will evolve with improved understanding of pellet physics and reactor design (particularly with respect to operating wavelength and efficiency) they provide a useful basis for assessing alternative laser systems.

Reactor laser systems may possibly use the active gain media of demonstrated fusion research lasers, but reconfigured in amplifier and systems architectures amenable to high average power operation. Alternatively, they may use entirely new active gain and pumping techniques specifically identified and developed for the fusion application. This chapter discusses several new laser media/concepts falling in the latter category.

1.2 Scope of the chapter

To place the material of this chapter in perspective, we begin in § 2 with brief summaries of the principles of inertial confinement fusion, the reactor laser systems performance requirements and the characteristics of laser systems currently used in fusion research. We follow in § 3 with an analysis of the systems performance requirements in terms of the physics and technological parameters of potentially useful laser media; power amplifier and systems architectures are discussed together with related issues of energy extraction, beam propagation and operation at high average power. On the basis of these analyses, four new laser media/concepts of potential utility in fusion applications have been identified and are singled-out for review here: (1) Group VI atoms, (2) rare earth solid-state hybrids, (3) rare earth molecular vapors and (4) backward wave Raman pulse compressors. In §§ 4 to 7 we discuss these systems, giving the basic principles of operation, the physics and technical characteristics of the gain medium and excitation sources, and conceptual designs of large-scale power amplifier devices. We conclude in § 8 with a brief summary of additional new laser media and techniques of interest for fusion applications being pursued concurrently in the laser community.

2 Laser requirements for fusion applications

2.1 Introduction

In its simplest form a laser fusion reactor (Lubin and Fraas 1971, Nuckolls et al. 1973, Emmett et al. 1974) consists of a combustion chamber to contain the fusion microexplosion, a pellet factory to make and project spherical liquid DT pellets into the center of the chamber at a rate of several per second and a laser system to suitably irradiate the pellets. Energy released in the fusion reaction is carried by 14 MeV neutrons, charged particles and electromagnetic radiation. To produce electrical energy, the released fusion energy is absorbed in a liquid lithium blanket surrounding the implosion volume which is circulated to drive a conventional steam–thermal conversion cycle. Upon irradiation of an injected pellet, the following sequence of events will occur (see fig. 2).

First a low density atmosphere, extending to several pellet radii, is created by ablating the pellet surface with a laser prepulse. This atmosphere is then irradiated more or less uniformly from all sides by a second, much more intense laser pulse. Absorption of the laser light in the outer atmosphere heats the electrons. The atmosphere and the pellet surface are heated by electron diffusion and transport. As the electrons move inward through the atmosphere, scattering and solid angle effects greatly increase the spherical symmetry. Violent ablation

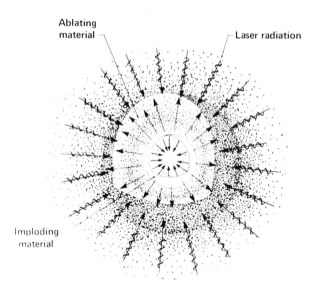

Fig. 2. Laser implosion of a pellet. The atmosphere extends to several pellet radii and is formed before the main laser pulse by a prepulse that ablates some of the pellet surface. Absorption of the laser light in the outer atmosphere generates hot electrons which move inward, heating the atmosphere and pellet surface; scattering and solid-angle effects greatly increasing the spherical symmetry. Violent ablation and blow-off of the pellet surface generate the pressure that implodes the pellet. The pellet core then undergoes thermonuclear burn.

and the blow-off of the pellet surface generate the pressure which implodes the pellet. The laser pulse is shaped in time to achieve ultrahigh compression of the pellet. Fusion yields several hundred times larger than the input laser energy for laser energies on the order of 1 MJ have been predicted by sophisticated computer simulation calculations (Nuckolls et al. 1972, Brueckner and Jorna 1974, Clarke et al. 1973, Fraley et al. 1974, Nuckolls 1974). In these calculations a portion of the pellet mass is first isentropically compressed to a high density Fermi degenerate state and thermonuclear burn is initiated in the central region. A thermonuclear burn front then propagates radially outward from the central region heating and igniting the dense fuel.

In § 2, the laser subsystem requirements for a fusion reactor will be presented following elementary considerations of the thermonuclear reaction cycle, the basic physical processes occurring in fusion pellets and the reactor blanket energy conversion characteristics. This material is provided as background information on the fusion process itself. Those readers interested only in the laser system requirements should proceed directly to § 2.4.

2.2 Thermonuclear reaction cycles and confinement

Controlled TN fusion will make use of one or more of the following nuclear reactions (Glasstone and Lovberg 1960, Ribe 1975, Steiner 1975, Conn 1976)

1. *Deuterium–Tritium (D–T) cycle*

$$_1D^2 + {}_1T^3 \rightarrow {}_2He^4(3.5 \text{ MeV}) + {}_0n^1(14.1 \text{ MeV}); \quad Q = 17.6 \text{ MeV}.$$

Tritium generation in blanket

$$_0n^1 + {}_3Li^6 \rightarrow {}_2He^4(2.0 \text{ MeV}) + {}_1T^3(2.6 \text{ MeV}); \quad Q = 4.6 \text{ MeV}.$$

2. *Deuterium–Deuterium (D–D) cycle*

$$_1D^2 + {}_1D^2 \begin{cases} \rightarrow {}_1T^3(1.0 \text{ MeV}) + {}_1H^1(3.0 \text{ MeV}); & Q = 4.0 \text{ MeV} \\ \rightarrow {}_2He^3(0.92 \text{ MeV}) + {}_0n^1(2.4 \text{ MeV}); & Q = 3.3 \text{ MeV}. \end{cases}$$

3. *Deuterium–Helium Three (D–He³) cycle*

$$_1D^2 + {}_2He^3 \rightarrow {}_2He^4(3.7 \text{ MeV}) + {}_1H^1(14.7 \text{ MeV}); \quad Q = 18.4 \text{ MeV},$$

where Q is the total energy released by the nuclear reaction. Both the D–D and D–T cycles involve the release of an energetic neutron. Fusion reactors based on these cycles will therefore require a blanket capable of slowing down and extracting neutron kinetic energy. If $_3Li^6$ is used in the D–T cycle tritium can be bred with a net energy yield or enhancement of 4.6 MeV, making 22.2 MeV the total yield for the D–T cycle. A great potential advantage of the D–He³ cycle is that the reaction products are charged particles which, in addition to offering the potential of direct conversion, will have neutrons produced only as a result of concomitant D–D fusion reactions. This neutron production level will be much lower than in the D–T cycle and this will alleviate many of the neutron related technology problems, such as materials radiation damage and induced radioactivity.

The choice between these potential fuel cycles for the D–T reaction in the near term becomes clear by examining the reaction rates $\langle \sigma v \rangle$ for the various fusion reactions shown in fig. 3. The reaction rate is derived by averaging the appropriate reaction cross section σ over a Maxwellian distribution of speeds for both incident particles. It is clear that the D–T cycle offers the greatest advantages since its reaction rate is largest and peaks at the lowest temperature of 70 keV. It thus appears likely that the first fusion reactor will employ the D–T cycle. It is also apparent from fig. 3 that the reaction rate for the fusion reaction becomes large only at high energies, and thus it is necessary to heat the reacting species to temperatures where the fusion rate becomes reasonably large. For the D–T cycle, this means temperatures of order 10 keV.

There are several general figures of merit that have been used to characterize the condition of a reacting plasma: D–T gain, Lawson condition and ignition condition. The thermal energy of an equilibrated reacting D–T plasma is $3n\theta$, where θ is the product of temperature T times Boltzmann's constant. Owing to charge neutrality the electron density n is twice the equal deuteron and triton densities. In this plasma, fusion energy is released at a rate $(\frac{1}{2}n)^2\langle \sigma v \rangle Q$. If the thermal energy is confined for a characteristic time τ, then the D–T or fuel gain is

$$G_F = (\tfrac{1}{2}n)^2 \langle \sigma v \rangle Q / (3n\theta/\tau) = n\tau (\langle \sigma v \rangle / 12\theta) Q, \tag{1}$$

or

$$n\tau = G_F(12\theta/\langle \sigma v \rangle Q). \tag{2}$$

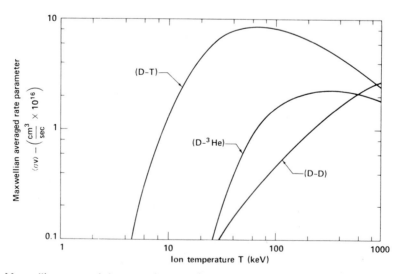

Fig. 3. Maxwellian-averaged thermonuclear reaction rates for deuterium–tritium, deuterium–helium-3 and deuterium–deuterium reactants plotted as as a function of kinetic temperature.

The D–T gain defined by eq. (1) is the ratio of the thermonuclear energy released by the plasma to the thermal energy invested in the D–T fuel. Curves of $n\tau$ versus T are plotted in fig. 4 for different values of D–T gain. For a given gain the minimum product of plasma density and the plasma energy confinement time occurs at approximately 25 keV and it increases sharply for $T < 10$ keV.

The requirement for a fusion plasma to approximately achieve energy breakeven is often given as the Lawson criterion (Lawson 1957). To derive this condition plasma loss mechanisms must be accounted for. Lawson assumed that bremsstrahlung radiation was the main loss mechanism. This energy loss rate is given (Glasstone and Lovberg 1960) for hydrogenic plasma by $P_b[\text{W/cm}^3] = C_b n^2[\text{cm}^{-3}]\theta^{1/2}[\text{keV}]$, where $C_b = 4.8 \times 10^{-31}$ W cm^3/keV$^{1/2}$. Assuming that the fusion reaction power plus the bremsstrahlung and plasma thermal power are available for conversion to electricity at an overall efficiency ϵ, the Lawson condition arises from the following power balance on a unit volume of the plasma

$$3n\theta/\tau + P_b = \epsilon\{(\tfrac{1}{2}n)^2\langle\sigma v\rangle Q + P_b + 3n\theta/\tau\}, \tag{3}$$

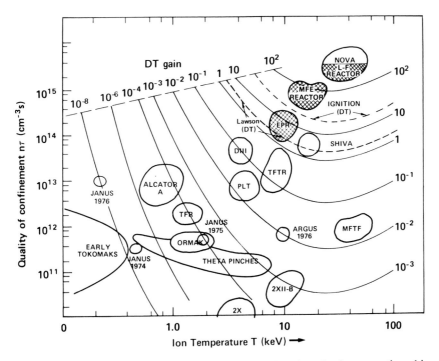

Fig. 4. Map of thermonuclear plasma conditions achieved and projected using magnetic and inertial confinement facilities. Plasma conditions are characterized by density–confinement time product and by ion temperature. Curves of constant DT plasma gain are indicated, together with the Lawson and ignition conditions.

or solving for $n\tau$,

$$n\tau = \frac{3\theta(1-\epsilon)}{\epsilon[\frac{1}{4}\langle\sigma v\rangle Q + C_b\theta^{1/2}] - C_b\theta^{1/2}}, \tag{4}$$

which is a function of plasma temperature only. Lawson assumed $\epsilon = \frac{1}{3}$, and a curve of $n\tau$ versus temperature for the D–T cycle is shown in fig. 4 for this value of ϵ. The minimum $n\tau$ product required to meet the Lawson criterion is approximately $5 \times 10^{13}\,\mathrm{cm}^{-3}\,\mathrm{s}$ at 25 keV; however, it is usually quoted at a temperature of 10 keV where it attains the value of $10^{14}\,\mathrm{cm}^{-3}\,\mathrm{s}$. In general, the Lawson criterion is too optimistic since the choice of ϵ is quite large. As will be shown later, this is especially true in the case of inertial confinement fusion owing to the relatively low efficiency of implosions. In addition, a plasma that meets the Lawson condition is not necessarily ignited and self-sustaining because the analysis does not include the possibility of direct deposition in the plasma of some or all of the energy of the fusion reaction products. The condition which characterizes an ignited and self-sustaining plasma is thus another important general criterion.

The ignition condition is obtained by balancing the power lost from the plasma due to bremsstrahlung and thermal losses in a time τ against the power deposited in the plasma from the slowing down of fusion reaction products. Letting $\bar{f}$ denote the fraction of fusion energy deposited in the plasma the balance is

$$(\tfrac{1}{2}n)^2\langle\sigma v\rangle\bar{f}Q = 3n\theta/\tau + P_b, \tag{5}$$

and the ignition condition on $n\tau$ becomes

$$n\tau = 3\theta/(\tfrac{1}{4}\langle\sigma v\rangle\bar{f}Q - C_b\theta^{1/2}). \tag{6}$$

For the D–T cycle, $\bar{f}Q \approx 3.5\,\mathrm{MeV}$, the energy of the α-particle which would be contained in the plasma. This criterion is also shown in fig. 4 and the minimum $n\tau$ in this case is approximately $2 \times 10^{14}\,\mathrm{cm}^{-3}$. Thus, it is more difficult to meet the ignition condition in a D–T mixture than to meet the Lawson criterion. We note that the temperature at which the energy deposited by nuclear fusion within the reacting system just exceeds the energy lost by bremsstrahlung is often referred to as the ideal ignition temperature (Glasstone and Lovberg 1960). For the D–T and D–D cycles these temperatures are 4 and 36 keV respectively, and represent the lowest possible operating temperature for a self-sustaining thermonuclear plasma (fig. 4). For temperatures lower than the ideal ignition values, the bremsstrahlung losses exceed the rate of thermonuclear energy deposition by charged particles in the reacting system. To compare the above criteria with what has been achieved experimentally and is planned in magnetic and inertial confinement, fig. 4 also shows the regime of operation of several well-known thermonuclear facilities. The remarkable flexibility of a laser irradiation system is illustrated by the diverse plasma conditions achieved with JANUS (Ahlstrom and Nuckolls 1977). Utilizing the JANUS spherical illumination system, temperature was traded for density in the imploded fuel and densities of approximately five times liquid density ($0.2\,\mathrm{g/cm^3}$ for D–T) were achieved at ion temperatures of a couple of tenths of a kilovolt compared to early JANUS results in which approximately 2 keV plasma was produced at densities near $\frac{1}{2}$ liquid density. The ARGUS results show a DT ion temperature of the order of 10 keV at liquid density and a DT gain of 1%. Projections of plasma conditions achievable with the SHIVA and SHIVA–NOVA Nd:glass irradiation facilities are also shown in fig. 4.

2.3 Pellet gain

A description of the relationship between the performance of a complex fusion pellet and laser irradiation characteristics (power, energy, wavelength, etc) requires large-scale computer calculations. The conclusions drawn from a limited number of such calculations are presented at the end of this section, expressed in terms of laser system requirements. However, to provide some feel for the complexity of the problem and some direct insight into the important physical processes in fusion pellets, we discuss next a model of a simple pellet (Kidder 1972, 1974, 1975, 1976, Brueckner and Jorna 1974).

We have seen that breakeven, ignition, self-heating and appreciable fuel burn-up conditions can be related to constraints on $n\tau$. In inertial confinement fusion, however, it is conventional (Nuckolls et al. 1972, Nuckolls 1974) to replace the confinement time τ by $R_p/4c_s$ and the number density n by the mass density ρ. The quantity τ is then a measure of the time required for a rarefaction wave traveling at the isothermal sound speed c_s to significantly disassemble a compressed spherical pellet of radius R_p. The factor of four arises because in a sphere approximately half the mass is beyond 80% of the radius. Thus for inertially confined thermonuclear microexplosions the confinement parameter is $\rho R_p = (4m_i c_s)n\tau$, where m_i is the average ion mass of the fuel.

In the spherical compression of a fixed mass M_p the density increases as R_p^{-3} and thus ρR_p increases as $(3M_p\rho^2/4\pi)^{1/3}$. The resulting increase in the thermonuclear reaction rate compared to disassembly rate leads to a corresponding increase in the burn efficiency (Nuckolls 1974)

$$\phi_p = \rho R_p/[(\rho R_p)^* + \rho R_p], \tag{7}$$

where $(\rho R_p)^* = 8m_i c_s/\langle\sigma v\rangle$ is nearly constant in the 20 to 70 keV temperature range characteristic of efficient D–T microexplosions. In this range $(\rho R_p)^*$ is approximately 6 and it increases sharply below 20 keV. If ρR_p is sufficiently large the reaction products may be trapped by the burning fuel. The range of 14 MeV neutrons in D–T is $\rho\Lambda_n \simeq 4.6$ g/cm². Similarly, the range of 3.5 MeV α-particles depends on plasma properties (Fraley et al. 1974) and $\rho\Lambda_\alpha \simeq 0.3$ to 0.5 g/cm² in D–T at 5 to 10 keV when ρ is in the 10^3 to 10^4 g/cm³ range. Consequently under efficient burn-up conditions ($\phi_p > 1/3$, $\rho R_p > 3$ g/cm²), although the fuel is ignited at temperatures $\leqslant 10$ keV, trapping of α-particle reaction products rapidly heats the reaction zone to temperatures in the 20 to 70 keV range. The yield from 1μg of D–T fully burned ($\phi_p = 1$) is 326 kJ (Fraley et al. 1974) and consequently when

$\phi_p = 1/3$ the yield is $Y_{DT} = 109 \, kJ/\mu g$. This yield does not include blanket multiplication.

If the parameter $\bar{\beta}$ is introduced to account for α-particle self-heating and thermonuclear propagation, then the average thermal energy per particle required for ignition may be written approximately as $3\theta_i/2\bar{\beta}$, where θ_i is the ignition temperature. In magnetic confinement fusion (§ 2.2) $\bar{\beta}$ is approximately one; however, in laser fusion, by exploiting central fuel ignition and thermonuclear propagation, values much greater than one may be achieved. The ignition temperature is usually $\sim 10 \, keV$ in order to avoid the relatively long time to self-heat from the ideal ignition temperature (§ 2.2). If the compressed fuel is such that $\rho R_p \geqslant 0.3 \, g/cm^2$, then only about $0.3 \, g/cm^2$ in the central region need be heated to approximately $10 \, keV$ in order to initiate a radially propagating burn front which ignites the entire pellet. In this case initially $\phi_p \approx 0.05$ and a yield $Y_{DT} = 16 \, kJ/\mu g$ of fusion energy will be released from the critical region, about one-fifth of this energy is in α-particles, sufficient to heat three times more D–T to $10 \, keV$. This central ignition can be achieved by compression shock convergence controlled by laser pulse shaping during implosion. For example, if the D–T temperature just prior to ignition is θ_i from the center to Λ_α and falls off as R_p^{-2} beyond, then $\bar{\beta}^{-1} \sim (\Lambda_\alpha/R_p)^2$ (Nuckolls et al. 1972, Nuckolls 1974). In this case practical limitations on implosion symmetry set a minimum on $\bar{\beta}^{-1}$ of approximately 0.03, which occurs for $\rho R_p \leqslant 3 \, g/cm^2$. Other models of central spark ignition (Afanasev et al. 1975, Kidder 1975, Guskov et al. 1976, Lindl 1977, 1978) suggest, however, that even smaller values of $\bar{\beta}^{-1}$ may be achieved. In the present example the average ignition energy is $34 \, J/\mu g$. This is also the minimum compressional energy of D–T at a density of $10^3 \, g/cm^3$ and this minimum is $\frac{3}{5}\epsilon_F$ when the electrons are Fermi degenerate, i.e., $\theta \ll \epsilon_F$, where ϵ_F is the Fermi energy $(h^2/8m_e)(3n/\pi)^{2/3}$. For a density of $10^3 \, g/cm^3$ ($n \approx 2.5 \times 10^{26} \, cm^{-3}$), $\epsilon_F \sim 1 \, keV$, and the D–T electrons are degenerate if the implosion is carried out, so that $\theta \ll 1 \, keV$ in the fuel, except in the central region where ignition occurs. In general, the ideal minimum cannot be achieved owing to electron preheat or the approximately 1 M bar initial shock produced by the laser prepulse. Under these circumstances the specific (per electron) compressional energy $3\alpha_F\eta^{2/3}$ in electron volts may be referenced to the minimum value by using a parameter $\alpha_F > 1$ which corrects for nonideal Fermi gas behavior ($\alpha_F = 1$). The quantity η measures the compression density ρ relative to liquid density ρ_L.

The total internal energy of the compressed pellet of radius R_p is approximately the sum of the ignition and compression terms and may be written

$$E_F = (3n\theta_i/\bar{\beta}_e)(\tfrac{4}{3}\pi R_p^3),\tag{8}$$

where $\bar{\beta}_e = \bar{\beta}[1 + (\bar{\beta}\alpha_F\eta^{2/3}/\theta_i)]^{-1}$ represents the effective reduction in $\bar{\beta}$ due the energy of degenerate compression. The corresponding thermonuclear energy yield can be written as

$$E_{TN} = \tfrac{1}{2}nQ\phi_p(\tfrac{4}{3}\pi R_p^3).\tag{9}$$

Then, if ϵ_{LF} is the fraction of the laser energy converted into internal energy of the fuel, the pellet gain can be written, using eqs. (8) and (9),

$$G_p = \epsilon_{LF} E_{TN}/E_F = \bar{\beta}_e \epsilon_{LF} (Q/6\theta_i)\phi_p. \tag{10}$$

The conversion fraction ϵ_{LF} can be written as a product $\epsilon_{LA} \epsilon_{AF}$, where ϵ_{LA} is the fraction of the laser light absorbed by the pellet and ϵ_{AF} is the fraction of the absorbed laser energy that is converted into internal energy of the compressed thermonuclear fuel. The laser light absorption fraction ϵ_{LA} depends on the details of the target design and on laser–plasma interaction mechanisms, and may achieve values of ~0.5. However, of the laser light absorbed by the target, 90 to 95% is lost via the blow-off from the ablation process (Nuckolls et al. 1972, Nuckolls 1974, Bruckner and Jorna 1974) so that $\epsilon_{AF} \simeq 0.05$ to 0.10 and consequently $\epsilon_{LF} \simeq 0.025$ to 0.05. For the example above, when $\rho R_p = 3 \text{ g/cm}^2$ and $\bar{\beta}^{-1} \sim 0.03$, the minimum average energy of ignition and compression E_F/M_p is roughly 70 J/μg. Since the fusion energy produced at this ρR_p is approximately 110 kJ/μg the fuel gain is about 1600. Then, as a consequence of inefficiency of the laser–plasma interaction and ablation process, the pellet gain G_p will lie in the range of 40 to 80. Without self-heating and propagation ($\bar{\beta} = 1$) the corresponding gains would be between one and 2.5 in this model. Finally, recent calculations (Gitomer et al. 1977, McCrory and Morse 1977) suggest that for solid spherical targets the efficiency ϵ_{AF} may vary as λ^{-1}, where λ is the laser wavelength. This is because at shorter wavelengths the laser light is deposited closer to the ablation surface.

If the approximation is made that ρR_p is somewhat less than $(\rho R_p)^*$ we have

$$\rho R_p \simeq (6\theta_i/Q)(G_p/\bar{\beta}_e \epsilon_{LF})(\rho R_p)^* \tag{11}$$

and the corresponding confinement quality factor

$$n\tau \simeq (12\theta_i/Q\langle\sigma v\rangle)(G_p/\bar{\beta}_e \epsilon_{LF}), \tag{12}$$

which is analogous to eq. (2). From eq. (12) at an ignition temperature of 10 keV the Lawson condition, $n\tau = 10^{14} \text{ cm}^{-3}$ s, requires $G_p/\bar{\beta}_e \epsilon_{LF} \simeq 2$, and thus from eqs. (11) and (7) this corresponds to a fractional burn-up of approximately 0.3%. For 'optical breakeven' $G_p = 1$, and then if $\bar{\beta}_e \epsilon_{LF} \simeq 0.05$, the confinement parameters must be $n\tau \simeq 10^{15} \text{ cm}^{-3}$ s and $\rho R_p \simeq 0.16 \text{ g/cm}^2$, leading to a fractional burn-up of ~3%. Alternatively, if a self-heating factor $\bar{\beta}_e \gtrsim 10$ is achieved, then optical breakeven occurs at or below the Lawson condition. Clearly, these requirements can be reduced if self-heating and propagation can be exploited. Using eqs. (11) and (8) the laser energy requirement may be written

$$E_L = (4\pi\theta/m_i\rho_L^2)[(6\theta_i/Q)(\rho R_p)^*]^3 G_p^3/(\bar{\beta}_e \epsilon_{LF})^4 \eta^2. \tag{13}$$

For the D–T fuel cycle, Q can be 22.2 MeV if an Li^6 blanket is used, and at 10 keV eq. (13) can be simplified to (Kidder 1972)

$$E_L[\text{MJ}] \simeq (4G_p)^3/(\bar{\beta}_e \epsilon_{LF})^4 \eta^2. \tag{14}$$

Within the context of this simple model, the corresponding pellet mass and confinement time requirements are (Kidder 1972)

$$M_p[mg] \simeq (4G_p/\bar{\beta}_e\epsilon_{LF})/\eta^2 \tag{15}$$

and

$$\tau[ns] \simeq (G_p/\bar{\beta}_e\epsilon_{LF})/\eta. \tag{16}$$

Although this is a simple model of the laser fusion pellet performance it reveals the extreme sensitivity of laser requirements to laser light absorption fraction, ablation efficiency, fuel compression, self-heating and propagation and to pellet gain requirements. According to eqs. (13) and (14) the laser energy requirement decreases as the square of the compression and the fourth power of the laser light absorption fraction, self-heating factor and the energy coupling between absorbed laser energy and pellet core. Correspondingly, as shown in eqs. (15) and (16), the confinement time and fuel mass requirements decrease as the pellet coupling and compression increase.

Over a wide range of compressed pellet parameters ($10^2 < \rho[g/cm^3] < 10^3; \frac{1}{2} < \rho R_p < 2$) a more accurate analysis (Nuckolls et al. 1972, Nuckolls 1974) of eq. (10) has been performed (Nuckolls et al. 1972) leading to an expression for the laser light energy

$$E_L \propto \epsilon_{LF}^{-a}\eta^{-b} \tag{17}$$

at constant gain, where $a \simeq \frac{4}{3}$ to 2 and $b \simeq 2$, indicating that, compared to the simpler model above, the dependence on compression is at least as strong and generally stronger than the dependence on the absorption and compression efficiencies. Using these results fig. 5 shows the variation of gain with compression and laser light energy. The curves have been normalized to computer

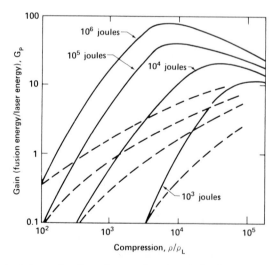

Fig. 5. Computer calculations of pellet gain as a function of density compression for various laser energies. Solid curves include effects of propagation; dashed curves assume uniform ignition (no propagation). The curves are computed for optimum pellet mass and laser pulse shape (Nuckolls et al. 1972).

simulations of implosion and burn. The solid curves include propagation, whereas the dashed curves assume uniform ignition and no propagation. The importance of self-heating and propagation is readily apparent. Gains approaching 100 are predicted for laser energies of 1 MJ. At compressions less than a thousand the gain increases strongly with compression because of increasing burn efficiency and self-heating of the fuel. The gain decreases with compressions greater than 10^4 because of depletion of the D–T, increased ablative energy losses and because the energy of compression (against degeneracy pressure) becomes dominant. Achievement of this performance requires (Nuckolls et al. 1973, Nuckolls 1974, Emmett et al. 1974) nearly isentropic compressions in which the laser wavelength and pulse shape may be varied.

Implosion of hollow (rather than solid) pellets containing one or more concentric shells makes possible order of magnitude reductions in laser power (Nuckolls et al. 1974, Brueckner and Jorna 1974, Afanasev et al. 1975). Recently it has been suggested (Afanasev et al. 1975, Lindl 1977, 1978) that pellet gains as great as 10^3 might be possible with such targets. Smaller laser powers are required to implode hollow pellets because a smaller implosion pressure acts over a larger volume to generate the required implosion energy. Consequently the focused laser light intensity is lower, offering the possibility of greater collisional (Nuckolls 1974, Lindl 1974) as opposed to anomalous absorption. Preheat effects (Morse and Nielson 1973) can be minimized by using various shell materials (Nuckolls et al. 1974, Lindl 1977, 1978). However, the symmetry and fluid stability problems (Shiau et al. 1974, Henderson et al. 1974, Brueckner and Jorna 1974, Bodner 1974, Lindl and Mead 1975, Afanasev et al. 1976) are more severe in hollow pellet implosions. Numerical simulations show that laser-driven implosions of moderately hollow pellets to the super dense Fermi degenerate state required to achieve large gains are feasible when the laser power history, a train of short pulses (Lindl and Mead 1975), is tailored to generate an optimum sequence of several weak shocks.

The implosion models presented above revealed that laser coupling efficiency is as important as compression in laser fusion implosions. Furthermore, this coupling efficiency is strongly dependent on the nature of the laser–plasma interaction process which is a function of laser intensity and wavelength. Although great progress has been made recently in understanding the laser–plasma interaction process much remains to be done before the dependence on laser light characteristics can be fully established.

Computer simulations and theoretical analyses (Dawson et al. 1969, Shearer 1970, Chen 1974, Liu et al. 1974, Estabrook et al. 1975, Forslund et al. 1975a, Max 1977) indicate that the laser–plasma interaction process may be characterized by the dimensionless parameter $\bar{\xi} = V_0/V_T$ which is the ratio of the electron oscillatory and thermal velocities, $\bar{\xi}^2 \propto \lambda^2 I/\theta$. The parameter $\bar{\xi}$ is a measure of the relative strengths of the ponderomotive force due to the laser beam and the plasma pressure gradient. The interaction is weak, and classical collisional inverse bremsstrahlung absorption is expected to dominate the interaction process when $V_0/V_T \ll 1$.

When this condition holds, the heated electron distribution tends toward a Maxwellian distribution. Such a distribution minimizes the number of high energy electrons that can preheat the central fuel region. If the density scale length is too long then the absorption process must compete with stimulated scattering processes and parametric instabilities (Dawson and Lin 1974) excited outside the critical layer. Even in the collisional absorption regime it is important that the plasma scale length not be too large.

When the interaction parameter is increased above about 0.1 (the regime in which all recent laser fusion experiments have been performed) nonlinear effects such as quivering of the electrons in the electric field of the laser (Shearer 1970), ponderomotive density gradient steepening (Kidder 1972, Estabrook et al. 1975, Forslund et al. 1976), filamentation (Langdon and Lasinski 1975, Kaw et al. 1973, Estabrook 1976) and stimulated scattering processes such as Brillouin scattering (Liu et al. 1974, Forslund et al. 1975b, Phillion et al. 1977a, b) all reduce the efficiency of collisional absorption. Owing to the ponderomotive force the density scale length in the vicinity of critical density diminishes and a density jump develops, greatly reducing the significance of parametric instability absorption at the critical layer. The dominant absorption and scattering mechanisms then become stimulated Brillouin and Raman scattering in the low density plateau region outside the critical density layer and resonance absorption at the steepened critical surface. The disadvantage of resonance absorption is that high energy non-maxwellian electrons are efficiently generated by the intense resonant field that develops at the steepened critical layer (Ginzburg 1964, Coffey 1971, Friedberg et al. 1972, Estabrook et al. 1975, Forslund et al. 1975a, 1976, Estabrook and Kruer 1977). As is well known (Morse and Nielson 1973, Brueckner and Jorna 1974) these high energy electrons lead to fuel preheat which may greatly increase the driving pressure required to achieve high values of ρR_p and efficient thermonuclear burn.

Recent interferometric (Attwood et al. 1978) and polarimetry (Phillion et al. 1977a, b) measurements of plasma density profiles have confirmed that density profile steepening does occur. In addition, laser light absorption and scattering measurements (Haas et al. 1977a, 1977b, Ripon 1977a, 1977b, Phillion et al. 1977a, b, Manes et al. 1977, Max 1977, Kruer et al. 1977) reveal that under these circumstances resonant absorption and stimulated Brillouin scattering play a dominant role and that collisional absorption is of secondary importance. Recent measurements of high energy X-ray emission from irradiated targets at both 1.06 and at $10.6\,\mu$m show the development of a strong nonthermal tail at high intensities. These tails, which are produced by resonance absorption, are nearly exponential in energy above several tens of kilovolts with an effective 'temperature' that scales approximately as $(\lambda^2 I)^{1/3}$ (Forslund et al. 1977, Estabrook and Kruer 1977). These results suggest that target preheat will be less at shorter wavelengths. Although short wavelength laser light produces softer heated electron distributions, if the laser pulse length and corresponding plasma scale length are long, stimulated scattering processes, such as Brillouin scattering, may act to reduce the amount of absorption (Phillion et al. 1977a, b). However, recent theoretical studies (Kruer et al. 1973, Thomson et al. 1974, Thomson and Karash 1974, Thomson, 1975) suggest that parametric instabilities such as stimulated Raman and Brillouin scattering can be controlled by using finite bandwidth ($\Delta\lambda/\lambda \geqslant 0.01$ to 0.10) laser radiation. Short wavelength fusion laser systems employing rare gas halide ($\Delta\lambda/\lambda \sim 0.01$) and metal vapor excimer ($\Delta\lambda/\lambda \sim 0.10$) lasers, to be discussed in §§ 4 to 8, may be used to achieve wide bandwidth target irradiation. Clearly, much remains to be learned about the interaction of laser radiation with plasmas, especially in the long pulse regime required for isentropic high density compressions.

2.4 Fusion reactor and laser systems requirements

The overall efficiency of a laser fusion power plant can be evaluated through a basic power flow diagram shown in fig. 6. The plant recirculating power fraction (P_{in}/P_g) is a function of laser efficiency ϵ_L, pellet gain G_p, thermal efficiency ϵ_T and blanket multiplication M, and is given by (Maniscalco 1975, 1976)

$$P_{in}/P_g = [\epsilon_L G_p \epsilon_T (0.8M + 0.2)]^{-1}. \tag{18}$$

The output power from the plant is related to the target or laser irradiation frequency ν_L, laser output energy and recirculating power fraction by

$$P_b = P_g - P_{in} = (\nu_L E_L/\epsilon_L)\{(P_{in}/P_g)^{-1} - 1\}. \tag{19}$$

The recirculating power fraction is a figure of merit for any power plant. In

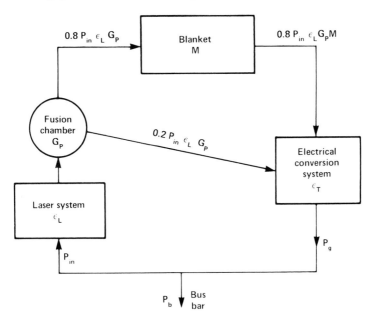

Fig. 6. Laser fusion reactor power flow diagram (DT).

general, plant capital costs scale as the gross electrical power P_g, while revenues scale with the net electrical power P_b. For reference purposes, fossil-fueled power plants operate with recirculating powers less than 5% while the majority of present nuclear plants operate at or slightly above this figure.

In the context of a power plant, laser system efficiency ϵ_L is a lumped parameter that includes power conditioning, medium regeneration, laser and optical transport efficiencies. It is defined as a ratio of the focusable beam energy to total energy input. Blanket energy multiplication M is defined as the ratio of blanket thermal energy to fusion neutron energy. For pure fusion systems it represents the slight gain ($M \leq 1.3$) in energy within the neutron absorbing blanket due to neutron multiplying and exothermic capture reactions such as $Li^6(n, \alpha)H^3$. Higher values of M are possible in fusion–fission hybrid systems in which fissionable material is placed in a blanket around the fusion micro-explosion chamber. The high energy neutrons produced in the fusion core boost the fission process, even in depleted uranium, producing large multiplications of the fusion energy. The overall system efficiency ϵ_s of a laser fusion power plant can also be related to the recirculating power fraction by

$$\epsilon_s = \epsilon_T\{1 - P_{in}/P_g\}. \tag{20}$$

From these expressions for recirculating power fraction it logically follows that the product of laser efficiency and pellet gain $\epsilon_L G_p$ provides an excellent figure of merit for gauging the prospects of efficiently generating electrical power with laser fusion (Maniscalco 1976). This product is called the fusion energy gain and represents the ratio of thermonuclear energy output to energy input to the laser driver. Fusion energy gain is an extremely revealing performance parameter because economical power production with laser fusion will fundamentally depend on the pellet gains that can be achieved and the overall efficiencies at which large pulsed laser drivers can operate.

Although present power plants have about 5% recirculating power fractions, most laser fusion reactor system studies have assumed that when such plants become available values of 25% will be competitive. Under these circumstances and if thermal conversion efficiencies are $\epsilon_T \simeq 0.4$, then plant efficiencies will be low, $\epsilon_s \simeq 0.3$. From eq. (18) fusion energy gains will have to be

$$\epsilon_L G_p \simeq 10/(0.8M + 0.2). \tag{21}$$

For pure fusion systems this becomes $\epsilon_L G_p \simeq 10$ and hence for 1 and 10% efficient lasers pellet gains of 10^3 and 10^2 will be required respectively. According to the discussion in § 2.3, achievement of these pellet gains will require 1 to 10 MJ lasers. In pure fusion systems the efficiency of the laser driver is of great importance. From eq. (21) it is also clear that laser and pellet requirements can be eased by using blankets with high multiplication M.

Fig. 7 gives the fusion system gain requirements for electrical breakeven and various recirculating power fractions for various values of M. As higher M values are selected different fission blanket materials are required, as shown in fig. 7. The figure dramatically shows that an order of magnitude decrease in $\epsilon_L G_p$

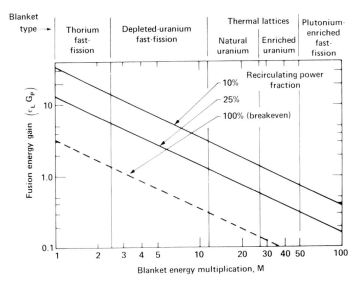

Fig. 7. Laser fusion core-gain requirements for hybrid fusion–fission power systems (Maniscalco et al. 1977).

can be achieved. For example, a fast-fission depleted uranium hybrid with 25% recirculating power requires an $\epsilon_L G_p$ of about one compared to a value of ten without a blanket (i.e., pure fusion reactor). Thus an order of magnitude reduction in either laser system efficiency or pellet gain is possible with a hybrid blanket. However, such a concept is not without limitations and liabilities since radiological hazards and environmental problems are reintroduced by hybrid concepts. For $M > 50$ the higher power densities and plutonium recovery problems associated with liquid metal fast breeder reactors must be considered. Therefore current fission–fusion concepts are under active investigation to determine their viability. In any event, this reactor analysis, when contrasted to the discussion of §2.3, clearly indicates that lasers with efficiencies greater than 1 to 10% will be required as reactor drivers. Early reactors may be hybrids to compensate for low laser efficiencies.

The laser system requirements for a fusion reactor depend on the thermonuclear reaction cycle, the dependence of pellet performance on laser energy, pulse shape and wavelength and the reactor blanket conversion characteristics. Unfortunately, little experimental evidence is available at the present time for the long pulse high peak power regime in which laser-driven high density implosions must operate. This current situation makes it difficult to accurately define reactor laser requirements. However, based on the previous discussion and the numerous numerical simulations of pellet performance for different laser assumptions that have been performed, it is possible to set down a range of requirements for the fusion laser, and these are listed in table 1. As discussed earlier, current theoretical understanding of high density, high gain laser-driven microexplosions indicates that a laser with a high efficiency, short wavelength,

Table 1
Laser system performance requirements for
fusion power plants.

Property	Value range
Laser wavelength	250–2000 nm
Pulse energy	1–3 MJ
Pulse duration	5–10 ns
Peak power	≥ 200 TW
Pulse repetition rage	a few Hz
Average power	≈ 10 MW
Overall efficiency	$>1\%$

high energy and high peak power capability will be required to achieve the high pellet gains required in reactor operation. If these performance capabilities are achieved laser fusion power plants will operate, according to eqs. (18) and (19), in the 1 GW range.

2.5 Laser systems for fusion research

As a point of reference for discussion of advanced laser media and concepts, we provide in this section a brief summary of the properties of demonstrated single-pulse lasers currently used in fusion research.

During the past half-decade, three types of laser systems have been developed with performance characteristics adequate to address the physics issues central to inertial confinement fusion (§ 2.3). Although using three different laser media (neodymium-doped glass, molecular carbon dioxide, atomic iodine), all three laser systems utilize the master oscillator–power amplifier (MOPA) systems architecture. Historically, Nd:glass was the medium of choice, particularly in the subnanosecond pulse regime, and has sustained the greatest development effort worldwide (Emmett et al. 1974). Recently, significant output power has also been demonstrated in CO_2 (Stratton 1976, Richardson et al. 1976) and in atomic iodine (Brederlow et al. 1977, Basov and Zuev 1976) laser systems. The important microscopic physical and kinetic properties of these three laser media are summarized in table 2, along with performance characteristics of the highest-power systems of each type operating at present. The systems characterized in table 2 are (1) the five-terawatt ARGUS Nd: glass laser system at the Lawrence Livermore Laboratory (Simmons et al. 1977), (2) the terawatt beam line of the eight-beam CO_2 laser system at the Los Alamos Scientific Laboratory (Boyer 1977) and (3) the one terawatt Asterix-III atomic iodine laser system at the Plasma Physics Laboratory of the Max Planck Institute (Brederlow et al. 1977).

The Nd:glass laser medium is characterized by a large energy-density storage (~ 500 J/ℓ), long population inversion time (300 μs), low stimulated emission cross

Table 2
Operating performance characteristics of large laser systems developed for fusion research.

	ARGUS 2-beam Nd:glass	DBM 1-beam CO_2	Asterix-II iodine beam
Medium properties			
Laser transition	$^4F_{3/2} \to {}^4I_{11/2}$	$00°1 \to 10°0$	$^2P_{1/2} \to {}^2P_{3/2}$
Level splittings	Stark	Rotational	Hyperfine
Wavelength [μm]	1.06	10.6	1.32
Radiative lifetime [ms]	0.4	4	130
Inversion lifetime [μs]	300	4	10
Emission cross section σ [10^{-19} cm^2]	0.3	20	20
Saturation fluence [J/cm^2]	5	0.2	0.5
System performance			
Pump source	Xe-flashlamp	E-beam/sustainer	Xe-flashlamp
Pump pulse width t_p [μs]	800	3	10
Mean stored energy den. [J/ℓ]	500	10	10
Peak power [TW]	5.	1.2	1.0
Laser pulse width t_L [ns] – nom.	0.2	1.0	0.5
Number of beam lines	2	1	1
Exit beam dia. [cm]	20	15	15
System efficiency [%]	~ 0.1	~ 2	~ 0.2

section $(3 \times 10^{-20}$ cm$^2)$ and a relatively short wavelength (1.06 μm). Optimized multi-terawatt laser chains are designed so as to minimize optically-induced, nonlinear refractive index effects which can produce small-scale laser beam breakup and loss of focusability (Trenholme 1976). A Nd:glass MOPA chain typically consists of a mode-locked Nd:YAG master oscillator, glass-rod pre-amplifiers, and a series of face-pumped disk amplifiers of increasing aperture. Blocks of amplifier gain are isolated from one another and from target back reflection by appropriately placed sets of Faraday rotators and thin-film polarizers. Optical beam stability throughout the system is maintained by the use of proper beam apodization (Johnson 1975) and imaging-spatial filters (Hunt et al. 1977b). Based on the current technology of low nonlinear refractive index materials and thin-film polarizers, multi-terawatt laser chains are peak-power-limited for pulse widths below 0.5 ns and energy-limited for pulse widths above 1.0 ns. In the peak-power-limited regime, the medium cannot be saturated and the laser efficiency is typically 0.1%. In the energy-limited regime, optical damage limits will permit saturation of the laser medium and achievement of an order of magnitude increase in systems efficiency.

As this chapter was being written, final integration of hardware into the SHIVA laser was completed and the system was successfully operated at a power level in excess of 10 TW and at an energy in excess of 10 kJ. Containing 20 ARGUS-like laser chains, this system will soon deliver 20 to 30 TW of laser radiation to fusion targets. This system is also distinguished by the fact that the pulse arrival time, pointing and focusing of each beam at the target are accomplished automatically. On the basis of improved glasses, better beam handling techniques and the demonstrated technological ability to control a large number of laser chains, the construction of a 200 TW laser system (SHIVA–NOVA) at the Lawrence Livermore Laboratory has been initiated. The SHIVA–NOVA baseline design is presently being completed, and project start is set for February 1978.

Large 10 kJ Nd:glass lasers are presently being constructed at the Lebedev Physical Institute, at the University of Rochester and at Osaka University. The Lebedev Del'fin system (Basov 1976) will consist of 216 beams utilizing all glass-rod amplifiers; the Lebedev UMI-35 system (Korobkin et al. 1976) will consist of 64 beams amplified in an array of large slab amplifiers. The OMEGA TEN System at the University of Rochester, New York (Lubin 1977) is designed with 24 beams of staged rod amplifiers. At Osaka University, the Gekko XII system (Yamanaka et al. 1977) will also utilize rod/disk amplifier chains, but the baseline design had not been completed at this writing.

Molecular CO_2 lasers operating in the nanosecond regime utilize CO_2–He–N_2 mixtures of typically $1:3:0.25$ with total pressures of 1 to 3 atm (Stratton 1976). Efficient excitation of large volumes of laser gases is achieved using e-beam-sustainer pumping (Fenstermacher et al. 1972, Reilly 1972). Under typical operating conditions (Strattan 1976), the CO_2 laser medium is characterized by a modest energy density storage ($10\,J/\ell$), population inversion lifetime ($4\,\mu s$) and stimulated emission cross section ($2 \times 10^{-18}\,cm^2$). In contrast with Nd:glass, the CO_2 laser provides an advantageous gaseous form, and relatively high efficiency at the expense of a factor of ten longer wavelength ($10.6\,\mu m$). In analogy with Nd:glass systems, high peak-power CO_2 laser chains typically utilize a mode-locked master oscillator, single-pulse switch-out, discharge preamplifiers and a series of e-beam-sustainer power amplifiers of increasing aperture. In the larger machines at LASL, multipassing of the output amplifier has also been used. Units of amplifier gain are isolated from one another and from target back reflection by appropriately placed saturable absorbers and retro-pulse break-down optical switches. Scaling to higher peak power per unit aperture is paced by the technical performance of the saturable absorbers available and by optical damage to thin-film-coated optics. At the present time, an eight-beam CO_2 system (EBS) designed to produce 10 kJ/1 ns pulses is in final assembly at the Los Alamos Scientific Laboratory. A beam-line of the dual-beam module of this system was recently operated at an energy of 1200 J in a 1 ns pulse and at an overall efficiency of 2%. This system is to be followed by the construction of a 100 kJ CO_2 laser system (Antares) at LASL with operation scheduled in 1982 (Boyer 1977).

High energy, nanosecond iodine lasers typically operate with a few Torr of C_3F_7I gas and with an atmosphere or more of argon buffer gas. The iodine laser medium combines the attractive features of the short wavelength of Nd:glass and the gaseous state of CO_2, while being characterized by moderate energy density storage ($10\,J/\ell$), cross section ($2 \times 10^{-18}\,cm^2$), and population inversion lifetime ($20\,\mu s$) under working conditions (Brederlow et al. 1977). In analogy with both Nd:glass and CO_2 systems, iodine MOPA systems consist of a mode-locked master oscillator, preamplifiers and a series of power amplifiers of increasing aperture. Amplifier gain and aperture parameters are chosen to maintain the beam energy density below optical damage limits while achieving significant gain saturation and energy extraction efficiency. Overall systems efficiencies of 0.2% have been achieved in large systems. As in all MOPA fusion lasers, units of amplifier gain are isolated from one another and from target back reflection by appropriately distributed sets of Faraday isolators and polarizers or of saturable absorbers. In the Asterix-III system, iodine amplifiers are pumped with xenon flashlamps of relatively short pulse duration ($\sim 10\,\mu s$). Exploding-wire-driven, shock heated plasma sources also have been used to pump large iodine power amplifiers. Amplifiers of this type will see use in an iodine system for fusion experiments at the Lebedev Physical Institute (Basov et al. 1977).

All three laser media described above can be classified as energy-storage media. That is, the upper laser level integrates energy from the pump source for times more than a thousand-fold longer than the extracted laser pulse width, greatly reducing the pump source power requirements. In contrast, the pulsed HF chemical laser is intrinsically a non-energy-storing system and its upper laser levels must be pumped at a rate approximately equal to the required extraction rate. This system is currently being examined for fusion applications (Moreno et al. 1977). To provide the exceptionally high pumping rates required, high pressure mixtures of H_2, F_2 and rare gas are induced to react rapidly by initiation with a pulsed relativistic electron beam. The laser electrical efficiency can be very high ($\sim 100\%$) and the overall laser efficiency including chemical regeneration may be usefully large. The major issues confronting the pulse HF chemical laser for use in a fusion context are the achievement of efficient multiline extraction and focusing of amplifier energy without excessive prepulse and ASE heating of targets, and the extraction of useful energy in the few nanosecond time scale for total energies in the megajoule range.

The question 'why can't the laser media used in current research lasers be used for high average power fusion applications?' naturally arises. The answer is that maybe they can be so used, but for each medium certain issues must be successfully redressed. For multinanosecond long pulses, the Nd:glass laser may achieve an efficiency of a few percent. However, simultaneous achievement of high average power, multi-Hertz outputs will be predicated on successful accommodation of the relatively low thermal conductivity of the solid-state Nd:glass medium. A conceptual approach to solving this design problem involves the use of arrays of thin, large-aperture disk slabs, pumped through their faces and cooled by helium gas flowing turbulently through the channels formed by the slab faces. In the case of iodine, current systems are already operated appreciably into saturation; achievement of greater than 1% efficiency will probably require a significant innovation in pumping technique, development of an efficient process for reforming spent laser medium and design of efficient gas flow systems which provide adequate output beam quality (Krupke and George 1976). By contrast, the CO_2 laser is intrinsically amenable to high average power operation at an efficiency in excess of a few percent. However, because fusion

pellet gains at constant peak power may be several times lower at 10 μm than at
1 μm, the CO_2 laser may have to operate at correspondingly higher efficiency to
be useful. Alternatively the CO_2 laser may be required to supply significantly
greater peak power than a laser at substantially shorter wavelength. It is
expected that an understanding of the wavelength dependence of pellet gain will
be obtained experimentally in the next few years. In the meantime, the potential
average power capabilities of the known and demonstrated media will continue
to be assessed. While significant progress can be expected toward the reduction
and/or elimination of many of the limiting aspects of the laser systems described
above, a directed program to identify and develop advanced lasers specifically
matched to fusion applications has been established within the Inertial
Confinement Fusion (ICF) Program of the Department of Energy. The analyses
and experimental activities described in this chapter are a part of this applied
program.

3 Fusion laser media/systems concepts

3.1 Introduction

An analysis of the laser system performance requirements for fusion power
plants set forth in table 1 can lead to an identification of the physical and
technological constraints that a candidate laser medium must satisfy. In general
terms, the candidate laser must operate in a wavelength regime in which high
optical power and energy transport can be accomplished with low loss; laser
media excitation and extraction techniques must be both efficient and scalable,
and efficient high average power operation must be possible. Finally, the overall
laser system must be reliable and economical. Although several gaseous and
solid-state media presently known provide some of these desired properties,
those which show promise of satisfying all requirements for a fusion reactor
laser are indeed quite rare.

In § 3 the general physical and technological characteristics required to
achieve the fusion laser performance specifications given in table 1 will be
discussed. In § 3.2, solid and gaseous media damage or breakdown limitations on
laser beam energy transport will be briefly reviewed. In § 3.3, electron-beam and
e-beam-augmented discharge excitation techniques will be discussed in relation
to fusion laser requirements for efficient and scalable pump sources. In § 3.4,
fusion laser systems architectures (cascaded, regenerative or multipass amplifiers) and amplifier excitation and extraction characteristics will be discussed.
Limitations imposed by optical transport, pump technology, depumping losses
such as fluorescence and parasitic self-oscillation, nonsaturable photoabsorption
losses and system architecture will be related to establish the parameter regimes
in which fusion laser amplifiers are likely to operate. In § 3.5 the propagation and

focusing of high power laser beams will be considered. Finally the special requirements for high average power laser operation will be described and related to the problems of total amplifier efficiency.

3.2 Optical materials damage constraints

The design of all high power, high energy laser systems will be strongly driven by the optical damage fluences of dielectric materials required to be in the beam path. Given the total energy output required of the laser the optical damage fluence will determine the total emitting area of the laser system at the output. This total area will have to be divided into a number of apertures determined by a combination of laser physics, laser pumping and cost constraints. For example, if the final stage optical damage limit is 1 J/cm^2 then a 1 MJ laser must have a radiating aperture of 10^6 cm^2. If each final amplifier could be made to have a $1 \text{ m} \times 1 \text{ m}$ output aperture then 100 such amplifier modules would be required to achieve the 1 MJ output. An increase in the damage threshold by a factor of ten for such a system would reduce the number of amplifiers to ten and probably result in much greater system reliability, lower cost and simplicity. Based on present knowledge, damage fluence levels will be at most a few tens of joules per square centimeter and consequently large-aperture ($>1 \text{ m}^2$) final amplifier modules will be required.

3.2.1 Solid media damage constraints

Several recent reviews (Smith 1978, Glass and Guenther 1973, 1974, 1975, 1976, 1977, Milam 1977, Milam and Bradbury 1973, Bliss 1971) have been written summarizing the various damage mechanisms in solids. Some mechanisms require imperfections in the material or depend on properties which are subject to control during the growing or melting process. Others are a result of more intrinsic properties and constitute fundamental limitations on the performance of the material. At present, accurate theoretical prediction of damage thresholds is not possible and therefore experimental measurements are required in all cases.

The most extensive measurements (Milam 1977) of optical damage to solid materials have been performed in the vicinity of $1 \mu m$ in connection with the development of short pulse Nd:glass lasers. Optical damage thresholds were found to vary among optical component types and to depend on pulse length. As an example, Milam's measurements for 125 ps glass laser pulses can be summarized as follows. Surface damage thresholds for bare glass surfaces fabricated by normal polishing using nonabsorbing (at $1 \mu m$) abrasives, ranged from 8 to 13 J/cm^2. Special surface preparation was ineffective in increasing these damage thresholds and obvious surface scratches seemed to play only a minor role in the damage process. Threshold damage invariably resulted in micropits, indicating that failure was due to isolated damage-susceptible areas of an unidentified nature. Anti-reflection dielectric thin films damaged at $3.9 \pm 2.0 \, \text{J/cm}^2$, while both the average threshold and the range of observed thresholds for highly reflective dielectric film structures were somewhat larger. Thin-film polarizers damaged at levels comparable to those for anti-reflection films. Diamond turned copper and silver surfaces withstood 1 to $4 \, \text{J/cm}^2$, polished copper damaged at 0.15 to $1.5 \, \text{J/cm}^2$ and the damage thresholds of thin inconel films were found to vary with film thickness from 0.02 to $0.05 \, \text{J/cm}^2$. Damage to dielectric surfaces or films was sharply defined on a given sample, exhibiting no readily observable statistical features. Damage in metal surfaces was much less easily diagnosed and was frequently quite subtle in nature. Details of these observations and their implications for current damage theories have been discussed by Milam (1977). Recent damage measurements at $1 \, \mu\text{m}$ (Milam 1977) suggest that the damage thresholds for bare, polished fused silica and dielectric thin film reflectors increase approximately as the square root of the laser pulse duration in the 0.1 to 1.0 ns regime. These results suggest that surface roughness and absorption may become increasingly detrimental as the pulse duration increases. Earlier ruby laser (694.3 nm) measurements (Milam and Bradbury 1973) showed that the damage thresholds for high-quality electron gun deposited thin films also increased as the square root of the pulse length in the 0.2 to 23 ns regime. These observations were interpreted as due to the heating of included metallic or other high absorbing particles (Bliss 1971, Bliss et al. 1972, Hopper and Uhlmann 1972).

At wavelengths very much shorter than $1 \, \mu\text{m}$ damage thresholds decrease. However, the magnitude and wavelength dependence of this variation has not been well documented experimentally. All optical materials are opaque at wavelengths less than about 100 nm and most convenient materials absorb strongly at 150 nm (Ballard et al. 1972), so wavelengths this short are not suitable for high energy lasers unless a windowless configuration is developed. For high intensity near-ultraviolet lasers two-photon absorption becomes an important process (Smith 1978). In table 3 the band gap or two-photon absorption threshold for several optical materials is tabulated, together with the two-photon energy for a number of short wavelength lasers. When the two-photon energy exceeds the band gap energy, electrons can be excited into the conduction band and if the laser intensity is high enough, avalanche breakdown can occur. A crude estimate (Murray and Hoff 1972) indicates that two-photon absorption of 172 nm Xe$_2^*$ excimer laser radiation in crystals such as LiF and MgF_2 might reach $3 \, \text{cm}^{-1}$ at an intensity of $10^{10} \, \text{W/cm}^2$. Clearly, detailed measurements at short wavelengths are needed to establish the intensity and wavelength dependence of this potentially serious phenomenon.

The difficulties associated with the transmission characteristics of solid optical components at short wavelengths and high fluxes can be circumvented in principle by using subsonic or supersonic flow windows (Parmentier and Greenberg 1973). In the subsonic case the boundary between co-flowing streams of different gases moving at the same pressure and speed to avoid Kelvin–Helmholtz instability (Landau and Lifshitz 1959) of the interface can be used

Table 3
Laser two-photon energy and two-photon optical
material absorption thresholds.

Optical material	Threshold [eV]	Laser
	19.6[a]	Ar$_2$
	17.0[a]	Kr$_2$
	14.4[a]	Xe$_2$
	14.3[a]	ArCl
	13.0[a]	ArF
LiF	13.0	
	11.2[a]	KrCl
MgF$_2$	11.0	
KF	10.9	
NaF	10.5	
RbF	10.4	
	10.1[a]	KrF
CsF	10.0	
LiCl	10.0	
CaF$_2$	10.0	
SrF$_2$	9.0	
BaF$_2$	9.0	
	8.9[a]	XeBr
NaCl	8.6	
KCl	8.5	
LiBr	8.5	
Al$_2$O$_3$	8.3	
RbCl	8.2	
	8.1[a]	XeCl
KBr	7.8	
NaBr	7.7	
SiO$_2$	7.7	
	7.4[a]	N$_2$
MgO	7.3	
	7.1[a]	XeF

[a]Energy of two laser photons – for a given laser two
photon absorption can occur in optical materials with
thresholds lower than this energy.

instead of a solid window. If the gases have a different density and temperature, the interface must also be oriented vertically or parallel to the gravitational field vector in order to avoid Rayleigh–Taylor instability (Landau and Lifshitz 1959). Generally, significant disruption of the subsonic flow window by diffusive mixing processes is not important. Subsonic flow windows are promising in applications involving ultraviolet fluorescence pumping sources (Monsler 1977). Supersonic flow or aerodynamic windows (Parmentier and Greenberg 1973) use the momentum of a supersonic jet to support a pressure difference between, for example, the laser cavity and the ambient atmosphere or a low pressure region. Several designs using expansion, compression or shock-expansion waves to support the pressure difference have been proposed (Parmentier and Greenberg 1973). Unfortunately, supersonic flow windows require large mass flows and high flow pressure drops, and therefore for large apertures consume substantial amounts of power (≥ 100 times the power to condition the laser gas), and so do not presently appear to be practical for average power large aperture fusion laser applications.

3.2.2 Gaseous media damage constraints

There are generally two mechanisms that can lead to the interruption of the propagation of a high intensity laser beam through gaseous media. One is a multiphoton process and the other is a cascade or collision-induced absorption. Both lead to breakdown or ionization of the gas. Direct, single-photon processes are only important in the ultraviolet where photoionization of excited states may lead to gas breakdown. Many excellent reviews (De Michelis 1969, Kroll and Watson 1972, Grey Morgan 1975, Smith and Meyerand 1976) of gas breakdown have been published and only a brief summary is given here.

The multiphoton mechanism involves the simultaneous absorption of the number of photons n required to equal the ionization potential of the gas. Multiphoton ionization probabilities are generally low, except when accidental resonances occur, and scale as I^n, where I is the laser beam intensity. For visible laser radiation sources it is observed experimentally that the pulse duration must be as short as tens of picoseconds for atmospheric pressure gases or the pressure must be less than 1 Torr for pulses of tens of nanoseconds duration in order for multiphoton ionization to dominate over cascade ionization. For example, the calculated (Beeb and Gold 1966, Keldysh 1965) and experimental (Krasyuk et al. 1969, 1970) multiphoton ionization thresholds of the rare gases and air for ruby laser (694.3 nm) light pulses are approximately 10^{13} W/cm^2 for argon, 10^{14} W/cm^2 for air and are generally in the 10^{12} to 10^{14} W/cm^2 range. Multiphoton ionization breakdown processes may become very important in ultraviolet lasers and much remains to be done to quantify this issue for the excimer lasers.

In the cascade breakdown process laser radiation is absorbed by free electrons in the gas during their collisions with atoms and ions until sufficient energy is gained to ionize an atom by an inelastic collision. This mechanism is dominant at long wavelengths and competes with the multiphoton process at short wavelengths. A very simplified model (Smith and Meyerand 1976) shows that the threshold intensity for cascade breakdown for large-area beams can be written as

$$I_{th} \simeq \frac{\epsilon_0 c m_e \epsilon_i \omega^2}{e^2 \nu_c} \left\{ \frac{\ln(N/n_0)}{t_L} + \frac{1}{\epsilon_i} \left(\frac{d\bar{\mathscr{E}}}{dt} \right) + \nu_a \right\}, \tag{22}$$

where the physical constants are defined in the usual way, ϵ_i is the ionization energy, ω the radian frequency of the laser light whose pulse length is t_L, ν_c is the electron momentum transfer collision frequency, N/n_0 is the ratio of the final to initial electron density, $d\bar{\mathscr{E}}/dt$ is the electron collision energy loss rate and ν_a is the attachment rate. In fusion laser applications when the pulse length is less than approximately 10 ns, the last two terms in eq. (22) are not important. Then, from eq. (22), the breakdown threshold is fluence limiting and is proportional to $\omega^2\epsilon_i/p$, where p is the gas pressure. This behavior has been well established experimentally for long wavelength visible and infrared laser radiation. Typical breakdown fluences for Nd:glass (1.06 μm) and ruby (694.3 nm) laser pulses are in the 10^2 to 10^3 J/cm^2 range, and depend on the presence of particulates and other impurity concentrations. When the pulse length exceeds 10 ns the last two terms in eq. (22) become important and the breakdown threshold becomes power dependent. For additional results on gas breakdown the reviews cited earlier should be consulted. Although gas breakdown is a potentially serious problem, especially for infrared and ultraviolet lasers, at the present time transport through solid optical components generally sets the lowest fluence limits (~ 10 J/cm^2) on visible and ultraviolet laser beam propagation.

3.3 Media excitation techniques

The lasers generally considered for fusion applications are those pumped with electrical prime power. For each laser medium, the prime electrical power must be suitably conditioned and coupled to the medium to produce a useful gain. This coupling may involve exciting the medium directly with an electron beam or with an electron-beam-augmented discharge (Schlitt and Bradley 1975, Bradley 1975, Daugherty 1976). Alternatively the electrical energy may be converted into light (either coherent or incoherent) which is then used to pump the fusion laser medium.

The simplest and one of the most efficient techniques for electrically exciting large volumes of laser media uses a high energy electron beam. When high energy electrons enter a gaseous medium, electron collision processes ionize and excite the medium through the production of secondary, tertiary and higher-order electrons. Taking argon as an example, detailed calculations (Peterson and Allen 1972) indicate that approximately 65% of the electron-beam energy, for energies greater than several hundred electron volts, goes into ionization, as shown in fig. 8. The remainder of the electron-beam energy produces electronic excitation and thermal heating of the gas. In e-beam excited laser media, ionic recombination processes usually dominate the formation kinetics of the laser molecule (Lorents 1976, Werner and George 1976, Daugherty 1976). Assuming that each ionization event leads to a laser photon, the excitation efficiency ϵ_{eb} can be as high as $h\nu/W_i$, where W_i is the e-beam energy loss per ion pair formed. Values of W_i are typically on the order of twice the gas ionization potential and in the range of 20 to 35 eV. Consequently for a green laser photon ϵ_{eb} may be on the order of 10%, whereas for photons in the near ultraviolet values of 40% and more are possible. Therefore, the highest overall excitation efficiency for direct

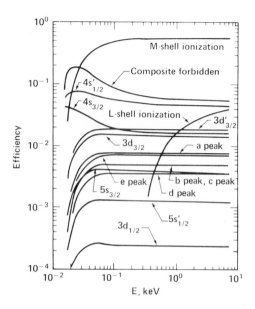

Fig. 8. Production efficiency of each state of argon as a function of the incident electron-beam energy (Peterson and Allen 1972).

e-beam pumping occurs for laser transitions in the visible or near ultraviolet and not in the infrared.

The excitation efficiency limit is generally not achieved in practice owing to collisional quenching processes and inefficiencies in the e-beam deposition process. In general, the value of the primary electron kinetic energy plays a minor role in the pumping kinetics. This is because excitation and ionization cross sections vary slowly for electron energies in the 10^4 to 10^7 eV range. The required energy of the beam electrons is therefore determined by other considerations. It must be high enough (> 200 keV) so that the electrons efficiently penetrate (Seltzer and Berger 1974) the foil separating the electron gun and the gas. If the electron-beam energy or the atomic number Z of the gas medium is too high, bremsstrahlung radiation and backscatter energy losses can reduce the e-beam energy deposition efficiency in the laser medium. In fig. 9 the fractional energy loss due to bremsstrahlung and backscatter of a 1 MeV electron beam is plotted as a function of atomic number of the gas medium (Schlitt and Bradley 1975). The results are not sensitive to variations in electron energy or medium density but only to the average Z of the medium. Clearly if an efficient system is desired, the primary constituent of the medium will have to have $Z < 10$. At higher voltages, bremsstrahlung X-ray production becomes particularly important. For example, a 10 MeV beam pumping xenon loses 25% of its energy to X-rays, thus reducing excitation efficiency and increasing system costs owing to shielding requirements. Therefore, practical working voltages fall in the range of 200 keV to 5 MeV.

Within this range the prime consideration in determining gun voltage is that

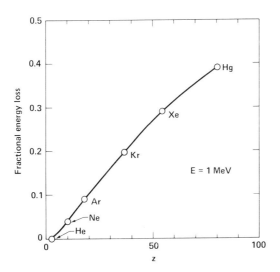

Fig. 9. Electron-beam energy loss due to backscatter and bremsstrahlung emission plotted as a function of the atomic number Z of the gas medium (Schlitt and Bradley 1975).

the electrons must be sufficiently energetic to uniformly pump the gas volume of interest. The condition for large volume excitation can be readily estimated. As the beam travels in the gas it spreads out because the electrons are scattered by the gas atoms; the power density of the beam diminishes the farther it goes. Therefore, the dimension W of the laser medium parallel to the e-beam propagation direction will be approximately the electron range. The electron range R_e in a gas without externally applied electric or magnetic fields is given approximately by (Schlitt and Bradley 1975)

$$R_e \simeq \frac{V_{eb}}{\rho(dE/dx)},$$ (23)

where ρ is the gas density, V_{eb} is the electron-beam voltage and dE/dx is the deposition function (Berger and Seltzer 1964). For electron-beam energies in excess of 250 keV the deposition function is approximately 2 MeV/[gm/cm²] for most gases (hydrogen is an exception). The depth of penetration of the electrons into the gas is less than the electron range because of scattering. Then W is set at some fraction F of the electron range, depending on the desired pump uniformity. SANDYL (Colbert 1973) calculations (Schlitt and Bradley 1975) for the superposition of oppositely directed beams suggest that F is approximately 0.4 and depends in detail on the particular geometry and gas. The mean volumetric power deposition rate is

$$\frac{d\mathscr{E}_{eb}}{dt} = \frac{1}{F}\left(\frac{J_{eb}}{e}\right)\rho\frac{dE}{dx}$$ (24)

if foil and foil holder losses are not included. Therefore, for a constant e-beam pump pulse exerted over a time t_p, the integrated deposited energy density is

$$\mathcal{E}_{eb} \simeq \frac{1}{F}\left(\frac{J_{eb}}{e}\right) t_p \rho \frac{dE}{dx}. \tag{25}$$

If ϵ_j is the fraction of the primary electron-beam energy which is deposited in an atomic or molecular state j whose threshold is U_j (see fig. 8), then the production rate of state j is

$$S_j = \frac{\epsilon_j}{F}\left(\frac{J_{eb}}{e}\right) \rho \frac{dE}{dx}. \tag{26}$$

For example, the ionization efficiency is $\epsilon_i = U_i/W_i$ (Peterson and Allen 1972, Bass and Green 1973), where U_i is the ionization potential of the ground state of the gas.

Several generally useful formulas can be derived from the previous results. The electron-beam energy required to excite a gas whose depth is W is

$$V_{eb}[\text{MeV}] = 8.4 \times 10^{-5} Ap[\text{atm}] \frac{W}{F}[\text{cm}] \left[\frac{1}{2}\frac{dE}{dx}\left[\frac{\text{MeV}}{\text{g/cm}^2}\right]\right], \tag{27}$$

where A is the atomic weight of the gas. The specific energy density deposited in the medium by a constant electron-beam pulse is

$$\frac{\mathcal{E}_{eb}[\text{J}/\ell]}{p[\text{atm}]} \simeq J_{eb}[\text{A/cm}^2] \frac{t_p[\mu\text{s}]}{F}\left(\frac{A}{12}\right)\left[\frac{1}{2}\frac{dE}{dx}\left[\frac{\text{MeV}}{\text{g/cm}^2}\right]\right]. \tag{28}$$

To illustrate the last results, if a laser amplifier contains mostly argon at 1 atm pressure and is 1 m across, then the e-beam energy required to excite this amplifier is approximately 0.59 MeV. If an electron beam operating at 10 A/cm^2 for 1 μs is used, then the specific energy density deposited in the medium will be 58 J/ℓ atm.

The previous approximate results indicate that large volumes of laser gas can be excited at relatively high specific energy densities with an electron beam. Detailed analyses of the uniformity and efficiency of e-beam excitation of a bounded volume require a full three-dimensional simulation by a numerical code such as SANDYL (Colbert 1973), or alternatively require extensive experimental studies to determine optimum conditions. A typical spatial distribution of deposited energy, calculated (Schlitt and Bradley 1975) using the SANDYL electron transport code, is shown in fig. 10. Two 1.5 MeV electron beam each 2 m × 5 m in area are assumed incident from opposite sides on a laser cell 1.7 m thick. Neon gas at a pressure of 4 atm was assumed with a 25 μm thick nickel foil. Contours of equal energy deposition/unit volume are plotted. The contour interval is 10% of the peak deposition density. The contours indicate that 41% of the incident energy is deposited in the region with energy deposition density 0.9 to 1.0 of the peak value. Thus even though foil losses and backscatter are small (90% of the incident energy is deposited in the gas), a requirement of $\leqslant 10\%$ uniformity in energy deposition greatly limits the overall deposition efficiency. Thus, in practice for e-beam-excited media there will generally be a trade-off between energy deposition efficiency and uniformity. Externally applied magnetic fields can be used to improve energy deposition uniformity (Daugherty 1976).

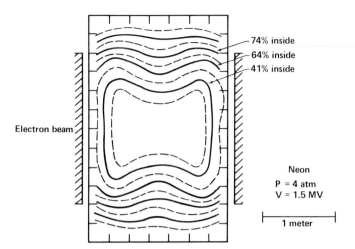

Fig. 10. Spatial distribution of deposited energy for two opposed electron beams injected into neon gas. Contours of equal energy deposition/unit volume are plotted. The contour interval is 10% of the peak deposition density (Schlitt 1976).

For the electron-beam-pumping technique, the power conditioning system consists of several elements. Low power electrical energy from commercial power lines is transformed to high voltage and stored. The energy-storage reservoir integrates the low power input for delivery on the shorter time scale of the electron-beam pulse. A pulse-forming network then shapes the electrical pulse applied to an electron gun. The electron-beam generated by the gun is then directed into the laser medium.

In general there are two ways to form pulses, using either distributed or lumped pulse-forming networks. The former may be as simple as a section of charged cable and the latter may be the lumped equivalent of this cable. Which circuit may be used is determined by the rise time (L/R time constant) of the component inductances and the load resistance, and by the $(LC)^{1/2}$ time constant of the circuit. Nominally the distributed circuit is required for very short or very powerful pulses. Distributed circuits are pulse charged at high voltage, and during charge act as a capacitor. Examples of the pulse charger are the Marx (Fitch 1971), LC generator (Bradley 1975), transformer (Abramyan 1968) and Van de Graaff. Alternate techniques exist to produce similar results, such as charging distributed or lumped lines (Fitch and Howell 1964) at low voltage and erecting them to high voltage.

Of the many types of electron guns, such as thermionic (Fink and Schumacher 1974), glow (Holliday and Isaacs 1971), cold cathode (Parker 1974), field emission (Dyke 1964), photo emission (Passner 1972), multipaction (Liska 1971) and plasma (Iremashvili 1973), only the first three have found widespread use for pumping lasers. The first two types of guns are typically used for low voltage (0.1 to 0.3 MeV), low current density ($\leqslant 10$ A/cm^2) and long pulse ($\geqslant 1$ μs to continuous) applications. The third gun type covers this range and extends to high voltage, greater current density and shorter pulse widths.

The thermionic and glow guns are generally operated at long pulse widths necessary to produce reasonable energy density in the gas at their intrinsically limited current density. The cold cathode gun is limited to pulsed operation by formation of a conducting plasma which shorts out the accelerating field (diode closure). Additionally, the thermionic and glow guns may be grid controlled whereas the cold cathode gun may not be so controlled, again because of plasma closure. The operating regimes for these guns overlap; post-acceleration techniques may be used to obtain higher voltage or focusing to increase the current density of thermionic and glow guns.

In order to generate the electron beam it is necessary to have the high voltage gun electrodes in a high vacuum varying from about 10^{-3} Torr for cold cathode devices to about 10^{-6} Torr for thermionic emitters. Operationally, foils backed by a support structure are used as penetrable barriers to separate the electron-gun vacuum region from the laser medium, which is usually near atmospheric pressure. These foils are usually made of materials of low average atomic number; for example, ~ 0.7 mils (5 mg/cm^2) of aluminum, to maximize transmission (Bradley 1975, Daugherty 1976).

The ionizing power or excitation flux required by the laser medium can result in severe foil heating. The foil temperature rise must be limited to a certain value in order to avoid structural failure. This places a limit on the electron-beam energy fluence that can be transmitted through the foil at a given pulse repetition frequency (Daugherty 1976). For an e-beam operating in the 1 to 10 pps regime, fig. 11 shows that the foil damage electron-beam energy fluence limit is ap-

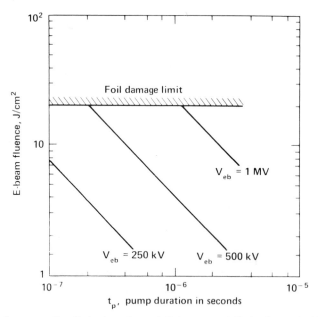

Fig. 11. E-beam fluence scaling limitations due to foil damage and diode closure (cold cathode diode) plotted as a function of pump duration and electron-beam voltage (Monsler 1977).

proximately $20 \, \text{J/cm}^2$ (Daugherty 1976). However, as noted above, for e-beam current densities greater than 1 to $10 \, \text{A/cm}^2$, cold cathode guns are generally used. In these guns, the phenomena of plasma or diode closure (Daugherty 1976) limits the pulse length at a given current density and electron-beam voltage. An estimate of the constraint imposed by diode closure is plotted in fig. 11 for 250 keV, 500 keV and 1 MeV electron beam voltages. Fig. 11 indicates that for cold cathode electron guns, diode closure is a more severe limit than foil heating for current densities less than several hundred amperes per square centimeter and that thermionic emitting guns operating in the $10 \, \text{A/cm}^2$ regime offer potentially significant pump fluence advantages at the low pulse repetition rates of interest in fusion reactor lasers. Alternatively, development of techniques to control plasma closure in cold cathode guns may eliminate this difference.

The scaling of e-beam pumps is also determined by vacuum and insulator breakdown limitations and e-beam pinch due to self-generated magnetic fields. In the pinch effect, the self magnetic field of the beam overcomes the repulsive space charge field and causes the beam to collapse inward. The self magnetic field is dependent on the detailed current distribution in the beam and the degree of collapse is proportional to the time spent in the field, i.e., to the length of the electron-beam trajectory between cathode and anode. The pinch effect is therefore geometry, current density and e-beam-voltage dependent. For a high aspect ratio rectangular e-beam, the maximum cathode module width H_{max} before significant beam pinch occurs is approximately (Schlitt and Bradley 1975)

$$H_{max}(\text{cm}) \sim 24[V_{eb}(\text{kV})]^{1/4}/[J_{eb}(\text{A/cm}^2)]^{1/2} \qquad (29)$$

for $V_{eb} \geq 250 \, \text{keV}$. Therefore if a laser amplifier of width $H > H_{max}$ is to be pumped, the e-beam must be modularized with each module having a width less than H_{max}, or an external magnetic field must be applied parallel to the direction of current flow in the diode. For the example cited earlier in which a $1 \, \text{m} \times 1 \, \text{m}$ amplifier cross section containing mostly argon was excited by a 0.59 MeV, $10 \, \text{A/cm}^2$ e-beam for $1 \, \mu\text{s}$, eq. (29) indicates that $H_{max} \sim 37 \, \text{cm}$. Therefore, avoidance of beam pinch and achievement of uniform excitation requires two opposed e-beam systems each composed of three adjacent modules. Alternatively, if an external magnetic field is used which exceeds the e-beam self magnetic field by a sufficient amount, the electron trajectories will not collapse and the pinch effect will be controlled. The cost, efficiency and scalability of the magnetic field technique may represent serious limitations. However, although e-beam modularization is a promising technique for exciting large amplifier volumes on the time scales of interest in laser fusion, significant problems associated with module return current control and pulse-forming line design must be overcome.

Although large volumes of laser medium can be ionized by high energy electron beams, in many cases the excited laser species can be more efficiently produced by use of an electric discharge (Daugherty 1976). In a self-sustained discharge, low energy (1 to 5 eV) electrons are produced and heated by an electric field. These electrons then transfer their energy by inelastic collision processes to atoms and molecules. In some circumstances this technique offers the possibility of a significant increase in excitation efficiency. For example, in excimer lasers the lasing molecule can form by collisional quenching of the first excited electronic state of an atomic species present in the medium. If ΔE is the inelastic electron energy loss associated with production of this lowest electronic state, then the discharge excitation efficiency is approximately $h\nu/\Delta E \gtrsim 2\epsilon_{eb}$, since ΔE is less than the ionization energy and so is over a factor of two less than W_i. Although other inelastic and elastic electron energy transfer processes occur in a discharge, it has been shown that as much as 80% of the electron energy can reach the intended excited state under carefully controlled discharge conditions. Thus the overall discharge electrical efficiency can be significantly greater than that obtained by e-beam excitation.

To be useful in fusion laser applications the discharge must be capable of exciting large volumes for a time of order a microsecond. Two phenomena limit the simultaneous achievement of these conditions: discharge instabilities and the electromagnetic skin depth effect. Generally, self-sustained discharges are limited to low pressure operation owing to the occurrence of a variety of plasma instabilities (Haas 1973, Nighan 1976, Daugherty 1976). As the pd product of the discharge is increased above a few tens of Torr-centimeters by increasing the pressure p or transverse discharge dimension d, a fairly uniform low-pressure discharge constricts into an arc. To achieve large volume excitation, $d \simeq 100\,\mathrm{cm}$ requires operating at low pressure (<0.1 to $1\,\mathrm{Torr}$) and consequently at unacceptably low excited state densities. The scalability of short pulse high conductivity discharges is also limited by the electromagnetic skin depth effect (Daugherty 1976). For example, in nonstorage excimer laser applications discharge plasma conductivities σ of 10 to $50\,\mathrm{mho/m}$ must be established for pump times on the order of 0.1 to $10\,\mu\mathrm{s}$. The electromagnetic skin depth, $\delta \sim (2t_p/\pi\mu_0\sigma)^{1/2}$, under these circumstances varies from approximately $3\,\mathrm{cm}$ for a $100\,\mathrm{ns}$ discharge whose conductivity is $50\,\mathrm{mho/m}$ to nearly $70\,\mathrm{cm}$ for a $10\,\mu\mathrm{s}$ discharge with a $10\,\mathrm{mho/m}$ conductivity. To maintain discharge uniformity, one dimension of the discharge must be constrained to be less than the skin depth. Thus even if discharge stabilization techniques can be developed, the transverse scale of high conductivity discharge lasers will be significantly restricted by the field penetration effect. To circumvent these scaling limitations, several novel electron-beam-ionized or -augmented discharge techniques have been developed. These techniques include the following types of discharges: (1) the e-beam ionizer sustainer discharge (Daugherty et al. 1971, Fenstermacher et al. 1972) (2) the e-beam-augmented discharge (Mangano and Jacob 1975, Daugherty et al. 1976) (3) the beam-plasma instability discharge (AVCO Everett Res. Lab. 1975) and (4) the e-beam return current discharge (Hsia et al. 1977). The first two have achieved some success in circumventing discharge instabilities that affect device scalability, whereas the latter two are directed at eliminating the field penetration skin depth constraint in low pressure ($<100\,\mathrm{Torr}$) visible laser devices. The latter two techniques have been discussed in detail elsewhere (Daugherty 1976) and will not be considered further here. However, it should be noted that although the skin depth effect is circumvented in the return current discharge, the appearance of a large return current electric field will lead to instabilities similar to those found in the e-beam-augmented discharge.

In molecular lasers, the e-beam ionizer sustainer technique (Daugherty et al. 1971, Fenstermacher et al. 1972) has been very effective in controlling discharge instabilities. Indeed, in the past few years stable discharges have been scaled to *pd* products in excess of 10^4 Torr cm by this method for both the CO and CO_2 infrared lasers. Since the average electron energy (1 to 2 eV) is small compared to the ionization energy ($\gtrsim 12$ eV) in these lasers, the electron beam is used to produce the electrons, whereas the applied electric field heats them. In this way discharge instabilities (Haas 1973, Nighan 1976) are suppressed by controlling the ionization mechanisms. The electron-beam power required is typically $\lesssim 1\%$ of the joule heating discharge power provided by the applied electric field.

In electronic transition lasers the e-beam ionizer sustainer technique is no longer so effective. The average electron energy required to produce high electronically excited state densities is much higher (3 to 5 eV). The fraction of the total excitation power required to operate the e-beam under these circumstances is large, and the excitation and ionization processes are not fully separated as in the e-beam ionizer sustainer discharge (Mangano and Jacob 1975, Daugherty 1976, Daugherty et al. 1976). Consequently, this new type of discharge is referred to as an e-beam-augmented as opposed to sustained discharge.

It has been pointed out (Daugherty et al. 1976) that in e-beam-augmented discharges, an ionization instability can occur when

$$\bar{k}_i = \partial \ln k_i / \partial \ln n_e > (m - 1) + m(S_i/n_e n_g k_i), \tag{30}$$

where the electron density n_e dependence of the effective ionization rate k_i is due to ionization of excited states and electron–electron collisions. The electron density dependence of the loss rate $n_e^m \alpha_e$ is characterized by the exponent m; for attachment $m = 1$, for dissociative recombination $m = 2$, etc. S_i denotes the electron production rate due to the e-beam [eq. (26)]. Generally $\bar{k}_i \lesssim S_e$, where S_e is the number of electronic states participating in the ionization kinetics. The stability of the discharge depends on the number of excited electronic states that participate in the ionization process, the electron density dependence of the loss process and the ratio of the e-beam ionization rate to the effective ionization rate due to the applied electric field. In the case of rare gas halogen lasers Daugherty et al. (1976) have pointed out that attachment ($m = 1$) is the dominant loss and that multistep ionization of excited states of the rare gases dominates the ionization process under some circumstances. According to eq. (30) these circumstances lead to instability (Daugherty et al. 1976) unless $S_i > n_e n_g k_i$, or equivalently, using the steady-state condition,

$$n_e n_g k_i + S_i - n_e^m \alpha_e = 0 \tag{31}$$

unless $\alpha_e > 2 n_g k_i$, where $\alpha_e = \nu_a$ is the attachment coefficient. The discharge is stabilized when the e-beam ionization rate exceeds the ionization rate due to the applied electric field. This type of instability is also expected to occur in metal vapor discharge laser systems (§ 8) where multistep ionization processes are also very important. This instability leads to a collapse or filamentation of the discharge and thus limits the sustainer discharge enhancement factors (the ratio of the power deposited in the laser mixture by the discharge to the power deposited by the e-beam) that can be obtained for these systems. For the rare gas halide lasers enhancement factors of 5 to 10 are possible (Daugherty et al. 1976).

In some media it is not possible to produce the lasing species directly by electrical means owing to unfavorable excitation cross sections, collisional kinetic processes or the nature of the matter such as solid-state media. Under these circumstances optical excitation techniques have often been useful. The optical radiation must be generated by electrically exciting another medium and therefore the potential efficiency of this technique is generally not as high as the direct electrical methods. Although there are many ways to optically excite a medium in laser fusion applications, only those capable of efficient large volume excitation are of interest. The medium pumped must be able to efficiently absorb and store or integrate the pump radiation for the duration of the pumping process (0.1 to 10 μs). Several optical pumping techniques are described in §§ 4 to 8.

3.4 Energy extraction considerations

According to table 1, a fusion reactor laser system must deliver a multi-hundred-terawatt pulse of short wavelength radiation to each fusion pellet. Design of such a laser system requires careful consideration of the energy and power extraction characteristics of each of its amplifier elements and the staging or systems architecture within which these elements must function. In addition to the physics and technology issues associated with the development of large-scale amplifier elements, economic and reliability constraints will require the development of innovative systems architectures. In current Nd:glass laser designs the systems architecture consists of a cascaded chain of amplifiers (fig. 12a) driven by a master oscillator in which the final amplifier stores most of the energy but accounts for perhaps 20 to 30% of the total cost of the laser system. This suggests that architectures based on large-scale regenerative (fig. 12b) or multipass, (fig. 12c) amplifier configurations may be employed to achieve significant economic and reliability benefits. In these systems a large amplifier is used to build up a low fluence or intensity input pulse to the level at which efficient extraction occurs. The 'front end' on such a system can be simple, reliable and account for only a small fraction of the overall system cost. Currently the technology of large-aperture, fast ($\lesssim 1 \mu$s) electro-optical switches is such that only 5 to 10 cm apertures are practical without segmentation; thus large-scale regenerative systems do not appear compelling at this time. However, for many media it is possible to build large-aperture, high energy and power gain amplifiers suitable for multipass operation. In these systems, and after only a

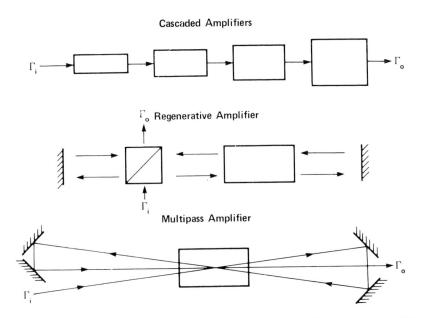

Fig. 12. Advanced laser system architectures: (a) cascaded amplifiers; (b) regenerative amplifier; (c) multipass amplifier.

few passes, say, $\leqslant 5$, a small signal can be efficiently amplified to high intensity. These systems are particularly promising because they are simple, potentially more reliable, relatively inexpensive and potentially more efficient, since the effective extraction time may be extended to allow lower level relaxation and gain recovery. In the most general case, we consider a large-scale amplifier that is pumped one or more times and through which an extraction pulse is repetitively passed to remove the available energy. The optimum pump-extraction scenario will depend on the character of the laser medium.

Laser media and concepts capable or service in the fusion application are conveniently discussed in terms of two generic types, energy-storage and nonstorage media. Energy-storage media are those in which the population inversion is radiatively and collisionally stable for times of order a microsecond. These kinetic properties allow for a relatively slow pumping rate and ease the burden on the pump technologies. These media must operate at high excited state densities (10^{16} to $10^{17}\,\mathrm{cm}^{-3}$) and possess relatively large energy-storage densities (10 to $100\,\mathrm{J}/\ell$). Such requirements place severe constraints on the stability of the medium during the excitation and extraction sequence. Non-storage media are those in which the upper laser level radiates on a time scale less than a few tens of nanoseconds. These systems have typically low excited state densities (10^{14} to $10^{15}\,\mathrm{cm}^{-3}$) and corresponding energy storage densities (0.1 to $1\,\mathrm{J}/\ell$). Nonstorage systems demand higher pumping rates but offer a wide range of systems options. For example, high output power and fluence can be achieved by multiple pulse excitation of an amplifier through which an extraction pulse is passed several times, once for each excitation cycle. The disadvantage of this approach is that many passes are required, a situation comparable with currently nonscalable regenerative systems architectures. A second approach employs these media in the long pulse mode as optical pumps for energy-storage laser media. In this case quasi-continuous operation is used to achieve sizeable pump fluences with currently available or reasonably attainable pulsed power technologies. Yet a third approach uses nonstorage media in the quasi-continuous excitation mode and short pulses are generated by sequential temporal pulse stacking (Glass 1976). In this technique, several short pulses of, say, 5 to 10 ns duration would be propagated sequentially, temporally back to back, at slightly different angles through a large nonstorage power amplifier that is pumped continuously for 50 to 100 ns. After amplification the short pulses would be combined to produce a single high intensity 5 to 10 ns pulse for target irradiation. Alternatively, pulse compression (Glass 1976) of a single long pulse may be achieved by using a nonlinear process. Such a scheme using backward wave Raman scattering (Maier et al. 1966, 1969, Glass 1967, Kaiser and Maier 1972, Shen 1976) has been proposed (Murray et al. 1978). In the following, the microscopic properties and concomitant technological implications of these two generic types of laser media are developed further by consideration of their power amplification and energy extraction characteristics.

3.4.1 Storage laser extraction characteristics

In energy-storage laser media both the characteristic pump time and the upper laser level lifetimes are long compared to the extraction time. The laser radiation transport and the population inversion rate equation therefore take simple forms:

$$\frac{1}{c}\frac{\partial I}{\partial t} + \frac{\partial I}{\partial z} = \sigma \Delta N I - \gamma I \tag{32}$$

and

$$\frac{\partial}{\partial t}\Delta N = -\frac{\Delta N I}{\Gamma_s}, \tag{33}$$

where it is assumed that the lower laser level does not relax during extraction. Under these circumstances $\Delta N = N_u - (g_u/g_L)N_L$, where N_u (N_L) and $g_u(g_L)$ are the upper and lower laser level populations and degeneracies respectively. The saturation fluence is given by $\Gamma_s = [g_L/(g_L + g_u)]h\nu/\sigma$, where $\sigma \sim \lambda^2/4\pi^2\Delta\nu\tau_R$ is the stimulated emission cross section for the laser transition whose fluorescence linewidth and radiative lifetime are $\Delta\nu$ and τ_R respectively. The quantity γ accounts for any nonsaturable losses which may be present. If the lower laser level is not bottlenecked during extraction, the present formulation can be used provided the limit $g_u/g_L \to 0$ is imposed. Eqs. (32) and (33) have been studied in some detail (Frantz and Nodvik 1963, Bellman et al. 1963, Kryukov and

Letokhov 1969, Trenholme and Manes 1972). By suitable combination of these equations, an equation for the pulse energy fluence

$$\Gamma(z) = \int_{-\infty}^{\infty} I(z, t)\, dt \tag{34}$$

can be constructed (Trenholme and Manes 1972)

$$d\Gamma/dz = \alpha_0 \Gamma_s[1 - e^{-\Gamma/\Gamma_s}] - \gamma\Gamma, \tag{35}$$

where the small-signal gain α_0 is related to the population inversion immediately prior to extraction, ΔN_0, by $\alpha_0 = \sigma \Delta N_0$. In the analysis of short pulse amplifier extraction dynamics, eq. (35) is particularly valuable since it governs the energetics of the amplifier.

In the absence of loss, eq. (35) can be integrated analytically giving

$$\Gamma(z) = \Gamma_s \ln\{e^{\langle\alpha_0\rangle z}[e^{\Gamma_I/\Gamma_s} - 1] + 1\}, \tag{36}$$

where $\langle\alpha_0\rangle z = \int_0^z \alpha_0\, dz'$ is the integrated small-signal gain. The laser pulse energy fluence at any point in the amplifier depends only on the input fluence Γ_I, the saturation fluence and the small-signal gain integrated from the input. The spatial distribution of the gain is important only insofar as it affects the magnitude of the small-signal gain integral. Since the stored energy density in the amplifier is $N_{u_0} h\nu$, the extraction efficiency is

$$\epsilon_{ext}(z) = (\Gamma(z) - \Gamma_I)/N_{u_0} h\nu z. \tag{37}$$

When Γ is much less than the saturation fluence, eq. (36) indicates that the pulse energy fluence increases exponentially along the amplifier. Under these circumstances the extraction efficiency is very low. However, when Γ is greater than the saturation fluence the pulse fluence increases linearly with distance down the amplifier and the extraction efficiency approaches the maximum value $g_L/(g_L + g_u)$ if the initial population of the lower laser level is zero. Bottlenecking of the lower level limits the extraction efficiency. If there is no bottlenecking during extraction or if $g_L \gg g_u$, then all the available energy stored in the upper laser level can be removed, i.e., $\epsilon_{ext} = 1$. If the lower laser level can be relaxed between passes in bottlenecked multipass amplifiers, extraction efficiencies exceeding $g_L/(g_L + g_u)$ can be achieved.

In the absence of loss it is also possible to solve eqs. (32) and (33) analytically (Frantz and Nodvik 1963, Bellman et al. 1963) and to obtain an expression for the evolution of the pulse intensity envelope. This result has been used (Trenholme and Manes 1972) to investigate pulse distortion in saturated amplifiers. It was found that significant pulse temporal distortion occurs if the output fluence exceeds 5 to 10 saturation fluences. Above approximately $10\Gamma_s$ the details of the output pulse shape depend on very low-level portions of the input pulse shape and are subject to distortion from noise. Therefore, an amplifier designed to run well saturated must be driven with a very exact pulse shape or some form of pulse stacking (Thomas and Siebert 1976, Martin and Milam 1976, Martin et al.

1977) must be employed to construct the desired pulse shape. Saturation effects can also produce significant distortion of the spatial profile of the pulse (Trenholme and Manes 1972). As a pulse goes deeper into saturation its weaker portions experience more growth, since they do not deplete the available gain as much as the strong central beam. However, techniques have been developed (Hunt et al. 1977a, b) to control the transverse profile of a laser beam.

The optimum microscopic physical and kinetic parameters for an energy-storage, short wavelength laser amplifier are determined by several constraints. The maximum excited state line density along the propagation axis is determined by the requirement that the output fluence be less than the optical damage fluence:

$$(\alpha_0 L)\Gamma_s/h\nu \le N_{u_0}L < \Gamma_{\text{damage}}/h\nu = 5 \times 10^{18} \, \text{cm}^{-2}\Gamma_{\text{damage}}[\text{J}/\text{cm}^2]\lambda \, [\mu\text{m}]. \qquad (38)$$

Within this constraint, achievement of a high extraction efficiency using a multipass amplifier requires that the saturation fluence be less than or comparable to the damage fluence and that $\alpha_0 L$ be as large as possible and still satisfy eq. (38). This latter condition minimizes the number of passes and the concomitant number of costly optical components. For a 500 nm wavelength laser amplifier on the order of a meter in length, operation at a damage limit of 10 J/cm^2 requires an excited state density of about 2.5×10^{16} cm^{-3}.

Efficient pumping of a storage laser amplifier requires that the collisional and radiative lifetime of the upper laser level must exceed the pump time. At high excited state densities the collisional self-quenching rate constant k_q of the upper level must be such that $k_q < (t_p N_{u_0})^{-1}$. For an energy-storage time of order a microsecond k_q must be less than 10^{-11} cm^3/s, a value significantly less than gas-kinetic. At high energy densities the radiative lifetime of the upper laser level is generally determined by fluorescence depumping and parasitic oscillation limitations. Fluorescence depumping can occur when the radiation spontaneously emitted by the excited upper laser level is amplified to the extent that it increases the effective fluorescence loss rate. The magnitude of this effect depends on the size, shape, gain, refractive index and line profile of the laser medium. For example, at the edge of a spherical volume with spatially uniform small-signal gain the instantaneous ratio A of the spontaneous to stimulated lifetimes is (Trenholme 1972)

$$\mathscr{R} = \frac{3}{2\beta_f}\left[\frac{2e^{\beta_f}}{\beta_f}\left(1 - \frac{1}{\beta_f}\right) + \frac{2}{\beta_f^2} - 1\right] - 1 \qquad (39)$$

for a single frequency and $\beta_f = \alpha_0 D$, where D is the diameter of the sphere. This fluorescence amplification problem is especially severe in large laser systems where $\alpha_0 L$ may be large. In actual laser materials the fluorescence and gain are distributed in wavelength according to some line profile. The effective values of the spontaneous-stimulated lifetime ratio $\mathscr{R}$ are then found by averaging the single frequency values over the line. Table 4 lists results of the calculation of $\mathscr{R}$ for several fluorescence line profiles, flat top, gaussian and lorentzian. From the point of view of fluorescence amplification, the worst possible case would be a

Table 4

Fluorescence depumping parameter[a] $\mathscr{R}$ for a spherical volume.

$\alpha_0 D$	$\mathscr{R}_F{}^b$	$\mathscr{R}_G{}^b$	$\mathscr{R}_L{}^b$
1.	0.5	0.33	0.23
2.	1.4	0.88	0.60
5.	13.	6.5	4.1
10.	590.	200.	120.
20.	3.5×10^6	8.1×10^5	4.7×10^5
50.	6.1×10^{18}	8.8×10^{15}	5.0×10^{17}
100.	8.0×10^{39}	8.0×10^{38}	4.6×10^{38}

[a]Trenholme (1972).
[b]Fluorescence line profile: Flat (F), Gaussian (G), Lorentzian (L).

flat-topped or rectangular line profile, since in this case the line average is equal to the peak value. Since the effective lifetime τ_L of the inversion in the presence of fluorescence depumping is $\tau_L = \tau_R/\mathscr{R}$ the constraint imposed by finite pump time requires that the stimulated emission cross section satisfy

$$\sigma < \frac{\lambda^3}{4\pi^2 c t_p \mathscr{R}} \left(\frac{\lambda}{\Delta\lambda}\right) = 2.7 \times 10^{-18} \, \text{cm}^2 \frac{(\lambda[\mu m])^3}{t_p[\mu s]\mathscr{R}} \left(\frac{\lambda}{\Delta\lambda}\right). \tag{40}$$

As noted above, pumping of large volumes in times less than approximately 100 ns will be technologically very difficult. Therefore, achievement of these conditions requires atoms and molecules with metastable electronic levels connected to lower lying states only by electric-dipole-forbidden transitions (magnetic dipole; electric quadrupole, forced electric dipole, etc.). These metastable levels must also possess a high degree of collisional stability.

In general, fluorescence amplification sets no definite upper limit on stored energy, but instead makes pumping more and more difficult as the energy density rises. Parasitic oscillation, on the other hand, sets an abrupt upper limit to stored energy at the oscillation threshold. Generally for storage lasers with large radiative lifetimes ($\gtrsim 10 \, \mu s$), the limits set by parasitic oscillation are more severe.

Parasitic oscillation (Trenholme 1972, Glaze et al. 1974) takes place when a medium with optical gain has within it a light path which returns on itself. Under this condition, the material will break into oscillation when the gain is large enough to overcome the path losses. If the mode of oscillation fills an appreciable fraction of the laser volume, it sets a sharp upper limit to the stored energy and, in general, the threshold for parasitic oscillation in an amplifier depends on geometry, size, gain and the reflection characteristics of the bounding surfaces. In gas laser amplifiers parasitics may be controlled by using absorbing or diffusely reflecting surfaces (Chester 1973) or by addition of saturable absorber gases to the laser medium. Diffuse reflectors suppress parasitics because reflected radiation is continually dispersed in many directions. Approximate solutions (Chester 1973) of the diffuse parasitic oscillation problem suggest that for a diffuse reflectance on the order of 1% $\alpha_0 L$ should be less than 3 to 5 for rectangular boundary conditions encountered in flowing gas lasers.

The scaling limitations set by parasitics can be established by noting that avoidance of this phenomenon requires that

$$\sigma N_{u_0} L_j < (\alpha_0 L)_{max}, \tag{41}$$

where L_j is the amplifier dimension in the jth direction. This condition favors square amplifier modules. Along the propagation axis of a laser amplifier eqs. (38) and (40) require that $N_{u_0} L$ must be less than the smaller of $\Gamma_{damage}/h\nu$ and $(\alpha_0 L)_{max}/\sigma$. Therefore the amplifier is damage fluence limited if $\sigma > \sigma_0 = (\alpha_0 L)_{max} h\nu/\Gamma_{damage}$. For a damage fluence of $10\,J/cm^2$ and $(\alpha_0 L)_{max} \sim 4$ the optimum cross section is $\sigma_0[cm^2] \sim 8 \times 10^{-20}/\lambda\,[\mu m]$ or $3 \times 10^{-19}\,cm^2$ at $0.25\,\mu m$ and $8 \times 10^{-20}\,cm^2$ at $1\,\mu m$. Of course, for larger damage fluences, smaller cross sections are optimal. For cross sections less than optimal, smaller gain $\alpha_0 L$ must be used to avoid damage and more passes are required for a given input fluence. In practical multipass amplifiers there will be a lower limit on $\alpha_0 L$ of near unity. Correspondingly, if the cross section is greater than σ_0, the output fluence will be less than the damage fluence and the optics will not be used as efficiently as possible. The output aperture of the machine must be larger than desirable to achieve a given total output energy.

Using the previous results, the maximum energy that can be extracted from an amplifier module becomes

$$E_L < E_{L_{max}} = \Gamma_{damage}(\alpha_0 L)^2_{max}/(\sigma N_{u_0})^2. \tag{42}$$

This result shows the advantage of operating at high damage fluences, low small-signal gains and large gain length products or large amplifier lengths. For the previous example, if $\alpha_0 \sim 10^{-2}\,cm^{-1}$, corresponding to an amplifier length of 4 m, then the maximum amplifier energy capacity is about 1 MJ. In practice it is generally not technically feasible to excite the volume or build optical components for a 4 m × 4 m aperture and so smaller apertures will be used.

Eq. (41) indicates that the smaller $\alpha_0 = \sigma N_{u_0}$ is, the larger the module energy that can be achieved. However, efficient extraction in the presence of nonsaturable losses ($\gamma \neq 0$) sets a limit on the minimum value of α_0. When nonsaturable loss is present, efficient extraction and fluence gain are affected. As the fluence is increased to extract the available energy, the gain decreases but the loss remains constant. Therefore once saturation sets in, the energy can no longer grow indefinitely as it did in the lossless case. In fact, it cannot rise above a maximum energy fluence Γ_m which is determined by setting eq. (35) equal to zero (Trenholme and Manes 1972). When α_0/γ is near unity the maximum fluence is small and approaches $2\Gamma_s(1 - \gamma/\alpha_0)$. However, when α_0/γ exceeds approximately 4, the ratio Γ_m/Γ_s is very nearly equal to α_0/γ. Clearly, if α_0 is less than γ the beam is attenuated and decays steadily to zero. If the beam enters the amplifier above the maximum flux, it decays until it reaches this flux. The limit imposed by nonsaturable loss is significant in two ways. In energy-storage lasers, gains are generally relatively low, $\leq 1\%$ cm; and since absorption losses can have values on the order of 0.1%/cm in visible and ultraviolet lasers (Hawryluk et al. 1977), values of $(\alpha_0/\gamma) > 5$ to 10 may be difficult to achieve. Since efficient extraction in a single pass requires $\Gamma_1 > \Gamma_s$, the potential efficiency of the device can be impaired. The second limitation involves a limitation on power or fluence gain.

The equation for energy transport in the presence of loss [eq. (35)] must be integrated numerically. Fig. 13 shows the results of such calculations where all curves are plotted as a function of $(\alpha_0 - \gamma)z$. The significance of finite values of α_0/γ is clearly evident. For instance, if the maximum output fluence is limited by damage so that eq. (38) holds, then efficient extraction in the presence of nonsaturable loss requires $\alpha_0/\gamma \gg (\alpha_0 L)$ or $\gamma L \ll 1$ where the inequality symbol implies a factor of 5 to 10. For the previous example this requires $\gamma < 0.025\%$/cm. Recent measurements of nonsaturable loss rates in visible and ultraviolet lasers have yielded values of $\sim 0.1\%$/cm (Hawryluk et al. 1977). These results suggest that nonsaturable losses present a nontrivial amplifier design problem. The impact of this constraint is to require operation at higher small-signal gains and shorter amplifier lengths, thereby significantly reducing the maximum energy that can be generated by a single amplifier module.

Based on the foregoing discussion we have summarized the required microscopic kinetic and laser properties of potentially useful energy-storage laser media in table 5. Because of the stringent requirements placed on several microscopic parameters, it is perhaps not surprising then that the number of atomic and molecular species satisfying simultaneously all these conditions is relatively small. It has been found that at least two generic species appear to manifest the required kinetic and structural properties: (1) the Group VIa atoms – oxygen, sulfur, selenium and tellurium; (2) the trivalent rare earth solids and molecular vapors. The trivalent rare earth ions are, of course, well-known active centers in energy-storage solid-state lasers currently used in fusion research. Given these atomic and molecular species which manifest suitable microscopic structural and kinetic properties, a successful laser system can only be achieved if methods are found for pumping the laser medium which are both efficient and energetically

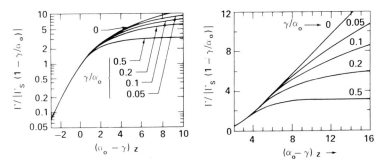

Fig. 13. The growth of pulse energy fluence Γ in a saturating storage laser amplifier with a fixed nonsaturating loss coefficient γ (Trenholme and Manes 1972).

Table 5
Microscopic physical and kinetic parameters for energy-
storage short wavelength laser medium.

Property	Symbol	Value
Saturation fluence	$\Gamma_s = h\nu/\sigma$	$<10\,\text{J/cm}^2$
Laser cross-section	σ	$>5 \times 10^{-20}\,\text{cm}^2$
Gain coefficient	$\alpha_0 = \sigma\Delta N$	10^{-3}–$10^{-2}\,\text{cm}^{-1}$
Inversion density	$\Delta N = \alpha_0/\sigma$	10^{16}–$10^{17}\,\text{cm}^{-3}$
Inversion lifetime	τ_L	$\sim 1\,\mu\text{s}$
Max. quenching rate	$k_q < 1/\tau_L \Delta N$	$< 10^{-11}\,\text{cm}^3\text{s}^{-1}$

scalable with 'tractable' technologies. High energy fusion amplifiers based on these active species are discussed in §§ 4to 8.

3.4.2 *Nonstorage laser long pulse extraction characteristics*

Most atoms and molecules will not possess energy levels and transitions appropriate for energy-storage lasers operating at a short wavelength. On the other hand, many atoms and molecules will be capable of sustaining short wavelength laser action on electric-dipole-allowed transitions with characteristic upper laser level lifetimes of typically 10 ns. Good examples of this are the rare gas monohalide excimer lasers (KrF, ArF, XeF, etc.) pumped by electron beam (Ewing and Brau 1975, Brau and Ewing 1975) and electron-beam-augmented discharge techniques (Mangano and Jacob 1975). These lasers may exhibit efficiencies in excess of 10% and are volumetrically scalable to high energies for pulse widths of ~ 100 ns or longer. Unfortunately, such pulse lengths are more than an order of magnitude too long for driving fusion implosions (table 1). Nevertheless, the high efficiencies and scalability of these short wavelength lasers compels one to consider the pumping and pulse compression techniques described earlier. Technical extraction issues involved in implementing long pulse nonstorage lasers are discussed next.

In nonstorage laser media the characteristic pumping and extraction times are equal and both are long compared to the lifetime of the upper laser level. Therefore, the pump rate into the upper laser level equals the stimulated emission rate plus the spontaneous and collisional loss rate. The radiation transport equation for a homogeneously broadened line is

$$\mathrm{d}I/\mathrm{d}z = \alpha_0 I/(1 + I/I_s) - \gamma I, \qquad (43)$$

where if the lower laser level is dissociative or of negligible density, $\alpha_0 = \sigma N_{u_0}$ is the small-signal gain and γ is the nonsaturable loss coefficient. The saturation intensity is related to the lifetime τ_L of the upper laser level by $I_s = h\nu/\sigma\tau_L$. The extraction efficiency may be written

$$\epsilon_{\text{ext}}(z) = (I - I_1)/(\alpha_0 z)I_s.$$

In the absence of loss the extraction efficiency approaches unity if $I_1 > I_s$.

If nonsaturable loss (Hawryluk et al. 1977) is present in the system eq. (43) indicates that as the intensity increases above the saturation intensity the gain saturates and eventually equals the loss when the intensity reaches

$$I_m = I_s[\alpha_0/\gamma - 1]. \tag{45}$$

The intensity I_m is the maximum intensity that can be extracted from the amplifier. As this intensity is approached the extraction efficiency of the amplifier decreases to zero. The performance characteristic of a nonstorage amplifier with nonsaturable loss is illustrated in fig. 14. For a value of $\alpha_0/\gamma = 19$ the extraction efficiency is plotted as a function of I_1/I_s for several values of γL. In fig. 14a the extraction efficiency increases as I_1/I_s increases, then reaches a maximum and ultimately decreases to zero. In the absence of loss the extraction efficiency would approach unity for large values of I_1/I_s. It is clear that nonsaturable loss limits extraction efficiency and amplifier length. In fig. 14b the maximum attainable extraction efficiency is plotted as a function of $(\alpha_0/\gamma - 1)$ for several values of γL. Large values of α_0/γ are clearly desirable. As an example of the importance of these results, in the atmospheric pressure electron-beam-excited KrF laser (Ewing and Brau 1975) the production efficiency of the excimer KrF* has a maximum of approximately 25%. The saturation intensity is near $1\,MW/cm^2$ (Mangano et al. 1977). For short pulse ($\sim 100\,ns$) excitation, modeling calculations indicate that α_0/γ will be less than or comparable to 30 and therefore the maximum flux obtainable is less than $30\,MW/cm^2$. From fig. 14b the maximum extraction efficiency for such a device where $\gamma L \sim 1$, is $\sim 50\%$, which leads to a medium efficiency of approximately 12%. Only half of the available energy is extracted owing to nonsaturable loss.

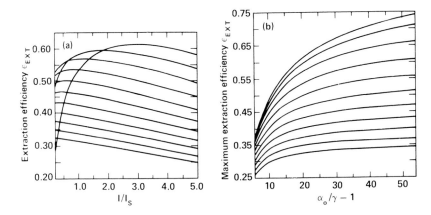

Fig. 14. Extraction efficiency characteristics of nonstorage laser amplifiers plotted parametric in the nonsaturable loss coefficient–amplifier length product γL from 0.1 (top) to (2.6) bottom) in increments of 0.25 (Eimerl 1977).

3.5 Propagation of laser energy

Laser light focusing requirements for target irradiation place constraints on the
optical quality of media in which the laser light is generated and through which
the laser light must pass on its way to the target. Early Nd:glass laser development experiments revealed that nonlinear phase and amplitude distortion
produced by small-scale self-focusing (Bliss et al. 1974) could cause serious
defocusing at high peak power. Since then the development of laser glasses with
much lower nonlinear indices (Milam and Weber 1976) and the development of
spatial filtering (Simmons et al. 1975, Glaze 1976) and optical relay beam-
transport techniques (Hunt et al. 1977a, b) have greatly reduced this problem. In
§ 3.5 sources of phase and amplitude distortion in gaseous and solid media are
identified and compared to establish their relative importance. Focusing of
beams and the constraints imposed by beam aberrations are then considered.

3.5.1 Phase and amplitude distortion processes

The propagation of high intensity laser light through gaseous and solid-state
media in a laser system is degraded by phase and amplitude distortions due to
several mechanisms. In gaseous media some of the more important ones are
likely to be gas flow turbulence (Prokhorov et al. 1974) in amplifiers and beam
transport tubes, density disturbances generated by volumetric energy deposition
(Feinberg et al. 1975, Aushermann et al. 1978), saturation (Booth and Troup
1969, Close 1967) and shock wave generation by hot electrodes or e-beam foil
energy loading (Pugh et al. 1974, Culick et al. 1976).

   The index of refraction for a gas is close to unity and can be written
$n = 1 + \beta(\rho/\rho_s)$, where ρ_s is the gas density at standard conditions (1 atm, 273 K)
and β is the Gladstone–Dale constant, which depends only weakly on
wavelength in the visible. Representative values of β at 589.3 nm are listed in
table 6. It is important to note that β is approximately an order of magnitude
smaller for helium and neon compared to other common gases.

   From the point of view of geometric optics (Born and Wolf 1970), nonuni-
formities in the refractive index across the beam, $\Delta n_\perp = \beta \Delta \rho_\perp / \rho_s$, can produce
average phase variations

$$\Delta\phi_\perp = k\beta \frac{\rho_0}{\rho_s} \int_0^L \frac{(\Delta\rho)_\perp}{\rho_0} \, dz = k\beta \frac{\rho_0}{\rho_s} \langle (\Delta\rho/\rho_0) \rangle L \tag{46}$$

and average angular deflections

$$\Delta\theta_\perp = \beta \frac{\rho_0}{\rho_s} \int_0^L \frac{|\nabla_\perp \rho|}{\rho_0} \, dz = \beta \frac{\rho_0}{\rho_s} (\langle |\nabla_\perp \rho| \rangle / \rho_0) L, \tag{47}$$

where ρ_0 is the density of the medium, $\perp$ denotes displacement normal to the

Table 6

Gladstone–Dale refractive index coefficients[a] for 589.3 nm light reduced to
1 atm and 20°C.

Gas	$\beta \times 10^4$	Gas	$\beta \times 10^4$
Air	2.92	Hydrogen	1.32
Ammonia	3.76	Krypton	4.27
Argon	2.81	Mercury	9.33
Benzene	17.62	Methane	4.44
Bromine	1.32	Neon	0.67
Carbon dioxide	4.51	Nitric oxide	2.97
Carbon monoxide	3.38	Nitrogen	2.97
Carbon tetrachloride	17.68	Nitrous oxide	5.16
Chlorine	7.73	Oxygen	2.72
Fluorine	1.95	Water vapor	2.54
Helium	0.36	Xenon	7.02

[a]Kaye and Laby (1973).

axis of propagation, k is the laser light wavenumber and L is the propagation
distance in the medium. The mean quantities $\Delta\phi_\perp$ and $\Delta\theta_\perp$ depend on the
character of the density fluctuations and are nonzero only if the density
fluctuations are not random or fully turbulent. In high average power lasers this
is often the case because acoustic disturbances set up characteristic density
variations that depend on specific boundary conditions and are therefore not
fully turbulent. If the medium is turbulent then the average phase and angular
deflection of a ray vanishes but its mean square value does not. As an example,
if the density (refractive index) of the turbulence can be characterized by a
gaussian correlation function with scale length l then the square root mean
square phase and angular deflection quantities (Chernov 1960, Tatarski 1961) are
approximately

$$[\langle(\Delta\phi_\perp)^2\rangle]^{1/2} = (1/\sqrt{2})\pi^{1/4}(lL)^{1/2}k\beta(\rho_0/\rho_s)[\langle(\Delta\rho_1/\rho_0)^2\rangle]^{1/2} \tag{48}$$

and

$$[\langle(\Delta\theta_\perp)^2\rangle]^{1/2} = 2\pi^{1/4}(L/l)^{1/2}\beta(\rho_0/\rho_s)[\langle(\Delta\rho_\perp/\rho_0)^2\rangle]^{1/2}. \tag{49}$$

These results, when compared to eqs. (46) and (47), indicate that turbulent
fluctuations produce less distortion than nonturbulent correlated inhomogeneities of comparable scale length. In general, the density fluctuations
generated in an average power gas laser amplifier will be neither fully correlated
or turbulent and will have properties specific to the scale size of the device and
the acoustic damping devices. Therefore, eqs. (46) through (48) are only useful in
estimating the bounds that distortions can have and the approximate conditions
that must be achieved in order to have adequate beam quality. For example, for
density fluctuations of the same magnitude, small scale length turbulence ($l \ll L$)
produces larger deflections than large scale ($L \sim l$) or well-correlated dis-

turbances. Phase fluctuations are larger for smaller wavelength whereas average deflections are virtually independent of wavelength. The significance of these results will become clear later. Laser operation at a density ρ_0 less than or equal to standard density with a gas mixture composed primarily of helium or neon is highly desirable. For example, density fluctuations of the same character and magnitude in xenon at 10 atm produce distortions that are 200 times those produced in helium at 1 atm. For a 500 nm laser amplifier of length 4 m containing helium (argon) at 1 atm, if $(\Delta\rho_\perp/\rho_0) \sim 10^{-3}$ the relative phase variations [eq. (44)] will be $(\Delta\phi_\perp/2\pi \sim 0.28(2.3)$ respectively. Recent experimental and theoretical studies of average power lasers (§ 3.6) indicate that $\Delta\rho_\perp/\rho_0$ values in the 10^{-4} to 10^{-3} range can be achieved.

In solid media the index of refraction can be written $n = n_0 + \gamma_2 I$, where $n_0\gamma_2[\text{cm}^2/\text{W}] = 4.19 \times 10^{-3} n_2[\text{esu}]$ and I is the local laser intensity. The values of n_0 and n_2 for several laser materials are listed in table 7. The low power or linear index of refraction is n_0, and n_2 is the nonlinear refractive index. Nonuniform temperature T and mechanical stress distributions σ_{ij} in solids can produce variations in media length L and refractive index n_0 (Quelle 1966, Welling and Bickart 1966, Born and Wolf 1970, Jorner et al. 1977). Stress induced birefringence (Born and Wolf 1970) is also a potentially serious problem since laser disks oriented at Brewster's angle act as polarizer/analyser combinations. From the point of view

Table 7
Refractive index coefficients for 1.06 μm light[a].

Material	Type	n_0	$\nu_D{}^d$	$n_2[10^{-13}$ esu]
Optical glass				
Borosilicate	BK-7	1.517	64	1.24 ± 0.12
Fused silica	4000	1.458	68	0.96 ± 0.10
Flurosilicate	FK-5	1.487	70	1.07 ± 0.11
Fluorophosphate	FK-51	1.487	84	0.69 ± 0.07
Nd laser glass				
Silicate	ED-2	1.567	54	1.41 ± 0.14
Phosphate	LHG-5	1.541	63	1.16 ± 0.12
Phosphate	LHG-6	1.532	66	1.01 ± 0.10
Phosphate	EV-1	1.507	69	0.91 ± 0.09
Fluorophosphate		1.492	81	0.71 ± 0.14
Faraday rotator glass				
Tb-silicate	FR-5	1.686	52	2.1 ± 0.4
Crystals				
CaF_2	Cubic	1.434	95	0.57 ± 0.15
LiF	Cubic	1.392	99	0.35 ± 0.10
MgF_2	Tetragonal	1.378 (o)[b]	108	0.30 ± 0.10
		1.389 (e)[c]	98	
KH_2PO_4	Tetragonal	1.468 (e)	70	1.0 ± 0.3
		(1.509(o)	57	

[a]Milam and Weber (1976).
[b]The abbreviation 'o' is for ordinary waves.
[c]The abbreviation 'e' is for extraordinary waves.
[d]Abbé number.

of geometric optics these nonuniformities in the refractive index produce average phase variations

$$\Delta\phi_{\perp} = k \sum_{ij} \int_0^L \Delta n_{\perp}(T, \sigma_{ij}, I)\, dz + k\langle n(T, \sigma_{ij}, I)\rangle_{\perp}\Delta L_{\perp}(T, \sigma_{ij}) \tag{50}$$

and angular deflections

$$\Delta\theta_{\perp} = \sum_{ij} \int_0^L |\nabla_{\perp} n(T, \sigma_{ij}, I)|n_0^{-1}\, dz, \tag{51}$$

where $\langle n \rangle$ denotes the average value between the two observation points. If mechanical and thermally induced stresses are not important, then eq. (50) can be approximated by

$$\Delta\phi_{\perp} = kn_0 L\{n_0^{-1}(\partial n_0/\partial T)_{\sigma=0} + L^{-1}(\partial L/\partial T)_{\sigma=0}\}\Delta T_{\perp} + \Delta B_{\perp}, \tag{52}$$

where

$$B = k\gamma_2 \int_0^L I\, dz \tag{53}$$

is the nonlinear phase change through the material. For most solid media $(\partial n_0/\partial T)/n_0$ and $(\partial L/\partial T)/L$ have values in the 10^{-6} to $10^{-5}/°C$ range; in addition, usually $\partial n_0/\partial T < 0$ whereas $\partial L/\partial T > 0$. The possibility exists that solid media may be developed for average power applications by designing the system so that these contributions effectively cancel. Control of mechanical and thermally induced stresses would then be the limiting consideration. At a wavelength of 1 μm, a 4cm thick CaF_2 laser window would contribute a $(\Delta\phi_{\perp}/2\pi) \approx 0.7$ for $\Delta T_{\perp} \simeq 1°C$ and would contribute a $(\Delta B_{\perp}/2\pi) \sim 0.08$ for an intensity of 10^{10} W/cm^2. In this example, thermal effects would dominate. In any case, it is interesting to note in general that the solid and gaseous contributions to laser beam distortions are comparable for solid-state and gas lasers.

In solid-state media, an instability (Bespalov and Talanov 1966) in the laser light propagation occurs owing to the nonlinear contribution to the refractive index. It develops from intensity modulations on the beam due to noise sources (scratches, dust, etc.) in the laser and is known as small scale self-focusing (Bespalov and Talanov 1966, Akhmanov et al. 1968, Bliss et al. 1974, Suydam 1975). The instability develops in the following way. In a high intensity region of the beam the index is higher and therefore more radiation bends into this region, increasing the local intensity (index). Because of this resulting intensification of the irregularity the rate of concentration of the light is increased. Intensity ripples of less than 1% can be amplified, producing a 100% modulation of the beam. The local growth rate of the instability depends on the beam intensity ripple size (Bespalov and Talanov 1966). Therefore the actual growth of intensity irregularities in a laser system is controlled by a complicated interplay of nonlinear focusing, natural diffraction, gain in laser media, apertures and noise sources. Experimental and theoretical studies (Bliss et al. 1974, Hunt et al. 1976, Glaze 1976) associated with Nd:glass lasers show that beam degradation due to ripple growth is well described by the *B*-integral [eq. (53)]. It is found that when *B* accumulates much above 2 the focusability of the laser beam rapidly diminishes. If left unchecked these distortions eventually become unmanageable and lead to component damage, breakdown or a substantial decrease in the focusable power that can be extracted from the clear aperture of the final amplifiers of the system. Recently it has been suggested (Hunt et al. 1977a, b) that the low pass filtering and imaging properties of spatial filters can be used to minimize the detrimental effects of self-focusing and other laser beam aberrations. The filtering property can be used to remove small-scale spatial irregularities from the beam before they can grow to significant power content. The imaging properties are used to increase the extraction efficiency and reduce whole-beam self-focusing (Hunt et al. 1976) and diffraction effects by continuously reimaging more uniform beam intensity profiles.

3.5.2 Focusing of aberrated laser beams and laser media optical quality constraints

Target irradiation requirements specify the focusability constraints that the output beam of a fusion laser system must satisfy. Several criteria and formulations which relate focusability requirements to laser medium homogeneity have been proposed. However, as will be seen here, some of these criteria are inadequate and in some cases misleading.

The simplest criterion is to require the laser spot size to be equal to or somewhat smaller than the pellet cross section. Fusion reactor pellet sizes are on the order of a few millimeters. For a near-diffraction-limited system the Straehl focal plane intensity ratio (Born and Wolf 1970) has been used to estimate (Holmes and Avizonis 1976) the ratio of the diffraction-limited spot size to the aberrated spot size. For a gaussian beam this gives the radius

$$r_t \geq 1.22(\lambda f_L/D)\left[1 - \sum_j (\Delta\phi_j)^2\right]^{-1/2}. \tag{54}$$

The quantity in the square root is the Straehl ratio, i.e., the ratio of the intensity on-axis in the focal plane to the intensity for a diffraction-limited spot size. Since this expression is not rigorous and only approximately valid for $\sum_j (\Delta\phi_j)^2 \ll 1$ it is of relatively little value here, because fusion lasers operating in the visible and ultraviolet will likely have $\sum_j (\Delta\phi_j)^2 > 1$. It should also be noted that misapplication of eq. (54) can lead to erroneous conclusions, particularly with regard to the media and laser wavelength dependence of the focusability constraint.

Continuing along the line that a useful criterion is the requirement that the focused laser beam spot size be less than the target diameter, a more accurate approach than eq. (54) is to use the results of Talanov's moment theory (Vlasov et al. 1971) for the propagation of aberrated laser beams. Talanov's theory is completely rigorous and very powerful. According to this theory there are several integral invariants or moments of the intensity distribution of a propagating laser beam which remain invariant throughout the propagation of the beam.

Application of this theory (Vlasov et al. 1971) indicates that after passage through a focusing lens of focal length f_L the minimum mean square radius of an aberrated laser beam is

$$\langle r^2 \rangle_{min} \simeq f_L^2 < (\nabla_\perp \phi_0/k)^2 - r_\perp \nabla_\perp \phi_0/f_L k + (\nabla_\perp \bar{E}_0/k\bar{E}_0)^2 \rangle, \tag{55}$$

where $\phi_0(r_\perp)$ and $\bar{E}_0(r_\perp)$ are the phase and electric field amplitude distributions respectively of the beam across the input to the lens. The quantity $\langle f \rangle$ denotes the intensity weighted average of the function f across the beam profile at the input to the lens. Expressions similar to eq. (55) can also be derived (Vlasov et al. 1971) for the focal length and depth of focus of the aberrated laser beam. If the input phase distribution ϕ_0 is due primarily to linear sources, then the geometrical optics term $\nabla_\perp \phi_0/k$ is essentially independent of laser wavelength and only the diffraction term $(\nabla_\perp \bar{E}_0/k\bar{E}_0)^2$ contributes a wavelength dependence. The implications of this result will become clear shortly. Using eq. (55) the focusability condition becomes $r_t^2 \geq \langle r^2 \rangle_{min}$. To make an estimate of the constraint imposed by this condition suppose that $L_\perp$ is the average characteristic transverse dimension of the coherent density fluctuation producing the phase distortion ϕ_0, then the focusability condition is approximately

$$r_t^2 \geq f_L^2 \{\beta^2 (\rho_0/\rho_s)^2 (\langle \Delta\rho_\perp/\rho_0 \rangle)^2 (L/L_\perp)^2 + (\lambda/D)^2\}. \tag{56}$$

If the density fluctuations are sufficiently large that the diffraction contribution is small then the condition eq. (56) becomes a constraint on the relative density fluctuations in the laser medium:

$$\langle (\Delta\rho_\perp/\rho_0) \rangle \lesssim (r_t/f_L)(1/\beta)(\rho_s/\rho_0)(L_\perp/L). \tag{57}$$

From this simple criterion the medium quality constraint is reduced by minimizing the focal length of the focusing lens, the Gladstone–Dale constant β of the working medium, the operating density ρ_0 relative to standard density ρ_s and maximizing the transverse scale length of the characteristic density fluctuations of the medium. As an example, suppose the working medium is helium at 1 atm, that mirror damage requires a stand-off distance of $f_L = 100$ m and that the target specification is $r_t = 1$ mm. If the density disturbance in the amplifier is such that $L/L_\perp = 10$, then eq. (56) requires that the average density fluctuations in the medium satisfy $\langle \Delta\rho_\perp/\rho_0 \rangle \lesssim 2.9 \times 10^{-2}$. This constraint is not trivial; however, in high average power CO_2 lasers, values of $\langle \Delta\rho_\perp/\rho_0 \rangle$ in the range 10^{-4} to 10^{-3} have been achieved (Feinberg et al. 1975, Aushermann et al. 1978). Clearly if the working gas were xenon at 1 atm this constraint would require the present state-of-the-art in average power laser medium optical quality control.

Although the condition $r_t^2 \geq \langle r^2 \rangle_{min}$ is useful since it uses a rigorous result from Talanov's Moment Theorems, it is still not precise enough since it does not specify the fraction of the incident laser energy that is contained within the specified radius r_t. For example, it is easy to construct intensity distributions in which a significant fraction of the energy in the beam lies outside the mean square radius $\langle r^2 \rangle_{min}$. Therefore, a more accurate and useful criterion is that the fraction of the total laser power incident on the target that lies outside the radius r_t, $\Delta P_{r_t}/P_0$, be less than a certain value, say 0.01, i.e., $\Delta P_{r_t}/P_0 \leq 0.01$.

Although this calculation has not been done for the general case of arbitrary phase and amplitude distortion, a simple example gives insight into the possible character of the general result. If the refractive index fluctuations of the medium are statistically random with a Gaussian correlation function whose scale length is l and if diffraction effects are not important (Chernov 1960, Tatarski 1961), then

$$\Delta P_{r_t}/P_0 = \exp(-r_t^2/r_{eff}^2), \tag{58}$$

where

$$r_{eff} = [\langle r^2 \rangle]^{1/2} = 2\pi^{1/4} f_L (L/l)^{1/2} \beta (\rho_0/\rho_s)[\langle(\Delta\rho/\rho_0)^2\rangle]^{1/2}. \tag{59}$$

Here it was assumed that the principal contribution to the root mean square refractive index fluctuation comes from gas density fluctuations. It is interesting to note that this expression for the mean square radius [eq. (59)] can be derived from the Talanov theory by using $\langle r^2 \rangle_{min} = f_L^2 \langle(\nabla_\perp \phi_0/k)^2\rangle$ and the approximations of Tatarski (1961) and Chernov (1960). Using eqs. (58) to (59), the radius outside of which 0.01 of the total beam power resides is

$$r_t = 2.2 r_{eff} = 5.7 f_L (L/l)^{1/2} \beta (\rho_0/\rho_s)[\langle(\Delta\rho/\rho_0)^2\rangle]^{1/2}. \tag{60}$$

Again, owing to the large aberration geometrical optics limit, the criterion is essentially independent of wavelength. We note that the primary difference between the present criterion and that developed approximately from the Talanov theory applied to coherent density fluctuations [eq. (56)] is the ratio of the medium length to the characteristic inhomogeneity scale length entered as the square root for the turbulence and linearly in the simple approximation. Random scattering tends to reduce the refractive spreading. From eq. (60) the medium quality condition becomes

$$[\langle(\Delta\rho/\rho_0)^2\rangle]^{1/2} \leq (1/5.7)(r_t/f_L)(1/\beta)(\rho_s/\rho_0)(l/L)^{1/2}. \tag{61}$$

The constraints imposed by eq. (60) are plotted in fig. 15 for several gases. The results indicate the potentially significant advantages associated with the use of low β gases such as helium or neon, operation at low pressure and elimination of small-scale turbulence or density inhomogeneities.

Although these results are somewhat qualitative they indicate the type of analysis that must be performed to relate visible and ultraviolet laser media optical quality requirements to fusion reactor target irradiation constraints. Estimates given in fig. 15 show that in the case of gaseous media, control of density fluctuations in large-scale average power fusion laser amplifiers will be an important technological problem.

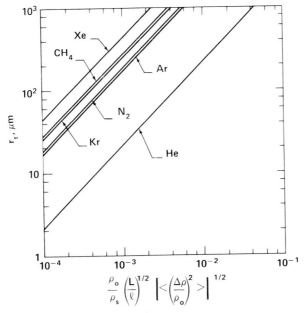

Fig. 15. Medium homogeneity constraint on laser beam focusability for an amplifier of length L containing statistically random density fluctuations ($\Delta\rho/\rho_0$) with a Gaussian correlation function whose scale length is l (Chernov 1960, Tatarski 1961). The quantity r_t is the focal plane radius outside of which 0.01 of the total beam power is focused with a 100 m focal length lens; Eq. (60).

3.6 Average power considerations

During optical (Zuzak and Ahlborn 1969, Ahlborn and Zuzak 1969, Ishii and Ahlborn 1976) and electrical (Pugh et al. 1974, Feinberg et al. 1975 Culick et al. 1976, Aushermann et al. 1978) laser excitation energy is deposited in the medium and its boundaries creating thermal or entropy (Kovasnay 1953, Chu and Kovasnay 1958) and acoustic disturbances which perturb the optical quality of the laser medium. Gasdynamic time scales are typically much longer than the excitation pulse length and therefore these disturbances do not significantly disturb the optical quality of the laser medium during excitation. In the case of photolytically pumped media (§ 4), this condition requires that bleaching or photodissociation waves propagate supersonically with respect to the local sound speed (Ishii and Ahlborn 1976). Therefore formation of density disturbances (§ 3.5) and chemical decomposition place limits on the maximum repetition rate, energy per pulse and output power obtainable. In addition, a certain amount of the power supplied to the laser system must be expended to recondition the laser medium between pulses. If this flow conditioning power becomes comparable to the laser power the overall efficiency of the laser system may be degraded. In the following these gas dynamic problems associated with high power gas lasers are reviewed.

3.6.1 Average power laser medium control

When volumetric heating dominates, a convenient analytical model of average power laser medium dynamics is to consider pressure waves as having emanated from a slug of gas which is instantaneously heated during laser excitation. The instantaneous heating of the gas results in a pressure Δp_L and temperature ΔT_L rise in the laser cavity

$$\Delta p_L/p_L = \Delta T_L/T_L = (\gamma - 1)\mathscr{E}_T/p_L = (\gamma - 1)10^{-2}\mathscr{E}_T[J/\ell]/p_L[atm], \qquad (62)$$

where $\mathscr{E}_T$ is the thermal energy per unit volume added to the laser medium. According to eq. (62), specific thermal energy loadings $\mathscr{E}_T/p_L \gtrsim 10^2\,J/\ell$ atm produce strong pressure and temperature disturbances. In high energy pulsed chemical laser systems (Aushermann et al. 1978) operating in this regime about one-half of the initial thermal translational energy released upon lasing is transferred into shock, compression and rarefaction waves, and the remaining energy is convected out of the system with the flowing purge gas. For specific thermal loadings $\lesssim 10\,J/\ell$ atm, the disturbance generated is an acoustic wave which propagates at nearly the local sound speed. Compressional heating is small and the added thermal energy is removed by convection.

In the design of high average power lasers several schemes or techniques have been developed to control density and refractive index inhomogeneities. Repetitively pulsed electric lasers (Feinberg et al. 1975) of the 'humdinger' category and high energy pulsed chemical lasers (Aushermann et al. 1978) have typically employed a high pressure drop flow wall and a vented duct or a duct lined with sound absorbing material such as is illustrated in fig. 16a. The purpose of the flow wall is twofold: the first is to reflect upstream traveling disturbances and send them downstream into the attenuators, the second is to isolate the gas flowing into the optical cavity from acoustic effects created by excitation. When acoustic disturbances interact with these choked orifices entropy disturbances or hot spots (Kovasnay 1953, Chu and Kovasnay 1958) can be generated which must then be convected through the laser excitation region. Flow wall/absorbing or vented duct configurations have achieved low density fluctuations at high repetition rates ($\Delta \rho/\rho < 10^{-3}$, $\nu_L > 10\,Hz$) in small aperture high specific thermal energy deposition ($\mathscr{E}_T/p_L > 10^2\,J/\ell$ atm) laser systems. For large aperture laser systems absorbing or vented ducts will not be as effective since reverberation times will be longer.

Recently it has been demonstrated (Feinberg et al. 1975) that high quality gas flow can be established with a low pressure loss flow wall such as the cascaded vane system illustrated in fig. 16b. In this system attenuation is accomplished by multiple internal reflections and dissipation in the porous vane material as the acoustic waves propagate through the cascade. The curved vanes produce a gradual turning of the flow, minimizing flow pressure losses. Other techniques using muffler, baffle, through-flow and center body anechoic chamber concepts have also been investigated (Feinberg et al. 1975, Aushermann et al. 1978). To control disturbances and achieve high gas mass utilization and high repetition

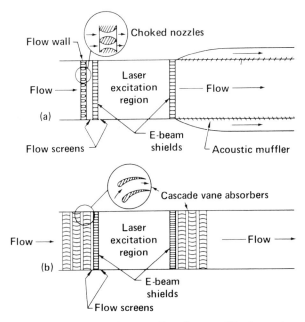

Fig. 16. High average power laser medium flow conditioning systems.

rates and therefore high flow efficiency, the low pressure loss system (fig. 16) includes upstream and downstream absorbers, flow straighteners and flow screens (Schubauer et al. 1950) and other obstructions such as electrodes or e-beam shields which result in flow losses. The primary purpose of the flow straighteners and screens is the removal of velocity variations (shear) across the entrance of the laser cavity. In addition to upstream propagation of the acoustic waves, temporary flow reversal is possible, thereby increasing the time to clear the heated gas. Of the low pressure drop systems that have been investigated, the cascaded vane scheme, (fig. 16b) has been found to yield a low loss coefficient ($\bar{\omega} = \Delta p / \frac{1}{2}\rho u^2 = 0.1$ to 0.2/stage (nozzle–diffuser vane combination)), high attenuation (~ 17 dB/stage) and if several stages are employed a reflection coefficient in the 0.01 to 0.1 range. According to Feinberg et al. (1975) cascaded vane systems offer acoustic damping performance comparable to high pressure loss flow wall systems. The final acoustic design of an average power fusion laser amplifier will depend on the specific thermal loading $\mathscr{E}_T/p_L$, the medium optical quality or laser beam focusing requirements and the flow efficiency that must be achieved.

3.6.2 Total laser system efficiency

In § 2 the total system efficiency ϵ_L of a fusion reactor laser was defined as the laser power p_L delivered to the target divided by the total power (electrical P_{EC}

and flow conditioning P_{FC}) required to operate the laser:

$$\epsilon_L = P_L/(P_{EC} + P_{FC}) = (P_L/P_O)/(P_{EC}/P_O + P_{FC}/P_O) = \epsilon_0/(\epsilon_{EC}^{-1} + \epsilon_{FC}^{-1}). \qquad (63)$$

P_O is the laser output power. In eq. (63), $\epsilon_0 = P_L/P_O$ is the laser energy transport efficiency, $\epsilon_{EC} = P_O/P_{EC}$ and $\epsilon_{FC} = P_O/P_{FC}$ are the electrical and flow conditioning efficiencies. The laser energy transport efficiency ϵ_0 takes account of laser beam power losses incurred by the beam as it travels from the output of the final amplifier through isolators that prevent return of reflected light, through a port to the final focusing element, and through the reactor chamber beackground gas onto the pellet. It also includes loss of beam energy at the target due to a lack of focusability arising from phase and amplitude distortions of the beam generated in the laser medium. In this way, ϵ_0 is parametrically coupled to the flow conditioning efficiency, ϵ_{FC}. In a well-designed optical transport system ϵ_0 might be in the 90 to 95% range (Monsler 1977). The electrical conditioning efficiency is the ratio of laser radiation power to electrical power provided to operate the laser. For an electron-beam-pumped laser it is the product of the component efficiencies for the e-beam electrical pulse-forming network (80 to 90%), e-beam formation and deposition (75 to 80%), laser medium excitation and laser power extraction efficiencies; $\epsilon_{EC} = \epsilon_{PF}\epsilon_{EB}\epsilon_{LM}\epsilon_{ext}$. The flow conditioning efficiency accounts for the power required to circulate the gas, adequately damp acoustic disturbances, remove waste heat and chemically regenerate the laser gas.

From eq. (63) it can be seen that the laser system efficiency ϵ_L is effectively the product of the optical transport efficiency and the smaller of the electrical or flow subsystem efficiencies. Usually most of the power invested in the laser system will be employed to operate a large final amplifier where most of the laser energy will be generated. Consequently, the efficiency of the laser will be largely determined the performance characteristics of this amplifier. The average output power of this final laser amplifier can be written

$$P_O = \mathscr{E}_0 L_F A_L \nu_L, \qquad (64)$$

where $\mathscr{E}_0 = E_L/\epsilon_0 V$ is the laser output energy per pulse per unit volume. The volume of the final amplifier is $V = L_F A_L$, where L_F is the amplifier's length in the gas flow direction and A_L is its corresponding cross-sectional area. The laser system produces pulses at a repetition frequency ν_L.

Neglecting chemical regeneration, the flow conditioning power for this final amplifier is proportional to the mass flow rate $\dot{m}$ and the stagnation temperature of the gas, T_0:

$$P_{FC} = \xi \dot{m} C_p T_0 = [\gamma/(\gamma - 1)]\xi p_L U_L A_L(T_0/T_L), \qquad (65)$$

where C_p is the specific heat at constant pressure, U_L is the speed of the flowing laser gas, p_L is the pressure of the laser medium and ξ is the fraction of the circulating power required to return the gas to its initial state on each round trip. If the laser flow system is treated as an ideal wind tunnel (Liepmann and Roshko 1957) and the energy added to the laser medium due to excitation is not considered then

$$\xi = \{[1 - \Delta p/p_0]^{(1-\gamma)/\gamma} - 1\}, \tag{66}$$

where γ is the gas specific heat ratio, Δp is the pressure drop required to maintain the gas flow against losses from acoustic damping devices, etc., and p_0 is the reservoir or stagnation pressure. Subsonic flows are required to minimize the generation of entropy (thermal) or acoustic disturbances; furthermore, if low pressure drop acoustic damping schemes are employed the expression for ξ given by eq. (66) simplifies considerably because the reservoir and laser sections are at nearly the same temperature and pressure and $\xi = (\gamma - 1)\Delta p/\gamma p_L$. The flow-conditioning power is then simply $P_{FC} \simeq \Delta p u_L A_L$.

The flow conditioning efficiency can be obtained by combining eqs. (64) and (65):

$$\epsilon_{FC} = (\mathscr{E}_0/p_L)/(\gamma/\gamma - 1)f\xi, \tag{67}$$

where $f = u_L/L_F \nu_L$ is the laser medium exchange factor (the ratio of the distance the gas flows between pulses to the optical cavity size in the flow direction). Laser medium exchange factors of approximately 1 to 2 are typical of subsonic gas lasers. From eq. (67) it can be seen that there are three dimensionless parameters which determine the flow efficiency; $(\mathscr{E}_0/p_L)$, f and ξ. To achieve high flow system efficiency the medium exchange factor and the pressure drop and flow Mach number must be kept as low as possible. In addition, the laser output energy per unit volume must be as large as possible compared to the laser medium pressure which is equal to two-thirds of the gas thermal energy density. In energy density units $p_L = 10^2 \text{J}/\ell$ at 1 atm. The ratio $\mathscr{E}_0/p_L$ is sometimes referred to as the volumetric efficiency or the specific laser output energy of the laser amplifier. The significance of the volumetric efficiency of a laser amplifier is illustrated in fig. 17 (Monsler 1977), where total laser system efficiency is plotted as a function of laser electrical conditioning efficiency for representative high pressure drop (fig. 16a) and low pressure drop (fig. 16b) flow conditioning systems. The energy transport efficiency ϵ_0 was taken to be 95%. The volumetric efficiency is taken as a parameter. In the high pressure drop system (fig. 16a) the nozzle and flow screen arrangement require a substantial pressure drop. For a total pressure drop around the flow conditioning circuit of 30 psi a pressure ratio of 3 is required for the pumps. A medium exchange factor of 1.5 was also assumed to be required. From fig. 17 for this case the importance of achieving a high volumetric efficiency in the laser medium is clear. For example, if the laser system electrical efficiency is 4% and if the laser medium can only operate at volumetric efficiency of $1 \text{J}/\ell$ atm, then the total laser system efficiency will be 0.4%, an order of magnitude lower than the electrical efficiency. To achieve its full electrical efficiency this laser would have to operate at greater than $50 \text{J}/\ell$ atm. The potential advantage of a low pressure drop system is illustrated by further consideration of fig. 17. Suppose it is possible to use a low pressure drop system such as illustrated in fig. 16b with a 5 psi pressure drop, comparable to an ordinary wind tunnel, and a medium exchange factor of 1.5. Again, if the

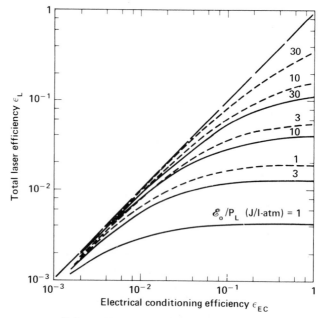

Fig. 17. Total laser system efficiency plotted as a function of laser electrical conditioning efficiency for different laser specific output energies, $\mathscr{E}_0/p_L$, and flow loop pressure losses; low pressure drop $\xi = 0.12$ (---) and high pressure drop $\xi = 0.55$ (———) flow systems.

laser system electrical efficiency is 4% and the laser medium operates with a volumetric efficiency of 1 J/ℓ atm, then the total laser system efficiency will be 1%. The flow penalty diminishes as the volumentric efficiency is increased. At 30 J/ℓ atm the total laser system efficiency is nearly equal to its electrical efficiency of 4%. Fig. 17 clearly shows the profound implications that flow conditioning penalties can have for a laser system.

Fig. 17 suggests that in order to achieve its full electrical efficiency a laser medium must operate at as high a volumetric efficiency as possible, preferably >10 J/ℓ atm, with a low pressure drop acoustic damping system. In some laser media it can be anticipated that high electrical efficiencies will be achievable only at low volumetric efficiencies. For instance, operation at high volumetric efficiencies may be precluded by excited–excited state kinetic losses involving the laser species. Also operation at high volumetric efficiency or specific laser output may produce large specific thermal loadings and pressure pulses which must then be dissipated. For example, in the case of the 4% efficient laser operating at a specific output of 10 J/ℓ atm, if 40% of the electric energy appears as specific thermal energy, 100 J/ℓ atm, then very strong acoustic disturbances will be generated. In this case a high pressure drop system may be required to achieve the required acoustic damping and optical medium quality. On the other hand if the laser medium is highly radiative so that only 4%, or 10 J/ℓ atm, of the electrical energy appears as thermal energy then a low pressure drop system may be sufficient to achieve high medium optical quality. This example illustrates the potentially strong interplay between total laser system efficiency requirement, optical transport and laser kinetics. In general, we see that gaseous laser media that are radiatively efficient and that operate at volumetric efficiencies in the 10 to 30 J/ℓ atm range have the greatest promise of achieving the highest average power total laser system efficiencies.

4 Photolytic group VI atomic lasers

4.1 Introduction

The microscopic properties required for an energy-storage fusion laser medium have been discussed in § 3.4. Murray and Rhodes (1973, 1976) identified the column four and column six atoms (those with p^2 and p^4 ground electronic configurations) from the periodic table as potentially viable candidates which meet these requirements. Electric dipole transitions among the levels of the ground configuration are forbidden by parity and angular momentum selection rules, and the energy required to excite an electron to the next higher shell is large compared to the separation between states of the ground configuration. Studies to date have not been directed towards atoms with the p^2 configuration because efficient methods for producing the excited state population have not been identified. Fig. 18 shows the energy levels of the p^4 electron configuration of these atoms. The highest energy state, the maximally symmetric $^1S_0$, is resistant to collisional deactivation by a wide variety of other atoms and molecules. In contrast, the lower lying $^1D_2$ and the excited levels of the ^3P ground state are characteristically much more reactive. The properties of these

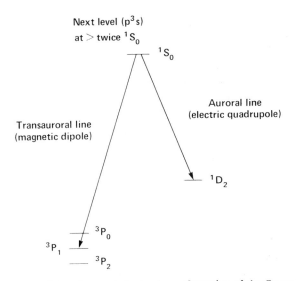

Fig. 18. Energy levels of the p^4 ground electronic configuration of the Group VIa elements.

states in the oxygen atom have been studied in more detail than any of the other atoms with these configurations, because the oxygen atom and its forbidden transitions among these states are very important in upper-atmosphere phenomena (Chamberlain 1961, Omholt 1971).

Both the ^1S and ^1D states are radiatively metastable in the dipole approximation but weak higher-order moments are allowed. These are shown in fig. 18 and have acquired the common name shown. The auroral green line is in fact one of the most intense features of atmospheric emission and is the source of the green color of aurora. For oxygen the radiative lifetime of the auroral line at 577.7 nm is of the order of 0.8 s. The transauroral lines at 297.2 and 295.8 nm have radiative lifetimes on the order of 15 s (Corney and Williams 1972, Kernahan and Pang 1975). By way of example, the Doppler-broadened auroral transition of oxygen at 300 K has a peak optical cross section of 8.7×10^{-20} cm^2 and a saturation energy fluence of 4 J/cm^2, so that a laser operating on the auroral line has energy storage and extraction properties within the region of most interest for a high energy amplifier.

A variety of techniques have been used in studies on the production of the upper ^1S laser level. Optically pumped lasers can have significant advantages for high energy storage devices if one can identify efficient pump sources spectrally well-matched to appropriate pump transitions. Excitation is usually more specific in optically pumped devices, which avoids producing undesirable absorbing or deactivating species and a consequent loss of efficiency and storage time. The medium may be free of electrons and ions which will be present in discharge- or e-beam-pumped systems and which can lead to further deactivation losses. In addition the total particle density in optically pumped storage lasers can be much lower than for high voltage direct electron-beam-excited devices. This can reduce the magnitude of index of refraction gradients arising from density fluctuations and may allow the propagation of high intensity beams of higher quality in optically pumped gas lasers than in comparable electron-beam or discharge pump devices.

High intensity laser pump sources such as the rare gas excimers, rare gas monohalides and dihalogens offer a potential advantage over conventional flashlamps for optical pumping since the efficient coupling of pump radiation to an absorbing medium is greatly simplified for a well-collimated pump beam. A large excimer pump source, operating at a relatively low excited state density to enhance fluorescence efficiency, can thereby be converted to a beam of very high brightness for optical pumping, avoiding the inevitable losses of complex reflective light collecting optics.

In their analyses Murray and Rhodes (1976) point out that one very attractive technique for the production of ^1S excited state species is the photolytic decomposition of a diatomic or triatomic molecule. Such a scheme is shown in fig. 19. Here N_2O is photolyzed using the fluorescence from an argon excimer radiation. As can be seen in this rough schematic, the peak of the Ar$_2^*$ fluorescence band lies very near the peak of the N_2O absorption band. Upon absorption of a photon at 1300 Å, the N_2O molecule is promoted to a repulsive $^1\Sigma^+$ state and

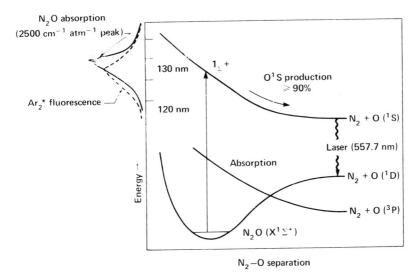

N$_2$O absorption
(2500 cm^{-1} atm^{-1} peak)→

Ar$_2^*$ fluorescence —

130 nm

120 nm

1Σ^+

O^{1}S production
≥ 90%

N$_2$ + O (^{1}S)

Laser (557.7 nm)

Absorption

N$_2$ + O (^{1}D)

N$_2$ + O (^{3}P)

Energy →

N$_2$O (X$^1\Sigma^+$)

N$_2$–O separation

Fig. 19. Schematic representation of the photolysis of N$_2$O into a N$_2$ ground state molecule and an oxygen atom in the ^{1}S state using uv radiation emitted by Ar$_2$ excimers.

is subsequently dissociated, leaving as fragments N$_2$ in its ground electronic state and oxygen in the excited ^{1}S state. Transitions then can occur between the ^{1}S state of oxygen and the lower lying ^{1}D and ^{3}P states.

4.2 Parent molecules and their properties

The absorption of a photon by a molecule causes a specific energy change unlike electronic or collision processes. In addition there are fewer particles leaving the reaction complex after the photolytic reaction, so the number of allowed channels for the reaction products is reduced. Photolytic reactions are therefore more specific than electronic or molecular collisional reactions, and reactions with a high yield of ^{1}S states are more easily identified and studied. Photolytic reactions have a disadvantage, however; photons, rather than easily produced energetic electrons or molecules, must be generated to carry the energy to the photolytic fuel. A discussion is given here of some of the photolytic reactions in which the production of O, S, Se atoms in the ^{1}S state has been demonstrated and some of the collisional deactivation processes which are present in these systems.

For the case of N$_2$O the singlet sigma ground state dissociates adiabatically to ground state N$_2$(X) plus the metastable O(^{1}D). The strongest electronic transition in the molecular spectrum is the $X^1\Sigma^+ \rightarrow 2^1\Sigma^+$ bonding to anti-bonding transition which causes the molecule to dissociate yielding ground state N$_2$(X) and O(^{1}S). This transition is fairly well isolated in energy from other allowed transitions, a property which simplifies selective excitation. Fig. 20 shows the absorption

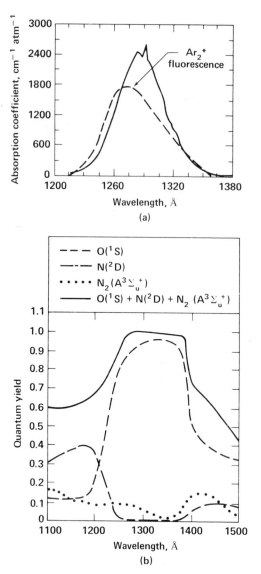

Fig. 20. Photolysis of N_2O: (a) vacuum ultraviolet absorption spectrum of N_2O (Zelikoff et al. 1970) with the Ar_2 excimer fluorescence emission superposed; (b) yield of $N_s(X) + O(^1S)$, $NO(X) + N(^2D)$ and $N_2(A) + O(^3P)$ (Black et al. 1975a).

spectrum of N_2O (fig. 20a) (Zelikoff et al. 1970) and the yield of $O^1(S)$ measured per absorbed photon in this region (fig. 20b) (Black et al. 1975a). The absolute yield of $O^1(S)$ is near unity, making the N_2O molecule an obvious choice as a fuel for $O(^1S)$ production in a laser system. A similar yield of $O(^1S)$ (Kayano et al. 1975) is obtained in the photolysis of CO_2 at a wavelength of 1100 Å. Unfortunately, the yield of $O(^1S)$ in the photolysis of O_2 peaks at 1060 Å and is everywhere $\lesssim 0.1$ (Lawrence and McEwan 1973). The low yields in O_2 are presumably due to the great multiplicity of overlapping electronic states, bound and unbound, in the appropriate spectral region. Most of these correlate to other asymptotes.

Fig. 21a shows the absorption spectrum of the molecule OCS in the vacuum ultraviolet spectral region. Fig. 21b shows the yield of $S(^1S)$ (Black et al. 1975b) produced by the photolysis of OCS near 1460 Å. The qualitative behavior of the $S(^1S)$ yield is much like the yield of $O(^1S)$ from the analogous states of N_2O. The emission of the Kr_2^* rare gas excimer is also shown in this figure, suggesting that Kr_2^* photolysis of OCS may prove to be an effective source of $S(^1S)$. The absorption spectrum of OCS in the $2^1\Sigma^+$ region shows considerable structure, which indicates quasi-bound or predissociative states in this region rather than a purely dissociative absorption. This structure is of some importance in a practical laser system since a quasi-bound predissociative state implies a lifetime in the $2^1\Sigma^+$ level of at least a few vibrational periods. Internal conversion in a

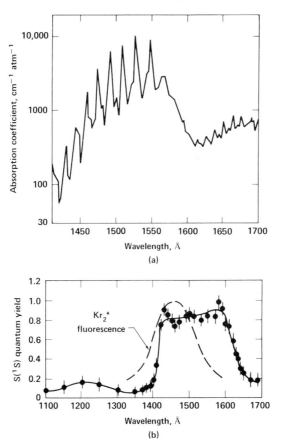

Fig. 21. Photolysis of OCS: (a) vacuum ultraviolet absorption spectrum of OCS (Rabelais et al. 1971); (b) yield of $S(^1S)$ (Black et al. 1975b) with the Kr_2 excimer fluorescence emission superposed.

collision during this lifetime could take the molecule to a different energy surface and reduce the S(^{1}S) yield. This, in turn, might lead to a restriction of the operating pressure in an S(^{1}S) laser. The N$_2$O absorption spectrum has a shape characteristic of a purely repulsive upper state and would not be expected to show significant internal conversion. This conclusion is supported by the successful demonstration of O(^{1}S) production from N$_2$O photolysis in an argon buffer at pressures up to 70 atm (Murray et al. 1974, Hughes et al. 1976).

Fig. 22 shows the absorption spectrum (Finn and King 1975) and quantum yield (Black et al. 1976) for Se(^{1}S) in the vacuum ultraviolet for the molecule OCSe. Also shown is the Xe$^*_2$ rare gas excimer fluorescence which has a peak at a wavelength of 1720 Å. The photodissociation spectrum for producing selenium in the ^{1}S state shows a similar structure to that seen in OCS. As can be seen, the quantum yield at 1720 Å is near unity.

The design of large-scale photolytic Group VI atomic lasers will be largely driven by the properties of the OCS and OCSe molecular 'fuels'. The peak

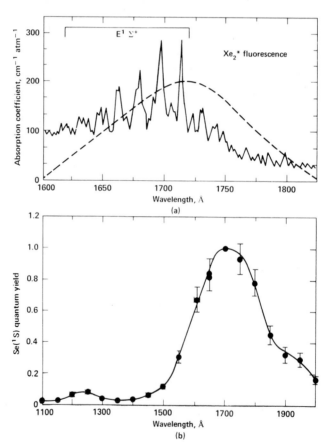

Fig. 22. Photolysis of OCSe: (a) vacuum ultraviolet absorption spectrum of OCSe (Finn et al. 1975) with the Xe$_2$ excimer fluorescence emission superposed; (b) yield of Se(^{1}S) (Black et al. 1976).

photodissociation cross sections of OCS and OCSe molecules at wavelengths matching their respective excimer pumps are $\sim 10^{-16}$ cm^2. Since the fuel molecular number density in fusion amplifiers must be of order 10^{16} cm^{-3} to achieve a useful gain coefficient and stored energy density, the unphotolyzed laser medium will be optically thick. However, a condition of uniform gain throughout the amplifier volume can be achieved by 'bleaching' the photolytic fuel with excimer pump radiation (see § 4.4) whose intensity is in excess of the 'bleaching' fluence.

Table 8 summarizes the characteristic parameters of the photolytically produced Group VIa atoms and the excimer pumps, including the bleaching fluence requirements and laser stimulated emission cross section (Murray and Rhodes 1976). For the highly efficient rare gas excimer pumps (Ar$_2^*$, Kr$_2^*$ and Xe$_2^*$), fuels are available which have high quantum yields into ^{1}S states. The bleaching fluences are the order of 0.1 to 1 J/cm^2, and are well matched to the available fluences from the excimer pumps. The most attractive Group VI atom laser transitions are the auroral transition in oxygen at 557.7 nm, the auroral transition in sulfur at 772.5 nm and the transauroral transition in selenium at 488.7 nm. The transauroral transition in sulfur at 458.9 nm is considerably less attractive for two reasons: (1) its saturation fluence is high (14 J/cm^2) and (2) for a given ^{1}S density in sulfur the gain on that transition is 15 times lower than that on the auroral transition. This means that the gain on the sulfur auroral transition must be suppressed in order to operate the short wavelength transauroral transition.

Electrons, heavy particles and photons can interact with the ^{1}S or the excited parent molecules in such a way as to reduce the actual yield in these systems. The photon processes discussed earlier are operative in OCS and OCSe, and place a limit on the intensity of the pump radiation. In addition to direct photoionization of the ^{1}S state by the pump radiation, the following processes are important in the collisional deactivation of the ^{1}S species. In these equations, X represents the Group VI atom, oxygen, sulfur or selenium, and M represents a heavy particle.

$$X(^1S) + X(^1S) \rightarrow 2X(^3P, \, ^1D), \qquad (68)$$

$$X(^1S) + X(^1S) \rightarrow X_2^+ + e, \qquad (69)$$

$$X(^1S) + e(slow) \rightarrow X(^3P, \, ^1D) + e(fast), \qquad (70)$$

$$X(^1S) + e \rightarrow X^+ + 2e, \qquad (71)$$

$$X(^1S) + X(^3P) \rightarrow 2X(^3P), \qquad (72)$$

$$X(^1S) + M \rightarrow X(^3P) + M. \qquad (73)$$

Eq. (68) represents the process of self-quenching whereby two ^{1}S atoms collide producing either ground state atoms or ^{1}D atoms which will rapidly collisionally collapse to the ground state. The second process [eq. (69)] is commonly called associative ionization; because of potential curve crossings, two ^{1}S atoms upon collision can form the molecular ion and associated electron. Cross sections for

Table 8
Characteristics of photolytically produced Group VIa atoms.

Excimer pump	λ_p [Å]	Laser fuel	Bleaching fluence [J/cm^2]	Terminal level-energy [cm^{-1}]		λ [Å]	σ_L [10^{-20} cm^2]	Γ_s [J/cm^2]
Ar$_2^*$	1265	N$_2$O	0.04	O(^{1}D$_2$)	15 868	5577	8.7	4.0
NeF*	1200	CO$_2$	0.05	O(^{1}D$_2$)	15 868	5577	8.7	4.0
Kr$_2^*$	1460	OCS	0.10	S(^{1}D$_2$)	9 239	7725	45.0	0.6
				S(^{3}P$_1$)	397	4589	3.0	14.0
Xe$_2^*$	1720	OCSe	~1.0	Se(^{1}D$_2$)	9 576	7768	48.0	0.5
				Se(^{3}P$_1$)	1 989	4887	40.0	1.0

this process are not presently known for the Group VI atoms but the cor-responding cross sections for the rare gases are very large. Note that the electron which is produced in associative ionization can deactivate other ^{1}S atoms to ^{1}D and ^{3}P states via superelastic electron collisions [eq. (70)]. It is well known that the cross sections for this electron deactivation process are very large. If sufficiently energetic electrons are present, they can ionize [eq. (71)] or promote ^{1}S atoms to a higher level of excitation. By analogy with the rare gases, cross sections for these processes are expected to be large. Eq. (72) represents the collisional deactivation of the ^{1}S state by ground state atoms. Experiment-ally, the rate coefficient for this process is only known for oxygen, $7.5 \times$

Table 9
Rate coefficients for the collisional quenching of Group VI metastables by parent molecules and fragments.

Group VI spec	Collision partner	Rate coefficient [cm^3 s^{-1}]
O(^{1}S)	CO$_2$, O$_2$	3.6 $(-13)^a$
	CO	$\leqslant$4.5 (-15)
	O	2 (-11)
	N$_2$O	1.1 (-11)
	N$_2$	4 (-18)
	NO	8 (-11)
	N	1.8 (-11)
S(^{1}S)	OCS	4 (-13)
	CO	$\leqslant$3.5 (-16)
	CS, S, O	—
Se(^{1}S)	OCSm	1.6 (-10)
	CO	$\leqslant$1.6 (-16)
	Cse, O, Se	—

^{a}The numbers in parenthesis represent the power of ten multiplicative factor (e.g., 3.6 $\times$ 10^{-13}).

10^{-12} cm^{-3} s^{-1}. Eq. (73) represents the collisional deactivation of the ^{1}S state by a heavy particle. The heavy particle can be either the parent molecule itself or its fragments. Table 9 contains some of the quenching rate coefficients for Group VI metastables by parent molecules and by fragments (Murray and Rhodes 1976). Note that the rate coefficients for N_2O on $O(^1S)$ and for OCSe on $Se(^1S)$ are near gas-kinetic. In order to avoid significant loss of ^{1}S excited states by this loss channel, the parent N_2O or OCSe molecules must be entirely dissociated on a time scale shorter than the collisional quenching time. We shall return to a discussion of this in § 4.4 when we examine photolytic pumping schemes.

4.3 Recent experimental results

An extensive review of the experimental data on the photolytic production of Group VI atoms has recently been given by Murray and Rhodes (1976). In this subsection recent results that have been obtained at our Laboratory by H. Powell since that publication will be given. In these experiments a 50 ns relativistic electron beam (Powell and Murray 1975) was used to excite xenon or krypton gas at pressures in excess of 100 psi in a cell 10 cm high, 50 cm long and 2 cm in width (the actual radiating width was determined by the effective range of the high energy electrons and was about 2 cm). The excimer radiation emitted along the long axis of the cell (50 cm) was transported through a magnesium fluoride window into the photolytic cell which was 10 cm in length. Because of the amplified spontaneous emission in the excimer cell, the excimer radiation pattern at the entrance to the photolytic cell had a cross-sectional area of approximately 10 cm by 1 cm. It was found that 2.1 J of xenon excimer radiation at 1720 Å and 1.3 J of krypton excimer radiation at 1460 Å could be deposited in the photolytic cell in 50 ns.

The radiative decay rate for the ^{1}S state is on the order of 0.1 to 1 s; thus the intensity of the radiation which emanates from transitions from the ^{1}S is very small. Because of this the measurements on the spectral and temporal charac-teristics of the auroral and transauroral transitions are quite difficult, particularly so because the relativistic electron beam generates a large amount of electrical noise. It is possible to increase the radiative decay rate of the ^{1}S state with the addition of a sufficient quantity of a rare gas into the photolytic cell. The rare gases argon, krypton and xenon will serve to increase the effective matrix element for the $^1S \rightarrow {^1D}$ transition by collisionally inducing a quadrupole moment in the lower ^{1}D state. The presence of a diluent has an additional advantage in that it serves to distribute the kinetic energy of the photodissociation products (typically 1 to 2 eV per ^{1}S atom) among the background gas atoms, thereby reducing the overall temperature rise in the photolytic cell and minimizing the deleterious effects of hot atom reactions.

Fig. 23 shows the time history of the radiation in the 100 Å band centered at 7700 Å from 0.5 Torr of OCS photolyzed at 1460 Å. Shown are the oscilloscope traces for various krypton diluent pressures in the photolytic cell. For these data

0.5 torr OCS photolyzed at 1460 Å
30 mJ/cm²

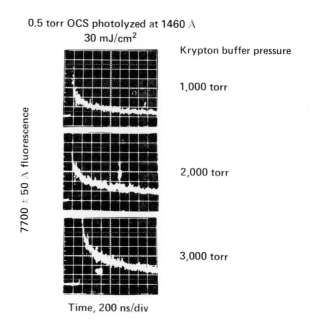

Fig. 23. Temporal evolution of the collision induced sulfur S(¹S) emission. The signals shown represent radiation emitted within a 100 Å bandwidth, centered at 770 nm. Krypton is used to collision induce the S(¹S) emission, and, as shown, serves to increase the strength of the signal. The rapid decay rate is thought to be due to electron deactivation processes.

the input pump fluence at 1460 Å was 30 mJ/cm², a value approximately one-third the pump saturation fluence (table 8). The presence of the krypton diluent does not affect the decay kinetics, which are still collision dominated, but merely serves to increase the intensity of the detected signal. Similar data have been obtained at input pump fluences up to 57 mJ/cm², and the decay characteristics for times longer than a few hundred nanoseconds remain basically the same. From such data one can conclude that the effective lifetime of the S(¹S) state is of the order of 1 μs for the conditions of this particular experiment; however, one distrubing and as yet unexplained aspect of these recent measurements is the fact that at higher pump fluences the absolute yield of S(¹S) is less than for lower pump fluences. Similar results have been obtained for the absolute yield of Se(¹S) from the photolysis of OCSe. These results are contrary to what was expected; namely, it had been thought that when the fuel (OCS, OCSe) was completely photolyzed the kinetic lifetime of the ¹S state would increase. The precise mechanism for this ¹S deactivation process is not presently known; it is thought, however (Powell 1978), to be due to the creation, heating and subsequent multiplication of electrons. That is, if while producing large yields of ¹S atoms some electrons are formed via photoionization of either the fuel (two step) or ¹S atoms (one step), then superelastic electron collisions [eq. (70)] will not only deactivate the ¹S state but will also heat the electron gas. These hotter electrons can then ionize other ¹S atoms, further reducing the absolute yield as

well as increasing the electron population. The process then continues until the
^1S population comes into a quasi-static local equilibrium with other excited state
populations.

Indirect evidence which tends to support this hypothesis comes from experiments where a buffer gas which cools the low energy (a few electron volts) electrons (CO for example) is added to the photolytic fuel (OCS or OCSe). When ~50 Torr of CO is added to 1 Torr of OCS the kinetic lifetime at high excited state densities of the S(^1S) state increases from ~200 ns to several microseconds. As expected for this proposed quenching mechanism, the addition of a similar quantity of a rare gas has no effect on the lifetime. Work is presently underway in an attempt to quantify this deactivation mechanism.

Powell and Ewing (1978) have recently observed laser action in a mixture of OCSe (1 torr) and CO (50 torr) gases photolyzed by xenon excimer radiation. Laser oscillation was observed on both the auroral and the transauroral transitions of Se. Stimulated emission on the auroral transition of S has also been observed by Powell and co-workers.

In summary then, recent experiments on the photolysis of OCS and OCSe via krypton and xenon excimer radiation have yielded several important results: (1) OCS and OCSe densities greater than 10^{16} have been bleached, (2) collisional quenching processes appear to be significant and (3) laser action has been observed in both Se and S atoms.

4.4 Photolytic pumping schemes – Incoherent and coherent

For the systems of interest, the effective absorption cross sections σ_P for photolyzing the parent molecules into the desired final states lie in the range of 10^{-17} to 10^{-16} cm^2. Typical ^1S excited state densities for nominal sized amplifier modules will lie in the range 10^{16} to 10^{17} cm^{-3}, requiring therefore a parent molecule pressure in the 1 to 3 Torr range. The confluence of these facts indicates that the characteristic absorption length for the photolytic pump radiation, defined as $\Lambda = (N_A \sigma_p)^{-1}$, will lie in the range 0.1 to 1 cm. At low input pump fluences and for no ^1S collisional deactivation processes operative, both the excited state density and the input pump fluence will exponentially decay with a scale length of Λ. For input fluences greater than the bleaching fluence given by $\Gamma_b = h\nu_p/\sigma_p$, the energy in the input pump beam is sufficient to fully consume the parent (photolytic) molecules. Under these conditions the input pump beam will propagate through the media fully consuming the parent molecules, until such a point is reached where the pump fluence at that position lies below the pump bleaching fluence. Such a phenomenon has been called 'bleaching wave excitation' owing to the fact that, as a consequence of the photodissociation process, the medium is rendered optically thin and that the spatial position where this bleaching process takes place propagates as a wave through the photolytic media.

It is useful to consider in a simple fashion the spatial and temporal charac-teristics of both the pump beam and the photolyzed parent molecules. Consider a reaction of the form

$$h\nu_p + A \rightarrow B + C^*. \tag{74}$$

In this equation, A represents the parent molecule, for example OCS, B represents one of the fragments such as CO, and C^* represents the ^1S state, in this case sulfur ^1S. Taking the z-axis to be the axis of propagation of the pump beam and assuming that the input pump beam is both uniform in time and in the x, y dimension, one can write (Haas 1977) that the pump intensity at any position z and any time t is given by the following expression:

$$I_p(z, t) = I_p(0)\{1 + \exp[-(V_b t - z)\sigma_p N_A(0)]$$

$$- \exp[-(t - z/c)\sigma_p I_p(0)/h\nu_p]\}^{-1}, \tag{75}$$

where $I_p(0)$ and $N_A(0)$ are the input pump intensity and initial parent molecule density, and V_b is the asymptotic bleaching wave velocity given by

$$V_b = (I_p(0)/h\nu_p)/[N_A(0) + I_p(0)/ch\nu_p] \simeq I_p(0)/N_A(0)h\nu_p. \tag{76}$$

Here the approximation has been made that the photon density [the second term in the denominator of eq. (76)] is much less than the initial parent-molecule density, which is generally the case for the systems of interest here. For parent-molecule pressures in the Torr range and pump intensities in the megawatt to $100 \, MW/cm^2$ range, eq. (76) indicates that the bleaching wave velocities are much less than the speed of light and lie typically in the range 10^6 to $10^7 \, cm/s$. Such slow velocities will have a significant effect on the maximum size amplifier devices may assume, particularly when kinetic quenching proces-ses are operative. Under conditions typical for Group VI atoms, one must trade-off the device size in the direction of bleaching wave propagation with parent molecule density and pump intensity in such a way as to minimize the effects of kinetic quenching.

One can also write an expression giving the density of parent or precursor molecules (or for that matter the density of excited atoms) in the absence of any collisional quenching processes. The expression for the spatial and temporal dependencies of the precursor density is given by

$$N_A(z, t) = N_A(0) - N_{C^*}(z, t)$$

$$= N_A(0)\{1 - \exp[(V_b t - z)\sigma_p N_A(0)] - \exp[-\sigma_p N_A(0)z]\}^{-1}. \tag{77}$$

It is useful to estimate the pump intensity required to assure that quenching of the ^1S state by collisions with the parent molecules does not dominate the kinetic processes. Clearly this quenching process is operative only where both the parent molecules and ^1S states co-exist; namely, at the bleaching wave front. Denoting by k_q the rate coefficient for the kinetic quenching process of interest, and recognizing that the width of the bleaching front is of order Λ, one has the following relationship for the required pump intensity:

$$I_p \gg (N_A h\nu_p/\sigma_p)k_q. \tag{78}$$

This, simply stated, is a requirement that the bleaching time $(h\nu_p/I_p\sigma_p)$ be much less than the quenching time $(1/N_A k_q)$. For the worst possible case of OCSe with a quenching rate constant of $k_q = 2 \times 10^{-10}\,\text{cm}^3\,\text{s}^{-1}$ and a bleaching fluence equal to $3 \times 10^{-2}\,\text{J/cm}^2$ at the ArF wavelength, a flux $I_p > 3\,\text{MW/cm}^2$ is required for a number density of $10^{17}\,\text{cm}^{-3}$, and $I_p > 300\,\text{kW/cm}^2$ for a number density of $10^{16}\,\text{cm}^{-3}$. These are reasonably modest minimum flux requirements. The number densities are ~ 10 times more favorable for Kr_2^* photolysis of OCS. As we have seen, eq. (78) gives the minimum pump intensity required to compete with collisional deactivation processes. The duration of the pump pulse is also set in a similar way by collisional deactivation processes. It was mentioned above that sulfur ^1S densities in the range $10^{16}\,\text{cm}^{-3}$ have been produced which last for times on the order of a microsecond. Therefore, these experimental data show that the pump times for efficient utilization of the media must be $\sim 1\,\mu s$. These two conditions set the characteristic intensity and pulse width for the photolytic pumping system.

While the foregoing analyses were fairly idealized in that uniform pump intensities in the transverse dimensions were assumed, they were useful in that they allowed the identification of the parameter range of interest. In designing actual laser devices, it is important to simulate as closely as possible the actual device under study. A computer simulation of a photolytic Group VIa laser device using fluorescent excimer pumping of the photolytic medium is shown in fig. 24. Electron-beam excitation of broadband excimer fluorescence is first calculated; this radiation is then transmitted through either a solid or gaseous window into the photolytic region. The fluorescence intensity is such that a bleaching wave is launched in this medium. In this simulation, double-sided excitation is used to achieve a more uniform deposition of the broadband excimer fluorescence in the photolytic medium. Fig. 25 shows the results of a computer calculation for one-sided excitation of OCSe using Xe_2^* excimer radiation. The lower portion of this figure is the electron-beam deposition area

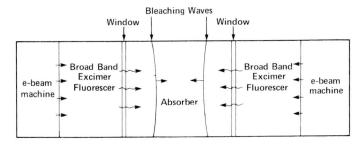

Fig. 24. Schematic representation of a double-sided fluorescently pumped photolytic laser. Relativistic electron beams impinge upon a rare gas which radiates broadband excimer radiation. These emissions are transported through a window into the photolytic media launching bleaching waves.

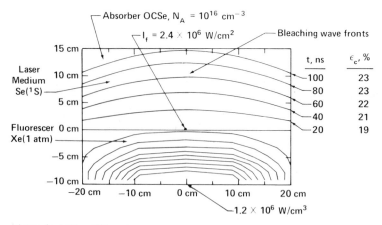

e-beam deposition 10% contours

Fig. 25. Two-dimensional computer results for a single-sided fluorescently pumped photolytic laser. Within the xenon fluorescer region is shown the e-beam deposition contours. This radiation is transported into the OCSe photolytic medium and lines of constant fluorescer intensity and the temporal evolution of the bleaching wave front computed. The ^1S excited state production efficiency ϵ_c is also shown.

and the lines represent iso-intensity e-beam-deposition contours. These contours are obtained using a three-dimensional Monte Carlo electron-beam-deposition code (SANDYL). The fluorescence cell contains Xe at 1 atm and is irradiated for 100 ns by a 300 kV, 100 A/cm^2 electron beam. The peak electron-beam deposition is 1.2×10^6 W/cm^3 and the peak fluorescence intensity in the laser medium cell is 2.4×10^6 W/cm^2. The e-beam deposition is assumed in this case to be linearly related to the xenon fluorescer radiation via an efficiency factor of 0.4. However, one does not need to place this restriction on the xenon fluorescer e-beam-deposition relationship. A sophisticated rare gas excimer kinetics code exists which can treat the range where linearity between input and output does not occur owing to excited state–excited state interactions. To calculate the transport of the fluorescence radiation into the photolytic medium, we assume that the intensity at any point in the absorber region is the sum of the intensity contributions from all regions of the fluorescer source. Using a formalism similar to that used in eqs. (75) and (77) we can compute the local ^1S density and bleaching wave intensity at any position of the absorber for any time. The upper part of fig. 25 shows the result of such a calculation. For example, at a time of 100 ns (top curve) the profile of the bleaching wave front as a function of distance is shown. Also shown are the trajectories of the bleaching wave fronts. The figures in the column labeled ϵ_c represent the efficiency for converting the excimer-fluorescence photons into ^1S atoms in the volume that is bounded by the deposition of the bleaching wave and the window separating the fluorescer cell and absorber media. At 100 ns, we see that 23% of the photons that were emitted from the fluorescer region produced ^1S atoms. To get the actual energy efficiency one must include the photon defect (the ratio of the ^1S energy to fluorescer photon energy) as well as the ratio of the e-beam deposition energy to fluorescer energy. As stated earlier, the latter ratio is in the range 40 to 50% for rare gases (Turner 1975, 1977). In the calculation presented in fig. 25 the ratio of e-beam deposition energy to fluorescer energy was taken as 0.4. It is important to point out that, for these examples of 2-D bleaching wave propagation simulations, kinetic quenching processes have not been included. If the bleaching times were such that kinetic processes were beginning to limit the storage of ^{1}S density, then the actual conversion efficiencies would be substantially less. For small devices operating with an absorber density in the Torr range, this is not an important consideration.

The numerical examples shown in figs. 24 and 25 are for an incoherently pumped photolytic medium. For this scheme the rare gas excimer fluorescent radiation can be created with an intrinsic efficiency (i.e., deposited electrical energy to vuv radiation) approaching 50%; the major issue for fluorescence-pumped devices has to do with the development of a suitable device geometry which maximizes the efficiency of the transport of the fluorescent pump radiation into the photolytic medium. With coherent pumping of the photolytic medium the same formalism used above can be applied. However, the need for detailed transport codes is minimal since the coherent pump radiation will be propagated axially or transversely into the photolytic medium. Kinetic processes will affect the spatial character of the excited state density profile and must be calculated.

The major issue with coherent pumping has to do with the overall efficiency of the pump source (the radiation transport efficiency in general being large, >50%). To quantitatively assess the operating characteristics of a rare gas excimer coherent amplifier, a detailed kinetics and radiation model was developed (Werner and George 1976, Werner et al. 1977). The intrinsic extraction efficiency characteristics as a function of the local laser field for a xenon excimer amplifier are shown in fig. 26. These results represent local quantities

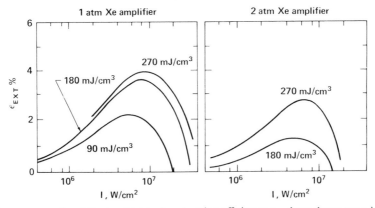

Fig. 26. Computer results of the local intrinsic extraction efficiency ϵ_{ext} dependence upon input laser intensity for a xenon amplifier. The various curves shown are for differing pumping rates (1 μs pulse length).

and not device characteristics. As seen, the intrinsic extraction efficiency ϵ_{ext} increases with increasing laser intensity I, as it should, and then saturates and actually decreases for I values in excess of 10^7 W/cm^2. The decrease in ϵ_{ext} is due to the quasi-nonsaturating character of the photoionization optical loss of the $^3\Sigma$ excimer state (Werner 1975). As can be seen for the cases chosen in this figure, the maximum intrinsic extraction efficiency is about 4%. Analysis and modeling studies show that a xenon excimer laser will most probably have a device efficiency less than 10%. Higher xenon device efficiencies may be possible if Ar–Xe mixtures and sustainer-enhanced pumping at low excited state densities are employed. Current modeling studies are underway to explore this operating regime.

4.5 *Extraction characteristics and efficiency estimates*

The critical assessment of the extraction efficiency of a fusion laser amplifier module must include in a self-consistent manner the spatial dependences of the gain media and short pulse input beam. Fig. 27 illustrates such a calculation for the transverse fluorescence pumping of an amplifier module containing OCSe by xenon excimer radiation. Shown in fig. 27a are the extraction efficiencies and input fluence characteristics for a multipass amplifier configuration (see § 3.4). The laser transition chosen is the selenium transauroral line at 488.7 nm and the lower level is assumed to be fully bottle-necked; under these conditions $\epsilon_{ext} = 0.75$. The short pulse input fluence is taken to be a fifth-power exponential, exp

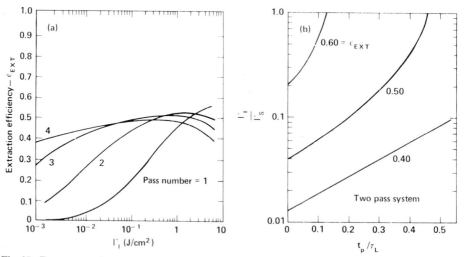

Fig. 27. For a xenon fluorescently pumped OCSe photolytic laser operating on the transauroral line: (a) the calculated extraction efficiency in a multipass amplifier geometry as a function of input fluence, with number of amplifier passes as a parameter; (b) the calculated normalized input fluence Γ_1/Γ_S as a function of the pumping time to lifetime ratio t_p/τ_1 for several extraction efficiency (ϵ) values.

Table 10
Component efficiencies for a Group VI laser in selenium.

Efficiency components	System	Xe$_2^*$ laser OCSe	Windowless fluorescer OCSe
		Percentage efficiency[a,b]	
Pulse forming deposition and fill factors			65
Pump medium		5–10 [20]	45
Pump after flow costs		2.8–5.6 [10]	20
Transport/coupling		90	20–40 [60]
Fill factors		80	
Extraction X quantum		22	22
Laser electrical		0.4–0.8 (1.6)	0.7–1.4 (2.1)
Total laser after flow in laser		0.4–0.7 (1.4)	0.6–1.1 (1.5)

[a] Brackets indicate percentage with possible break-throughs.
[b] Parentheses indicate total with breakthrough increment.
[c] Requires cleaner pump medium.

$-(R/R_0)^5$, where R_0 is on the order of the amplifier radius R_A. The other relevant parameters are transverse pumping time 400 ns, kinetic lifetime 1 μs, optical transit time 100 ns, optical loss per pass 0.05 and maximum small-signal gain $(\alpha_0 L) = 4$. Under these conditions the maximum extraction efficiency is seen to approach 0.5. This lower value (compared to 0.75) is a direct consequence of the finite pump time, the optical fill factor and transport losses, and the short lifetime (1 μs) associated with the excited state population.

Shown also in this figure (fig. 27b) is the effect of varying the kinetic lifetime on the input fluence requirement for a two-pass amplifier configuration for several desired extraction efficiencies. As anticipated, the longer the kinetic lifetimes relative to the pumping time, the lower is the required input fluence for a given extraction efficiency. In addition, for a given input fluence the device extraction efficiency decreases as the kinetic lifetime decreases (pump time remains constant).

Table 10 summarizes the component and overall efficiencies for both a coherent and incoherently pumped photolytic laser in selenium (transauroral line). The first line describes the conversion of raw electrical energy into deposited electron-beam energy within the pump medium. The next line gives the intrinsic pump efficiency; as discussed earlier the fluorescence efficiency can be large (45%), whereas because of stimulated loss processes the laser efficiency is expected to be much lower (~10%). The third line summarizes the total pump efficiency including flow and pulsed power losses; fluorescence light is produced with higher efficiency than laser light; however, when radiation transport and coupling losses are included the overall system efficiencies for incoherent and coherent pumping are comparable and lie in the range 0.4 to 1.5%.

A conceptual design of a large-scale ($400 \times 100 \times 50 \, cm^3$) selenium power amplifier (50 kJ) which incorporates the experimental findings, kinetic analysis and systems modeling discussed in the previous subsections is shown in fig. 28. The laser 'fuel' is a mixture of OCSe (several Torr) and helium diluent (an atmosphere). A subsonic, laminar flow of the fuel is maintained through the central amplifier cavity in a vertical direction. In this fluorescence-pumped design, two parallel flows of Xe fluorescer gas are maintained parallel to the fuel flow with matching velocities and pressures. Large-area electron beams (1 MeV, 10 A/cm², 1 μs) are fired into the Xe gas regions on each side, producing copious spontaneous emission at the photolysis wavelength. A short (nanosecond) extraction pulse of light at the transauroral wavelength 488.7 nm is propagated through the amplifier cavity transverse to the flow direction along the long dimension of the amplifier (measuring several meters in length). The gas flow rate is adjusted to essentially replace the fuel between pulses (0.1 to 1 s).

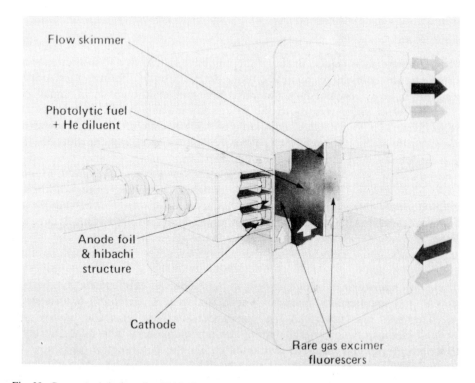

Fig. 28. Conceptual design of a 50 kJ, fluorescently pumped Group VI laser. Double-sided pumping of the rare gas using large area electron beams is shown. The excimer radiation is transported into the photolytic media through a flowing gas (gas curtain) window.

5 Rare earth solid-state hybrid lasers

5.1 Introduction

The trivalent rare earth ions (RE^{3+}), when placed in a suitable ligand environment, manifest microscopic properties suitable for energy-storage lasers in the short wavelength spectral regions. As noted earlier, the successful development of neodymium (Nd^{3+}) glass lasers exploited the long storage time ($>300\ \mu s$), high energy-density storage ($>500\ J/\ell$), near-optimum stimulated-emission cross section ($\sim 3 \times 10^{-20}\ cm^2$) and saturation fluence ($\Gamma_s \sim 4\ J/cm^2$) to produce multi-terawatt optical beams of nanosecond time duration.

In view of these near-ideal microscopic laser properties it was deemed useful to explore methods by which they can be exploited in efficient, high average power laser systems. The simplest approach to this end might be to configure the Nd:glass laser medium in a geometry which can accommodate the thermal loading of the glass host resulting from broadband flashlamp pumping of the Nd^{3+} ions. As a point of reference, approximately $1.7\ J/cm^3$ of heat must be removed from the Nd:glass medium for every J/cm^3 stored in the upper laser. Because of the relatively low thermal conductivity of laser glasses, the gain medium must be arrayed in the form of thin, large-aperture slabs which can be faced-pumped and cooled with helium flowing in channels formed by the slab faces. This 'axial gradient' cooling technique was first demonstrated in the 1960's (Young 1969) on small-aperture devices in which optical distortions were almost completely dominated by edge effects. This cooling technique is more nearly optimal for use in the fusion application because of the requirement for amplifiers with large apertures. At the same time, such amplifiers have become more feasible because of the recent advances in the development of new glasses with very low nonlinear refractive indices (Weber et al. 1978). Of course, one is constrained to utilize neither a broadband flashlamp as a pump nor Nd^{3+} ions as active centers; a logical extension of the flow-cooled Nd-slab amplifier concept (to substantially reduce the specific thermal loading of the solid host material and to minimize problems with cyclical stress) is to utilize a RE^{3+} ion whose upper laser level can be resonantly excited with an efficient narrowband pump such as a rare gas monohalide laser (XeF, KrF, ArF, etc.), and whose energy level structure minimizes the conversion of electronic to lattice thermal energy. We have designated this type of laser (gas-laser-pumped solid-state laser) as a rare earth 'hybrid' laser. As an example, the thulium (Tm^{3+}) ion dispersed in an appropriate solid and pumped with the XeF laser appears to have the required properties for such a hybrid laser and is discussed in this section.

5.2 Dynamics of Tm^{3+} solid-state media

The relevant energy levels and transitions of the Tm^{3+} ion are shown in fig. 29. The Tm^{3+} ion exhibits a resonance absorption at a wavelength of 353 nm,

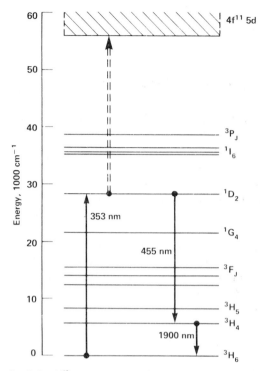

Fig. 29. Electronic levels of the $4f^{12}$ ground electronic configuration of Tm^{3+} showing the resonant pump transition at 353 nm, possible (dashed) excited $4f \rightarrow 5d$ transition at the pump wavelength, and the laser transition at 455 nm.

terminating on the $^1D_2$ electronic level and matching the output wavelength of the XeF excimer laser (Brau and Ewing 1975). The conversion of input pump energy either into thermal energy in the solid-state host via phonon emission or into stimulated or spontaneous emission of light will depend on the strength and character of the Tm^{3+} ion–lattice coupling. For the fusion application, the host must meet the following basic requirements: (1) exhibit a low ($<0.7 \times 10^{-13}$ esu) nonlinear index of refraction, (2) be scalable in transverse aperture (~ 100 cm) with good optical properties, (3) provide RE^{3+} ion sites lacking inversion symmetry, (4) exhibit a reasonably low energy cut-off ($<1\,200$ cm^{-1}) of the optical phonon spectrum and (5) exhibit as large a coefficient of thermal conductivity as possible. The fluorophosphate and fluoroberyllate glasses appear to meet all of these criteria, although thermal conductivities higher than 0.01 W/cm°C would be desirable. Crystalline calcium fluoride and KY_3F_{10} are characterized by sufficiently low nonlinear refractive indices ($\sim 0.6 \times 10^{-13}$ esu) (Milam et al. 1977) and relatively high thermal conductivities (0.1 W/cm°C). Scaling doped crystals to large sizes may be possible, in view of their cubic structures, by hot-forging of macrocrystals. The KY_3F_{10} host with a trivalent rare earth site has a clear advantage over the CaF_2 host with its regular divalent sites. The KY_3F_{10} host will likely not be plagued with clustering of dopant RE^{3+} ions as observed for the

CaF$_2$ host (Tallant et al. 1976). While these or other host materials may prove useful, we will continue our discussions here using simple glasses as the host material.

The $^1D_2$ level in silicate, phosphate and fluoride glasses will exhibit a radiative lifetime of several tens of microseconds (Reisfeld and Eckstein 1975). Because the 6000 cm^{-1} energy gap to the next lower lying level ($^1G_4$) is large compared to the 1200 cm^{-1} cut-off energy of the optical phonons (Layne et al. 1977), the $^1D_2$ level will have a very low probability for decay by multiphonon emission with an attendant thermal heating of the host solid. The radiative decay of the $^1D_2$ upper laser level strongly favors emission to the $^3H_4$ electronic level (Spector et al. 1977) resulting in stimulated emission cross sections estimated to be greater than 5×10^{-20} cm^2 at a wavelength of 455 nm. Once in the $^3H_4$ level, the Tm^{3+} ion will decay radiatively at 1 900 nm with high probability, again owing to the 6 000 cm^{-1} energy gap to the ground $^3H_6$ electronic level. Aside from these single-ion multiphonon lattice heating processes, which appear to be quite small in this case, several other $^1D_2$ quenching channels must be considered in lasers of this type. Under conditions of strong pumping and $^1D_2$ population densities appropriate for laser amplification (e.g.,) 10^{18}/cm^3), it is important that Tm^{3+}($^1D_2$) ions not be further excited by the pump to levels of the 4f^{11}5d electronic configuration via parity-allowed transitions (indicated in fig. 29 by a dotted arrow). Such doubly-excited ions will rapidly relax nonradiatively to high lying levels of the ground configuration with a substantial amount of lattice heating. The energy locations of 4f^{11}5d electronic levels (near 74 000 cm^{-1} in the free ion but depressed to lower energy in the solid) will be sensitive to the ligand fields of the host. The need to avoid this quenching channel may therefore limit the host materials suitable for this application. Analysis of the electronic structure of the Tm^{3+} ion also indicates that quenching of $^1D_2$ population may occur via near-resonant interactions with unexcited Tm^{3+} ions of the form

$$Tm(^1D_2) + Tm(^3H_6) \rightarrow Tm(^1G_4) + Tm(^3H_4) + \bar{Q}[600 \text{ cm}^{-1}], \tag{79}$$

where $\bar{Q}$ represents the energy difference in electronic energies involved in the process. This energy appears as lattice heating. Further quenching of the Tm($^1G_4$) ion produced in this process is possible by a near-resonant interaction with a ground state Tm ion. The rate of these quenching processes can be controlled by varying the Tm doping level in the glass. However, in these quenching interactions the excitation persists primarily as electronic energy which ultimately will be fluoresced away without appreciable thermal heating of the host. For host materials of interest, estimates of the partitioning of excitation energy into re-emitted light and thermal heating indicate that a specific thermal loading of 0.2 J/cm^3 for every J/cm^3 of available output energy can be achieved, or roughly an order of magnitude smaller than in flashlamp-pumped Nd:glass.

Preliminary fluorescence kinetics studies have been performed (Jacobs 1977) on Tm-doped ED-4 silicate laser glass using intense pumping with a XeF laser. Fig. 30 indicates that $^1D_2$ excited state densities of 10^{18}/cm^3 can be achieved without evidence of two-step excitation of 5d electronic levels. Kinetic storage

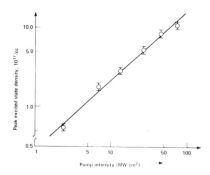

Fig. 30. Peak fluorescence intensity of the $^1D_2 \to {}^3H_4$ transition at 455 nm, expressed as the peak $^1D_2$ number density, as a function of pump intensity (Jacobs 1977).

times of order a microsecond appear achievable with a Tm doping of about 1 weight %, appropriate for large-scale devices; small-scale XeF pump Tm:glass laser experiments are now in progress (Jacobs 1977).

5.3 A Tm:glass hybrid power amplifier

It is instructive to consider some of the principal issues and the characteristic parameters for a hybrid laser of some scale, say 10 kJ. A schematic representation of such a power amplifier is shown in fig. 31. A number of square Tm:glass slabs of thickness t and transverse dimension W are held in a parallel arrangement with a channel width s between adjacent slabs. Helium gas at a nominal pressure of an atmosphere is flowed through the channels under turbulent flow conditions at a Mach number M, Reynolds number Re and Nusselt number Nu. In this schematic design, XeF pump radiation is propagated through the slabs at the Brewster's angle and coaxially with an input extraction

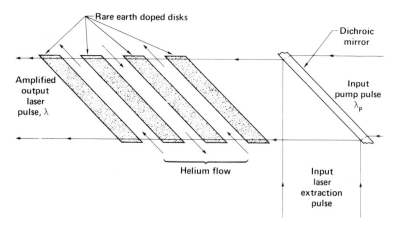

Fig. 31. Schematic layout of a rare earth, solid-state hybrid laser. Rare earth ions doped in transparent dielectric slabs of high aspect ratio are resonantly pumped by a high brightness pump source of wavelength λ_p. Slabs are cooled by turbulent flow of helium gas.

pulse with a wavelength of 455 nm. In more practical designs, one might expect to pass the extraction beam through the amplifier several times at small angles to the axial symmetry axis. Pump beams tailored in intensity in the transverse plane would be propagated in the amplifier along the axial symmetry axis.

An optimum hybrid laser design must take into account the interdependence of a great number of physical parameters: Tm ion concentration, upper laser level lifetime, time interval between passes through the amplifier, single-pass gain, pump and output fluences compared to optical damage fluence, B-integral value, etc. Such optimization procedures have not yet been fully carried out, but some feeling for characteristic values can be derived in the following example. Table 11 lists a nominal set of values for Tm:glass. We have assumed a total optical thickness of the slabs of $L = 20$ cm. For the selected Tm ion density (0.12 mol%), the fluorescence lifetime of the $^1D_2$ upper laser level is 4μs (Jacobs 1977) and the absorption coefficient is $\alpha_p = 0.1$ cm^{-1}. If we assume an excited state density of 2×10^{18} cm^{-3} (800 J/ℓ), the small-signal gain coefficient is 0.1 cm^{-1} and the effective saturation fluence is 5 J/cm^2, a value which takes into account level degeneracies and the fact that the system will bottleneck in the terminal laser level (this level is radiative with a multi-millisecond lifetime).

The extraction properties of such a device have been calculated assuming a pump pulse time of 0.5 μs, a single-pass loss of 5%, a single-pass transit time of 100 ns, a fifth-power exponential spatial pulse shape and a nonlinear refractive index of 0.7×10^{-13} esu. Fig. 32 shows the calculated extraction efficiency as a function of input fluence, with number of passes as a parameter. Fig. 33 shows the corresponding accumulation of B-integral (§ 3.5) parametrically with the number of passes (3 ns long extracted pulse). These figures indicate that for an input fluence of a few hundred millijoules per square centimeter, an extraction efficiency of 40% can be achieved with a B-value of five and with five passes. The B-value could in principle be lowered with spatial filtering and optical relaying between transits, but with some additional loss per pass. Alternatively, at an input fluence of 1 J/cm^2, the extraction efficiency is again near 40% in four passes, but with a B-value of only three. The energy stored in such a device depends on the maximum transverse aperture at which the gain be stably stored. For a maximum gain–length product of 5, the maximum transverse dimension (without segmenting of the slabs) is $W = 40$ cm. The output energy of a device of

Table 11
Characteristic parameters for Tm:glass medium.

Property	Value
Pump cross-section σ_p[cm^2]	1.5×10^{-21}
Laser cross-section σ[cm^2]	5×10^{-20}
Saturation fluence Γ_s[J cm^{-2}]	5
Tm-ion concentration N[ions cm^{-3}]	7×10^{19}
$^1D_2$ level lifetime τ_L[μs]	4
Pump absorption coeff. α_p[cm^{-1}]	0.1

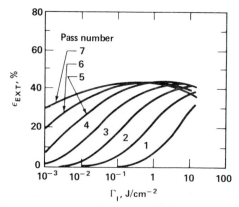

Fig. 32. Calculated extraction efficiency of Tm³⁺:glass hybrid laser in a multipass amplifier geometry as a function of input fluence, with number of amplifier passes as a parameter.

this sort can be of order 10 kJ. The pump and output beam fluences, summarized in table 12, are tractable values in terms of optical damage limits and the nominal output fluence from an XeF pump laser. Assuming an XeF pump laser efficiency of 5 to 10%, we estimate an overall hybrid laser system efficiency of 0.7 to 2%, including flow cooling, which is treated below.

For reasons of simplicity, mechanical stability and minimum optical losses, it is desirable to divide the gain medium into the fewest number of slabs. However, for given specific thermal deposition and cooling rates, a thicker slab will suffer greater temperature rise and surface stress than a thin slab. Some simple estimates can be made for the mechanical power consumption required to cool the laser for a typical slab thickness t. In the first approximation, we assume the slab to be thermally loaded uniformly with a specific thermal power $\mathcal{E}_T \nu_L [\text{W/cm}^3]$, where $\mathcal{E}_T$ is the specific thermal energy deposited in the glass and ν_L is the pulse repetition rate. The combination of uniform thermal loading and surface cooling will set up a parabolic thermal gradient normal to the slab

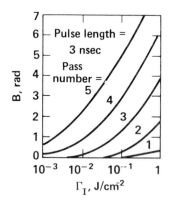

Fig. 33. Calculated 'B-integral' for Tm³⁺:glass hybrid laser in a multipass amplifier geometry, as a function of input fluence, with number of amplifier passes as a parameter.

Table 12
Characteristic parameters for a 10 kJ
Tm:glass amplifier.

Property	Value
Slab width w[cm]	40
Slab thickness t[cm]	2
Number of slabs	10
Gain length L[cm]	20
Pump pulse width t_p[μs]	0.5
Pump fluence Γ_p[J cm^{-2}]	6
Pump intensity I_p[MW cm^{-2}]	12
Excited state density N^*[cm^{-3}]	2×10^{18}
Stored energy density [J/ℓ]	800
Gain coefficient α_0[cm^{-1}]	0.1
Extraction efficiency ϵ_{ext}[%]	40
Laser output fluence Γ_0[J cm^{-2}]	10
B-integral	5
Laser pulse width t_L[ns]	3
Output energy E_0[J]	8000
Total beam power P_0[TW]	3

surface with temperature T_c at the mid-plane of the slab and T_f at the slab surface. The slab will be in compression at the mid-plane and in tension at the surface (Boley and Werner 1960). We assume cooling is achieved by flowing helium at about 1 atm through the cooling channels of width s at a velocity u. If the Reynolds number of the flow is greater than 10^4 the gas flow is turbulent, and generally the boundary layer will be a fraction of the channel thickness. The Reynolds number Re is related to the gas density ρ and viscosity μ by $Re = \rho us/\mu$. If the slab is cooled symmetrically from two sides, the mid-plane temperature T_c is related to the face temperature T_f and the cooling gas temperature T_g by the expressions

$$\Delta T_c \equiv T_c - T_f = \mathscr{E}_T \nu_L t^2/8\kappa \tag{80}$$

and

$$T_c = T_g + (\mathscr{E}_T \nu_L t^2/8\kappa)[1 + 4\kappa/ht], \tag{81}$$

where κ is the thermal conductivity (Thornton et al. 1969) of the laser medium (~ 0.01 W/cm °C for glass) and h is the heat transfer coefficient (Bird et al. 1960). The latter parameter is related to the thermal conductivity of the helium cooling gas κ(He) $\sim 1.44 \times 10^{-3}$ W/cm °C, the Nusselt number Nu and the channel thickness s by the expression

$$h = \kappa(He)Nu/s. \tag{82}$$

In the range of interest to us, it can be shown (Bird et al. 1960) that $Nu \simeq 1.5 \times 10^{-2}(Re)^{0.8}$, so that eq. (81) becomes

$$T_c = T_g + 14.5 \mathscr{E}_T \nu_L t^2[1 + 0.168(s)^{0.2}/t(M)^{0.8}], \tag{83}$$

where M is the cooling gas Mach number.

For the Tm:glass amplifier considered above, and for a 1 Hz pulse repetition rate, we have from eq. (80), $\Delta T_c = 1.9t^2$ for t in cm and T_c in °C. For $t = 2$ cm (e.g., 10 slabs), $\Delta T_c = 7.5$°C. This temperature difference will give rise to a surface stress (tension) of about 600 lb/in² (Boley and Weiner 1960). Note that this stress is isoplanar in the central portion of the slab and will cause no depolarization of a linearly polarized extraction pulse properly incident to the slab at the Brewster angle. Of course, edge effects, not considered here, will show up within a distance $2t$ from the edge of the slab. But because of spatial roll-off of the extraction beam profile in this region, edge effects will be of greatly reduced importance. For the slab dimensions assumed in the example, and assuming a flow velocity of 100 m/s (0.1 Mach number) and a channel width of $s = 1$ cm, eq. (83) shows that $T_c = T_g + 13.3$°C. It is also easy to calculate that the temperature rise of the cooling gas is only a few degrees in this case. The selection of channel thickness and flow velocity will largely be determined by balancing the rise in T_c against the rise in the power to drive the flow. Preliminary estimates based on reasonable assumptions of ducting losses, heat exchanger losses and pump efficiency show that mechanical cooling power will be a few tenths of a watt per square centimeter of slab surface for a flow velocity of order 100 m/s. These values are small compared to the electrical power per unit area of glass required of the XeF pump laser, so that it appears that flow cooling will not dominate the overall systems efficiency.

However, it is clear from this simple analysis that materials with higher thermal conductivity and lower specific thermal loading would be advantageous. Such materials would allow for a fewer number of slabs, thicker slabs and reduced flow cooling power. Efforts to identify such materials with suitable active ions and matching pumps are continuing. In any event it will be necessary to extend the simple analysis outlined here to include nonuniform thermal loading and a detailed treatment of the optical stress field.

6 Rare earth molecular gas lasers

6.1 Introduction

The rare earth 'hybrid' solid-state-laser concept described in § 5 has been proposed as one means of conserving the attractive microscopic properties of RE^{3+} ions for use in an energy-storage, high-average-power laser system. The conceptual thrust there was to minimize the specific thermal loading of selected rare-earth-doped solid-state materials by using resonant, narrowband pumping with a scalable, efficient gas laser. Yet another approach for conserving the attractive microscopic properties of RE^{3+} ions is the use of molecular gases containing these ions and which are directly amenable to average power scaling through convective gas flow (Krupke 1974a, 1976b, c). By appealing to the body of data gathered on the spectral-kinetic properties of RE^{3+} ions in solids and liquids, it is possible by analogy to identify specific combinations of RE^{3+} ions and molecular ligands of interest. In the following subsections, we will discuss the physical, chemical, and spectral-kinetic properties of suitable RE molecular gases, the options for laser pumping, and conclude with some estimates of the parameters of large energy power amplifiers.

6.2 RE molecular gas laser media

The physical, chemical and electronic properties of RE molecular gases suitable for use in the fusion application can be derived on heuristic grounds (Krupke 1976a). These molecular gases must be reasonably volatile (10 Torr at $T_g <$ 1000°C), thermally stable at the working temperature T_g, provide a RE^{3+} ion site lacking inversion symmetry, be free of molecular electronic transitions in the visible and near-ultraviolet spectral regions, and possess low energy fundamental vibrational modes. At least three RE molecular gas species appear to meet these criteria: (1) RE trihalides, (2) RE-Group III halide complexes and (3) RE-thd chelates.

The thermodynamic and spectral absorption properties of several RE trihalides have been studied (Gruen and DeKock 1966, et al. 1967, Gruen 1971). These molecules exhibit exceptionally low energy vibrational fundamentals (Wells et al. 1977), are transparent to about $28\,000\,cm^{-1}$ and occur in a planar geometry with D_{3h} point group symmetry at the RE^{3+} ion. Although vapor pressures of these molecules are marginal (1 Torr, $T_g \sim 1000$°C), they appear to be nearly ideal in all other properties.

RE molecules with considerably higher volatility often can be obtained when a RE trihalide salt (REX_3) is heated in the presence of an atmosphere or more of volatile Al_2Cl_6 (Øye and Gruen 1969). The latter studies indicate the formation of RE complex molecules according to the reactions

$$NdCl_3(s) + 1.5Al_2Cl_6(g) \leftrightarrow NdAl_3Cl_{12}(g) \tag{84}$$

and

$$NdCl_3(s) + 2Al_2Cl_6(g) \leftrightarrow NdAl_4Cl_{15}(g), \tag{85}$$

which are assumed to be in simultaneous equilibrium. Here 's' and 'g' refer to solid and gaseous phases. Note that the partial pressure of the Al_2Cl_6 complexing agent will be typically 2 to 4 atm at RE molecular number densities of interest and that the dynamical and technological impact of this agent must be considered in laser design. The spectral studies of Øye and Gruen (1969) indicate that the RE molecular complexes formed are free of molecular electronic transitions below $43\,000\,cm^{-1}$ and that the RE^{3+} ion is situated in a nonsymmetric site. These vapor complexes exhibit fundamental vibrational mode energies only up to $\sim 600\,cm^{-1}$, a value estimated to be sufficiently low for our purposes. While nearly ideal in most respects, the vapor complexes suffer from the relatively high pressure of the Al_2Cl_6 complexing agent, the sensitive thermodynamic equilibrium between solid and vapor components, and from a high chemical activity –

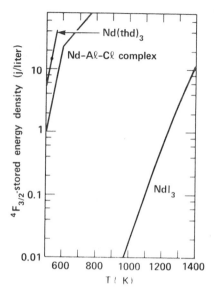

Fig. 34. Energy density stored in the $^4F_3$ J-level of neodymium-bearing vapors as a function of equilibrium temperature. Half of the vapor phase molecules are assumed excited.

all factors which place significant constraints on the physical design of a large laser amplifier.

The RE-thd chelate molecules (2,2,6,6-tetramethyl-3,5-heptanedione) are noted for their relatively high volatility (20 Torr at $T_g \sim 250°C$) and their high thermal stability (Sicre et al. 1969). However, these molecules do exhibit molecular electronic transitions near $28\,000\,cm^{-1}$ (Gruen et al. 1967) and possess fundamental vibrational mode energies up to $1500\,cm^{-1}$. Because of the latter properties, we expect that only a very few RE^{3+} ions possess electronic level structures suitable for laser action in this ligand environment. Nonetheless, the RE-thd chelates are of great interest because of their single-component nature, their chemical passivity and their relatively high vapor pressure.

The volatilities of the three RE molecular species discussed above are illustrated in fig. 34 in the context of a 1060 nm transition Nd^{3+} laser. Under the assumption that one-half of the neodymium-bearing molecules per unit volume are excited to the $^4F_{3/2}$ upper laser level, we can plot the stored energy density in joules per liter as a function of the laser medium temperature. Note that both the metal-complex and thd-chelate vapors can provide stored energy densities in excess of $20\,J/\ell$ for temperatures below 300°C. During the remainder of the discussion here, we will focus on the latter two molecular systems.

6.3 Rare earth molecular kinetics

For reasons developed below, the neodymium (Nd^{3+}) and terbium (Tb^{3+}) ions are most likely of all the RE ions to be useful in molecular gas lasers. The relevant

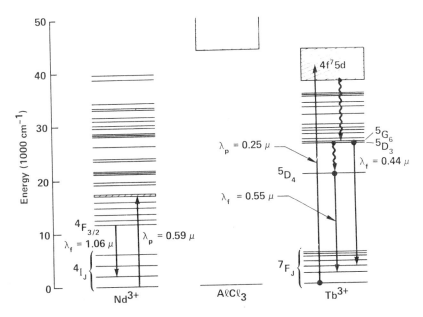

Fig. 35. Energy level diagrams for Nd^{3+} and Tb^{3+} ions in relation to the lowest lying electronic molecular level of Al_2Cl_6 vapor. Principle pump and fluorescence transitions for Nd^{3+} and Tb^{3+} are indicated.

electronic levels and transitions of these two ions are shown in fig. 35. In analogy with both liquid and solid-state neodymium lasers, we expect the gas phase Nd^{3+} laser transition to occur between $^4F_{3/2}$ and $^4I_{11/2}$ J-manifolds at a wavelength of 1060 nm. In analogy with liquid and solid-state terbium lasers, we expect the gas phase Tb^{3+} laser transition to occur between $^5D_4$ and $^7F_5$ J-manifolds at a wavelength of 545 nm. Some possibility also exists, dependent on the molecular ligand selected, for terbium laser action to occur between $^5D_3$ and $^7F_4$ J-manifolds at a wavelength of 435 nm.

To help assess the suitability of these ions for use in a large-scale, energy-storage laser amplifier, the radiative properties of the relevant electronic levels (radiative lifetime, fluorescence branching ratios, stimulated emission cross section, etc.) need to be estimated. These quantities can be calculated within the framework of the Judd (1962) – Ofelt (1962) model of 'forced' electric dipole transitions demonstrated to be quantitatively accurate in rare-earth-doped crystals (Krupke 1966, Weber et al. 1973), glasses (Krupke 1974b) and liquids (Carnall et al. 1965). According to the Judd–Ofelt model, the expectation value of the square of the electric dipole operator S between initial level $\langle(S, L)J|$ and final level $|(S', L')J'\rangle$ is given by the sum

$$S = \sum_{t=2,4,6} \Omega_t |\langle(SL)J\|u^{(t)}\|(S'L')J'\rangle|^2, \tag{86}$$

where the quantities $|\langle\|u^{(t)}\|\rangle|^2$ are doubly-reduced unit tensor matrices which are independent of the environment and depend only on the quantum numbers of initial and final states of a given RE^{3+} ion. Computer codes have been written by Caird (1973), by Crosswhite (1976), and by Rajnak (1976) to calculate these matrices between all pairs of levels of all the RE^{3+} ions. The three parameters $\Omega_2, \Omega_4, \Omega_6$ are independent of the quantum numbers of the initial and final levels of a given RE^{3+} ion, depend on the ligand environment and may be regarded as phenomenological parameters characterizing *all* radiative transition probabilities (f-numbers) for a given ion in a given ligand environment. The Ω values for a specific RE vapor are usually determined by measuring the absolute absorbances of several transitions originating on the ground state and adjusting the Ω parameters in magnitude to reproduce the measured intensities. Once the three Ω parameters have been determined in this manner, the Einstein transition probability A between any two excited states may be directly calculated.

Estimates of the intensity parameters for neodymium and terbium vapors are given in table 13, together with Ω values found for neodymium laser glass (Krupke 1974b). The neodymium vapor parameters were derived from the published absorption spectra and are probably good to a factor of two; accuracy is limited by an uncertainty in vapor number density and the limited number of transitions available. For the terbium vapors Ω values cannot be determined from absorption spectra since there is almost a total lack of useful transitions. Parameters for TbI_3 and Tb-thd were estimated by extrapolation of parameters for other rare earth ions (Gruen et al. 1967) and parameters for Tb–Al–Cl were inferred from fluorescence lifetime measurements (Hessler et al. 1977a). It is interesting to note that the Ω_4 and Ω_6 parameters for all systems are similar (to within a factor of three), whereas the Ω_2 values for the vapors can be as much as a hundred times greater than for solids. The large variability of Ω_2 parameters among species is a well-known effect expressed through transitions for which $\Delta J = 2$, the so-called hypersensitive transitions. In terms of laser applications, 'hypersensitive' transitions can provide relatively strong stimulated emission cross sections and pump transition cross sections ($\sim 10^{-19} \text{ cm}^2$) which will be needed to offset the relatively low molecular number densities ($\sim 10^{18} \text{ cm}^{-3}$) dictated by vapor volatilities.

Table 13

Estimated Judd–Ofelt intensity parameters for several neodymium and terbium-bearing vapors.[a]

Species	Ω_2	Ω_4	Ω_6	References
Nd–thd	$1.5E-18$	$9E-20$	$9E-20$	Gruen et al. (1967)
NdI_3	$2.7E-18$	$9E-20$	$9E-20$	Gruen and DeKock (1966)
Nd–Al–Cl	$1.8E-19$	$4.8E-20$	$3.9E-20$	Øye and Gruen (1969)
Nd:ED-2	$3.3E-20$	$4.7E-20$	$5.2E-20$	Krupke (1974b)
TbI_3	$1.5E-18$	$5E-20$	$5E-20$	Gruen and DeKock (1966)
Tb–Al–Cl	$2.8E-18$	$5E-20$	$5E-20$	Hessler et al. (1977a,b)
Tb–thd	$2.5E-18$	$5E-20$	$5E-20$	Gruen et al. (1967)

[a] Ω in units of cm^2.

Table 14
Calculated radiative transition probabilities and *S*-values
for the neodymium ion.

Species	$A(^4F_{3/2} \rightarrow ^4I_{11/2})$ $[s^{-1}]$	$\tau_{rad}(^4F_{3/2})$ $[\mu s]$	$S(10^{-20})$ $[cm^2]$
Nd-thd$_3$	740	650	4.9
NdI$_3$	740	650	4.9
Nd–Al–Cl	340	1350	2.3
Nd:ED-2	1380	370	2.8

The calculated radiative transition probabilities for the neodymium vapors and Nd : glass are listed in table 14. The calculated radiative lifetime of the Nd–Al–Cl complex is 1.35 ms and that of the chelate is 650 μs. The *S*-values for the $^4F_{3/2} \rightarrow ^4I_{11/2}$ transition in Nd–Al–Cl vapor and Nd : glass are essentially identical; since the fluorescence linewidths in neodymium vapors and Nd : glass are nominally the same (~ 350 cm^{-1}), the vapors should exhibit stimulated emission cross sections of 2 to 4×10^{-20} cm^2. Probabilities for the $^5D_4 \rightarrow ^7F_5$ and $^5D_3 \rightarrow ^7F_4$ transitions of Tb^{3+} and metastable level lifetimes are given in table 15. Radiative lifetimes of several hundred microseconds are predicted with stimulated emission cross sections likely to lie in the range 3 to 5×10^{-20} cm^2.

The projected radiative properties of the neodymium and terbium vapors are quite attractive; these vapors may serve as energy-storage gain media provided nonradiative quenching processes do not dominate excited state kinetics. The possible nonradiative quenching channels for Nd^{3+} and Tb^{3+} ions in the complex and chelate vapors have been considered by Krupke 1976a; these include: (1) excitation of molecular electronic levels which are relaxed nonradiatively, (2) direct excitation of multiple quanta of molecular vibration and (3) RE–RE bimolecular collisions involving near-resonant $4f \rightarrow 4f$ electronic transitions.

Table 15
Calculated radiative transition probabilities and *S*-values
for the terbium ion.

Species	$A(^5D_4 \rightarrow ^7F_5)$ $[s^{-1}]$	$\tau_{rad}(^5D_4)$ $[\mu s]$	$S(^5D_4 \rightarrow ^7F_5)$ $[10^{-20}$ cm$^2]$
TbI$_3$	1130	690	2.2
Tb–Al–Cl	1700	460	3.4
Tb-thd	1500	500	3.1

Species	$A(^5D_3 \rightarrow ^7F_4)$ $[s^{-1}]$	$\tau_{rad}(^5D_3)$ $[\mu s]$	$S(^5D_3 \rightarrow ^7F_4)$ $[10^{-20}$ cm$^2]$
TbI$_3$	1290	440	1.0
Tb–Al–Cl	1900	310	1.5
Tb-thd	1700	340	1.4

Some specific examples of these processes for the neodymium and terbium ions are as follows:

Multi-quantum Vibrational Relaxation

$$Nd(^4F_{3/2}, n) \rightarrow Nd(^4I_{15/2}, m),$$ (87)

$$Tb(^5D_3, n) \rightarrow Tb(^5D_4, m),$$ (88)

where n, m, indicate the number of highest energy vibrational quanta excited in the molecule. The values of n and m are related by the expression

$$(n - m)\hbar\omega_{max} \simeq \Delta E,$$ (89)

where $\hbar\omega_{max}$ is the energy of the highest energy vibrational mode of the molecule and ΔE is the energy difference between the initial and final rare earth 4f electronic states. In analogy with solid-state (Reisberg and Moos 1968) and liquid phase (Haas and Stein 1972) studies of multiphonon relaxation of rare earth ions, we would expect the rate for this type of process to vary as $\exp(-\Delta E)$.

RE–RE Bimolecular Collisions

$$Nd(^4F_{3/2}) + Nd(^4I_{9/2}) \rightarrow 2Nd(^4I_{15/2}) + \bar{Q},$$ (90)

$$Tb(^5D_3) + Tb(^7F_6) \rightarrow Tb(^5D_4) + Tb(^7F_0) + \bar{Q},$$ (91)

where $\bar{Q}$ is the energy defect of 4f electronic energies involved.

Because the complex molecules possess no molecular electronic levels below $\sim 43\,000\,cm^{-1}$ and have relatively low energy vibrational modes, it is expected that only the RE–RE bimolecular collision quenching channel is likely to be of importance for the $Nd(^4F_{3/2})$, $Tb(^5D_4)$ and $Tb(^5D_3)$ electronic levels in the vapor complex. In contrast, the chelate molecules possess molecular electronic levels near $28\,000\,cm^{-1}$ and vibrational energies up to $\sim 1500\,cm^{-1}$, so that one or more of the nonradiative quenching channels might be anticipated in the chelate vapors.

6.4 Recent experimental results

The kinetics of Nd–Al–Cl vapor have been studied experimentally by Jacobs et al. (1977) and Jacobs and Krupke (1977a). Fluorescence emission originating on the $^4F_{3/2}$ level was excited using 532 nm radiation from a Chromatix doubled Nd:YAG laser. Excitation occurred on a 100 ns time scale and the $^4F_{3/2}$ fluorescence was prompt, indicating that the relaxation from higher lying levels to the $^4F_{3/2}$ level is fast, as expected. To obtain kinetic data under conditions of high excited state density, Jacobs and Krupke (1977a) used pulse dye laser excitation of the prominent Nd–Al–Cl 4f → 4f absorption band at 587 nm.

The important regimes of $^4F_{3/2}$ population kinetics are illustrated in fig. 36 by the 1.06 μm fluorescence decay curves recorded under differing conditions of ground state Nd–Al–Cl molecular density N_0 and $^4F_{3/2}$ excited state density N^*.

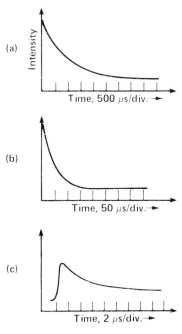

Fig. 36. Decay of fluorescence emission from the $^4F_{3/2}$ J-level of Nd–Al–Cl vapor complex under various conditions of temperature (vapor number density) and excited state number density N^*: (a) $T = 483$ K. $N_0 \lesssim 3 \times 10^{16}$ cm^{-3}, $N^* \sim 0$, $\tau_f \sim \tau_{rad} \sim 800$ μs; (b) $T \sim 800$ K, $N_0 = 10^{18}$ cm^{-3}, $N^* \sim 0$; $\tau_f \sim 46$ μs; (c) $T \sim 800$ K, $N_0 = 10^{18}$ cm^{-3}, $N^* \sim 5 \times 10^{17}$ cm^{-3}, $\tau_f \sim 4$ μs.

A fluorescence lifetime of ~ 800 μs was observed for low N_0 and N^* values (fig. 36a). This lifetime is quite comparable to the radiative lifetime of 1350 μs estimated using the Judd–Ofelt model. As N_0 is increased by raising the temperature, the fluorescence is increasingly quenched: for $N_0 \sim 10^{18}$/cm^3 and a low pumping rate (fig. 36b), the fluorescence lifetime decreased to ~ 46 μs. This quenching is attributed to a near-resonant, multipolar energy exchange process between excited and ground state Nd molecules [eq. (90)]. The effective rate coefficient for this process is found to be $k_g = 5 \times 10^{-14}$ cm^3/s. This relatively low rate, corresponding to a transfer probability per collision $\sim 10^{-4}$, results from the weak forced electric dipole moments of the participating transitions. When N^* is increased by strong pumping to be comparable to N_0 ($N^* \sim 5 \times 10^{17}$/cm^3), the fluorescence decay time is reduced by an order of magnitude to several microseconds (fig. 36c). The energies of Nd^{3+} electronic levels are spaced such that a large number of near-resonant energy-exchange channels are available in the collision of two molecules excited to the $^4F_{3/2}$ J-level. The transition strengths involved in these collisions of this type are 50 times larger than those involved in ground state quenching; the excited state–excited state quenching coefficient $k_q = 5 \times 10^{-13}$ cm^3/s is therefore in qualitative agreement with the simple scaling of transition strengths. In as much as we primarily envision large-scale RE vapor amplifiers being pumped with excimer lasers which are energetic and efficient for output pulse widths of order 1 to 2 μs, a fluorescence lifetime of several microseconds is acceptably long. It is significant to note also that the peak fluorescence intensity in these experiments was a linear function of pump intensity, indicating that deleterious multiphoton absorption or dissociation processes are of little importance.

Under strong pumping conditions, optical amplification of a 1.06 μm probe beam has been observed (Jacobs and Krupke 1977b) with a time signature of the $^4F_{3/2}$ level fluorescence. Under pump conditions for which half of the Nd–Al–Cl molecules were excited ($2 \times 10^{17}/cm^3$) a single-pass gain of 4 to 5% was measured in a 18 cm long cell, corresponding to an average small-signal gain coefficient ~ 0.25 cm^{-1}. Using estimates for the transition cross section ($\sim 2 \times 10^{-20}$ cm^2) and N^* ($\sim 2 \times 10^{17}$ cm^{-3}), the predicted small-signal gain coefficient $\sim 0.4\%$ cm^{-1} is in reasonable agreement with the observed value. This excited state density corresponds to a stored energy density ~ 35 J/ℓ, a value which compares favorably with both CO_2 and iodine short pulse laser amplifier media.

The fluorescence kinetics of the $^5D_4$ and $^5D_3$ J-levels of Tb–Al–Cl complex molecules have been studied by Hessler et al. (1977a, b) under conditions of low excited state densities. Using a xenon flashlamp, Tb molecules were excited via 4f → 5d allowed transitions lying near 42 000 cm^{-1} (see fig. 35), followed by vibrational relaxation to the $^5D_3$ and $^5D_4$ J-manifolds. The fluorescence spectrum corresponding to the $^5D_4 \rightarrow ^7F_J$ transitions (Hessler et al. 1977a) is shown in fig. 37. The strong 545 nm transition is quite prominent, as suggested by Judd–Ofelt calculations. At low temperature ($T < 200°C$) the $^5D_4$ fluorescence decay was exponential with a lifetime of 2.3 ms. As the temperature was raised a double exponential decay signature was observed with lifetimes of 2.3 and 0.46 ms. This

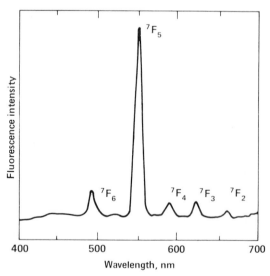

Fig. 37. Visible fluorescence of Tb–Al–Cl complex showing $^5D_4 \rightarrow ^7F_J$ transition array (Hessler et al. 1977a).

behavior was attributed to the existence of Tb molecules with two different molecular structures whose equilibrium density shifted with temperature. These fluorescence decay times were taken as the radiative lifetimes, because none of the nonradiative quenching channels cited earlier are expected to be competitive (Hessler et al. 1977a).

Under certain conditions of Tb–Al–Cl molecular ground state density and temperature, $^5D_3 \rightarrow {}^7F_J$ fluorescence transitions were also observed (Hessler et al. 1977b) with prominent appearance of the 435 nm band. Measurements of the $^5D_3$ fluorescence decay and the subsequent rise of $^5D_4$ fluorescence indicated the presence of the RE–RE bimolecular quenching channel described by eq. (91). In related experiments, Krupke and Jacobs (1977) studied the fluorescence kinetics of $^5D_4$ and $^5D_3$ J-manifolds under conditions of high excited state densities appropriate for laser amplification. In these experiments, an intense ($5 \, MW/cm^2$) KrF laser operating at 249 nm was used to excite Tb molecules via $4f \rightarrow 5d$ transitions. The fluorescence kinetics of $^5D_3$ and $^5D_4$ J-levels was found to be a complex function of excitation indensity (excited state density). The kinetics data are currently being analyzed; at this time it is concluded that excited-state–excited-state collisions are significant, perhaps involving $4f \rightarrow 5d$ transitions. In any event, the exceptionally long fluorescence lifetimes ($\sim 100 \, \mu s$) at high excited state densities ($N^* > 10^{16} \, cm^{-3}$) attest to the unique collisional properties of the Tb–Al–Cl molecular vapor and support further consideration of this molecular species for laser fusion amplifiers. Experiments to observe optical amplification in the Tb–Al–Cl vapor system are in progress.

Let us now turn to the kinetic behavior of Nd-thd and Tb-thd chelate. The first fluorescence kinetics measurements on the Tb-thd chelate were performed by Jacobs et al. (1975). In these experiments the $^5D_4$ fluorescence was excited by pumping the molecular ligands near 340 nm followed by ligand to RE ion transfer, and fluorescence decay was measured as functions of temperature and molecular number density N_0 under conditions of low excited state density. The $^5D_4$ lifetime was found to be independent of molecular number density and a strong decreasing function of increasing temperature ($t_L \sim 1 \, \mu s$; $T_g \sim 200°C$). The observed temperature dependence of decay rate was well characterized by an Arrhenius expression, indicating energy transfer to the lowest lying triplet level of the chelate, followed by rapid relaxation to the molecular electronic ground state. No fluorescence was observed from the $^5D_3$, indicating either no excitation of this level or very rapid vibrational relaxation to lower lying Tb levels.

More recently, Jacobs (1977) has measured Tb-thd fluorescence under conditions of intense pumping ($> 10 \, MW/cm^2$) using a KrF laser operating at 249 nm. Again, some evidence of $Tb(^5D_4) + Tb(^5D_4)$ quenching was found as expressed by a shortening of the fluorescence lifetime from $\sim 1 \, \mu s$ to a few hundred nanoseconds for estimated excited state densities $N^* \sim 10^{16} \, cm^{-3}$. Uncertainties in excited state density are due to lack of experimental information on the quantum efficiency of ligand to $^5D_4$ level transfer.

In the case of Nd-thd fluorescence kinetics (Krupke and Jacobs 1977), excitation of the $F_{3/2}$ J-level was accomplished by direct pumping of a $4f \rightarrow 4f$

transition at 587 nm with a powerful dye laser (~ 10 MW/cm^2; 1 μs). The fluorescence lifetime was measured to be $\sim 0.5 \, \mu$s and was found to be independent of temperature and molecular number density N_0. This lifetime is much less than the estimated radiative lifetime of 1 ms; fluorescence quenching is attributed to excitation of a few high energy molecular vibrations, according to eq. (87). In these experiments, it was observed that the peak fluorescence intensity varied linearly with the pump intensity up to 10 MW/cm^2, indicating the absence of significant multi- or successive photon absorption processes, and of significant molecular dissociation. Furthermore, the fluorescence lifetime was found to be independent of excited state density for $N^* \sim 10^{17}$ cm^{-3}, a population value appropriate for a gas phase power amplifier.

6.5 Laser pumping options

Four generic types of pumping of RE molecular vapor lasers have been considered (Krupke 1976a): (1) resonant pumping of the relatively weak $4f \rightarrow 4f$ absorption bands, with emphasis on 'hypersensitive' absorption transitions, (2) optical pumping of the relatively strong, electric-dipole-allowed $4f \rightarrow 5d$ transitions, (3) optical pumping of the lowest lying molecular ligand or charge transfer bands, followed by internal conversion of energy to 4f levels and (4) direct electrical excitation of 5d levels of RE ions or of molecular ligand transitions.

For optical pumping methods to be efficient, pump sources must be energetically scalable, intrinsically efficient and spectrally matched to RE molecular pump transitions. To identify specific RE–pump combinations for experimental study, the efficient rare gas halogen lasers operating at discrete wavelengths in the ultraviolet and near-visible spectral regions were considered as likely pump sources. Particular attention was given to the more efficient and scalable of these nonenergy storing lasers: KrF – 249 nm, ArF – 193 nm and XeF – 351 nm. It is believed that these lasers will be able to produce tens of kilojoule output pulses on a microsecond time scale with overall efficiencies approaching 10%.

In the case of Nd–Al–Cl vapor only $4f^3$ electronic levels of the Nd^{3+} ion will be spectrally accessible. Pumping must therefore proceed through $4f \rightarrow 4f$ transitions such as the hypersensitive transition at a wavelength of $\lambda_p = 587$ nm and with a cross section of 7×10^{-20} cm^2. No existing excimer laser operates directly at this wavelength and one must appeal in this case to wavelength shifting techniques of scalable excimer sources. One possibility here is to take advantage of efficient (80%) conversion of XeF laser radiation at 353 nm to 582 nm radiation of electronic Raman scattering in varium vapor, as demonstrated by Djeu and Burnham (1977).

In the case of Tb–Al–Cl vapor, strong metastable excitation can be obtained by KrF laser pumping of the $4f \rightarrow 5d$ transitions near 40 000 cm^{-1}. Although the 5d orbital may be somewhat delocalized from the Tb^{3+} center, it is speculated that excitation will convert internally into the 4f shell sufficiently rapidly, so that the overall pump conversion efficiency will be high. Experiments are currently in progress to establish this conversion efficiency.

At the present time, very little is known about the electronic structure of RE–Al–Cl vapors. It is possible that energies of molecular transitions (ligand excitation and charge transfer excitation) may shift below 45 000 cm^{-1} for different rare earth ions. The additional possibility then arises of pumping the molecular transitions followed by energy transfer to the RE ion; Carnall et al. (1978) have observed some evidence of a red shift in the uv absorption edge of Er–Al–Cl vapor relative to the uv edge of Al$_2$Cl$_6$ vapor which may represent a charge-transfer band of the Er–Al–Cl molecule.

In the case of the thd-chelates, weak absorption due to molecular singlet transitions begins near 25 000 cm^{-1}, followed by intense absorption at 35 000 cm^{-1}. The molecular excitations obscure high lying 4f and 5d levels of the RE^{3+} ions. As observed in liquid phase RE chelates, molecular excitation can lead to efficient internal conversion to the 4f shell in some cases (Lempicki et al. 1965). In this eventuality for vapor chelates, both XeF(353 nm) and KrF(249 nm) represent excellent pump sources. Copious fluorescence emission has been observed from Tb- and Nd-thd chelates under excitation by these excimer pumps, but quantitative photon yields have not yet been measured.

The possibility of electrically pumping a RE molecular vapor laser is an intriguing one. Because of the paucity of information on cross sections for relevant electron scattering processes in the vapors of interest here, only very general comments can be offered at present. First, it is unlikely that a stable, self-sustained discharge can be maintained in any of the vapor systems at the required pressures and for the relatively long times commensurate with inversion storage lifetimes. This is particularly true for the metal complexes where Al$_2$Cl$_6$ working pressure will be several atmospheres. Rather, it is much more likely that electrical excitation would have to proceed through ionization channels using relativistic electron beams. In the case of the metal complexes, the high ratio of Al$_2$Cl$_6$ pressure to RE species pressure will likely render e-beam pumping inefficient. In the case of single-component vapors, such as RE halogens or RE chelates, laser mixtures would presumably utilize appreciable pressures (> 1 atm) of a rare gas whose ionization would dominate electron energy deposition. One would have to rely on rare gas excimer formation followed by collisional transfer of energy to RE molecular species before excimer emission. Of course, collisional ionization and/or dissociation of RE molecules would also dissipate energy. It has been shown that complex organic molecules such as POPOP vapor dye molecules can be excited to emit characteristic fluorescence when mixed with xenon buffer gas and excited with an e-beam (Cordray et al. 1977). Fragmentation of fluorescing molecules did not appear to be a major electron energy dissipation channel. While these observations are encouraging, experiments would have to be performed on the specific species of interest in order to assess the potential for direct electrical pumping of RE molecular vapors.

6.6 Multi-kilojoule Tb-thd fusion amplifier

A laser amplifier using RE molecular gases is most easily visualized in terms of the Tb-thd chelate vapor because of its single-component chemical nature, its chemical passivity and its relatively low operating temperature ($\sim 250°C$). The laser medium would likely consist of a Tb-thd chelate vapor at a number density of 2×10^{17} (225°C) buffered with helium gas at, say, an atmosphere. For a device with ~ 50 kJ of energy available at 545 nm, the medium is envisaged to flow through a volume measuring 400 cm long, 100 cm high and 30 cm wide. This volume would be pumped optically with KrF lasers through the large-area opposing quartz side windows. The optical extraction pulse would propagate through the medium several times, nominally along the 400 cm dimension and perpendicular to the orthogonal pump and medium flow directions. The microscopic parameters of the RE-thd laser medium are summarized in table 16 and a set of nominal values for a large-scale power amplifier is given in table 17. Note that the laser medium is optically thick at the KrF pump wavelength but exhibits a bleaching fluence of only 0.08 J cm^{-2}. Under the pump conditions chosen for this example, the medium will be fully bleached (but not dissociated) in about 0.2 μs, a time less than the population inversion storage time of 0.4 μs. This example shows that for a maximum gain–length product $\alpha_0 L = 4$, 100 kJ of energy can be stored in the $^5D_4$ level for times of order 0.4 μs, and that consistent with level degeneracies, approximately half of this energy (50 kJ) can be extracted. Moderation of the medium temperature rise after excitation can be achieved with increased helium buffer but with some degradation of output beam quality. This attractive device performance is predicated on our assumption of a near-unity quantum conversion of pump photons to Tb molecules in the $^5D_4$ level; the effacacy of this laser medium depends critically on the validation of a near-unity efficiency in the small-scale experiments now in progress. Based on an assumed KrF pump laser efficiency of 6 to 12%, we estimate an overall rare

Table 16

Microscopic parameters for a Tb-thd chelate laser amplifier.

Parameter	Value[a]
Pump wavelength	249 nm(KrF)
Pump cross section	1×10^{-17} cm^2
Pump bleaching fluence	0.08 J cm^{-2}
Laser wavelength	545 nm($^5D_4 \rightarrow {^7F_5}$)
Laser cross section	(5×10^{-20} cm^2)
Laser saturation fluence	5 J cm^{-2}
Quantum defect	0.46
Internal conv. eff.	(1)

[a] Values in parentheses are estimated and are subject to confirmation by direct measurement.

Table 17
Device parameters for a 50 kJ Tb-thd chelate
laser amplifier.

Parameter	Value
Tb-thd₃ density	2×10^{17} cm^{-3}
Medium temp.	225°C
Inversion density	2×10^{17} cm^{-3}
Inversion lifetime	$0.4\,\mu$s
Stored energy density	$80\,J/\ell$
Available energy density	$40\,J/\ell$
Small-signal gain coeff.	1 m^{-1}
Max. gain–length product	⩽4
Maximum dimension	= 400 cm
Buffer pressure	~1 atm(He)
Available energy/ℓ atm	$40\,J\,\ell^{-1}\,atm^{-1}$
Total stored energy	100 kJ
Total available energy	50 kJ
Pump pulse width	$0.2\,\mu$s
Pump fluence	2.5 J/cm^2
Pump flux	12 MW/cm^2
Bleaching velocity	7.5×10^7 cm/s
Bleaching time	$0.2\,\mu$s
Height	100 cm
Width	30 cm

earth vapor laser system efficiency of 1 to 2%, taking into account pump transport, fill factors, extraction and flow cooling inefficiencies.

7 Raman pulse compressors

7.1 Introduction

Pulse compression by backward stimulated Raman scattering is not a new concept. It was first proposed by Maier et al. in 1966 and has since been reconsidered for applications several times. One might reasonably begin the present discussion with direct quote from the conclusions of a paper by Glass (1967): 'In the design of Raman lasers, beam trapping dictates the use of a gaseous medium for high power densities. The most flexible configuration consists of a mode-controlled oscillator followed by an amplifier which brings the pulse up to the saturation level and then by a backward-traveling amplifier to shorten and intensify the output pulse. Intensities of the order of 10^{10} W/cm^{-2} seem possible ...'. This very well describes the systems considered here. Other studies include experiments on and computer modeling of gas phase Raman oscillators and amplifiers by Kachen (1975) and by Kachen and Lowdermilk (1977); theoretical work (Carman 1975) and experiments (Lowdermilk and

Kachen, 1976) on two-photon extraction of energy stored in excited states of laser media.

Fig. 38 illustrates the principle involved in pulse compression using a backward wave Raman amplifier. A long laser pulse which we will call the pump pulse propagates from right to left into a Raman-active interaction zone, producing gain at a frequency given by $\nu_p - \nu_R = \nu_S$, where ν_p is the pump frequency and ν_R is the frequency of the Raman transition which is excited. A Stokes input pulse at ν_S is injected into the interaction zone at the time when the pump pulse has just reached the exit window of the interaction zone. As the Stokes pulse propagates from left to right and the pump pulse from right to left, the Stokes pulse at ν_S is amplified at the expense of the pump pulse. After a time $\frac{1}{2}t_p$, the Stokes pulse emerges as an amplified pulse traveling in the backward direction with respect to the pump pulse. The Stokes pulse can be much shorter in time than the pump pulse, as shown in fig. 38, so the net result is to compress a long laser pulse at ν_p into a short pulse at ν_S. There is a net loss of energy since $\nu_S < \nu_p$. One can also think of the backward Raman amplifier as an energy-storage laser (§ 3.4) in which the energy is stored in the form of a collimated beam of photons at ν_p and which is then extracted with a short pulse at ν_S.

The recent availability of efficient ultraviolet laser pump sources may be the necessary development to convert the Raman pulse compressor from a good idea to a practical one (Murray and Szoke 1977). Comparing KrF and Nd:glass lasers as possible pump sources for a compressor, we note that a Raman amplifier saturating at the same energy fluence requires about a factor of 12 lower gas pressure for the KrF pump than for the Nd:glass pump. For Nd:glass,

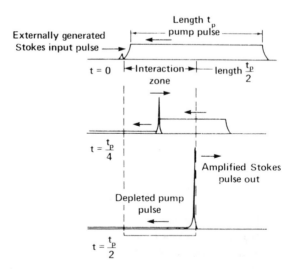

Fig. 38. Schematic representation of the principle of operation of a backward wave Raman compressor laser. A long pump pulse is propagated counter to a short Stokes pulse through a Raman active medium. As the two pulses pass through the interaction region, energy is transferred from the pump pulse to the Stokes pulse.

the required operation of the Raman amplifier at these higher pressures also introduces more competition from Brillouin scattering, self-focusing and other nonlinear effects. Short wavelengths are desirable for laser fusion drivers and a KrF pumped device will emit in the 250 to 300 nm range rather than in the 1500 to 2000 nm range for a Nd:glass pumped device. Electron-beam-pumped KrF lasers can probably be constructed with good efficiency at pulse lengths of 50 to 100 ns, so that pulse compression ratios of 50 to 10 are required to achieve 1 to 10 ns output pulses.

7.2 The backward wave Raman process

Atomic or molecular Raman scatterers can be employed in the backward wave amplifier. The selection of an optimum scatterer involves trade-offs between physics and technology issues, some of which will become evident below. Generally speaking, however, molecular scatterers are often in the gaseous state at room temperature with near-atmospheric pressures; atomic scatterers typically require high temperatures to attain even modest pressure. Because of the low densities available for atomic vapors, the choice of a particular scatterer will be based largely on the realization of a near-resonant Raman scattering of a specific pump laser. In the present work we will treat the important issues associated with molecular scatterers rather than atomic scatterers, since molecular scatterers appear at this time to be most promising.

Fig. 39 gives the energy level scheme for the Raman effect. For the molecular scatterers of interest, ν_R represents a Raman-active vibrational mode frequency

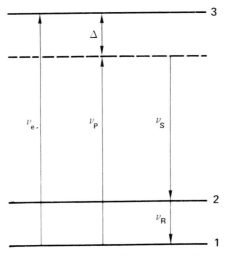

Fig. 39. Schematic diagram of the energy levels of a molecular Raman scatterer. Level 1 is the vibrationless ground electronic state; level 2 is a Raman-active vibrational level of the ground electronic state; level 3 is an excited electronic state of parity opposite to that of level 1. The energy Δ is the energy defect between the pump energy $h\nu_p$ and the energy of level 3.

ν_e, the frequency of the optically allowed transition between levels 1 and 3 and Δ the frequency defect. The gain at the first Stokes frequency ν_S for a stimulated Raman amplifier can be written in a variety of ways. For molecular scatterers where $\nu_p \ll \nu_e$ (i.e., Δ large) the gain can be written (Lallemand 1971) in terms of the differential scattering cross section $\partial\sigma/\partial\Omega$ as

$$g = (\lambda_S^2/h\nu_p)\,h(\nu - \nu_S)(\partial\sigma/\partial\Omega)I_p\,NL. \tag{92}$$

where L is the length of the Raman amplifier, N the number density of scatterers, $h(\nu)$ the normalized line shape function and I_p is the intensity of the pump radiation. In this expression the population in level 2 (i.e., the first vibrational level $V = 1$) is assumed to be negligible at all times. This gain expression may be rearranged and cast into a more conventional amplifier form:

$$\begin{aligned} g &= (\lambda_S^2 ch(\nu - \nu_S)(\partial\sigma/\partial\Omega)N)\,(I_p/h\nu_p c)L \\ &\equiv \gamma_R I_p L \\ &\equiv \sigma_R \Delta n_p L. \end{aligned} \tag{93}$$

The effective Raman stimulated-emission cross section σ_R expresses the strength of the interaction between the Stokes field and a population inversion density Δn_p which is the density of pump photons per unit volume. The quantity γ_R is the specific Raman gain. From this latter expression we see that, conceptually, the backward wave Raman amplifier is an energy-storage laser, the energy being stored in the optical pump field and the Raman media merely serving as a means for extracting this energy in a short pulse. To convert the inversion density in pump photons per cm^{-3} to more conventional units, note that a pump flux of $10^8\,\mathrm{W\,cm^{-2}}$ at 249 nm corresponds to an inversion density of $4 \times 10^{15}\,\mathrm{cm^{-3}}$ or to an energy storage of $3.3\,\mathrm{J\,\ell^{-1}}$ at 249 nm. The energy available for extraction by backward scattering is $6.6\,\mathrm{J\,\ell^{-1}}$.

Backward wave scattering has an additional property which confuses this representation of Raman gain. In fig. 38, note that at a fixed position and with a pump pulse of length t_p, one has a photon density Δn_p and gain coefficient $\sigma_R \Delta n_p$ for a time t_p. A backward Stokes pulse, however, passes through the pump pulse with a relative velocity of $2c$ and is subjected to an effective photon density of $2\Delta n_p$ for a time $\frac{1}{2}t_p$, giving the same net gain but an effective inversion density equal to twice the photon density in the pump pulse. A Stokes pulse propagating in the forward direction samples an inversion Δn_p for all times (neglecting dispersion and assuming the scattering medium is infinite) but it can only grow to an intensity corresponding to the local photon density Δn_p rather than extracting energy from the enitre length of the pump pulse. The consequence of this is that the saturation fluence Γ_S for the forward Stokes pulse is $h\nu_S/\sigma_R^f$, whereas for the backward Stokes pulse it is given by $2h\nu_S/\sigma_R^b$. Here σ_R^f and σ_R^b refer to the simulated Raman cross section for forward and backward scattering respectively. In general $\sigma_R^f > \sigma_R^b$ (see § 7.2).

We now discuss several specific systems for both molecular and atomic Raman scatterers and give their properties.

7.3 *Molecular Raman scatterers and their properties*

The effective Raman stimulated emission cross section σ_R must now be estimated for excitation by an efficient rare gas halide source such as KrF at 249 nm. Unfortunately differential scattering cross sections have not been measured at this wavelength for any scatterers, but they may be inferred from measurements at other wavelengths using any of several models which have been developed for the frequency dependence of $\partial\sigma/\partial\Omega$. The simplest of these is the model of Placzek (1959) (Behringer and Brandmuller 1956) which gives a frequency dependence of the form

$$\partial\sigma/\partial\Omega \propto \nu_s^4/(\nu_e - \nu_p)^2(\nu_e + \nu_p)^2, \tag{94}$$

where ν_e is the frequency of an electronic intermediate state through which the Raman transition is induced. More complex models can sometimes give a better fit to the data (Behringer and Brandmuller 1956, Albrecht 1961), but Placzek's model will be used here as a good first approximation.

Raman intensities for gas phase scatterers have been reduced to absolute values using several different techniques. It is customary to determine an absolute differential scattering cross section for N_2 by these procedures and to use this as a working standard to determine the cross section for other gases. Values (Fenner et al. 1973) of $\partial\sigma/\partial\Omega$ for N_2 at a 90° scattering angle range from 5.5 to 3.3×10^{-31} cm^2 sr^{-1} at 488 nm. For this analysis the low value of 3.3×10^{-31} cm^2 sr^{-1} shall be assumed. If a higher value proves correct, the pressure in the scattering cell may be reduced proportionally to produce the desired σ_R. The differential scattering cross section has an angular dependence. For the highly polarized Q-transitions of interest here, it is constant and equal to $\partial\sigma/\partial\Omega$ (90°) for any direction in the plane perpendicular to the assumed linear polarization of the pump wave, including the forward and backward scattering cases of interest here.

The gases chosen for preliminary analysis as scatterers exhibit several important special features for use in a Raman amplifier. The two molecules chosen, N_2 and CH_4, have such close Q-line spacings that the transitions of interest can be described by the envelope of the Q-branch at any pressure of interest (here ≥ 1 atm). The line shapes of these overlapped transitions are essentially independent of pressure and are isotropic in scattering angle, although at high pressures (≥ 100 atm) they can show significant narrowing. The N_2 line shape is fairly well established, while CH_4 has not been studied in great detail. Nitrogen is chosen as an inert gas which is also used as a standard for Raman intensities. Methane has a very high Raman cross section with a narrow Q-branch for the ν_1 vibration (carbon stationary with four hydrogens moving in phase with equal amplitude) and represents the class of similar molecules such as SF_6, SiH_4, CF_4, etc.

In the present work, we will evaluate a Raman amplifier using a KrF laser operating at 249 nm as a pump source. If XeF or ArF laser pumps were to be used, the relative desirability of various scatterers might change. Ammonia has

Table 18
Properties of several candidate Raman scatterers.

Property	N_2Q_{01} branch	$CH_2\nu_1Q_{01}$ branch
Shift [cm^{-1}]	2327	2914
Linewidth [cm^{-1}]	1.5	1.5 (± 0.5)
Assumed intermediate state ν_e [cm^{-1}]	100 000	$\sim$70 000
$\left.\frac{\partial\sigma}{\partial\Omega}\right\|249 \Big/ \left.\frac{\partial\sigma}{\partial\Omega}\right\|488$ (Placzek model)	24	38
$\left.\frac{\partial\sigma}{\partial\Omega}\right\|488$ *referred to N_2	3.3×10^{-31}	$*19.8 \times 10^{-31}$
$\left.\frac{\partial\sigma}{\partial\Omega}\right\|249$ *calculated*	8.1×10^{-30}	7.5×10^{-29}
ν_s [cm^{-1}]	37 834	37 247
λ_s (nm)	264	268
$\Delta\nu_p = 0$ σ_r [cm^2] Γ_s [J cm^{-2}]	2.9×10^{-20} p 26/P	2.8×10^{-19} p 0.5/P
$\Delta\nu_p = 5$ cm^{-1} σ_r [cm^2] Γ_s [J cm^{-2}]	6.7×10^{-21} p 113/p	6.4×10^{-20} 11/p

not been considered, for example, although it has a high cross section and will show significant resonance enhancement in $\partial\sigma/\partial\Omega$ for KrF and XeF pump wavelengths. The linewidth of the ν_1 Q-branch of NH_3 is not well established, and the presence of near-resonant scattering effects may also introduce undesirable nonlinearities for a KrF pump. In general, cross sections for ArF will be nominally a factor of four larger, and for XeF a factor of four smaller than those for KrF (neglecting resonance enhancement). Optical problems for ArF pumping (193 nm) will be much more severe. Table 18 summarizes the important parameters for KrF Raman scattering from nitrogen and methane.

7.4 Recent experimental results

This subsection is devoted to the KrF Raman scattering experiments in methane by Murray et al. (1978) at the Lawrence Livermore Laboratory. The purpose of their experiments was to verify the theoretical predictions for the backward wave Raman gain as a function of linewidth and pressure and in particular to measure the gain asymmetry between the forward and backward propagating Stokes pulses. It is predicted theoretically that the small signal stimulated Raman gain for forward scattering (co-propagating Stokes and pump radiation) and for backward scattering (counter-propagating Stokes and pump radiation) can differ significantly for several reasons, one of the most important of which is the effect of pump laser linewidth. The gain for backward scattering is proportional to the convolution of the laser line profile and the scatterer line profile, which for lorentzian lines is proportional to $(\Delta\nu_p + \Delta\nu_S)^{-1}$, where $\Delta\nu_p$ and $\Delta\nu_S$ are the linewidths of the laser and scatterer respectively. As discussed above, methane displays a homogeneous, quasi-lorentzian line of width $\Delta\nu_S \sim 0.3\,\mathrm{cm}^{-1}$, which is nearly independent of density at moderate pressures. If the medium has negligible dispersion, however, the gain in the forward direction becomes independent of pump laser linewidth and proportional only to $(\Delta\nu_S)^{-1}$. This enhanced gain is produced by a parametric mixing process in which different pump frequencies with their matched Stokes frequencies can couple to the same phased array of Raman scatterers in the forward direction. The high forward gain is present only for a Stokes frequency and phase distribution which matches the pump distribution exactly, such as a Stokes wave produced by superfluorescence from the pump wave.

Fig. 40 shows the experimental layout used for backward Raman gain measurements. The pump source is a KrF TEA laser operated in an unstable resonator cavity and injection-locked by a separate etalon-controlled KrF oscillator which determines its output linewidth and spatial quality. The 15 ns long, 70 mJ output pulse from this device is split into two equal parts, one of which is focused softly by a beam-contracting telescope through a 2m long Raman oscillator cell containing methane gas at 10 to 30 atm pressure. The equivalent focal length of this lens arrangement is about 10 m. The pump at 249 nm induces

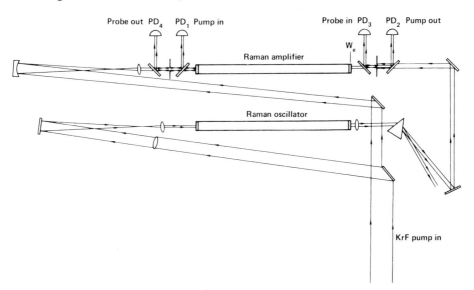

Fig. 40. Experimental arrangement for measuring the backward wave Stokes gain for Raman scattering of KrF radiation from methane (Murray et al. 1978).

superfluorescent output at the first Stokes wavelength of 268 nm, corresponding to the 2916 cm^{-1} Q-branch discussed above. The output from the oscillator passes through a prism monochromator which selects the first Stokes component and directs it as a 268 nm probe pulse through a 2 m Raman amplifier cell maintaining the same CH$_4$ pressure as the oscillator, and placed between two 1.5 mm collimating apertures. The other half of the KrF pump pulse passes through a beam-contracting telescope and is directed through the same collimating apertures as the probe pulse, but in the opposite direction. The transit time through the optics is arranged so that the leading edges of the probe pulse and the KrF pump pulse meet at the exit window of the amplifier cell W_e, and the transit time (length) of the Raman oscillator cell is selected to be 7.5 ns so that the leading edge of the probe pulse sweeps through the entire 15 ns of gain produced by the pump pulse. Beam splitters and biplanar photodiodes are used to record the pump and probe pulses before and after transit through the amplifier, with appropriate neutral density and bandpass filters. The intensity calibration of the diodes is derived from comparison with a calorimeter. The KrF laser linewidth is monitored by a camera photographing the ring pattern of a Fabry–Perot interferometer.

Fig. 41 shows typical oscilloscope traces of the pump and probe signals in the experiment, demonstrating small-signal gain in the backward direction for a probe pulse intersecting a pump pulse of linear integrated intensity $\int I_p \, dz = 3 \times 10^9$ W/cm^9 and with a linewidth of 0.6 cm^{-1} in methane gas at a density of 25 Amagat. The absence of distortion in the pump pulse in fig. 41 shows the absence of pump depletion by either backward Raman amplification or by superfluorescence in the amplifier cell of the first Stokes component in the forward direction, effects which have been seen at higher intensities. The probe pulse appears distorted after amplification because only the leading edge of the pulse experiences the full gain produced by the pump pulse. Later parts of the probe pulse enter the amplifier cell after part of the pump pulse has already excited, and they therefore experience smaller gain.

Fig. 42 shows a more complete plot of backward gain data obtained with this apparatus as a function of methane pressure for two pump laser linewidths,

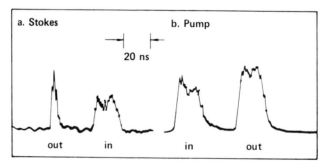

Fig. 41. Input and output signatures for KrF pump pulse and backward wave Stokes pulse for a pump pulse spectral linewidth of 0.6 cm^{-1} and a methane density of 25 Amagat (Murray et al. 1978).

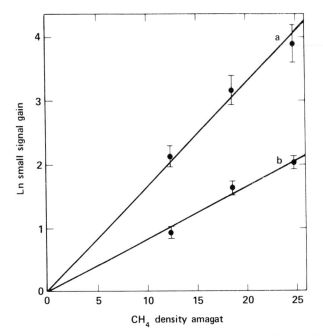

Fig. 42. Measured backward wave Stokes gain as a function of methane density for two values of the spectral linewidth of the KrF pump pulse (Murray et al. 1978).

obtained by varying the etalons in the KrF injection oscillator. The gain is seen to be proportional to pressure for each linewidth (to within the precision of this experiment), which implies that any change in the linewidth of the methane Raman transition over this pressure range is small compared to the laser linewidth used in the experiment. The gain also shows the predicted $(\Delta\nu_S + \Delta\nu_p)^{-1}$ dependence to within the precision of the experiment.

An absolute determination of the methane Raman gain coefficient requires an accurate knowledge of the pump intensity in the amplifier cell as well as precise lineshape information, and cannot at present be stated with high precision. Using the data of fig. 42 the backward Raman gain is given by $\exp[2 \pm 1 \times 10^{-11} \rho(\Delta\nu_p + 0.3)^{-1} \int I_p \, dz]$, where $\int I_p \, dz$ for the experimental conditions is $3 \pm 1.5 \times 10^9 \, \mathrm{cm} \, \mathrm{W}^{-1}$, ρ is the CH_4 density in Amagat and $\Delta\nu_p$ is in wavenumbers. The measured backward gain agrees with the theoretical prediction given above to within the accuracy of the measurement.

Forward gain for superfluorescence produced from the collimated pump beam in the amplifier cell has been inferred by observation of the intensity of forward superfluorescence in the amplifier at higher pump intensities. The intensity of this superfluorescence is substantially independent of laser linewidth over the range studied, including also a few measurements at linewidths as high as 30 to 50 cm^{-1}, achieved by removing all etalons from the injection oscillator. This is also in agreement with theory.

In addition to these gain measurements, the limitations imposed upon the pump and Stokes fluences due to multiphoton dissociation and ionization of several molecular Raman scatterers have been determined. It was found that for methane at atmospheric pressure, pump fluences less than $100\,\mathrm{J\ cm^{-2}}$ could be propagated through the cell without difficulty.

7.5 Extraction characteristics and efficiency estimates

In this subsection we discuss in a quantitative fashion the conversion efficiency of pump radiation into the backward Raman wave for a multipass off-axis amplifier. While other device architectures have been studied, the underlying physical processes operative are common to all. An analysis of the KrF pump source is also made here followed by an assessment of the overall laser system efficiency.

There are several physical processes which limit the performance of the pulse compressor; the most serious is the parasitic growth of first and second Stokes radiation. The forward-scattered first Stokes wave at the frequency ν_S will have a gain of $(R\sigma_R^b \Delta n_p L)$ [See eq. (93)] for no pump depletion, where σ_R^b is the backward stimulated Raman cross section and R is the ratio of the forward-to-backward stimulated Raman cross sections, namely,

$$R \equiv \sigma_F^f / \sigma_R^b \simeq (\Delta\nu_S/(\Delta\nu_P + \Delta\nu_S))^{-1}. \tag{95}$$

This expression is valid for pressure-broadened methane. Design studies indicate that a device with a backward Raman gain of five (that is, a small-signal gain of exp 5) can be stably operated. Since this parasitic process grows from noise it is only necessary to have its small-signal gain coefficient be less than about 20. This implies that $R < 4$, in order to ensure that little conversion of pump energy into superfluorescent forward scattered Stokes radiation takes place. For the case of methane, experiments show that R values from 1 to 2 are possible, so that this process is not of overriding significance.

The Raman parasitic process of lowest order and highest gain, and therefore of most concern in an amplifier design, is the generation of a second Stokes pulse propagating coincident with the 268 nm backward first Stokes pulse. The 268 nm pulse injected at the left-hand side of the amplifier in fig. 43 produces gain and some fluorescence at 291 nm, corresponding to a shift of $2914\,\mathrm{cm^{-1}}$ from 268 nm. As can be seen, that part of the fluorescence or parametrically generated 291 nm radiation which propagates coincident with the 268 nm pulse undergoes gain for the entire transit time of the amplifier. Since this second Stokes radiation is amplified spontaneous emission, it may not be useful for focusing onto a target and thus represents a loss of the optical energy.

Before proceeding to discuss the results of a computer analysis of this problem, it is useful to elucidate the scaling relationship of the second Stokes gain by considering the propagation characteristics of the Stokes pulse. Making the assumption that the Raman cross section for scattering a first Stokes photon into the second Stokes wave $\sigma_{R_{2S}} \simeq \sigma_R^f$ (or $\gamma_{R_{2S}} \simeq \gamma^f$, a valid approximation for

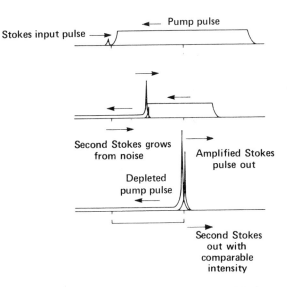

Stokes input pulse →

← Pump pulse

Second Stokes grows
from noise

Amplified Stokes
pulse out

Depleted
pump pulse

Second Stokes
out with
comparable
intensity

Fig. 43. Schematic representation of the development of parasitic second Stokes wave generation in the
backward direction (e.g., propagating in the same direction as the injected first Stokes pulse).

nonresonant Raman scattering), the limitation to the backward traveling wave
Raman amplifier set by second Stokes production driven by the amplified
backward Stokes wave is

$$g_{2S} = \gamma^f \int I_S \, dz < g_{2S_{max}} \sim 20, \tag{96}$$

where g_{2S} is the second Stokes gain and I_S is the intensity of the Stokes pulse. As
shown by Maier et al. (1969), when the Stokes pulse width is much less than the
pump pulse width the theory of Frantz and Nodvik (1963) and Bellman et al
(1963) can be extended to treat Stokes pulse amplification inside the Raman cell.
According to this theory the Stokes pulse intensity envelope satisfies

$$I_S(z, \tau)[1 + U^{-1}(z, \tau)] = I_S(0, \tau)[1 + U^{-1}(0, \tau)], \tag{97}$$

where $\tau = t - z/c$, and the auxiliary function $U(z, \tau)$ is related to the Stokes pulse
energy fluence Γ_1 up to time τ by

$$U(z, \tau) = -1 + \exp[\Gamma_1(z, \tau)/\Gamma_S] \tag{98}$$

and satisfies

$$U(z, \tau) = U(0, \tau) \exp \gamma^b I_p z. \tag{99}$$

The Stokes pulse saturation fluence Γ_S is $2h\nu_S/\sigma_R^b = 2/\gamma^b c$. The total Stokes pulse
energy fluence, defined by eq. (34), $\Gamma_1(z) = \Gamma_1(z, \tau \to \infty)$ is governed by eq. (36).
Eq. (97) also applies to pulse propagation in storage laser media (§ 3.4) and
indicates that the quantity $I_S[1 + U^{-1}]$ is 'conserved' on passing through the
amplifier. When the auxiliary saturation function U becomes much larger than 1,

the left side of eq. (97) reduces to the pulse envelope $I(z, \tau)$ and eq. (97) becomes an especially convenient way to estimate pulse distortion. Because of this analogy between backward Stokes pulse propagation and pulse propagation in storage laser amplifiers, the pulse distortion results (Trenholme and Manes 1972) discussed in § 3.4 also apply to backward traveling wave Raman amplifiers. Therefore when the Stokes pulse output fluence exceeds $\sim 10 \, \Gamma_S$ significant pulse distortion can be anticipated (§ 3.4). As indicated earlier for fusion applications the Stokes output fluence will be approximately 2 to 5 saturation fluences and therefore pulse distortion will not be significant. The condition [eq. (96)] that the parasitic second Stokes wave does not appear can now be evaluated by substituting eqs. (97) to (99) into eq. (96). For a gaussian input Stokes pulse intensity envelope, eq. (96) becomes (Eimerl 1977)

$$g_{2S} = g_{1S} R \kappa \epsilon_{ext}^2 \chi < g_{2S_{max}} \sim 20, \tag{100}$$

where $g_{1S} = \gamma^b I_p L$ is the gain coefficient for the Stokes wave, and χ is a slowly varying function

$$\chi = \left(\frac{\nu_{2S}}{\nu_S}\right) \frac{I_S(\tau_f)}{I_S(0)} \psi [1 - \exp(-f\epsilon_{ext}\psi)]^{-1} \tag{101}$$

and

$$\psi = 1 + \frac{1}{\epsilon_{ext} g_{1S}} \ln \left\{ \frac{1 - \exp(-g_{1S})}{1 - \exp[-g_{1S}(1 - \epsilon_{ext})]} \right\}. \tag{102}$$

The extraction efficiency for the Stokes pulse can be written

$$\epsilon_{ext} = (\Gamma_{1_0} - \Gamma_{1_1})/\Gamma_p, \tag{103}$$

where the Stokes output fluence Γ_{1_0} can be evaluated using eq. (34) and where Γ_p is the pump laser fluence. In eq. (100), the pulse compression is

$$\kappa = (4 \ln 2/\pi)^{1/2}(t_p/t_S), \tag{104}$$

where t_S is the full width at half maximum pulse width of the input Stokes pulse. The quantity f is the fraction of the pulse that has passed at the peak of the gain g_{2S}, namely: $\Gamma_1(\tau_f) = f\Gamma_1(L) = f\Gamma_1(L, \tau \to \infty)$.

Eq. (100) defines the trade-off that must be made between the pulse compression ratio κ, the amplifier extraction efficiency ϵ_{ext} and the small-signal gain g_{1S}, since the value of g_{2S} must not exceed ~ 20 in order to minimize the deleterious effects of superfluorescent second Stokes emission. As an example, for $R = 2$, $\kappa = 10$, $g_{1S} = 5$, eq. (100) gives $\epsilon_{ext} = 55\%$.

It is now apparent that a large single-stage compression at high efficiency using a standard pulse compressor architecture can only be achieved if some means can be found for suppressing the second Stokes radiation. A selective absorption medium could be employed or near-resonant Raman scatterers, where $\sigma_{2S}^f \ll \sigma_R^f$ could also be used. Several near-resonant Raman scatterers in atomic vapor systems have been identified and experimental studies are presently underway.

An assessment of the design criteria for an amplifier requires a more refined analysis than that presented above. To accomplish this task a detailed space–time computer model (Eimerl 1977) was developed which contains pump depletion, forward first and backward second Stokes losses as well as the space–time evolution of the backward first Stokes pulse. Fig. 44 illustrates the results of this model for the case where $\kappa = 15$, $R = 4$ and $g_{1S} = \gamma^b I_p L = 5$. In fig. 44a the first Stokes input intensity was sufficiently small enough so that the growth of the second Stokes pulse was not important. For these parameters 2% of the pump radiation was converted into forward Stokes emission and 19% was usefully converted into the backward Stokes wave. The temporal pump depletion is evident in this figure. Increasing the first Stokes input intensity in an attempt to increase the extraction efficiency has an adverse effect on the output, as shown in fig. 44b. Now the gain at the second Stokes frequency is more than sufficient to allow it to grow and begin to saturate and extract energy from the first Stokes pulse, 'eating' out its center. It is important to note that only a small

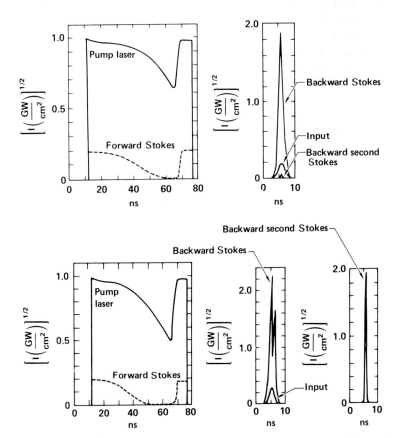

Fig. 44. Computer calculation of the temporal behavior of pump, forward first Stokes, and backward first and second Stokes pulses in a Raman compressor system with $\kappa = 15$, $R = 4$, and $g_{1s} = \gamma^b I_p L = 5$: (a) backward first Stokes input amplitude = 0.2 $[GW/cm^2]^{1/2}$; (b) backward first Stokes input amplitude = 0.3 $[GW/cm^2]^{1/2}$ (Eimerl 1977).

change in input intensity (less than a factor of two) caused this dramatic effect, indicating that the boundary in parameter space for second Stokes growth is quite sharp. In fig. 44b the conversion efficiency to forward Stokes was 1.5%, to backward Stokes 23% and to second Stokes 6%.

The regions in parameter space where the parasitic growth of either the forward first Stokes or backward second Stokes is small enough so as not to degrade the amplifier performance characteristics are shown in fig. 45. These results [eq. (100)] were obtained by integrating the space–time Frantz–Nodvik equations (Maier et al. 1969) under the assumption that the two major parasitic waves, the forward first Stokes wave and the backward second Stokes wave, do not grow to the point where they distort the waveforms of the waves that feed them. For a compression ratio of 10, fig. 45 shows the boundaries where the forward first Stokes gain coefficient is equal to 20 (vertical lines derived from the previous analyses) and where the backward second Stokes gain coefficient is equal to 20 (shallow, horizontal curves) as a function of the extraction efficiency [eq. (99)] and backward gain coefficient, for several values of the forward-to-backward cross-section ratio R. For a given R value, stable operation is possible below and to the left of the boundary. It is anticipated that, for methane, R will lie between 1 and 3, therefore the extraction efficiency should lie in the range between 0.5 and 0.75. The second Stokes boundaries are parametric in the product of R and κ [eq. (100); thus the curve for $R = 2$, $\kappa = 10$ is equivalent to the one for $R = 1$, $\kappa = 20$.

Using the Frantz–Nodvik analysis developed in § 3.4, the variation of the extraction efficiency and fluence gain with the normalized input fluence can be

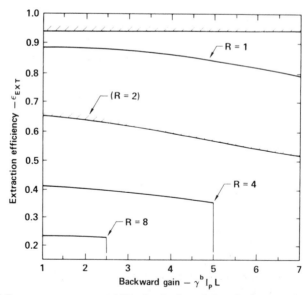

Fig. 45. Map of Raman compressor stability for backward first Stokes amplification for various values of the forward to backward gain ratio R (Eimerl 1977).

determined. These functional dependences are shown in fig. 46, together with the parasitic boundaries taken from fig. 45. For a given $R\kappa$ product, stable operation occurs below the boundary in fig. 46a and to the left of the boundary in fig. 46b. Such curves can be used to design a Raman amplifier; for the example discussed throughout this section ($K = 10$, $R = 2$, $\gamma^b I_p L = 5$), the extraction efficiency is ~ 0.6 for an input fluence of $0.1\Gamma_s$. The fluence gain for this situation will be 25.

A comprehensive efficiency analysis of a fusion laser system entails both the extraction characteristics of the Raman amplifier and the kinetic analysis of the

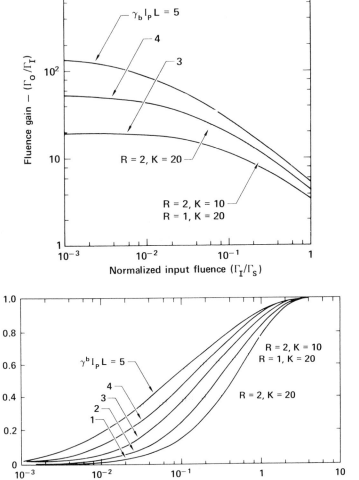

Fig. 46. Raman conversion efficiency (a) and fluence gain (b) as a function of normalized input fluence for several values of $g_{1s} = \gamma^b I_p L$. Limits for stable backward first Stokes growth are shown for various product values of $R\kappa$.

KrF pump laser. A discussion of the kinetic process operative in the KrF laser is given elsewhere in this book. It suffices here to discuss only the salient features of a KrF amplifier. Quantitative amplifier performance results have been obtained from a computer model which treats the microscopic kinetic processes using rate equations and solutions to the appropriate Boltzmann equation for the electron gas. The amplifier code contains nonsaturable loss processes (principally molecular fluorine absorption) and models the dynamic characteristics of the system under the action of a stimulating field. The presence of nonsaturable losses limits the maximum laser intensity to a value given by

$$I_m = I_s(\alpha_0/\gamma - 1), \tag{105}$$

where I_s is the saturation intensity ($\sim 1 \, \text{MW/cm}^2$ at 1 atm for KrF), α_0 the small-signal gain coefficient and γ the nonsaturable loss coefficient. In an amplifier, as the intensity approaches I_m, the intrinsic device efficiency (defined as the ratio of the extracted energy to the deposited electrical energy) approaches zero. The intrinsic efficiency ϵ_I is the product of the laser medium excitation and extraction efficiencies (§ 3.6) $\epsilon_I = \epsilon_{LM}\epsilon_{ext}$. Thus I is limited to practical values less than I_m. The results from the computer analysis for several specific cases are given in figs. 47 and 48. For large I, the intensity begins to saturate approaching a value given by eq. (101) and the efficiency is a decreasing function of increasing intensity. From such an analysis, a KrF pump laser delivering $20 \, \text{MW cm}^{-2}$ at an efficiency of 12% appears feasible.

In summary, the Raman pulse compressor appears to be a very attractive system for fusion applications both because of the potentially high system efficiencies and because of the 'rich' variety of system architectures possible. In addition several other features of this scheme are worth mentioning. Because the fusion target is not optically coupled directly to the radiating KrF laser pump, preheating of the target due to any KrF amplified spontaneous emission (ASE) will be minimal. Another advantage suggested theoretically is that the KrF pump laser need not be near-diffraction-limited; indeed it may not need to be less than

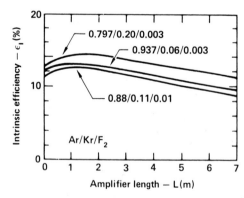

Fig. 47. Calculated extraction efficiency of a KrF amplifier as a function of amplifier length for several laser medium mixtures and for an input intensity of 2 MW/cm².

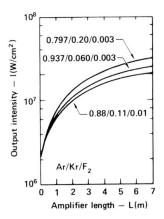

Fig. 48. Calculated output intensity of a KrF amplifier as a function of amplifier length for several laser mixtures and for an input intensity of 2 MW/cm².

100 times diffraction-limited – but it must be reasonably uniform in intensity across the beam. This possibility arises because the Stokes wave integrates the gain along the entire length of the Raman cell, tending to average out variations of the gain. If this prediction proves valid, it will greatly relax medium homogeneity requirements in the KrF pump with substantial energy savings in flow cooling. It also opens up the possibility of using multiple KrF pump beams to produce Raman gain in a single cell. These features are currently under investigation to aid in the design of energy efficient systems. We have seen that the detailed design and optimization of Raman compressor systems must take into account nonlinear parasitic processes. The lowest-order process will be generation and amplification of a second Stokes pulse propagating coincident with the 268 nm first Stokes pulse. If the intensity of the first Stokes pulse grows too large (under conditions of a high compression ratio) the second Stokes pulse may become sufficiently large to begin to deplete the first Stokes pulse. Analysis shows that pulse compression ratios of ten or less will permit stable generation on the first Stokes pulse. High compression ratios could be attained if a means could be found for suppressing the second Stokes gain. Several suggestions to accomplish this objective have been put forward but they have not yet been explored experimentally.

Finally, table 19 summarizes the projected component efficiencies for a Raman pulse compressor laser system for fusion applications. The total laser efficiency after flow is expected to lie in the range of 2 to 6% making this system one of the most efficient of those studied to date. The highest efficiency probably requires the use of potentially highly efficient e-beam-augmented discharges (Mangano and Jacob 1975, Daugherty 1976).

Table 19
Component efficiencies for a multiple KrF
Raman pulse compressor.

System components	Percentage efficiency
Pulse forming, deposition and fill factors	65
Pump medium	10–15 [25][a]
Pump after flow costs	6–8 (12)[b]
Transport/coupling	80
Fill factors	80
Extraction * quantum	55–75[c]
Laser electrical	2.1–4
Total laser after flow in laser	2–4 (6)[b]

[a]Brackets indicate percentage with possible breakthroughs.
[b]Parentheses indicate total with break-through increment.
[c]Second Stokes gain suppression.

8 Other new laser media

8.1 Metal-vapor excimer lasers

In addition to the short wavelength laser media discussed in some detail above, several other media are also being researched as part of DOEs Advanced Laser Program (Hoff 1978). Included among these are the metal vapor excimers, with particular emphasis on mercury (Hg_2) and mercury–cadmium (CdHg) molecules. Houtermans (1960) was the first to suggest the possibility of obtaining laser action in the 485 nm molecular band of Hg_2; this possibility was analyzed and addressed experimentally by Carbone and Litvak (1968) using a pulsed d.c. electric discharge in mercury vapor at 96 Torr. Stimulated radiation was not observed. The 485 nm molecular band has traditionally been attributed to collision-induced transitions from the stable excimer $^3O_u^-$ level ($^3P_0 + {}^1S_0$) to the repulsive $^1\Sigma_g^+$ ground state ($^1S_0 + {}^1S_0$). Measurements of negative gain (i.e., loss) in the 485 nm band were subsequently reported by Hill et al. (1973), using electron-beam excitation of the mercury medium. Based on more recent spectral-kinetic measurements by Drullinger et al. (1977) and on an extensive theoretical analysis by Smith et al. (1977), it is now believed that the 485 nm emission band arises from an excited mercury trimer complex. Despite the negative laser experiments with the mercury green band, it was noted by Schlie et al. (1976) that the intensity of the 335 nm mercury excimer band, attributed to transition from the ^31_u level ($^3P_1 + {}^1S_0$) to the repulsive $^1\Sigma_g^+$ ground level ($^1S_0 + {}^1S_0$), grows at the expense of the 485 nm band as the temperature of the vapor is increased above the saturation temperature. Using e-beam excitation of the mercury medium, Schlie et al. (1976) were able to demonstrate the presence of positive optical gain in the 335 nm band. The time behavior of the gain signal was exponential with a time constant of about 220 ns, which if taken as the radiative lifetime would imply a stimulated emission cross section of about 10^{-17} cm^2. The spectral analysis of Smith et al. (1977) predicts an effective stimulated emission cross section of about 10^{-18} cm^2, and an effective radiative lifetime of about 1 μs. Komine and Byer (1977a, b) have conducted extensive spectral kinetic-studies in molecular mercury by direct optical pumping of the weakly bound ground state molecules with intense radiation at 266 nm. These authors demonstrated that under their excitation conditions, the medium exhibited a significant net loss due to excited state absorption at 325 and 442 nm. They were able to estimate an absorption cross section of 10^{-17} cm^2 at 325 nm. A nonexponential decay of the metastable excimer levels was interpreted in terms of bimolecular-excimer quenching with a measured rate coefficient of $\sim 2 \times 10^{-10}$ cm^3 s^{-1} mol^{-1}. Komine and Byer (1977a) point out that this quenching channel would limit the excimer density to about 10^{17}/cc or 50 J/ℓ for a 100 ns storage time.

Interest in the metal vapor excimers in the context of energy-storage fusion lasers is predicated on the efficient excitation of molecules states via inelastic electron scattering. Judd (1976) has examined the importance of some of these electron processes in a computational study. Electron impact excitation rates for the lowest lying electronic states and rates for ionization from these states were determined through use of the Boltzmann transport equation for the electric field/electron density ratio varying between 5×10^{-17} and 10^{-15} V cm^{-2}. Ultraviolet preionized discharge experiments were conducted by York and Carbone (1976) in mercury vapor at a pressure of 2 atm. They observed volumetric energy depositions of 14 J/ℓ at an electric field/electron density ratio of 2×10^{-16} V cm^{-2} and for a discharge pulse width of 2 μs. In more recent measurements (Judd 1977), the discharge-excited medium has been probed for optical gain in the 325 nm band of the excimer. Under all conditions of excitation, only a net loss was observed, casting considerable doubt on the ultimate success of this system.

The hetero-nuclear metal excimer CdHg* may not suffer from an excessive amount of excited state absorption and prove to be an efficient energy-storage laser operating at a wavelength of 470 nm. Calculations of Fournier and McGeoch (1975 and 1978) indicated efficient discharge excitation of Cd(5^{3}P) electronic levels could be achieved in Cd–Hg vapor mixtures. Subsequently, McGeoch et al. (1976) observed 470 nm excimer emission from Cd–Hg vapor irradiated with a dye laser tuned to the ^{3}P$_1$ transition of the Cd atoms. Time resolved fluorescence emission studies indicated the existence of two radiating CdHg* species with radiative decay times of 2.8 and 4.0 μs; estimated peak stimulated emission cross sections were 2.0 and 2.9×10^{-20} cm^2. Some evidence of positive gain at 470 nm in CdHg* vapor has been reported by McGeoch and Fournier (1978) using flashlamp pumping. A small-signal gain coefficient of about 2×10^{-3} cm^{-1} was reported. As in the case of Hg$_2$, the CdHg* laser will be of practical interest only if large volumes of CdHg* vapor can be efficiently excited by electric discharge to achieve excited state densities of order $10^{16}/cm^3$ for times of order a micro-second. Under these conditions, discharge ionization stability will be determined by multistep ionization processes involving these excited states (§ 3.3). Therefore e-beam-augmented discharges will probably be required to achieve large volume excitation. Maximizing the electrical efficiency of these lasers will then require maximizing the discharge enhancement ratio (§ 3.3). In general, the efficiency and scalability of these metal vapor laser systems will be strongly determined by field penetration and discharge stability considerations. Efficient extraction of energy will also require a sufficiently small excited state absorption coefficient and favorable saturation kinetics (small nonsaturable losses) (§ 3.4). Laser kinetics aside, the extraction of laser radiation with sufficiently high beam quality appears to be a challenging problem in the metal vapor excimer systems because the strong, near-ultraviolet atomic transitions may significantly contribute to the index of refraction in the visible, even for relatively low densities. Control of the gas density fluctuations of the laser medium will therefore be at a premium as will operation of the medium at the highest possible small-signal gain coefficient. Analysis and experimental investigations of Hg_2 and $CdHg^*$ laser media are currently in progress at Mathematical Sciences Northwest (MSNW) and at the AVCO Everett Research Laboratory, respectively.

9 Summary and conclusions

The principles of operation, characteristic parameters, experimental data and design concepts for several possible advanced laser systems useful for fusion applications have been described. Exploratory research, including parallel experimental, theoretical, computational and systems-modeling efforts have begun to provide for a quantitative, comparative assessment of the utility of these systems. Preliminary analyses indicate that each of the systems described in some detail here can be scaled to high total energies using realizable, although undeveloped, technologies. Projected efficiencies for these systems range from about 1% to a high of about 5% (KrF/CH_4 Raman lasers), subject to the validation of several physics and technical assumptions. It is expected that on-going experiments and analyses will favorably resolve many of these remaining issues providing a solid basis for quantitatively assessing the scalability of these systems. It appears at this time that several laser systems may be scaled to meet the minimum technical requirements for a laser fusion reactor. Should this prove to be the case, the selection of the best system for use must also take into account difficulties in engineering development, capital cost, and systems maintenance and reliability.

Acknowledgments

The work report here derives in large measure from our colleagues at the Lawrence Livermore Laboratory. With respect to specific laser media, it is a pleasure to acknowledge major contributions from the following individuals. Group VI (H. Powell, J.R. Murray, J. Ewing), Rare Earth Ion Lasers (R.R. Jacobs), and Raman Compressors (J.R. Murray, A. Szoke, J. Goldhar, D. Eimerl). Significant contributions to systems modeling, design and theoretical calculations were made by R. Franklin, J. Ewing, L. Pleasance, L. Schlitt, C. Bender, A. Hazi, N. Winter and T. Rescigno. More extensive descriptions of their technical contributions can be found in LLL laser Program Annual Reports, UCRL-50021-74,-75,-76, available from the National Technical Information Service, Springfield, VA 22161, U.S.A. The authors are indebted to Ms. Judith Johnson for skill and dedication in the production of the manuscript.

References

Abramyan, E.A. (1968), Nucl. Instr. Meth. **59**, 22.

Afanasev, Yu.Y., N.G. Basov, P.P. Volosevich, E.G. Gamalii, O.N. Kroklin, S.P. Kurdyumov, E.I. Levanov, V.B. Rosanov, A.A. Samarskii and A.N. Tikhonov (1975), JETP Letters **21**, 68.

Afanasev, Yu.F., N.G. Basov, E.G. Gamalii, O.N. Krokhin and V.B. Rozanov (1976), JETP Letters **23**, 566.

Ahlborn, B. and W.W. Zuzak (1969), Can. J. Phys. **47**, 1709.

Ahlstrom, H.G. and J.H. Nuckolls (1977), Laser Fusion Implosion and Target Interaction Physics, University of California Report, UCRL-79540, Presented at the 1977 IEEE/OSA Conference on Laser Engineering and Application, June 1-3, 1977, Washington, D.C.

Akhmanov, S.A., A.R. Sukhorukov and R.V. Khokhlov (1968), Usp. Fiz. Nauk **93**, 609.

Albrecht, A. (1961), J. Chem. Phys. **34**, 1476.

Attwood, D.T., D.W. Sweeney, J.M. Auerbach and P.H.Y. Lee (1978), Phys. Rev. Letters **40**, 184.

Aushermann, D.R., I.E. Alker and E. Baum (1978), Acoustic Suppression in a Pulsed Chemical Laser, AIAA Paper No. 78-237, Presented at the AIAA 16th Aerospace Sciences Meeting, Huntsville, Alabama, Jan. 1978.

AVCO Everett Research Lab. Inc. (1975), Advanced Laser Research and Development Final Tech. Rept., Contract DAAHO1-72-C-0995, August, 1975.

Ballard, S.S., J.S. Browder and J.F. Ebersole (1972), Transmission and Absorption of Special Crystals and Certain Glasses, in: American Institute of Physics Handbook, 1972 (D.E. Gray ed.; McGraw-Hill; New York), pp. 6–58 to 6–94.

Basov, N.G. (1976), Powerful Laser Thermonuclear Installation Del'fin (to be published).

Basov, N.G. and V.S. Zuev (1976), Opt. Commun. **18**, 167.

Basov, N.G.,Yu.S. Ivanov, O.N. Krokhin, Yu.A. Mikhailov, G.V. Sklizkov and S.I. Fedotov, (1972), JETP Letters **15**, 417.

Bass, J.N. and A.E.S. Green (1973), J. Appl. Phys. **44**, 3726.

Beeb, H.B. and A. Gold (1966), Multiphoton Ionization of Rare Gas and Hydrogen Atoms, in Physics of Quantum Electrons (P.L. Kelley, B. Lax and P.E. Tannenwald, eds.; McGraw-Hill: New York).

Behringer, J. and J. Brandmuller (1956), Z. Electrochem. **60**, 643.

Bellman, R., G. Birnbaum and W.G. Wagner (1963), J. Appl. Phys., **34**, 780.

Berger, M.J. and S.M. Seltzer (1964), Tables of Energy Losses and Ranges of Electrons and Positrons, NASA SP-3012.

Bespalov, V.I. and V.I. Talanov (1966), JETP Letters **3**, 307.

Bird, R.B., W.E. Stewart and E.N. Lightfoot (1960), Transport Phenomena (Wiley; New York), pp. 396–407.

Black, G., R.L. Sharpless, T.G. Slanger and D.C. Lorents (1975a), J. Chem. Phys. **62**, 4266.

Black, G., R.L. Sharpless, T.G. Slanger and D.C. Lorents (1975b), J. Chem. Phys. **62**, 4274.

Black, G., R.L. Sharpless and T.G. Slanger (1976), J. Chem. Phys. **64**, 3785.

Bliss, E.S. (1971), Opto-Electronics, **3**, 99.

Bliss, E.S., D. Milam and R.A. Bradbury (1972), Appl. Optics **12**, 602.

Bliss, E.S., D.R. Speck, J.F. Holzrichter, J.H. Erkilla and A.J. Glass (1974), Appl. Phys. Letters **25**, 448.

Bodner, S.E. (1974), Phys. Rev. Letters **33**, 761.

Boley, B.A. and J.H. Werner (1960), Theory of Thermal Stresses (Wiley; New York), Chapt. 9.

Booth, D.J. and G.J. Troup (1969), IEEE J. Quantum Electronics **QE-5**, 547.

Born, M. and E. Wolf (1970), Principles of Optics (Pergamon Press; New York).

Boyer, K. (1977), Proc. of the U.S.–Japan Seminar on Laser Interaction with Matter, Univ. of Rochester, Nov. 1976. Edited by C. Yamanaka, Inst. of Laser Eng., Osaka University.

Bradley, L.P. (1975), Electron Beam Pumping of Visible and Ultraviolet Gas Lasers, University of California Report UCRL-77213.

Brau, C.A. and J.J. Ewing (1975), Appl. Phys. Letters **27**, 435.

Brederlow, G., K.J. Witte, E. Fill, K. Hohla and R. Volk (1977), Digest of Technical Papers, Conference on Laser Engineering and Applications, p. 61.

Brueckner, K.A. and S. Jorna (1974), Rev. Mod. Phys. **46**, 325.

Caird, J. (1973), Air Force Avionics Laboratory Report, AFAL-TR-73-323, Part I, Rare Earth Laser Engineering Program, Contract No. F33615-72-C-2038. Hughes Aircraft Co.

Carbone, R.J. and M.M. Litvak (1968), J. Appl. Phys. **39**, 2413.

Carman, R.L. (1975), Phys. Rev. A **12**, 1048.

Carnall, W.T., P.R. Fields and B.G. Wybourne (1965), J. Chem. Phys. **42**, 3797.

Carnall, W.T., J.P. Hessler, H.R. Hoekstra and C.W. Williams (1978), J. Chem. Phys. **68**, 4304.

Chamberlain, J.W. (1961), Physics of the Aurora and Airglow (Academic Press; New York).

Chen, F.F. (1974), Physical Mechanisms for Laser–Plasma Parametric Instabilities, in: Laser Interaction and Related Phenomena (H. Schwarz and H. Hora, eds.; Plenum Press; New York), vol. 3B, pp. 291–313.

Chernov, L.A. (1960), Wave Propagation in a Random Medium (McGraw-Hill; New York).

Chester, A.N. (1973), Appl. Optics **12**, 2139.

Chu, B.T. and L.G. Kovasznay (1958), J. Fluid Mech. **3**, 494.

Clarke, J.S., H.N. Fisher and R.J. Mason (1973), Phys. Rev. Letters **30**, 89.

Close, D.H. (1967), Phys. Rev. **153**, 360.

Coffey, T.P. (1971), Phys. Fluids **14**, 1402.

Colbert, H.M. (1973), SANDYL, A Computer Program for Calculating Combined Photon-Electron Transport in Complex Systems, Sandia Livermore Laboratory Rept. SCL-DR-720109.

Conn, R.W. (1976), Two Lectures on Fusion Reactors, Fusion Technology Program Nuclear Engineering Dept. Univ. of Wisconsin, Madison, Wisconsin.

Cordray, R., F.K. Tittel and W.L..Wilson (1977), Appl. Phys. **12**, 245.

Corney, A. and O.M. Williams (1972), J. Phys. B **5**, 686.

Crosswhite, H. (1976), Unpublished Computer Code, Chemistry Division, Argonne Natl. Laboratory, Argonne, Ill. 60439, U.S.A.

Culick, F.E.C., P.I. Shen and W.S. Griffin (1976), IEEE J. Quantum Electronics **QE-12**, 566.

Daugherty, J.D. (1976), Electron Beam Ionized Lasers, in: Principles of Laser Plasmas (G. Bekefi ed.; Wiley; New York), pp. 369–419.

Daugherty, J.D., E.R. Pugh and D.H. Douglas-Hamilton (1971), Bull. Amer. Phys. Soc. **116**, 399.

Daugherty, J.D., J.A. Mangano and J.H. Jacob (1976), Appl. Phys. Letters **28**, 581.

Dawson, J., P. Kaw and B. Green (1969), Phys. Fluids, **12**, 875.

Dawson, J.M. and A.T. Lin (1974), University of California, Los Angeles, Report PPG-191.

De Michelis, C. (1969), IEEE J. Quantum Electronics **QE-5**, 188.

Djeu, N. and R. Burnham (1977), Appl. Phys. Letters **30**, 473.

Drullinger, R.E., M.M. Hessel and E.W. Smith (1977), J. Chem. Phys. **67**, 5656.

Dyke, W.P. (1964), Sci. Amer. 108.

Eimerl, D. (1977), private communication.

Emmett, J.L., J. Nuckolls and L. Wood (1974), Sci. Amer. **230**, No. 6, 24.

Estabrook, K.G. (1976), Phys. Fluids, **19**, 1733.

Estabrook, K.G. and W.L. Kruer (1977), Properties of Resonantly Heated Electron Distributions, University of California Report, UCRL-79617.

Estabrook, K.G., E.J. Valeo and W.L. Kruer (1975), Phys. Fluids, **18**, 1151.

Ewing, J.J. and C.A. Brau (1975), Appl. Phys. Letters **27**, 350.

Feinberg, R.M., R.S. Lowder and O.L. Zappa (1975), Low Pressure Loss Cavity for Repetitively Pulsed Electric Discharge Lasers, AFWL-TR-75-99, February 1975.

Fenner, W.R., H.A. Hyatt, J.M. Kellam and S.P.S. Porto (1973), J. Opt. Soc. Amer. **63**, 73.

Fenstermacher, C.A., M.J. Nutter, W.T. Lelend and K. Boyer (1972), Appl. Phys. Letters **20**, 56.

Fink, J.H., and B.W. Schumacher (1974), Optik **39**, 543.

Finn, E.J. and G.W. King (1975), J. Molec. Spectros. **56**, 52.

Fitch, F.A. (1971), IEEE Trans. Nucl. Sci. **18**, 190.

Fitch, F.A. and V.T.S. Howell (1964), Proc. IEEE, **111**, 849.

Forslund, D.W., J.M. Kindel, K. Lee, E.L. Lindman and R.L. Morse (1975a), Phys. Rev. A **11**, 679.

Forslund, D.W., J.M. Kindel and E.L. Lindman (1975b), Phys. Fluids **18**, 1002.

Forslund, D.W., J.M. Kindel, K. Lee and E.L. Lindman (1976), Phys. Rev. Letters **36**, 35.

Forslund, D.W., J.M. Kindel and K. Lee (1977), Phys. Rev. Letters **39**, 284.

Fournier, G.R. and M.W. McGeoch (1975), Proc. Inst. Phys. 2nd. Natl. Quantum Electronics Conf., Oxford.

Fournier, G.R. and M.W. McGeoch (1978), J. Appl. Phys. **49**, 2651.

Fraley, G.S., E.J. Linnebur, R.J. Mason and R.L. Morse (1974), Phys. Fluids **17**, 474.

Frantz, L.M. and J.S. Nodvik (1963), J. Appl. Phys. **34**, 2349.

Friedberg, J.P., R.W. Mitchell, R.L. Morse and L.I. Rudsinski (1972), Phys. Rev. Letters **28**, 795.

Ginzburg, V.L. (1964), The Propagation of Electromagnetic Waves in Plasmas (Pergamon Press; New York).

Gitomer, S.J., R.L. Morse and B.S. Newberger (1977), Phys. Fluids **20**, 234.

Glass, A.J. (1967), IEEE J. Quantum Electronics **3**, 516.

Glass, A.J. (1976), Optical Pulse Compression, in: Energy Storage, Compression and Switching, (W.H. Bostick, V. Nardi and O.S.F. Zucker, eds.; Plenum Press; New York), pp. 399–404.

Glass, A.J. and A.H. Guenther (1973), Appl. Optics **12**, 637.

Glass, A.J. and A.H. Guenther (1974), Appl. Optics **13**, 74.

Glass, A.J. and A.H. Guenther, eds. (1975), Laser Induced Damage in Optical Materials: 1974, NBS Special Publication 414.

Glass, A.J. and A.H. Guenther, eds. (1976), Laser Induced Damage in Optical Materials, 1975, NBS Special Publication 435.

Glass, A.J. and A.H. Guenther, eds. (1977), Laser Induced Damage in Optical Materials, 1976, NBS Special Publication 462.

Glasstone, S. and R.H. Lovberg (1960), Controlled Thermonuclear Reactions (Van Nostrand; Princeton, New Jersey), Chapter 2.

Glaze, J.A. (1976), Opt. Eng. **15**, 136.

Glaze, J.A., S. Guch and J.B. Trenholme (1974), Appl. Optics **13**, 2808.

Grey Morgan C. (1975), Rep. Prog. Phys. **38**, 621.

Gruen, D.M. (1971), Progress in Inorganic Chemistry, (S. Lippard, ed.; Wiley; New York), Vol. 14, p. 119.

Gruen, D.M. and C.W. DeKock (1966), J. Chem. Phys. **45**, 455.

Gruen, D.M., C.W. DeKock and R.L. McBeth (1967), Amer. Chem. Soc. Advances in Chemistry Series, No. 71, 102.

Gus'kov, S.Yu, O.N. Krokhin and V.B. Rozanov (1976), Nucl. Fusion **16**, 957.

Haas, R.A. (1973), Phys. Rev. A **8**, 1017.

Haas, R.A. (1977), private communication.

Haas, R.A., W.C. Mead, W.L. Kruer, D.W. Phillion, H.N. Kornblum, J.D. Lindl, D. MacQuigg, V.C. Rupert and K.G. Tirsell (1977a), Phys. Fluids **20**, 322.

Haas, R.A., H.D. Shay, W.L. Kruer, M.J. Boyle, D.W. Phillion, F. Rainer, V.C. Rupert and H.N. Kornblum (1977b), Phys. Rev. Letters **39**, 1533.

Haas, Y. and G. Stein (1972), Chem. Phys. Letters **15**, 12.

Hawryluk, A.M., J.A. Mangano and J.H. Jacob (1977), Appl. Phys. Letters **31**, 164.

Henderson, D.B., R.L. McCrory and R.L. Morse (1974), Phys. Rev. Letters **33**, 205.

Hessler, J.P., F. Wagner, Jr., C.W. Williams and W.T. Carnall (1977a), J. Appl. Phys. **48**, 3260.

Hessler, J., J. Caird, W. Carnall, F. Wagner, Jr. and C. Williams (1977b), 13th Rare Earth Research Conference Record, Oglebay Park, West Virginia, Oct.

Hill, R.M., D.J. Eckstrom, D.C. Lorents and H.H. Nakano (1973), Appl. Phys. Letters **23**, 373.

Hoff, P.W. (1978), Digest of Technical Papers, Inertial Confinement Fusion Meeting, San Deigo, Feb 7–9, ThA6-1, Opt. Soc. Amer.

Holliday, J.H. and G.G. Isaacs (1971), J. Vac. Sci. Technol. **38**, 15.

Holmes, D.A. and P.V. Avizonis (1976), Appl. Optics **15**, 1075.

Holzrichter, J. (1975), Laser Program Annual Report – 1974, Lawrence Livermore Laboratory, UCRL-50021-74, p. 441.

Holzrichter, J.F. (1976), Laser Program Annual Report, Lawrence Livermore Laboratory, UCRL-50021-75, pp. 61–129.

Hopper, R.W. and D.R. Uhlmann (1972), J. Appl. Phys. **41**, 4023.

Houtermans, F.G. (1960), Helv. Phys. Acta **33**, 933.

Hsia, J., J.A. Mangano and J.H. Jacob (1977), Plasma Return Current Discharge in Proc. U.S. – Japan Joint Seminar on the Glow Discharge, and Its Fundamental Processes, Joint Institute for Laboratory Astrophysics, Univ. of Colo. and U.S. Nat. Bur. of Stand., Boulder, Colo., July, 1977.

Hughes, W.M., N.T. Olson and R. Hunter (1976), Appl. Phys. Letters **28**, 81.

Hunt, J.T., P.A. Renard and R.G. Nelson (1976), Appl. Optics **15**, 1458.

Hunt, J.T., P.A. Renard and W.W. Simmons (1977a), Appl. Optics **16**, 779.

Hunt, J.T., J.A. Glaze, W.W. Simmons and P.A. Renard (1977b), Suppression of Self-Focusing Through Relay Imaging and Low Pass Spatial Filtering, University of California Report UCRL-79904.

Iremashvili, D.V. (1973), Zh. Eksp. Teor. Fiz. Pio'ma Red. **17**, 11.

Ishii, S. and B. Ahlborn (1976), J. Appl. Phys. **47** 1076.

Jacobs, R.R. (1977), Lawrence Livermore Laboratory, private communication.

Jacobs, R.R. and W.F. Krupke (1977a), Electronic Transition Lasers II (L.E. Wilson et al. eds.; MIT Press; Cambridge, Mass.), p. 247.

Jacobs, R.R. and W.F. Krupke (1977b), IEEE. J. Quantum Electronics **13**, 103D.

Jacobs, R.R., M. J. Weber and R.K. Pearson (1975), Chem. Phys. Letters **34**, 80.

Jacobs, R.R., W.F. Krupke, J.P. Hessler and W.T. Carnall (1977), Opt. Commun. **21**, 395.

Johnson, B.C. (1975), Laser Program Report 1974, Lawrence Livermore Laboratory UCRL-50021-74, p. 135.

Johnson, R.R. (1974), Bull. Amer. Phys. Soc. **19**, 886 (A).

Jorner, R.E., J. Marburger and W.H. Steier (1977), Appl. Phys. Letters **30**, 485.

Judd, B.R. (1962), Phys. Rev. **127**, 750.

Judd, O. (1976), J. Appl. Phys. **47**, 5297.

Judd, O. (1977), 7th. Winter Colloquium on Quantum Electronics, Park City, Utah.

Kachen, G.I. (1975), Lawrence Livermore Laboratory Report UCRL-51753.

Kachen, G.I. and H. Lowdermilk (1977), Phys. Rev. **16**, 1657.

Kaiser, W. and M. Maier (1972), Stimulated Rayleigh, Brillouin and Raman Spectroscopy, in: Laser Handbook (F.T. Arecchi and E.O. Schulz-DuBois, eds.; North-Holland; Amsterdam), Vol. 2, pp. 1078–1150.

Kayano, I., T.S. Wauchop and K.H. Welge (1975), J. Chem. Phys. **63**, 110.

Kaye, G.W.C. and T.H. Laby (1973), Tables of Physical and Chemical Constants (Longman; London).

Kaw, P., G. Schmidt and T. Wilcox (1973), Phys. Fluids **16**, 1522.

Keldysh, L.V. (1965), Sov. Phys. – JETP **20**, 1307.

Kernahan, J.A. and P.H.L. Pang (1975), Can. J. Phys. **53**, 455.

Kidder, R.E. (1972), Interaction of Intense Photon Beams with Plasmas – II in: Fundamental and Applied Laser Physics (M.S. Feld and N.A. Kurnit, eds.; Wiley; New York), p. 107.

Kidder, R.E. (1974), Nucl. Fusion **14**, 797.

Kidder, R.E. (1975), Nucl. Fusion **15**, 405.

Kidder, R.E. (1976), Nucl. Fusion **16**, 3.

Komine, H. and R.L. Byer (1977a), J. Chem. Phys. **67**, 2536.

Komine, H. and R.L. Byer (1977b), J. Appl. Phys. **48**, 2505.

Korobkin, V.V., V.M. Ovchinnikov, P.P. Pashinin, Yu.A. Pirogor, A.M. Prokhorov and R.V. Serov (1976), VIII All Union Conference on Coherent and Nonlinear Optics, Tbilisi, Georgia, U.S.S.R., May 1976. p. 243.

Kovasnay, L.G. (1953), J. Aeron. Sci. **20**, 657.

Krasyuk, I.K., P.P. Pashinin and A.M. Prokhorov (1969), JETP Letters **9**, 354.

Krasyuk, I.K., P.P. Pashinin and A.M. Prokhorov (1970), Sov. Phys. – JETP **31**, 860.

Kroll, N. and K.M. Watson (1972), Phys. Rev. A **5**, 1883.

Kruer, W.L., K.G. Estabrook and K.H. Sinz (1973), Nucl. Fusion **13**, 952.

Kruer, W.L., R.A. Haas, W.C. Mead, D.W. Phillion and V.C. Rupert (1977), Collective Behavior in Recent Laser-Plasma Experiments, in: Plasma Physics – Nonlinear Theory and Experiments (H. Wilhelmsson, ed.; Plenum Press; New York), pp. 64–81.

Krupke, W.F. (1966), Phys. Rev. **145**, 325.

Krupke, W.F. (1974a), Lawrence Livermore Laboratory Report, UCID-16620, September 11, 1974, Prospects for Gaseous Rare Earth Lasers.

Krupke, W.F. (1974b), IEEE J. Quantum Electronics **QE-10**, 450.

Krupke, W.F. (1976a), Lawrence Livermore Laboratory Report, UCID-16993, January 6, 1976, Dynamics of Trivalent Rare Earth Molecular Vapor Lasers.

Krupke, W.F. (1976b), 12th Rare Earth Research Conference Record, Vail, Colorado, July. Vol. II. p. 1034.

Krupke, W.F. and E.V. George (1976), SPIE, **86**, 122.

Krupke, W.F. and R.R. Jacobs (1977), 13th. Rare Earth Research Conference Record, Oglebay Part, West Virginia, Oct.

Kryukov, P.G. and V.S. Letokhov (1969), Usp. Fiz. Nauk **99**, 169.

Lallemand, P. (1971), The Raman Effect (A. Anderson, ed.; Dekker; New York), Vol. I, Chapter 5.

Landau, L.D. and E.M. Lifshitz (1959), Fluid Mechanics (Addison-Wesley; Reading, Mass.).

Langdon, A.B. and B.F. Lasinski (1975), Phys. Rev. Letters **34**, 834.

Lawrence, G.M. and M.J. McEwan (1973), J. Geophys. Res. **78**, 8314.

Lawson, J.D. (1957), Proc. Phys. Soc. (London) **B70**, 6.

Layne, C.B., W.H. Lowdermilk and M.J. Weber (1977), Phys. Rev. **16B**, 10.

Lempicki, A., H. Samelson and C. Brecher (1965), Appl. Optics Suppl. 2, Chemical Lasers, 205.

Lerche, R.A., L.W. Coleman, J.W. Houghton, D.R. Speck and E.K. Storm (1977), Appl. Phys. Letters **31**, 644.

Liepmann, H.W. and A. Roshko (1957), Elements of Gas Dynamics (Wiley; New York).

Lindl, J.D. (1977), Bull. Amer. Phys. Soc. **22**, 1078.

Lindl, J.D. (1978), Low Aspect Ratio Double Shells for High Density and High Gain, in: Technical Digest for Topical Meeting on Inertial Confinement Fusion, Feb. 7–9, San Diego, CA, Sponsored by Optical Soc. of America, pp. ThC7-1 to -4.

Lindl, J.D. and W.C. Mead (1975), Phys. Rev. Letters **34**, 1273.

Liska, D.J. (1971), Proc. IEEE **118**, 1253.

Liu, C.S., M.N. Rosenbluth and R.B. White (1974), Phys. Fluids **17**, 1211.

Lorents, D.C. (1976), Physica **82C**, 19.

Lowdermilk, H. and G.I. Kachen (1976), Phys. Rev. A **14** 1472.

Lubin, M.J. (1977), Proc. of the U.S.–Japan Seminar on Laser Interaction with Matter, Univ. of Tochester, Nov. 1976. Edited by C. Yamanaka, Inst. of Laser Eng, Osaka University.

Lubin, M.J. and A.P. Fraas (1971), Sci. Amer. **224**, 21.

Maier, M., W. Kaiser and J.A. Giordmaine (1966), Phys. Rev. Letters **17**, 1275.

Maier, M., W. Kaiser and J.A. Giordmaine (1969), Phys. Rev. **177**, 580.

Manes, K.R., H.G. Ahlstrom, R.A. Haas and J.F. Holzrichter (1977), J. Opt. Soc. Amer. **67**, 717.

Mangano, J.A. and J.H. Jacob (1975), Appl. Phys. Letters **27**, 495.

Mangano, J.A., J.H. Jacob, M. Rokni and A. Hawryluk (1977), Appl. Phys. Letters **31**, 26.

Maniscalco, J.A. (1975), Laser Fusion Systems Studies, in Laser Program Annual Report – 1975, Lawrence Livermore Laboratory, Rept. UCRL-50021-75.

Maniscalco, J.A. (1976), Nucl. Technol. **28**, 98.

Maniscalco, J.A., W.R. Meier and M.J. Monsler (1977), Amer. Inst. Chem. Eng., Proc. of the 70th. Annual Meeting, Nov.

Martin, W.E. and D. Milam (1976), Appl. Optics **15**, 3054.

Martin, W.E., B.C. Johnson, K.R. Guinn and W.H. Lowdermilk (1977), Shaping Pulses in Fusion, Laser Focus, 44–50.

Max, C.E. (1977), Comparison of Theory and Simulations With Recent Laser Plasma Experiments, University of California, Report UCRL-79859.

McCory, R.L. and R.L. Morse (1977), Phys. Rev. Letters **38** 544.

McGeoch, M.W. and G.R. Fournier (1978), J. Appl. Phys. **49**, 2659.

McGeoch, M.W., G.R. Fournier and P. Ewart (1976), J. Phys. B **9**, L121.

McMahon, J.M., J.L. Emmett, J.F. Holzrichter and J.B. Trenholme (1972), IEEE J. Quantum Electronics **QE-9**, 992.

Mead, W. (1970), Phys. Fluids **13**, 1510.

Milam, D. (1977), Appl. Optics **16**, 1204.

Milam, D. and R.A. Bradbury (1973), Laser Damage in Dielectric Films, Laser Focus, Dec., 41–45.

Milam, D. and M.J. Weber (1976), J. Appl. Phys. **47**, 2497.

Milam, D., M.J. Weber and A.J. Glass (1977), Appl. Phys. Letters **31**, 822.

Monsler, M.J. (1977), private communication.

Moreno, J.B., G.A. Fisk and J.M. Hoffman (1977), J. Appl. Phys. **48**, 238.

Morse, R.L. and C.W. Nielson (1973), Phys. Fluids **16**, 909.

Murray, J.R. and P.W. Hoff (1972), Lasers for Fusion in: High Energy Lasers and Their Applications (S. Jacobs, M. Sargent III and M.O. Scully, eds.; Addison-Wesley, New York).

Murray, J.R. and C.K. Rhodes (1973), Lawrence Livermore Laboratory Report UCRL-51455.

Murray, J.R. and C.K. Rhodes (1976), J. Appl. Phys. **47**, 5041.

Murray, J.R. and A. Szoke (1977), 7th Winter Colloquium on Quantum Electronics, Part City, Utah.

Murray, J.R., H.T. Powell and C. K. Rhodes (1974), IEEE J. Quantum Electronics **QE-10**, 781.

Murray, J.R., J. Goldhar and A. Szoke (1978), Appl. Phys. Letters **33**, 399.

Nighan, W.L. (1976), Stability of High-Power Molecular Laser Discharges, in: Principles of Laser Plasmas (G. Bekefi, ed.; Wiley), pp. 257–314.

Nuckolls, J.H. (1974), Laser Induced Implosion and Thermonuclear Burn, in: Laser Interaction and Related Phenomena (H. Schwarz and H. Hora, eds.; Plenum Press; New York), Vol. 3B, pp. 399–425.

Nuckolls, J., L. Wood, A. Thiessen and G. Zimmerman (1972), Nature **239**, 139.

Nuckolls, J., J. Emmett and L. Wood (1973), Physics Today, 1.

Nuckolls, J., J. Lindl, W. Mead, A. Thiessen, L. Wood and G. Zimmerman (1974), Laser Driven Implosion of Hollow Pellets, in: Plasma Physics and Controlled Nuclear Fusion Research (International Atomic Energy Agency; Vienna), Vol. II, pp. 535–542.

Ofelt, G.S. (1962), J. Chem. Phys. **37**, 511.

Omholt, A. (1971), The Optical Aurora (Springer-Verlag; Berlin).

Oye, H.A. and D.M. Gruen (1969), J. Amer. Chem. Soc., **91**, 2229.

Parker, R.K. (1974), J. Appl. Phys. **45**, 2463.

Parmentier, E.M. and R.A. Greenberg (1973), AIAA J., **11**, 943.

Passner, A. (1972), Rev. Sci. Instr. **43**, 1640.

Peterson, L.R. and J.A. Allen, Jr. (1972), J. Chem. Phys. **12**, 6068.

Phillion, D.W., R.A. Lerche, V.C. Rupert, R.A. Haas and M.J. Boyle (1977a), Phys. Fluids **20**, 1892.

Phillion, D.W., W.L. Kruer and V.C. Rupert (1977b), Phys. Rev. Letters **39**, 1529.

Placzek, G. (1959), Rayleigh and Raman Effect, English translation UCR-Trans. -56(L).

Powell, H.T. and A.U. Hazi (1978), Chem. Phys. Lett. **59**, 71.

Powell, H.T. and J.J. Ewing (1978), Appl. Phys. Lett., **33**, 165.

Powell, H. and J. Murray (1975), Lawrence Livermore Laboratory, Laser Fusion Annual Report.

Prokhorov, A.M., F.V. Bunkin, K.S. Gochelashvilli and V.I. Shishov (1974), Vsp. Fiz. Nauk **114**, 415.

Pugh, E.R., J. Wallace, J.H. Jacob, D.B. Northam and J.D. Daugherty (1974), Appl. Optics **13**, 2512.

Quelle, F.W. (1966), Appl. Optics **5**, 633.

Rajnak, K. (1976), Unpublished Computer Code. Laser Division, Lawrence Livermore Laboratory, Livermore, California 94550, U.S.A.

Reilly, J.P. (1972), J. Appl. Phys. **43**, 3411.

Reisberg, L.A. and H.W. Moos (1968), Phys. Rev. **174**, 429.

Reisfeld, R. and Y. Eckstein (1975), J. Chem. Phys. **63**, 4001.

Ribe, F.L. (1975), Rev. Mod. Phys. **47**, 7.

Richardson, M.C., N.H. Burnett, G. Enright, P. Burtyn and K. Leopold (1976), Opt. Commun. **18**, 168.

Ripon, B.H. (1977a), Appl. Phys. Letters **30**, 134.

Ripon, B.H. (1977b), Absorption of Laser Light in Laser Fusion Plasmas, NRL Memorandum Report 3684.

Schlie, L.A., B.D. Guenther and R.D. Rathge (1976), Appl. Phys. Letters **28**, 393.

Schlitt, L.G. (1976), private communication.

Schlitt, L.G. and L.P. Bradley (1975), The Scaling of Electron Bream Sources for Laser Fusion Applications, University of California Report UCID-16864.

Schubauer, G.B., W.G. Spangenberg and P.S. Klebanoff (1950), Aerodynamic Characteristics of Damping Screens, NACA-TN-2001, January, 1950.

Seltzer, S.M. and M.J. Berger (1974), Nucl. Instr. and Meth., **119**, 157.

Shearer, J.W. (1970), A Survey of the Physics of Plasma Heating by Laser Light, University of California Report, UCID-15745.

Shen, Y.R. (1976), Rev. Mod. Phys., **48**, 1.

Shiau, J.N., E.B. Goldman and C.I. Weng (1974), Phys. Rev. Letters **32**, 352.

Sicre, J.E., J.T. Subois, K.J. Eisentraut and R.E. Sievers (1969), J. Amer. Chem. Soc. **91**, 3476.

Simmons, W., S.S. Guch, Jr., F. Rainer and J.E. Murray (1975), IEEE J. Quantum Electronics **QE-11**, 31D.

Simmons, W.W., D.R. Speck and J.T. Hunt (1977), Digest of Technical Papers, IEEE/OSA Conf. on Laser Eng. and Appl., p. 50.

Slivinsky, V.W., H.G. Ahlstrom, K.G. Tirsell, J. Larsen, S. Glaros, G. Zimmerman and H. Shay (1975), Phys. Rev. Letters **35**, 1083.

Smith, D.C. and R.G. Meyerand (1976), Laser Radiation Induced Gas Breakdown, in: Principles of Laser Plasmas (G. Bekefi, ed.; Wiley; New York), pp. 457–507.

Smith, E.W., R.E. Drullinger, M.M. Hessel and J. Cooper (1977), J. Chem. Phys. **66**, 5667.

Smith, W.L. (1978), Laser-Induced Breakdown in Optical Materials, University of California Report UCRL-80956 (to be published in *Optical Engineering*).

Spector, N., R. Reisfeld and L. Boehm (1977), Chem. Phys. Letters **49**, 49.

Steiner, D. (1975), Proc. IEEE **63**, 1968.

Stratton, T.F. (1976), High Power Gas Lasers, 1975, Inst. Phys. Conf. Ser. No. 29. (E.R. Pike, ed.; Inst. of Physics; Bristol and London), p. 284.

Suydam, B.R. (1975), IEEE J. Quantum Electronics **QE-11**, 225.

Swain, J.E. (1969), J. Appl. Phys. **40**, 3973.

Tallant, D.R., M.P. Miller and J.C. Wright (1976), J. Chem. Phys. **65**, 510.

Tatarski, V.I. (1961), Wave Propagation in a Turbulent Medium (McGraw-Hill; New York).

Thomas, E.E. and L.D. Siebert (1976), Appl. Optics **15**, 462.

Thomson, J.J. (1975), Nucl. Fusion **15**, 237.

Thomson, J.J. and J.I. Karash (1974), Phys. Fluids, **17**, 1608.

Thomson, J.J., W.L. Kruer, S.E. Bodner and J.S. DeGroot (1974), Phys. Fluids, **17**, 849.

Thornton, J.R., W.D. Gountain, G.W. Flint and T.G. Crow (1969), Appl. Optics **8**, 1087.

Trenholme, J.B. (1976), Laser Program Annual Report 1975, Lawrence Livermore Laboratory, UCRL-50021-75, p. 237.

Trenholme, J.B. (1972), Fluorescence Amplification and Parasitic Oscillation Limitations in Disk Laser, NRL Memorandum Rep. 2480.

Trenholme, J.B. and K.R. Manes (1972), A Simple Approach to Laser Amplifiers, University of California Report UCRL-51413.

Turner, C.E. (1975), Lawrence Livermore Laboratory, Laser Fusion Annual Report.

Turner, C.E. (1977), Appl. Phys. Letters **31**, 659.

Vlasov, S.N., V.A. Petrischev and V.I. Talanov (1971), Izv. VUZ Radiofiz. **14**, 1353.

Weber, M.J., T.E. Varitimos and B.H. Matsinger (1973), Phys. Rev. B **8**, 47.

Weber, M.J., C.F. Cline, W.L. Smith, D. Milan, D. Heiman and R.W. Hellwarth (1978), Appl. Phys. Letters **32**, 403.

Welling, H. and C.J. Bickart (1966), J. Opt. Soc. Amer. **56**, 611.

Wells, J.C., J.B. Gruber and M. Lewis (1977), Chem. Phys. **24**, 391.

Werner, C.W. (1975), Ph.D. Dissertation, Massachusetts Institute of Technology Cambridge 1975 (unpublished).

Werner, C.W. and E.V. George (1976), Principals of Excimer Lasers in: Principles of Laser Plasmas (G. Bekefi, ed.; Wiley; New York).

Werner, C.W., E.V. George, P.W. Hoff and C.K. Rhodes (1977), IEEE J. Quantum Electronics **QE-13**, 769.

Yamanaka, T., S. Nakai, Y. Kato, T. Sasaki, M. Matoba, K. Yoskida, Y. Mizumoto and C. Yamaka (1977), Proc. of the U.S.–Japan Seminar on Laser Interaction with Matter, Univ. of Rochester, Nov. 1976, Edited by C. Yamanaka, Inst. of Laser Eng., Osaka University.

York, G.W. and R.J. Carbone (1976), Proc. Third Summer Colloquium on Electronic Transition Lasers, Snowmass, Colo., Sept. 7–10, pp. 230–235.

Young, C.G. (1969), Laser Focus, November, 37.

Zelikoff, M., K. Watanabe and E.C.Y. Inn (1970), J. Chem. Phys. **21**, 1953.

Zuzak, W.W. and B. Ahlborn (1969), Can. J. Phys., **47**, 2667.

B6 | Continuous Picosecond Spectroscopy of Dyes

H.E. LESSING and A. VON JENA

Abt. Chemische Physik,
Universität Ulm,
F.R. Germany

Contents

© *North-Holland Publishing Company, 1979*
Laser Handbook, edited by M.L. Stitch

Abstract

This chapter deals with the photophysical and transport properties of dyes in solution that are relevant to their application as dye lasers, saturable absorbers or structural labels in biophysics. The necessary theoretical background on induced photodichroism, rotational diffusion and population kinetics is outlined. Experimental results are reviewed with emphasis on cw absorption-relaxation measurements. Theoretical tools are provided to deal with the coherent-coupling effect in the sample.

1 Introduction

The application to dyes occurred at the very beginning of picosecond spectroscopy. Shortly after passive mode-locking was reported by Mocker and Collins (1965) and DeMaria et al. (1966) interest turned to the transmission versus time characteristics of the dye in the passive absorber cell within the laser resonator. In the first published experiment by Shelton and Armstrong (1967) the output beam of the passively mode-locked Nd:glass laser was attenuated, successively delayed and redirected sideways through the absorber cell inside the resonator to probe the momentary transmission of the Eastman 9740 dye. In the next published experiment by Scarlet et al. (1968) the sample cell was placed external to the laser resonator so that dyes not working as saturable absorber could be measured also. To this end the laser pulse stream was primarily used to bleach the sample, whereas an attenuated pulse stream, that was derived from the laser output by a beam splitter, probed the transmission at successive delays. This method has persisted – with minor modifications – until today. Its inherent sampling principle allows one to take full advantage of the picosecond pulse width available from mode-locked lasers, whereas the fluorescence-decay methods used thus far were limited by the nanosecond response of the photomultipliers. Soon photochemists discovered that picosecond pulses would hold the promise to study photophysical processes in organic molecules. Single-shot lasers were installed in chemistry departments and their output frequency doubled, tripled or quadrupled to reach the interesting ultraviolet absorption bands of the molecules. With their time-consuming alignment procedures, erratic operation and tedious measurement procedures, mode-locked single-shot lasers are no easy research tool. In addition, in fluorescent molecules intensity-dependent lifetime shortening by stimulated fluorescence is an ever-present artifact, as shown early by Lessing et al. (1970).

The necessity to decrease excitation intensity for sensible results led to continuous picosecond spectroscopy. Runge (1971) showed that a continuous He–Ne laser could be mode-locked by dyes if the intra-resonator power was focused into the absorber cell. Since the dyes were lasing simultaneously, the interesting conclusion was that enough molecules must have been excited to be detectable also in transmission. Thus the power loss of five orders of magnitude as compared to single-shot lasers could be compensated for by focusing in order to arrive at useful intensities. Development along these lines resulted in two groups publishing continuous picosecond transmission measurements; Shank and Ippen (1975) using their continuous mode-locked dye laser and Lessing et al. (1975) using a mode-locked Ar laser. Instead of a hedge of error bars or noisy densitograms continuous picosecond spectroscopy delivers smooth transmission versus time curves revealing details not seen in the single-shot measurements.

Continuous mode-locked lasers were also used rather early on by Merkelo et al. (1969) for fluorescence-decay measurements limited by photomultiplier response. Only recently the feasibility of picosecond fluorescence-decay measurements with a sum-frequency light gate has been demonstrated by Hirsch et al. (1976) using a continuous mode-locked dye laser.

Continuous picosecond spectroscopy, then, is on the verge of becoming a routine method in photochemistry. This chapter will give the quantitative description of dye relaxation in solutions in some detail with emphasis on the transient-transmission measurement with identical wavelengths. This is not only because this mode of picosecond spectroscopy is the simplest and commonest at present, but also because it gives more information than the fluorescence-decay method, and from nonfluorescent molecules, too. Each step in theory is followed by a discussion of the relevant measurements which are compared only sporadically with single-shot results. While the treatment should hold for the electronic transitions of all organic molecules, actual measurements are predominantly on dyes with their absorption bands in the visible, where intense mode-locked pulses are available without frequency doubling. Finally an attempt is undertaken to compare the experimental arrangements published thus far.

2 Principles and methods

2.1 Photochemical primary processes

The light absorption and emission characteristics of organic molecules are governed by transitions between electronic levels that are characterized by excitations of the loosely bound π electrons. The vibrational and rotational substructure of these complex molecules is so dense that it appears as a smooth broadening of the electronic levels as a rule. As with atoms there is a singlet and triplet ladder with the spins antiparallel and parallel respectively. The ground state is a singlet S_0, the lowest triplet state is T_1. Except for a lowering of the triplet ladder due to exchange interaction, the successions of higher states S_n and T_n closely parallel each other. As a consequence and owing to the broadness of the levels there is a good chance that a molecule once excited by a photon to S_1 or T_1 can absorb another photon of the same energy and arrive in a S_n or T_n state respectively. In addition, there may be hidden levels, the transitions to which are parity forbidden from the ground state but not from S_1.

The kinetic aspects of the various paths of activation and deactivation are discussed using a Perrin scheme (fig. 1), where a single molecule is thought to experience successively all kinds of transitions. The individual lifetimes shown give, when averaged over all molecules, the average lifetimes indicated. The

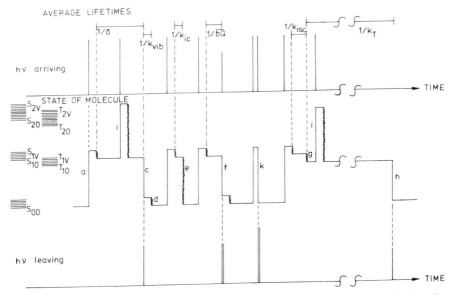

Fig. 1. Perrin scheme of photochemical primary processes in an organic molecule. The lettering is explained in the text.

observable continuous relaxation curves result as a summation over all random elementary acts weighted by the number of molecules undergoing them.

(a) *Absorption* $S_{00} \rightarrow S_{1v}$. The absorption process from the zero vibrational level of the ground state, in which most molecules reside owing to the Boltzmann distribution, ends usually in a higher vibrational level v of the first excited singlet state. The electron jump time is estimated to be 10^{-15} s (time for a wavelength to pass the molecule). The absorption rate for monochromatic light at frequency ν_e with isotropic polarization directions is given by $h\nu_e b\bar{Q}N_0(t)$ or $\sigma_1 I N_0(t)$, where b is the spectral Einstein coefficient for absorption (for the excitation frequency), $\bar{Q}$ the frequency-integrated photon density in the sample, I the total photon flux where $I = c\bar{Q}$ (c is the light velocity), and N_0 the population density of S_{00}. The connection between b, the absorption cross section σ_1 and the photochemist's decadic extinction coefficient ϵ can be read from Beer's law of absorption for the small-signal case and isotropic samples:

$$I(x) = I(0) \exp(-h\nu_e b N_0 x/c) = I(0) \exp(-\sigma_1 N_0 x) = I(0)10^{-\epsilon Cx}, \qquad (1)$$

where x is the thickness of the sample and C the concentration in mol ℓ^{-1}.

The Einstein coefficient integrated over the absorption band is given from quantum-mechanical perturbation theory for isotropic light polarization by

$$\bar{b} \equiv \int_{abs} b(\nu) \, d\nu = \frac{8\pi^3}{3h^2c} |p_{01}|^2, \qquad (2)$$

where p_{01} is the electronic transition moment between the electron densities of

S_0 and S_1 described by their electronic wavefunctions ψ_0^{el} and ψ_1^{el} respectively:

$$p_{01} = \langle \psi_0^{el} | e\mathbf{r} | \psi_1^{el} \rangle. \tag{3}$$

(b) *Vibrational relaxation* $S_{1v} \leadsto S_{10}$. It is an experimental fact that radiation is emitted only from the S_{10} level independent of the wavelength of excitation (Kasha rule). Consequently there must be a fast relaxation $S_{1v} \leadsto S_{10}$ within the vibrational manifold. It could be allowed for by a heuristic rate constant k_{vib}, but since it is estimated to be of the order of 10^{12} to $10^{13} \, s^{-1}$, the quasi-steady-state approximation is used when dealing with slower phenomena. Thus $N_{1v} \approx 0$, and every excitation finally ends up in S_{10}. Nevertheless it is the vibrational relaxation that warrants the four-level treatment of the absorption and emission kinetics of dyes (see § 3.1).

(c) *Spontaneous fluorescence* $S_{10} \rightarrow S_{0v}$. The total spontaneous fluorescence rate from S_{10} is given by

$$\bar{a} N_1(t), \text{ where } \bar{a} \equiv \int_{fluor} a(\nu) \, d\nu.$$

$\bar{a}$ is the Einstein coefficient for spontaneous emission integrated over the fluorescence band and N_1 the population density of S_{10}. The reciprocal $1/\bar{a} = \tau_0$ is called the natural or radiative lifetime that would be observed if all processes from S_{10} were radiative. The fluorescence band shape enters via $a(\nu)$. $a(\nu)$ can be calculated from $b(\nu)$ at the same frequency using the Einstein relation

$$a(\nu) = 8\pi (h^3/c^3) \nu^3 b(\nu). \tag{4}$$

Thus the radiative lifetime can be calculated from the absorption band $S_0 \rightarrow S_1$:

$$\frac{1}{\tau_0} = \bar{a} = 8\pi \frac{h^2}{c^2} \int_{abs} \nu^2 \sigma_1(\nu) \, d\nu, \tag{5}$$

which is of the order of $10^8 \, s^{-1}$.

(d) *Vibrational relaxation* $S_{0v} \leadsto S_{00}$. Essentially the same holds here as for step (b). There is a measurement by Laubereau et al. (1975) in Coumarin 6 giving a vibrational relaxation time in the ground state of 1.4 ps. Since radiationless energy is dissipated twice, the energy of the emitted photons is generally lower than the energy of photons suitable for absorption. As a consequence the fluorescence is shifted to the red as compared to the longest-wavelength absorption band (Stokes shift). There remains a mirror relationship between the fluorescence band and this absorption band relative to the resonant 0–0 transition $S_{00} \leftrightarrows S_{10}$ (Levshin rule).

(e) *Internal conversion*. Internal conversion can be thought of as an isoenergetic transition from the S_{10} vibronic level to a high S_{0v} vibronic level of the S_0 vibrational manifold with ensuing vibrational relaxation. It is taken into account by a heuristic rate constant k_{ic} so that the loss rate from S_1 is given by $k_{ic} N_1$.

(f) Stimulated fluorescence $S_{10} \rightarrow S_{0v}$. With strong excitation the stimulated emission process becomes appreciable where a fluorescence photon hits an already excited molecule. For this process alone the deactivation rate would be $\overline{h\nu bQ}N_1$, where b is the Einstein coefficient for stimulated emission (being equal to that for absorption at the corresponding mirror frequency) and Q the spectral fluorescence photon density, the bar indicating integration over the fluorescence band.

(g) Intersystem crossing. The radiative transition between singlet and triplet states is spin forbidden. The radiationless transition is accounted for by a intersystem-crossing rate constant k_{isc}, and the rate is $k_{isc}N_1$.

With the radiative and nonradiative transitions competing for the S_{10} population the observed fluorescence decay time τ in the small-signal case is determined by

$$\tau = (\bar{a} + k_{ic} + k_{isc})^{-1} \tag{6}$$

and the fluorescence quantum yield η by

$$\eta = \bar{a}/(\bar{a} + k_{ic} + k_{isc}) = \tau/\tau_0. \tag{7}$$

Also a triplet yield ϕ is defined as

$$\phi = k_{isc}/(\bar{a} + k_{ic} + k_{isc}) = k_{isc}\tau. \tag{8}$$

(h) Phosphorescence $T_{10} \rightarrow S_{0v}$. The phosphorescence photons are of still lower energy than the fluorescence photons, which means that the phosphorescence band is shifted to the red. More often than not room-temperature phosphorescence is suppressed by competing processes such as the quenching by oxygen in aerated solutions. For our purposes only the triplet depleting rate due to all processes combined, which is accounted for by $k_T N_3$, where N_3 is the population density of T_{10}, is of interest.

(i) Transitions to higher states $S_{10} \rightarrow S_{nv}$ and $T_{10} \rightarrow T_{nv}$. Kasha's rule and similar reasoning for the triplet ladder suggest a nonradiative recovery of these excitations. $T_1 \rightarrow T_n$ spectra are known from classical and $S_1 \rightarrow S_n$ spectra from nanosecond flash spectroscopy. The rates of these excitations are $\sigma_2 I N_1$ and $\sigma_3 I N_3$ respectively, σ_2 and σ_3 being the relevant absorption cross sections.

(j) Resonant stimulated emission $S_{1v} \rightarrow S_{00}$. This process at the frequency of the exciting radiation is far less important at medium intensities than the stimulated fluorescence $S_{10} \rightarrow S_{0v}$. Only at very high intensities may it compete with the fast vibrational relaxation.

In addition, a basic insight into molecular photochemistry investigations of primary photoprocesses gives important clues for the application of dyes as fluorescence brighteners, organic lasers, saturable absorbers and structural labels in biophysics. For the laser dyes the requirements are, in order of their importance: high extinction coefficient, low triplet yield, no absorption to higher states, and good quantum yield. For further details see Drexhage (1973) and Schäfer (1973).

For mode-locking saturable absorbers the requirements are similar, if the contribution of stimulated fluorescence to recovery shortening is kept in mind. For nonfluorescing absorbers the requirement is a short excited-state lifetime (see Lowdermilk 1979).

2.2 Transport properties

In dye solutions there is a continuous brownian motion of the individual dye molecules. Since the transition moment is connected rigidly to the molecular structure the optical properties are definitely influenced by the transport properties.

2.2.1 Rotational diffusion and libration

To derive the time course of the mean square of angular displacement one starts with the angle θ of rotation around a molecular axis (see, e.g., Becker 1966):

$$\overline{\theta^2} = \frac{1}{N} \sum_{i=1}^{N} \{\theta_i(t) - \theta_i(0)\}^2,$$

where N the number of molecules and θ_i the angle of rotation of molecule i. To describe rotational diffusion, the change in time of the mean square of the rotational displacement has to be considered:

$$\frac{d\overline{\theta^2}}{dt} = \frac{d}{dt} \left\{ \frac{1}{N} \sum_{i=1}^{N} \theta_i^2(t) \right\} - \frac{2}{N} \sum_{i=1}^{N} \dot{\theta}_i(t)\theta_i(0)$$

$$= 2(\overline{\theta\dot{\theta}}) - \frac{2}{N} \sum_{i=1}^{N} \dot{\theta}_i(t)\theta_i(0).$$

The second term has to be zero, since for every direction $\theta(0)$ the average angular velocity must vanish for all times t (isotropic directions). The first term can be related to the rotational diffusion constant D for the rotation around the appropriate axis. The latter is introduced via the angular flux density $j_\theta = n(\theta, t)\dot{\theta}$: $j_\theta = -D\partial n(\theta, t)/\partial\theta$, where $n(\theta, t)$ is the distribution of molecular orientations θ. Multiplying both sides by θ and integrating, one arrives at

$$(\overline{\theta\dot{\theta}}) = D \text{ or } d\overline{\theta^2}/dt = 2D. \tag{9}$$

Thus the rotational diffusion constant D can be related to the averaged molecular motion

$$D = \overline{\theta^2}/2t. \tag{10}$$

Since $\overline{\theta^2}$ is not amenable to measurement one needs an equation of motion. According to Langevin one can start from the angular analog of Newton's law

$$J\dot{\omega} = T(t) = -\omega/B + M(t), \tag{11}$$

where J is the moment of inertia, $\omega \equiv \dot{\theta}$ angular velocity and $T(t)$ is the torque. This was split by Langevin into a constant frictional force (B angular mobility) and a fluctuating part $M(t)$ that warrants fulfillment of the equipartition theorem $\frac{1}{2}J\omega^2 = \frac{1}{2}kT$. Averaging again, one obtains the required equation

$$(d/dt)(\overline{\theta\dot{\theta}}) + \frac{1}{BJ}(\overline{\theta\dot{\theta}}) = kT/J$$

with the solution

$$\overline{\theta\dot{\theta}} = BkT + C \exp(-t/(BJ)).$$

Comparing this with eq. (9) for times $t \gg BJ$ provides the familiar Einstein relation

$$D = BkT. \tag{12}$$

Since rotation of a molecule is not so much impeded by neighbouring molecules as translation, the rotational diffusion times $\tau_R = 1/(6D)$ may reach well into the picosecond range for low viscosity solvents. The general solution to the Langevin equation [eq. (11)] with initial condition $\theta(0) = 0$ is

$$\overline{\theta^2} = 2D(t - BJ\{1 - \exp(-t/(BJ))\}), \tag{13}$$

which contains eq. (10) as a special case. For small times t one obtains

$$\overline{\theta^2} = 2D\left(t - BJ\left\{\frac{t}{BJ} - \frac{t^2}{2B^2J^2}\right\}\right) = \frac{D}{BJ}t^2 = \frac{kT}{J}t^2. \tag{14}$$

We find here at short times a practically free rotation (libration), the angular velocity of which is determined by thermal energy $\omega = (kT/J)^{1/2}$. There is a difficulty in deciding when libration ends and rotational diffusion definitely takes over. A possible convention is to describe the rotation up to $t = BJ$ as libration [eq. (14)] and then as rotational diffusion. In fig. 2 this convention is depicted as the uninterrupted line, whereas the dashed curve gives the general solution [eq. (13)].

2.2.2 Lateral diffusion

Lateral diffusion is described analogously by the relation for the mean square of the displacement

$$\overline{x^2} = 2D_\ell t, \tag{15}$$

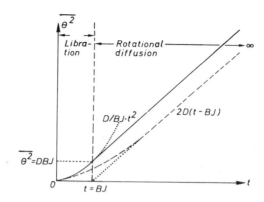

Fig. 2. Time range of libration and rotational diffusion.

where D_ℓ is the lateral diffusion constant. For Coumarin 6 it has been measured to be $5.1 \times 10^{-4} \text{mm}^2\text{s}^{-1}$ in methanol and $2.2 \times 10^{-5} \text{mm}^2\text{s}^{-1}$ in ethylene glycol (see Lessing 1977). The only effect of lateral diffusion could be that, after pulsed excitation, excited molecules diffuse out of or unexcited ones into the illuminated volume. Even for focused excitation (30 nm diameter) this effect is rather slow; for Coumarin 6 one calculates a diffusion time of 100 ms and 2.3 s respectively.

2.3 Experimental methods

Whereas the fluorescence-decay method has been in use for half a century, the transient-transmission method had to await the advent of the laser. Fig. 3 compares the gross features of both methods in the simplest case of singlet kinetics. The inability to achieve picosecond resolution in fluorescence decay has only recently been overcome by the continuous photogating method of Hirsch et al. (1976). Since the conventional fluorescence decay method is covered in several review papers (see, e.g., Ware 1972), in what follows only the transient-transmission method will be elaborated further.

2.3.1 Modified Lambert–Beer law

The bleaching by strong excitation pulses can be understood starting from the Lambert–Beer absorption law

$$\Delta I(x)/\Delta I(0) = \exp(-\sigma N x), \tag{16}$$

which holds only for the small-signal approximation. For a higher bleaching intensity $I(x, t)$ we have instead,

$$\partial I(x, t)/\partial x = -\sigma I(x, t)(N - \Delta N\{I(x, t), t\})$$

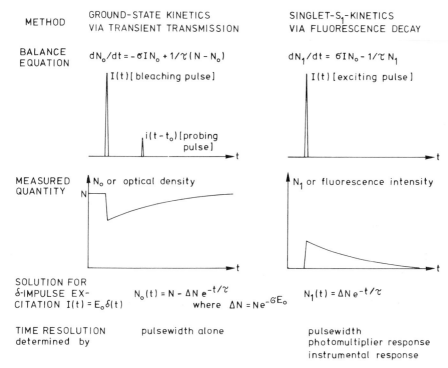

Fig. 3. Comparison of transient-transmission and fluorescence-decay methods.

where ΔN is the intensity-dependent ground-state depletion. The simplest scheme of picosecond spectroscopy is the one in which a weak probing waveform $i(x, t)$, which is split off the bleaching beam by a beam splitter and delayed by t_0, monitors the ground-state population without changing the latter appreciably (see fig. 3):

$$\partial i(x, t - t_0)/\partial x = -\sigma i(x, t - t_0)(N - \Delta N\{I(x, t),t\}).$$

After integration one has for the transmission of the weak probing waveform

$$i(t - t_0) = i_0(t - t_0) \exp(-\sigma d\{N - \langle \Delta N(t)\rangle_x\}),$$

with

$$\langle \Delta N(t)\rangle_x = \frac{1}{d} \int_0^d \Delta N\{I(x, t),t\}\, dx, \tag{17}$$

being the ground-state depletion averaged over the cell length d. Of course, only $\langle \Delta N(t \approx t_0)\rangle_x$ is being probed. Assuming δ-impulse waveforms with the delay t_0 we arrive at

$$i = i_0 \exp(-\sigma N d) \exp\{\sigma \langle \Delta N(t_0)\rangle_x d\}.$$

In the small-signal regime (ΔN small) the second exponential can now be expanded:

$$i = i_0 \exp(-\sigma Nd)[1 + \sigma\langle \Delta N(t_0)\rangle_x d] = i'[1 + \sigma\langle \Delta N(t_0)\rangle_x d], \tag{18}$$

where i' is just the transmitted intensity without bleaching. In view of this a signal Y proportional to the sampled quantity ΔN alone suggests itself:

$$Y(t_0) = (i - i')/i' = (T_1(t_0) - T)/T = \Delta T/T = \sigma\langle \Delta N(t_0)\rangle_x d \tag{19}$$

$(T_1(t_0) =$ transmission *with* excitation at delay t_0, and $T =$ transmission without excitation) that can be shaped easily in continuous picosecond spectroscopy by chopping of the bleaching beam and synchronous detection of the probing beam (see § 6).

2.3.2 Time resolution

What has been said so far holds for a pulse width that is large compared to the transit time $\tau_d = d/c$ through the cell. At the other extreme, pulse width small compared to transit time, the case of co-running pulses has to be distinguished from the case of counter-running pulses. For counter-running pulses we have

$$Y(t_0) = \frac{\Delta T(t_0)}{T} = \sigma d\sigma N \frac{1}{\tau_d} \int_{-\tau_d/2}^{+\tau_d/2} E\{c(t' + \tau_d/2)\}a(t_0 - t')\, dt', \tag{20}$$

where E is the pulse photon content per area unit and $a(t - t')$ is the normalized response function $a(t - t') = \theta(t - t')\{\exp - (t - t')/\tau\}$ containing the Heavyside step function $\theta(t - t')$. Evidently the molecular response is smeared out by averaging the differential contributions over the sample depth. Since convenient sample depths are of the order of 1 mm the time resolution then is 5 ps. For co-running pulses we arrive at

$$Y(t_0) = \Delta T(t_0)/T = \sigma d\sigma N \frac{1}{d} \int_0^d E(x)\, dx \int_{-\infty}^{x} dt f(t - t_0) \int_{x}^{t} dt' a(t - t'), \tag{21}$$

where $f(t)$ is the pulse shape. Now it is not the sample depth that limits the time resolution but the pulse width. In view of the better separability of bleaching and probing light and of optimal focus overlap with the counter-running geometry, one has to check carefully whether the theoretical time resolving potential of the co-running geometry can be reached in a projected set-up.

2.3.3 Beam focusing

In continuous picosecond spectroscopy peak powers available from cw mode-locked lasers range from one hundred watts to kilowatts, which is some five orders of magnitude weaker than available from single-shot lasers. Since the amount of bleaching is proportional to photon flux per unit area, this handicap can be compensated by extreme focusing, taking full advantage of the optical

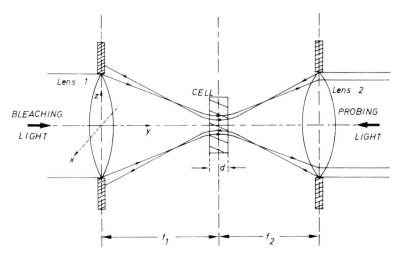

Fig. 4. Focusing geometry for counter-running pulses.

quality of the TEM_{00} laser beam. Fig. 4 shows the sample geometry when working with counter-running pulses. Lens 1 for the bleaching beam is chosen with a larger focal length than lens 2 for the probing beam in order to provide some allowance for alignment. The diffraction-limited spot size for a beam diameter d_0 and focal length f is given by $w_0 = 2\lambda f/\pi d_0$, where λ is the wavelength of light. In order to avoid strong variations of beam cross section over the sample, one has to bear in mind the criterion that the Rayleigh range $Y_R = \pi w_0^2/\lambda$ equals the sample depth d, so that extreme focusing is limited. Under these circumstances the optimal spot size for a sample of thickness d turns out to be $w_0 = 0.54(\lambda d/\pi)^{1/2}$ from which, together with w_0 from above, the proper f-number of the focusing lens can be estimated. The amount of population change that can be achieved in a typical continuous experiment is estimated here based on the following data: peak power $P = 50\,W$, average beam cross section $F = \bar{w}^2\pi = 1.25 \times 10^3\,\mu m$, energy of a photon $E_p = 4 \times 10^{-19}\,J$, pulse duration $\Delta t = 3 \times 10^{-10}\,s$, and absorption cross section $\sigma = 10^{-16}\,cm^2$.

Assuming no relaxation during the excitation process we can take from § 2.1, $\dot{N}_0(t) = -\sigma I N_0(t)$, which after integration over the pulse duration Δt yields

$$N(\Delta t) = N \exp\left\{-\sigma \int_0^{\Delta t} I(t)\,dt\right\},$$

and insertion of the numbers gives $N(\Delta t) \approx N(1 - 0.25)$. With 25% of the molecules being excited there is no longer the small-signal case. Therefore the beam is usually attenuated, leaving a reserve for molecules with weaker absorption.

2.3.4 Photoselection

By assuming light with isotropic polarization directions in § 2, the influence of polarization and molecular orientation did not show up in the transition probabilities discussed there. In fact, with linearly polarized light represented by the electric field component E_z,

$$w = (\pi^2/h)|E \cdot p_{12}|^2 = (\pi^2/h)|E_z|^2|p_{12}|^2 \cos^2 \theta,$$

or written with photon flux density I, we have

$$w(\theta) = 3\sigma I \cos^2 \theta \qquad (22)$$

instead of σI for isotropic polarization directions. Even with naturally polarized light there is an orientational dependence of absorption probability. As a consequence we have in both cases immediately after excitation some preferential orientation of S_1 molecules, a phenomenon for which the term photoselection has been coined by Albrecht (1961). To be more specific, the transition moments of the S_1 molecules have a $\cos^2 \theta$ distribution around the polarization direction of the bleaching light (see fig. 5). The ground-state molecules show a complementary distribution $(N/4\pi) - N_1(\theta, \phi)$. When the probing pulse arrives by the time t_0 later, S_1 molecules have rotated out of its polarization direction or relaxed to the ground state and S_0 molecules have rotated into this direction. Evidently, rotational diffusion as well as level kinetics contribute to the overall relaxation of transmission. This induced photodichroism, as it is sometimes called, has been observed already by Neporent and Stolbova (1961) on a slow time scale. Lombardi et al. (1964) analogously observed rotation of a photoproduct in a viscous matrix. The first picosecond experiment on induced photodichroism was by Eisenthal and Drexhage (1969). Since they used a special 45° geometry it remained unnoticed by other workers in the field that light polarization and rotational diffusion contribute virtually to any transient-transmission experiment on dyes in solution. It was only recently that a rotation-free transient-transmission measurement could be demonstrated (Lessing and von Jena 1976). For fluorescence-decay measurements it has long been known that by choosing special propagation directions for excitation and emission the isotropic level kinetics can be extracted (see, e.g., Fleming et al. 1976).

Linear photodichroism of dye-labeled biopolymers is turning into an important structural method in biophysics.

2.3.5 Coherent wave superposition

On closer inspection the various designs for delaying the probing pulse after the bleaching pulse turn out to be actually interferometers in the classical sense. Instead of continuous beams we have the interference of wave packets that are coherent because of their common origin from the same laser pulse. Of course, the overlap takes place only near zero delay. With an absorbing sample at the

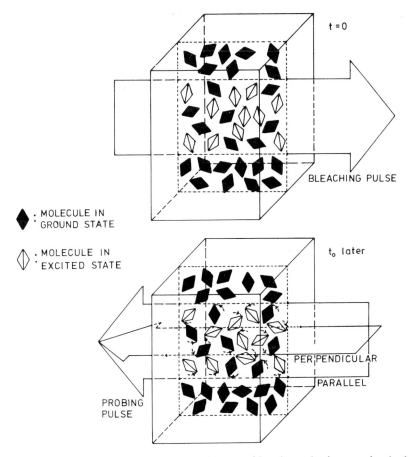

Fig. 5. Principle of transient-transmission method governed by photoselection, rotational relaxation and ground-state recovery.

position of overlap there is an interesting effect, coherent beam coupling. By way of interference a spatial hole-burning is established within the sample. This serves as an amplitude grating that diffracts light from the bleaching beam into the probing beam. In this way it appears that the sample has a higher transmission at $t_0 = 0$ than one would expect from the excitation intensity.

The first hint of this effect can be seen in a measurement by Shank and Auston (1975) in the solid state. Meanwhile it is reproducibly established by continuous picosecond spectroscopy; in fact, its presence to the predicted extent is a good criterion for optimal alignment of bleaching and probing beams. On the other hand, it can be regarded as an artifact that has distorted some conclusions drawn from the initial portion of the transient-transmission curve in the past. The effect can be seen even with perpendicular polarizations of bleaching and probing beams, where it depends on rotational diffusion of the dye molecule (see von Jena and Lessing 1979a). A quantitative understanding of the effect is useful for the evaluation of transient-transmission measurements. It disappears if bleaching and probing beam have different wavelengths. The coherent-coupling effect can be used also as an experimental method for the determination of level and rotational kinetics, as demonstrated by Phillion et al. (1975). In their set-up the signal is the intensity of a diffracted beam (see § 6).

3 Rotation-free relaxation

It has already been discussed in § 2.3.4 that linearly or even naturally polarized light leaves a preferentially oriented set of ground-state molecules that randomizes via rotational diffusion in liquid solution. By choosing the magic angle of 54.7° between the polarization directions of probing and bleaching beam the contributions of photoselection and orientation are cancelled (see § 4.2). The signal then appears as if the sample had been excited using isotropic polarization directions. It is only owing to this singular angle that rotation-free level kinetics can be observed with the linearly polarized light from mode-locked lasers.

3.1 Introduction of balance equations

With the transition probalities and nomenclature introduced in § 2.1 the population kinetics in the three lowest levels of the molecules can be described by balance equations with an underlying four-level scheme. The S_{0v} and S_{1v} levels are assumed to be swiftly depopulated and thus are practically empty, whereas the S_{00} and S_{10} levels have appreciable populations and occupation times. For this reason the resonant stimulated emission $S_{1v} \rightarrow S_{00}$ is neglected. For the resonant 0–0 transitions $S_{00} \rightarrow S_{10}$ and $S_{10} \rightarrow S_{00}$ a two-level scheme should be adopted, but its contribution is neglected here. The T_{10} level will be populated solely by intersystem crossing and depopulated by a (usually nonradiative) rate constant k_T. With these assumptions and monochromatic excitation at frequency ν_0 the balance for the S_0, S_1 and T_1 population density and the particle conservation condition are as follows:

$$dN_0(t)/dt = -\sigma(\nu_0)I(t)N_0(t) + (k_f + k_{ic})N_1(t) + k_T N_3(t),$$
$$dN_1(t)/dt = \sigma(\nu_0)I(t)N_0(t) - k_1 N_1(t),$$
$$dN_3(t)/dt = k_{isc}N_1(t) - k_T N_3(t),$$
$$N = N_0(t) + N_1(t) + N_3(t). \tag{23}$$

Here k_f has been written for $\bar{a}$. Recovery from S_1 to S_0 is governed by $k_f + k_{ic}$ only, since intersystem crossing at k_{isc} does not contribute, whereas the disappearance of S_1 is governed by $k_1 = k_f + k_{ic} + k_{isc}$.

3.2 Influence of stimulated fluorescence

Since dye molecules act as a four-level system, stimulated fluorescence $S_{10} \rightarrow S_{0v}$ becomes important at rather low excitation. The kinetic treatment of stimulated emission goes back to Statz and deMars (1960) who introduced the view that besides the population-density balance for the molecular levels a photon-density balance has to be fulfilled simultaneously. In the following adaptation by Lessing et al. (1970) again the simplified four-level scheme is assumed and the triplet contribution neglected for simplicity. For reasons of compatibility with the photon density Q the population-density balance is formulated with Einstein coefficients a and b instead of k_F and σ respectively:

$$dN_0(t)/dt = -\overline{h\nu bq}\, N_0 + \overline{h\nu bQ}\, N_1 + \bar{a}N_1/\eta, \qquad (24)$$

$$dQ(\nu, t)/dt = +\overline{h\nu bQ}\, N_1 + aN_1 - pQ, \qquad (25)$$

where $q(\nu, t)$ is the exciting photon density, $Q(\nu, t)$ the fluorescence photon density, η the small-signal fluorescence quantum yield and the bar means frequency integration over the respective band. The complication due to reabsorption in the overlapping region of the longest-wavelength absorption band and fluorescence band has been neglected here, since stimulated emission occurs preferably on the longer-wavelength side of the fluorescence band. p is a photon-dissipation rate constant, which is known for the common laser resonator of length L and mirror reflectivity R to be $(c/L)\ln(1/R)$. For a spherical sample volume V of diameter d without reflecting walls an expression for p has been given, derived on the assumption of spatially constant population and photon density:

$$p = 9c/2d. \qquad (26)$$

A derivation by L. White (unpublished) as quoted by Fleming et al. (1977) arrives at a slightly different value $p = 8c/3d$.

Since pQ is the only loss term in the photon balance equation, it presents the fluorescence signal that can be measured outside the sample

$$pQ(\nu, t)V \text{ photons s}^{-1}\text{Hz}^{-1}.$$

Eqs. (24) and (25) form a pair of coupled nonlinear integro-differential equations solvable only numerically. Fig. 6 shows the solution for δ-impulse excitation of Rhodamine 6G with increasing percentages of the dye molecules having been excited. The normalized decay of $N_1(t)$ that can be detected in transmission, since $N_0(t) = N - N_1(t)$, already shows for 1% of the molecules excited some deviation from the small-signal exponential. For 12.5% the effect is evident. As a puzzling consequence of the nonlinearity of stimulated emission the fluorescence decay pQV is shortened even more, i.e., no longer parallels the decay of $N_1(t)$ as in the small-signal case. Systematic investigations of this lifetime shortening by stimulated fluorescence using single-shot mode-locked lasers have been done for fluorescence decay by Lessing et al. (1970) and Fleming et al. (1977), and for transient transmission by Müller and Willenbring (1974). There are many more involuntary examples in the single-shot picosecond literature, where usually high excitation intensities are used to get large transmission changes or high fluorescence intensities that can be handled by photogating. The

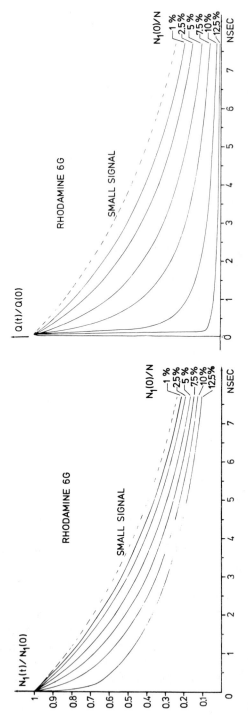

Fig. 6. Excitation-dependent lifetime shortening effect in normalized population density N_1 (left) and photon density Q (right) calculated for Rhodamine 6G. From Lessing et al. (1970).

perturbing influence of stimulated fluorescence that makes comparison of large-signal measurements without computer analysis illusory has only recently been realized by other workers in the field (see Busch et al. 1975). Illustrative examples of large-signal single-shot measurements are contrasted with small-signal measurements in tables 1 and 2.

On the other hand, recovery shortening due to stimulated fluorescence is instrumental for the mode-locking process in saturable absorbers (Lessing et al. 1975). This became evident when Runge (1971) passively mode-locked a He–Ne laser for which the absorber was working simultaneously as a dye laser. The finding that laser dyes are better mode-lockers than nonlasing ones was applied by Yasa and Teschke (1975) to the mode-locking of dye lasers. They called it double mode-locking, since the absorber fluorescence appears as a longer-wavelength pulse too. Stimulated fluorescence is probably helpful in mode-locking the ruby laser with cryptocyanine or DDI. The absorber fluorescence pulse should appear at 738 nm, but since it has not yet been observed among the ruby laser pulses at 693 nm, the term double mode-locking has not been applied here.

Although the examples in tables 1 and 2 seem to prove the contrary, transient transmission is less prone to shortening by stimulated fluorescence than fluorescence decay in the same molecule (see fig. 7). In this respect the transient-transmission method is superior to the fluorescence-decay method. However, when dealing with strong excitation the small-signal approximation [eq. (18)] breaks down for the transient-transmission signal. Instead we now have

$$Y(t_0) = \Delta T/T = \exp\{\sigma_1 \langle \Delta N(t_0) \rangle_x d\} - 1, \tag{27}$$

where σ_1 corresponds to absorption of the probing light and $\Delta N = N_1 + N_3$. The relative change in transmission is no longer linearly related to the population density change of S_0. If a molecule with zero triplet yield is assumed for

Table 1.

Fluorescence decay (1/e)-time of 1, 1′-diethyl-4, 4′-carbocyanine iodide (cryptoncyanine) and 1, 1′-diethyl-2, 2′-dicarbocyanine iodide (DDI).

Excitation	Cryptocyanine	DDI	Reference
8 (GW cm^{-2})	22 ps	14 ps	Duguay and Hansen (1969)
Phase fluorimeter	270 ps	150 ps	Bonch-Bruevich et al. (1969)

Table 2.

Absorption recovery (1/e)-time of 3, 3′-diethyloxa-dicarbocyanine iodide (DODCI).

Excitation	Recovery	Reference
4 (GW cm^{-2})	10 ps	Busch et al. (1975)
(MW cm^{-2})	1.2 ns	Shank and Ippen (1975)

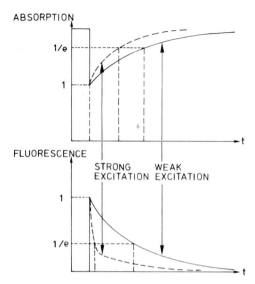

Fig. 7. 1/e times for transient transmission (top) and fluorescence decay (bottom) illustrating different lifetime shortening by stimulated fluorescence.

simplicity, we have

$$Y(t_0) = \Delta T/T = \exp\{\sigma_1 \langle N_1(t_0)\rangle_x d\} - 1, \tag{28}$$

with now appreciable population N_1 in S_1. For $\sigma_1 \langle N_1(t_0)\rangle_x d = 1$ the measured signal $Y = \exp(1) - 1 = 1.72$ deviates considerably; for $\sigma_1 \langle N_1(t_0)\rangle_x d \leq 0.1$ we have a tolerable deviation, $Y \leq 0.105$. Taking $\{Y(t_0)\}_{max} < 0.1$ as the necessary condition, one must keep the energy of the bleaching pulse sufficiently low, while the extinction $\sigma_1 N d$ should be in the range 1 to 3. The large-signal case can be solved by a numerical treatment of eqs. (23). Fig. 8 (left) demonstrates that with increasing excitation saturation of the $S_0 \rightarrow S_1$ absorption leads to a steeper rise in the S_1-population. Fig. 8 (right) shows in addition that owing to eq. (28) the

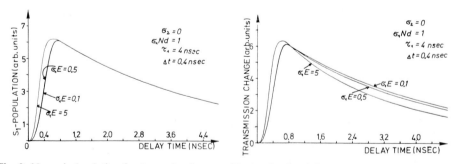

Fig. 8. Numerical solution for large-signal case without stimulated fluorescence or triplet involvement. Left: calculated S_1 population density versus time. Right: relative transmission change signal calculated from S_1. Both normalized to same ordinate at $t = 2\Delta t$, where the exciting pulse is over (coherence-free).

recovery of the corresponding transmission change versus time curve becomes faster if the S_0-population is strongly bleached. In a continuous picosecond spectrometer this large-signal 'shortening' of the transient-transmission signal is comparable to lifetime shortening by stimulated fluorescence. It is for this reason that the appearance of stimulated-fluorescence shortening cannot conclusively be determined. To this end an absolute picosecond spectrometer would have to be built (see § 6.2).

3.3 Small-signal solutions for medium-fast processes

If the stimulated-emission term $\overline{h\nu bQ}$ is negligible in eqs. (24) and (25) we have uncoupled linear differential equations that can be solved analytically. Assuming a quasi-steady state for the photon-density balance results in the fluorescence outside the sample being

$$pQV = p(\bar{a}/p)N_1V = \bar{a}N_1V \text{ or } k_fN_1V,$$

expressing the familiar proportionality of fluorescence to S_1 population. The solutions of the system of eqs. (23) for monochromatic δ-impulse excitation $I(t) = E\delta(t)$ bleaching the sample at $t = 0$ are

$$N_1(t) = \sigma EN \exp(-k_1t), \tag{29}$$

$$N_3(t) = \sigma EN \frac{k_{isc}}{k_1 - k_T}(\exp(-k_Tt) - \exp(-k_1t)). \tag{30}$$

Thus

$$N_0(t) = N - N_1(t) - N_3(t)$$

$$= N - \sigma EN \left(\exp(-k_1t) + \frac{k_{isc}}{k_1 - k_T}\{\exp(-k_Tt) - \exp(-k_1t)\}\right). \tag{31}$$

Introducing the triplet yield $\phi = k_{isc}/k_1$ the change in ground-state population is

$$\Delta N(t) = N - N_0(t) = N_1(t) + N_3(t) = \sigma EN((1 - \phi)\exp(-k_1t) + \phi\exp(-k_Tt)), \tag{32}$$

whereas the fluorescence response to δ-impulse excitation,

$$F(t) = k_fN_1(t)V = k_fV\sigma EN \exp(-k_1t), \tag{33}$$

records only the singlet S_1 population. The rotation-free transient-transmission signal

$$Y(t) = \Delta T/T = \sigma d\sigma EN((1 - \phi)\exp(-k_1t) + \phi\exp(-k_Tt)) \tag{34}$$

gives the kinetics of the singlet S_1 as well as the triplet T_1. Because of this advantage it is believed that continuous picosecond spectroscopy has a future as a routine method for photochemists. Fig. 9 compares a fluorescence-decay measurement by Cramer and Spears (1978a) on Rose bengal with a transient-transmission measurement on the same molecule. Both measurements have been

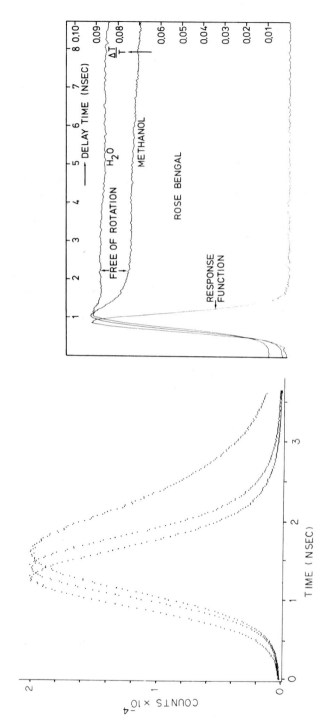

Fig. 9. Instrumental response and fluorescence signal of Rose bengal (left) giving, after deconvolution, $\tau_1 = 115$ ps in water and 540 ps in Methanol. After Spears and Cramer (1978a). Instrumental response and transient transmission of Rose bengal (right).

done with pulses from cw mode-locked lasers, but only the fluorescence measurement is determined by photomultiplier response. The transient transmission clearly shows a peak due to the combined populations of singlet and triplet, then the singlet population decays to the ground state with rate constant k_1 leaving behind the plateau due to triplet population, which practically does not decay within the nanosecond range of measurement.

3.4 Participation of higher levels: inverse effect

If higher singlet and triplet levels are reached from S_{10} and T_{10} by absorption of photons from the bleaching pulse (see fig. 10) the transmission of the probing pulse is now dependent on N_1 and N_3 too:

$$i(t_0) = i_0(t_0) \exp(-\sigma_1 d\{N - \langle \Delta N(t_0)\rangle_x\} - \sigma_2 d\langle N_1(t_0)\rangle_x - \sigma_3 d\langle N_3(t_0)\rangle_x), \qquad (35)$$

σ_2, σ_3 being the relevant absorption cross sections. Using the particle conservation $N_0(t) + N_1(t) + N_3(t) = N$ one arrives at

$$i(t_0) = i_0(t_0) \exp(-\sigma_1 dN + (\sigma_1 - \sigma_2)\langle N_1(t_0)\rangle_x d + (\sigma_1 - \sigma_3)\langle N_3(t_0)\rangle_x d). \qquad (36)$$

For $\sigma_1 = \sigma_2 = \sigma_3$ no transmission change would be observed with arbitrary excitations. While eq. (34) still holds to first order, even with higher transitions, the measurable signal is now

$$Y(t_0) = \sigma_1 ENd((\sigma_1 - \sigma_2)(1 - \phi) \exp(-k_1 t_0) + (\sigma_1 - \sigma_3)\phi \exp(-k_T t_0)), \qquad (37)$$

i.e., the usual transient-transmission curves are now modified by the differences in absorption cross section. This is illustrated in fig. 10 for an $S_1 \to S_n$ transition

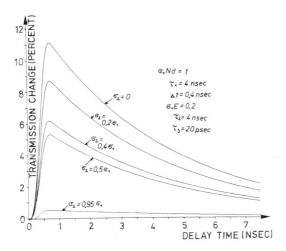

Fig. 10. Influence of $S_1 \to S_n$ absorption (σ_2) on transient transmission signal for moderate excitation $\sigma_1 E = 0.2$ (coherence-free).

only, where for S_n the lifetime has been assumed to be 20 ps. At the chosen excitation $\sigma_1 E = 0.2$ the small-signal case is accomplished where the achieved N_1 is so low that the time course of S_2 does not show up even in the $\sigma_2 = 0.95\sigma_1$ case.

Only with very high excitation $\sigma_1 E$ does a bump come out of the signal that reproduces closely the pulse shape (see fig. 11). The S_2 relaxation time would have to be extracted by deconvolution. The conditions for observing this bump are $\sigma_2 \simeq \sigma_1$ and very high excitation, e.g., $\sigma_1 E = 10$, leading to N_1 equalling 80% of N in Fig. 11 (right).

Finally, if $\sigma_2 > \sigma_1$ we have the inverse effect where the sample turns optically denser upon irradiation. This has been observed in two Oxazine dyes, Nile blue and Cresyl violet, where the 514 nm excitation is far away from the absorption band maximum in the red. Fig. 12 shows a measurement on Nile blue, where the rotation-free signal indeed shows negative transmission, i.e., increased absorption. In fact the curve started from a small positive value due to residual triplet population from excitation 10 ns earlier, went through zero and should rise again beyond the range of measurement to reach the low triplet plateau. Neglecting $T_1 \rightarrow T_n$ absorption this can be described by

$$Y(t_0) = (\sigma_1 - \sigma_2)d\langle N_1(t_0)\rangle_x + \sigma_1 d\langle N_3(t_0)\rangle_x.$$

Save for the inverse effect, the time course in itself of singlet relaxation is not influenced by $S_1 \rightarrow S_n$ absorption in the small-signal case. Only by comparing the measured amount of maximum transmission change with an estimate of the expected one, as in § 2.3.3, could the presence or absence of $S_1 \rightarrow S_n$ absorption be judged.

Using bleaching and probing pulses of different wavelength the time course of singlet S_1 population can be followed separately, or a change of the absorption cross section σ_2 in time can be detected for shorter times, where the S_1 population is known to remain practically constant. An example of the latter case is the recent measurement by Teschke et al. (1977) on the cis–trans isomerization in the excited state of stilbene. Fig. 13 shows the level energies versus twisting angle and two excited-state absorption measurements on trans- and cis-stilbene. The frequency-doubled pulse of 308 nm is used to excite the molecule to the singlet S_1. The attenuated dye laser pulse of 615 nm is not absorbed by the ground state S_0, but by the excited state S_1 with an unknown cross section σ_2. While the excited-state lifetime is known to be of the order of 1 ns, any changes in the picosecond range must reflect a change of the absorption cross section σ_2 with time. The different relaxations for trans- and cis-stilbene in fig. 13 are explained by internal twisting to different equilibrium positions in the excited state with concomitant scanning of the absorption profile $\sigma_2(\nu)$ by the 615 nm pulse. The relaxation was not observable when trans-stilbene molecules were imbedded in a rigid matrix of lucite (not shown).

If it happens that fluorescence takes place from the S_2-level, as is well-known for Azulene, one can use successive two-photon excitation to measure the S_1-lifetime (Ippen et al. 1976). With the response function $a(t - t') =$

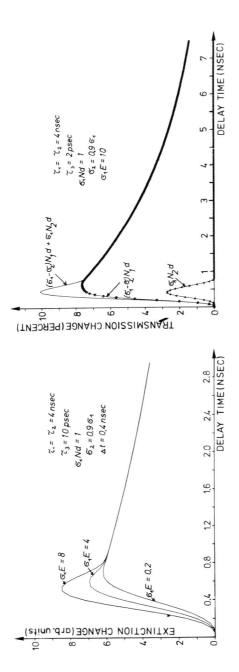

Fig. 11. Influence of $S_1 \rightarrow S_n$ absorption on transient-transmission signal for increasing excitations $\sigma_1 E$. Left: S_n radiationless decay constant $\tau_3 = 10$ ps. Right: $\tau_3 = 2$ ps (coherence-free).

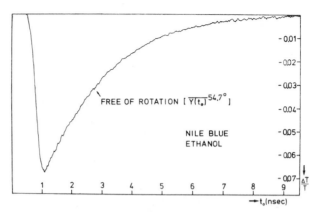

Fig. 12. Transient-transmission measurement of inverse effect in Nile blue. (coherence-free). From Lessing and von Jena (1978).

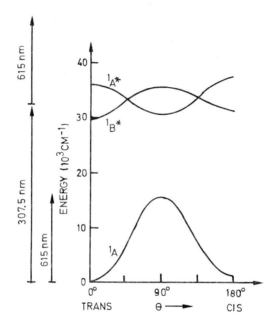

Fig. 13. Electronic levels of cis- and trans-stilbene.

$\exp(-(t-t')/\tau)$ for the S_1 relaxation, one gets for the S_1-population

$$N_1(t) = \sigma_1 N \int_{-\infty}^{t} a(t-t')i(t')\,dt', \tag{38}$$

where $i(t')$ is the intensity inside the sample. With the light wavelength properly chosen, absorption due to the transition $S_1 \rightarrow S_2$ is possible, leading to a S_2 population (assuming $\tau_{S_2} \gg \Delta t$, where

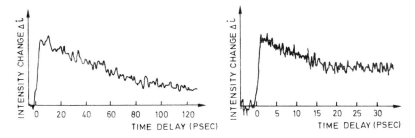

Fig. 14. Rotation-free excited-state absorption of trans-stilbene (left) and cis-stilbene (right). After Teschke et al. (1977).

$\Delta t \equiv$ pulse width)

$$N_2(t) = \sigma_2 \int\limits_{-\infty}^{t} N_1(t - t')i(t')\,dt', \tag{39}$$

which is detectable in fluorescence emission. Since the level of S_2 is arranged to be populated only via $S_1 \rightarrow S_2$ transitions, fluorescence follows the time course of S_1 population if $i(t')$ consists of two pulses separated from each other by an increasing time delay t_0.

For co-running pulses, where one is modulated by a light chopper (Ippen et al. 1976), one gets as the resultant intensity

$$\begin{aligned} i(t, t_0) &= (\epsilon_0/\mu_0)^{1/2}|E_0(t) + E_{mod}(t + t_0)|^2 \\ &= i_0(t) + i_{mod}(t + t_0) + 2(\epsilon_0/\mu_0)^{1/2}E_0(t)E_{mod}(t + t_0)\cos\phi, \end{aligned} \tag{40}$$

where $E_0(t)$ resp. $E_{mod}(t + t_0)$ are the envelopes of the electric fields from the modulated and the unmodulated light pulses respectively and ϕ is the phase difference between the two light waves. Evidently an interference term contributes to the fluorescence intensity near $t_0 = 0$, even for S_1 relaxation times $\tau \gg \Delta t$ (pulse width). This can be seen as follows by inserting eqs. (40) into eqs. (39) and (38) and combining the resulting equations. Since the measured quantity is the integrated fluorescence intensity $F(t_0)$ excited by the two time-separated pulses, one has to calculate the time-integrated S_2 population which is proportional to $F(t_0)$:

$$\begin{aligned} F(t_0) &\sim \int\limits_{-\infty}^{+\infty} N_2(t, t_0)\,dt = \sigma_2 N \int\limits_{-\infty}^{+\infty} dt'i(t', t_0)\left\{\sigma_1 \int\limits_{-\infty}^{t} dt''a(t' - t'')i(t'', t_0)\right\} \\ &= \sigma_1\sigma_2 N \int\limits_{-\infty}^{+\infty} dt'\{i_0(t') + i_{mod}(t' + t_0) + 2(\epsilon_0/\mu_0)^{1/2}E_0(t') \\ &\quad \times E_{mod}(t' + t_0)\cos\phi\}\int\limits_{-\infty}^{t'} dt''a(t' - t'')\{i_0(t'') + i_{mod}(t'' + t_0) \\ &\quad + 2(\epsilon_0/\mu_0)^{1/2}E_0(t'')E_{mod}(t'' + t_0)\cos\phi\}. \end{aligned}$$

Since only the modulated part of $F(t_0)$ is detected, one has to pick only terms containing a modulated term. Furthermore, the signal variations caused by continuously changing ϕ between 0 and 2π when increasing the time delay average out, since they occur over a delay path in the order of the light wavelength. What remains is

$$
\begin{aligned}
F(t_0) \sim \sigma_2\sigma_1 N \Bigg\{ & \int_{-\infty}^{+\infty} dt' i_{\mathrm{mod}}(t'+t_0) \int_{-\infty}^{t'} dt'' a(t'-t'') i_0(t'') \\
& + \int_{-\infty}^{+\infty} dt' i_0(t') \int_{-\infty}^{t'} dt'' \, a(t'-t'') i_{\mathrm{mod}}(t''+t_0) \\
& + \int_{-\infty}^{+\infty} dt' i_{\mathrm{mod}}(t'+t_0) \int_{-\infty}^{t'} dt'' a(t'-t'') i_{\mathrm{mod}}(t''+t_0) \\
& + 2(\epsilon_0/\mu_0)^{1/2} \int_{-\infty}^{+\infty} dt' E_0(t') E_{\mathrm{mod}}(t'+t_0) \\
& \times \int_{-\infty}^{t'} dt'' a(t'-t'') E_0(t'') E_{\mathrm{mod}}(t''+t_0) \Bigg\}.
\end{aligned}
\tag{41}
$$

The first two terms correspond to a probing of S_1 population excited by an i_0-pulse with delayed i_{mod}-pulses and vice versa. The third one corresponds to probing S_1 at constant delay, thus containing little information on S_1 relaxation. The last term is caused by interference between the coherent pulses i_0 and i_{mod}. It diminishes if t_0 exceeds the pulse width, regardless of the S_1 lifetime. By choosing $i_{\mathrm{mod}} \ll i_0$ the third term can be neglected. For $t_0 = 0$ one has

$$
F(t_0 = 0) \sim 4\sigma_1\sigma_2 N \int_{-\infty}^{+\infty} dt' i_{\mathrm{mod}}(t') \int_{-\infty}^{t'} dt'' a(t'-t'') i_0(t''),
$$

while for $|t_0| \gg \Delta t$

$$
F(t_0) \sim \sigma_1\sigma_2 N \int_{-\infty}^{+\infty} dt' i_{\mathrm{mod}}(t'+|t_0|) \int_{-\infty}^{t'} dt'' a(t'-t'') i_0(t'')
$$

which means that for a long-lived S_1 population the spike height quadruples the signal for nonoverlapping (and noninterfering) pulses.

Ippen et al. (1977) show with their apparatus that from the time course of the noncoherent fluorescence signal the S_1 level of azulene has a lifetime $\tau \approx 1.9 \pm 0.2$ ps.

3.5 Triplet yield determination

On closer inspection of fig. 9 and eq. (34),

$$Y(t_0) = \Delta T(t_0)/T = \sigma_1 d\sigma_1 EN ((1 - \phi) \exp(-k_1 t_0) + \phi \exp(-k_T t_0)), \tag{42}$$

one finds a peak of transient transmission representing the initial excitation to S_1 followed by a plateau due to triplet T_1 population that remains nearly constant in time within the limited range of measurement. With these two features transient transmission is able to provide an elegant method for the absolute determination of the triplet yield (Lessing et al. 1976). The principle can be seen from the above expression for δ-impulse excitation, but a more practical derivation is given afterwards. For the peak at $t_0 = 0$ we have, peak $= \sigma_1 d\sigma_1 EN$, and for the plateau for times $t_0 \gg 1/k_1$, but $t_0 \ll 1/k_T$, we get, plateau $= \sigma_1 d\sigma_1 EN\phi$.

The number of excitations $\sigma_1 EN$ – this is usually difficult to estimate in conventional flash photolysis (compare, e.g., Wilkinson 1975) – cancels when the triplet yield is formed by the ratio plateau/peak $= \phi$.

For a bleaching pulse of finite width $I(t) = Ef(t)$ the time course of the ground-state population is given by

$$N_0(t) = \int_{-\infty}^{t} f(t)N_0(t - t') \, dt'$$

$$= N - \sigma N \int_{-\infty}^{t} I(t')(\exp[-k_1(t - t')]$$

$$+ \frac{k_{\mathrm{isc}}}{k_1 - k_T} \{\exp\{-k_T(t - t')\} - \exp[-k_1(t - t')]\}) \, dt.$$

This can be solved analytically for a $\sin^2$-pulse of half-width Δt with further simplification, when $\Delta t \ll 1/k_1$ is assumed. The values for the peak and plateau are then determined by $\Delta N(2\Delta t)$ and $\Delta N(t \gg \tau_1)$ respectively. We obtain

$$\Delta N(2\Delta t) = \sigma EN \left[\frac{1 - k_{\mathrm{isc}}/k_1}{1 - k_1 \Delta t} + \frac{k_{\mathrm{isc}}}{k_1} \right], \tag{43}$$

$$\Delta N(t \gg \tau_1) = \sigma EN \, k_{\mathrm{isc}}/k_1. \tag{44}$$

If we now form the quantity A from experiment

$$A = \frac{\text{peak} - \text{plateau}}{\text{plateau}} = \frac{\Delta N(2\Delta t) - \Delta N(t \gg \tau_1)}{\Delta N(t \gg \tau_1)} \tag{45}$$

we obtain, with eqs. (43) and (44),

$$\phi = [1 + A(1 + k_1 \Delta t)]^{-1}. \tag{46}$$

The correction $k_1 \Delta t$ accounts for singlet decay that already occurs during excitation by the bleaching pulse. Thus even for a practical pulse shape we have a quick determination of triplet yield from a single transient-transmission curve.

For $\Delta t \to 0$, i.e. no S_1 relaxation during the pulse, eq. (46) reduces to

$$\phi = 1/(1 + A) = \text{plateau/peak.}$$

For S_1 lifetimes shorter than the pulse width there is no quick method of arriving at the triplet yield. Instead the general solution (42) has to be fitted to the curve with ϕ as a parameter. As an illustration of triplet-yield determinations the heavy-atom effect in substituted derivatives of Uranine has been measured in fig. 15. For pure states intersystem crossing from S_1 to T_1 is a spin-forbidden transition with very low transition probability. According to perturbation theory the spin–orbit interaction between singlet and triplet states gives rise to a mixing of all singlet and triplet states. Because of this one has nonvanishing transition matrix elements and consequently triplet population can occur at all after singlet excitation. Since the magnitude of ℓs coupling increases for atoms with increasing atomic number Z, one expects for a dye with heavy-atom substituents a larger triplet yield than for the same dye with light substituents. Table 3 gives the evaluation of the curves. Evidently, the X substituents close to the chromophore have more influence on the triplet yield than the more distant Y substituents. For Rose bengal ϕ is even smaller than for Erythrosine B, although the Y substituent is heavier. Further triplet yield data collected from transient-transmission measurements are given in table 4, together with the singlet lifetimes.

An estimate of the triplet lifetime τ_T is possible from the residual signal S_{resid} at $t_0 < 0$, just before the transient transmission rises. This residual amount of triplet population stems from the preceding bleaching event, whereupon the triplet population has decayed during the interpulse interval. Knowing the pulse repetition rate ν_{rep} and the signal S_{max} of the transient-transmission curve, at maximum delay (approx. 10 n sec), one finds

$$\tau_T = [\nu_{\text{rep}} \ln(S_{\text{max}}/S_{\text{resid}})]^{-1}.$$

Of course, in flow cells the flow should be slow enough not to influence this estimate.

Fig. 16 shows a molecule where the S_1 decay is much faster than the pulse width with concomitant reduction of the transmission maximum. Thus the triplet yield could be determined with a large error only. But there is the feasibility of estimating the inter-system-crossing rate constant k_{isc} by making another approximation in eqs. (23). For $\tau_1 \ll \Delta t$ we get a S_{10} population density that follows the bleaching pulse shape quasi-stationarily: $N_1(t) = \tau_1 \sigma I(t) N$.

Via the triplet balance we obtain for a $\sin^2$-pulse the maximum transmission change $\Delta N(t = \Delta t) = \sigma E N (\tau_1/\Delta t + \frac{1}{2} k_{\text{isc}} \tau_1)$, and for the plateau, i.e., for $t > 2\Delta t$, but $t \ll 1/k_T$, $\Delta N(t > 2\Delta t) = \sigma E N k_{\text{isc}} \tau_1$.

Taking from experiment the quantity

$$B = [\Delta N(t = \Delta t) - \tfrac{1}{2}\Delta N(t > 2\Delta t)]/\Delta N(t > 2\Delta t)$$

one obtains $B = 1/k_{\text{isc}}\Delta t$.

In our case we obtain for $B \simeq 3$ and $\Delta t \simeq 400 \, \text{ps}$ a $k_{\text{isc}} \simeq 10^9 \, \text{s}^{-1}$, but τ_1 could not

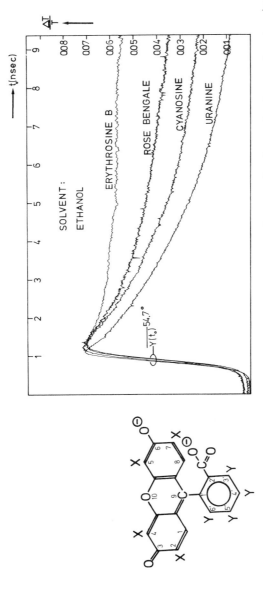

Fig. 15. Heavy-atom effect in substituted xanthene dyes. Left: structural formula. Right: rotation-free and coherence-free transient-transmission measurement in ethanol. From Lessing et al. (1976).

Table 3.
Heavy-atom effect for Uranine derivatives.

Compound	Uranine	Cyanosine	Rose bengal	Erythrosine B
X =	H	Br	J	J
Y =	H	Cl	Cl	H
ϕ	0.02	0.33	0.61	0.83
τ_1	3.7	3.3	1.1	0.5

Table 4.
Triplet yields ϕ and singlet lifetimes τ_1 as well as $\tau = \tau_1(1 - \phi)$ from transient transmission measurements.

A. Solvent: Ethanol

Dye		τ[ns]	τ_1[ns]	Φ_{exp}	τ_1[ns] (lit.)	ϕ (lit.)
Rhodamine B	(X)	3.1 ± 0.2	3.1 ± 0.2	0.00 ± 0.02		
Pyronine G	(X)	2.3 ± 0.2	2.3 ± 0.2	0.00 ± 0.02		
Uranine	(X)	3.7 ± 0.2	3.6 ± 0.2	0.02 ± 0.02	3.6[a]	0.03[b]
Rhodamine 6G	(X)	3.9 ± 0.1	3.8 ± 0.1	0.02 ± 0.02	3.4[c]	
Fluorescein 27	(X)	4.0 ± 0.1	3.9 ± 0.1	0.02 ± 0.02		
Phloxine	(X)	3.7 ± 0.3	3.0 ± 0.2	0.19 ± 0.03		
Cyanosine	(X)	4.3 ± 0.3	3.3 ± 0.2	0.22 ± 0.03		
Eosin Y	(X)	3.9 ± 0.3	2.6 ± 0.2	0.33 ± 0.03		0.42[d] 0.44[e]
Rose bengal	(X)	2.8 ± 0.5	1.1 ± 0.2	0.61 ± 0.05	0.74[f]	
Erythrosine B	(X)	3.1 ± 1.0	0.5 ± 0.2	0.83 ± 0.06		1.05[d]
Neutral red	(Az)	1.1 ± 0.1	1.0 ± 0.1	0.08 ± 0.02		
Safranin T	(Az)	3.2 ± 0.4	1.6 ± 0.2	0.50 ± 0.05		
Cresyl violet	(Ox)	3.1 ± 0.2	2.9 ± 0.2	0.04 ± 0.02		
Nile blue	(Ox)	2.1 ± 0.2	1.9 ± 0.2	0.04 ± 0.02		
Acridine orange	(Ac)	2.8 ± 0.2	2.8 ± 0.2	0.00 ± 0.02	4.4[g]	0.30[e]

B. Solvent: Methanol

Dye		τ[ns]	τ_1[ns]	Φ_{exp}	τ_1[ns] (lit.)	ϕ (lit.)
Rhodamine B	(X)	2.0 ± 0.2	1.9 ± 0.2	0.05 ± 0.02		
Pyronine G	(X)	2.1 ± 0.2	2.1 ± 0.2	0.00 ± 0.02		
Rhodamine 6G	(X)	3.9 ± 0.2	3.8 ± 0.2	0.02 ± 0.02	3.4[c]	
Fluorescein 27	(X)	4.1 ± 0.2	4.1 ± 0.2	0.00 ± 0.02		
Phloxine	(X)	4.2 ± 0.3	3.2 ± 0.2	0.23 ± 0.03		
Cyanosine	(X)	4.2 ± 0.3	2.9 ± 0.2	0.30 ± 0.03		
Rose bengal	(X)	3.1 ± 0.7	1.2 ± 0.2	0.61 ± 0.05	0.54[f]	0.76[h]
Erythrosine B	(X)	4.0 ± 2.0	0.3 ± 0.1	0.94 ± 0.06	0.42[h]	0.60[h]
Neutral red	(Az)	0.8 ± 0.1	0.7 ± 0.1	0.05 ± 0.02		
Cresyl Violet	(Ox)	2.8 ± 0.2	2.7 ± 0.2	0.03 ± 0.02		
Acridine orange	(Ac)	2.7 ± 0.2	2.6 ± 0.2	0.02 ± 0.02		

$X \equiv$ Xanthene, $Az \equiv$ Azine, $Ox \equiv$ Oxazine, $Ac \equiv$ Acridine.
[a]From Seybold et al. (1969).
[b]From Soep et al. (1972).
[c]From Porter et al. (1977).
[d]From Nemoto et al. (1969a).
[e]From Nemoto et al. (1969b).
[f]From Spears and Cramer (1978).
[g]From Rammensee and Zanker (1960).
[h]From Gollnick and Schenck (1964).

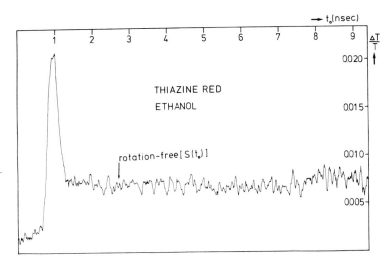

Fig. 16. Rotation-free and coherence-free transmission of a molecule with fast S_1 relaxation and slower intersystem crossing.

be determined. Whereas in some aromatic hydrocarbons the radiationless portion of the singlet deactivation occurs by intersystem crossing exclusively, we have here a case of definite $S_1 \leadsto S_0$ internal conversion. The transient-transmission method supplies this information from a single measurement. In closing this paragraph the necessity of rotation-free and coherence-free measurements for the absolute triplet-yield determination has to be stressed. Its precision depends critically on the transient-transmission maximum not being distorted by rotational diffusion or the coherent-coupling effect.

3.6 Vibrational relaxation

In § 3.1 vibrational relaxation was dealt with by the approximation of quasi-stationarity for medium-fast processes leading to the four-level scheme for dye kinetics. Investigations in the picosecond to subpicosecond range intended to measure the vibrational relaxation constant k_{vib} require consideration of the kinetics of the vibronic level S_{1v} into which the molecules are primarily excited by photons of energy $E_{00,1v} = h\nu_0$. For the population density N_{1v} of S_{1v} and N_1 of S_{10} we have the balance equations (ignoring triplet population)

$$\dot{N}_{1v}(t) = \sigma(\nu_0)I(t)N_0(t) - k_{vib}N_{1v}(t) - \sigma(\nu_0)I(t)N_{1v}(t) - k_{fv}N_{1v}(t),$$

$$\dot{N}_1(t) = k_{vib}N_{1v}(t) - k_1N_1(t). \tag{47}$$

The third term describes stimulated emission $S_{1v} \to S_{00}$ that competes with vibrational relaxation in this two-level scheme. Due to the equality of the Einstein coefficients for absorption and stimulated emission at the same frequency this resonant stimulated emission is governed by the same cross section as the excitation. The inclusion of a resonant spontaneous fluorescence from S_{1v} is an open question. It will certainly not change the time course of the S_{1v} population appreciably if it is of the same order of magnitude as k_f, i.e., much smaller than k_{vib}. To that extent Kasha's rule is not violated. We will see that some experimental methods inherently rely on a change of k_{fv} or the corresponding cross section σ_{em} for stimulated emission during vibrational relaxation. Assuming the small-signal case $N_0(t) \simeq N = \text{const}$ and δ-impulse excitation $I(t) = E\delta(t)$ one obtains, neglecting spontaneous emission (in close formal analogy to the singlet-triplet kinetics),

$$N_{1v}(t) = \sigma(\nu_0)EN \exp(-k_{vib}t),$$

$$N_1(t) = \sigma(\nu_0)EN\{\exp(-k_1t) - \exp(-k_{vib}t)\}, \tag{48}$$

where E is again $\int I(t)\,dt$ in photons per time and unit area.

To follow the time course of the vibronic state S_{1v} and thus measure the vibrational rate constant k_{vib} there are essentially three principles of measurement that have been tried in single-shot measurements and are now profiting from the higher precision of continuous picosecond spectroscopy: (a) transient transmission, bleaching and probing at identical wavelengths; (b) transient gain, bleaching the absorption band and probing in the fluorescence band; and (c) rise of fluorescence, bleaching the absorption band and detecting the emission.

For the transient transmission, when bleaching and probing with photons of identical energy $h\nu_0$ the generalized Beer's law is now (for isotropic polarization directions)

$$i(t) = i_0 \exp(-\sigma(\nu_0)d\{N_0(t) - N_{1v}(t)\}). \tag{49}$$

The second term affects gain contributed by resonant stimulated emission $S_{1v} \to S_{00}$. The spontaneous contribution into the small solid angle of the probing beam is ignored, as usual. The transient-transmission signal is then

$$Y(t_0) = \Delta T/T = \sigma(\nu_0)d(\Delta N_0(t_0) + N_{1v}(t_0)) = \sigma(\nu_0)d\,(N_1(t_0) + 2N_{1v}(t_0)), \tag{50}$$

or with the small-signal solutions [eq. (48)],

$$Y(t_0) = \Delta T/T = \sigma(\nu_0)d\{\exp(-k_1t_0) + \exp(-k_{vib}t_0)\}\sigma(\nu_0)EN.$$

Thus vibrational relaxation appears as an additional bleaching on top of the slower decay due to $N_1(t_0)$ of S_{10}. Rotational relaxation is frozen at these short times and will therefore not contribute to the time course. Save for the shorter time scale the signal looks like the combined singlet and triplet kinetics for the special case of $\phi = 0.5$.

With identical frequencies of bleaching and probing light, however, the coherent coupling effect (see § 5) usually cannot be avoided. Using a result that will be derived later in § 5, the signal amplitude at $t_0 = 0$ due to coherent coupling will be in the rotation-free measurement

$$S(0) = \langle Y(0)\rangle_\parallel^{\text{incoh}} + \tfrac{1}{3}\langle Y(0)\rangle_\perp^{\text{incoh}}, \tag{51}$$

which reduces in the case of slow rotational motion, that applies to the present situation, to

$$S(0) = 2\langle Y(0)\rangle_{54.7°}. \tag{52}$$

In other words, the combined signal reaches double the height of the undistorted signal at delay time $t_0 = 0$. The half-width Δt of the coherent-coupling spike is determined by either the pulse width or the coherence length of the laser pulses (see § 5). With this usually unavoidable artifact a rotation-free transient-transmission measurement of vibrational relaxation should have the appearance of fig. 17 (right). Evidently, the coherent spike limits the time resolution for detecting vibrational relaxation, since it hinders signal deconvolution for $1/k_{vib} \ll \Delta t$. In fact, if improperly fitted, it may make one believe in a nonexistent vibrational relaxation with the spike width Δt simulating the relaxation time $1/k_{vib}$. Proper fitting means not only including the coherent coupling (with error sources due to pulse width and coherence length) but also keeping in mind that the relative heights of slow and fast decays are not arbitrary. Rather there is a one-to-one ratio as given by eq. (50).

A continuous picosecond measurement on Cresyl violet has been performed by Ippen and Shank (1975). By graphical subtraction of an assembled curve containing $N_1(t_0)$ and a coherent-coupling contribution, the authors obtain a residual signal with a time constant of 2.8 ps attributed to vibrational relaxation [see fig. 18 (right)]. The assembled curve does not show the coherent-coupling spike to be expected (compare Shank et al. 1976), if the rise of $N_1(t_0)$ and the shape of the spike are based on the same bleaching pulse shape. With a bleaching pulse width of $\Delta t \approx 2$ ps the observed signal could be explained by the coherent-coupling effect alone.

The same situation is demonstrated for an analog experiment on a slower

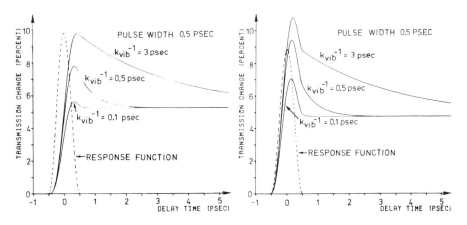

Fig. 17. Appearance of vibrational relaxation $S_{1v} \rightsquigarrow S_{10}$ in transient transmission (identical wavelengths). Left: coherence-free. Right: including coherent-coupling effect.

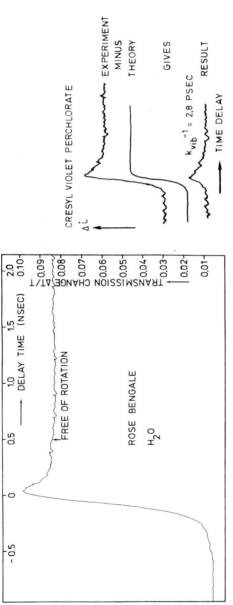

Fig. 18. Left: rotation-free transient transmission with subnanosecond resolution of Rose bengale showing triplet population. The initial bump is due to the coherent spike. Right: signal extraction from a transient-transmission measurement with picosecond resolution showing singlet population. After Ippen and Shank (1975).

time scale in fig. 18 (left). The dye shown is known to have a pure triplet population, i.e., no fast relaxation $S_1 \leadsto S_0$. The decay seen on top of the triplet plateau is not S_1 relaxation, but in fact entirely due to the coherent-coupling effect.

In the transient-gain method of Ricard et al. (1972) the probing pulse has a light frequency ν_p suitable to induce stimulated fluorescence downwards from S_1 within the fluorescence band. As a consequence the transmission may become larger than one, i.e., amplification occurs with gain

$$g(t_0) \equiv i(t_0)/i_0 = \exp\{\sigma_{em}(\nu_p)N_{10}(t_0)d\}. \tag{52'}$$

This formulation implies the assumption that $S_1 \rightarrow S_0$ fluorescence can exit only from S_{10}, σ_{em} is the cross section for stimulated fluorescence which is connected to the radiative lifetime $1/k_f$ for fluorescence via eq. (4). The time course of $N_1(t_0)$ is the one given in eq. (48). At $t_0 = 0$ there is no S_{10} population, i.e., $N_1(0) = 0$. As a consequence $g(t_0 = 0) = 1$, with subsequent increase due to the rise in S_{10} population through vibrational relaxation. Fig. 19 (left) shows transient-gain curves calculated for realistic parameters. Owing to the short times there is no contribution of rotational diffusion and owing to the different light frequencies no coherent-coupling spike, either. This experimental advantage is offset by the special assumption that stimulated emission exits from S_{10} exclusively, which is neither theoretically nor experimentally justified. Generally all sublevels between S_{1v} and S_{10} may contribute to the measurable gain according to

$$g(t_0) = \exp\left\{d \sum_{n=0}^{v} \sigma_{em}(E_{1n}, \nu_p)N_{1n}(t_0)\right\},$$

where E_{1n} is the energy of the S_{1n}-level ($n = 0, \ldots, v$). If for instance $\sigma_{em}(E_{1n}, \nu_p)$

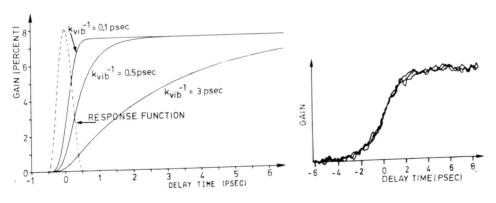

Fig. 19. Left: calculated transient-gain curves. Right: transient-gain curves showing the rise of singlet population of Rhodamine 6G and B in methanol and glycerol. After Shank et al. (1977).

is nearly constant for all states between S_{1v} and S_{10}, that means

$$\sigma_{em}(E_{1n}, \nu_p) = \sigma_{em}(\nu_p):$$

the gain is given by

$$g(t_0) = \exp\left\{\sigma_{em}(\nu_p)d \sum_{n=0}^{v} N_{1n}(t_0)\right\}.$$

Since

$$\sum_{n=0}^{v} N_{1n}(t_0) = \Delta N_0(t_0)$$

one gets an amplification

$$g(t_0) = \exp\{d\sigma_{em}(\nu_p)\Delta N_0(t_0)\} = \exp\{d\sigma_{em}(\nu_p)\sigma(\nu_0)EN \exp(-k_1t_0)\},$$

which is completely insensitive to the vibrational relaxation. How far this behaviour may be included in the gain curves of Shank et al. (1977) cannot be stated, since data on $\sigma_{em}(E_{1n}, \nu_p)$ are not available [see fig. 19 (right)]. Obviously the rise characteristics of the signals in fig. 19 (right) do not show a contribution due to vibrational relaxation, which of course may be due to the fact that it is extremely fast ($k_{vib} > 10^{-13}$ s^{-1}). As shown here, the gain method using bleaching and probing pulses with different wavelengths indeed avoids the unwanted coherent-coupling effect, but delivers no absolutely certain information on vibrational relaxation.

Essentially the same conclusions hold for the fluorescence-risetime method, where the spontaneous-fluorescence rate constants k_{fn} govern the measured signal

$$F(t_0) \sim \sum_{n=0}^{v} k_{fn}N_{1n}(t_0).$$

If these constants are all identical, $k_{fn} = k_f$, and one ends up again with a signal that is insensitive to vibrational relaxation:

$$F(t_0) \sim k_f\sigma(\nu_0)EN \exp(-k_1t_0). \tag{53}$$

Only for the special assumption $k_{fn} = \delta_{n0}k_f$ we have

$$F(t_0) \sim k_fN_1(t_0) = k_f\sigma(\nu_0)EN(\exp(-k_1t_0) - \exp(-k_{vib}t_0)). \tag{54}$$

Single-shot experiments evaluated with this philosophy are due to Mourou and Malley (1975). It cannot be decided conclusively if nonobservation of a vibrational contribution is due to eq. (53) or due to eq. (54) with a large k_{vib}.

A single-shot experiment on the onset of stimulated fluorescence by Rentzepis et al. (1970) interprets a delay of 6 ps as vibrational relaxation. The onset of stimulated fluorescence is governed by the system of eq. (24) and (25), which has to be solved numerically. A rough estimate of the onset time is obtained, if nevertheless the photon density balance [eq. (25)] is solved separately:

$$\dot{Q} = -pQ + h\nu bQN_1 + aN_1.$$

Taking $N_1 \simeq N_1(0) = \sigma E N$ one arrives at

$$F(t) = pQ(t)V = pa\sigma ENV/(h\nu b\sigma EN - p)(\exp\{(h\nu b\sigma EN - p)t\} - 1).$$

Obviously, there is an onset time $(h\nu b\sigma EN - p)$ that poses a lower limit to the rise time of stimulated fluorescence regardless of vibrational relaxation. Summarizing, one must conclude that the ordinary transmission method using pulses of identical wavelength–even though somewhat disturbed by coherent coupling–offers the best chance for detecting vibrational relaxation. It is the only method which definitely measures the time course of $N_{1v}(t)$, while the results of the other techniques depend on a change of the emission probabilities for molecules relaxing through the whole vibronic ladder.

4 Induced photodichroism

4.1 Photoselected initial values

In dealing with photoselection the theoretical description of molecular relaxation has to be modified to include molecular orientation and light polarization. Molecular orientation is included using population densities $N_i(t, \theta, \varphi)$ in a system of polar coordinates θ, φ. These are defined as the number of molecules per unit volume in state i, the transition moments of which point at time t into a solid-angle element $d\varphi \sin \theta \, d\theta$ centered around direction θ, φ. The balance equations [eqs. (23)] can be rewritten in terms of these quantities if the polarization of light is included in the cross sections by replacing the isotropic σ by orientation-dependent quantities $\sigma m^i(\theta, \varphi)$. The following expressions hold for light polarized in the z-direction, light polarized in the y-direction and naturally polarized light respectively:

$$\sigma m^z(\theta, \varphi) = 3\sigma \cos^2 \theta,$$
$$\sigma m^y(\theta, \varphi) = 3\sigma \sin^2 \theta \sin^2 \varphi, \tag{55}$$
$$\sigma m^{nat}(\theta, \varphi) = \tfrac{3}{2}\sigma(\cos^2 \theta + \sin^2 \theta \sin^2 \varphi).$$

The initial populations created by δ-impulse excitation are solutions of the modified balance equations, when orientational relaxation is neglected, and can be expanded in the small-signal case $(\sigma E \ll 1)$:

$$N_0^z(0, \theta, \varphi) = \frac{N}{4\pi} \exp(-3\sigma E \cos^2 \theta) \simeq \frac{N}{4\pi}(1 - 3\sigma E \cos^2 \theta),$$

$$N_0^y(0, \theta, \varphi) = \frac{N}{4\pi} \exp(-3\sigma E \sin^2 \theta \sin^2 \varphi) \simeq \frac{N}{4\pi}(1 - 3\sigma E \sin^2 \theta \sin^2 \varphi), \tag{56}$$

$$N_0^{nat}(0, \theta, \varphi) = \frac{N}{4\pi} \exp(-\tfrac{3}{2}\sigma E\{\cos^2 \theta + \sin^2 \theta \sin^2 \varphi\})$$
$$\simeq \tfrac{1}{2}N_0^z(0, \theta, \varphi) + \tfrac{1}{2}N_0^y(0, \theta, \varphi).$$

If these photoselected populations are probed by light with isotropic polarizations, which is hard to realize experimentally, one has to sum over all orientations (θ, φ), e.g.,

$$\langle N_0^z(0)\rangle_{\theta,\varphi} = \frac{N}{4\pi} \int\limits_0^{2\pi} d\varphi \int\limits_0^{\pi} d\theta \sin\theta(1 - 3\sigma E\cos^2\theta) = N - \sigma EN,$$

and one obtains the same result for all three cases. In other words, photoselection cannot be detected when probing with light of isotropic polarizations.

With polarized light, however, an anisotropic population $N_0(t, \theta, \varphi)$ can be
detected. The modified version of the absorption law of § 2.3.1 for bleaching
pulse I and probing pulse i is

$$\frac{\partial^2}{\partial\varphi\sin\theta\partial\theta}\frac{\partial i(x, t - t_0)}{\partial x} = -\sigma m^j(\theta, \varphi)i(x, t - t_0)$$
$$\times\left(\frac{N}{4\pi} - \Delta N\{I(x, t), t, \theta, \varphi\}\right)$$

describing the contribution of molecular transition moments pointing into the
θ, φ-direction to the intensity change of the probing light, the polarization of
which enters via $m^j(\theta, \varphi)$. The distribution $N_0(t, \theta, \varphi)$ is replaced by the isotropic
ground-state population $N/4\pi$ before excitation minus the induced anisotropic
change ΔN. The latter contains the polarization of the bleaching light. Since only
total intensities can be measured, the absorption law has to be integrated over all
directions θ, φ. Assuming the probing light to be polarized in the z-direction we
call the 'parallel case' ($\parallel$) the situation when the bleaching light is z-polarized
also, and the 'perpendicular case' ($\perp$) the situation when the bleaching light is
y-polarized. With the initial distributions [eq. (56)] we obtain the following initial
values of the transient-transmission signal:

$$\langle Y(0)\rangle_\parallel = \sigma d\langle \Delta N_0(0)\rangle_\parallel = \sigma d \int\limits_0^{2\pi} d\varphi \int\limits_0^{\pi} \sin\theta\, d\theta\, 3\cos^2\theta\, \frac{N}{4\pi}\, 3\sigma E\cos^2\theta$$
$$= \tfrac{9}{5}\sigma^2 ENd. \tag{57}$$

$$\langle Y(0)\rangle_\perp = \sigma d\langle \Delta N_0(0)\rangle_\perp = \tfrac{3}{5}\sigma^2 ENd. \tag{58}$$

For isotropic bleaching and isotropic probing one has

$$\langle Y(0)\rangle_{\text{iso}} = \sigma^2 ENd,$$

and for bleaching and probing with natural polarization one obtains

$$\langle Y(0)\rangle_{\text{nat,nat}} = \tfrac{6}{5}\sigma^2 ENd \neq \langle Y(0)\rangle_{\text{iso}}.$$

Thus it is no use putting depolarizers into bleaching and probing beams, since
even with naturally polarized light the isotropic case cannot be established.

The averaging over sample depth for the focusing geometry of the continuous
transient-transmission set-up, as maintained in § 2 and 3, is omitted here for
simplicity.

4.2 Rotation-free detection

The sequence

$$\langle Y \rangle_\| > \langle Y \rangle_{\text{iso}} > \langle Y \rangle_\perp$$

suggests that there might be an angle ψ between bleaching and probing polarization such that photoselection is evened up and the transmission signal reflects the isotropic case. This magic angle is of great practical importance, since it allows one to detect the pure level kinetics without any contributions from orientational relaxation (Lessing and von Jena 1976).

If the probing light field is split into components parallel and perpendicular to the polarization direction of the bleaching light one has for the intensity

$$i(t_0) = i_\| + i_\perp = i \cos^2 \psi + i \sin^2 \psi,$$

and for the transient-transmission signal

$$\langle Y(t_0) \rangle_\psi = \langle Y(t_0) \rangle_\| \cos^2 \psi + \langle Y(t_0) \rangle_\perp \sin^2 \psi. \tag{59}$$

The condition for ψ is therefore

$$\langle Y(t_0) \rangle_{\text{iso}} \overset{!}{=} \langle Y(t_0) \rangle_\| \cos^2 \psi + \langle Y(t_0) \rangle_\perp \sin^2 \psi,$$

which can be written for the corresponding changes of population ΔN_0,

$$\langle \Delta N_0(t_0) \rangle_{\text{iso}} - \langle \Delta N_0(t_0) \rangle_\| \cos^2 \psi - \langle \Delta N_0(t_0) \rangle_\perp \sin^2 \psi \overset{!}{=} 0.$$

Assuming ΔN_0 to be independent of φ, i.e., to be rotationally symmetric about the bleaching-light propagation, one can integrate over φ obtaining

$$2\pi \int_0^\pi \sin \theta \, d\theta \Delta N_0(\theta, t_0)(1 - \tfrac{3}{2} \sin^2 \theta \sin^2 \psi - 3 \cos^2 \theta \cos^2 \psi) \overset{!}{=} 0$$

which is satisfied, if for arbitrary values of θ,

$$1 - \tfrac{3}{2} \sin^2 \psi + (\tfrac{3}{2} \sin^2 \psi - 3 \cos^2 \psi) \cos^2 \theta = 0.$$

The rotation-free signal can therefore be detected if

$$\tan^2 \psi = 2 \quad \text{or} \quad \psi = 54.7°. \tag{60}$$

Since the relative phase of the parallel and perpendicular components of the probing-light field vectors does not matter, a 90° phase is allowed, too. Therefore, instead of the 54.7° linear polarization arrangement, an elliptically polarized probing light with ellipticity

$$i_\perp / i_\| = 2 \tag{61}$$

does equally well.

Finally, one can choose to process the parallel and perpendicular signals $\langle Y \rangle_\|$ and $\langle Y \rangle_\perp$ outside of the sample in order to arrive at the rotation-free signal. Assuming again ΔN_0 to be independent of φ one obtains for the distance of the parallel and perpendicular branches from the isotropic one

$$\langle Y(t_0)\rangle_{\text{iso}} - \langle Y(t_0)\rangle_\perp = 2\pi\sigma d \int_0^\pi d\theta \sin\theta \Delta N_0(\theta, t_0)(1 - \tfrac{3}{2}\sin^2\theta)$$

and

$$\langle Y(t_0)\rangle_\| - \langle Y(t_0)\rangle_{\text{iso}} = 2\pi\sigma d \int_0^\pi d\theta \sin\theta \Delta N_0(\theta, t_0)(3\cos^2\theta - 1).$$

Because $3\cos^2\theta - 1 = 2 - 3\sin^2\theta$, one arrives at

$$\langle Y(t_0)\rangle_\| - (Y(t_0)\rangle_{\text{iso}} = 2(\langle Y(t_0)\rangle_{\text{iso}} - \langle Y(t_0)\rangle_\perp), \tag{62}$$

which is consistent with eq. (60). The rotation-free signal is then obtained by the arithmetic operation

$$\langle Y(t_0)\rangle_{\text{iso}} = \tfrac{1}{3}\langle Y(t_0)\rangle_\| + \tfrac{2}{3}\langle Y(t_0)\rangle_\perp.$$

4.3 Induced polarization rotation

By virtue of the different initial transmissions [eqs. (57) and (58)] linearly polarized probing light with an arbitrary angle ψ leaves the sample with its polarization direction rotated by $\Delta\psi$ (Shank and Ippen 1975). Our derivation following is based on the situation of fig. 20. The field components of the probing light are, *with* bleaching light I,

$$e_\|^1 = \{(\mu_0/\epsilon_0)^{1/2} i_\| \cos^2\psi\}^{1/2} = (\mu_0/\epsilon_0)^{1/4}|\cos\psi|\sqrt{i_0}\exp\{-\sigma d(N - \langle\Delta N\rangle_\|)/2\},$$

$$e_\perp^1 = \{(\mu_0/\epsilon_0)^{1/2} i_\perp \sin^2\psi\}^{1/2} = (\mu_0/\epsilon_0)^{1/4}|\sin\psi|\sqrt{i_0}\exp\{-\sigma d(N - \langle\Delta N\rangle_\perp)/2\}.$$

Without bleaching one has

$$e_\|^0 = (\mu_0/\epsilon_0)^{1/4}|\cos\psi|\sqrt{i_0}\exp\{-\sigma dN/2\},$$

$$e_\perp^0 = (\mu_0/\epsilon_0)^{1/4}|\sin\psi|\sqrt{i_0}\exp\{-\sigma dN/2\}.$$

From fig. 20 it is seen that

$$\tan\psi = e_\perp^0/e_\|^0,$$

$$\tan(\psi + \Delta\psi) = e_\perp^1/e_\|^1 = (e_\perp^0/e_\|^0)\exp\{-\sigma d(\langle\Delta N\rangle_\| - \langle\Delta N\rangle_\perp)/2\}$$

$$= \tan\psi \exp\{-\sigma d(\langle\Delta N\rangle_\| - \langle\Delta N_\perp\rangle)/2\}.$$

With the trigonometric relation for $\tan(\psi + \Delta\psi)$ one obtains for the angle of polarization rotation $\Delta\psi$,

$$\tan\Delta\psi = \tan\psi \frac{\exp\{-\sigma d(\langle\Delta N\rangle_\| - \langle\Delta N\rangle_\perp)/2\} - 1}{1 + \tan^2\psi \exp\{-\sigma d(\langle\Delta N\rangle_\| - \langle\Delta N\rangle_\perp)/2\}}.$$

For the small-signal case $\sigma d\Delta N \ll 1$, one can expand the exponential in the

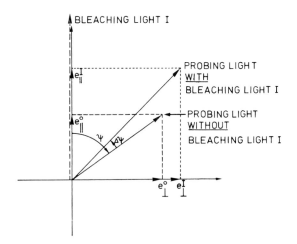

Fig. 20. Induced polarization rotation.

numerator and approximate the denominator by $1 + \tan^2 \psi$ with the result

$$\Delta\psi(t_0) = -[\tan\psi/(1 + \tan^2\psi)]\sigma d(\langle\Delta N(t_0)\rangle_\| - \langle\Delta N(t_0)\rangle_\perp)/2. \tag{63}$$

The probing-light polarization is thus rotated always towards the bleaching-light polarization direction. For the $\|$ case $\psi = 0$, and for the $\perp$ case $\psi = \pi/2$ there is no polarization rotation. To obtain the angle ψ for maximum $\Delta\psi$ the derivative of (63) yields the condition $\cos\psi = \sin\psi$, i.e., $\psi = \pi/4$.

With δ-impulse excitation the maximum angle of polarization rotation is $\Delta\psi_{max} = -0.3\sigma d\sigma EN = 1.7°$, the numerical result being calculated for extinction $\sigma dN = 1$ and a population change $\Delta N/N = \sigma E = 10\%$.

The angle of rotation of the transmitted probe light $i_{45°}$ is measured by Shank and Ippen (1975) by a polarizer adjusted for extinction without bleaching. For this arrangement the intensity falling on the photodetector is the measured signal. We obtain

$$i(\Delta\psi(t_0)) = Ki(t_0)\cos^2\{\pi/2 + \Delta\psi(t_0)\} = Ki(t_0)\sin^2\{\Delta\psi(t_0)\}$$

$$= \tfrac{1}{16}Ki(t_0)(\sigma d\{\langle\Delta N(t_0)\rangle_\| - \Delta N(t_0)\rangle_\perp\})^2, \tag{64}$$

where K is the polarizer transmission in the noncrossed case and $i(t_0)$ is the probing-light intensity after the sample, but before the polarizer

$$i(t_0) = Ci_0\exp\{-\sigma d\langle\Delta N(t_0)\rangle_{45°}\} \simeq Ci_0(1 + \sigma d\langle\Delta N(t_0)\rangle_{nat}). \tag{65}$$

For an evaluation of data according to eq. (64) with bleaching of, say, more than 1%, it would be advantageous to exclude the displacement of eq. (65) by a suitable normalization. If this is done the transient polarizer transmission $i\{\Delta\psi(t_0)\}/i(t_0)$ is proportional to the square of the rotational relaxation function according to eq. (64).

For the orientation-independent case $\psi = 54.7°$ discussed previously we find a polarization rotation

$$\Delta \psi_{54.7^\circ} = -\sqrt{\tfrac{1}{6}}\sigma d\{\langle \Delta N \rangle_\| - \langle \Delta N \rangle_\perp\} \neq 0, \tag{66}$$

i.e., one still has orientational relaxation in this geometry. Nevertheless we can give an orientation θ of the polarizer, for which the isotropic level kinetics can be detected in the small-signal case. Combining eqs. (64) and (65) we have

$$i(\Delta\psi(t_0)) = KCi_0\{\cos^2\theta + \sigma d(\langle\Delta N(t_0)\rangle_\|(\cos^2\theta - \cos\theta\sin\theta)/2$$

$$+ \langle\Delta N(t_0)\rangle_\perp(\cos^2\theta + \cos\theta\sin\theta)/2)\}. \tag{67}$$

This comes down to the condition for a special polarizer position θ_R,

$$\tan\theta_R = \tfrac{1}{3} \quad\text{or}\quad \theta_R = 18.4^\circ, \tag{68}$$

for which the detector intensity depends on $\langle\Delta N(t_0)\rangle_{iso}$ alone:

$$i[\Delta\psi(t_0)]_{\theta_R} = KCi_0\{0.9 + 0.3\sigma d(\langle\Delta N(t_0)\rangle_\| + 2\langle\Delta N(t_0)\rangle_\perp)\} \tag{69}$$

$$= 0.9 KCi_0\{1 + \sigma d\langle\Delta N(t_0)\rangle_{iso}\}.$$

With some arithmetical manipulation of the measured signal the quantity of interest can be singled out:

$$\{i(\Delta\psi(t_0))_{18.4} - i(0)_{18.4}\}/i(0)_{18.4} = \sigma d\langle\Delta N(t_0)\rangle_{iso},$$

$$\{i(\Delta\psi(t_0))_{\pi/4} - i(0)_{\pi/4}\}/i(0)_{\pi/4} = \sigma d\langle\Delta N(t_0)\rangle_\perp, \tag{70}$$

$$\{i(\Delta\psi(t_0))_{-\pi/4} - i(0)_{-\pi/4}\}/i(0)_{-\pi/4} = \sigma d\langle\Delta N(t_0)\rangle_\|.$$

4.4 Orientational relaxation

Whereas the initial values of transient transmission can be regarded as due to photoselection alone, the complete time course is determined by the rotational relaxation of the molecular transition moments.

4.4.1 Diffusional equation of rotation

The orientation of a molecule is most generally described by Eulerian angles α, β, γ which give the position of the molecular system with axes $i = 1, 2, 3$ relative to the laboratory system. The orientational distribution of the molecules is then given by a population density $N(\alpha, \beta, \gamma, t)$. In analogy to lateral diffusion the coefficients D of rotational diffusion are introduced, connecting the angular flux density

$$j_i = N(\alpha, \beta, \gamma, t)\omega_i$$

(ω_i is the angular velocity around axis i) with the gradient $\partial N(\alpha, \beta, \gamma, t)/\partial\phi_i$

around axis i by setting

$$j_i = -D \frac{\partial}{\partial \phi_i} N(\alpha, \beta, \gamma, t).$$

To cover all conceivable cases one can admit contributions from gradients around axes $j \neq i$ to the flux density around i:

$$j_i = -\sum_{j=1}^{3} D_{ij} \frac{\partial}{\partial \phi_j} N(\alpha, \beta, \gamma, t),$$

where D_{ij} is the tensor of rotational diffusion. With the continuity equation

$$\operatorname{div} \mathbf{j} + \partial N / \partial t = 0, \quad \text{where } \mathbf{j} = N\boldsymbol{\omega},$$

one obtains for the rotational diffusion equation

$$\frac{\partial N(\alpha, \beta, \gamma, t)}{\partial t} = \sum_{i=1}^{3} \sum_{j=1}^{3} D_{ij} \frac{\partial^2}{\partial \phi_i \partial \phi_j} N(\alpha, \beta, \gamma, t). \tag{71}$$

On the other hand, the quantum-mechanical operator of angular momentum for rotation around axis i has the shape $l_i = (\hbar/i)\partial/\partial\phi_i$, so that a formal analogy between rotational diffusion and the quantum mechanics of the asymmetric top can be established with the coefficients $\hbar^2/2D_{ij}$ corresponding to the components of the molecular tensor of inertia. If one refers to the molecular main-axis system for simplicity where the 'deviation moments' $\hbar^2/2D_{ij}$ with $i \neq j$ disappear, one has the diffusion equation formulated as the quantum-mechanical analog

$$\partial N(\alpha, \beta, \gamma, t)/\partial t = -\sum_{i=1}^{3} (D_i/\hbar^2) l_i^2 N(\alpha, \beta, \gamma, t). \tag{72}$$

The rotational diffusion coefficients D_i have to be explained by detailed molecular models of solute and solvent (see § 4.7).

4.4.2 Isotropic rotator

With isotropic rotational diffusion the distinction between molecular system and laboratory system (defined by the directions of polarization) is not necessary. This simplifies the problem considerably, since the orientational distribution of the transition moments can be given in spherical coordinates (θ, ϕ) in the laboratory system. With the Laplace operator in spherical coordinates

$$\Delta_{\theta,\phi} = \frac{1}{\sin \theta} \frac{\partial}{\partial \theta} \left(\sin \theta \frac{\partial}{\partial \theta} \right) + \frac{1}{\sin^2 \theta} \frac{\partial^2}{\partial \phi^2}$$

the system of balance equations (23) including rotational diffusion is now

$$\partial N_0(\theta, \phi, t)/\partial t = -\sigma I(t) m^i(\theta, \phi) N/4\pi + k_f N_1(\theta, \phi, t)$$
$$+ k_T N_3(\theta, \phi, t) + D\Delta_{\theta,\phi} N_0(\theta, \phi, t),$$

$$\partial N_1(\theta, \phi, t)/\partial t = \sigma I(t) m^j(\theta, \phi) N/4\pi - k_1 N_1(\theta, \phi, t)$$
$$+ D\Delta_{\theta,\phi} N_1(\theta, \phi, t), \tag{73}$$

$$\partial N_3(\theta, \phi, t)/\partial t = k_{isc} N_1(\theta, \phi, t) - k_T N_3(\theta, \phi, t)$$
$$+ D\Delta_{\theta,\phi} N_3(\theta, \phi, t),$$

$$N/4\pi = N_0(\theta, \phi, t) + N_1(\theta, \phi, t) + N_3(\theta, \phi, t).$$

Here use has been made of the fact that owing to ground-state probing only the orientation (θ, ϕ) of transition moments $S_{00} \to S_{1v}$ is important, even for excited molecules. The polarization of the bleaching light enters via the $m^j(\theta, \phi)$ [compare eqs. (55)].

What can be measured is a weighted average

$$\langle \Delta N_0 \rangle_{\parallel,\perp} = \int_0^{2\pi} d\phi \int_0^{\pi} d\theta \, m^{z,y}(\theta, \phi) \{ N/4\pi - N_1(\theta, \phi, t, I(t))$$
$$- N_3(\theta, \phi, t, I(t)) \}. \tag{74}$$

With these averages a considerable simplification of eqs. (73) can be brought about:

$$d\langle N_0(t) \rangle/dt = - B^{\parallel,\perp} \sigma I(t) N + kf\langle N_1(t) \rangle + k_T\langle N_3(t) \rangle + 6D(N_0(t) - \langle N_0(t) \rangle),$$
$$d\langle N_1(t) \rangle/dt = B^{\parallel,\perp} \sigma I(t) N - k_1\langle N_1(t) \rangle + 6D(N_1(t) - \langle N_1(t) \rangle),$$
$$d\langle N_3(t) \rangle/dt = k_{isc}\langle N_1(t) \rangle - k_T\langle N_3(t) \rangle + 6D(N_3(t) - \langle N_3(t) \rangle), \tag{75}$$
$$N = \langle N_0(t) \rangle + \langle N_1(t) \rangle + \langle N_3(t) \rangle, \quad B^{\parallel} = \tfrac{9}{5}, \quad B^{\perp} = \tfrac{3}{5},$$

where only the parallel or perpendicular average has to be used throughout. It is then sufficient to solve this system of equations once and to use the initial values (57), (58) and the isotropic solutions (29), (30) to construct the special polarization-dependent solution.

For δ-impulse bleaching $I(t) = E\delta(t)$, the solutions are

$$\langle N_1(t) \rangle_{\parallel,\perp} = \sigma E N \{ \exp(-k_1 t) \} + A_{\parallel,\perp} \exp\{ -(k_1 + 6D)t \},$$

$$\langle N_3(t) \rangle_{\parallel,\perp} = \sigma E N \frac{k_{isc}}{k_1 - k_T} (\exp(-k_T t) - \exp(-k_1 t)$$
$$+ A_{\parallel,\perp} \{ \exp\{ -(k_T + 6D)t \} - \exp\{ -(k_1 + 6D)t \} \}),$$

and $\langle \ldots \rangle_{nat}$ and $\langle \ldots \rangle_{iso}$ can be formed accordingly. The A-factors have the following values: parallel case, $A_{\parallel} = \tfrac{4}{5}$; perpendicular case, $A_{\perp} = -\tfrac{2}{5}$; natural case, $A_{nat} = \tfrac{1}{5}$; and isotropic case, $A_{iso} = 0$.

The polarization-weighted ground-state population is then

$$\langle N_0(t) \rangle_{\parallel,\perp} = N - \sigma E N \left\{ \left(1 - \frac{k_{isc}}{k_1 - k_T} \right) \exp(-k_1 t) \right.$$
$$+ A_{\parallel,\perp} \left(1 - \frac{k_{isc}}{k_1 - k_T} \right) \exp\{ -(k_1 + 6D)t \}$$
$$\left. + \frac{k_{isc}}{k_1 - k_T} \exp(-k_T t) + A_{\parallel,\perp} \frac{k_{isc}}{k_1 - k_T} \exp\{ -(k_T + 6D)t \} \right\}, \tag{76}$$

where again $\langle N_0(t)\rangle_{\text{nat}}$ and $\langle N_0(t)\rangle_{\text{iso}}$ can be formed consistently. It is immediately apparent that the orientational relaxation time $\tau_R = 1/(6D)$ is always combined with lifetimes $\tau_1 = 1/k_1$ or $\tau_3 = 1/k_T$; in other words, reorientation is detectable only as long as molecules are excited. But in contrast to fluorescence depolarization the transient-transmission method allows one to study rotational diffusion during triplet decay. Thus the rotation of nonfluorescent molecules can be studied, with the additional advantage that, owing to the longevity of the T_{10}-state the signal is determined by the orientational relaxation alone. The transient-transmission signals actually measured in the parallel and perpendicular case are

$$\langle Y(t_0)\rangle_{\parallel,\perp} = \Delta T_{\parallel,\perp}(t_0)/T = \sigma d(N - \langle N_0(t_0)\rangle_{\parallel,\perp}).$$

These have been calculated for different values of τ_R using eq. (76) and are plotted in fig. 21 (right). Evidently, the larger τ_R is, the later the $\parallel$ and $\perp$ branches merge into the isotropic curve. The measurement of fig. 21 (left) is on a molecule with less triplet yield. The isotropic curve divides the difference between $\parallel$ branch and $\perp$ branch in the ratio $2:1$, as it should according to eq. (60).

Because of $A_\parallel > 0$ and $A_\perp < 0$ the parallel-case signal is always larger and the perpendicular-case signal always smaller than the isotropic-case signal, because, e.g., in the latter case bleached molecules still have to rotate into the probing polarization direction. The isotropic part of the signal can be removed altogether when forming the difference of parallel and perpendicular cases which we call the 'rotational signal':

$$R(t_0) = \langle Y(t_0)\rangle_\parallel - \langle Y(t_0)\rangle_\perp = \sigma d(\langle \Delta N_0(t_0)\rangle_\parallel - \langle \Delta N_0(t_0)\rangle_\perp)$$

$$= \sigma d\sigma EN \tfrac{6}{5}\{(1-\phi)\exp\{-(k_1+6D)t_0\} + \phi\exp\{-(k_T+6D)t_0\}\}. \tag{77}$$

This turns into a single exponential if the triplet yield is either 0 or 1, since we have then either pure singlet or pure triplet decay respectively:

$$R(t_0) = \sigma d\sigma EN \tfrac{6}{5}\exp\{-(k_i+6D)t_0\}, \text{ with } k_i = k_1 \text{ or } k_T.$$

This can be evaluated easily by plotting logarithmically

$$\ln R(t_0) = -(k_i+6D)t_0 + \text{const}$$

to obtain the rotational relaxation time $\tau_R = 1/(6D)$, if k_i is known (e.g., from the rotation-free measurement). Since for the triplet decay $k_T \ll 6D$, the slope of the straight line is determined by $6D$ alone. On the other hand, the isotropic part of the signal is singled out when forming the sum

$$S(t_0) = \tfrac{1}{3}(\langle Y(t_0)\rangle_\parallel + 2\langle Y(t_0)\rangle_\perp) = \langle Y(t_0)\rangle_{54.7°}, \tag{78}$$

giving the pure population kinetics and its rate constants k_i. Obviously, either the signals $\langle Y(t_0)\rangle_\parallel$ and $\langle Y(t_0)\rangle_\perp$, or the signals $R(t_0)$ and $S(t_0)$, have to be measured simultaneously for a complete analysis of transient transmission. Fig. 22 shows calculated curves for the rotation-free signal $S(t_0)$ and rotational signals $R(t_0)$ for different values of τ_R. Fig. 23 gives examples of measurements for both modes of evaluation. The simultaneous measurement of the $\parallel$ and $\perp$ cases during a

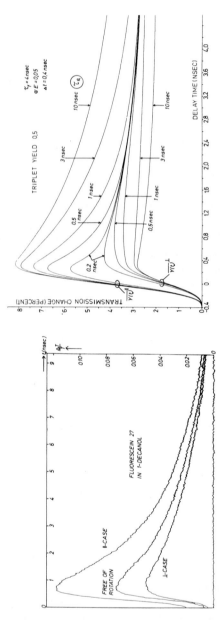

Fig. 21. Right: calculated coherence-free transient-transmission curves for a pulse width of 0.4 ns. Left: coherence-free transient transmission of fluorescein 27. Zero line written with bleaching beam blocked. From Lessing and von Jena (1976).

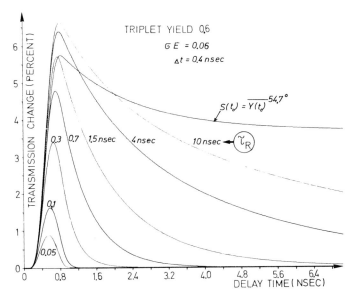

Fig. 22. Calculated coherence-free transient-transmission curves showing the rotation-free signal $S(t_0)$ as well as rotational signals $R(t_0)$ with τ_R as a parameter.

single scan is enabled by switching the bleaching-beam polarization between both cases. The slower rotational relaxation in the more viscous solvents is evident, as well as the larger difference between the initial values.

Besides rotational diffusion, intermolecular energy transfer via the Förster mechanism is able to randomize anisotropic molecular distributions created by photoselection, a phenomenon known in fluorescence spectroscopy as concentration depolarization (see Förster 1951). By means of a dipole–dipole interaction an initially excited molecule A′ is deactivated and a neighbouring, initially unexcited molecule A excited to S_1. In contrast to randomization by rotational motion, energy transfer occurs between photoselected molecules A′ and isotropically distributed molecules A. The dipole–dipole interaction is feasible, even if the A transition moment has a direction differing from the A′ transition moment. Since this process occurs isoenergetically, the energy transfer starting from the S_{10} or T_{10} states of molecules A′ can reach only those states of molecules A that are still covered by the absorption band $S_{00} \rightarrow S_{1v}$. Thus it is the short-wavelength wing of the fluorescence band overlapping the absorption band that contributes to singlet-energy transfer between like molecules. The transition probability $k_{\text{Fö}}$ contains therefore besides the dipole–dipole energy an overlap integral between absorption and emission band

$$k_{\text{Fö}}(\Theta_{A'}, \Theta_A, \Theta_{A'A}) = \frac{3c^2}{32\pi^4\epsilon_0} \frac{[\cos\Theta_{A'A} - 3\cos\Theta_{A'}\cos\Theta_A]^2}{R_{A'A}^6}$$

$$\times \int\limits_{-\infty}^{+\infty} \frac{d\nu}{\nu^2}\, \sigma_{\text{ab}}(\nu)\sigma_{\text{em}}(\nu),$$

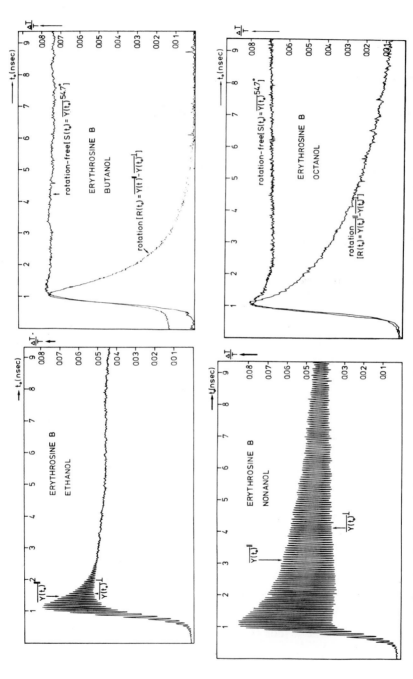

Fig. 23. Coherence-free transient-transmission measurements. Left: measurement alternating between ∥ and ⊥ case represented by the envelope curves. Right: rotation-free measurement S and rotational signal R; top row, lower viscosity; bottom row, higher viscosity.

where $\Theta_{A'A}$ is the angle between the emitting transition moment of A′ and the absorbing one of A, $\Theta_{A'}$ and Θ_A are the angles of the respective moments relative to the connecting line between A and A′, and $R_{A'A}$ is the average distance between A and A′. σ_{ab} is the absorption cross section for transitions $S_{00} \to S_{1v}$ and σ_{em} is the emission cross section for transitions $S_{10} \to S_{0v}$.

A completely different picture results if donor and acceptor molecules belong to species with different absorption and fluorescence bands. Here the energy-transfer rate may increase considerably if the absorption band of the acceptor A overlaps well with the fluorescence band of the donor D. Also, if probing light absorption is determined solely by the donor, $\sigma_A(\nu_{probe}) \ll \sigma_D(\nu_{probe})$, energy transfer does not influence the orientational distribution seen by the probing light. Rather there is a shortening contribution to level kinetics.

Thus energy transfer between unlike molecules appears in the rotation-free transient-transmission signal, whereas energy transfer between like molecules does not. As a consequence there is no necessity to measure energy transfer between unlike molecules in the rotational signal $R(t_0)$ (see Rehm and Eisenthal 1971), where a rotation-free measurement would have been less complicated.

4.5 *Influence of* $S_1 \to S_n$ *transition-moment orientation*

If higher transitions $S_1 \to S_n$ or $T_1 \to T_n$ cannot be neglected one has the following transmission law for linearly polarized light:

$$i(t_0) = i_0(t_0) \exp\{d(-\sigma_1 N + \sigma_1 \langle N_1(t_0) \rangle_{\|,\perp} - \langle \sigma_2 N_1(t_0) \rangle_{\|,\perp}$$
$$+ \sigma_1 \langle N_3(t_0) \rangle_{\|,\perp} - \langle \sigma_3 N_3(t_0) \rangle_{\|,\perp})\}, \tag{79}$$

where the inclusion of σ_2 and σ_3 into the averaging indicates that the polarization of the corresponding transitions does not necessarily coincide with that of σ_1, but includes an angle Θ_S and Θ_T respectively. Starting from the initial distribution of the σ_1-transition moments $N_1(\theta, \phi) = 3/(4\pi)\sigma EN \cos^2 \theta$, one can calculate for every orientation θ the distribution of the σ_2- and σ_3-transition moments, which point into solid angles with angular aperture Θ_S and Θ_T respectively, using the transformation $\cos \alpha = \cos \theta' \cos \theta + \sin \theta' \sin \phi' \sin \theta$. Thus one obtains after integration over the primed angles

$$g_{\sigma_2,\sigma_3}(\theta, \Theta_{S,T}) = \frac{1}{2} \sin^2 \Theta_{S,T} \sin^2 \theta + [\cos^2 \Theta_{S,T}] \cos^2 \theta.$$

With these distributions g the polarization dependent contribution due to σ_2 results:

$$\langle \sigma_2 N_1(t_0) \rangle_{\|,\perp} d = \sigma_2 d\sigma_1 EN \{\exp(-k_1 t_0)$$

$$+ A_{\|,\perp}(\cos^2 \Theta_S - \frac{1}{2} \sin^2 \Theta_S) \exp\{-(k_1 + 6D)t_0\}\},$$

as well as a similar expression for $\langle \sigma_3 N_3(t) \rangle_{\|,\perp}$. The transient-transmission signal probing higher transitions is therefore

$$\langle Y(t_0)\rangle_{\parallel,\perp} = \sigma_1 ENd\{(\sigma_1 - \sigma_2)\exp(-k_1 t_0) + (\sigma_1 - \sigma_3)\phi[\exp(-k_T t_0)$$

$$-\exp(-k_1(t_0)] + A_{\parallel,\perp}(\sigma_1 - \sigma_2(\cos^2\Theta_S - \frac{1}{2}\sin^2\Theta_S))$$

$$\times \exp\{-(k_1 + 6D)t_0\} + \phi A_{\parallel,\perp}(\sigma_1 - \sigma_3(\cos^2\Theta_T - \frac{1}{2}\sin^2\Theta_T))$$

$$\times [\exp\{-(k_T + 6D)t_0\} - \exp\{-(k_1 + 6D)t_0\}]\}.$$

Fig. 24 shows illustrative examples for two special cases. For the higher transition moments oriented perpendicular to $S_0 \rightarrow S_1$, $\sigma_2 = 2\sigma_1$, and no triplet contribution (left): one has of course, the inverse effect in the rotation-free signal, but this is mitigated in the $\parallel$ case by the fact that the absorbing $S_1 \rightarrow S_n$ moments are initially turned out of the probing polarization. Thus $\langle Y(t_0)\rangle_\parallel$ starts with positive values, crosses over the zero line and finally approaches the negative $\langle Y(t_0)\rangle_{54.7°}$. On the other hand, with $\sigma_1 = \sigma_2$ and triplet participation [fig. 24 (right)] the rotation-free signal is determined by a pure triplet population rise. However, owing to the perpendicular transition moments the $\perp$ case starts with negative values of transmission. Note the drastic difference between $\parallel$ and $\perp$ branches.

In fig. 12 a rotation-free measurement of the inverse effect in Nile blue was presented. The induced photodichroism of the same dye and of Cresyl violet is presented in fig. 25. The shown computer fit was optimized with Θ_S as a parameter, which turned out to be 0° within experimental accuracy.

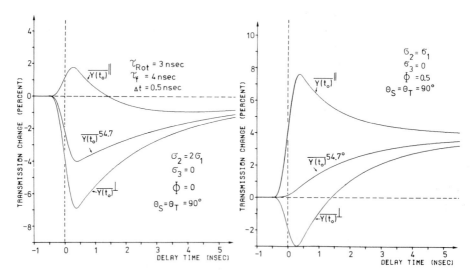

Fig. 24. Calculated transient-transmission curves with higher transitions (coherence-free). $S_1 \rightarrow S_n$ transition moment perpendicular to $S_0 \rightarrow S_1$. Left: σ_2 twice σ_1, no triplet. Right: equal cross sections, triplet yield 0.5. From Lessing and von Jena (1978).

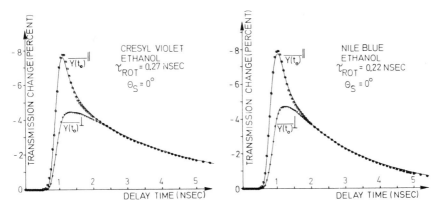

Fig. 25. Computer fit to induced photodichroism of molecules showing inverse effect (tucked up in the graph). Fitting parameter was θ_s, the angle between $S_0 \to S_1$ and $S_1 \to S_n$ transitions, which turned out to be $0°$ for the optimal fit within experimental accuracy. From Lessing and von Jena (1978).

4.6 Symmetric rotator

A more thorough discussion of induced photodichroism has to consider aniso-tropic rotational diffusion. While the most general solution has been discussed in the literature (see Favro 1960), the symmetric rotator with $D_1 = D_2 \neq D_3$ will be treated here, since it is presumably more within experimental reach. The diffusion equation describes the motion around the principal axes:

$$\partial N(\Omega_e, t)/\partial t = -\{(D/\hbar^2)(l_1^2 + l_2^2) + (D_3/\hbar^2)l_3^2\}N(\Omega_e, t),$$

where Ω_e is the set of Eulerian angles describing the orientation of the system of principal axes and $N(\Omega_e, t)$ is their orientational distribution. The orientation of the principal axes can be expanded using the eigenfunctions $\Phi_{lmk}(\Omega_e)$ of the quantum-mechanical symmetric rotator

$$N(\Omega_e, t) = \sum_{l=0}^{\infty} \sum_{m=-l}^{+l} \sum_{k=-l}^{+l} a_{lm}^{(k)} \Phi_{lmk}(\Omega_e) \exp(-E_l^{(k)}t),$$

where $a_{lm}^{(k)}$ are time-independent expansion coefficients and $E_l^{(k)}$ eigenvalues that are $(2l + 1)$-fold degenerate.

The expansion coefficients $a_{lm}^{(k)}$ are given by

$$a_{lm}^{(k)} = \int d\Omega_e N(\Omega_e, t = 0) \Phi_{lmk}(\Omega_e),$$

where $N(\Omega_e, t = 0)$ is the initial distribution of the molecular principal axes. From (56) one knows that the initial distribution of the transition moments is

$$N_1(\theta, \phi) = 3/(4\pi)\sigma EN \cos^2 \theta.$$

This has to be transformed into an initial distribution of principal axes expressed by Eulerian angles using the transformation

$$Y_{lm}(\theta, \phi) = [4\pi/(2l+1)]^{1/2} \sum_{m'=-l}^{+l} \Phi_{lmm'}(\Omega_e) Y_{lm}(\theta', \phi').$$

Using the identity

$$\cos^2 \theta = \frac{1}{3}\{2(4\pi/5)^{1/2} Y_{20} + (4\pi)^{1/2} Y_{00}\}$$

one obtains for the time-dependent distribution

$$N(\Omega_e, t) = \sum_{k=-2}^{+2} a_{20}^{(k)} \Phi_{20k}(\Omega_e) \exp(-E_2^{(k)}t) + a_{00}(0) \Phi_{000}(\Omega_e),$$

where

$$a_{20}^{(k)} = \frac{2}{5}\sigma EN Y_{2k}(\theta', \phi'), \quad a_{00}^{(0)} = \sigma EN Y_{00}(\theta', \phi').$$

This time-dependent distribution is sampled by linearly polarized probing light. The measured signals $\langle Y(t_0)\rangle_{\|,\perp}$ again depend on the weighted average

$$\langle N(t)\rangle_{\|,\perp} = \int d\Omega_e m_{\|,\perp}(\Omega_e) N(\Omega_e, t).$$

The weight function $m_{\|,\perp}(\Omega_e)$ contains the angular part of the transition probability written in Eulerian coordinates. Relative to the laboratory system

$$m_{\|}(\theta, \phi) = 3 \cos^2 \theta, \quad m_{\perp}(\theta, \phi) = 3 \sin^2 \theta \sin^2 \phi.$$

Rewritten in Eulerian angles one has

$$m_{\|}(\Omega_e) = (8\pi/5) \sum_{m'=-2}^{+2} \Phi_{20m'}(\Omega_e) Y_{2m'}(\theta', \phi') + 4\pi \Phi_{000}(\Omega_e) Y_{00}(\theta', \phi')$$

and a more complicated expression for $m_{\perp}(\Omega_e)$. Using the eigenvalues $E_2^{(k)} = 6D_3 + (D - D_3)k^2$ one has the solution

$$\langle N(t)\rangle_{\|,\perp} = EN \left\{ \begin{matrix} (16/25\pi) \\ -(8/25\pi) \end{matrix} \right\} \sum_{k=0}^{2} |Y_{2k}(\theta', \phi')|^2 (2 - \delta_{k0}) \exp(-E_2^{(k)}t)$$
$$+ 4\pi |Y_{00}(\theta', \phi')|^2. \tag{81}$$

If one uses this solution in the system of balance equations [eqs. (23)] one obtains the following transient-transmission signals in the small-signal case:

$$\langle Y(t_0)\rangle_{\|,\perp} = \sigma d \{\text{eq. (81)}\} \left(\left(1 - \frac{k_{isc}}{k_1 - k_T}\right) \exp(-k_1 t_0) \right.$$
$$\left. + \frac{k_{isc}}{k_1 - k_T} \exp(-k_T t_0) \right).$$

The measured signals thus contain three rotational time constants $\tau_{R_1} = 1/6D$, $k = 0$, $\tau_{R_2} = 1/(D_3 + 5D)$, $k = 1$, and $\tau_{R_3} = 1/(4D_3 + 2D)$, $k = 2$, which influence the time course, depending on the orientation (θ', ϕ') of the transition moment relative to the system of principal axes besides the singlet $(\tau_1 = 1/k_1)$ and triplet $(\tau_T = 1/k_T)$ lifetimes.

The rotational signal turns out to be

$$
\begin{aligned}
R(t_0) &= \langle Y(t_0) \rangle_\parallel - \langle Y(t_0) \rangle_\perp \\
&= \sigma d\sigma E N \{ 0.9 \sin^4 \theta' \exp(-t_0/\tau_{R_3}) \\
&\quad + 3.6 \sin^2 \theta' \cos^2 \theta' \exp(-t_0/\tau_{R_2}) \\
&\quad + 0.3(3 \cos^2 \theta' - 1)^2 \exp(-t_0/\tau_{R_1}) \} \\
&\quad \times \left\{ \left(1 - \frac{k_{\text{isc}}}{k_1 - k_T} \right) \exp(-k_1 t_0) + \frac{k_{\text{isc}}}{k_1 - k_T} \exp(-k_T t_0) \right\}.
\end{aligned}
$$

In a plane molecule the transition moment will be within the molecule's plane. For approximately identical longitudinal and transversal dimensions the symmetric rotator case can be assumed, with the transition moment oriented at $\theta' = \pi/2$ relative to the principal axes' ellipsoid. In this case the rotational signal contains *two* different rotational terms with different decay. One has to imagine therefrom that, after z-polarized excitation, randomization occurs by rotation around the symmetry axis (D_3) as well as around one of the axes in the molecular plane:

$$
\begin{aligned}
R(t_0)_{\theta'=\pi/2} &= \sigma d\sigma E N \{ 0.9 \exp(-t_0/\tau_{R_3}) + 0.3 \exp(-t_0/\tau_{R_1}) \} \\
&\quad \times \left\{ \left(1 - \frac{k_{\text{isc}}}{k_1 - k_T} \right) \exp(-k_1 t_0) + \frac{k_{\text{isc}}}{k_1 - k_T} \exp(-k_T t_0) \right\}.
\end{aligned}
\tag{82}
$$

For an oblong molecular one can think of the case $\theta' = 0$ as well as $\theta' = \pi/2$. With $\theta' = \pi/2$, again $R(t_0)$ has the shape of eq. (82) with two rotational time constants according to the shape of the molecule. For $\theta' = 0$ one finds only one rotation-

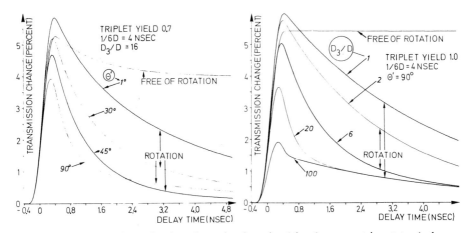

Fig. 26. Calculated rotational signals and rotation-free signal for the symmetric rotator (coherence-free). From von Jena and Lessing (1979b). Left: for different angles θ' between transition moment and symmetry axis. Right: for $\theta' = 90°$ and different ratios D_3/D of the diffusion constants.

dependent term in $R(t_0)$:

$$R(t_0)_{\theta'=0} = 1.2\sigma d\sigma EN \exp(-t_0/\tau_{R_1})\left\{\left(1 - \frac{k_{isc}}{k_1 - k_T}\right)\exp(-k_1 t_0)\right.$$

$$\left. + \frac{k_{isc}}{k_1 - k_T}\exp(-k_T t_0)\right\}.$$

The time dependence of this signal cannot be distinguished from isotropic rotational diffusion. With the transition moment along the longitudinal axis, as assumed, the rotation around the symmetry axis cannot contribute to randomization.

Transient-transmission signals calculated for prolate molecule for different orientations θ' of the transition moment show only slight deviations from single exponential decay, except for $\theta' = \pi/2$ [see fig. 26 (left)]. For $\theta' = 54.7°$ the slow term disappears altogether. Fig. 26 (right) shows for $\theta' = \pi/2$ how the ratio D_3/D influences the nonexponential shape of the signal. If the fast contribution of D_3 rotation cannot be resolved experimentally, it can at least be detected by the initial transmission being incompatible with the experimental decay constant (see also § 4.8).

4.7 Hydrodynamic models of rotational diffusion

In § 2.2.1 the Einstein relation

$$D = BkT$$

was introduced relating the formal rotational-diffusion constant D to the rotational mobility B of the molecule. Taking the molecule as a sphere of radius r that rotates with angular velocity $\omega = d\theta/dt$ around an arbitrary axis within a viscous liquid one has, via the Navier–Stokes equations for the torque acting against rotation, $T = 8\pi\eta r^3\omega$. The underlying assumption is that the particles of the viscous liquid stick to the surface of the sphere (stick boundary condition), that is to say,

$$v_{liquid} = \omega \times r \tag{83}$$

for all space vectors pointing to the surface of the sphere. Comparison with eq. (11) gives the rotational mobility $B = 1/8\pi\eta r^3$, wherefrom the rotational relaxation time τ_R is

$$\tau_R = \tfrac{4}{3}\pi\eta r^3/kT = \eta V/kT, \tag{84}$$

where V is the sphere volume. This is the relation used also in the Perrin formula for steady-state fluorescence depolarization. The next step of sophistication should be the symmetric rotator with symmetry axis 3. Its rotational diffusion constants are conveniently represented using the functions $\Phi_x = \Phi_y$ and Φ_z introduced by Memming (1961) and depicted in fig. 27 as a function of the ratio $\gamma = b/a$ of the axes of the rotational ellipsoid

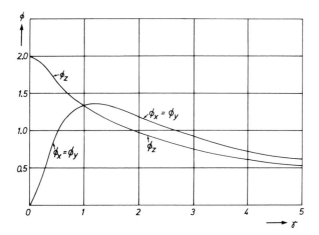

Fig. 27. Dependence of rotational diffusion constants from the ratio $\gamma = a/b$ of the axes of the symmetric rotator. After Memming (1961).

$$D_1 = D_2 = \frac{1}{8}kT/\eta V\Phi_x, \quad D_3 = \frac{1}{8}kT/\eta V\Phi_z, \tag{85}$$

where the molecular volume is now, of course, $V = \frac{4}{3}\pi ab^2$. Considering the shape of the curves in fig. 27, it can be seen that the rotational-diffusion times are distinctly different only for prolate molecules. For oblate molecules, because of $D_3 \simeq D$, eq. (85) and fig. 27 predict rotational relaxation with a single relaxation time. Rotation around the symmetry axis displaces only minimal amounts of liquid, but the stick boundary condition (83) produces a large braking torque. That rotation around an axis perpendicular to the symmetry axis produces somewhat less torque, mainly owing to liquid displacement, comes as a surprise with this model.

The stick boundary condition (83) postulates strong attractive forces between solvent and solute molecules. As an alternative, this is replaced by the slip boundary condition due to Hu and Zwanig (1974). Their model assumes that the interaction between the solute molecule and its environment is restricted to displacement of the liquid, the internal-friction work of which impedes rotation. Since rotation of a sphere or rotation of a rotational ellipsoid around the symmetry axis displaces no solvent, these cases are entirely unhindered in this model $(D_3 \to \infty)$. This is illustrated in fig. 28. Owing to $D_3 \to \infty$ the rotational signal of § 4.6 reduces to

$$R(t_0) = \sigma d\sigma EN\, 0.3(3\cos^2\theta' - 1)^2 \exp(-t_0/\tau_R^{slip})\left\{\left(1 - \frac{k_{isc}}{k_1 - k_T}\right)\exp(-k_1 t_0)\right.$$

$$\left. + \frac{k_{isc}}{k_1 - k_T}\exp(-k_T t_0)\right\}$$

for these cases of the slip model, where $\tau_R^{slip} = 1/(6D^{slip})$. Thus a quasi-isotropic behaviour results with initial values that may have arbitrary values between 0 and $1.2\sigma d\sigma EN$ depending on the orientation θ' of the transition moment. The

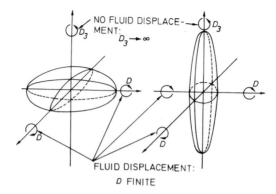

Fig. 28. Visualization of mechanical hindrance of the motion of the symmetric rotator.

ratio $\tau_R^{slip}/\tau_R^{stick}$, where $\tau_R^{stick} = 1/(6D^{stick})$ with $D^{stick} = D_1 = D_2$, has been calculated by Hu and Zwanig as a function of the ratio $\gamma = b/a$ of the axes of the rotational ellipsoid and is depicted in fig. 29. One can see that only for strong deviations from the spherical shape do the observable rotation-relaxation times around axis 1 or 2 become equal, i.e., $\tau_R^{slip}/\tau_R^{stick} = 1$ for the two models. As the ratio $\tau_R^{slip}/\tau_R^{stick}$ is smaller than one except for $\gamma = 0$ the slip boundary condition implies always a faster decay of the contribution $\exp(-6Dt_0)$ to the signal than the stick boundary condition. Moreover, the initial values depending on transition-moment orientation θ' are different for the two models except for $\theta' = 0$. For this case–transition moment aligned with the symmetry axis–the different decays of the rotational signal give the only criterion for deciding between the two models.

Fig. 30 shows two examples of comparisons between theory and experiment. Although the two structurally different molecules are far from spherical in shape, the evaluation assuming a single rotational-relaxation time apparently suffices. For Rhodamine 6G as an oblate molecule this is to be expected from the above discussion of the symmetric rotator in terms of the stick model. For

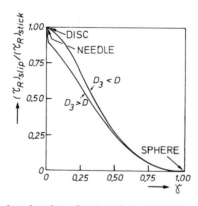

Fig. 29. Relative rotational relaxation times for the 'slip' and 'stick' models as a function $\gamma = a/b$ of the axes of the symmetric rotator. After Hu and Zwanig (1974).

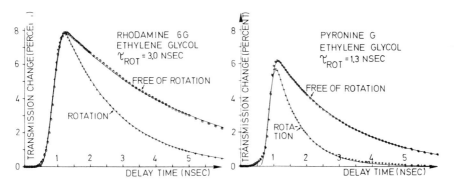

Fig. 30. Experimental (dots) and calculated (lines) rotational signal plus rotation-free signal (coherence-free). Left: oblate molecule. Right: oblong molecule. From Lessing and von Jena (1978).

Pyronine G as an oblong molecule there should be three distinct relaxation times according to the stick model, unless $\theta' = 0°$ (compare § 4.6). Thus the lack of anisotropic rotation requires that the transition moment must lie in the symmetry axis of the molecule.

Table 5 compares rotational-relaxation times evaluated from transient transmission with the parameters of the stick and the slip models that have been calculated from atomic Van-der-Waals increments. A single time constant was sufficient to fit the experimental data. Evidently the stick condition provides more realistic relaxation times from molecular dimensions than the slip condition, but more than one relaxation time is not observed, for the reasons discussed above.

Thus far rotational diffusion has been discussed in terms of molecular volume of the solute and solvent viscosity alone, whereas specific intermolecular interactions were excluded from the hydrodynamic model. Recent measurements by Fleming et al. (1976), Spears and Cramer (1978b) and von Jena and Lessing (1979c) show that specific solute–solvent interactions can strongly affect the orientational relaxation. Fig. 31 demonstrates that Rose bengal, which is a dianion dye, and Rhodamine 6G, a cation dye, rotate with almost the same relaxation time in formamide ($\eta = 4.1$ cP)–as expected from their only slightly differing volumes. In pentanol, however, which is as viscous as formamide, the orientational relaxation is drastically slowed down for dianion dyes and moderately for Rhodamine 6G (see table 6). Table 6 compares furthermore the rotational relaxation in dimethyl sulfoxide (DMSO) and propanol (both $\eta = 2.2$ cP). While Rhodamine 6G rotates equally fast in both solvents, the dianion dyes Fluorescein, Rose bengal, Eosin Y and Erythrosine rotate much slower in propanol than in DMSO.

Table 6 clearly shows that the dianion dyes interact more strongly with alcohol molecules than Rhodamine 6G does. The increase in orientational relaxation times for the alcohols can only be due to filling the space above and below the disc-shaped dye molecules by solvent molecules up to a sphere, the diameter of which corresponds to the longest axis of the dye molecule. The first

Table 5.

Rotational-relaxation times evaluated from transient transmission and calculated from atomic Van-der-Waals increments. From von Jena and Lessing (1979c).

Structure I, Structure II		τ_R exp. [ps]	τ_R lit. [ps]	$(\tau_R)_1$ stick [ps]	$(\tau_R)_2$ stick [ps]	$(\tau_R)_3$ stick [ps]	$(\tau_R)_1$ slip [ps]
A. Solvent: Acetone ($\eta = 0.4$ cP)							
Acridine orange	(I)	50 ± 10		44	34	20	16
Rhodamine 6G	(II)	90 ± 15		51	52	57	18
Erythrosine B	(II)	80 ± 15		48	49	54	17
Phloxine	(II)	60 ± 15		48	49	54	17
Cyanosine	(II)	80 ± 15		49	50	55	18
B. Solvent: Methanol ($\eta = 0.6$ cP)							
Acridine orange	(I)	110 ± 10		66	50	30	24
Neutral red	(I)	100 ± 10		60	44	26	21
Pyronine G	(I)	90 ± 15		65	49	29	23
Rhodamine 6G	(II)	170 ± 20	{100[i] 120[j]	76	78	85	27
Rose bengal	(II)	150 ± 15	190[f]	82	85	92	29
Erythrosine B	(II)	200 ± 20		72	75	81	26
Phloxine	(II)	180 ± 20		72	75	81	26
Cyanosine	(II)	190 ± 20		75	78	85	27
Rhodamine B	(II)	110 ± 15		72	75	81	26
Fluorescein	(II)	200 ± 20		53	55	60	20
DODCI		95		from Shank and Ippen (1975)			
C. Solvent: Ethanol ($\eta = 1.2$ cP)							
Acridine orange	(I)	130 ± 15		133	99	59	48
Neutral red	(I)	180 ± 15		120	88	51	43
Pyronine G	(I)	100 ± 15		130	97	58	47
Rhodamine 6G	(II)	310 ± 20	{290[i] 290[j] 270[l]	152	157	170	55
Rose bengal	(II)	420 ± 30	450[f]	165	171	185	59
Erythrosine B	(II)	380 ± 30		145	150	162	52
Phloxine	(II)	420 ± 30		145	150	162	52
Cyanosine	(II)	420 ± 30		151	156	169	54
Rhodamine B	(II)	260 ± 20	270[l]	145	150	162	52
Fluorescein	(II)	330 ± 20		106	110	121	40
DODCI		200		from Shank and Ippen (1975)			
D. Solvent: Propanol ($\eta = 2.2$ cP)							
Acridine orange	(I)	200 ± 20		243	181	108	88
Pyronine G	(I)	160 ± 20		238	178	106	85
Neutral red	(I)	340 ± 20	{440[j] 450[i]	220	161	94	79
Rhodamine 6G	(II)	490 ± 30		279	288	312	101
Rose bengal	(II)	830 ± 50	840[f]	303	314	339	108
Erythrosine B	(II)	760 ± 50		266	275	292	96

Table 5 (*continued*).

Structure I, Structure II		τ_R exp. [ps]	τ_R lit. [ps]	$(\tau_R)_1$ stick [ps]	$(\tau_R)_2$ stick [ps]	$(\tau_R)_3$ stick [ps]	$(\tau_R)_1$ slip [ps]
Phloxine	(II)	810 ± 60		266	275	297	96
Cyanosine	(II)	830 ± 60		277	285	307	100
Fluorescein	(II)	750 ± 50		195	202	222	71
E. Solvent: Decanol ($\eta = 13.5$ cP)							
Acridine orange	(I)	1200 ± 100		1490	1110	662	535
Neutral red	(I)	2200 ± 200		1350	988	577	485
Rhodamine 6G	(II)	3400 ± 250	2300[i]	1710	1770	1910	615
Erythrosine B	(II)	5800 ± 300		1630	1690	1820	585
Phloxine	(II)	6200 ± 300		1630	1690	1820	585
Cyanosine	(II)	6700 ± 350		1700	1750	1890	610
Fluorescein	(II)	4300 ± 300		1200	1240	1360	440
Safranine T	(II)	3200 ± 250		1300	1350	1480	470
F. Solvent: Ethylene glycol ($\eta = 19.5$ cP)							
Acridine orange	(I)	1200 ± 100		2150	1600	955	770
Pyronine G	(I)	1300 ± 100		2100	1570	935	755
Neutral red	(I)	1600 ± 100		1950	1430	835	700
Rhodamine 6G	(II)	3000 ± 200	2200[i] 2000[j] 2300[m]	2470	2560	2760	890
Rose bengal	(II)	4100 ± 200		2710	2810	2980	970
Erythrosine B	(II)	3900 ± 200		2350	2440	2630	845
Phloxine	(II)	3800 ± 250		2350	2440	2630	845
Cyanosine	(II)	3600 ± 250		2450	2540	2730	880
Fluorescein	(II)	3000 ± 250		1730	1790	1960	630
Rhodamine B	(II)	2800 ± 250		2360	2450	2640	850
Safranine T	(II)	2200 ± 150		1880	1950	2140	675
DODCI		3000 ± 1000		from Shank and Ippen (1975)			

[f]From Spears and Cramer (1978).
[i]From Chuang and Eisenthal (1971).
[k]From Fleming et al. (1976).
[l]From Penzkofer and Falkenstein (1976).
[m]From Heiss et al. (1975).

attempt to explain this space-filling effect can be made by assuming hydrogen bonding between solute and solvent molecules. The dianion dyes have four oxygen atoms with unshared electron pairs, thus being good hydrogen acceptors. Since the DMSO-molecule is a pure H-acceptor too, no strong interaction can be expected, which is in good agreement with experiment. Because of their hydroxyl groups alcohols are good H-donors, leading to strong coupling with H-accepting dyes. Even though formamide contains both acceptor ($C = O$) and donor groups (NH_2) it evidently couples only weakly to the di-anion dyes. The rotational behaviour of Rhodamine 6G seems to be determined by the fact that it contains acceptor as well as donor groups like formamide. So the less suggestive

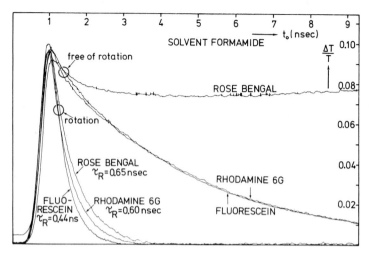

Fig. 31. Rotational and rotation-free signals (coherence-free) of different dyes in a hydrogen-bonding solvent.

Table 6.
Rotational relaxation times [ps] in solvents with different hydrogen-bonding capabilities.

Solvent	Rhodamine 6G	Uranine	Rose bengale	Erythrosine	Eosin Y
Dimethyl sulfoxide	480 ± 30	220 ± 25	340 ± 30 420 ± 40^f	430 ± 30	250 ± 25
Propanol	490 ± 30	750 ± 40	830 ± 40 840 ± 70^f	760 ± 40	780 ± 40
Formamide	600 ± 30	440 ± 30	650 ± 35	590 ± 30	
Pentanol	1090 ± 50	1620 ± 70	1780 ± 100	1560 ± 60	

^{f}From Spears and Cramer (1978).

differences of its rotation times for DMSO and propanol, and formamide and pentanol respectively, as compared to those of the di-anion dyes, are not surprising. The experimental value for the rotation in pentanol shows, however, that the almost pure donor properties of the alcohols result in stronger solvent attachment than the donor *and* acceptor abilities of formamide together. This is at variance with a former single-shot measurement of Chuang and Eisenthal (1971), which states that the rotation in formamide and pentanol is identical. Of course, the volumes of the pentanol and the formamide molecule are not the same. So it may be possible that to a certain degree the differences in the rotation times for formamide and pentanol are due to different degrees of space filling.

It is not clear whether the position of donor or acceptor groups in a dye molecule plays a role. Since the maximum effective hydrodynamic volume can only be a sphere whose diameter is given by the longest molecular axis, one may assume that the active groups near the center of the molecules are the most important. Bonding to them leads to the required space filling, while outstanding active groups like the ones at the ends of the dye's chromophore group seem just to help in satisfying the stick boundary condition.

The obvious differences in the orientational relaxation times for different dyes of the di-anion type (see especially DMSO in table 6) which have almost exactly the same volume (and shape) show that interactions other than hydrogen bonding influence rotational diffusion. No satisfactory explanation for such effects is available at present.

Summarizing, one must say that the Debye–Einstein model is quite a good base for the description of rotational diffusion. Deviations due to solute–solvent interactions can be taken into account only in a qualitative way.

4.8 Solubilizer effects

If a dye does not dissolve well in water, a solubilizing tenside may be added to increase the solute concentration in the mixture. This may be accompanied by a marked change of the orientational behaviour. Fig. 32 (left) shows the effect for various amounts of added tenside. Already with the addition of 0.1 volume % of the tenside a slowly decaying component appears in the rotational signal in addition to the fast one characterizing the dye in pure water. Its amplitude increases with further addition of tenside, until there is no further increase at around 5% by volume. That the effect may not appear for a dye of the same class is demonstrated in fig. 32 (right).

The explanation of the effect relies on the fact that the chain molecules of the tenside form micelles, with the hydrophobic tails forming the interior and the hydrophilic heads the exterior. The oil-like interior dissolves some dye molecules better than water, and as a consequence part of the dye molecules enter the highly viscous, oil-like interior of the micelle, as depicted schematically in fig. 33. The ability to enter the micelle appears to be independent of the charge on the dye, since both anions and cations enter the micelles equally well.

The experimental evidence in fig. 32 shows that the dye molecules are distributed between the micellar and the aqueous phase. Moreover, the dyes in the micelles experience the rotation of the micelle as a whole ('external rotation') as well as their own rotation inside the micelle ('internal rotation'). We have therefore the rotational signal

$$R(t_0) = x \exp\{-(6D^{int} + 6D^{ext} + k_1)t_0\}$$
$$+ (1 - x) \exp\{-(6D^{H_2O} + k_1)t_0\}, \tag{87}$$

with

$$x = \frac{\text{number of molecules inside micelles}}{\text{total numbers of solute molecules}}$$

and D^{H_2O} representing the diffusion coefficient of the dye in aqueous solution. After fitting the expression to the experimental data, the slow decay, which is

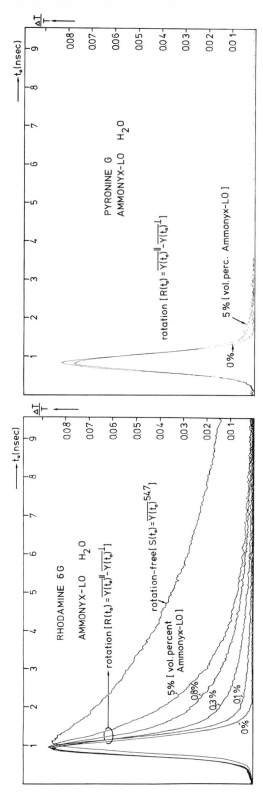

Fig. 32. Rotational signals from transient transmission of dyes in micellar solutions (coherence-free). Left: dye molecule that is distributed between aqueous and micellas phase. Right: dye molecule that remains in aqueous phase. From Lessing and von Jena (1979).

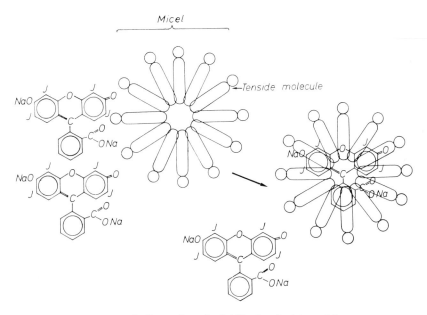

Fig. 33. Schematic illustration of solubilization (not to scale).

the first term in (87), can be evaluated. Table 7 gives rotational relaxation times 1 $(6D^{int} + 6D^{ext})$ of the slow component for several dye molecules in two solubilizer–water systems. Evidently the dye molecules do not serve simply as a label to the otherwise identical micelles, which would imply $D^{int} \to \infty$. Rather the experimental relaxation times have the same dependence on the molecular structure as in the pure solvent, i.e., Acridine orange and Pyronine G rotate faster than Rhodamine 6G and Erythrosine B. The external rotational diffusion constant D^{ext} can be singled out by making comparison measurements using two dyes, 1 and 2, with distinctly different internal diffusion constants D^{int}, such as Rhodamine 6G and Erythrosine B. For each measurement one has

$$D_{1,2} = D_{1,2}^{int} + D_{1,2}^{ext}.$$

Table 7.
Slower rotational relaxation time in water plus solubilizer [in ns]. From Lessing and von Jena (1979).

	Ammonyx LO	Triton X-100
Acridine orange	2.4 ± 0.2	2.7 ± 0.25
Pyronine G	—	2.4 ± 0.25
Cresyl violet	2.6 ± 0.2	4.1 ± 0.35
Rhodamine 6G	2.6 ± 0.2	6.2 ± 0.5
Cyanosine	3.6 ± 0.25	7.5 ± 0.3
Rose bengal	3.7 ± 0.2	8.7 ± 0.3
Erythrosine B	4.3 ± 0.2	9.9 ± 0.3

Since the external rotation depends on the micelle volume only,

$$D_1^{ext} = D_2^{ext} = D^{ext}.$$

Introducing the ratio $D_2^{int}/D_1^{int} = w_{int}$ for the relative diffusion constants inside the micelle and the ratio $D_2/D_1 = w$ from the comparison experiment, one has finally

$$D_{ext} = [(w_{int} - w)/(w_{int} - 1)]D_1.$$

If w comes close to w_{int}, then D^{ext} must be very small, i.e., external rotation contributes negligibly to the observed rotational relaxation. If, however, w approaches one, then $D_1 = D_{ext} = D_2$ and a orientational relaxation determined by external rotation alone is observed.

The ratio w_{int} would be obtained in the simplest way from another experiment in pure viscous solvent. But one cannot be sure that this ratio follows just the ratio of the hydrodynamic volumina as predicted by the Debye–Einstein model, also inside the micelle. A better way to get information about the real situation in the micellar microenvironment is to follow the temperature variation, since the microviscosity inside the micelle shows a stronger variation than the viscosity of water. Thus in the high temperature case the faster internal rotation dominates the signal giving w_{int}, whereas in the low temperature case the more rigid interior results in $D_1 = D_2 = D^{ext}$. This has been studied, with the result $\tau_R^{ext} = (13 \pm 2)$ ns for Ammonyx LO and (22 ± 4) ns for Triton X-100. Assuming a spherical shape for the micelles the Einstein relation (84) leads to a radius of (2.9 ± 0.5) and (3.5 ± 0.6) nm respectively, the latter being in good agreement with a value of 3.2 nm obtained from dynamic Rayleigh scattering by Corti and Degiorgio (1975). Our results are at variance with a view introduced by Phillion et al. (1974) to explain the change of the orientation-dependent dye-laser threshold with concentration of tenside. For the explanation it is assumed implicitly that all dye molecules are dissolved in micelles. The change of rotational behaviour with the amount of solubilizer is then attributed to a change of the micellar volume. With the distribution of molecules between micellar and aqueous phase that depends on the amount of solubilizer, a changing micellar volume need not to be invoked. In fact, the relaxation time of the slow signal component for Rhodamine 6G dissolved in Triton X-100 shows no dependence on solubilizer amount within experimental error (see table 8). Even if there were   dependence, the rotational behaviour of the dye in the micellar phase is–as we have seen–determined mostly by internal rotation which is independent of the micellar volume.

Solubilizer experiments have yielded, moreover, the most distinct hint to libration. Fig. 34 shows two measurements of induced photodichroism performed with identical excitation intensity. As can be seen, the separation between ∥ and ⊥ maxima is smaller for the micellar solution of Fluorescein 27 than for the decanol solution.

Starting from § 2.2.1 one can calculate the contribution of a libration to the signal as follows: transition moments that have pointed in the same direction during δ-pulse excitation are imagined to be distributed by libration on a cone, the apex angle $\Theta_0(t)$ of which increases quadratically with time according to eq.

Table 8.

Relaxation time of slow signal component versus amount of tenside. From Lessing and von Jena (1979).

τ_R(micelle) [ns]	Amount of tenside [vol. %]
6.8 ± 1.0	0.2
6.3 ± 0.7	1.0
6.7 ± 0.5	5.0

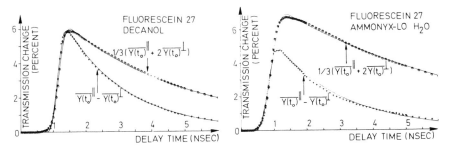

Fig. 34. Experimental evidence of libration from transient transmission (coherence-free): rotational-signal maximum lower than 1.2 times rotation-free maximum. Left: dye molecule in pure solvent $R_{max}/S_{max} = 0.99$. Right: dye molecule in solubilizer–water mixture $R_{max}/S_{max} = 0.73$.

(14). An initial distribution at $t = 0$,

$$\Delta N_0(\theta, 0) = (N/4\pi)3\sigma E \cos^2 \theta,$$

thus changes into

$$\Delta N_0(\theta, t) = \left(\frac{N}{4\pi}\right) 3\sigma E\{\tfrac{1}{2} \sin^2 \Theta_0(t) \sin^2 \theta + \cos^2 \Theta_0(t) \cos^2 \theta\}.$$

The weighted distributions are therefore

$$\langle \Delta N_0(t) \rangle_{\|,\perp} = \sigma E N \left(\begin{Bmatrix} 3/5 \\ 6/5 \end{Bmatrix} \sin^2 \Theta_0(t) + \begin{Bmatrix} 9/5 \\ 3/5 \end{Bmatrix} \cos^2 \Theta_0(t) \right)$$

$$\approx \sigma E N \left(\begin{Bmatrix} 9/5 \\ 3/5 \end{Bmatrix} + \begin{Bmatrix} 6/5 \\ 3/5 \end{Bmatrix} \Theta_0^2(t) \right)$$

after expansion for small $\Theta_0(t)$. Using the solution (14) the rotational signal due to libration is

$$R(t_0) = \sigma d(\langle \Delta N_0(t_0) \rangle_{\|} - \langle \Delta N_0(t_0) \rangle_{\perp}) = \sigma d\sigma E N (\tfrac{6}{5} - \tfrac{9}{5} \Theta_0^2(t_0))$$
$$= \sigma d\sigma E N (\tfrac{6}{5})(1 - \tfrac{3}{2}kT/Jt_0^2), \tag{88}$$

where population relaxation is omitted for simplicity. This can be compared with the solution for rotational diffusion

$$R(t_0) = \sigma d\sigma E N \tfrac{6}{5} \exp(-6Dt_0).$$

For times $t_0 > BJ$ the signal (88) is continued by

$$R(t_0) = \sigma d\sigma EN \tfrac{6}{5}(1 - \tfrac{3}{2}DBJ)\exp(-6D_0 t).$$

The hydronamic model gives a negligible angle of libration $\Theta_0(t_0 = BJ) \simeq 10^{-3}/\eta$ for a dye molecule of molecular weight of 500 and a radius of 5 Å. For comparable size of solute and solvent molecules the Langevin model fails. One has to assume here that the solute molecule makes relatively large unhindered rotations θ_L. The initial value of $R(t_0)$ then is reduced accordingly by the factor $(1 - \tfrac{3}{2}\theta_L^2)$.

In principle the reduction of $R(t_0)_{max}$ could also be explained by means of the symmetric-rotator model and the slip condition, assuming a certain angle θ' between transition moment and axis of symmetry (see § 4.6). The rotational signal for this case is, from eq. (86),

$$R(t_0) = \langle Y(t_0)\rangle_\parallel - \langle Y(t_0)\rangle_\perp$$
$$= 0.3(3\cos^2\theta' - 1)^2 \sigma d\sigma EN \exp(-6Dt_0)$$

with a single time constant $\tau_R(\text{slip}) = 1/(6D)$. Depending on the orientation θ' of the transition moment, arbitrary initial values between 0 and $1.2\sigma d\sigma EN$ could be adjusted. Against this approach consider not only the discrepancy between calculated $\tau_R(\text{slip})$ and experimental τ_R values in table 4, but also the argument that for flat rotational ellipsoids like Fluorescein 27 the transition moment can stick out of the molecular plane very little, i.e., $\theta' = 90°$ and $R(t_0)_{max} \simeq 0.3 S(t_0)_{max}$. This would be far too small when compared with experimental findings.

Another interfering effect is saturation due to large-signal conditions where the expansion of the exponential as in eqs. (56) is no longer feasible. For δ-impulse excitation we have

$$\langle N_0(0)\rangle_\parallel = \int_0^{2\pi} d\phi \int_0^\pi d\theta \sin\theta\, 3\cos^2\theta\, N/4\pi \exp\{-\sigma E3\cos^2\theta\}$$
$$= N\{1 - \tfrac{9}{5}\sigma E + (27/14)(\sigma E)^2 + O(\sigma^3 E^3)\},$$
$$\langle N_0(0)\rangle_\perp = N\{1 - \tfrac{3}{5}\sigma E + \tfrac{1}{5}(27/14)(\sigma E)^2 \pm O(\sigma^3 E^3)\}.$$

The correction term for the perpendicular case is smaller, since saturation occurs primarily in the bleaching polarization, whereby the perpendicular probing polarization is less affected. The rotation-free initial value $S(0)$ can now be given:

$$S(0) = \langle Y(0)\rangle_{iso} = \sigma d\{\tfrac{1}{3}\langle \Delta N_0(0)\rangle_\parallel + \tfrac{2}{3}\langle \Delta N_0(0)\rangle_\perp\}$$
$$= \sigma dN\{\sigma E - 0.9(\sigma E)^2\}.$$

The distance between the $\parallel$ and $\perp$ initial values relative to the initial value of the rotation-free signal is given in table 9 for increasing excitation, i.e.,

$$\frac{\langle Y(0)\rangle_\parallel - \langle Y(0)\rangle_\perp}{S(0)} = \frac{R(0)}{S(0)} = \frac{\tfrac{6}{5} - (54/35)\sigma E}{1 - 0.9\sigma E}.$$

Table 9.
Reduction of the distance between ∥ and ⊥ maxima relative to
the rotation-free maximum with
increasing excitation.

$R(0)/S(0)$	$\sigma_1 E$
1.200	≪ 1
1.176	0.05
1.149	0.10
1.087	0.20
1.010	0.30

The results for the initial values hold, of course, for the maxima of the experimental convolved signals. To arrive at an appreciable deviation from the relative small-signal distance 1.200, strong excitation $\sigma E_1 \geqslant 0.3$ would be required, as can be seen from table 8.

Nevertheless, fig. 34 shows two cases where the rotational-signal maximum falls below 1.2 times the rotation-free maximum at rather low excitation. For the dye in the pure solvent decanol the fitting procedure gave the ratio $R_{max}/S_{max} = 0.99$. In the aqueous dye–solubilizer mixture, which showed practically no fast rotational decay due to Fluorescein in water, the ratio turned out to be 0.73. Apparently, we have here libration of the dye dissolved in the micellar core, but one has to think of libration also in pure solvents like decanol.

5 Coherent coupling effects

Thus far the experimental transient-transmission signals have been marked as coherence-free, which is to say, that by careful adjusting, different cross sections of the bleaching and probing beams came into bearing within the sample volume. Owing to a peculiarity of the laser system employed this meant slightly different frequencies of bleaching and probing light. In this way the coherent wave superposition introduced in § 2.3.5, that is present in any transient-transmission experiment with beams of identical frequency, could be avoided. In what follows the theory of the coherent coupling effect is given in terms of transient transmission and rotational relaxation, and includes the polarization dependence (von Jena and Lessing 1979a).

5.1 Modulated population changes

To study the coherent interactions the electric field strengths $E_A(x, t)$ of the bleaching light and $E_M(x, t)$ of the probing light have to be considered. Their superposition within the sample generates a spatially periodic population change that influences in turn the propagation of the bleaching and probing light (see Sargent 1976).

Considering at first the case of parallel polarization directions ($\psi = 0$) one has, introducing the relative phase ϕ of bleaching and probing wave,

$$E_A(x, t) = E_A \exp\{i(\omega t - kx + \phi)\},$$

$$E_M(x, t) = E_M \exp\{i(\omega t + kx)\},$$

and the superposition

$$
\begin{aligned}
E_A(x, t) + E_M(x, t) &= (E_A - E_M) \exp\{i(\omega t - kx + \phi)\} \\
&\quad + E_M \exp\{i(\omega t + \phi/2)\} \\
&\quad \times \{\exp\{-i(kx - \phi/2)\} \\
&\quad + \exp\{i(kx - \phi/2)\}\} \\
&= (E_A - E_M) \exp\{i(\omega t - kx + \phi)\} \\
&\quad + E_M \exp\{i(\omega t + \phi/2)\} 2 \cos(kx - \phi/2).
\end{aligned}
$$

The resulting field turns out to be composed of a wave running in the direction of the original bleaching wave with amplitude $(E_A - E_M)$ and a standing wave with amplitude $2E_M$. The corresponding intensity is

$$
\begin{aligned}
I(x, t) &= (\epsilon_0/\mu_0)^{1/2} |E_A(x, t) + E_M(x, t)|^2 \\
&= (\epsilon_0/\mu_0)^{1/2} \{E_A^2 + E_M^2 + 2E_A E_M \cos(2kx - \phi)\},
\end{aligned}
$$

with a periodically space-dependent part that is proportional to the field strengths of probing *plus* bleaching light. The time dependence in $I(x, t)$ relates to the pulse envelope only. If the frequencies of the two fields differ by $\Delta\omega$ ($\ll \omega$), there is a modulated intensity part $I_{mod}(x, t) = (\epsilon_0/\mu_0)^{1/2} 2E_A E_M \cos\{2(kx + \Delta\omega t) + \phi\}$ that runs with phase velocity $\Delta\omega/k$ through the sample.

In the case of perpendicular polarization directions ($\psi = \pi/2$) the superposition has to be written with field vectors

$$\boldsymbol{E}_A(x, t) = E_A \exp\{i(\omega t - kx + \phi)\}\boldsymbol{e}_z,$$

$$\boldsymbol{E}_M(x, t) = E_M \exp\{i(\omega t + kx)\}\boldsymbol{e}_y,$$

giving

$$
\begin{aligned}
\boldsymbol{E}_A(x, t) + \boldsymbol{E}_M(x, t) &= (E_A - E_M) \exp\{i(\omega t - kx + \phi)\}\boldsymbol{e}_z \\
&\quad + E_M \exp\{i(\omega t + \phi/2)\}\{\exp\{-i(kx - \phi/2)\}\boldsymbol{e}_z \\
&\quad + \exp\{i(kx - \phi/2)\}\boldsymbol{e}_y\}.
\end{aligned}
$$

If the superposition field is projected on two axes z_1 and z_2 at 45° to $\boldsymbol{e}_z$ and $\boldsymbol{e}_y$ the corresponding intensities are

$$
\begin{aligned}
I_{z_1}(x, t) &= (\epsilon_0/\mu_0)^{1/2} |\{\boldsymbol{E}_A(x, t) + \boldsymbol{E}_M(x, t)\}\boldsymbol{e}_{z_1}|^2 \\
&= \tfrac{1}{2}(\epsilon_0/\mu_0)^{1/2} \{E_A^2 + E_M^2 + 2E_A E_M \cos(2kx - \phi)\},
\end{aligned}
$$

$$I_{z_2}(x, t) = \tfrac{1}{2}(\epsilon_0/\mu_0)^{1/2} \{E_A^2 + E_M^2 - 2E_A E_M \cos(2kx - \phi)\}.$$

Thus the intensities polarized along z_1 and z_2 respectively contain modulated terms that are antiphase. From the intensities given thus far, the following changes of population result for the simplest case of a pure singlet population, isotropic rotational diffusion and δ-impulse excitation:

$$\langle \Delta N(x, t)\rangle_{\parallel} = \{\tfrac{4}{5}\exp(-6Dt) + 1\}\exp(-t/\tau)\sigma N/(h\nu)(\epsilon_0\mu_0)^{1/2}$$
$$\times [(E_A^2 + E_M^2) + 2E_A E_M \cos(2kx - \phi)]. \tag{89}$$

To formulate the perpendicular case the modulated population changes for the probe-light components along z_1 and z_2 are required:

$$\langle \Delta N(x, t)\rangle_{z_1}^{mod} = \tfrac{1}{2}\exp(-t/\tau)\sigma N/(h\nu)(\epsilon_0/\mu_0)^{1/2}2E_A E_M \cos(2kx - \phi)$$
$$\times [\{\tfrac{4}{5}\exp(-6Dt) + 1\} - \{-\tfrac{2}{5}\exp(-6Dt) + 1\}], \tag{90}$$

and the same expression for $\langle \Delta N(x, t)\rangle_{z_2}^{mod}$, but with the factors $-\tfrac{2}{5}$ and $\tfrac{4}{5}$ exchanged. Additionally the unmodulated population change must be included:

$$\langle \Delta N(x, t)\rangle_{\perp} = \{-\tfrac{2}{5}\exp(-6Dt) + 1\}\exp(-t/\tau)\sigma N/(h\nu)(\epsilon_0/\mu_0)^{1/2}(E_A^2 + E_M^2). \tag{91}$$

5.2 Wave propagation

From the calculated population changes the propagation of an electromagnetic wave can be predicted according to the wave equation in one dimension:

$$\epsilon_0\epsilon\mu_0\partial^2 E(x, t)/\partial t^2 - \partial^2 E(x, t)/\partial x^2 = 0.$$

The plane waves $E_{\pm}(x, t) = E_0\exp\{i(\omega t \pm kx)\}$ are solutions with wavenumber $k = (\epsilon_0\epsilon/\mu_0)^{1/2}\omega$.

In an absorbing medium the dielectric constant ϵ and the wavenumber k are complex: $\epsilon = \epsilon' + i\epsilon''$, $k = k' + ik''$. The solution of the wave equation

$$E(x, t) = E\exp\{i(\omega t + k'x)\exp(-k''x)\}$$

now contains a damping term that determines also the spatial dependence of the corresponding intensity:

$$I(x) = I(0)\exp(-2k''x).$$

Comparison with the Lambert–Beer law identifies $k'' = \tfrac{1}{2}\sigma N$, or in the case of population changes $\langle \Delta N\rangle_{\parallel, \perp}\}$ accordingly

$$k'' = \tfrac{1}{2}\sigma\{N - \langle \Delta N\rangle_{\parallel, \perp}\}.$$

Returning to the wave equation and splitting of the fast time dependence due to the frequency ω of light, there remains

$$\partial^2 E(x, t)/\partial x^2 + k^2 E(x, t) = 0. \tag{92}$$

$E(x, t)$ now being the slow amplitude variation of the wave. Obviously, the population changes $\langle \Delta N\rangle_{\parallel, \perp}$ influence via k'' in k the wave propagation. One puts for the total field $E(x, t)$ due to the counter-running bleaching and probing fields with delay t_0, analogously to Kogelnik and Shank (1972)

$$E(x, t - t_0) = E_A(x, t)\exp\{-i(k'x - \phi)\} + E_M(x, t - t_0)\exp(ik'x)$$

and obtains, inserting this into eq. (92) and neglecting terms with k''^2 ($\ll k'^2$),

$$- 2ik' \exp\{-i(k'x - \phi)\}\, \partial E_A(x, t)/\partial x + 2ik' \exp(ik'x)\partial E_M(x, t - t_0)/\partial x$$
$$+ \exp\{-i(k'x - \phi)\}\partial^2 E_A(x, t)/\partial x^2 + \exp(ik'x)\partial^2 E_M(x, t - t_0)/\partial x^2$$
$$+ ik'\sigma\{N - \langle \Delta N\rangle_{\|,\perp}\}\{E_A(x, t) \exp[-i(k'x - \phi)]$$
$$+ E_M(x, t - t_0) \exp(ik'x)\} = 0.$$

5.2.1 Parallel orientation

It is convenient to distinguish the nonmodulation part $\langle \Delta N\rangle$ and the modulated part $\langle \Delta N\rangle^{\mathrm{mod}}$ of $\langle \Delta N\rangle_{\|,\perp}$. Convolving the δ-impulse response of (89) with the bleaching pulse shape, one has for the unmodulated part for parallel orientation

$$\langle \Delta N(x, t)\rangle_\| = A \int_{-\infty}^{t} dt' E_A^2(x, t')f_1(t - t'), \tag{93}$$

with

$$f_1(t - t') = \{\tfrac{4}{5}\exp[-6D(t - t')] + 1\} \exp\{-(t - t')/\tau\}, \quad A = \frac{\sigma N}{h\nu}\,(\epsilon_0\mu_0)^{1/2},$$

and for the modulated part

$$\langle \Delta N(x, t)\rangle_\|^{\mathrm{mod}} = A \int_{-\infty}^{t} dt' E_M(x, t' - t_0)E_A(x, t')f_2(t - t'), \tag{94}$$

where

$$f_2(t - t') = f_1(t - t')\, 2 \cos(2k'x - \phi).$$

Eq. (94) shows that for a modulated population change to appear a certain amount of overlap between the two counter-running pulses is required within the sample.

To sketch the further derivation, eqs. (93) and (94) are inserted into the wave equation and use is made of the Euler relation for $2 \cos(2k'x - \phi)$. The result is a coupling between bleaching and probing, such that bleaching light appears with direction and phase identical to the probing light and thus reaches the detector. Moreover, probing light appears with the direction and phase of the bleaching light, but this is of little concern, since $E_M(x, t - t_0) \ll E_A(x, t)$. The ensuing differential equation for the wave in the original probe-light direction can be simplified owing to the identical bleaching and probing pulse shapes (Ippen and Shank 1977):

$$E_M(x, t' - t_0) = a(x)E_A(x, t' - t_0),$$

with a space-dependent factor $a(x)$. Using this one arrives at the final differential equation

$$\frac{\partial^2 E_M(x, t - t_0)}{\partial x^2} + 2ik' \frac{\partial E_M(x, t - t_0)}{\partial x}$$

$$+ ik'\sigma\left\{ N - A \int\limits_{-\infty}^{t} dt' E_A^2(x, t') f_1(t - t') \right.$$

$$- A \int\limits_{-\infty}^{t} dt' E_A(x, t') E_A(x, t' - t_0) f_1(t - t')$$

$$\times \left. \frac{E_A(x, t)}{E_A(x, t - t_0)} \right\} E_M(x, t - t_0) = 0.$$

The ansatz for the solution

$$E_M(x, t - t_0) = E_M(0, t - t_0) \exp(-\lambda x),$$

where λ depends on x only slightly for small population changes, leads to

$$\lambda(t, t_0) = \tfrac{1}{2}\sigma\{N - \langle\langle\Delta N(x, t)\rangle_x\rangle_\parallel^{\text{incoh}}\} - \lambda_\parallel^{\text{coh}},$$

where

$$\langle\langle\Delta N(x, t)\rangle_x\rangle_\parallel^{\text{incoh}} = (1/x) A \int\limits_{0}^{x} dx' \int\limits_{-\infty}^{t} dt' E_A^2(x', t') f_1(t - t'),$$

$$\lambda_\parallel^{\text{coh}} = \frac{\sigma}{2} (1/x) A \int\limits_{0}^{x} dx' [E_A(x', t)/E_A(x', t - t_0)] \int\limits_{-\infty}^{t} dt'$$

$$\times E_A(x', t' - t_0) E_A(x', t') f_1(t - t').$$

The probing intensity is

$$i(x, t - t_0) = i_0(t - t_0) \exp(-2\lambda x),$$

where λ contains the usual partially bleached ground-state population $N - \langle\Delta N\rangle^{\text{incoh}}$ and a new term $\lambda_\parallel^{\text{coh}}$ which describes the bleaching light intensity diffracted into the probing light direction due to the coherent interaction.

The resulting coherent signal contribution is given by

$$\langle Y(t_0)\rangle_\parallel^{\text{coh}} = \sigma d\sigma N \left\{ \int\limits_{-\infty}^{+\infty} dt\, i_0(t - t_0) \right\}^{-1} \left\{ \int\limits_{-\infty}^{+\infty} dt \left([i_0(t - t_0) i_0(t)]^{1/2} \right. \right.$$

$$\times \int\limits_{-\infty}^{t} dt' f_1(t - t') \left\{ (1/d) \int\limits_{0}^{d} dx [I_A(x, t' - t_0) \right.$$

$$\left. \left. \left. \times I_A(x, t')]^{1/2} \right\} \right) \right\}. \tag{95}$$

The incoherent signal is

$$\langle Y(t_0)\rangle_\parallel^{\text{incoh}} = \sigma d\sigma N \left\{ \int\limits_{-\infty}^{+\infty} dt\, i_0(t - t_0) \right\}^{-1}$$

$$\times \left\{ \left[\int\limits_{-\infty}^{+\infty} dt i_0(t - t_0) \right. \right.$$

$$\left. \left. \times \int\limits_{-\infty}^{t} dt' \left(f_1(t - t') \left\{ (1/d) \int\limits_{0}^{d} dx I_A(x, t') \right\} \right) \right] \right\}.$$

On closer inspection of eq. (95) one realizes that the coherent part depends two-fold on the temporal overlap of the counter-running pulses. For incomplete overlap $t_0 \neq 0$ the diminished modulated part of the population change [eq. (94)] can turn around only bleaching intensity from the pulse edges into the probing-beam direction. For $t_0 = 0$ there is the remarkable fact that

$$\langle Y(0) \rangle_{\parallel}^{\text{coh}} = \sigma d\sigma N \left\{ \left[\int\limits_{-\infty}^{+\infty} dt\, i_0(t) \right]^{-1} \left\{ \int\limits_{-\infty}^{+\infty} dt \left(i_0(t) \int\limits_{-\infty}^{t} dt' f_1(t - t') \right. \right. \right.$$

$$\left. \left. \left. \times \left\{ (1/d) \int\limits_{0}^{d} dx I_A(x, t') \right\} \right) \right\} \right\} = \langle Y(0) \rangle_{\parallel}^{\text{incoh}}. \tag{96}$$

Completely independent from the relaxation mechanisms involved, the signal $\langle Y(0) \rangle_{\parallel}^{\text{incoh}}$ is just doubled at $t_0 = 0$, which is consistent with steady-state saturation spectroscopy (compare Keilmann 1977).

5.2.2 Perpendicular orientation

For perpendicular polarization direction of bleaching and probing light one considers wave propagation of the projections on the z_1 and z_2 axes. The scalar wave equation (92) governs each of the projections of the bleaching and probing fields. The population changes involved are given by (90).

The ansatz for the waves must consider the relative phase of the projections

$$E_{z_1}(x, t, t_0) = (1/\sqrt{2})[E_M(x, t - t_0) \exp(ik'x) + E_A(x, t) \exp\{-i(k'x - \phi)\}].$$

$$E_{z_2}(x, t, t_0) = (1/\sqrt{2})[E_M(x, t - t_0) \exp(ik'x) - E_A(x, t) \exp\{-i(k'x - \phi)\}]. \tag{97}$$

The decisive minus sign appears since the projection of E_A onto Z_2 is in antiphase to that onto z_1 because of the perpendicular polarization directions. The z_1-wave propagation can be treated analogously as in § 5.2.1, if the following substitutions are made:

$$f_1(t - t') \rightarrow f_3(t - t') = [-\tfrac{2}{5} \exp\{-6D(t - t')\} + 1] \exp\{-(t - t')/\tau\},$$

$$f_2(t - t') \rightarrow f_4(t - t') = \tfrac{1}{2}[\tfrac{6}{5} \exp\{-6D(t - t')\}] \exp\{-(t - t')/\tau\} 2 \cos(2k'x - \phi)$$

$$= f_4'(t - t') 2 \cos(2k'x - \phi),$$

and for the z_2-wave propagation accordingly

$$f_1(t - t') \rightarrow f_3(t - t'), \qquad f_2(t - t') \rightarrow -f_4(t - t'),$$

and moreover $E_A \rightarrow -E_A$.

For the z_1-projection the results from § 5.2.1 can be adopted directly with the above substitutions, whereas for the z_2-projection the coupling term has to be re-evaluated. The result for the perpendicular signal is

$$\langle Y(t_0) \rangle_\perp = \langle Y(t_0) \rangle_\perp^{\text{incoh}} + \langle Y(t_0) \rangle_\perp^{\text{coh}} = \sigma d\sigma N \left\{ \int_{-\infty}^{+\infty} dt\, i_0(t - t_0) \right\}^{-1}$$

$$\times \left\{ \int_{-\infty}^{+\infty} dt \left(i_0(t - t_0) \int_{-\infty}^{t} dt' f_3(t - t') \right. \right.$$

$$\times \left\{ (1/d) \int_0^d dx I_A(x, t') \right\} \Bigg)$$

$$+ \int_{-\infty}^{+\infty} dt \left([i_0(t - t_0) i_0(t)]^{1/2} \right.$$

$$\times \int_{-\infty}^{t} dt' f_4'(t - t') \Big\{ (1/d) \int_0^d dx [I_A(x, t')$$

$$\times I_A(x, t' - t_0)]^{1/2} \Big\} \Bigg) \Bigg\}. \tag{98}$$

Eq. (98) reveals the fact that, even for the perpendicular case, coherent coupling is present which had not been observed before. Moreover, the rotational diffusion now has an overwhelming influence on the $\perp$ effect in contrast to the $\parallel$ effect. This can be seen from the case of complete overlap $t_0 = 0$:

$$\langle Y(0) \rangle_\perp^{\text{coh}} = \sigma d\sigma N \left\{ \left\{ \int_{-\infty}^{+\infty} dt\, i_0(t) \right\}^{-1} \int_{-\infty}^{+\infty} dt \left(i_0(t) \int_{-\infty}^{t} dt' \right. \right.$$

$$\times \tfrac{1}{2}(\tfrac{6}{5} \exp\{-6D(t - t')\}) \exp\{-(t - t')/\tau\} \Big\{ (1/d) \int_0^d dx I_A(x, t') \Big\} \Bigg) \Bigg\}.$$

The rotational signal at zero delay is

$$R(0)^{\text{inc}} = \sigma d\sigma N \left\{ \left\{ \int_{-\infty}^{+\infty} dt\, i_0(t) \right\}^{-1} \int_{-\infty}^{+\infty} dt \left(i_0(t) \int_{-\infty}^{t} dt' \right. \right.$$

$$\tfrac{6}{5} \exp\{-(6D + 1/\tau)(t - t')\} \Big\{ (1/d) \int_0^d dx I_A(x, t') \Big\} \Bigg) \Bigg\}.$$

We thus have

$$\langle Y(0)\rangle_{\perp}^{coh} = \tfrac{1}{2}R(0)^{incoh},\tag{99}$$

i.e., in contrast to the parallel case the coherent signal part does not equal the incoherent one. Only for very slow rotations ($D \to 0$) do $\langle Y(0)\rangle_{\perp}^{coh}$ and $\langle Y(0)\rangle_{\perp}^{inc}$ become identical, since the functions f_3 and f'_4 then show the same time course

$$f_3(t - t') = \{-\tfrac{2}{5}\exp\{-6D(t - t')\} + 1\}\exp\{-(t - t')/\tau\} \to \tfrac{3}{5}\exp\{-(t - t')/\tau\},$$

$$f'_4(t - t') = \tfrac{3}{5}\exp\{-(6D + 1/\tau)(t - t')\} \to \tfrac{3}{5}\exp\{-(t - t')/\tau\}.$$

For very fast rotational diffusion, i.e., $1/6D \ll$ pulsewidth, $\langle Y(t_0)\rangle_{\perp}^{coh}$ vanishes. This can be explained by the antiphase of the two modulated population changes $\langle \Delta N\rangle_{z_1}$ and $\langle \Delta N\rangle_{z_2}$. Transition moments that have left the direction z_1 during excitation just replenish the deficiency due to the modulated population change in the z_2 direction and vice versa. Only in highly viscous liquids does the distribution of directions remain constant, and therefore the two complementary modulated populations are preserved during the pulse duration.

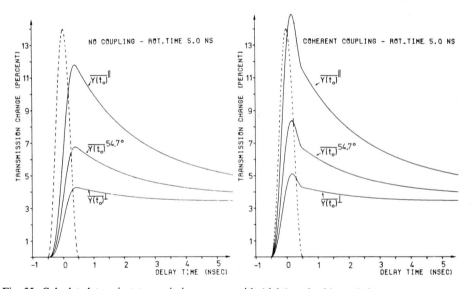

Fig. 35. Calculated transient-transmission curves with (right) and without (left) coherent-coupling effect. Dashed line: self-convolution of laser pulse. Rotational relaxation time comparable to singlet lifetime. From von Jena and Lessing (1979a).

Fig. 35 compares calculated transient-transmission curves with and without the coherent-coupling effect. An additional bump appears on the $\parallel$ and $\perp$ branches as well as on the rotation-free branch. As a consequence the maximum of each branch is shifted towards $t_0 = 0$ as compared to the maxima of the curves without the effect. The relation (96) is fulfilled, i.e., the $\parallel$ signal at $t_0 = 0$ with bump is double the ordinate of the $\parallel$ branch without this effect. Relation (99) is also fulfilled, i.e., the $\perp$ signal at $t_0 = 0$ with bump equals the ordinate of the $\perp$ branch plus half the difference between $\parallel$ and $\perp$ branches of the signal without the effect. Even the rotation-free signal has a bump, the ordinate at $t_0 = 0$ of which is given below. Evidently, any curve fitting ignoring the coherent effect will lead to discrepancies or to the belief that there are two relaxations involved. It is true that the coherent-coupling bump does not exceed a small interval around t_0 given essentially by the self-convolution of the pulse shape, but the extraction of fast rotational diffusion from the signals is impeded. Fig. 36 gives an alternating measurement of transient transmission that clearly shows the features of coherent coupling.

The rotation-free amplitude $S(0) = \langle Y(0)\rangle_{54.7^\circ}$ and that of the rotational signal $R(0) = \langle Y(0)\rangle_\parallel - \langle Y(0)\rangle_\perp$ can be calculated from (96) and (99):

$$S(0) = \tfrac{2}{3}\langle Y(0)\rangle_\parallel^{\text{incoh}} + \tfrac{2}{3}\langle Y(0)\rangle_\perp^{\text{coh}} + \tfrac{1}{3}R(0)^{\text{incoh}}$$
$$= \langle Y(0)\rangle_\parallel^{\text{incoh}} + \tfrac{1}{3}\langle Y(0)\rangle_\perp^{\text{incoh}},$$

$$R(0) = 2\langle Y(0)\rangle_\parallel^{\text{incoh}} - \langle Y(0)\rangle_\perp^{\text{coh}} - \tfrac{1}{2}R(0)^{\text{incoh}}$$
$$= \tfrac{3}{2}\langle Y(0)\rangle_\parallel^{\text{incoh}} - \tfrac{1}{2}\langle Y(0)\rangle_\perp^{\text{incoh}}.$$

In the case of slow rotational diffusion one has

$$S(0) = 2S(0)^{\text{incoh}}, \qquad R(0) = 2R(0)^{\text{incoh}},$$

i.e., here the coherent-coupling effect is equal for both signals. For fast rotational diffusion, i.e., $R(0)^{\text{incoh}} \approx 0$, one has

$$S(0) = \tfrac{4}{3}S(0)^{\text{incoh}}, \qquad R(0) \equiv R(0)^{\text{coh}} = S(0)^{\text{incoh}},$$

i.e., the coherent coupling effect is able to simulate a rotational signal where there is none.

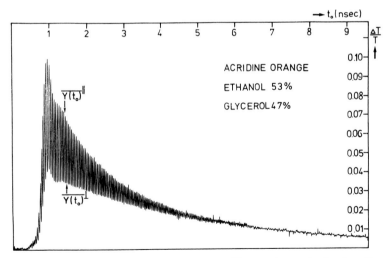

Fig. 36. Alternating transient-transmission measurement. The envelopes clearly show the coherent-coupling effect. From von Jena and Lessing (1979a).

Fig. 37 shows how the height of the coherent-coupling bump differs for the different signal branches if the rotational relaxation is faster (compare fig. 35). One may not be confused from the more distinctly separated coherent signals in fig. 37–this effect is due to the use of a shorter coherence length in the calculations (see 5.3). Whereas the bump heights in the ∥ case are comparable, the bump on the ⊥ branch is now lower than for the case of slower rotational relaxation, as expressed by eq. (96). This dependence on rotational relaxation and therefore on viscosity is illustrated in the experimental examples of fig. 38. Whereas the bump on the ∥ envelope is essentially the same for the three solvents of increasing viscosity, the effect on the ⊥ envelope is barely visible as a shoulder in acetone, which becomes more distinct the wider the alternating stripe (i.e., the rotational signal) as predicted by eq. (99).

According to eqs. (96) and (98) the relative proportions of the coherent and incoherent parts of the total signal are independent of the bleaching-pulse energy, i.e., they vary proportionally, if the pulse energy is varied. This has been checked on the rotation-free signal in fig. 39. The signal with ten-times-reduced bleaching-pulse energy was brought to the same height as the incoherent-signal maximum by increasing the sensitivity of the detection electronics. Apart from the stronger noise both coherent bumps reach the same height. This test allows one to distinguish the coherent-coupling effect from the similar appearance of contributions of higher transitions (compare fig. 11 and § 3.4).

A final verification of the coherent-coupling theory for the parallel and perpendicular cases is found in the inverse effect of § 3.4. Eqs. (96) and (99) hold for this case equally well, and fig. 40 shows calculated curves of the inverse effect with and without the coherent-coupling effect. Since excitation leads here

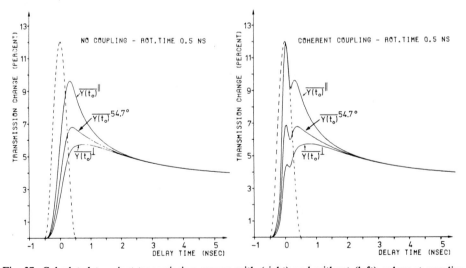

Fig. 37. Calculated transient-transmission curves with (right) and without (left) coherent-coupling effect. Rotational relaxation time faster than singlet lifetime. Coherence length shorter than spatial pulse extension. Dashed line: self-convolution of laser pulse. From von Jena and Lessing (1979a).

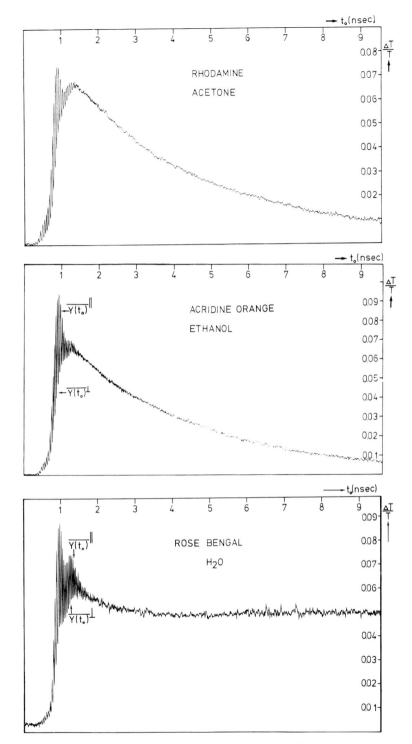

Fig. 38. Alternating transient-transmission measurements for increasing (top to bottom) rotational diffusion times in low-viscosity solvents. Top: Rhodamine 6G (acetone). Center: Acridine orange (ethanol). Bottom: Rose bengal (water). From von Jena and Lessing (1979a).

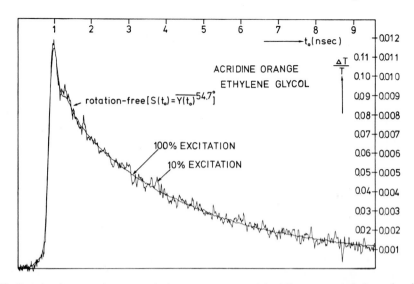

Fig. 39. Rotation-free transient-transmission measurement of Acridine orange (ethylene glycol) with different bleaching-pulse energies. From von Jena and Lessing (1979a).

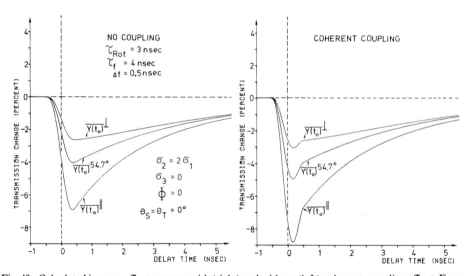

Fig. 40. Calculated inverse-effects curves with (right) and without (left) coherent-coupling effect. From von Jena and Lessing (1979a).

to a decrease in transmission, the probing light is also weakened by the coherent-coupling effect. This conforms to the wave propagation formalism derived above, since for a change of sign of the modulated population changes [eqs. (89) and (90)] the wave equation (92) supplies additional waves running in the direction of the probing waves that are antiphase to the latter ($-\Delta N^{mod} = \exp(i\pi)\Delta N^{mod}$). Instead of constructive interference in dyes showing bleaching

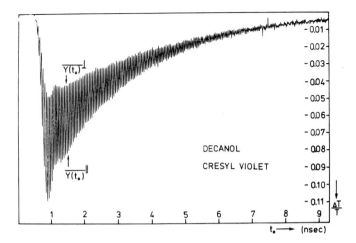

Fig. 41. Alternating transient-transmission measurement showing inverse effect of Cresyl violet (decanol). From von Jena and Lessing (1979a).

we have therefore a destructive interference of part of the measuring light. The alternating measurement on Cresyl violet (decanol) in fig. 41 clearly demonstrates this behaviour.

5.3 Influence of source parameters

Up to now the question of the coherence length has not been considered. It was tacitly assumed that there are no phase jumps during the laser pulse, in other words the coherence length l_c should be larger than the spatial pulse extend $s = 2c\Delta t$, where Δt is the pulse half-width in the time domain. If, however, $l_c < s$, shifts of the phase $\phi(t)$ are feasible between the counter-running waves. If the phase shifts occur very frequently the modulated intensity part vanishes, according to

$$\langle I_{\mathrm{mod}}(x, t)\rangle_t = (\epsilon_0/\mu_0)^{1/2} 2 E_M E_A (1/2\pi) \int_0^{2\pi} d\phi \, \cos(2kx - \phi) = 0.$$

In other words, in spite of the temporal overlap of the pulses there will be no periodic population change and hence no coherent-coupling effect in this case. As long as the coherence length is larger than the sample depth d, however, the effect is still there at delay $t_0 = 0$, since here the phase jumps occur simultaneously in both waves. Hence the relative phase ϕ of the waves stays constant. Finally, if l_c is even smaller than d, the spatial extent of the modulated intensity part is reduced and the coherent bump diminished by l_c/d.

Outside of the zero delay position (corresponding delay path $l = ct_0$) constant-phase superposition is feasible, as long as the condition $l < l_c$ is fulfilled, and the coherent bump exists. For $l > l_c$ the coherent contributions on top of the signal disappear. Therefore the half-width of the coherent bump can be considered as a measure of the coherence length of the laser pulses. Stated in another way: if the coherence length is larger than the equivalent pulse width, the coherent-coupling bump will render the self-convolution of the pulse. If it is shorter, the bump will be narrower than the self-convolution of the pulse. The latter case can be found in the theoretical curve of fig. 37 (right), where the pulse self-convolution is clearly broader than the coherent-coupling bump. In the preceding experimental examples taken with 514 nm pulses the bump half-width is smaller than the pulse self-convolution and suggests a coherence length that is half the spatial extent s of the pulse. However, it turned out that the coherence length varies with the age of the argon laser tube. A more drastic relation of bump width to self-convolution width could be demonstrated when using 488 nm pulses that are 1.5 ns wide. Fig. 42 shows a rotation-free and an alternating transient-trans-mission measurement, where a rather narrow bump is visible in all three cases. Apparently, the relation of coherence length to pulse width is even smaller for the blue pulses than for the green pulses.

If bleaching and probing waves have frequencies differing by $\Delta\omega$, the modulated-intensity part is no longer stationary, but sweeps through the sample with velocity $\Delta\omega/k$. As a consequence the modulated population change can build up appreciably only if the spatial grating relaxes so fast that it does not smear out the build-up of gratings at new positions during this sweep. A simplified model with a single relaxation time and using rectangular pulses of duration Δt results in the following modulated population for $t_0 = 0$:

$$\Delta N(\tau, \Delta\omega) = (\epsilon_0/\mu_0)^{1/2}(\sigma N/h\nu)2E_M E_A \int_0^{\Delta t} dt \exp\{-(\Delta t - t)/\tau\}$$
$$\times \cos(2kx + 2\Delta\omega t).$$

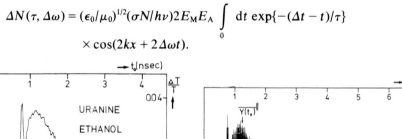
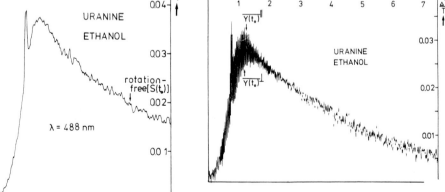

Fig. 42. Transient-transmission measurements of Uranine (ethanol) using blue laser pulses showing coherent-coupling effect. Left: rotation-free measurement. Right: alternating measurement. From von Jena and Lessing (1979a).

Assuming $\Delta t \ll \tau$ one obtains from this a Lorentzian dependence of ΔN on $\Delta \omega$:

$$\Delta N(\tau, \Delta \omega) \sim \frac{(1/\tau)^2}{(1/\tau)^2 + (2\Delta \omega)^2}.$$

For short lifetimes τ rather large frequency deviations are feasible before the coherent-coupling effect is weakened appreciably; for $\Delta \omega = 1/2\tau$ the effect is halved. On the other hand, for long lifetimes like those of the triplet, already small frequency differences suffice to eliminate the coherent-coupling effect.

This can be accomplished when using cavity-dumped mode-locked lasers. Fig. 43 shows three experiments written on the same graph. The alternating measurement was made with optimized alignment of bleaching and probing beams in the sample geometry. The envelopes representing the $\parallel$ and $\perp$ case clearly show the coherent coupling bump. By intentional selection of different areas within the cross sections of bleaching and probing beams using diaphragms, the bump can be made disappear, as the individually written $\parallel$ and $\perp$ branches vlearly show. It is believed that this is due to different frequencies $\omega + \Omega$ and $\omega - \Omega$, that arise from the laser frequency ω in the Bragg cell of the dumper, which is driven by an acoustic frequency Ω. One can imagine that the two diffracted beams in the folded laser cavity which form the output beam are separated at large distances due to their reflection on different sections of an intracavity spherical mirror, such that the beam cross section contains areas of different frequencies. One has thus the choice of either keeping the effect and accounting for it in curve fitting, or of tuning the sample geometry until it disappears. The coherent-coupling effect is certainly not restricted to continuous picosecond spectroscopy. It is only owing to the low reproducibility and discrete nature of single-shot measurements that it has not been considered earlier. On one

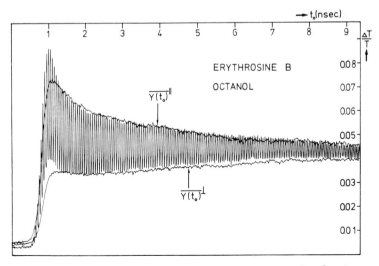

Fig. 43. Alternating transient-transmission measurement and coherence-free $\parallel$ and $\perp$ traces of the same sample, Erythrosine B (octanol). From von Jena and Lessing (1979a).

hand it gives a good criterion for optimal alignment of the probing and bleaching beams, on the other hand it complicates the evaluation of experiments. Whenever possible it should be avoided by using bleaching and probing light of nonidentical frequencies. If this is not feasible the quantitative theory now available should help to extract unequivocal information from transient-transmission experiments.

6 Experimental techniques

Although the field of picosecond spectroscopy is still dominated by the single-shot techniques (see Shapiro 1977), the continuous methods start are beginning to come to the fore owing to their sensitivity, reproducibility and ease of alignment.

6.1 Fluorescence apparatus

Mode-locked cw lasers appear attractive for decay measurements because of their higher intensity, shorter pulse-width and better stability when fed from a continuous discharge instead of the repetitive discharge lamps used thus far. The short pulse widths available make fluorescence decay time resolution no longer source limited but detector limited. Conventional photomultiplier tubes still have a trailing edge of the noise pulse of 1 to 2 ns, which determines time resolution. Recently developed crossed-field photomultipliers have 130 ps rise and fall times; however, the gain is still limited to 10^3. Processing of repetitive signals allows rather low fluorescence intensities to be detected, so that the small-signal regime is usually applicable.

The first fluorescence decay measurements using a cw mode-locked laser have been published by Merkelo et al. (1970). Their set-up is shown in fig. 44, together with the He–Ne laser pulse shape taken by a semiconductor photodiode. Time resolution of the fluorescence detection is limited by the response of the RCA 7102 photomultiplier, which had a response of 1 ns, but better resolution may be obtained by deconvolution. Alternation of the excitation polarization directions allowed the measurement of fluorescence depolarization times which were evaluated by a phase-shift method. In this way singlet lifetimes and rotational–relaxation times of Methylene blue and Chlorophyll 13 were determined as a function of temperature.

The apparatus of Harris et al. (1977) makes use of a synchronously-pumped cavity-dumped cw Rhodamine 6G laser that is frequency doubled, giving tunable uv output from 267 to 334 nm (see fig. 45). For their method of sampling, single-photon timing and signal averaging, a nearly gaussian transient response of 1 ns fwhm is reported (photomultiplier RCA 8850). The time resolution is thus detector limited, but can be improved by one order of magnitude with the use of a crossed-field photomultiplier. Fluorescence lifetimes of scintillator and label dyes have been measured. Time-correlated single-photon counting is used in the

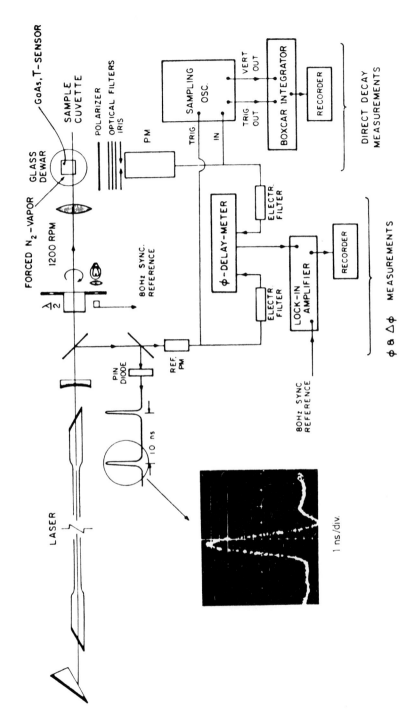

Fig. 44. Fluorescence-decay and depolarization apparatus of Merkelo et al. (1970).

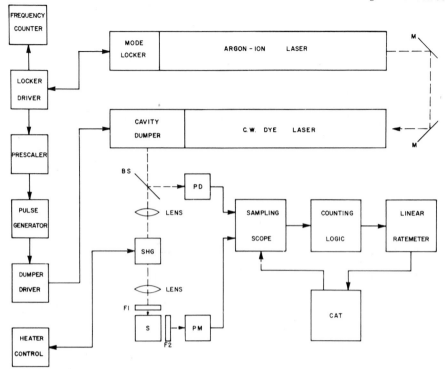

Fig. 45. Fluorescence-decay apparatus of Harris et al. (1977).

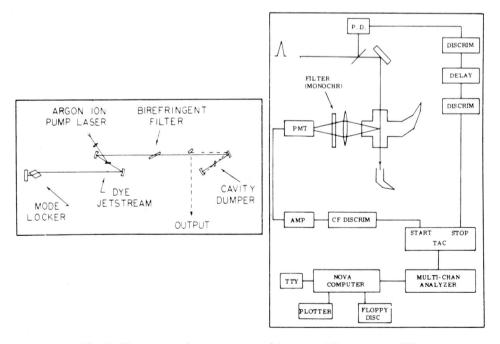

Fig. 46. Fluorescence-decay apparatus of Spears and El-Mangŭch (1977).

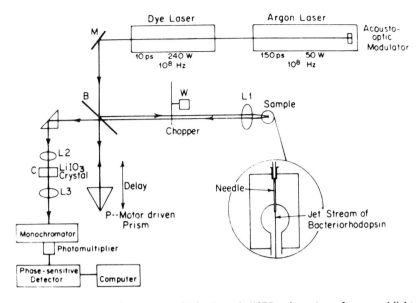

Fig. 47. Fluorescence-decay apparatus of Hirsch et al. (1976) using a 'sum-frequency' light gate.

system of Spears et al. (1978) with a mode-locked cavity-dumped cw dye laser that is frequency doubled (compare fig. 46). The time resolution is detector limited (photomultiplier RCA C31034). The instrumental response has a 0.9 ns wide fwhm (see fig. 9). Fluorescence decay and depolarization investigations have meanwhile been made with this equipment (see Spears and Cramer 1978).

An interesting way to circumvent the limitations due to detector response is the sum-frequency light-gate apparatus of Hirsch et al. (1976). The fluorescence to be sampled is mixed with the delayed laser pulse and focused into a nonlinear crystal that generates light at the sum frequency of laser output and fluorescence emission. Thus light at the sum frequency in the uv is generated only as long as the delayed laser pulse is present. In this way, the crystal serves as a gate for up-conversion of the fluorescence to the uv light. The response of detection then is essentially given by the self-convolution of the laser pulse of reportedly 10 ps width. The efficiency of conversion from fluorescence to sum-frequency light was 0.2% with 100 W pump power and an angle-tuned ADP crystal. The equipment has been used to study the concentration dependence of the fluorescence lifetime of bacteriorhodopsin. The method promises to make continuous picosecond fluorimetry competitive with the single-shot techniques.

6.2 Transient-transmission apparatus

The extension of classical flash spectroscopy into the nanosecond to picosecond range is entirely due to the advent of the laser. The directionality of the laser beam allows the creation of time delays via optical path-lengths and thus makes it feasible to derive the bleaching and probing flashes from the same laser pulse.

The method of Phillion et al. (1975) is not a transient-transmission method, but is related, since it measures delayed probing light scattered from the transient grating that was set-up in the sample by two co-running bleaching pulses (see fig. 48). The measured signal is the probing light scattered into the first order of diffraction, which is, in the plane-wave approximation (see von Jena and Lessing 1979d),

$$i_M^{\pm 1}(d, t - t_0) = \tfrac{1}{4}(\sigma E)^2 f_{\parallel,\perp}(t)^2 T(1 - T)^2 i_M^0(0, t - t_0),$$

where

$$f_{\parallel,\perp}(t) = (1 - \phi)[\exp(-k_1 t) + A_{\parallel,\perp} \exp\{-(k_1 + 6D)t\}]$$
$$+ \phi[\exp(-k_T t) + A_{\parallel,\perp} \exp\{-(k_T + 6D)t\}],$$

and $i_M^0(0, t - t_0)$ is the probing-light intensity before the sample E is here the total number of photons per cm^2 for *one* excitation beam. The A factors are as given in § 4.4.2.

The advantage of the induced-grating method is that relaxation measurements are not restricted to rather strongly absorbing molecules. A weak diffracted probing light can be still detected while the corresponding transmission change may be already buried within the noisy steady-state signal of a photodetector measuring directly the probing light intensity. Furthermore the anisotropy of diffusion processes in anisotropic samples (i.e. crystals) can be investigated.

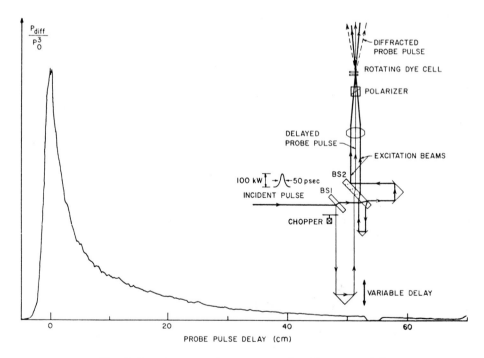

Fig. 48. Transient-scattering apparatus of Phillion et al. (1975).

Since the absorption of dyes suffices normally to get transmission changes in the 10% region, for dyes absorption experiments of the interferometer type are more appropriate due to their easier beam alignment.

A typical short-time experimental set-up for transient-transmission measurements, shown in fig. 49, has been reported by Shank and Ippen (1975). It works with co-running pulses, thus offering a time resolution limited by pulse width alone, but this implies rather poor overlap of probing and exciting beam in the probe. Since only the short-time range is investigated the variable delay line is kept short, and beam-diameter variation through the whole delay-time range need not be corrected for. Signal processing is for this set-up provided by synchronously demodulating the photomultiplier signal and data-storage in a multi-channel analyzer.

Greater delay-time ranges up to 10 ns require a different experimental approach to transient-transmission measurements. Now beam-diameter variations can be no longer neglected and beam alignment must be extremely stable over the whole delay path of approximately 1500 mm. These requirements can only be met by a set-up based on counter-running pulses, as shown in fig. 50 (see also von Jena and Lessing 1979b). The disadvantage of this method–time resolution restricted by probe length–is at present of minor importance, since a mode-locked argon laser with a pulse width of 300 ps is used. Moreover, for low-viscosity liquids a flow-cell depth of 0.1 mm is possible, which appears to be sufficient even for the shortest pulse widths now available from synchronously pumped cw dye lasers. On the other hand, relaxation times up to several nanoseconds can be measured with great precision. So, with the appropriate optical pulse generator, a time range of 10^4 can be covered by a set-up like that given in fig. 50.

The basics of the signal-processing method incorporated in the apparatus of fig. 50 are shown in fig. 51. The second and the third pulse row show the modulation of the bleaching beam. The bleaching pulses are switched off and on by a light chopper, and in addition experience periodic changes of their polarization state. When the probing pulses (fourth row) enter the sample they are modulated owing to the momentary intensity (off-on) and polarization state

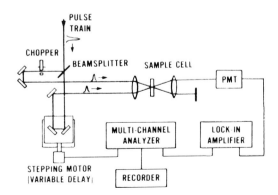

Fig. 49. Transient-transmission apparatus of Ippen et al. (1976).

(parallel-perpendicular) of the bleaching pulses, see fifth row. Integrating the detector signal by an RC circuit (see fig. 50) one gets a voltage signal containing ac and dc components. The ac component is detected by a lock-in amplifier, while the desired dc component (probing-light intensity with excitation off) is sampled by a boxcar integrator. Fig. 52 demonstrates on the upper trace a signal as provided for measurements of $Y_\parallel - Y_\perp$. It shows a double modulation due to the periodic changing of the polarization state and the additional, faster chopping rate. The lower trace rectangular signal defines the sampling point at which the dc voltage component is measured.

As discussed in § 2, chopping the bleaching beam allows measurements of transmission changes $\Delta T/T$ separately that are proportional to the change in population density for weak signals only. A proposed apparatus by which a direct measurements of $\Delta N(t)$ can be achieved is shown in fig. 53. It is based on a polarization angle of 45° between bleaching and probing light, and requires three lock-in amplifiers working at different frequencies. By proper setting of the attenuators for the bleaching beam blocked one can get simultaneously the rotation-free signal

$$S(t_0) = \ln\{(i_\parallel i_\perp^2/i_0^3)\} = 3\sigma d \langle \Delta N(t_0)\rangle_{54.7°}$$

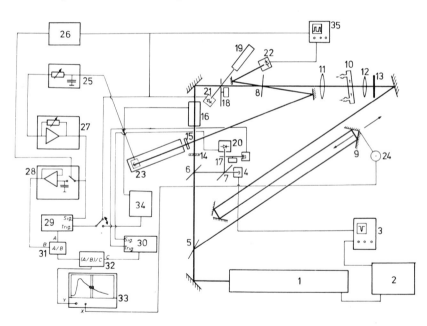

Fig. 50. Transient-transmission apparatus of Lessing et al. (1975). Legend: 1, mode-locked argon laser; 2, driver electronics; 3, sampling scope; 4, crossed-field photomultiplier; 5, 6, 7, 8, beam splitters; 9, retroreflector; 10, flow cell; 11, 12, lenses; 13, aperture; 14, attenuator; 15, interference filter; 16, Pockels cell; 17, 18, light chopper; 19, He–Ne laser; 20, 21, 22, photodiodes; 23, photomultiplier (1 P 28); 24, potentionmeter; 25, low-pass filter; 26, trigger delay; 27, current-to-voltage converter; 28, sample-and-hold amplifier; 29, 30, lock-in amplifier; 31, 32, ratio circuits; 33, XY-recorder; 34, square-wave generator; 35, oscilloscope.

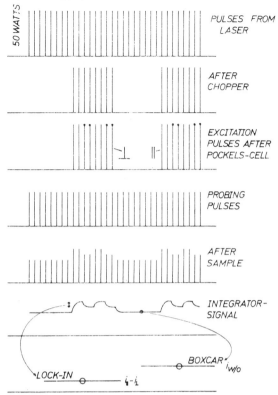

Fig. 51. Signal-detection scheme. Chop rate slower than polarization alternation.

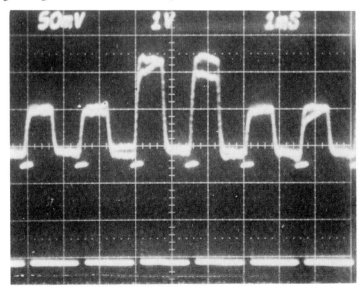

Fig. 52. Oscillogram of signal after current-to-voltage converter (27 in fig. 50). Chop rate faster than polarization alternation.

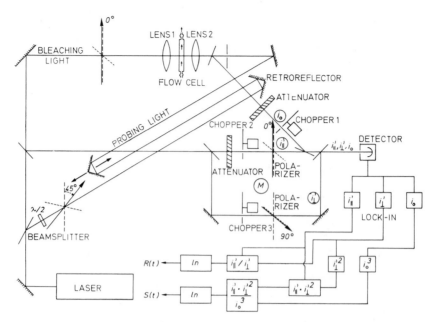

Fig. 53. Proposed absolute photometer for transient-transmission measurements.

and the rotational signal

$$R(t_0) = \ln(i_\parallel / i_\perp) = \sigma d [\langle \Delta N(t_0)\rangle_\parallel - \langle \Delta N(t_0)\rangle_\perp].$$

7 Final comment

The intention of this review was to record the present degree of understanding of picosecond measurements in dyes. The continuous methods help to avoid the disturbing stimulated fluorescence and to measure small-signal curves. These contain information on population relaxation and orientational relaxation, as well as on the triplet yield. Moreover, the contribution of higher transitions allows the determination of relative orientation of the relevant transition moments. Isotropic hydrodynamic models usually suffice to describe the rotational relaxation of dyes in solutions, except for the case of specific interactions.

References

Albrecht, A.C. (1961), J. Molec. Spectrosc. **6**, 84.
Alfano, R.R. and S.L. Shapiro (1972), Opt. Commun. **6**, 98.
Becker, R. (1966), Theory of Heat (Springer; New York).
Bonch-Bruevich, A.M., N.N. Zatsepina, T.K. Razumova, G.M. Rubanova, I.F. Tupitsin and V.N. Shuvalova (1969), Opt. Spectrosc. **26**, 51.

Busch, G.E., K.S. Greve, G.L. Olson, R.P. Jones and P.M. Rentzepis (1975), Chem. Phys. Letters 33, 417.

Chuang, T.J. and K.B. Eisenthal (1971), Chem. Phys. Letters 11, 368.

Corti, M. and V. Degiorgio (1975), Opt. Commun. 14, 358.

DeMaria et al. (1966), Appl. Phys. Letters 8, 174.

Drexhage (1973), Structure and Properties of Laser Dyes, in: Schäfer, F.P., ed., Dye Lasers, Topics in Applied Physics, Vol. 1, 1973 (Springer-Verlag, Heidelberg).

Duguay, M.A. and J.W. Hansen (1969), Opt. Commun. 1, 254.

Eisenthal, K.B. and K.H. Drexhage (1969), J. Chem. Phys. 51, 5720.

Favro, L.D. (1960), Phys. Rev. 119, 53.

Fleming, G.R., J.M. Morris and G.W. Robinson (1976), Chem. Phys. 17, 91.

Fleming, G.R., A.E.W. Knight, J.M. Morris, R.J. Robbins and G.W. Robinson (1977), Chem. Phys. 23, 61.

Förster, Th. (1951), Fluoreszenz Organischer Verbindungen (Vandenhoeck & Ruprecht; Göttingen).

Gollnick, K. and G.O. Schenck (1964), Pure Appl. Chem. 9, 507.

Harris, J.M., L.M. Gray, M.J. Pelletier and F.E. Lytle (1977), Molec. Photochem. 8, 161.

Heiss, A., F. Dörr and I. Kühn (1975), Ber. Bunsenges. 79, 294.

Hirsch M.D., M.A. Marcus, A. Lewis, H. Mahr and N. Frigo (1976), Biophys. J. 16, 1399.

Hu, C.M. and R. Zwanzig (1974), J. Chem. Phys. 60, 4354.

Ippen, E.P. and C.V. Shank (1975), Sub-Picosecond Spectroscopy with a Mode-Locked CW Dye Laser, in: Lasers in Physical Chemistry and Biophysics, Proceedings of the 27th International Meeting of the Societé de Chimie Physique, Thiois, 1975 (J. Joussot-Dubien, ed.; Elsevier; Amsterdam), pp. 293–302.

Ippen, E.P. and C.V. Shank (1977), Techniques for Measurement, in: Ultrashort Light Pulses, Topics in Applied Physics, (S.L. Shapiro, ed.; Springer-Verlag; Heidelberg), Vol. 18.

Ippen, E.P., C.V. Shank and A. Bergman (1976), Chem. Phys. Letters 38, 611.

Ippen, E.P., C.V. Shank and R.L. Woerner (1977), Chem. Phys. Letters 46, 20.

Keilmann, F. (1977), Appl. Phys. 14, 29.

Kogelnik, H. and C.V. Shank (1972), J. Appl. Phys. 43, 2327.

Laubereau, A., A. Seilmeier and W. Kaiser (1975), Chem. Phys. Letters 36, 232.

Lessing, H.E. (1977), Fluorescence Photon Correlation, in: Photon Correlation Spectroscopy and Velocimetry (H.Z. Cummins and E.R. Pike, eds.; New York), p. 526.

Lessing, H.E. and A. von Jena (1976), Chem. Phys. Letters 42, 213.

Lessing, H.E. and A. von Jena (1978), Chem. Phys. Letters 59, 249.

Lessing, H.E. and A. von Jena (1979), Chem. Phys. (to be published).

Lessing, H.E., E. Lippert and W. Rapp (1970), Chem. Phys. Letters 7, 247.

Lessing, H.E., A. von Jena and M. Reichert (1975), Chem. Phys. Letters 36, 517.

Lessing, H.E., A. von Jena and M. Reichert (1976), Chem. Phys. Letters 42, 218.

Lombardi, J.R., J.W. Raymonda and A.C. Albrecht (1964), J. Chem. Phys. 40, 1148.

Lowdermilk, W.H. (1979), this volume, chapt. B1.

Memming, R. (1961), Z. Phys. Chem. 28, 168.

Merkelo, H., S.R. Hartman, T. Mar and G.S. Singhal Govindjee (1969), Science 164, 301.

Merkelo, H., J.H. Hammond, S.R. Hartman and Z.I. Derzko (1970), J. Luminescence 1, 502.

Mocker, H.W. and R.J. Collins (1965), Appl. Phys. Letters 7, 270.

Mourou, G. and M.M. Malley (1975), Chem. Phys. Letters 32, 476.

Müller, A. and G.R. Willenbring (1974), Ber. Bunsenges. 78, 1153.

Nemoto, M., H. Kokubun and M. Koizumi (1969a), Bull. Chem. Soc. Japan 42, 1223.

Nemoto, M., H. Kokubun and M. Koizumi (1969b), Chem. Commun. 1969, 1095.

Neporent, B.S. and O. Stolbova (1961), Opt. Spectrosc. 10, 146.

Penzkofer, A. and W. Falkenstein (1976), Chem. Phys. Letters 44, 547.

Phillion, D.W., D.J. Kuizenga and A.E. Siegman (1974), J. Chem. Phys. 61, 3828.

Phillion, D.W., D.J. Kuizenga and A.E. Siegman (1975), Appl. Phys. Letters 27, 85.

Porter, G., P.J. Sadkowski and C.J. Tredwell (1977), Chem. Phys. Letters 49, 416.

Rammensee, H. and V. Zanker (1960), Z. angew. Physik 12, 237.

Rehm, D. and K.B. Eisenthal (1971), Chem. Phys. Letters 9, 387.

Rentzepis, P.M., M.R. Topp and R.P. Jones (1970), Phys. Rev. Letters 25, 1742.

Ricard, D., H. Lowdermilk, F. Ducuing (1972), Chem. Phys. Letters 16, 617.

Runge, P.K. (1971), Opt. Commun. 3, 434.

Sargent, M. (1976), Appl. Phys. 9, 127.

Scarlet, R.I., J.F. Figueira and H. Mahr (1968), Appl. Phys. Letters 13, 71.

Schäfer, F.P. (1973), Principles of Dye Laser Operation, in: Schäfer, F.P., ed., Dye Lasers, Topics in Applied Physics, Vol. 1, 1973 (Springer-Verlag, Heidelberg).

Seybold, P.G., M. Goutermann and J. Callis (1969), Photochem. Photobiol. 9, 229.

Shank, C.V. and E.P. Ippen (1975), Appl. Phys. Letters 26, 62.

Shank, C.V. and D.H. Auston (1975), Phys. Rev. Letters 34, 479.

Shank, C.V., E.P. Ippen and R. Bersohn (1976), Science 193, 50.

Shank, C.V., E.P. Ippen and O. Teschke (1977), Chem. Phys. Letters 45, 291.

Shapiro, S.L., ed. (1977), Ultrashort Light Pulses, Topics in Applied Physics (Springer-Verlag; Heidelberg 1977), Vol. 18.

Shelton, J.W. and J.A. Armstrong (1967), IEEE J. OE-3, 302.

Soep, B., A. Kellmann, M. Artin and L. Lindquist (1972), Chem. Phys. Letters 13, 241.

Spears, K.G. and L.E. Cramer (1978a), J. Am. Chem. Soc. 100, 221.

Spears, K.G. and L.E. Cramer (1978b), Chem. Phys. 30, 1.

Spears, K.G. and M. El-Manguch (1977) Chem. Phys. 24, 65.

Statz, H. and G. DeMars (1960), in: Quantum Electronics (Ch. Townes, ed.; Columbia Univ. Press; New York).

Teschke, O., E.P. Ippen and G.R. Holtom (1977), Chem. Phys. Letters 52, 233.

von Jena, A. and H.E. Lessing (1979a), Appl. Phys. (to be published).

von Jena, A. and H.E. Lessing (1979b), Ber. Bunsenges. (to be published).

von Jena, A. and H.E. Lessing (1979c), Chem. Phys. (to be published).

von Jena, A. and H.E. Lessing (1979d), Opt. Quant. Electr. (to be published).

Ware, W.R. (1972), in: Creation and Detection of the Excited State (A.A. Lamola, ed.; Marcel Dekker; New York), Vol. 1, p. 213.

Wilkinson, F. (1975), Triplet Quantum Yields and Singlet–Triplet Intersystem Crossing, in: Organic Molecular Photophysics (J.B. Birks, ed.; Wiley; London) Vol. 2.

Yasa, Z.A. and O. Teschke (1975), Opt. Commun. 15, 169.

Author Index

Subject Index